D0152533

ULSI Technology

McGraw-Hill Series in Electrical and Computer Engineering

ULSI Technology

EDITED BY

C. Y. Chang

Chair Professor, College of Electrical Engineering and Computer Science
National Chiao Tung University
Director, National Nano Device Laboratories
Hsinchu, Taiwan, ROC

S. M. Sze

UMC Chair Professor
Department of Electronics Engineering
Director, Microelectronics and Information Systems Research Center
National Chiao Tung University
Hsinchu, Taiwan, ROC

THE McGRAW-HILL COMPANIES, INC.

New York St. Louis San Francisco Auckland Bogotá Caracas Lisbon
London Madrid Mexico City Milan Montreal New Delhi
San Juan Singapore Sydney Tokyo Toronto

A Division of The McGraw-Hill Companies

ULSI TECHNOLOGY

Acknowledgments begin on page 724 and appear on this page by reference.

This book is printed on acid-free paper.

1 2 3 4 5 6 7 8 9 0 DOC DOC 9 0 9 8 7 6 5

ISBN 0-07-063062-3

This book was set in Times Roman by Publication Services, Inc.
The editors were Lynn Cox and John M. Morriss;
the production supervisor was Denise L. Puryear.
The cover was designed by Farenga Design Group.
Project supervision was done by Publication Services, Inc.
R. R. Donnelley & Sons Company was printer and binder.

Cover photo: Electron micrograph of contact holes filled with CVD tungsten plugs (see Chapter 8). The diameter is 0.25 micron. *Courtesy of the National Nano Device Laboratories, National Science Council, R.O.C.*

Library of Congress Catalog Card Number: 95-81366

ABOUT THE EDITORS

C. Y. CHANG is Chair professor at the National Chiao Tung University and Director of the National Nano Device Laboratories, Taiwan, R.O.C. He has been the Dean of the College of Engineering and the Dean of the College of Electrical Engineering and Computer Science. He has written a book, *GaAs High Speed Devices,* has published more than 200 technical papers in international journals, and holds eight patents in ULSI, VHIC, and optoelectronics. He has taught many electronics engineers who have contributed significantly to the rapid growth of ULSI industries in Taiwan. Dr. Chang was elected a Fellow of IEEE in 1988 for his "contributions to semiconductor development and to education."

S. M. SZE is UMC Chair Professor of the Electronics Engineering Dept. and Director of Microelectronics and Information Systems Research Center, the National Chiao Tung University, Taiwan, R.O.C. For many years he was a member of the technical staff at AT&T Bell Laboratories, Murray Hill, New Jersey. Dr. Sze has made pioneering contributions to semiconductor devices and processing technologies, including the invention of the nonvolatile memory (1967) and the fabrication of MOSFETs in the 0.1 μm regime (1982). Author or coauthor of more than 100 technical papers, Dr. Sze has written three books on semiconductor devices and edited six books on VLSI technology, high-speed devices, semiconductor sensors, and related topics. He has been elected Fellow of IEEE, member of the Academia Sinica, and member of the National Academy of Engineering.

To Our Colleagues and Students—
Past, Present, and Future
On the Centennial of Our University—
The National Chiao Tung University

CONTENTS

LIST OF CONTRIBUTORS

C. Y. Chang
National Chiao Tung University
Hsinchu, Taiwan, ROC

T. S. Chao
National Nano Device Laboratories
Hsinchu, Taiwan, ROC

H. C. Cheng
National Nano Device Laboratories
Hsinchu, Taiwan, ROC

R. B. Fair
Microfabrication Technology
Center for Microelectronic Systems Technologies
Microelectronics Center of North Carolina
Raleigh, North Carolina, USA

R. Jansen
Crystal Consulting
Eindhoven. The Netherlands

W. Y. Lee
Electronics Research and Service Organization
Hsinchu, Taiwan, ROC

Y. J. T. Lii
Motorola
Austin, Texas, USA

R. Liu
AT& T Bell Labs
Murray Hill, New Jersey, USA

C. Y. Lu
Vanguard International Semiconductor Corporation
Hsinchu, Taiwan, ROC

K. Nakamura
National Nano Device Laboratories
Hsinchu, Taiwan, ROC

T. F. Shao
Texas Instruments, Inc.
Dallas, Texas, USA

T. Tachikawa
Mitsubishi Electrical Corp.
Mizuhara Itami, Hyogo, Japan

H. P. Tseng
Vanguard International Semiconductor Corporation
Hsinchu, Taiwan, ROC

F. C. Wang
Texas Instruments, Inc.
Dallas, Texas, USA

P. J. Wang
National Nano Device Laboratories
Hsinchu, Taiwan, ROC

J. T. Yue
Advanced Micro Devices
Sunnyvale, California, USA

PREFACE

ULSI Technology describes the theoretical and practical aspects of the most advanced state of electronics technology—ultralarge-scale integration (ULSI), where an integrated circuit (IC) chip contains over 10 million semiconductor devices. With ULSI technology, the cost of electronics products will decrease while the system functionality and performance will increase. The ULSI chips will result in the realization of smart and brilliant electronic systems, and in the improvement of quality of life and global productivity.

To fabricate IC chips with such complexity, we have to employ the most sophisticated process equipment, to follow the most precise process steps, and to adopt the most stringent cleanroom specifications. The basic process steps for ICs were considered in *VLSI Technology, 2nd Edition* (McGraw-Hill, 1988). Because of their importance to ULSI circuits, topics such as cleanroom technology, wafer-cleaning technology, manufacturing technology, and the rapid thermal process, which were essentially not covered in the 1988 VLSI book, are extensively discussed in *ULSI Technology.* In addition, many key processes, such as lithography, etching, metallization, and process integration, have been totally revised and updated. However, because of space limitations, certain classic topics such as crystal growth, conventional thermal processes, analytical technologies, and yield are covered only briefly or not covered at all. We suggest that our readers consult *VLSI Technology, 2nd Edition* for details.

In *ULSI Technology,* each chapter has an introduction that provides a general discussion of a specific aspect of ULSI processing. Subsequent sections present the basic science underlying individual process steps, the necessity for particular steps in achieving required parameters, and the tradeoffs in optimizing device performance and manufacturability. The problems at the end of each chapter form an integral part of the development of the topic.

The book is intended as a textbook for senior undergraduate or first-year graduate students in applied physics, electronics engineering, and materials science; it assumes that the reader has already acquired an introductory understanding of the physics and technology of semiconductor devices. Because it elaborates on IC processing technology in a detailed and comprehensive manner, it can also serve as a reference for those actively involved in integrated circuit fabrication and process development.

In the course of writing this text, many people have assisted us and offered their support. First, we express our appreciation to the management of our industrial and academic institutions, without whose help this book could not have been written. We have benefited from suggestions made by our reviewers: Dr. K. M. Brown of Digital Equipment Corporation, Drs. P. Chang and I. D. Liu of United Microelectronics Corporation, Drs. J. Chen, N. S. Tsai, R. Tsai, and F. C. Tseng of Taiwan Semiconductor Manufacture Company, Dr. L. P. Chen of National Nano Device Laboratories, Dr. T. C. Chen of International Business Machines, Dr. P. Fang of Advanced Micro

Devices, Dr. B. J. Lin of Linovation Incorporated, Prof. S. Murarka of Rensselaer Polytechnic Institute, Prof. K. Ohtsuka of Meisei University, Dr. S. Okazaki of Hitachi Ltd, Prof. C. Osbum of North Carolina State University, Dr. J. Sung of AT&T Bell Laboratories, Dr. N. Tsai of MOSEL-Vitalic, Inc., Prof. C. Y. Yang of Santa Clara University, Dr. J. L. Yeh of HP Laboratories, Prof. G. Declerck of IMEC, and Prof. T. Ohmi of Tohoku University.

We are further indebted to Mr. N. Erdos of AT& T Bell Laboratories for technical editing of the manuscript. We also thank Mrs. Shenn-May Lee Chang and Ms. C. C. Chang for handling the correspondence with our contributors and reviewers, Mrs. T. W. Sze for preparing the Appendixes, and Ms. P. L. Huang, Ms. L. J. Chang, Ms. Y. M. Chen, Ms. F. F. Fang, Ms. S. L. Hsiau, Ms. S. Y. Teng, and Mr. S. Y. Wu for preparing the Index.

We wish to thank the Ministry of Education, ROC, the National Science Council, ROC, and the Spring Foundation of the National Chiao Tung University for their financial support. One of the editors (S. M. Sze) would especially thank the United Microelectronics Corporation (UMC), Taiwan, ROC, for the UMC Chair Professorship grant that provided the environment to work on this book.

C. Y. Chang
S. M. Sze

INTRODUCTION

GROWTH OF THE INDUSTRY

The United States has the largest electronics industry in the world, with a global market share of over 40%. Since 1958, the beginning of the integrated-circuit (IC) era, the factory sales of electronic products have increased by about thirty times [see Fig. 1, curve (*a*)[1,2]]. Electronics sales, which were $303 billion in 1993, are projected to increase at an average annual rate of 8.5% and reach a half-trillion-dollar level by the year 2000. In the same period, the IC market itself has increased at an even higher rate [see Fig. 1, curve (*b*)[1,2]].* IC sales in the United States were $28 billion in 1993 and are expected to grow by 13% annually, reaching $65 billion by the year 2000. The main impetuses for such phenomenal market growth are the intrinsic pervasiveness of electronic products and the continued technological breakthroughs in integrated circuits.

The world markets of electronics and semiconductor industries will grow at comparable rates. Figure 2 shows the 1993 world electronics industry with a global sales volume of $679.7 billion. Also shown are the market shares of the six major electronics applications: computer and peripherals equipment at 32.3%, consumer electronics at 21.2%, telecommunication equipment at 16.5%, industrial electronics at 14.3%, defense and space at 11.5%, and transportation at 4.2%. By the year 2000, the world electronics industry is projected to reach $1200 billion, which will surpass the automobile, chemical, and steel industries in sales volume.

Figure 3 shows the 1993 world semiconductor industry, with total sales of $85.6 billion. Only 14% is related to optoelectronics and discrete semiconductor devices. IC sales constitute 86% of the total volume, with the largest segment being memory ICs, followed by microprocessor and microcontroller units, logic ICs, and analog ICs. In 2000, the semiconductor industry is projected to reach $200 billion, with over $170 billion in integrated circuits.

Figure 4 shows the market shares of the three major IC groups: MOSFET, bipolar transistor, and ICs made from III–V compound semiconductors.[3] At the beginning of the IC era, the IC market was broadly based on bipolar transistors. However, because of the advantages in device miniaturization, low power consumption, and high yield, sales volume of MOS-based ICs has increased steadily and in 1993 amounted to 75% of the total IC market. By the year 2000, MOS ICs will capture the largest market share (88%) of all ICs sold. This book, therefore, emphasizes MOS-related ULSI technology.

---

*There were only two years in which the growths were negative: in 1974, due to the Middle East oil embargo, and in 1985, due to overproduction of personal computers.

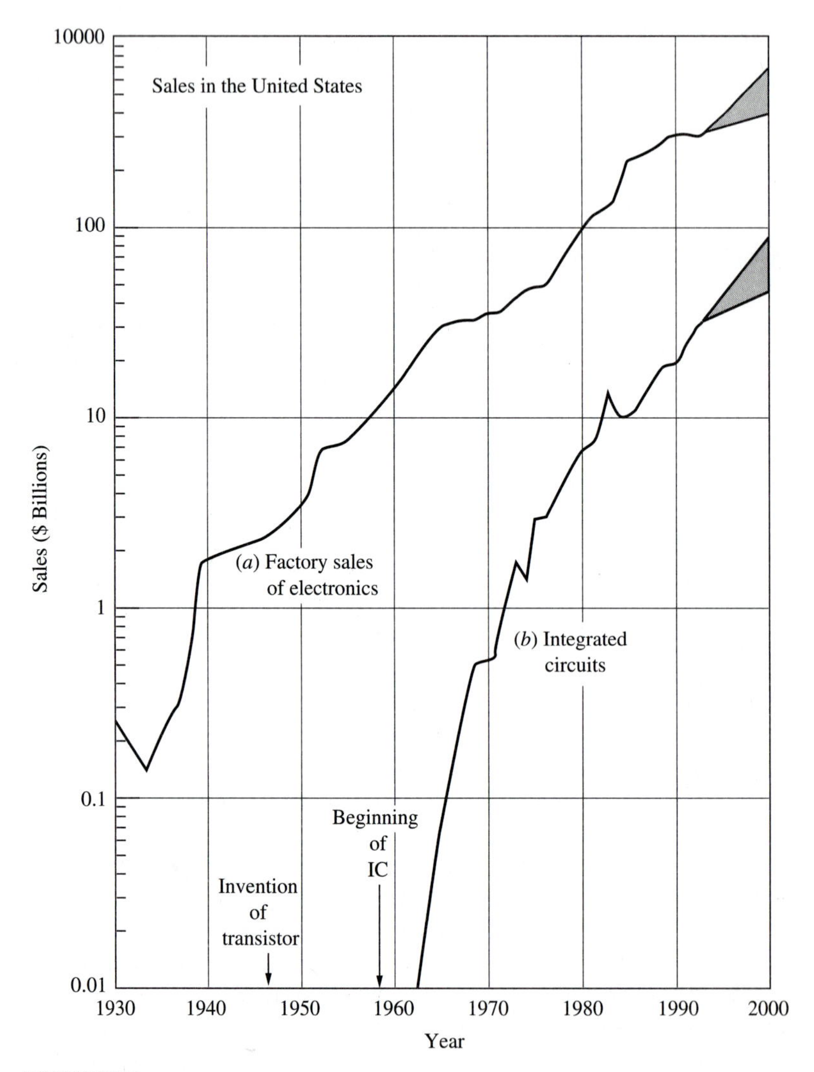

FIGURE 1
(*a*) Factory sales of electronics in the United States for the 64 years between 1930 and 1993 and projected to 2000. (*b*) Integrated circuit market in the United States for 32 years between 1962 and 1993 and projected to 2000. *(After Refs. 1 and 2.)*

DEVICE MINIATURIZATION

Figure 5, curve (*a*), shows the rapid growth in the number of components per MOS memory chip.[4,5] Note that the MOS IC complexity has advanced from small-scale integration (SSI), to medium-scale integration (MSI), to large-scale integration (LSI), to very-large-scale integration (VLSI), and finally to ultralarge-scale integration (ULSI), which has 10^7 or more components per chip. We note that since 1975 the growth has been maintained at a rate of about 40% annually; in other words, the

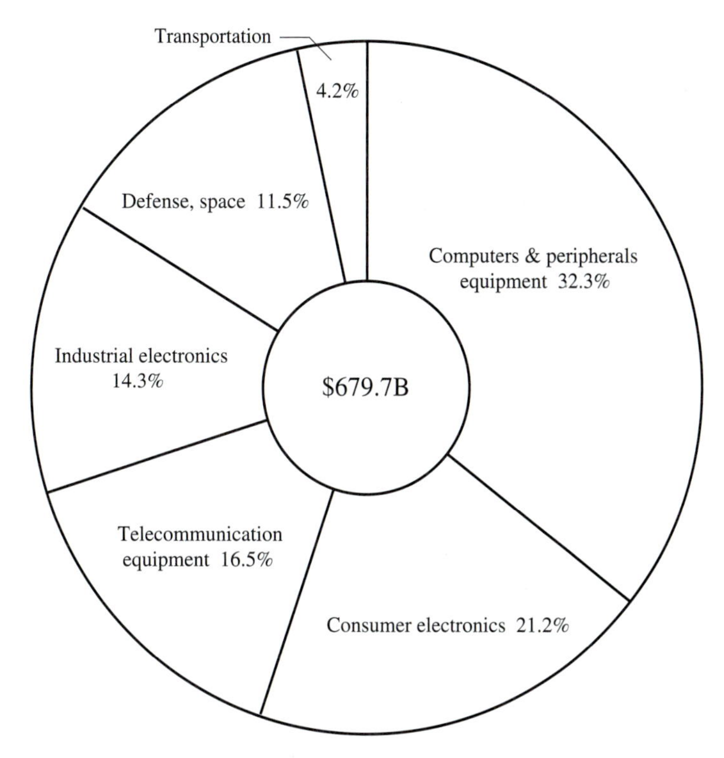

FIGURE 2
1993 world electronics industry. *(After Dataquest, 1994.)*

number of components has doubled every two years. At this rate, over 100 million components per chip will be available before the year 2000; in the early 21st century we will move into the gigabit range, with IC chips having more than one billion components.[6,7] Also shown in Fig. 5 is the growth of the number of components for bipolar, MESFET, and MODFET ICs. They are about two orders of magnitude lower in complexity compared with MOS-based ICs.

The most important factor in achieving the ULSI complexity is the continued reduction of the minimum device-feature length [see Fig. 6, curve (*a*)]. Since 1960, the annual rate of reduction has been 13%, which corresponds to a reduction by a factor of two every six years. At this rate, the minimum feature length will shrink from its present length of 0.5 μm to 0.2 μm in the year 2000. The junction depth of the source and drain junctions, and the gate oxide thickness are also being reduced at a similar rate as shown in curves (*b*) and (*c*) of Fig. 6, respectively.

The reduction of the device feature length and related dimensions has resulted in reduced overall device size and unit price per function. Figure 7 shows the relative price and size reductions.[8] In the past fifty years, prices have gone down by 100 million times, and the size has been reduced by a factor of one billion. By 2000 the price per bit is expected to be less than 0.1 millicent for a 64-megabit memory chip. Similar price reductions are expected for logic ICs. Additional benefits from device

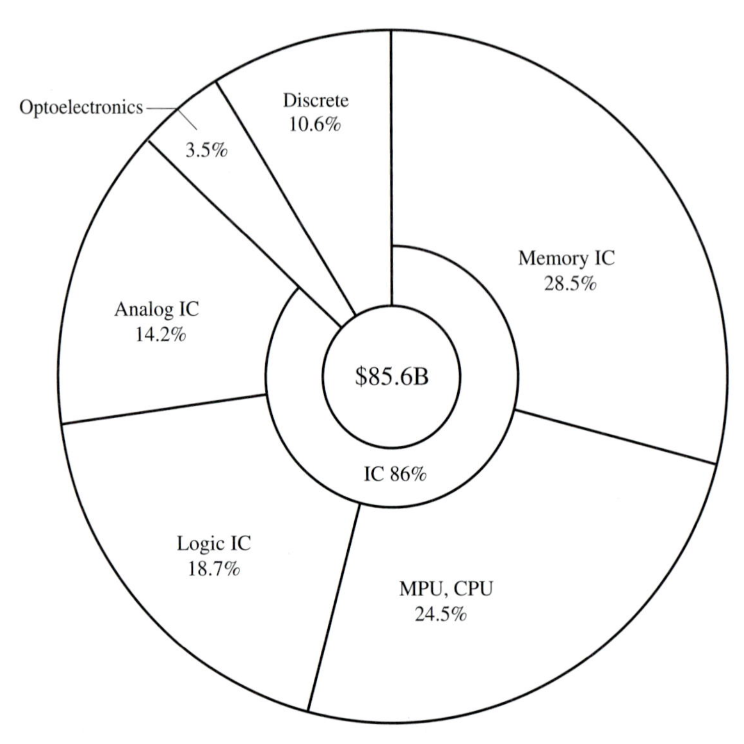

FIGURE 3
1993 world semiconductor industry. *(After Dataquest,1994.)*

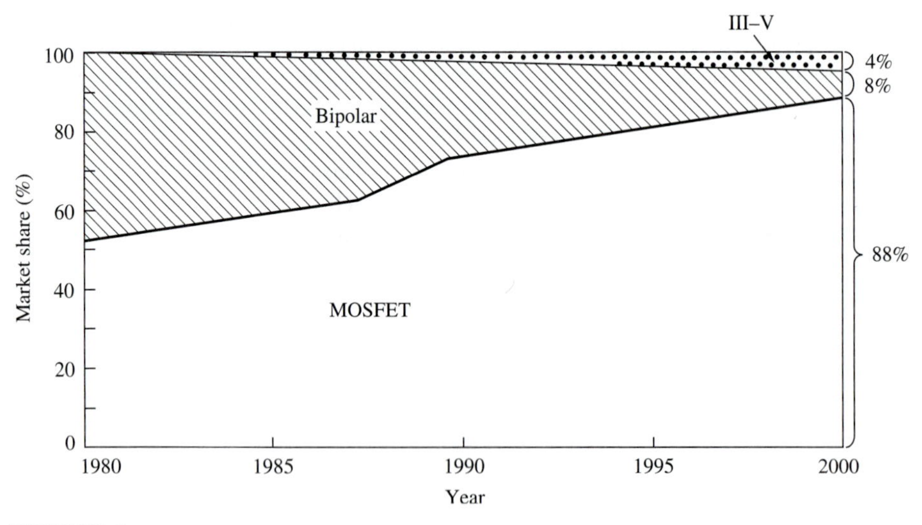

FIGURE 4
World IC market (1980–2000). *(After Zdebel, Ref.3.)*

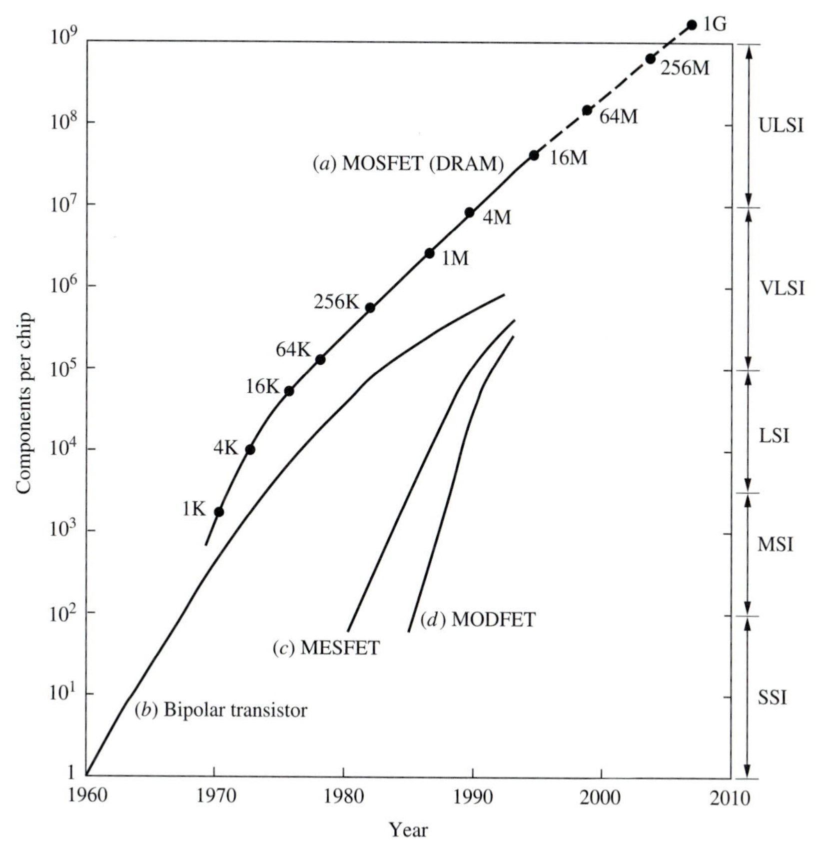

FIGURE 5
(*a*) Exponential growth of the number of components per MOS IC chip. *(After Moore, Ref. 4, and Myers, Ref. 5.)* (*b*), (*c*). and (*d*) Components per chip versus year for bipolar, MESFET, and MODFET ICs, respectively.

miniaturization include improvement of device speed (which varies inversely with the device feature length) and reduction of power consumption (which varies approximately with the square of the feature length). Higher speeds lead to expanded IC functional throughput rates, so that future ICs can perform data processing, numerical computation, and signal conditioning at 100 and higher gigabit-per-second rates.[9] Reduced power consumption results in lowering the energy required for each switching operation. Since 1960 the required energy, called the *power-delay product,* has decreased by six orders of magnitude.[10]

ORGANIZATION OF THE BOOK

Figure 8 shows how the 12 chapters of this book are organized. Chapter 1 considers cleanroom technology. The continued miniaturization in ULSI devices implies more

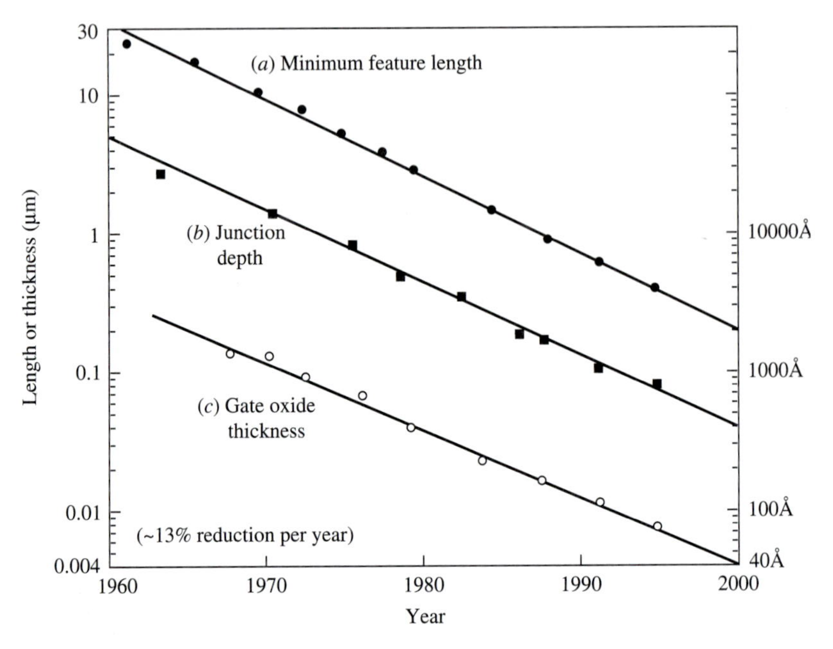

FIGURE 6
Exponential decrease of (*a*) minimum feature length, (*b*) junction depth, and (*c*) gate oxide thickness of MOSFET.

stringent requirements with respect to contamination control. Without an ultraclean processing environment, ULSI circuits simply cannot be realized.[11]

ULSI technology is synonymous with *silicon* ULSI technology. The unique combination of silicon's adequate bandgap, stable oxide, and abundance in nature ensures that in the foreseeable future no other semiconductor will seriously challenge its preeminent position in ULSI applications. Some important properties of silicon are listed in Appendix A.

Once the silicon wafers are in the cleanroom, we enter into the wafer-processing sequence, described in Chapters 2 through 8 and depicted in the wafer-shaped central circle of Fig. 8. Each of these chapters considers a specific process step. Of course, many processing steps are repeated many times in IC fabrication; for example, lithography and etching steps may be repeated 10 to 20 times. In ULSI technology the wafer-cleaning technology is as important as the cleanroom technology. Without a contamination-free wafer surface, the ICs will suffer from low yield and poor reliability. Because of limitations on the total length of the book, many classic topics, such as crystal growth, oxidation, diffusion, and ion implantation, are only briefly mentioned. The reader may consult textbooks on VLSI technology for details.[12]

The individual processing steps described in Chapters 2 through 8 are combined in Chapter 9 to form devices and integrated circuits. Chapter 9 considers the fundamental building process modules and four important IC families: CMOS (complementary MOSFET), bipolar ICs, BiCMOS (a combination of bipolar and CMOS),

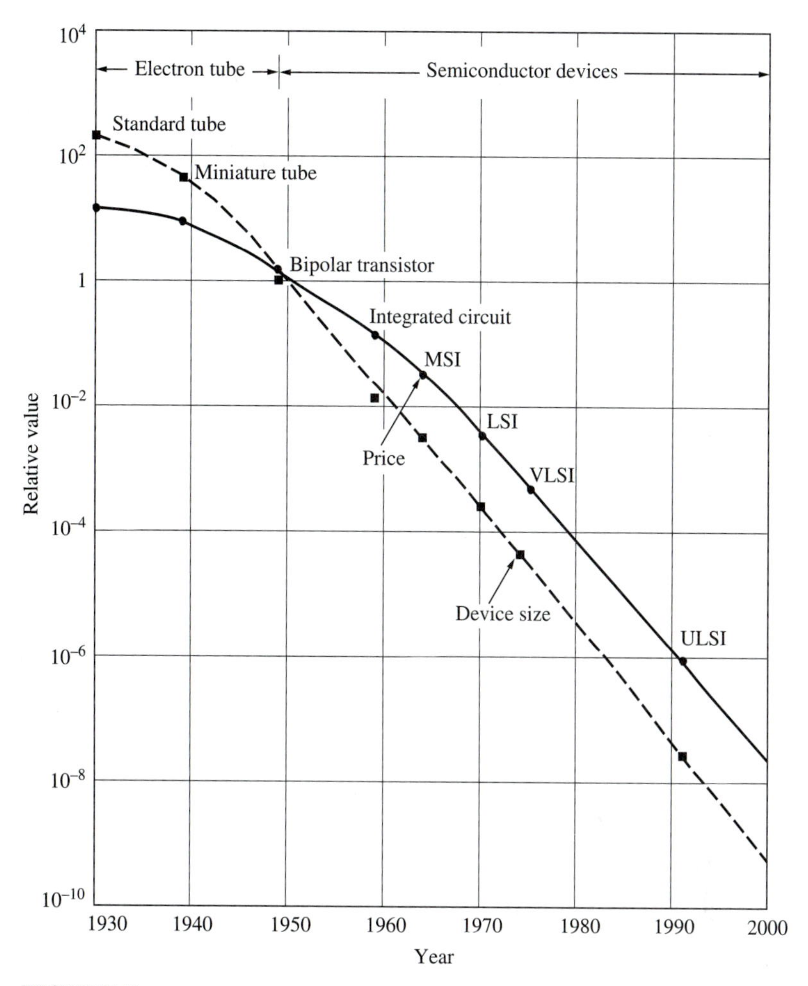

FIGURE 7
Price and size reduction of active electronic compents. *(After Shoda, Ref. 8.)*

and MOS memory ICs. After the completely processed wafers are tested, those chips that pass the tests are ready to be packaged. Chapter 10 describes the assembly and packaging of ULSI chips. Chapter 11 considers the manufacturing technology, that is, the strategy and logistics to implement various technologies to produce ULSI chips that meet customers' specifications in a timely fashion and to generate adequate return on investment for the IC manufacturer. Chapter 12 describes a multitude of reliability issues related to ULSI processes. As device dimensions move to the sub-half-micron and sub-quarter-micron regime, ULSI processing becomes more automated, resulting in tighter control of all processing parameters. At every step of production, from wafer cleaning to device packaging, numerous requirements are being imposed to improve the device performance and reliability.

To keep the notation simple in this book, we sometimes found it necessary to use a symbol more than once, with different meanings. However, within each chapter a

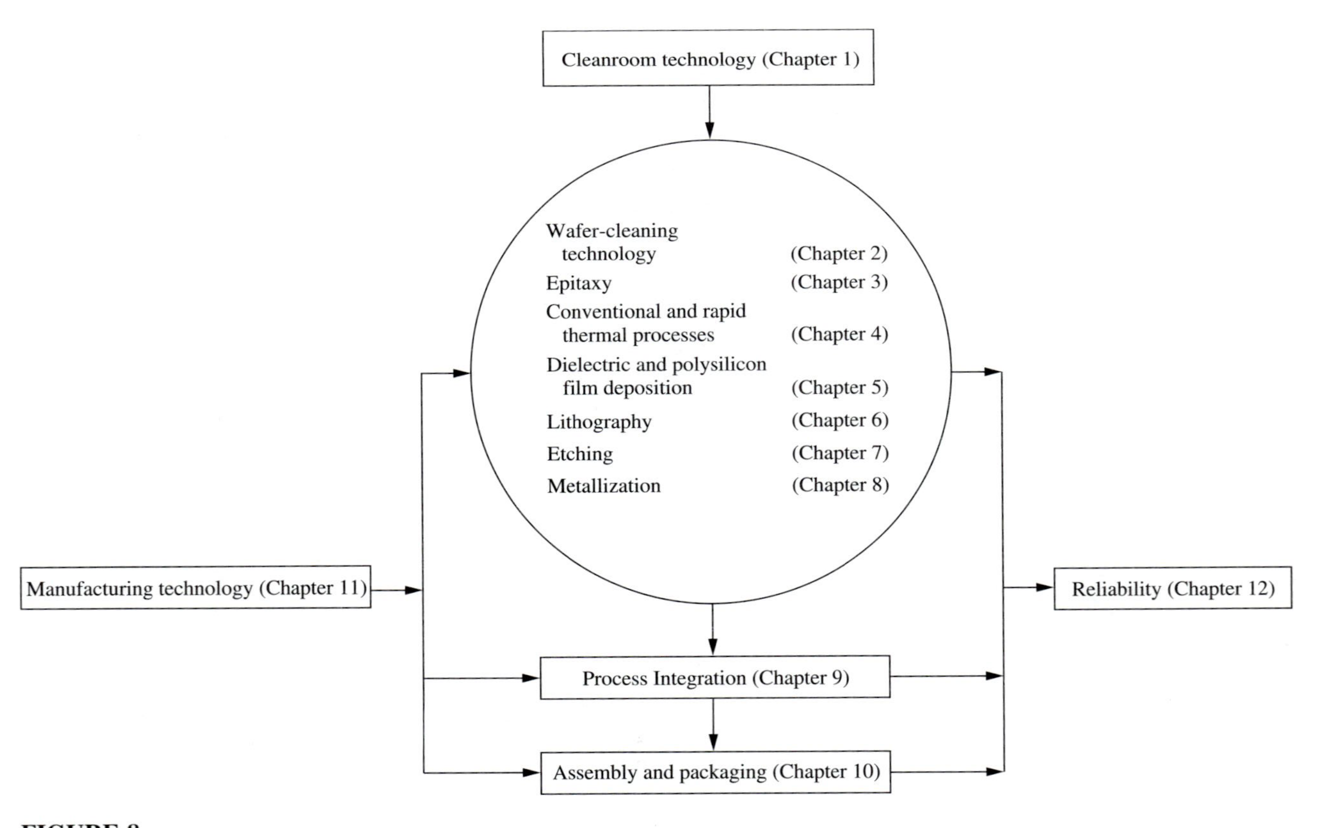

FIGURE 8
Organization of this book.

symbol has only one meaning and is defined the first time it appears. Many symbols do have the same or similar meanings consistently throughout this book; they are summarized in Appendix B.*

ULSI technology is presently moving at a rapid pace. The number of ULSI publications has doubled every year since 1990, the beginning of the ULSI era. Many topics, such as lithography, rapid thermal processing, and metallization, are still under intensive study. Their ultimate capabilities are not fully understood. The material presented in this book is intended to serve as a foundation. The references listed at the end of each chapter can supply more information.

REFERENCES

1. *1994 Electronic Market Data Book,* Electronic Industries Association, Washington, D.C., 1994.
2. *1994 Annual Report of Semiconductor Industry,* Industrial Technology Research Institute, Hsinchu, Taiwan, ROC, 1994.
3. P. J. Zdebel, "Current Status of High Performance Silicon Bipolar Technology," *14th Annual IEEE GaAs IC Symp. Tech. Digest,* 15 (1992).
4. G. Moore, "VLSI, What Does the Future Hold," *Electron Aust.,* **42,** 14 (1980).
5. W. Myers, "The Drive to the Year 2000," *IEEE Micro,* **11,** 10 (1991).
6. P. K. Chatterjee and G. B. Larrabee, "Gigabit Age Microelectronics and Their Manufacture," *IEEE Trans. VLSI Syst.* **1,** 7 (1993).
7. K. Mori, H. Yamada, and S. Takizawa, "System on Chip Age," *Proceedings of the International Symposium on VLSI Technology, Systems, and Applications,* k15 (1993).
8. K. Shoda, "Home Electronics in the 1990s," *Proceedings of the International Symposium on VLSI Technology, Systems, and Applications,* 1(1991).
9. H. Komiya, M. Yoshimoto, and H. Ishikura, "Future Technological and Economic Prospects for VLSI," *IEICE Trans. Electron.* **E76-C,** 1555 (1993).
10. R. W. Keyes, "Limitations of Small Devices and Large Systems," in N. G. Einspruch, Ed., *VLSI Electronics,* Academic, New York, 1981, Vol. 1, p. 186.
11. T. Ohmi, "ULSI Reliability through Ultraclean Processing," *Proc. IEEE,* **81,** 716 (1993).
12. For example, S. M. Sze, Ed., *VLSI Technology, 2nd Ed.,* McGraw-Hill, New York, 1988.

---

*Also included are the International System of Units (Appendix C) and Physical Constants (Appendix D).

CHAPTER 1

Cleanroom Technology

H. P. Tseng and R. Jansen

1.1 INTRODUCTION

Microtechnologies are developing in a way that makes both the production tool and the process environment increasingly critical for the manufacturing process. The issue is, above all, to prevent defects in the product or, in other words, to increase the yield. This is achieved by the control of the wafer environment. The environmental factors include the control of temperature, relative humidity, electrostatic discharges, airborne particles, chemical contamination, electromagnetic fields, oxygen, and vibration.

The continuous miniaturization in device technology implies more and more stringent requirements with respect to contamination control.[1] Table 1 lists the changes in requirements from 1980 projected to 2004. In the 1980s a cleanliness class of 100–1000 was sufficient for the feature size of 2 μm. Today, a cleanliness class of 0.1 is the requirement when addressing a feature size of 0.5 μm down to 0.25 μm. In the future, inert minienvironments may be the only applicable solution.[2,3] To eliminate microcontamination and reduce native oxide growth on silicon wafers, the wafer processing and loading/unloading sections of a process tool are enclosed in an extremely high cleanliness minienvironment flushed with ultrapure nitrogen containing no oxygen and moisture. Similar developments can be observed with respect to the purities of the process utilities: from 1 ppm in the 1980s to 1 ppb in the very near future.

Reducing defect density is the most important aspect of increasing yield. Defect density is defined as the number of defects per cm^2 wafer that may occur during the processing due to all kinds of contamination. Particle contamination is still responsible for over 80% of defects. Table 2 summarizes the defect density requirements with respect to different design rules. Critical particle sizes, above which killing defects may be generated on the wafer surfaces, follow technology developments,

TABLE 1
Evolution of IC processing features

Mass production started	1980	1984	1987	1990	1993	1996	1999	2004
Wafer size, mm	75	100	125	150	200	200	200	300
Technology (DRAM)	64K	256K	1M	4M	16M	64M	256M	1G
Chip size (cm^2)	0.3	0.4	0.5	0.9	1.4	2.0	3.0	4.5
Feature size(μm)	2.0	1.5	1.0	0.8	0.5	0.35	0.25	0.2–0.1
Process Steps	100	150	200	300	400	500	600	700–800
Cleanroom class (to be defined in Sec. 1.2)	1,000–100	100	10	1	0.1	0.1	0.1/ mini-environment	0.1/inert mini-environment
Utility impurity (ppb)	1,000	500	100	50	5	1	0.1	0.01

Source: Courtesy of KLA.

decreasing from 0.12 μm at a 0.8-μm-feature size to 0.03 μm at a 0.25-μm-feature size. As the process technology migrates from 0.8 μm 4M DRAM (dynamic random access memory) to 0.25 μm 256M DRAM over a time span of one decade, the maximum tolerable defect densities per critical layer measured at the defect size of 0.12 μm must be reduced from 0.28 D/cm^2 to 0.004 D/cm^2, representing an improvement factor of 65 times.

It is evident that the requirements for the control of the wafer environment are of great importance to the yield. In practice, the control of the wafer environment is a crucial aspect of the wafer manufacturing processes. It constitutes the basis for the design of the cleanroom systems and process utilities systems.

TABLE 2
Minimum defect density requirement with respect to design rule

DRAM process technology	Unit	4M	16M	64M	256M
Design rule	Micron	0.8	0.5	0.35	0.25
Critical layer	Each	9	10	11	13
Critical particle size	Micron	0.12	0.09	0.05	0.03
Killing defect size	Micron	0.27	0.18	0.1	0.06
Defect density measured at the killing defect size	D/cm^2	0.50	0.40	0.32	0.22
Defect density measured at 0.12 μm particle size	D/cm^2	2.53	0.90	0.22	0.055
Defect density/critical level measured at 0.12 μm	D/cm^2	0.28	0.09	0.02	0.004
Defect improvement factor		1	3	14	65

Source: Courtesy of KLA.

Besides the extreme high-quality demands, issues such as flexibility, safety, reliability, and an accountable cost level are equally important to the design. Moreover, rapid developments within the market and process technology call for swift actions. At first glance, a number of these actions seem to be contradictory.

In the past, a typical reaction of designers and engineers to this rather complex problem was to rule out, as much as possible, all conceivable risks. This attitude caused unnecessarily high investments. Cost-effectiveness is the only acceptable solution. Cost-effectiveness can be understood as the optimum integration of appropriate quality, quantity, safety, flexibility, reliability, and time to achieve a competitive cost level. To achieve such cost-effectiveness, four conditions are of utmost importance, namely, know-how, approach, organization, and control.

The available know-how determines whether a design is or is not cost-effective. Inadequate knowledge often leads to overkill, resulting in excessive cost.

The project approach, combined with the project organization, constitutes the most important control mechanism. Modifications and surprises introduced at a late stage are caused mostly by an improper project approach.

The project organization is extremely important for efficiency, information, communication, and chain of responsibility. If projects are not properly organized, they tend to drift away from their objectives.

Control focuses on furnishing continuous, up-to-date information about the costs, timing, and quality. Such information is vital as a control tool.

In the past decade, submicrometer technology has enormously enriched the know-how and experience in both technique and approach. The learning curve is vitally important to each individual who is involved with the cleanroom technology in any way.

In an attempt to give some practical insights, a global clarification of the most important technological developments will be provided. The following subjects will be covered:

Section 1.2, on cleanroom classification, defines cleanroom cleanliness classes according to the current standard and explains their application.

Section 1.3, on cleanroom design concept, deals with the considerations needed to define the optimum process layout, and the design of a cleanroom. In addition, the common cleanroom performance criteria are listed, such as cleanliness class, temperature and humidity control, air quality, etc. Principal technical data that are essential for sizing the cleanroom facilities of a 200-mm-wafer fab are also presented.

Section 1.4, on cleanroom installation, provides basic information concerning the cleanroom subsystems such as exhaust, make-up air, recirculation air, chilled water, hot water, steam, filter ceiling, partition walls, and raised floor. Moreover, the recommended project approach, an overall time schedule for the design and realization of a cleanroom installation project, as well as a cost analysis are presented.

Section 1.5, on cleanroom operation, explains common cleanroom gowning procedures and cleanroom maintenance practices.

Section 1.6, on automation, clarifies various levels of process equipment control, from shop-floor-control software to the application of equipment interfaces.

Section 1.7, on related facility systems, provides basic information about the design of the deionized water system, process chemicals and gases, and the importance of space-management design.

Finally, in Section 1.8, we present conclusions on all the main issues along with a look of the future trends.[4]

1.2 CLEANROOM CLASSIFICATION

The laminar-flow cleanroom concept was developed at Sandia National Laboratories, Albuquerque, New Mexico, in 1961 to provide a particle-free environment for thin-film deposition and the assembly of delicate mechanisms. This cleanroom concept was swiftly adopted by the then-burgeoning semiconductor industry.

TABLE 3
Evolution of Fed. Std 209 Series for specifying cleanliness of air

Date	Fed. Std.	Highlights of the original and revised contents
Dec. 1963	209	Cleanroom operation principles
Aug. 1966	209 A	Cleanroom design and testing methods Air flow pattern Laminar flow and turbulent flow Air velocity, 90 ± 20 ft/min Pressure, temperature, humidity, vibration Audio frequency noise, air exchange rate Air cleanliness classification specified as the number of particles at sizes larger than 0.5 μm per cubic foot; class 100, 10,000, and 100,000
Apr. 1973	209 B	Changed air velocity from 90 ± 20 ft/min to 90 ± 20% ft/min and changed humidity from 45% to 40 ± 5%
May 1977	209B amendment	Added cleanliness class 1,000
Oct. 1987	209C	Major revision of cleanroom classification and testing method Added classes 1 and 10 Extended the particle measurements from 5 μm and 0.5 μm down to 0.3 μm and 0.2 μm for class 100, and down to 0.3 μm, 0.2 μm, and 0.1 μm for class 10 and class 1 Clearly defined particulate sampling locations and numbers of sampling and measuring time
June 1988	209D	Corrected several typographical errors found in Fed. Std. 209C
Sep. 1992	209E	Adopted metric system Added descriptor to specify the maximum allowable number of ultrafine particles per cubic meter Added sequential airborne particle sampling plan to the single air sampling plan specified in Fed. Std. 209D

TABLE 4
Metric definition of airborne particulate cleanliness classes per Fed. Std. 209E

	Particles/m³				
Class	**0.1 μm**	**0.2 μm**	**0.3 μm**	**0.5 μm**	**5 μm**
M1	3.50×10^2	7.57×10^1	3.09×10^1	1.00×10^1	
M1.5	1.24×10^3	2.65×10^2	1.06×10^2	3.53×10^1	
M2	3.50×10^3	7.57×10^2	3.09×10^2	1.00×10^2	
M2.5	1.24×10^4	2.65×10^3	1.06×10^3	3.53×10^2	
M3	3.50×10^4	7.57×10^3	3.09×10^3	1.00×10^3	
M3.5		2.65×10^4	1.06×10^4	3.53×10^3	
M4		7.57×10^4	3.09×10^4	1.00×10^4	
M4.5				3.53×10^4	2.47×10^2
M5				1.00×10^5	6.18×10^2
M5.5				3.53×10^5	2.47×10^3
M6				1.00×10^6	6.18×10^3
M6.5				3.53×10^6	2.47×10^4
M7				1.00×10^7	6.18×10^4

In December 1963, the U.S. government issued Fed. Std. (Federal Standard) 209 to standardize cleanroom design and operation guidelines. Over the past three decades, U.S. Fed. Std. 209 has undergone several revisions as illustrated in Table 3. In the last revision, Fed. Std. 209E, the original title used for nearly 30 years, "Cleanroom and Work Station Requirements, Controlled Environment," had been changed to "Airborne Particulate Cleanliness Classes in Cleanrooms and Clean Zones."

The recent addition of metric system units ensures that the Fed. Std. 209 series will be used by the international technical community as a world standard technical document for the characterization of the cleanliness of air. The definition of the metric airborne particulate cleanliness classes and the relevant air sampling method specified in Fed. Std. 209E are quite different than those of Fed. Std. 209D and, hence, warrant further discussion.

Tables 4 and 5 illustrate air cleanliness classes expressed in SI and English units, respectively. The numerical designation of the class in SI units is taken from the

TABLE 5
English definition of airborne particulate cleanliness classes per Fed. Std. 209E

	Particles/ft³				
Class	**0.1 μm**	**0.2 μm**	**0.3 μm**	**0.5 μm**	**5 μm**
1	3.50×10	7.50	3.00	1.00	
10	3.50×10^2	7.50×10	3.00×10	1.00×10^1	
100		7.50×10^2	3.00×10^2	1.00×10^2	
1000				1.00×10^3	7.00
10,000				1.00×10^4	7.00×10
100,000				1.00×10^5	7.00×10^2

logarithm (base 10) of the maximum allowable number of particles, 0.5 μm and larger, per cubic meter. The numerical designation of the class in English units is taken from the maximum allowable number of particles, 0.5 μm and larger, per cubic foot. Note that the concentration of the particle, in general, is roughly inversely proportional to the particle size to a 2.2 power.

In addition to the metric definition of new and existing classes, standard nomenclature for the verification of a class was introduced in Fed. Std. 209E. It is no longer sufficient to state that air meets class 100 or class 1; Fed. Std. 209E requires that classes are expressed by using the format

Class X (at Y μm)

where X represents the numerical designation of the airborne particulate cleanliness class and Y represents the particle size.

In Fed. Std. 209E, a new air cleanliness definition, the U descriptor is added. The U descriptor specifies the maximum allowable numbers of ultrafine particles in a cubic meter of air in the size range from approximately 0.02 μm to the upper detectable limit of a discrete-particle counter. Without referring to a specific particle size, the U descriptor defines the cleanliness as

U(X)

where X is the maximum allowable number of ultrafine particles per cubic meter of air.

The description of air cleanliness is illustrated by the following examples:

"Class M 2.5 (at 0.3 μm)" describes air with no more than 1060 particles/m^3 with a particle size of 0.3 μm and larger.
"Class 100 (at 0.5 μm)" describes air with no more than 100 particles/ft^3 with a particle size of 0.5 μm and larger.
"U(20)" describes air with no more than 20 ultrafine particles/m^3.

Figure 1 shows the maximum allowable particle concentrations of each cleanliness class in U.S. Fed. Std. 209E as a function of particle size. The air cleanliness class in Japanese standard JIS B9920 rev. (revised) is shown in Fig. 2 for comparison.

Similar to Fed. Std. 209E, JIS B9920 rev. also defines cleanliness class by employing the logarithm of the maximum allowable particles per cubic meter. However, instead of specifying the cleanliness class at particle size of 0.5 μm, the JIS B9920 rev. specified the cleanliness class at particle size of 0.1 μm.

Because it uses metric units, the Fed. Std. 209E has made the necessary change in the specification of air sample location for verification of air cleanliness.[5,6] The sample location should be uniformly spaced throughout the clean zone operated in either a unidirectional (laminar) airflow or a nonunidirectional airflow environment. The minimum number of sample locations required for verification in a clean zone is specified below:

Unidirectional airflow (the lesser of (a) or (b)):

SI units: (a) $A/2.32$
(b) $A \times 64/(10^{M})^{0.5}$

English units: (a) A/25
(b) $A/(Nc)^{0.5}$

Nonunidirectional airflow:

SI units: $A \times 64/(10^{M})^{0.5}$

English units: $A/(Nc)^{0.5}$

where A is the floor area of the clean zone (or entrance plane) in ft^2 for English units and m^2 for SI units, and M and Nc are the numerical designations of the class in SI and English units, respectively.

Note that number of locations must always be rounded to the next higher integer.

One of the most useful new features of Fed. Std. 209E is the sequential sampling option.[7] The single sampling plan in Fed. Std. 209E requires, as it did in Fed. Std. 209D, that the air volume to be sampled should anticipate 20 particles estimated from the air cleanliness class to be verified. For example, the sampling air volumes required for the verification of a cleanroom of class 1 should be at least 20 ft^3. Using a standard air sampling rate of 1 ft^3/min, it will take 20 minutes to complete a class

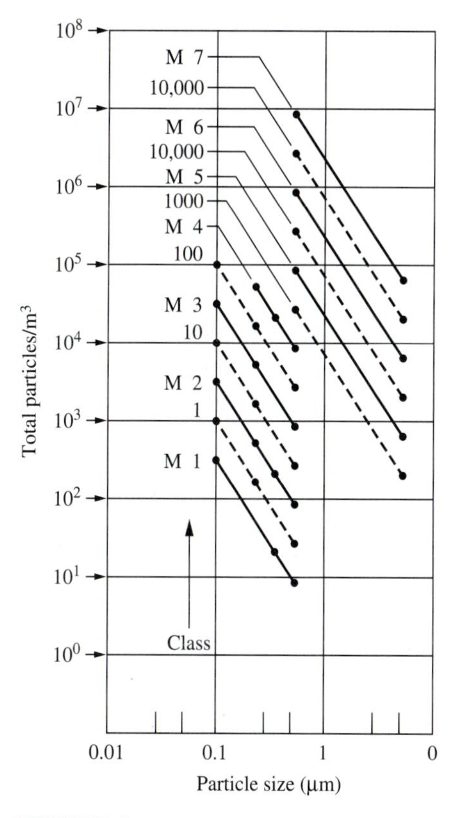

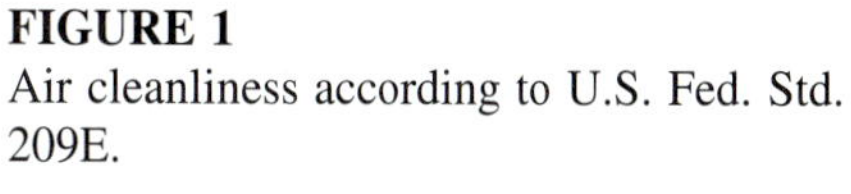
FIGURE 1
Air cleanliness according to U.S. Fed. Std. 209E.

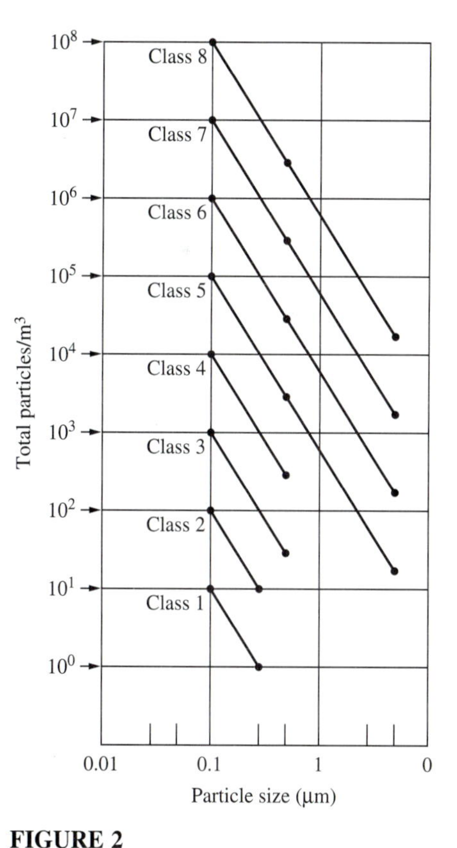

FIGURE 2
Air cleanliness according to Japanese Std. B9920 rev.

1 cleanliness verification test. For air that is very clean, this statistical requirement demands that air be sampled for a very long time at each location. Thus, measuring the cleanliness of air that contains very low concentrations of particles can be unacceptably expensive. However, if air at a location is sampled while a running count of particles is recorded as a function of the time, frequently one can obtain a clear-cut indication of whether the air at that location passes or fails in a very short time by applying statistical criteria to the pattern of the count. This procedure allows verification of an airborne-particulate cleanliness class using samples of shorter duration than is required by the original single-sampling plan.

To use the sequential-sampling plan, record the number of particles observed as a function of time. Compare the count, as sampling continues, with the upper and lower limits, which are calculated by the two equations

$$\text{Upper} = 3.96 + 1.03E \tag{1.1}$$

$$\text{Lower} = -3.96 + 1.03E \tag{1.2}$$

where E is the expected count estimated from the air cleanliness class.

If the cumulative observed count for the sample exceeds the upper limit, then sampling is stopped and the air is judged to have failed. If the cumulative observed count drops below the lower limit, then sampling is stopped and the air is judged to have passed. If the cumulative observed counts equal 20 or less at the end of the sample duration, which is calculated from the air cleanliness class to anticipate a maximum of 20 particles, the air is also judged to have met the requirement.

1.3 CLEANROOM DESIGN CONCEPT

The design process starts with the translation of the process technology and the production volume into the specifications for the architectural, electrical, HVAC (heating, ventilation, and air-conditioning), central, and process utilities systems. The process technology is an important standard for quality requirement. For example, the cleanroom relates strongly to the feature size (design rules) of the technology in question. The quantity requirements are derived from the production volume combined with the technology. As a matter of course, attention should be paid to future developments of process tools and their impact on the cleanroom design.

With the aid of a rough computation, a first process tool list is developed. The process utilities requirements are established on the basis of this list. Electrical, mechanical, exhaust, gas, chemical, and other demands are listed in a matrix according to equipment tool. An accurate estimation of the average consumption of each process utility is essential for sizing the utility capacity properly. The facilities are subdivided into the following main systems:

Site infrastructure, site facilities, and buildings
Power supply
Electrical and communication
Cleanrooms and HVAC

Central utilities
Process utilities
Environment and safety

The design criteria of each facility system and its subsystems are formulated, parallel to the definition of the specifications. These criteria include the quality and quantity characteristics as well as the schematic set-up. It should be clear that the process utilities requirements and the design criteria are determining factors for cost-effectiveness and that ignorance of these factors will almost always lead to overkill.

The design of a cleanroom normally starts with the optimization of cleanroom layout. Based on the targeted process technologies and the building architecture, a proper cleanroom system is selected and, subsequently, the relevant cleanroom performance criteria are defined.

1.3.1 Cleanroom Layout

Most cleanrooms have a rectangular shape with a central aisle. The central aisle serves as the corridor to each process tunnel extending sideways into the remaining portions of the cleanroom. The ideal width of the central aisle and process tunnel should be 2.4 meters to allow most process equipment to be transported in or out easily. The ideal height of the filter ceiling should be 3.4 to 3.6 meters above the raised floor to facilitate the transportation and installation of vertical furnaces, which are approximately 3 meters high.

Most 200-mm-wafer process equipment has the same throughput as the 150-mm-wafer equipment, except for steppers and high-current implanters. The footprints of the furnace, chemical stations, stepper, and photoresist coater/developer are also larger. Therefore a 200-mm-wafer fabrication plant requires 10% to 20% more cleanroom space than the 150-mm-wafer fab. A 200-mm, 16-Mbit, DRAM wafer fab with a monthly output capacity of 20,000 wafers requires approximately 5000 m^2 cleanroom area to accommodate the necessary process equipment and another 1000 m^2 cleanroom of lower classification to incorporate supporting facilities such as a gowning area, equipment-cleaning areas, and parts-cleaning areas, as well as a tube-cleaning area.

The entire IC wafer manufacturing process consists of 14 to 20 process modules. A typical process module, illustrated in Fig. 3, in general consists of 10 to 20 individual process steps. Within each process module, a layer of thin dielectric or conductive film is applied to the silicon wafer via oxidation, LPCVD (low-pressure chemical vapor deposition), APCVD (atmospheric-pressure chemical vapor deposition), PECVD (plasma-enhanced chemical vapor deposition), or metal-deposition techniques. Circuit patterns are then formed on a layer of photosensitive material in the photolithography area. Subsequently, the wafers are sent to either the etching or the ion implantation area. In the etching area, circuit patterns are etched into the thin dielectric or conductive film whereas in the implantation area ions such as As, P, or B are implanted into wafers to form devices. After stripping the photoresists, the implanted wafers are loaded into a diffusion furnace to anneal out the crystal defects

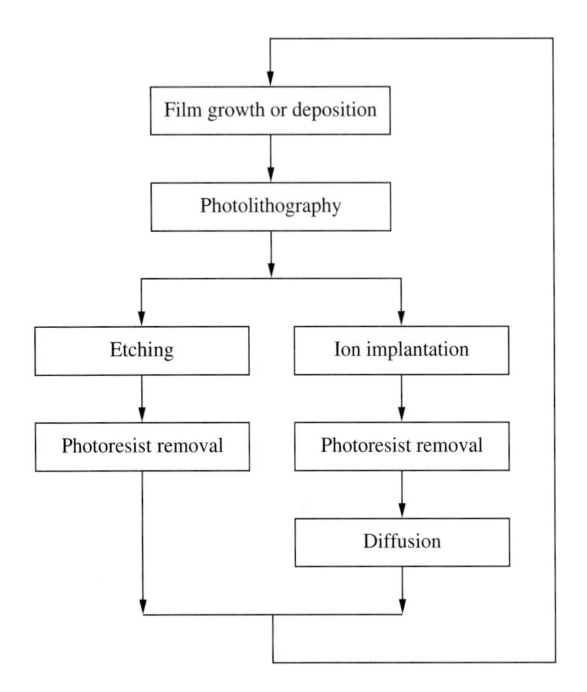

FIGURE 3
IC wafer manufacturing process module.

resulting from the ion implantation. The percentages of cleanroom space taken up by each process are summarized as follows:

Photolithography	25%
Diffusion and LPCVD	20%
Thin film	20%
Dry etching	15%
Implantation	10%
Wet process	10%

The following cleanroom layout principles are often used to optimize the IC production logistics.

Minimization of wafer transfer distance between each process area by putting photo, etching, and photo-resist removal close together.

Reduction of cross contamination by separating the back-end processes containing wafers exposed to metallic films from the front-end furnace and etching areas.

Flexibility for replacement and extension of process equipment.

Provision for interbay and intrabay automation and installation of equipment enclosures to encircle the wafer loading and unloading area in a highly classified environment (minienvironment).

Minimization of wafer transfer frequency between each process area by arranging process equipment in the following minimodule:

Furnace area:	Furnace, chemical station, and film measurement

Photo area:	Stepper, resist coater, developer, resist-baking hot plates, optical inspection, and SEM (scanning electron microscope)
Etching area:	Resist-baking hot plates, etcher, and optical inspection
Resist stripping area:	Resist asher, wet chemical station

An example of cleanroom layout is given in Fig. 4. The film growth and deposition areas are located upstream of the photo area, whereas the etching, ion implantation, and photo-resist strip areas are located downstream. All process areas are

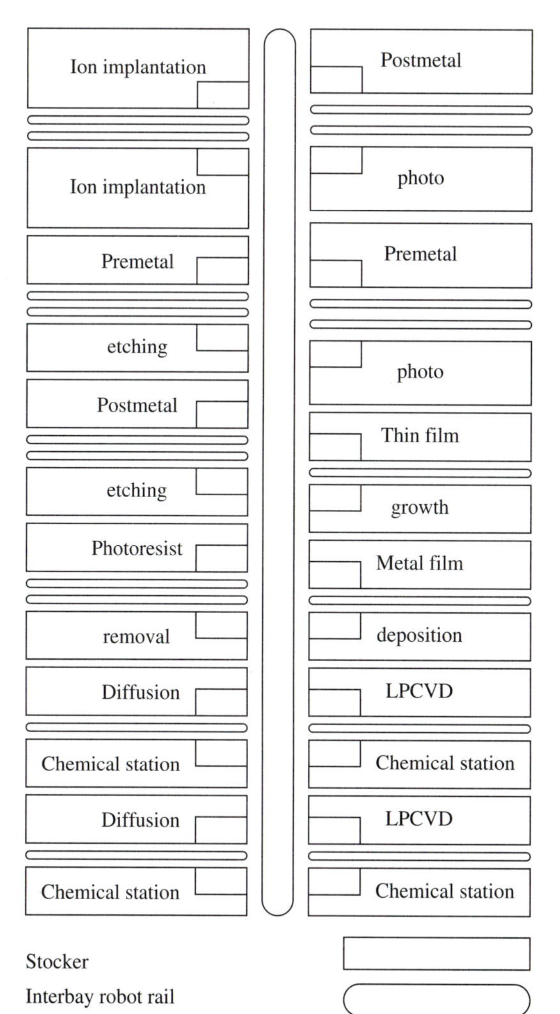

FIGURE 4
Cleanroom layout for a fully-automatic IC fab.

connected via an interbay wafer transfer system. Within each process area, wafers are retrieved from a stocker, which can store up to 150 wafer cassettes, and then transported to process equipment using intrabay automatically guided vehicles (AGVs).

1.3.2 Cleanroom Design[8–17]

Design can be developed on the basis of the formulated specifications and the design criteria. The prepared process layout is all-important. In principle, this layout is derived from the required process tools on the one hand and the logistic requirements on the other. In addition, the potential future developments within the process technologies have to be considered. This implies that the following trend must be taken into account.

In the 1970s, the manufacturing process required an environment that provided an overall cleanliness class 10,000 and a local class 100. The "ballroom" concept of one large room was created.

As a consequence of the developments of submicron technologies, a need for a highly classified area originated in the 1980s. This led to the introduction of the "tunnel" concept, in which a corridor separates the process area from the service area. To achieve the required air cleanliness, the majority of the equipment maintenance functions were accommodated in the low-classified service areas. At the same time, the costly high-classified process tunnels where the wafers were handled were reduced.

Today, the process technologies are focused on 16-Mbit and 64-Mbit DRAM products. The process environment conditions required by these technologies are so stringent that the enclosure of the process environment for each process tool is being considered, moving the process tool in the direction of "clean machines."

This requirement resulted in the creation of the "minienvironment" concept as shown in Fig. 5.[18–23] Within the enclosure of the minienvironment, an extremely high cleanliness class (0.1 at 0.1 μm) is realized, whereas the overall production area has a cleanliness class 1000. The wafer processing and loading/unloading sections of process equipment will be affected automatically by input/output devices. The transport of wafers outside the minienvironments is by means of SMIF pods (standard mechanical interfaces), a technique originally developed by Hewlett Packard.

In general, flexibility is a key issue within this framework. Modifications to the process layout or the process environment conditions should be easy to make at any time.

Another consideration in the initial design is the selection of the cleanroom system, since it is normative with respect to the building structure. The cleanroom principle is based on the isolation of the process environment from external influences using a laminar airstream. This airstream is filtered so that it contains only a few particles of a determined size.

In the current manufacturing environment, only 1 particle at 0.1 μm or larger per cubic foot air is regarded as the upper threshold value. In addition, the temperature, relative humidity, and speed of the airstream is controlled within very tight tolerances.

Side View

Class 1000 0.1 m/s

Class 1 0.4 m/s

Class 1000

Equipment enclosure

Stepper

Track

Robotic transfer system

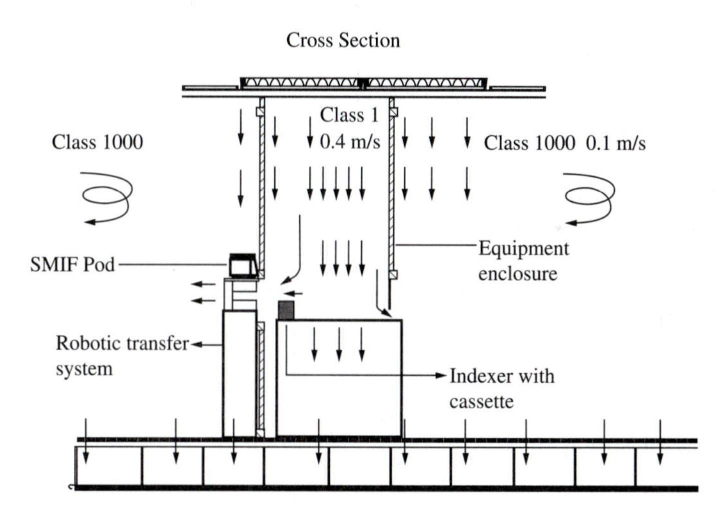

FIGURE 5
Creating an ultraclean minienvironment for a process tool using SMIF.

In the past, the common air speed was set at 0.45 m/sec. Currently, this value has been reduced to 0.35 m/sec based on experience. The reduced air speed offers a considerable cost reduction. In addition to the airflow, issues such as material selection, design details, and pressurization hierarchy are of great importance for the correct operation of the cleanroom.

A laminar airflow allows the creation of a cleanliness class 100, 10, 1, or even better. As a matter of course, the area in which such laminar airflow is established depends on the layout.

To guarantee a highly classified process environment, the conditions in the background (overall production area) have to be controlled. In general, a factor of 10 is adopted between the cleanliness class of the critical area (foreground) and the background. The cleanliness classes 100,000, 10,000, and 1000, up to 100, can be reached with a turbulent airflow. This means that air turbulence is acceptable, and, as a consequence, local air-supply diffusers can be applied and low air-speed values are permissible. The dividing line between laminar and turbulent flow, an important transition point with respect to the investment level, lies with cleanliness class 100. This is illustrated with the denomination of a turbulent or laminar class 100. Naturally, the generation of particles inside the cleanroom must be prevented as much as possible. For this reason, gowning procedures should be defined and implemented for persons entering the cleanroom and decontamination procedures must be followed for incoming material.

1.3.3 Cleanroom Concept

There are four basic concepts applicable to the cleanroom system.[24] All current designs can be derived from them. The first and least expensive concept is illustrated in Fig. 6. The supplied air is distributed in the room via a filter ceiling and is returned to the air-handling units via the lower sides of the walls. Both process and service functions are accommodated in one level. This approach has very low flexibility; layout modifications are difficult and, thus, relatively expensive. Consequently, this concept is seldom used. However, the introduction of filter fan units allows for a one-level concept with a certain degree of flexibility. The disadvantages with respect to the location of remote equipment and the connection to the utility distribution networks still remain.

The second, third, and fourth basic concepts depart from separated process and service levels. The process level is completely classified. Costs are reduced by accommodating only the primary process functions on this level. The service level has all piping, ductwork, and cabling systems, and it is also used for the installation of remote equipment. Moreover, the air will be returned via this level to the air-handling units, by which the process level receives a high cleanliness class without extra costs. All potential contamination sources should be enclosed and exhausted to prevent cross-contamination. All three concepts feature a high degree of flexibility. The difference between these concepts lies in the air-handling system, namely,

Option two, shown in Fig. 7, is based on the installation of centrifugal fan units on top of the process level. Air-return shafts are located at both sides of the building. An air-supply plenum is included between the fan units and the filter ceiling.

Option three, illustrated in Fig. 8, is based on the installations of axial fan units at both sides of the building. In this case the space between the roof and the filter ceiling serves as the air-supply plenum.

Option four, shown in Fig. 9, is based on the installation of filter fan units on top of the ceiling grid. Air-return shafts are located at both sides of the building. The space between the roof and the filter fan units serves as an air-supply

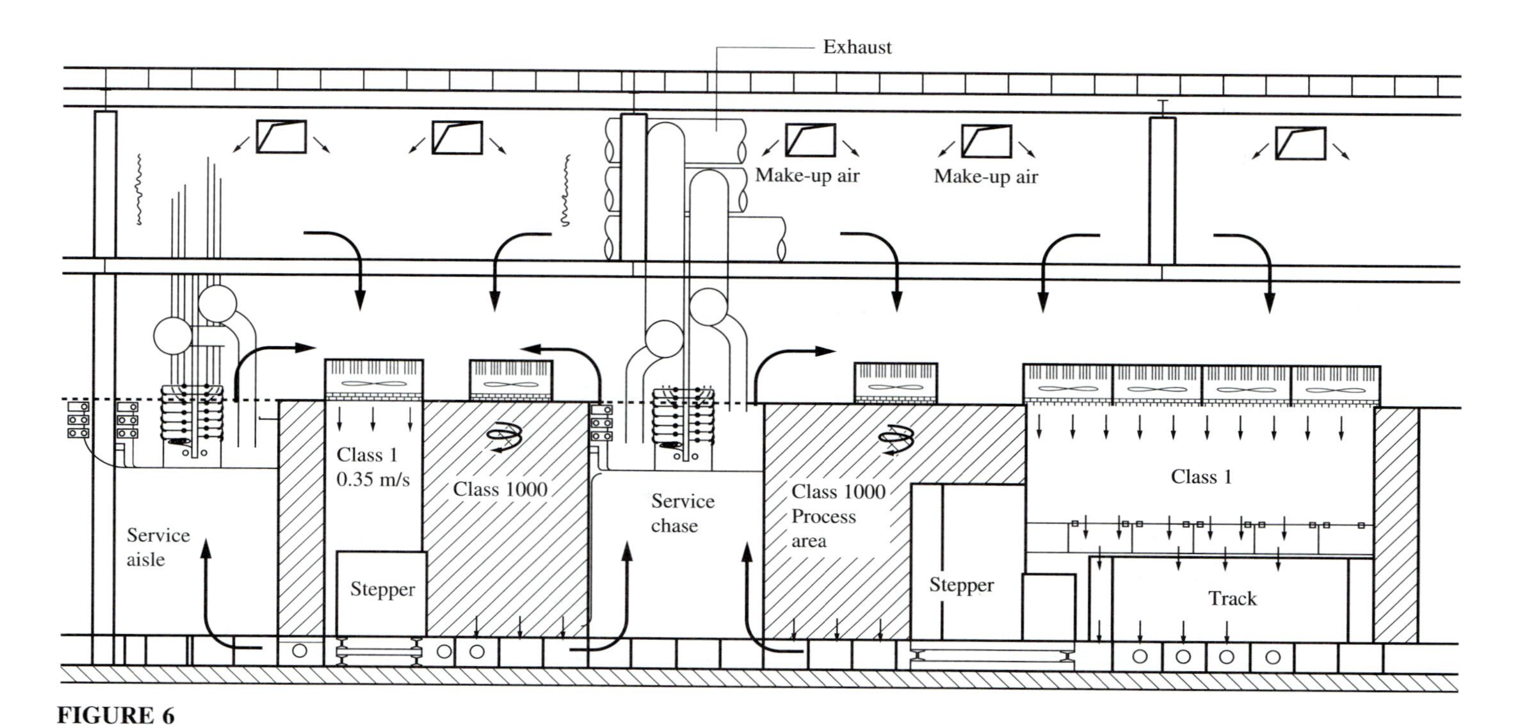

FIGURE 6
Ballroom-type cleanroom with process and service areas located on the same floor.

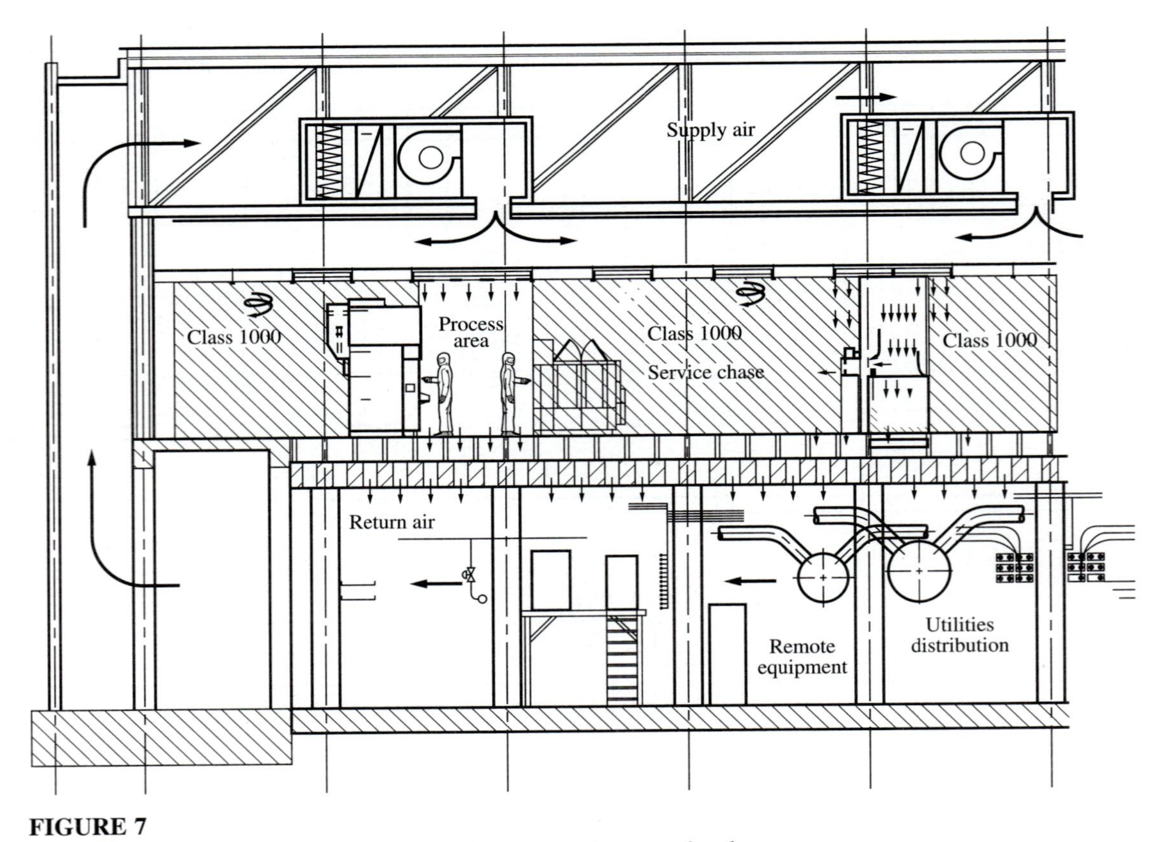

FIGURE 7
Cleanroom with centrifugal fan units installed on top of process level.

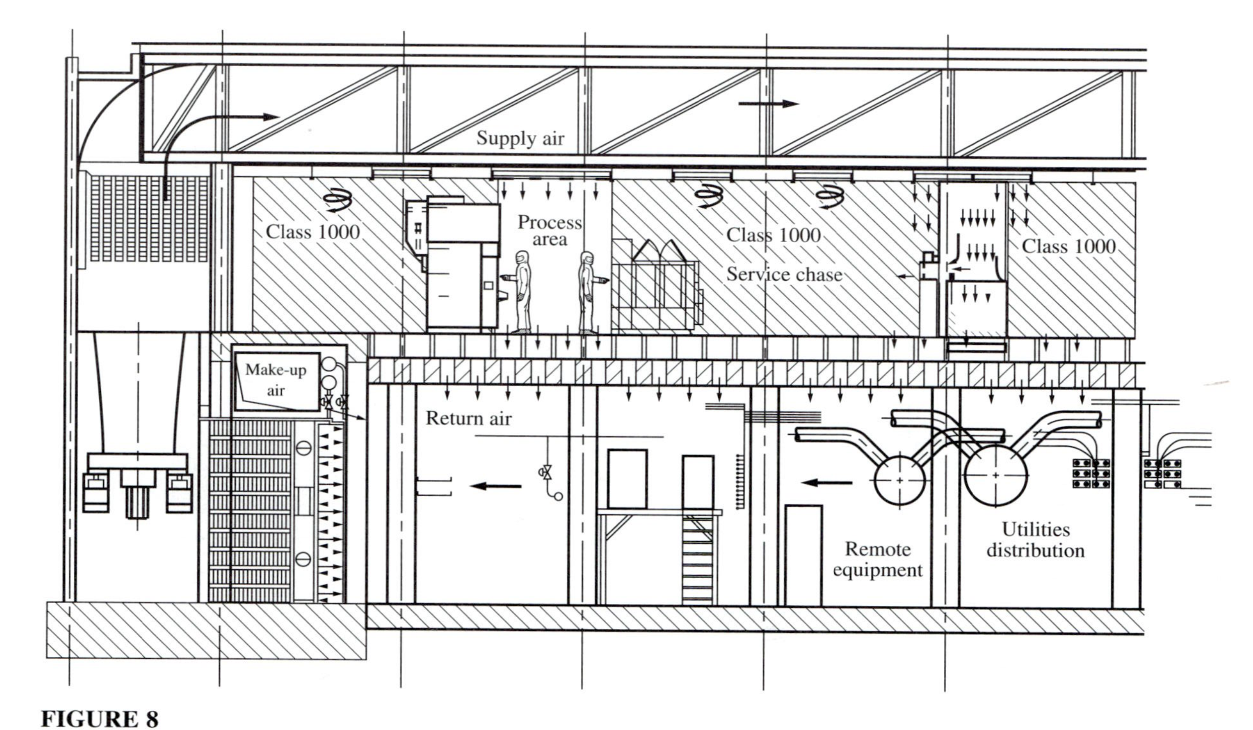

FIGURE 8
Cleanroom with axial fan units installed sideways connecting the air-supply plenum at the top and the air-return plenum at the bottom.

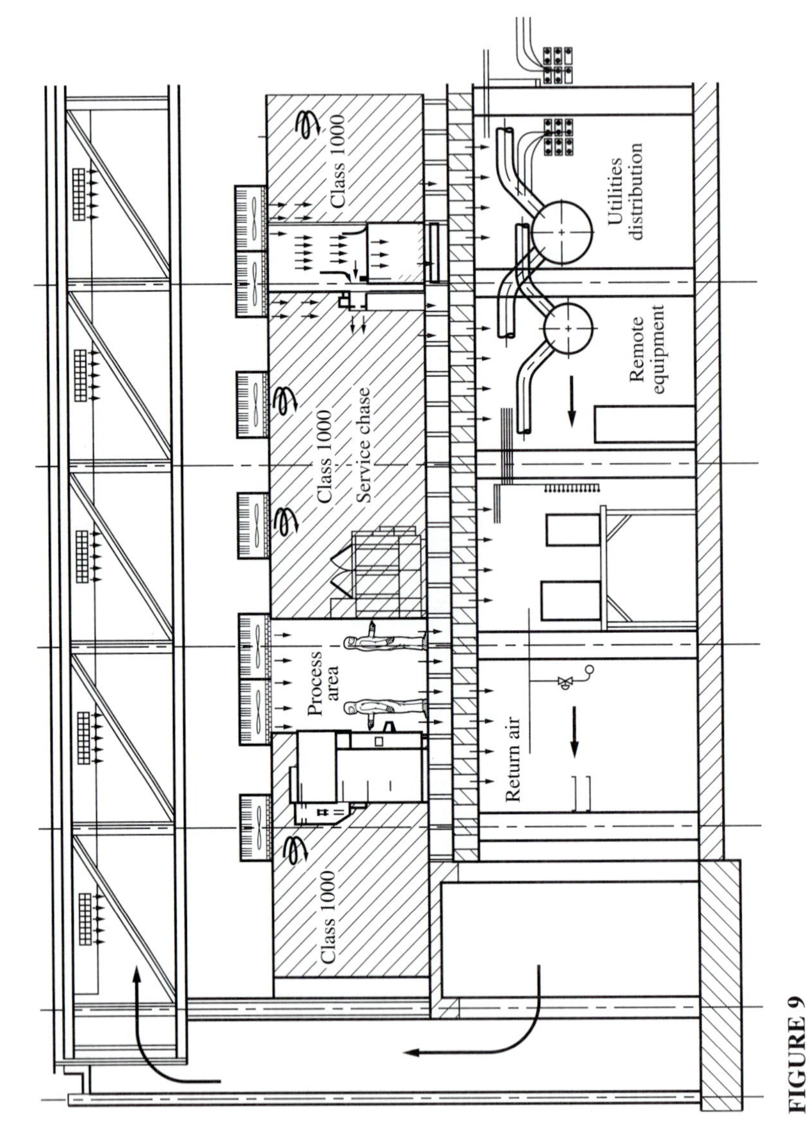

FIGURE 9
Cleanroom with filter fan units installed on the top of process area.

plenum. Since the air pressure inside the cleanroom is higher than that above the filter ceiling, airborne particles cannot infiltrate the cleanroom area. As a consequence, less expensive "dry" ceiling systems can be used. (Referred also to the filter ceiling system described in Section 1.4.1.)

1.3.4 Cleanroom Performance Criteria

Some common parameters[25,26] and values for the quality characteristics of a typical cleanroom are summarized in Table 6 and the details are elaborated in the following section.

Cleanliness class

The cleanest class (M-1) defined in the Fed. Std. 209E allows for

10 particles/m^3 or 0.28 particles/ft^3 at 0.5 μm
350 particles/m^3 or 9.91 particles/ft^3 at 0.12 μm

The actual technical goal is far beyond the definition of the Fed. Std. 209E. In practice, cleanliness class is often defined for the MPPS (most penetrating particle

TABLE 6
Principal cleanroom design criteria

Design parameters	Criteria
Cleanliness, pieces/ft^3	
Process area ≥ 0.12 μm	≤ 1
Service area ≥ 0.30 μm	≤ 1,000
Temperature control, °C	
Photolithography area	22 ± 0.1
Other cleanroom area	22 ± 0.5
Humidity control, %	
Photolithography area	43 ± 2
Other cleanroom area	43 ± 5
Air quality, ppb	
Total hydrocarbons (THC)	< 100
NO_x	< 0.5
SO_2	< 0.5
Cleanroom interior surface outgassing rate, torr-L/cm^2/sec	6.3×10^8
Cleanroom pressure, relative to outside ambient, Pa	> 30
Acoustical noise, dB	< 60
Vibration (8 to 100 Hz), μm/sec	< 3
Grounding resistance, Ω	< 1
Magnetic field variation, mG	< ±1
Charging voltage, V	< ± 50

size) of 0.10 μm to 0.12 μm. Figure 10[27] illustrates the filtration efficiencies of HEPA (high-efficiency particulate air) filters with respect to different particle sizes. Note that around 0.1 μm the filtration efficiency is the lowest.

The current cleanroom classification is, in general, defined as

Class 1	(at 0.12 μm) inside the process area
Class 1000	(at 0.3 μm) in the surrounding area

During cleanroom operation, in-line monitoring of particle density should be performed routinely. A practical goal is less than one particle/ft^3 with a size greater than 0.12 μm for critical process areas.

Temperature and humidity control

The recommended temperature control requirements are

metrology/lithography area	22 ± 0.1°C
other cleanroom area	22 ± 0.5°C

A relative humidity of 43% at 22°C can be readily achieved by using regular chilled-water systems with the chilled water inlet and outlet temperature controlled

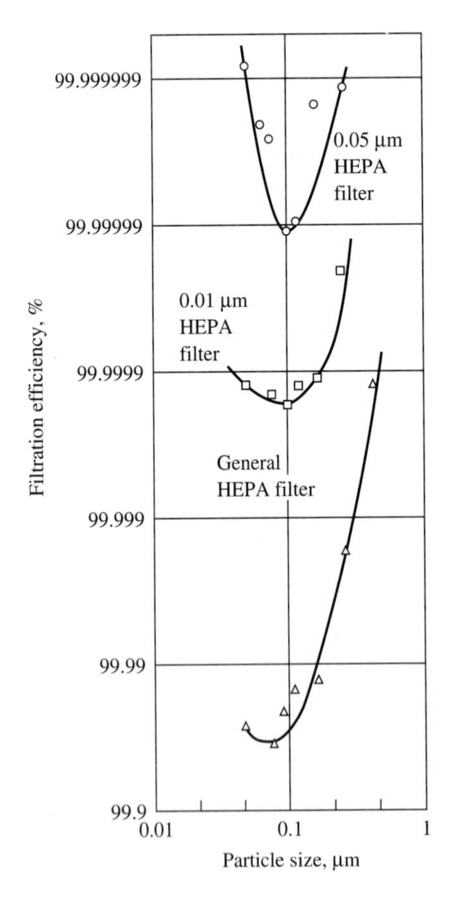

FIGURE 10
Filtration efficiencies of HEPA filters with respect to particle sizes.

at 5°C and 12°C, respectively. Tolerances of 2% at 22 ± 0.1°C are achievable via dew point control in the make-up air units. Typical relative humidity requirements are

metrology/lithography area	43% ± 2%
other cleanroom area	43% ± 5%

Air quality

Hydrocarbons comprise ethylene, benzene, methane, xylene, etc. The outside hydrocarbon contamination, typically in a range of 1.5 ppm to 50 ppm, can be reduced approximately 90% by use of active-carbon absorbers in the make-up air units. An approximate 98% reduction can be achieved with new active pleated filters. The inside hydrocarbon contamination can be reduced by the selection of low outgassing material finishes for the walls and floor. Chemically treated carbon composite filters located in the air recirculation can achieve sub-ppm THC (total hydrocarbon) results of 75 to 80 ppb.

Other external molecular contaminants are SO_2, SO_3, NO_2, halogen (CCl_4, BCl_3, Br_2, Cl_2), and sulfur compounds such as H_2S. The typical measured values of NO_x and SO_x in outside air, depending on local conditions and the location, are 5 to 100 and 15 to 50 ppb, respectively. The measured NO_x and SO_x levels inside a cleanroom using ion chromatography, in general, are less than 0.5 ppb.

Outgassing

The maximum outgassing rate of 6.3×10^8 torr-L/cm^2-sec is a criterion for the selection of all materials in contact with the airflow in the make-up air and the recirculation air system. Outgassing is the evolution of a volatile compound from cleanroom materials or process tools, which may condense on the wafer and create problems due to organic and ionic contamination.

Cleanroom pressurization

The cleanroom pressure should be positive (e.g., +30 Pa) relative to the outside pressure. Cleanroom pressure is automatically controlled by adjusting the make-up air volume via variable-frequency drives in the make-up air units.

The dynamic pressure (velocity pressure) of the air should be as low as possible to minimize pressure variation. The pressure variation and low-frequency sound waves of 0.2 to 8 Hz are caused by acceleration (air return) and deceleration (supply plenum) of airflow. Turbulent flow is reduced by eliminating sound attenuators. It is recommended that recirculation air velocity be kept below 2.5 m/sec.

Acoustical noise

The technical goal is to have a cleanroom in an "as-built" condition designed to meet the PNC (Pressure National Code) 60-dB sound levels as defined by the 1987 American Society of Heating Refrigerating and Air-Conditioning Engineers (ASHRAE) recommendation. A cleanroom in an "as-built" situation means no process tools are operating, but all fans and HVAC systems are running. The challenge is to meet PNC 60 sound levels without installing sound attenuators in the recirculation fan units.

Vibration isolation

Vibration criteria apply in both the vertical and horizontal directions at the top of the cleanroom waffle slab. Maximum vibration velocity should be no greater than 3 μm/sec, measured in 1/3-octave bands between 8 to 100 Hz. Below 8 Hz the criteria vary by 6 dB per octave. The criteria apply only at the photography and metrology area. Most steppers will perform with floor vibrations (waffle slab level) of 5 μm/sec, assuming that the criterion is met at the low end of the band at 8 and 10 Hz. Sensitive metrology equipment (such as the SEM) and defect measurement will determine vibration design criteria.

The entire clean production area should be provided with a waffle slab subdivided by joints. To meet the vibration limitations in the photolithography and metrology areas, the waffle slab is supported by 500 mm × 500 mm columns every 3.6 m. The column spacing in the other areas can be 7.2 m. The waffle slab typically has 500-mm-wide main ribs and 200-mm subribs with heights of 800 mm and 400 mm, respectively. The waffle grid is 600 mm × 1200 mm to match the raised floor and filter ceiling grids.

Grounding

Grounding is required for safety to protect people against a shock hazard. One overall grounding-grid system, buried approximately 400 mm below grade, should be installed. The grounding grids of the different buildings must be interconnected to form an equal potential grounding network.

The wye point (secondary side) of the transformers should be grounded via copper bars connected to the grounding grid. All major electrical equipment, such as power distribution panels and MCCs (motor control centers), must be grounded to the main grounding grid. Cable trays, pipe racks, metal tanks, handrails, supports, etc. should also be grounded.

A ground-grid bar is required for each power and distribution panel. All receptacles should be provided with a grounding pole. The grounding of the control system must be connected in a way that permits only a single path to the ground. The lightning-protection installation and the low-impedance grounding grid (see EMI) must also be connected to the grounding grid. The grounding-grid system should have a grounding resistance not exceeding 1Ω.

Electromagnetic interference (EMI)

Process tools are sensitive to electromagnetic interference reaching them through cables, pipes, or electrical and magnetic fields. At the same time, the tools generate conducted and radiated interference. Compatibility is achieved when the generated interference is kept below the sensitivity of process tools. The level of interference in the building is caused by external sources as well as internal sources. The effect of external sources can be limited by the building structure; the building should act as a shield against external sources of interference. This can be achieved by making the building a Faraday cage. Usually reinforced steel panels are installed over the roof and along the walls to provide the necessary shielding.

Doors and windows create leakage in the Faraday cage, so extra shielding is necessary. This can be achieved by creating other Faraday cages within our existing cage of the building. In this way, a number of zones can be provided to increase shielding.

The most important measure with respect to the floors, walls, and roofs is the use of planar grounds. A planar ground or low-impedance grounding grid should be installed under the raised floor in the clean production area.

Electrostatic discharge (ESD)

Static charges can develop on surfaces due to touching and rubbing (*triboelectricity*). The discharge of that electricity to machines and to people damages semiconductors, disturbs process tools, and may cause injury. ESD can be controlled by grounding all machines, by controlling the relative humidity, and by constructing walls and floor coverings of slightly conductive materials to route electrical charges to ground. The control of triboelectricity enhances dust and particulate contamination control. All metal racks, pipelines, cabinets, cable trays, and rails are grounded to an equal potential bar or planar ground. The metal pedestals of the raised floor connect to the planar ground under the raised floor. The metal framework of the cleanroom wall systems should be connected to this planar ground. To enhance ESD control, air ionization systems might be installed at certain locations in the process area.[28–33]

1.3.5 Principal Technical Data for Sizing the Cleanroom Facilities

Several key design parameters that are essential for sizing the cleanroom facilities of a 200-mm-wafer fab are summarized below. The dimensions of these parameters are all converted to units per cleanroom area.

Production capacity	85–120	mask × (wafer/month)/m^2
Exhaust	80	(m^3/hr)/m^2
Recirculation air	670	(m^3/hr)/m^2
Power for process equipment	800	W/m^2
Process cooling water	240	W/m^2
Heat load	450	W/m^2

The size of a cleanroom increases with process complexity and the photo mask level. Typically, the monthly production capacity of one square meter of cleanroom is around 85 to 120 mask steps, with DRAMs at the lower end of the range and ASICs (application-specific ICs) at the higher end.

The exhaust air rate is high; about 12% of the recirculation air is replaced by fresh make-up air. Approximately 65% of the exhaust rate is attributed to furnaces, chemical stations, and tube-cleaning equipment.

The average power requirement of process equipment is approximately 800 W/m^2. One-third of the heat generated is carried away by process-cooling water, and another third by the exhaust air. The rest of the heat dissipated by equipment, recirculating fans, and lighting is removed by the cooling coils installed in the recirculation air units.

The power consumptions of major facilities are listed below:

Facility	Consumption
Office building	6%
Deionized (DI) water system	3%
Process tool	30%
Testing equipment	10%
Utility equipment	15%
Support cleanroom	3%
Fab recirculation fans	7%
Boilers	8%
Chillers	18%

The power consumption of the office building is relatively small, whereas up to 40% of the power demand is used to run the process tools and testing equipment and the remaining 50% is used to support cleanroom operations, including pumps, recirculation fans, lighting, boilers, and chillers.

Other quantities of facilities required to complete the processing of a 200-mm, 16 Mbit DRAM wafer are as follows:

Chemicals	10 kg
Deionized water	4.5 ton
Compressed dry air	55 m^3
N_2	25 m^3
O_2	0.9 m^3
H_2	0.1 m^3
Power	470 Kwh

1.4 CLEANROOM INSTALLATION

The cleanroom installation section starts with an introduction of the basic information on the cleanroom subsystems followed by the discussion of the cleanroom installation approach. The investment cost required for constructing a semiconductor fab is analyzed last.

1.4.1 Cleanroom Subsystems[25, 34–36]

The air-handling principle of cleanrooms is based on the interrelated design of the process exhaust systems, the make-up air system, and the recirculation air system. Chilled water and hot water systems are required for temperature and relative humidity control. The filter ceiling system, the cleanroom partition system, and the raised floor system constitute the enclosing of a cleanroom area. The features of all these systems are described in the following paragraphs.

Process exhaust systems

The following main process exhaust systems can be identified:

General exhaust
Scrubbed exhaust
Solvent exhaust

The general exhaust and the solvent exhaust systems typically comprise ductworks, exhaust fans, bypass, and stacks. In addition, the scrubbed exhaust system includes scrubbers. The bypass allows entry of outside air if the pressure at the suction side of the operating fans exceeds a preset value.

The general exhaust system removes heat dissipated by the process equipment. This exhaust air should not contain acids, caustics, or solvents. The solvent exhaust system removes air containing solvents from the process equipment. The exhaust fans employed should be explosion-proof. The scrubbed exhaust system removes air containing acids and/or caustics from the process equipment. Ammonia, toxic, and silane exhaust systems are, in most cases, also connected to the scrubbed exhaust system upstream of the scrubber.

There are two types of scrubbers, namely, dry and wet scrubbers. A wet scrubber is an exhaust air treatment unit that removes the acids and caustics from the exhaust air by washing them out. The water used is typically city water. Waste water from the scrubber is routed to the neutralization plant of the waste treatment area.

A dry scrubber removes toxicants from the exhaust air by absorbing them into the scrubber material at a desired temperature. If the scrubber material is saturated, it has to be replaced.

Make-up air system

The make-up air system is designed to maintain the cleanroom pressure, to compensate for the air losses from building enclosure leaks, and to compensate for the process exhaust. Another function of the make-up air system is control of humidity in the cleanrooms, i.e., by humidification or dehumidification, depending on the process environment and ambient conditions. For exact adjustment of the make-up air volume with the exhaust rates, frequency convectors are incorporated in both process exhaust systems and the make-up air system.

A make-up air unit, illustrated in Fig. 11, withdraws outside air through an air intake grill and removes the majority of airborne particles via a prefilter. After passing through an air-cooling coil to remove excess moisture, the cold air is heated by an air heater prior to entering an air fan sandwiched between two sound attenuators. Fine particles are removed by a HEPA filter. The air is regulated by an air-volume controlling damper before it is fed into the make-up air ductwork, which distributes the make-up air to the suction side of the recirculation air units. During winter or dry season when the humidity is low, steam generated by a humidifier is added to the air stream to maintain the moisture at a designated level.

Recirculation air system

The recirculation air system serves three purposes:

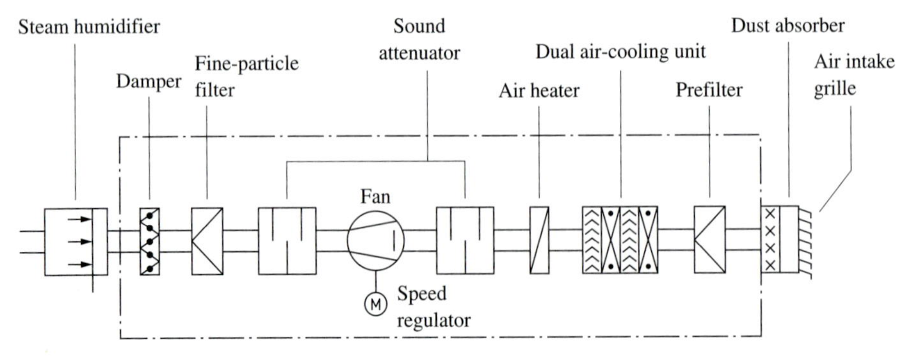

FIGURE 11
Schematic diagram of the make-up air unit.

Temperature control
Particle control
Air flow control

Depending on the cleanroom design principle, three types of recirculation air units can be used:

Filter fan units
Centrifugal fan units
Axial fan units

A filter fan unit consists of an enclosure with a fan and a final filter assembly. If required, the filter fan unit can be provided with a dry-cooling coil to control the temperature in the cleanroom area within very tight margins. Filter fan units can provide a high degree of flexibility; however, the cost for a given air-handling capacity is also the highest among the three types of recirculation units. The filter fan units are best suited for a small cleanroom without enough space to accommodate the big recirculation air units and the necessary ductwork.

A centrifugal fan unit recirculates air by using a centrifugal fan and removes particles and heat by passing air through air filters and a set of cooling coils. The entire air-handling unit is sandwiched between a pair of sound attenuators to cut down the sound level. The conditioned air is routed to the filters in the ceiling grid via either ductwork (more expensive solution) or a plenum.

In the case of an axial fan unit, the recirculating air is withdrawn through a series of components: a prefilter, a cooling coil, a sound attenuator, and an air-volume controlling damper, all located in the basement of a cleanroom for easy maintenance. The conditioned air is then routed through a vertical mounted axial fan and the second sound attenuator before it is distributed directly into the air supply plenum.

Both centrifugal and axial fan systems can be applied successfully within the current cleanroom performance criteria. Naturally, the applications of both systems have pros and cons when they are compared. This means that each application requires its dedicated measures. In general, the ultimate selection of centrifugal or axial fans is not determined by the comparison of advantages and disadvantages but by the chosen cleanroom concept.

The application of axial fans is most suitable when a fan bay concept is adopted. The selection of a cleanroom system with recirculation air units on top of the cleanroom leads to centrifugal fan systems and a separate air supply plenum. The key pros and cons of axial fans compared with centrifugal fans can be summarized as follows:

Higher investment level but lower running cost
Higher sound pressure levels at relatively high frequencies
More frequent and more comprehensive maintenance
Easier mounting due to compact sizes
Easier air flow adjustment
Directly driven by a motor instead of by particle-generating belts
Vibration generated at 19 to 30 Hz instead of below 17 Hz

Chilled water system[37]

The total cooling capacity of all make-up air units, recirculation air units, air coolers, ventilation units, and central and process utility systems is provided by a chilled water system. The dehumidification operation in the make-up air unit requires a 5 to 11°C temperature range to remove the excess moisture from the air (wet-cooling principle). On the other hand, the temperature of the cooling coils employed in the recirculation air units needs to be controlled above the dew point of 9°C to prevent any moisture from condensing so that the humidity inside the cleanroom can be maintained at a constant level (dry-cooling principle). By mixing the cooling water supply and return, a temperature of 14 to 18°C can be obtained.

A chiller consists of a condenser and an evaporator. In most cases, the chiller will be a water-cooled type. The required water cooling is provided by a cooling tower. Inside the cooling tower, the cooling water is sprayed downward to meet the uprising outside air drawn in by a fan mounted on top of the cooling tower. A small portion of water evaporates as the water travels through the cooling tower. The evaporation of water demands heat and, thus, heat is removed from the cooling water. The evaporated cooling water must be replaced.

In many cases, the cooling water required for process equipment will also be taken from the chilled water system using a heat exchanger. The primary side of this process cooling water system connects to the chilled water system, the secondary side is designed as an "open" system to keep the water pressure in the cooling water return lines as low as possible. For certain process tools such as sputters, the low return cooling water pressure is essential for preventing water from leaking into the process chamber.

In an open process cooling water system, the recirculating cooling water is returned to a holding tank opened to atmosphere. The level control sensors installed in the water holding tank allow a supply of deionized water to compensate for water losses due to evaporation. To avoid accumulation of particles in the systems, filter units are incorporated in the return lines.

Hot water system

The total heating capacity of all make-up air units, ventilation units, and certain process utility systems is provided by a hot water system. The hot boilers heat water to 90°C. The return temperature is, typically, 70°C.

Filter ceiling system

The filter ceiling system of the clean production area comprises ceiling grid framework, filters, blanks, and auxiliary steel structures fastened to the steel trusses or the concrete ceiling. The ceiling grid framework covers the entire clean production area and is suspended from the auxiliary steel structure by height-adjustable hanging rods to permit leveling. The grid is based on a 1200 × 600 mm system, i.e., on the size of the filters and blanks. The filter ceiling framework offers full flexibility with respect to the location of cleanroom partitions, filters, blanks, sprinkler heads, and lighting fixtures.

There are two kinds of filter ceiling systems, namely, the fluid filter ceiling system and the dry filter ceiling system. The designation of fluid or dry ceiling is related to the type of sealing. Figure 12 illustrates that the airtight sealing between filters and a fluid ceiling grid is provided by a self-healing, highly viscous, silicon-based sealing gel. In the case of a dry filter ceiling, the airtight sealing of filter assemblies and blanks onto the ceiling grid is provided by an elastic gasket. The filters and blanks are secured with fasteners.

The fluid ceiling system is typically applied in a cleanroom with an overpressure above the filter ceiling (air supply plenum) which must be kept absolutely airtight to prevent the unfiltered air from entering into the highly classified cleanroom. If the pressure inside the cleanroom area is higher than that above the filter ceiling, there is no risk of particulate infiltration, and a dry filter ceiling, which is not completely leaktight, can be applied. In the dry filter ceiling system, the pressurized air required for each filter is supplied either by a fan located right above the filter fan unit or by a flexible duct connected to each filter housing.

Although various manufacturers use different designations for filters, the names "HEPA" and "ULPA" (Ultra Low Penetration Air) are of the most frequent occurrence. In general, ULPA is used for filters having an efficiency of 99.9995% and more at particulate diameters > 0.12 μm. Filters with lower efficiencies are typically designated as HEPA.[38] For this reason, we recommend specifying a type of filter stating the efficiency and particle size. Details of HEPA filter construction are illustrated in Fig. 13.

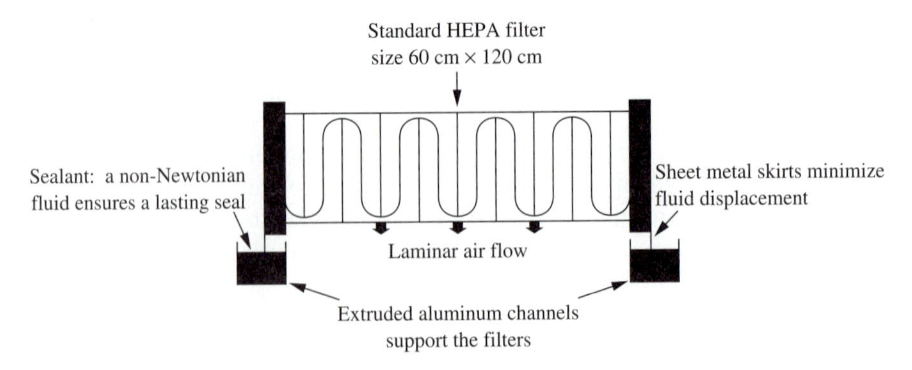

FIGURE 12
Positive fluid-sealing method for mounting HEPA filters.

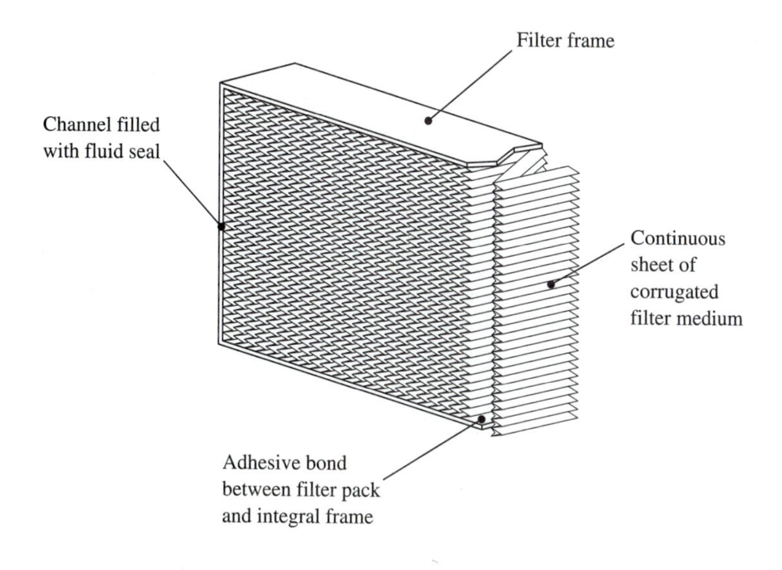

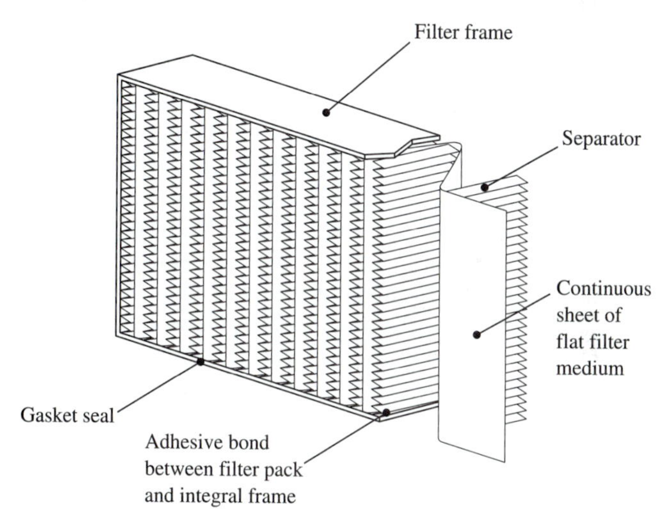

FIGURE 13
Details of HEPA filter construction.

The selection of filters depends on the level of cleanliness required. The cleanliness level in a cleanroom is controlled not only by the grade of air filter used but also by the amount of dust generated and the air flow pattern in the cleanroom. Therefore, filters of the same grade may result in different actual performance.

Cleanroom partition system

Partition walls separate the process area from the service area in the cleanroom. The entire cleanroom partition system should be grounded to meet ESD requirements. A typical cleanroom partition system consists of a framework, single-walled or double-walled panels, glass panels, and door units. The framework is constructed of extruded aluminum. The wall panels are typically metal sheet coated with a conductive paint. The glass panels are made of conductive clear glass or laminated plastic complete with associated sealings. The door units with glass have gaskets around the door posts and at floor side. The surface of the metal panels is smooth, cleanable, impact resistant, and does not generate or shed particles.

The partition wall system is constructed with flexible connections to minimize the transmission of vibrations between the filter ceiling grid and the partitions. All panels are easily dismountable and interchangeable to ensure optimum flexibility with respect to layout changes.

The cleanroom partition system has a total height equal to the distance between the raised floor and the filter ceiling (3.4 to 3.6 meters) and is 50 to 60 mm thick depending on the ceiling grid framework width. The cleanroom partition system, including all its components, must meet the isolation resistance requirement of 10^6 to $10^8 \Omega$.

Raised floor system

The raised floor should be installed 450 to 600 mm above the finished concrete waffle slab, and in general cover the entire clean production area. The grid of the raised floor is based on a 600×600 mm system and should be aligned with the centerlines of the filter ceiling grid. Some of the floor tiles should be perforated. The adjustment of the air-pressure drop and the balancing of the air flow in the clean production area can be achieved by selecting the floor tiles with proper perforating ratio.

The floor tiles should be static-dissipative and made of noncombustible vinyl that is chemical- and abrasion-resistant, impact-resistant, and smooth and meets the ESD isolation resistance requirement of approximately $10^6\ \Omega$.

The raised floor should be laterally stable in all directions with or without tiles present. Floor tiles are supported at each corner by height-adjustable pedestals. These pedestals are glued to or bolted on the finished concrete waffle slab. An insulation plate, placed on top of each pedestal, attenuates footstep sound and ensures electrical conductivity. To avoid horizontal vibrations, some of the pedestals are reinforced with bracing to increase the rigidity.

1.4.2 Construction Project Approach and Schedule

Cost effectiveness is the key starting point in the design of a semiconductor plant. This means that the final result in terms of quality, cost, and timing must be ensured during the entire course of the project.

The baseline-management method offers an excellent guideline. According to this method, the entire project is subdivided into clearly identifiable phases, from broad to detailed, and the progress in each phase is recorded. This provides a total picture of the project at any time to help project managers to make critical decisions.

A project can be subdivided into the following main phases: the initial design, the tendering, the elaboration (preparation for the execution, construction, documents, etc.), the realization, and the process-tool hookup.

The initial design constitutes the first and most important project step. During this phase, all starting points will be translated into an integrated building and services concept with the associated specifications, cost, and time indicated. The target project result is, thus, defined clearly at the beginning. All subsequent phases are a further elaboration of these initial concepts.

In general, the construction of an IC fab, from the start of the design activities to the first wafer out, extends over approximately two years. Design of facility systems and tendering of the relevant contracts takes 4 to 6 months. The construction of buildings and installation of all the utility systems take another 12 to 14 months. The final 4 to 6 months will be required to certify the facility systems, to install process equipment, and to test or tune processes.

The enclosed time schedule contains the timing of each project phase, subdivided into its essential parts. The development of site and buildings must receive the appropriate priority in the initial phase. When the fab-module shell is closed and the interior thoroughly cleaned, the installation of the cleanroom systems can start. In addition, the site infrastructure (roads) must be partly completed at a very early stage to reduce dust and to provide access to the fab.

The time schedule can be summarized by the following milestones:

Event	Cumulative months
Start initial design	0
Release conceptual design of the site/buildings	2
Release initial design document	3
Release preliminary design of the site/buildings	4
Receive approved building permit	6
Award contract for site/buildings	6
Release preliminary design of all facilities	6
Break ground	6
Start construction of buildings	7
Award contract for cleanroom/utilities	8
Award contract for mechanical/electrical work	8
Start cleanroom/utilities	15
Complete installation of cleanroom/utilities	20
Start equipment move-in/hookup	20
Complete testing/certification	21
Complete process-tool hookup	23
Complete functional testing	24
Start-up	24

1.4.3 Cost Analysis

After the specifications, the design criteria, the layout, and the conceptual designs of all subsystems are decided, the costs and time schedule can be determined accurately. Both are highly dependent on the dimensions and quality level. It is, therefore, incorrect to apply general characteristic values, e.g., the investment sum per square meters of cleanroom. Such standard figures are only applicable within a clearly defined frame of reference, e.g., IC fabs with a clean production area of some 3000 square meters projected for submicrometer technologies.

Approximately 75% of the investment is attributed to process equipment, whereas the remaining 25% is required for all the facilities including process utilities. The following table outlines the cost of each facility system:

Building	25%
Cleanroom systems	25%
Central utilities	10%
Process utilities	35%
Others	5%

A few years ago, the cleanroom systems constituted the most expensive item. To date the cost of the cleanroom systems has decreased mainly because of better knowledge, whereas the cost of process systems in the last years has increased substantially due to more stringent purity requirements and technology developments. Except for the building, the cleanroom, which houses process tools and test equipment, is the most expensive item. The average life span for the process equipment, the facility system, and the building is estimated at 5, 10, and 20 years, respectively. The depreciation costs can be calculated as follows:

Process equipment	88%
Facility system	10%
Building	2%

It is obvious that any significant reduction of facility costs, which may ultimately impair the yield of the wafers, has little effect on the manufacturing costs. It is, however, important to pursue realistic and soundly based expectations for quality and quantity and to exclude any unrealistic goals and overkilling, in order to make the facility systems cost-effective.

The facility systems should be designed to support at least three generations of process technology without major renovation. The following design principles are recommended:

The quality of cleanroom, process chemicals, DI water, and process gases should be of state-of-the-art.

The facility system should be highly flexibile with respect to future modification, tool change, and automation.

The facility system should be capable of continuous operation while the facilities are undergoing expansion or modification.

1.5 CLEANROOM OPERATIONS

Preventive maintenance of all the electrical and mechanical components is absolutely necessary to prevent unscheduled shutdown in the cleanroom. Continuous monitoring of particles, temperature, and humidity inside the cleanroom is required to alert process engineers to changes in the cleanroom environment so that steps may be taken to prevent temperature- and humidity-sensitive processes from drifting out of control. Proper gowning procedure and cleanroom maintenance practices are critical to prevent microcontamination.

1.5.1 Cleanroom Gowning

People are major sources of cleanroom contaminants. Not only do they generate a large number of contaminants but they are also in close proximity to the wafers at many stages of the IC manufacturing process. Therefore, an appropriate gowning procedure is necessary to minimize the exposure of hair, bare skin, and street clothes.

It has become a common practice for many IC manufacturers to require its personnel to change from street clothes and street shoes into company-provided clean undergowning and shoes before entering a cleanroom gowning area. In the gowning area, personnel are obliged to cover hair, nose, mouth, and even the eyes with a hair net, a face mask, and goggles. Gloves, a hood with an integral face mask, a cleanroom bunny suit, and booties are then worn to achieve as complete body coverage as possible.[39] Good cleanroom suit material is normally made of woven fabrics consisting of long synthetic fibers covered with a layer of PTFE (polytetrafluoroethylene) laminate material. This semiporous PTFE material can prevent particles from passing through but at the same time allow a high degree of vapor transmission.[40] The cleanroom suits and shoes should be washed regularly using deionized water and sodium-free detergent. The in-house or externally contracted laundry service should take appropriate measures to limit contamination while washing, packaging, transporting, and storing cleanroom garments.[41] All cleanroom clothing, including gloves and boots, should be electrically conductive to minimize the accumulation of electrical charges. Before entering a process area, a hand wash using DI water to remove particles from gloves may be necessary to prevent contamination.

1.5.2 Cleanroom Maintenance

Proper cleanroom maintenance is critical to minimize microcontamination. Some common industrial cleanroom maintenance practices are listed below.[42–47]

1. All personnel working in the cleanroom should receive special training in gowning, wafer handling, cleaning, and safety procedures.
2. Makeup and body powder is prohibited in the cleanroom.
3. Wool, fur collars, and jewelry should never be worn inside the cleanroom.
4. Pencils, clip ballpoint pens, noncleanroom paper, woods, aerosol spray cans, and boxes made of cardboard are prohibited in cleanroom areas.
5. Documents or photos that cannot be reproduced on cleanroom paper have to be laminated in a conductive PP (polypropylene) film before they are brought into the cleanroom.
6. Parts, boxes containing wafers, chemical bottles, and equipment should be thoroughly cleaned in a classified decontamination room adjacent to the cleanroom before they are brought into the cleanroom.

7. Ceiling grids and cleanroom walls should be mopped with DI water regularly. The floor should be washed with DI water several times each day. On average, one person is needed to clean each 1000 square meters of cleanroom.
8. Free-standing shield panels should be set up around process equipment during installation and maintenance to contain as much as possible any contamination that may be generated.
9. After equipment maintenance, dust or debris should be cleaned with a central vacuum cleaning system or a portable vacuum cleaner provided with a HEPA filter.
10. Maintenance hand tools should be cleaned in an ultrasonic bath to remove particles attached to their surfaces.
11. After they are detached from equipment, dirty parts should be covered before removal from the cleanroom.
12. Wafer cassettes or wafer boxes should be hand-carried one at a time. Movement of wafer cassettes by cart causes potential contamination problems due to vibration.
13. Pumps and equipment that have potential to generate contamination should be installed in a cabinet provided with an exhaust pipe connected to the general exhaust system.

1.6 AUTOMATION

With the ever-increasing process complexity, a process technology comprising some 20 mask levels and between 300 to 400 recipes becomes quite common. The number of WIPs (wafers in process) inside an IC fab routinely exceeds 30,000 pieces. It becomes critically important to implement automation to resolve the common IC manufacturing problems listed in Table 7.

TABLE 7
Common IC manufacturing problems and relevant solutions

Problem	Suggested actions
Wrong lot went to the equipment	Online lot validation
Operator cannot find cassette required	Automated stoker control
Operator selects wrong recipe	Real-time recipe download
Errors in manually recorded data	Data upload from equipment
Lot tracking not performed promptly	Real-time lot tracking
Equipment not correctly monitored	Real-time equipment monitoring
Schedule does not reflect fab condition	Real-time scheduling (just in time)
Inefficient recipe/bay setup	Recipe scheduling
Equipment capacity not fully utilized	Automation
Operators fail to report problem	Real-time process data management

Three levels of automation can be distinguished:[48,49]

1. *Shop-floor-control software.*[50] A shop-floor-control software resident in a computer is used to assist the operations to manually control process equipment and transfer wafer cassettes between stations.
2. *Computer Integrated Manufacturing (CIM).*[51] Automatic control of process equipment but manual transfer of wafer cassettes by operators.
3. *Mechanization.*[52–55] Automatic control of equipment and transfer of wafer cassettes directly to process tools by robots.

1.6.1 Shop-Floor-Control Software

The shop-floor-control software developed either in-house or purchased from outside vendors has been used extensively by almost every manufacturer to replace process run cards on which process steps and recipes were listed. This software usually has the following features:

1. Recipe management
2. Lot management

    WIP control
    Line balancing
    Exception lot handling (holds/releases, splits/merges, scraps)

3. Scheduling and dispatching

    Material priority handling and coordination
    Just-in-time scheduling and dispatching
    Dynamic rescheduling based on changing conditions

4. Data collection

    Equipment status
    Measurement data

5. Cost analysis
6. Analysis and reporting

    Lot history
    SPC (statistical process control) chart
    Production report

A good shop-floor-control program can significantly improve cycle time and reduce WIP by optimizing lot batching, scheduling, and dispatching.

1.6.2 Computer Integrated Manufacturing

To reduce human-induced errors and enhance the equipment productivity, many fabs built in the late 1980s started to link process equipment to a host computer. The following equipment control features have been implemented:

Bar code reader
Equipment setup
Recipe download
Process data acquisition and upload
Equipment state and status
Equipment alarm management
Equipment maintenance support

The semiautomatic fab operation mode can improve equipment utilization rate and fab yield considerably. Real-time equipment monitoring can minimize equipment idle time and provide early warning of potential equipment failures and problems. Automatic recipe download can improve wafer yield by eliminating recipe errors. It can also reduce equipment downtime resulting from incorrect operations.

1.6.3 Mechanization

The interbay automation system has been adopted by many IC manufacturers to automatically transfer and track wafer cassettes. Automatic stokers located in each process tunnel are linked by an overhead clean rail system. Wafers are transferred automatically from one process tunnel to the other on a vehicle driven by a linear motor. An interbay automation system can avoid confusion and save considerable time in tracking a wafer within a fab.

Intrabay automation is the last step to achieve a fully automatic operation mode. An AGV (automatic guided vehicle), shown in Fig. 14, is used to transfer wafer cassettes automatically from a stocker directly to process equipment. Intrabay automation can reduce the number of operators required. It can also significantly reduce wafer scratches and breakage by eliminating manual handling of the wafers.

However, the intrabay automation system has not been very popular outside Japan due to relatively high installation and maintenance costs. In the 300-mm wafer fab of the future, the wafer cassette might be so heavy that the intrabay automation system becomes absolutely necessary.

1.6.4 SECS Equipment Interface

The establishment of communication between a host computer and process equipment for downloading the processes recipe and monitoring of equipment status is a necessary step for factory automation. Programming the communication software, in general, follows the SECS (Semiconductor Equipment Communication Standard) protocol published by SEMI (Semiconductor Equipment Manufacturing Institute).

The SECS standard allows manufacturers to produce equipment that can be linked to any host computer without specific knowledge of the computer. The standard also allows host system designers to program a host computer and to plan a computer network without specific knowledge of the equipment functions. In addition, system designers can use the standard to connect different pieces of semiconductor equipment to each other.

FIGURE 14
Automatic wafer transfer to process tool using AGV. (*Courtesy of Murata of Japan.*)

There are two different sets of SECS standards: SECS-I and SECS-II. The SECS-I standard describes the physical connections, signal levels, data rate, and logical protocols used to exchange messages between a host computer and process equipment. The SECS-II standard defines the contents of the messages passed between the host computer and the process equipment.

The SECS-I stipulates that equipment should be connected to the host computer by a serial communication cable with an RS-232C connector. Through the RS-232C cable, data signals can be transmitted over a distance of more than 15 meters.

An SECS-II message consists of 13 to 257 bytes of 8-bit data. Every byte of data represents a character or a single-digit number. Each message block begins with a single byte giving the total length of the message block, followed by a 10-byte message header. The message header contains the ID number of a particular piece of equipment with which the host computer intends to establish communication. It also identifies the type of message data, such as an alarm code, equipment status code, etc. Up to a maximum of 244 bytes of message data follows immediately after the message header. Every SECS-II message ends with a 2-byte checksum.

The checksum computed by the host computer adds the binary value of all the preceding data bytes in an SECS-II message. After receiving the incoming SECS-II message, the equipment repeats the computation. If the data transmission is noise-free, the checksum computed by the equipment should match exactly that sent by the host computer.

1.6.5 Cost Analysis

For every $100 in equipment investment, an additional amount of approximately $15 is needed to implement a fully automatic system in a 200-mm fab. A cost breakdown of each level of automation is illustrated below:

Shop-floor-control software and hardware	10%
Interbay automation	20%
Semiautomatic process equipment interfacing	30%
Intrabay automation	40%

When an IC fab is operated in a fully automatic mode, the fab utilization rate can be increased up to 10%, the fab yield can be increased about 5% and the number of operators can be reduced by 75%. Therefore, the additional investment in fab automation is justified economically.

1.7 RELATED FACILITY SYSTEMS

Apart from the cleanroom subsystems described in Section 1.4, other process utility systems can significantly affect the wafer yield. The quality characteristics and the evolution of the specifications and the features are discussed in the following subsections.

1.7.1 DI Water[56–63]

Large quantities of DI (deionized) water are required to process IC wafers. The consumption of DI water increases with the wafer size. The DI water and chemical tanks of the earlier-generation 200-mm-wafer wet-benches are twice as large as those of a 150-mm-wafer wet-bench. As a result, the DI water and chemical consumption of a 200-mm-wafer IC fab are also doubled.

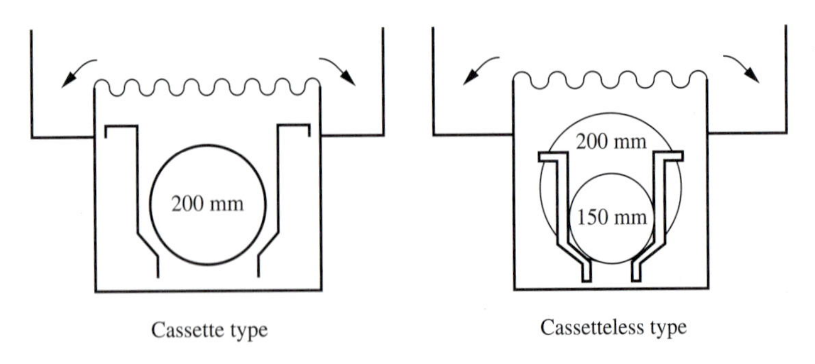

FIGURE 15
Comparison of the DI water tanks of cassette and cassetteless chemical stations.

Since the early 1990s, the cassetteless wet-benches have been developed. In these wet-benches, wafers are unloaded from wafer cassettes to a quartz boat, which is subsequently picked up by a robot arm. The robot arm dips the boat into a series of chemical and DI water tanks in a way similar to a conventional wet-bench. Figure 15 illustrates that the sizes of the DI water tanks of a 200-mm-wafer cassetteless wet-bench are about the same as a conventional 150-mm-wafer cassette type wet-bench. Table 8 also confirms that the consumption of DI water and chemicals of a 200-mm-wafer cassetteless wet-bench is close to that of a conventional 150-mm-wafer wet-bench. For each 200-mm, 16-Mbit, DRAM wafer produced, approximately 4.5 tons of DI water are required.

A state-of-the-art DI-water installation consists of two loops:

The make-up loop removes particles and colloids, total organic carbons, microorganisms, ionic impurities, and total dissolved solids from the raw water. In other words, it has a pretreatment section for removal of gross particles of sizes larger than 1 μm and a purifier section for removal of ionic impurities, bacteria, and dissolved gases.

The polishing loop removes the last traces of contaminants that originate in the construction materials or come from secondary microbial contamination.

There are many contaminants in raw water, such as particles, organic material, inorganic material, microorganisms, bacteria, and dissolved gases. As the raw water is fed through the DI water system, all of these contaminants are removed sequentially by a series of different types of filters, a high-vacuum degasifier, and an ion-exchange unit. Figure 16 shows the filtration spectrum of different types of filters employed in a DI water system.

The operating principles of multimedia filters, percoated filters, and microfilters in the make-up loop are similar. The filters let raw water pass through and trap the suspended solids and colloids in the filter media. When the pressure drop across the filter reaches a predefined threshold value, the dirty filtration media is regenerated

TABLE 8

Comparison between a 150-mm-wafer chemical station and a 200-mm-wafer chemical station

	Wafer Size			
	150 mm	200 mm		
Wafer transfer	Cassette	Cassette	Cassette	Cassetteless
Number of cassettes	Double	Single	Double	Double
Chemical tank, L	27.6	29.5	57.5	25
DI water tank, L	19	18	45	20
Particle				
DI water batch				
Density, pieces/L	30	30	30	10
Size, μm	0.3	0.3	0.3	0.16
Spin dry				
Density, pieces/L	10	10	10	10
Size, μm	0.3	0.3	0.3	0.16

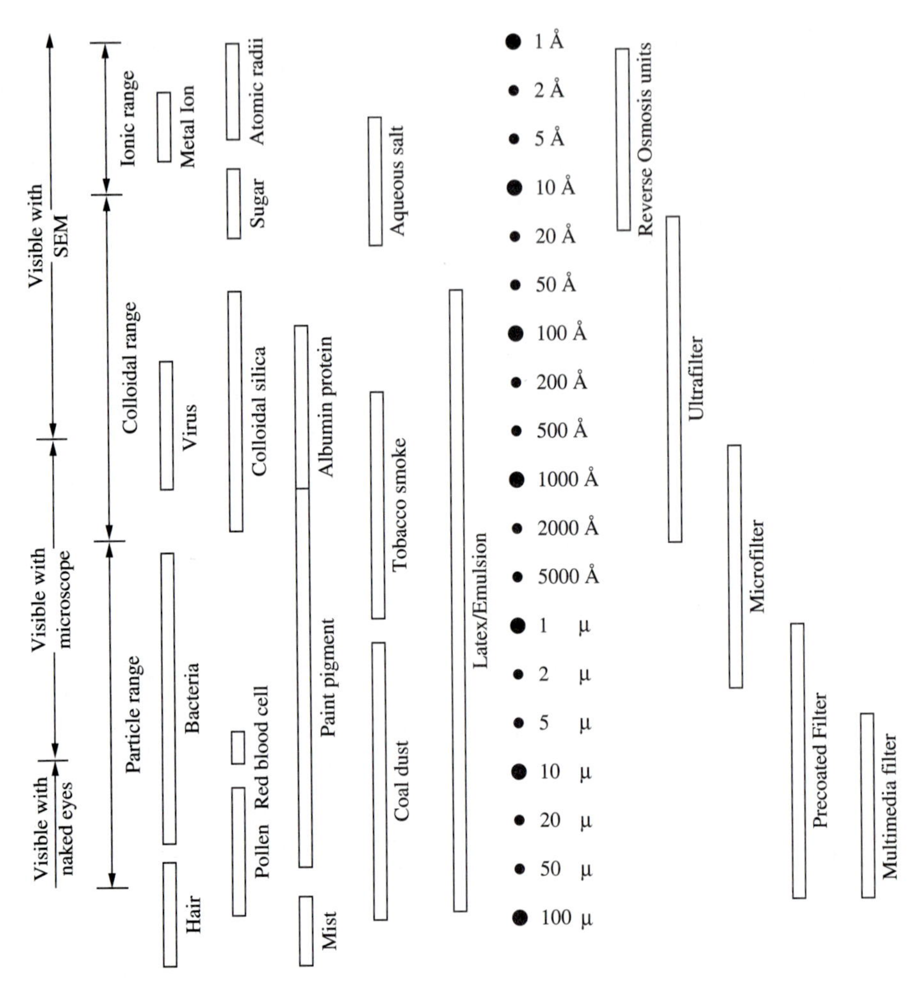

FIGURE 16
Filtration spectrum of different types of filters. (*Courtesy of Christ AG. Switzerland.*)

through a back-washing operation. Chemicals such as $FeCl_3$ are often injected at the filter inlet to enhance the agglomeration of suspended solids and thereby improve the filtration efficiency.

RO (reverse osmosis) units are used to remove smaller particles and metallic ions. The operating principle of RO units is illustrated in Fig. 17. As water enters the spiral-wound modules, water molecules diffuse through the porous membrane but microparticles, microorganisms, and ions are held back by the membrane. The rejected impurities are drained in the concentrated filtration effluent, and very few impurities are held back inside the membranes of the RO units. Conversely, multimedia filters and microfilters keep the captured particles in the filter media. RO units are regenerated by injecting RO permeate water on their feed side to wash away the sediments attached to the porous membrane.

Ultrafilters are used to remove submicrometer-size particles. During normal filtration mode, water is fed to the outside of a band of hollow fibers. As the water

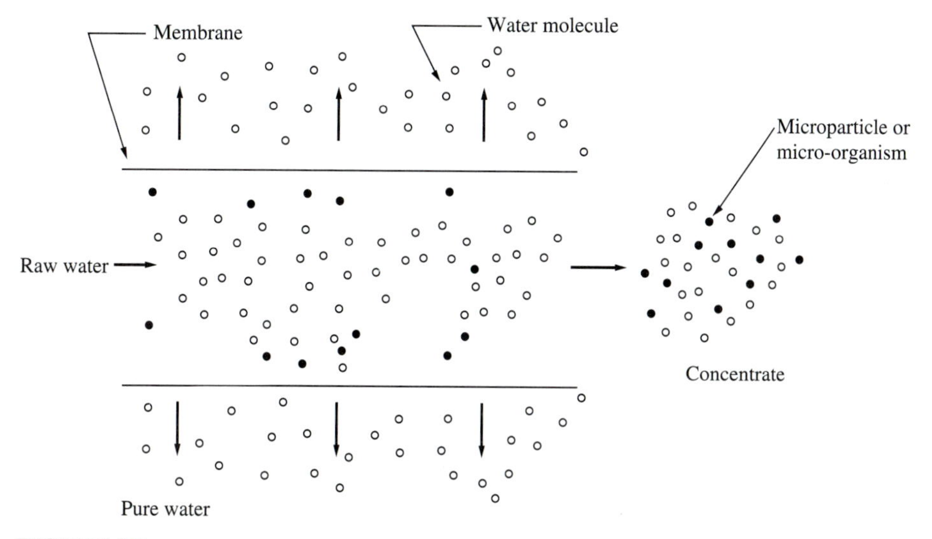

FIGURE 17
Operation principle of reverse osmosis filters.

penetrates through the porous fibers, particles are left behind on the exterior surface. During regeneration, the particles are washed away by the purified water fed in the reverse direction. A cross section of a hollow ultrafiltration fiber and a schematic diagram of an ultrafiltration cartridge are illustrated in Fig. 18.

The majority of TOC (total organic carbons) is removed in the make-up loop via the reverse osmosis units and the remaining portion of TOC is removed in the polishing loop by first ionizing it in the UV lamp modules and later capturing the ionized TOC in the mixed-bed polishing cartridges.

In the polishing loop, the DI water will be continuously circulated at a velocity greater than 1.5 m/sec from the process area back to the DI water installation to maintain low levels of contamination. All piping behind the polishing cartridge is constructed of PVDF (polyvinylidene fluoride) to minimize the release of particles and TOC.

PEEK (polyether ether ketone), a plastic tubing, and GOLDEP, an electropolished stainless steel piping passivated by Cr_2O_3 film, are reported to have much lower TOC- and metallic-ion dissolving characteristics compared to PVDF, and they have also been used in the polishing loop.

The evolution of DI water system design, shown in Fig. 19, can be characterized by the introduction of new techniques and materials.

For 256K DRAM devices, the DI water system includes the following features:

Pretreatment filters for removing particles in the micrometer range
RO units for removing ions, TOC, and particles
A vacuum degasifier for removing dissolving gases
Use of 254 nm UV lamps for bacterial sterilization
Use of mixed-bed ion exchange unit for removing residual ions
Use of membrane filters to capture particles

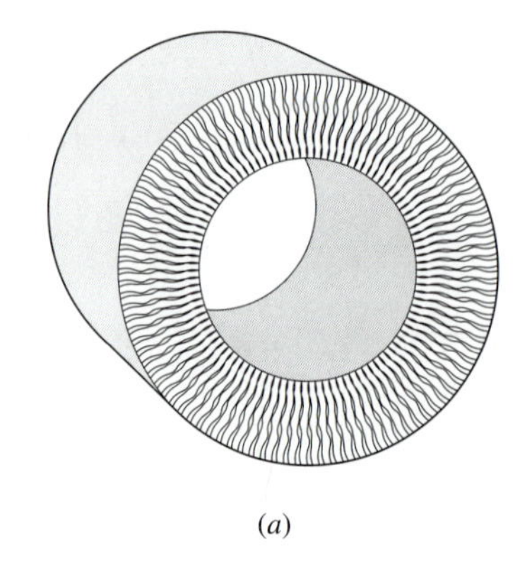

(*a*)

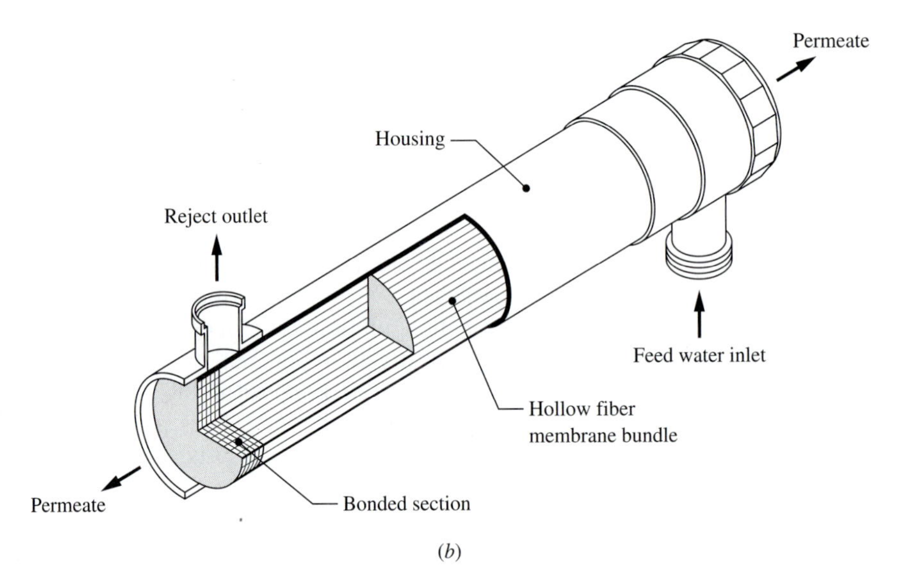

(*b*)

FIGURE 18
(*a*) Schematic diagram of the cross section of an ultrafiltration fiber. (*b*) Schematic diagram of an ultrafiltration cartridge. (*courtesy of Asahi Chemicals of Japan*)

For 1M DRAM devices, the following additional features are included:

Use of polishing cartridges with low TOC leachables
Use of ultrafiltration units as final filter
PVDF supply piping instead of PVC (polyvinyl chloride) or PP
Periodical sterilization of the distribution loop with ozone.

For 4M DRAM applications, the use of 185 nm UV lamp sterilization to reduce for TOC was adopted. For 16M DRAM, the DI water system process flow is similar to that for the 4M DRAM except that the RO and ultrafiltration filters are of better quality. For 64M DRAM applications, the use of secondary RO units in the make-up is forecast.

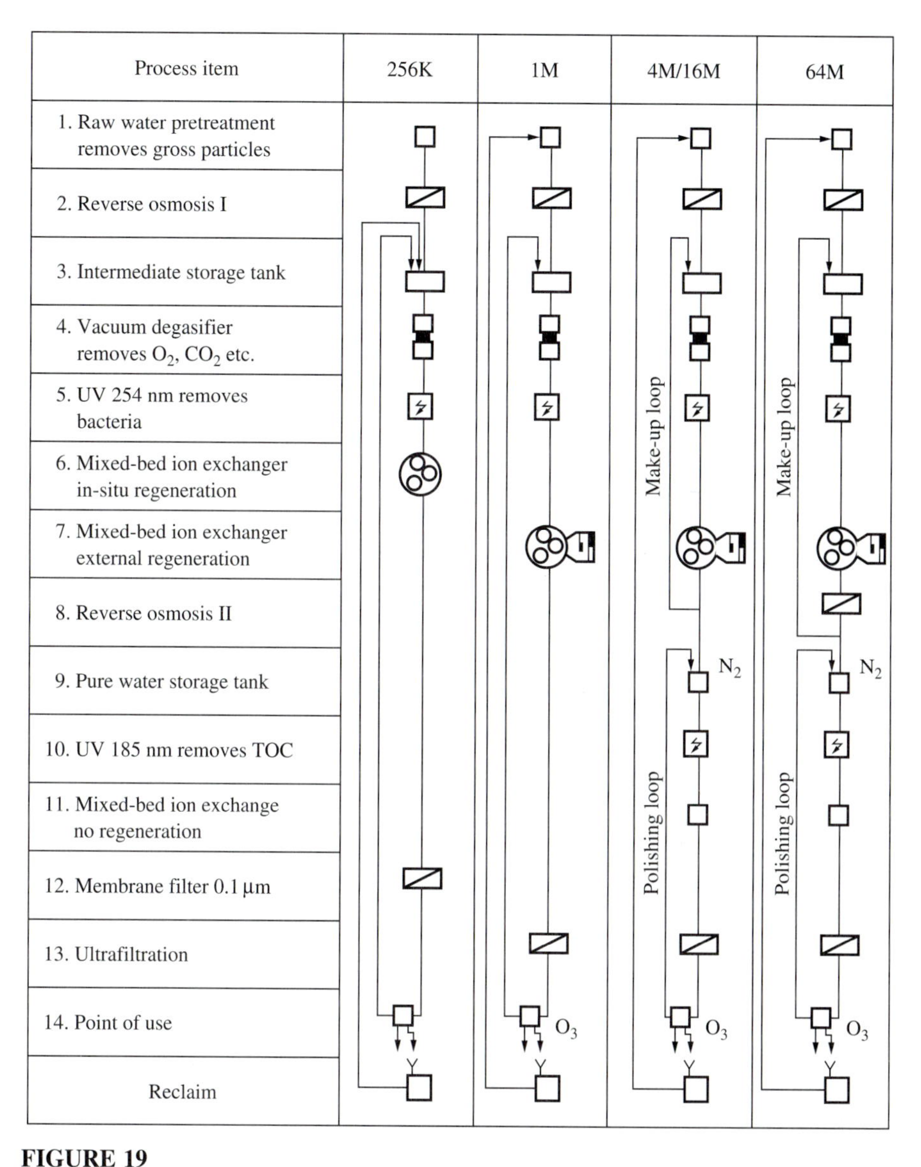

FIGURE 19
Evolution of ultrapure water system design. (*Courtesy of Christ AG. Switzerland.*)

Evolution of the DI water specifications is illustrated in Table 9. The specifications of most parameters, except for resistivity, have been tightened one order of magnitude for the 1-Mbit DRAM to the 16-Mbit DRAM.

One of the most critical impurities in DI water is the dissolved oxygen, which has to be kept as low as possible to prevent native oxide growth on bare silicon wafers. Much work has been done to improve the performance of vacuum degasifiers. There are three popular degasification processes: the hot-water process involves heating DI water to a temperature above 55°C; the nitrogen purging process entails injecting nitrogen counter flow into a DI water vessel; and the catalytic process involves

TABLE 9
Evolution of DI water specifications

	Process technology				
	256K DRAM	**1M DRAM**	**4M DRAM**	**16M DRAM**	**64M DRAM**
Design rule, μm	2.0	1.2	0.8	0.5	0.3
Resistivity, MΩ-cm	> 17.0	> 17.5	> 18.0	> 18.1	> 18.2
Particle (pieces/cc)					
> 0.2 μm	< 30	< 10			
> 0.1 μm	< 50	< 20	< 5	< 1	
> 0.085 μm			< 10	< 2	< 0.5
> 0.05 μm				< 5	< 1
Bacteria, unit/L	< 200	< 50	< 10	< 1	< 1
TOC, ppb	< 100	< 50	< 20	< 5	< 1
Oxygen, ppb	< 100	< 100	< 50	< 10	< 5
Silica, ppb	< 10	< 5	< 3	< 1	< 0.2
Na, ppb	1	< 1	< 0.1	< 0.05	< 0.01
Cl, ppb	1	< 1	< 0.1	< 0.05	< 0.01
Metal ion, ppb	1	< 1	< 0.1	< 0.05	< 0.01

Source: Courtesy of Christ AG. Switzerland.

feeding DI water through a vessel packed with palladium catalytic resin. These three degasifiers are expensive. Recently, a new degasifier operated at a room temperature of 23°C and employing neither nitrogen nor a catalyst has been developed. A vacuum level beyond 700 mm Hg is generated externally. The dissolved oxygen is removed by the boiling water vapor bubbling through the DI water. The dissolved oxygen level can be kept around 2 ppb.

Over 70% of the DI water supplied to the points-of-use can be recovered.[60] The reclaimed water can be fed directly to the raw water storage tank or pumped to an RO permeate storage tank after passing through various treatment steps such as activated carbon filters, a TOC UV lamp module, ion-exchange units, and RO units.

1.7.2 Chemical Supply System[64–65]

Close to 20 types of ultrapure chemicals are used in IC manufacturing processes. About two-thirds of them are acid and caustic and the remaining are organic solvents. These chemicals are used for the following processes:

Process	**Chemicals**
Prefurnace cleaning	NH_4OH, H_2SO_4, HCl, H_2O_2 and IPA
Thin film wet etching	HNO_3, H_3PO_4, HF, BHF (buffered HF)
Photoresist coating and developing	Photoresist, developer, acetone
Residual photoresist removal	H_2SO_4, H_2O_2, organic resist stripper
Postmetal process cleaning	Organic solvent

Particles and especially metallic impurities in the chemicals will significantly effect the wafer yield. The metallic impurities will cause degradation of gate oxide integrity and thereby result in early breakdown. The metallic impurities also enhance OSIF (oxidation induced stacking faults), resulting in junction leakage and a decrease of carrier lifetime.

TABLE 10
Evolution of chemicals specifications

	Process technology				
	256K DRAM	1M DRAM	4M DRAM	16M DRAM	64M DRAM
Design rule, μm	2.0	1.2	0.8	0.5	0.3
Particles (pcs/cc)					
> 0.5 μm	< 100				
> 0.3 μm		< 50	< 10		
> 0.2 μm		< 500	< 50	< 10	< 1
> 0.1 μm				< 100	< 10
Anion, ppb	< 2000	< 1000	< 500	< 100	< 50
Metal, ppb	< 100	< 50	< 10	< 1	< 0.2

Source: Courtesy of Merck-Kanto.

With the ever increasing chip size and process complexity, the continuing reduction in gate oxide thickness, and the ever tightening device leakage current requirement, the demand on the qualities of chemicals becomes very severe. Table 10 shows the evolution of the chemical specifications with respect to process technologies.

In submicron fabs, all bulk chemicals, except the highly viscous photoresist, are centrally delivered to points of use via chemical supply systems. A chemical supply system, as shown in Fig. 20,[65] consists of three parts: the chemical delivery modules normally located in a separate room adjacent to the cleanroom, the chemical delivery piping network, and the central controller.

There are two types of chemical supply modules. One is a big storage tank with a 1000 to 9000 L capacity equipped with external pumps and filters. This type of chemical supply module is used to supply chemicals consumed in large quantities, such as H_2SO_4, NH_4OH, H_2O, IPA, and photoresist developer. Chemicals are pumped from chemical drums supplied by the vendors into the big holding tank and then are delivered to the points of use as needed.

The other type of chemical supply module, used for delivering chemicals used in smaller volume, is a cabinet unit that houses two chemical drums, each with a 200 L capacity. One of the two chemical drums serves as a supply tank and the other as a standby. The chemical supply module automatically switches to a standby tank when the supply tank is empty. HEPA filters are often installed inside the chemical supply cabinet to provide a clean, laminar-flow environment.

It is a common practice to periodically recycle the chemicals inside the chemical drum through an external filter, when the tank is not delivering the chemicals, to reduce the particles. When the chemicals are delivered by a pneumatically driven bellows pump, particles trapped inside the filter may be pushed out due to the pulsating action of the pumps. To overcome this problem, some advanced chemical-supply modules employ a combination of the pumped and pressurized delivery techniques.[64] When the chemical supply module is in the supply mode, the chemical is delivered out of the holding tank by a constant nitrogen source, and when the chemical supply module is in the standby mode, the chemical is recycled through external pumps to filter out particles.

The piping networks delivering acid and caustic chemicals are made of PFA (perfluoroalkoxy) pipe enclosed in a seamless clear PVC tube that is connected to

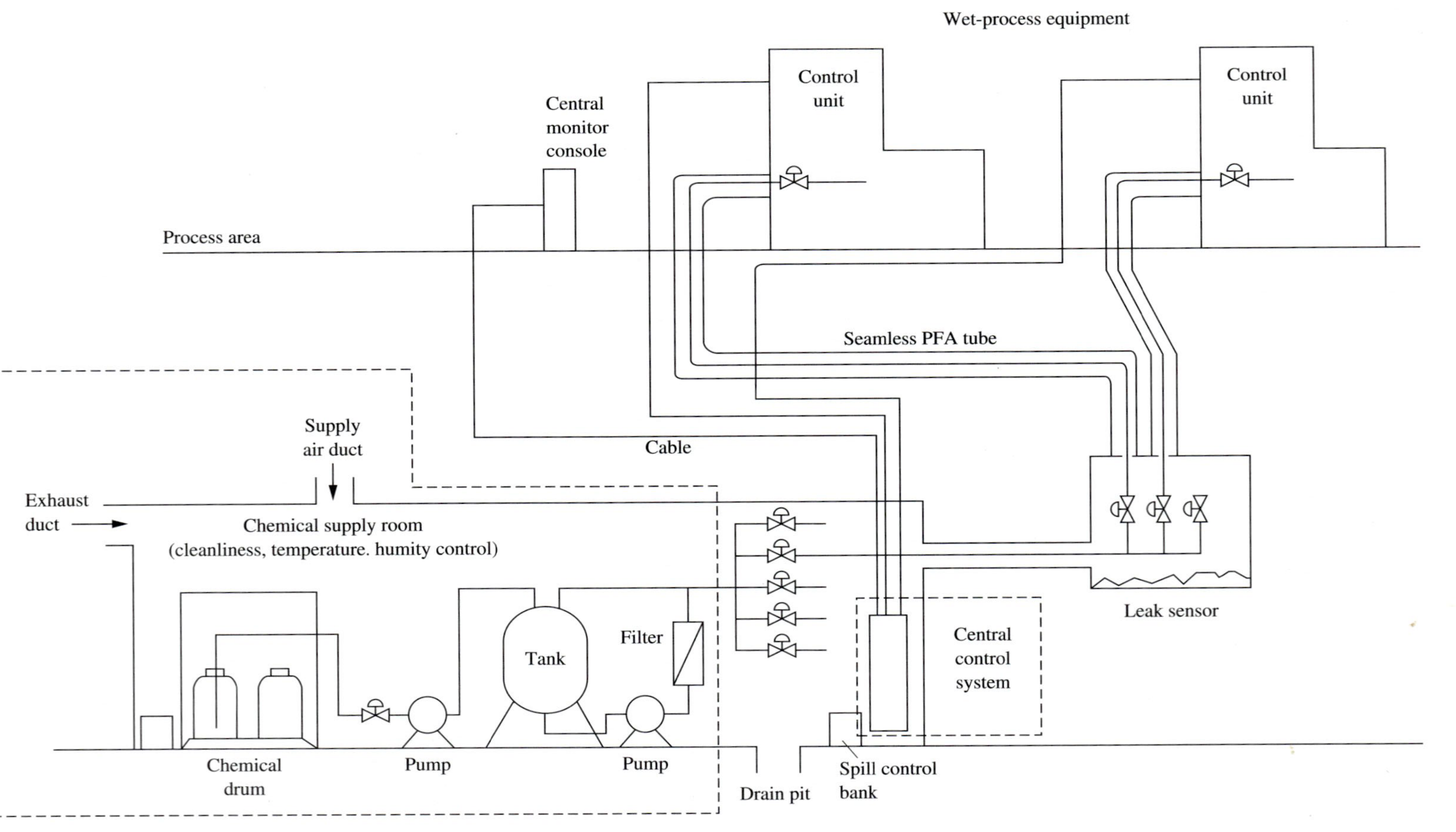

FIGURE 20
Schematic diagram of a chemical sypply system.

a valve box. Sensors are installed at the bottom of the valve box to detect chemical leakage. The piping for solvent is constructed of electropolished SS316L tubing.

Each chemical delivery module and all interface controllers, each connected to a point of use, are linked to a master controller that monitors the status of every delivery module and activates the chemicals supply function on receiving the request from any point of use.

1.7.3 Process Gas Piping System[66–73]

Over 20 different ultrapure gases are used in semiconductor manufacturing processes. Among these gases, N_2, O_2, H_2, and Ar are often called bulk gases because large quantities are used, whereas other gases, including He, AsH_3, PH_3, SiH_4, NH_3, and NF_3, are classified as special gases because only small quantities are needed.

In general, N_2 comes from three sources: an N_2 pipeline delivered from a remote air-separation plant, a liquefied-nitrogen storage tank located in the gas yard, and an on-site N_2 generation plant. O_2 and Ar are supplied from liquefied gas storage tanks. H_2 is delivered from either a liquefied-gas storage tank or a bank of high-pressure gas cylinders. These bulk gases pass through purifiers and gas filters to remove impurities and particles before entering a gas-distribution piping system located inside the cleanroom. The special gases are sent directly to process tools from gas cylinders located inside gas cabinets.

The gas cabinet is an exhausted safety enclosure that contains the gas cylinders and the associated gas-handling panels. The basic function of the gas cabinet is to allow purging and safe exchange of the cylinders. The gas panels incorporate all components for control and monitoring of the high-purity gases. The gas cabinet usually contains two process cylinders to allow automatic switch-over when one cylinder is empty. One nitrogen cylinder is also available for purging the piping line.

With a few exceptions, such as Cl_2 and SiH_2Cl_2, the supply pressure for bulk and special gases is normally kept around 5 kg/cm^2. At each point of use, the pressure of each gas has to be independently and locally controlled by a series of valves, pressure regulators, pressure sensors, and particle filters located inside a gas manifold box. Each gas is distributed at a special pressure dictated by the process tool. For each process tool there are one or more gas manifold boxes installed nearby for gas distribution and control.

The evolution of bulk-gas specifications with respect to process technology is shown in Table 11. From the 1M DRAM to the 4M DRAM, particles and impurities levels have to be reduced by one order of magnitude. To achieve the tight process-gas quality specifications required by a 16M DRAM process technology, the selection of gas components, piping-system design, and installation procedures becomes extremely critical.

Ultraclean gas components with the following features are used extensively in the submicrometer level process-gas piping systems:

Minimum dead space
Metal seal to prevent any leak to outside
Interior surface with submicrometer-level roughness

TABLE 11
Evolution of bulk gas specifications

	DRAM technology				
	256K	**1M**	**4M**	**16M**	**64M**
Design rule, μm	2.3	1.3	0.8	0.5	0.3
Particle level					
Size, μm	0.3	0.2	0.1	0.1	0.05
Density, pcs/cc	< 10	< 10	< 5	< 5	< 5
Impurity level, ppb					
O_2	< 100	< 50	< 10	< 5	< 1
CO			< 10	< 5	< 1
CO_2			< 10	< 5	< 1
CH_4			< 10	< 5	< 1
H_2O level					
Dew point, °C	< −76	< −80	< −90	< −100	< −120
Concentration, ppb	1,000	500	100	10	0.13
Metallic level, μg/m³	–	1	1	0.1	0.01

Source: Courtesy of BOC of England.

No thread seal and no moving parts
Fittings with bearing or welding applied to all joints

All of these features are aimed at minimizing particles and adsorbed moisture from the interior wall that makes contact with the ultrapure process gases, reducing the trapping of unwanted gas in the dead space, and preventing ambient moisture and oxygen from leaking into the piping systems. The performance characteristics of the ultraclean gas components can be illustrated by the following examples.

Figure 21 shows the schematic diagrams of a conventional branch piping method and an integrated gas valve.[67] The dead space of the former is 1.3 cc and of the latter

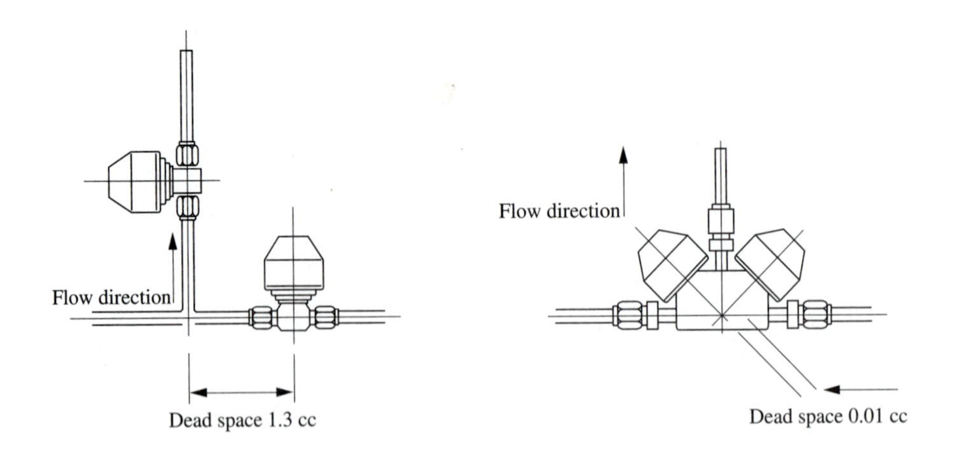

FIGURE 21
(*a*) Conventional branch piping method and (*b*) newly developed integrated gas valve.

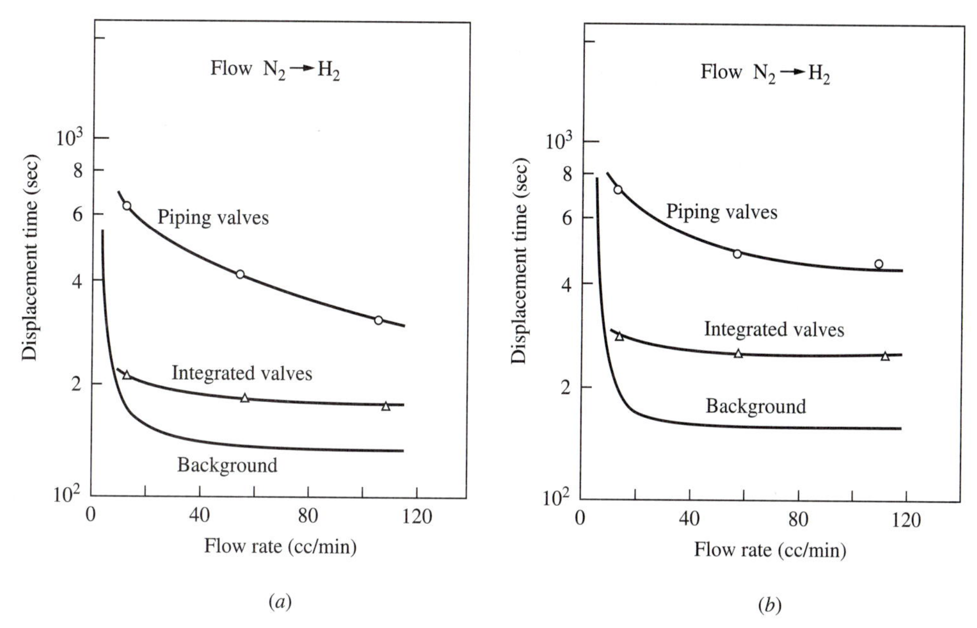

FIGURE 22
Gas displacement time (*a*) for branch valve and (*b*) for integrated valve.

is 0.01 cc. As illustrated in Fig. 22, the integrated valve, which has a dead space two orders of magnitude lower, offers much better gas displacement characteristics than the conventional branch piping method. Recently, block valves that have several integrated single valves in a single metal block have been used by some IC manufacturers in their process-gas distribution systems.

Hundreds of diaphragm valves and bellows valves are used in a process-gas piping system. Schematic diagrams of these two valves are shown in Fig. 23. Compared with the bellows valve, the diaphragm valve has much lower dead space, and therefore exhibits superior gas-displacement characteristics.[67]

In the past decade, gas components with an interior surface electropolished to a roughness below 1 μm have become a standard. The moisture outgassing characteristics of an EP (electropolished) surface can be improved further by growing a Cr_2O_3 passivation layer in an oxygen ambient. The oxygen-passivated EP piping has a golden color and is also known as GOLDEP piping. Figure 24 shows the moisture outgassing characteristics observed in the start-up phases of three different gas-piping systems.[71] Within a few hours of purging, the moisture level of the gas piping system employing GOLDEP tubing and gas components with all-metal seals (see case C) drops to a few ppb. It will take more than a month of continuous N_2 purging for the moisture level of the gas-piping system using conventional EP tubing and gas components with plastic seals to drop below 5 ppb. Figure 24 also illustrates the importance of purging continuously after the welding work is completed.

Table 12 summarizes the gas components and their relevant characteristics, such as interior surface roughness and external leakage rate, with respect to process technology ranging from 256K DRAM to 64M DRAM. The surface roughness of

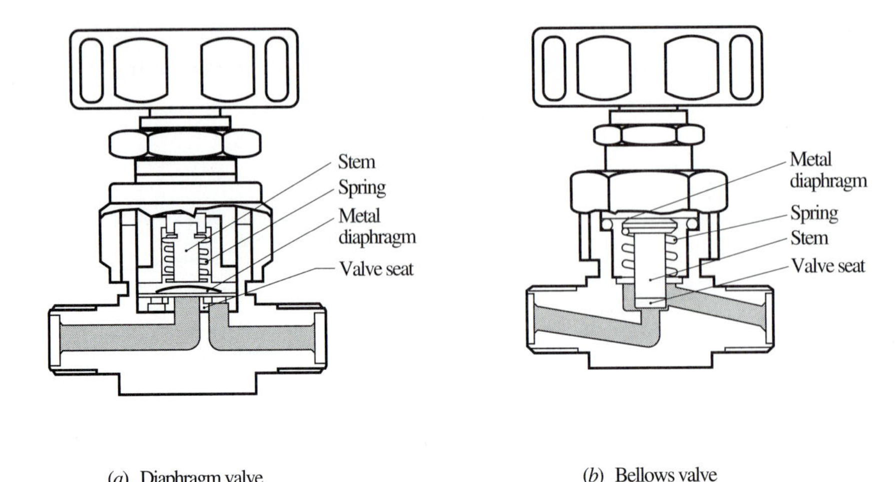

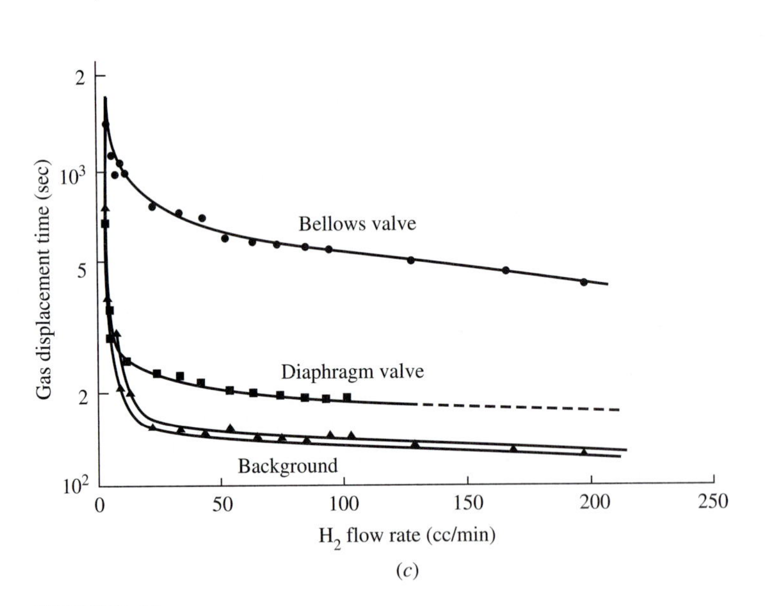

FIGURE 23
(*a*) Diaphragm valve, (*b*) bellows valve, (*c*) gas displacement characteristics.

gas piping and the external leakage of gas components has been improved by several orders of magnitude. The bellows valves have given way to diaphragm valves. The only plastic parts remaining in the gas-piping system are the membrane gas filters installed at the beginning of the main gas piping. Compared with the total gas flow rate in the main pipe, the outgassing rate of a membrane filter is negligibly small. The conventional Borden tube pressure gauge contains a spiral hollow coil that moves the pressure indicator needle by expanding and contracting as the pressure fluctuates. This type of spiral coil has a large dead space and also generates particles. The Borden tube pressure gauges are being replaced by flow-through type pressure

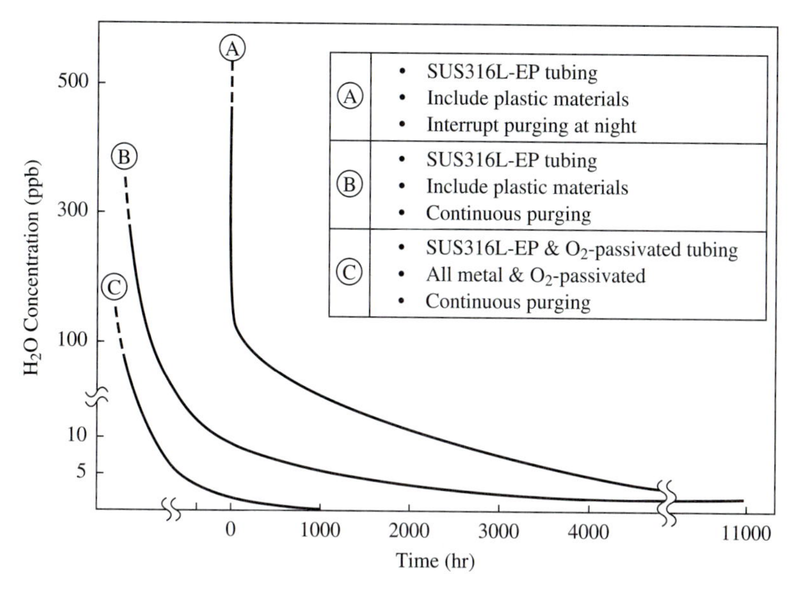

FIGURE 24
Moisture outgassing characteristics of three piping systems during the start-up stages.

TABLE 12
Evolution of process-gas components

DRAM technology	256K	1M	4M	16M	64M
Piping material	SS316LBA	SS316LBA	SS316LEP	SS316LEP	SS316LGEP
Internal surface	SCH-10S	SCH-10S	SCH-5S	SCH-5S	SCH-5S
Roughness, R_{max}, μm	< 3 − 5	< 1.0	< 0.7	< 0.3	< 0.1
Tube fitting	Swagelog	VCR	VCR	VCR	MCG/JSK
Ring			Metal	Metal	Metal
External leak, atm-cc/s	1 E−4	1 E−6	1 E−8	1 E−11	1 E−13
Valve					
Large	Bellows	Bellows	Bellows	Diaphragm	Diaphragm
Small	Bellows	Bellows and diaphragm	Diaphragm	All-metal diaphragm	All-metal diaphragm
Filter					
Cutoff size, μm	0.02	0.02	0.01	0.01	0.005
Large flow rate	Membrane	Membrane	Membrane	Membrane	Membrane
Small flow rate	Membrane	Membrane	Ceramic	Metal	Metal
Pressure regulator	Diaphragm	Diaphragm	Diaphragm	Diaphragm	Diaphragm
Pressure gauge	Borden tube	Borden tube	Borden tube	Flow-through pressure sensor	Flow-through pressure sensor

Source: Courtesy of BOC of England.

sensors that monitor gas pressure with a built-in metallic diaphragm attaching to the interior wall of the gas tubing.

Improvement has also been made to gas cylinder valves. The conventional CGA (Compressed Gas Association) cylinder valve has been replaced by a newly developed DISS (Diameter Index Safety System) cylinder valve. The DISS gas cylinder valve employs a metallic gasket to provide a better seal at the contact between the male gas cylinder outlet and the female connector attaching to the gas-distribution system. Special provisions are also made to minimize the torque force, and thereby the abrasion, generated at the surface of the gas cylinder outlet when the male and female DISS connectors are tightened.

The design and construction concepts of an ultraclean process-gas piping system is similar to the ultraclean gas components:

Leak free
Particle free
Dead space free
Outgas free
Corrosion free
Plastic material free

Purge valves are often installed at the ends of the subbranches extended sideways from the bulk-gas main piping. A small volume of gas is continuously purging out of the piping system to ensure that the process gas is not stagnant inside any section of the piping system.

In addition to the proper selection of gas components, the actual field installation procedures plays a decisive role in determining the final gas quality. All the components and tubing should be wrapped in a nitrogen-filled plastic package prior to shipment from the vendors. The on-site subassembly of gas components should be performed in a clean booth with a cleanliness better than class 100. The installation of the subassembled parts should also be in a clean environment, preferably after the installation of HEPA filters. During welding, Ar containing 3 to 5% H_2 is used as an arc gas and back-seal gas to prevent the hot welding spot from oxidizing with ambient oxygen and moisture. After welding, the piping should be purging continuously with N_2 to get rid of the adsorbed moisture. Helium leak tests are also applied to all the welding spots to ensure there will be no leakage due to defective welding.

1.7.4 Space Management

Space management refers to the arrangement of process utilities including make-up air and exhaust ductwork, process pipings, electrical cable tray and bus bars, etc. in both the utility building and the cleanroom area. It comprises the design of the process utilities distribution network and the design of the pipe rack on which the process utilities are installed. Space management, which is often overlooked during cleanroom design and construction, results in an optimum and flexible process utilities distribution.

Each process utility piping network can be subdivided into main pipings and branch pipings. Figure 25 illustrates the arrangement of main piping and branch

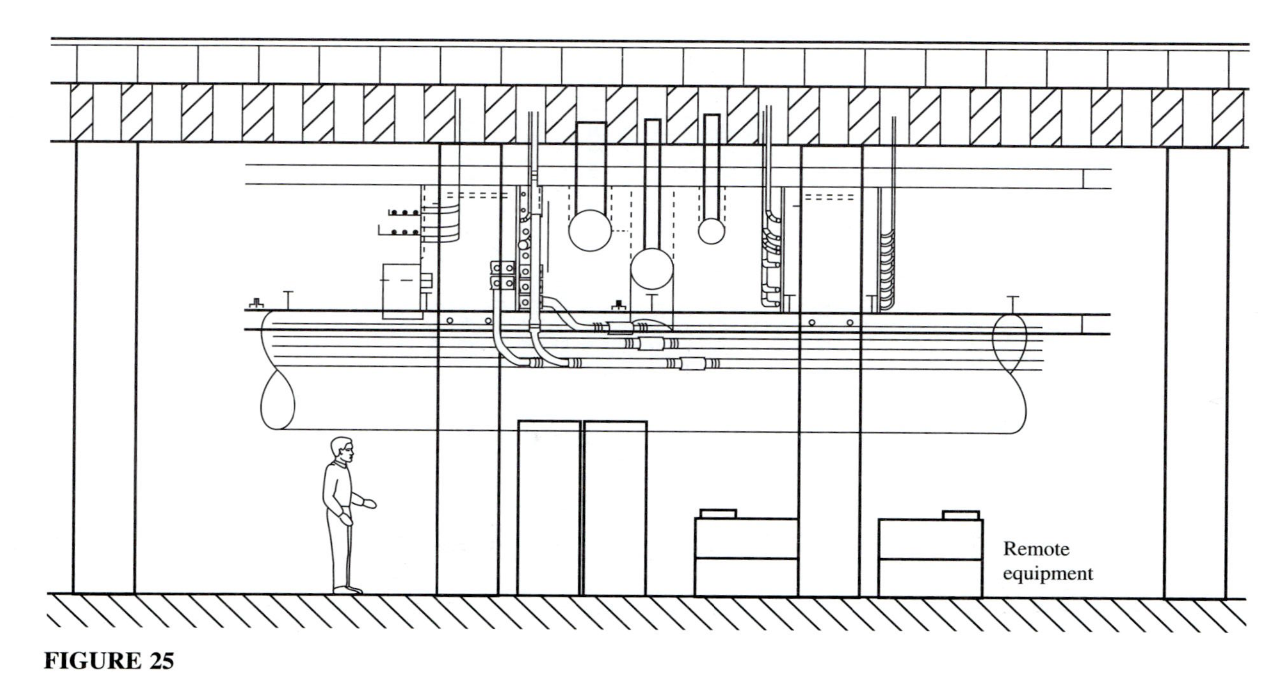

FIGURE 25
Example of the arrangement of the process utilities main and branch headers in the basement of a cleanroom.

headers in the basement of a cleanroom. The main headers of the process utilities are based on a horizontal distribution of piping, cabling, and exhaust ductwork. The branch headers of the process utilities, which crisscross over the main header, are based on a vertical distribution of piping, cabling, and exhaust ductwork. From the branch header, process utilities are connected upward to the process equipment located at the second floor or downward to the pumps and power cabinets located on the first floor.

The optimal height of the cleanroom basement, from the floor level to ceiling, should be approximately 6 meters to allow enough clearance for people to walk freely under the main utilities headers.

The pipe rack is required to support the main and branch headers of the process utilities and occasionally the central utilities, such as chilled water for the recirculation air units.

1.8 SUMMARY AND FUTURE TRENDS

Because devices are continuously miniaturized, the design of the cleanroom system and process utilities is increasingly critical for the IC manufacturing process. As feature sizes decrease and wafer sizes increase, purity requirements become more stringent. A low defect density demands a substantial improvement in the control of the wafer environment. The current cleanroom classification is based on 1 particle with a size equal to or larger than 0.1 μm per ft^3 in the wafer environment, obtained in a downward laminar air flow at a maximum speed of 0.35 m/sec.

The design of a cleanroom system must be flexible to allow any desired process layout. The two-level cleanroom concept is recommended, in which the upper level is the clean production area with the process equipment, and the lower level is a utilities distribution area and a return air plenum as well as a site for installing remote equipment. All cleanroom subsystems should be integrated for optimum performance. Material selection and basic setup for each subsystem must satisfy the requirements of the manufacturing process. Vibration, temperature, relative humidity, ESD and EMI, and particle and chemical contamination must all be controlled to ensure high quality cleanroom performance.

The baseline management method, which divides the entire project into clearly identifiable phases, is the recommended approach for a cleanroom construction project. The time span from beginning the design to starting up the process is estimated to be approximately 2 years, divided into a 6-month preparation period, a 14-month realization period, and a 4-month hookup and final-testing period. Of the total investment for a semiconductor fab, 75% purchases the process equipment and 25% pays for the facilities and utilities. Of the latter amount, approximately 35% pays for the process utilities and 25% finances the cleanroom system.

The generation of particles inside the cleanroom must be prevented as much as possible. Gowning procedures are imposed on persons entering the cleanroom and decontamination procedures must be followed for incoming materials. Sources of microcontamination inside the cleanroom are minimized by proper cleanroom maintenance.

Other related process utilities require the highest possible quality level. DI water, chemical supplies and process gases require attention and investment.

Total integration of process utilities requires space management—the optimal arrangement of the various utilities distribution networks and the supporting pipe rack. Trends in cleanroom design are toward minienvironments in which the wafer loading/unloading and processing areas are enclosed. Wafers are transferred automatically in and out of the minienvironment. Outside these minienvironments, the wafers are sealed inside a portable pod for transport. To eliminate microcontamination and reduce native oxide growth on silicon wafers, the nitrogen tunnel has been proposed for manufacturing 1G DRAM. In this future cleanroom, the wafers will be transported by robots, one piece at a time, inside a nitrogen tunnel that connects to each piece of process equipment. An alternative solution may be the use of cluster tools with dry HF etch chambers to clear the native oxide before a native-oxide-sensitive process is performed. In designing the cleanroom layout, enough space should be kept between process tools to accommodate the cluster tools, which usually have large footprints.

During the last three decades, the wafer size has increased from 75 mm to 200 mm, and the standard loading capacity of a wafer cassette has remained at 25 pieces. The lot size for the 300-mm wafer will become much smaller. Smaller lot size can significantly reduce the wafer processing cycle time; however, it can also dramatically increase the wafer handling frequency. Automatic transfer of the delicate wafers between process tools is the only applicable solution. Ultimately, the process tool manufacturing community must come to an agreement on standardizing the robotics interface so that total automation can be implemented smoothly.

Total automation is a starting point in the design of the process layout. This means that the required space and logistics must be planned. Three basic levels of automation can be distinguished: tool automation, intrabay automation, and interbay automation. Interbay automation is now a standard feature. Intrabay automation is becoming popular and will become absolutely necessary in the future 300-mm wafer fab.

REFERENCES

1. T. Ohmi, "Future Trends and Applications of Ultra-Clean Technology," *Tech Dig. IEDM,* 49 (1989).
2. T. Ohmi, "Ultraclean Cleanroom Environments: Closed System Essential for High-Quality Processing within Advanced Semiconductor Lines," *Microcontamination,* **8**(6), 27 (1990).
3. T. Ohmi and T. Shibata, "Developing a Fully Automated, Closed Wafer Manufacturing System," *Microcontamination,* **8**(7), 25 (1990).
4. T. Ohmi, "Breakthrough for Scientific Semiconductor Manufacturing in 2001. A Proposal from Tohoku University," Realize, Inc. (Japan) 1992.
5. D. W. Cooper, S. J. Grotzinger, L. R. Ackman, and V. Srinivasan, "Selecting Nearly Optimal Sampling Locations Throughout an Area. Application to Cleanrooms and Federal Standard 209," *Institute of Environmental Sciences, Annual Technical Meeting, New Orleans, LA,* 1990, p. 257.
6. S. J. Grotzinger and D. W. Cooper, "Selecting a Cost-Effective Number of Samples to Use at Preselected Locations," *Journal of the IES,* **35**(1), 41 (1992).

7. D. W. Cooper and D. C. Milholland, "Sequential Sampling Plan for Federal Standard 209," *Journal of the IES,* **33**(5), 28 (1990).
8. R. Simon, "Clean Room Technology for Semiconductor Manufacturing," (German), *Stroemungsmechanik und Stroemungsmaschinen,* **39** 139 (Nov., 1988).
9. K. Unno, "Cleanroom," *Denski Zairyo (Electronic Material) 1988 Sup. Vol.,* 134 (1988).
10. T. Kawamata, "Cleanroom," *Denski Zairyo (Electronic Material) 1989 Sup. Vol.,* 125 (1989).
11. N. Namiki, "Cleanroom," *Denski Zairyo (Electronic Material) 1990 Sup. Vol.,* 136 (1990).
12. B. Patel, J. Greiner, and T. R. Huffman, "Constructing a High-Performance, Energy-Efficient Cleanroom," *Microcontamination,* **9** (2), 29 (1991).
13. B. Newboe, "Successful Cleanroom Designs," *Semic. Intl.,* **14** (6), 106 (1991).
14. J. R. Weaver, "Cleanroom Reverse Design Principle," *Semic. Intl.,* **14** (10), 4 (1991).
15. R. W. Frick and M. C. Miller, "Designing NASA's Largest Cleanroom," *Microcontamination,* **10** (1), 37 (1992).
16. A. Saiki, "Cleanroom," *Denski Zairyo (Electronic Material) 1993 Sup. Vol.,* 143 (1993).
17. T. Kawamata, "Cleanroom," *Denski Zairyo (Electronic Material) 1994 Sup. Vol.,* 133 (1994).
18. R. A. Hughes, G. B. Moslehi, D. M. Campbell, K. J. Radigan, W. Lukaszek, and E. D. Castel, "Eliminating the Cleanroom: More Experiences with an Open-Area SMIF Isolation Site (OASIS)," *Microcontamination,* **8** (4), 35 (1990).
19. J. Inbody and V. E. Bradley, "Upgrading an Existing Wafer Fab Facility with SMIF Technology," *Microcontamination,* **8** (9), 25 (1990).
20. C. Y. Hsu and L. C. Tu, "Designing, Operating a Submicron Facility with Isolation Technology," *Microcontamination,* **10** (3), 29 (1992).
21. T. Baechle, G. Marvell, W. Fosnight, and M. Lynch, "Assessing the Capabilities of a Minienvironment to Meet Increasing Environmental Specifications," *Microcontamination,* **10** (5), 25 (1992).
22. T. Baechle, G. Marvell, and M. Lynch, "Evaluating the Capabilities of Mini-Environments Using Polished Silicon Monitor Wafers," *Microcontamination,* 10 (5), 35 (1992).
23. W. C. Grande, "Upgrading a Class 100 Fab through Use of Manual-Access Microenvironments," *Microcontamination,* **11** (1), 25 (1993).
24. A. Morizuki, "Cleanroom for Semiconductor Industries," *Denski Zairyo (Electronic Material),* 35 (Aug., 1990).
25. T. Ohmi, Y. Kasama, K. Sugiyama, Y. Mizuguchi, Y. Yagi, H. Inaba, and M. Kawakami, "Controlling Wafer Surface Contamination in Air Conditioning, Particle Removal Subsystems," *Microcontamination,* **8** (2), 45 (1990).
26. T. Ohmi, H. Inaba, "Cleanroom Special Edition — Environmental Control Technique," (Japanese), *Nikkei Microdevices,* 115, (Sep., 1989).
27. Suzuki, "Supercleanroom," (Japanese), *Hitachi Hyoron (Commentary),* **68** (9), 737 (1986).
28. M. J. Bader, "Ionizers for Clean Rooms, Eliminating Both Electrostatic Charging and Dust," (German), *Plastverarbeiter,* **39** (12), 44 (1988).
29. J. R. Turner, D. K. Liguras, and H. J. Fissan, "Clean Room Applications of Particle Deposition from Stagnation Flow: Electrostatic Effects," *Journal of Aerosol Science,* **20** (4), 403 (1989).
30. P. C. D. Hobbs, V. P. Gross, and K. D. Murray, "Suppression of Particle Generation in a Modified Clean Room Corona Air Ionizer," *J. Aerosol Science,* **21** (3), 463 (1990).
31. A. Steinman, "Evaluating Air Ionization Systems," *Evaluation Engineering,* **29** (4), (1990).

32. P. C. D. Hobbs, V. P. Gross, and K. D. Murray, "Reviewing Clean Corona Discharge: Laser-Produced Plasma Ionization Technologies," *Microcontamination,* **9** (6), 19 (1991).
33. T. Sebald, "Continuously Working Monitoring System for Clean Room Ionizers," *Environmental Engineering,* **5** (4), 16 (1992).
34. T. Takenami, T. Ohmi, and S. Fukuda, "Air Conditioning and Particle Filtration Systems for Energy Saving," *Sol. State Technol.,* **32** (4), 161 (1989).
35. P. Naughton, "HVAC Systems for Semiconductor Cleanrooms. Part 1: System Components," *ASHRAE Trans., Part 2,* 620 (1990).
36. P. Naughton, "HVAC Systems for Semiconductor Cleanrooms. Part 2: Total System Dynamics," *ASHRAE Trans., Part 2,* 620 (1990).
37. T. Ohmi, H. Inaba, and T. Takenami, "Using Water-Based Cooling Systems in Cleanroom Environments," *Microcontamination,* **7** (12), 27 (1989).
38. Y. Suzuki, S. Oikawa, and T. Sekiguchi, "Super Cleanroom Technology: A High-Tech Balancing Act," *Microcontamination,* **6** (9), 59 (1988).
39. B. Brandt and L. A. Wright, "Analyzing Particle Release of Cleanroom Headcoverings," *Microcontamination,* **8** (10), 53 (1990).
40. R. Spector, C. Berndt, and C. W. Berndt, "Reviewing Methods for Evaluating Cleanroom Garment Fabrics," *Microcontamination,* **11** (3), 31 (1993).
41. P. Ravis, G. H. Ranta, "How to Choose a Clean Room Laundry," *Institute of Environmental Sciences, Annual Technical Meeting, New Orleans, LA,* 1990, p. 355.
42. K. Skidmore, "Keep Your Clean Room Particle Free," *Semic. Intl.,* **12** (9), 94 (1989).
43. W. K. Kwok and J. T. Summers, "Characterization of Cleanroom Wipers: Particle Generation," *Institute of Environmental Sciences, Annual Technical Meeting, New Orleans, LA,* 365 (1990).
44. R. K. Schneider, "Developing and Implementing a Cleanroom Construction Protocol," *Microcontamination,* **8** (8), 35 (1990).
45. H. H. Schicht, "Contamination Control. An Indispensable Factor in High-Technology Manufacturing Tasks," *J. Aerosol Science,* **21,** Supp 1, 719 (1990).
46. J. Greiner and M. O'Halloran, "The Forgotten Functions: Support Facilities for the Modern Microenvironment," *Microcontamination,* **9** (3), 45 (1991).
47. A. M. Dixon, "Protesting Your High-Tech Investment through Sound Cleanroom Maintenence Practices," *Microcontamination,* **10** (11), 38 (1992).
48. T. Ohmi, "Proposal for Advanced Semiconductor Manufacturing Equipment: An Approach to Automated IC Manufacturing," *Proceedings of the 5th Symposium on Automated Integrated Circuits Manufacturing,* IEEE, 1989, p. 3.
49. T. Ohmi, T. Shibata, "Requirements of CAM in IC Technology," *Microelectron. Eng.,* **10** (3–4), 177 (1991).
50. P. K. John, "Optimal Partitions for Shop Floor Control in Semiconductor Wafer Fabrication," *Europ. J. Oper. Res.,* **59** (2), 294 (1992).
51. Y. Mizokami, "Total CIM System for Semiconductor Plants," IEEE/SEMI International Semiconductor Manufacturing Science Symposium—ISMSS '90, May 21–23, 1990.
52. R. S. Weiss, "Automation and Particulate Control within an IC Manufacturing Facility," *Proceedings of the 5th Symposium on Automated Integrated Circuits Manufacturing,* IEEE, 1989, p. 199.
53. H. Tabata, T. Yamashita, M. Murata, M. Onishi, and T. Tsubaki, "Autonomous Mobile Robot Transport System for Clean Rooms," (Japanese), *R&D, Research and Development, Kobe Steel, Ltd.,* **40** (3), 36 (1990).
54. I. Y. Wang, A. S. Li, and B. V. Ilene, "A Magnetic Levitation Transport Path," *IEEE Trans. Semic. Manuf.,* **4** (2), 145 (1991).

55. T. Araki, "Autotransfer System," Japanese, *Denski Zairyo (Electronic Material) 1994 Sup. Vol.,* 141 (1994).
56. H. Sato, M. Hashimoto, T. Shinoda, and Y. Hiratsuka, "Ultrapure Water System for 16M DRAM," Japanese, *Semiconductor World,* **8** (6), 54 (1989).
58. A. Houzuki and K. Ushigoe, "Ultrapure Water Manufacturing System," (Japanese), *Denski Zairyo (Electronic Material) 1989 Sup. Vol.,* 118 (1989).
59. M. Furuichi, "Ultrapure Water Manufacturing System," (Japanese), *Denski Zairyo (Electronic Material),* 69 (Aug., 1990).
60. M. Toto, "A Closed Ultrapure Water Manufacturing System," (Japanese), *Denski Zairyo (Electronic Material),* 59 (Aug., 1991).
61. K. Oda and T. Ozaki, "Ultrapure Water Manufacturing System," (Japanese), *Denski Zairyo (Electronic Material) 1991 Sup. Vol.,* 108 (1991).
62. Y. Yagi, T. Imaoka, Y. Kasama, and T. Ohmi, "Advanced Ultrapure Water Systems with Low Dissolved Oxygen for Native Oxide Free Wafer Processing," *IEEE Trans. Semic. Manuf.,* **5** (2), 7 (1992).
63. T. Doi, "Ultrapure Water Manufacturing System," (Japanese), *Denski Zairyo (Electronic Material) 1993 Sup. Vol.,* 137 (1993).
64. K. T. Pate, "Examining the Design, Capabilities, and Benefits of Bulk Chemical Delivery Systems," *Microcontamination,* **9** (10), 25 (1991).
65. K. Kobayashi, M. Tamura, T. Shimada, and H. Sakai, "Chemical Autodelivery System," *Denski Zairyo (Electronic Material) 1993 Sup. Vol.,* 157 (1993).
66. T. Ohmi, Y. Kasama, K. Sugiyama, Y. Mizuguchi, Y. Yagi, H. Inaba, and M. Kawakami, "Examining Performance of Ultra-High-Purity Gas, Water, and Chemical Delivery Subsystems," *Microcontamination,* **8** (3), 27 (1990).
67. T. Ohmi, K. Sugiyama, F. Nakahara, M. Tsuda, Y. Sugano, and N. Onaga, "The Reduction of Residual Gas After Purging by Eliminating the Dead Space in the Valve," (Japanese), *Nikkei Microdevices,* 126 (Jul., 1988).
68. Y. Mukagawa and K. Tamura, "The New Trend of Semiconductor Process and Supporting Facility Technologies," (Japanese), *Denski Zairyo (Electronic Material),* 22 (Aug., 1990).
69. K. Sugiyama, F. Nakahara, and T. Ohmi, "Designing a Gas Delivery System for Lower Submicron ULSI Processes," *Microcontamination,* **7** (7), 29 (1989).
70. D. H. Hope, R. J. Markle, T. F. Fisher, J. B. Goddard, J. Notaro, and R. D. Woodward, "Installing and Certifying SEMATECH'S Bulk-Gas Delivery Systems," *Microcontamination,* **8** (5), 31 (1990).
71. M. Nakamura, T. Ohmi, and K. Kawada, "All Metal and Oxygen Passivation Tubing Technology for Ultra Clean Gas Delivery System," *Institute of Environmental Science Technical Meeting,* San Diego, 605 (1991).
72. S. Kamoki, K. Sugiyama and M. Nakamura, "Gas Delivery System for Semiconductor Processes," (Japanese), *Electronic Material 1993 Sup. Vol.,* p. 151 (1993).
73. T. Aida, "Gas Delivery System for Semiconductor Processes," (Japanese), *Electronic Material 1994 Sup. Vol.,* 147 (1994).

PROBLEMS

1. Due to contamination that occurred in the cleanroom, the wafer defect density, measured at sizes above 0.3 μm, has increased fivefold from 0.2 D/cm^2 to 1.0 D/cm^2. Use the following equation along with the data provided in Tables 1 and 2 to estimate the yield loss of a 4M DRAM and a 16M DRAM wafer.

$$Y = e^{-DA}$$

where D is the defect density and A is the chip area. Note that the density of defects is roughly inversely proportional to the defect size to the second power.

2. The class 1 according to Japanese Std. B9920 rev. describes air cleanliness with no more than 10 particle/m^3 of a size 0.1 μm or larger. Calculate the equivalent air cleanliness class according to U.S. Fed. Std. 209E.

3. Estimate the numbers of sampling locations and the total air sampling time required to certify a submicrometer fab having 2000 m^2 class M1 process area and a 2000 m^2 class M3 service area. Use the single sampling plan and sample the air at 28.3 L/min.

4. An accidental spillage of an IPA bottle raised the cleanliness level of a 10 m^2 cleanroom from class 1000 to class 100,000. Assuming the air exchange rate of the cleanroom is 60 times per hour, estimate the time required for the air cleanliness to recover to class 1000.

5. The cleanliness of outside ambient air is around class M7 according to Fed. Std. 209E. Estimate the cleanliness of the air leaving the make-up air unit having a HEPA filter of 99.97% filtration efficiency. What is the minimum filtration efficiency of the ULPA filters required for the process area to reach a class M1 (at 0.1 μm) cleanliness?

6. The cleanliness of a cleanroom is achieved by recirculating air through HEPA filters. Estimate the annual electricity requirement for maintaining the air cleanliness of a 4000 m^2 submicrometer cleanroom. The HEPA filter coverage of the entire cleanroom is 60%. The average air speed leaving HEPA filters is 0.35 m/sec. The average pressure drop across the HEPA filter is 120 Pa. The efficiency of the axial fan unit is 85%.

7. By adopting the minienvironment concept, the HEPA coverage of the conventional cleanroom described in Problem 6 can be reduced by half. The total investment of implementing SMIF is around $25 million. Estimate the annual saving in electricity and compare that with the SMIF investment cost. The cost of electricity is $0.1 per kWh. The amortization period of SMIF investment is 7 years and the annual interest of the capital is 6%.

8. To maintain a constant air velocity of 0.35m/sec, the pressure drop across the HEPA filter has to be increased over time. Use the data provided in Problem 6 to estimate the annual increment in electricity cost for a HEPA filter having an initial pressure drop of 100 Pa and an annual pressure increment of 10 Pa.

9. To control the relative humidity level of a cleanroom at 43%, the outside air has to be cooled to 9°C to remove the excess moisture before being heated to the operating temperature of 22°C. Since a large amount of heat is released by the process tools located inside the cleanroom, significant energy saving on chillers and boilers can be realized by lowering the outlet temperature of the make-up air to, for example, 14°C instead of 22°C. Use the data provided in Section 1.3.5 to estimate the annual energy saving achieved in a submicrometer fab with a monthly production capacity of 15,000 wafers. Assume the average mask level is 20 and the average heat removal efficiency of a chiller is 5, i.e., 5 kW heat removed per kW of electricity required. The cost of electricity is $0.1 per kWh.

10. Estimate the gas displacement time of a 1/4-in. bellows valve at the flow rate of 50 cc/min. Repeat the same calculations for a diaphragm valve. The dead space of the bellows valve and the diaphragm is 2 cc and 0.2 cc, respectively. The gas displacement time is defined as the time required for the displaced gas to reach 0.01% of its initial concentration.

CHAPTER 2

Wafer-Cleaning Technology

C. Y. Chang and T. S. Chao

2.1 INTRODUCTION

ULSI technology requires more stringent and reliable means to control the surface smoothness and to get rid of particles and contamination, such as metallic and organic residues on the silicon wafer surface, than does VLSI technology.

The well-known RCA wet clean[1] processes have been used extensively since the 1970s. The basic mechanisms of cleaning will be discussed in the following section. This discussion is followed by a presentation of recent developments that modify the standard RCA process to meet the more severe requirements and various applications of ULSI processes. These processes include thinner (< 5 nm) gate oxide formation, metal contact, and high-crystal-quality epitaxy. Next, we will address dry cleaning processes that meet the future requirements of integrated cluster-processing systems. Recent developments in surface electronic-state configurations, surface termination (e.g., H-terminated or F-terminated), surface kinetics, and reactions will be discussed in order to understand the surface-cleaning processes.

Supporting technologies, such as O_2-content-free deionized water, pure gas delivery systems, special designs for avoiding particulate generation in the process chamber, cleanroom design, etc., that are crucial to avoid surface contamination of the processing wafer have been discussed in detail in Chapter 1 and will be readdressed where they are needed.

2.2 BASIC CONCEPTS OF WAFER CLEANING

The wafer-cleaning process has been the important and critical step in semiconductor manufacturing for over 30 years. The most common process used today is the RCA wet cleaning process.[1] The RCA cleaning process was first developed by Kern and

Puotinen in 1960 at RCA and was published in 1970. Two sequential cleaning solutions, NH_4OH–H_2O_2–H_2O (called *standard cleaning 1* (SC-1), a composition of 1 : 1 : 5 to 1 : 2 : 7 at 70 to 80°C) and HCl–H_2O_2–H_2O (called *standard cleaning 2* (SC-2), a composition of 1 : 1 : 6 to 1 : 2 : 8 at 70 to 80°C), were used. Both of these cleaning processes are based on hydrogen peroxide. At a high pH value SC-1 can effectively remove organic contamination and particles by oxidation. At a low pH value SC-2 can desorb metal contamination by forming a soluble complex. These basic concepts to remove contamination have not changed since they were developed. However, as the dimensions of devices have shrunk to submicrometer scale, additional requirements have been proposed for an ultraclean wafer surface in VLSI and ULSI. For complete fulfillment of an ultraclean wafer surface, it should be free from particles, organic contamination, metal contamination, surface microroughness, and native oxide.

Particle, organic, and metal contamination have been reduced with a controlled, super-cleanroom technology and improved wet cleaning technology. To improve the integrity of ultrathin gate oxide and to grow an epi-layer in a low-temperature process, the surface must be free of native oxide, or perfectly hydrogen-terminated. Another issue is the microroughness. This issue is not so serious when the thickness of the gate oxide is larger than 200 Å using 0.7 μm technology. However, as the thickness is reduced to less than 100 Å for 0.35 μm or smaller technology, surface microroughness at the SiO_2/Si interface becomes a prominent influence on channel electron mobility as well as on ultrathin gate oxide quality. The sources and the related effects of the contaminations are shown in Table 1. Different contaminants cause different effects on device reliability, which will be discussed in the next section. All these problems are summarized in Fig. 1 from the point of view of

TABLE 1
Sources and related effects of various contaminations

Contamination	Possible source	Effects
Particles	Equipment, ambient, gas, deionized (DI) water, chemical	Low oxide breakdown Poly-Si and metal bridging–induced low yield
Metal	Equipment, chemical, reactive ion etching (RIE), implantation ashing	Low breakdown field Junction leakage Reduced minority lifetime V_t shift
Organic	Vapor in room, residue of photoresist, storage containers, chemical	Change in oxidation rate
Microroughness	Initial wafer material, chemical	Low oxide breakdown field Low mobility of carrier
Native oxide	Ambient moisture, DI water rinse	Degraded gate oxide Low quality of epi-layer High contact resistance Poor silicide formation

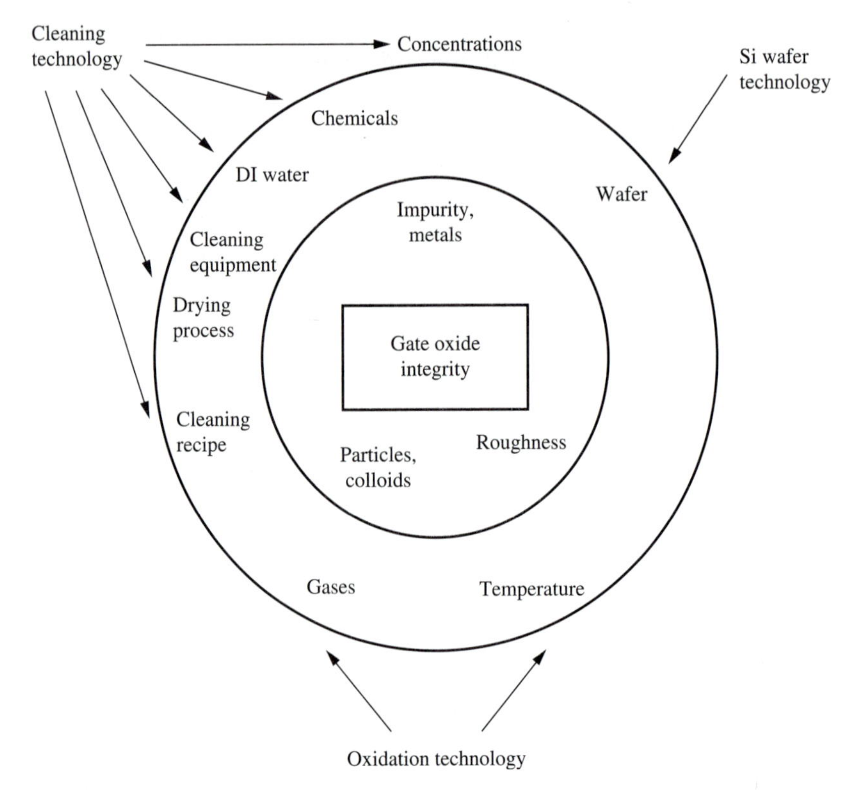

FIGURE 1
Interrelation of different factors on the gate oxide. (*After Verhaverbeke, Ref. 2.*)

gate oxide integrity.[2] These problems are important because gate oxide quality is one of the critical steps that determine the yield, reliability, and performance of a ULSI circuit. Problems due to roughness, impurity, and particles can be the result of cleaning technology (concentrations, chemicals, water, cleaning equipment, dry process, and cleaning recipe), oxidation technology (gases, temperature), and Si wafer technology (initial wafer). Because chemicals, deionized (DI) water, and Si wafers are available in high purity, the sources of most contaminations are tools and processes rather than the environment and operators. Although many cleaning processes have been developed to remove contaminants, the most important thing is to avoid contamination rather than to clean it up during processing. The basic concepts of each issue are discussed in the following section. The reliability problems related to each contaminant will also be discussed.

2.2.1 Mechanisms for Removing Particles and Contaminations

The mechanisms of removing particles and contaminations are addressed in the following paragraphs.

TABLE 2
Particle concentration (number/mL) in ULSI-grade semiconductor chemicals[6]

	≥ 0.2 μm	≥ 0.5 μm
NH_4OH	130–240	15–30
H_2O_2	20–100	5–20
HF	0–1	0
HCl	2–7	1–2
H_2SO_4	180–1150	10–80

Particles

Particle adhesion to a silicon surface usually occurs during the process from the equipment, ambient, gas, chemicals, and DI water. The particle contamination in ULSI-grade semiconductor chemicals is shown in Table 2. H_2SO_4 has the highest number of particles and HF the lowest. Figure 2 shows that the yield of gate oxide is reduced as the number of particles is increased.[3] The requirement of the particle size is scaled down with the minimum feature size of the technology.[4] The general guide for particle size is one-tenth of the device feature size. By the year 2000, 0.2-μm device features will require a particle size smaller than 0.02 μm.

To control particles, it is necessary to understand the mechanisms of adhesion and removal of particles. Particle adhesion is dominated by several mechanisms: forces due to static charge on the particle or to van der Waals forces, forces due to the formation of an electrical double layer, forces due to capillary action around the particle, and a chemical bond between the particle and the surface. Particle removal mechanisms can be classified into four types:

1. Dissolution
2. Oxidizing degradation and dissolution
3. Liftoff by slight etching of the wafer surface
4. Electric repulsion between particles and the wafer surface

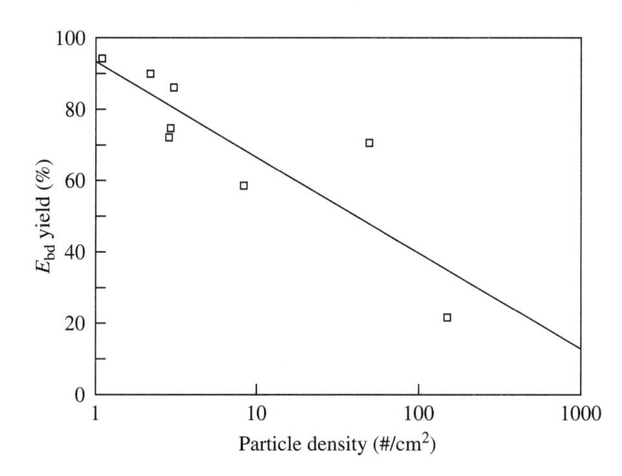

FIGURE 2
Dependence of E_{bd} yield on particle density. (*After Werkhoven et al., Ref. 3.*)

The oxidation and electrical repulsion mechanisms are illustrated in Figs. 3*a* and *b*.[5] Figure 3*a* shows the oxidation of a particle to make it soluble. Figure 3*b* shows the particle being removed from the surface by the electric repulsion. The SC-1 chemistry has both mechanisms. Hydrogen peroxide can oxidize the silicon surface, and OH^- from the NH_4OH can provide the negative charges on the silicon surface and the particle.

The deposition of the particle is a strong function of the pH value of the solution. Figure 4 shows the number of particles deposited as a function of pH.[6] Increasing the pH to 10 results in low particle deposition. Clearly, maximum particle deposition is observed in a highly acidic solution. A comparison of the particle-removal efficiency of various cleaning processes is shown in Table 3; the SC-1 was found to have the highest removal efficiency.

Some methods have been developed to improve the efficiency of removing particles. The most effective commercial method is the *megasonic* cleaning process.[7] SC-1 in conjunction with the megasonic cleaning process removes organic and inorganic particles from the surface at a temperature of less than 40°C. Figure 5*a* shows the detailed schematic of a static megasonic cleaning tank, which is noncontact and brushless, for removing particles ($< 1\ \mu m$) simultaneously from both sides of a silicon wafer. The model of the megasonic action on a particle on a silicon wafer surface is shown in Fig. 5*b*. Sonic energy from the transducer is directed parallel to the silicon surfaces, which are immersed in the SC-1 solution. The high-power (up to 300 W) and high-frequency (850 to 900 kHz) sonic pressure waves wet the particle first, and the solvent diffuses into the interface. Finally, the particle is totally wetted and is removed from the surface as a floating free particle. This particle-cleaning process is very useful in cleaning after the newly developed chemical-mechanical polishing (CMP) process to remove the slurry on the surface. In addition to particle removal, the megasonic process with SC-1 is reported to remove copper contamination effectively.[7] This is because the metal exchanges ions readily with the ammonia and therefore can be rinsed off.

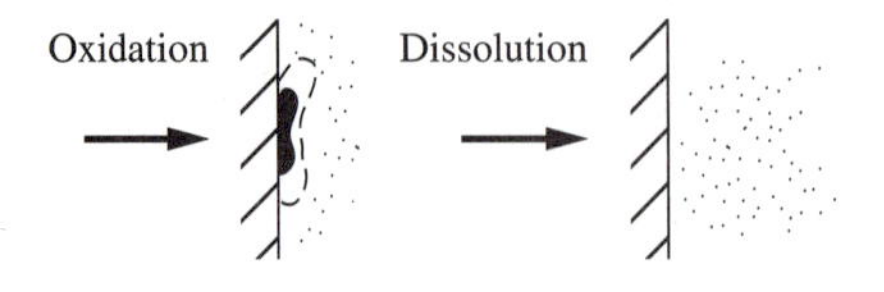

(*a*) in a powerful oxidizing agent

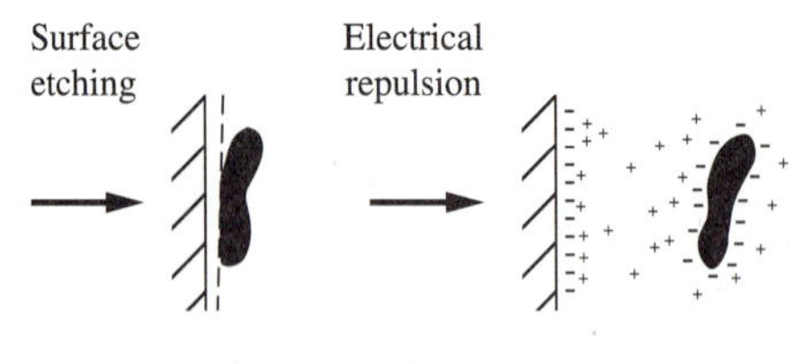

(*b*) in an alkaline solution

FIGURE 3
Acidic and alkaline wet chemical mechanisms to remove particles, organics, and metallics: (*a*) in a powerful oxidizing agent, (*b*) in an alkaline solution. (*After Singer, Ref. 5.*)

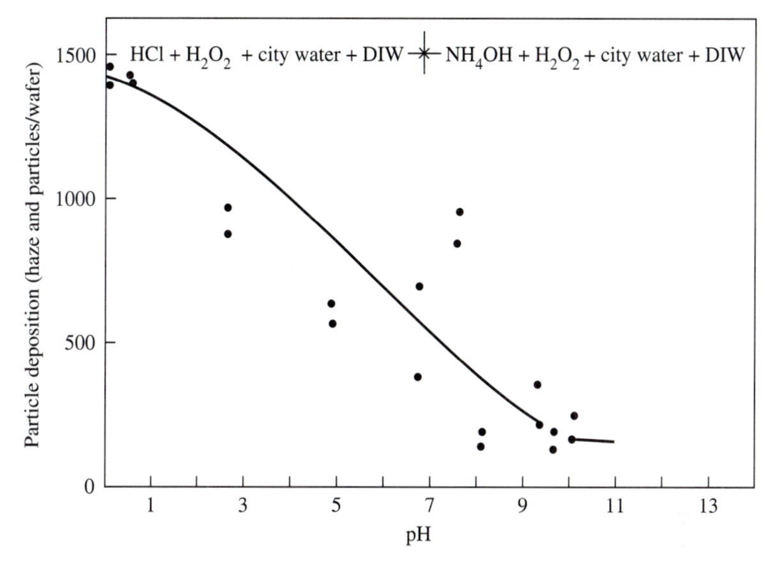

FIGURE 4
Deposition of tap water particles onto wafer surfaces as a function of seed solution pH. 10 percent by volume of tap water was mixed with ultrapure water. Acidity/basicity was controlled by the addition of either HCl with H_2O_2 or NH_4OH with H_2O_2. (*After Itano et al., Ref. 6.*)

Two points must be considered in using this process. One is that enough energy must be imparted to the particle so that it moves far enough from the surface not to readhere; this problem can be reduced by adding a continuous filtration system to trap the free particles from the surface into the solution. The other problem is that the concentration of the H_2O_2 will be reduced after long operation. Consequently, the smoothness of the silicon surface will be degraded by etching by the NH_4OH. This will be discussed in the microroughness section.

TABLE 3
Particle contamination removal efficiencies for various cleaning processes[6]

	Cleaning			
Pollution source	**SPM**[†]	**SC-1**[‡]	**SC-2**[§]	**PM**[¶]
City water	98.4	98.9	86.0	42.4
SiO_2	83.3	98.4	97.1	94.7
PSL[††]	91.7	99.2	55.2	7.2
Environmental air	95.8	96.3	86.9	0.

[†] SPM (Sulfuric–peroxide mixture): H_2SO_4: H_2O_2: (4:1) 5 min.
[‡] SC-1: NH_4OH: H_2O_2: H_2O (0.1:1.5) 10 min at 80–90°C.
[§] SC-2: HCl: H_2O_2: H_2O (1:1:6) 10 min at 80–90°C.
[¶] PM (peroxide mixture): H_2O_2: H_2O (1:5) 10 min at 80–90°C.
[††] Polystyrene latex sphere.

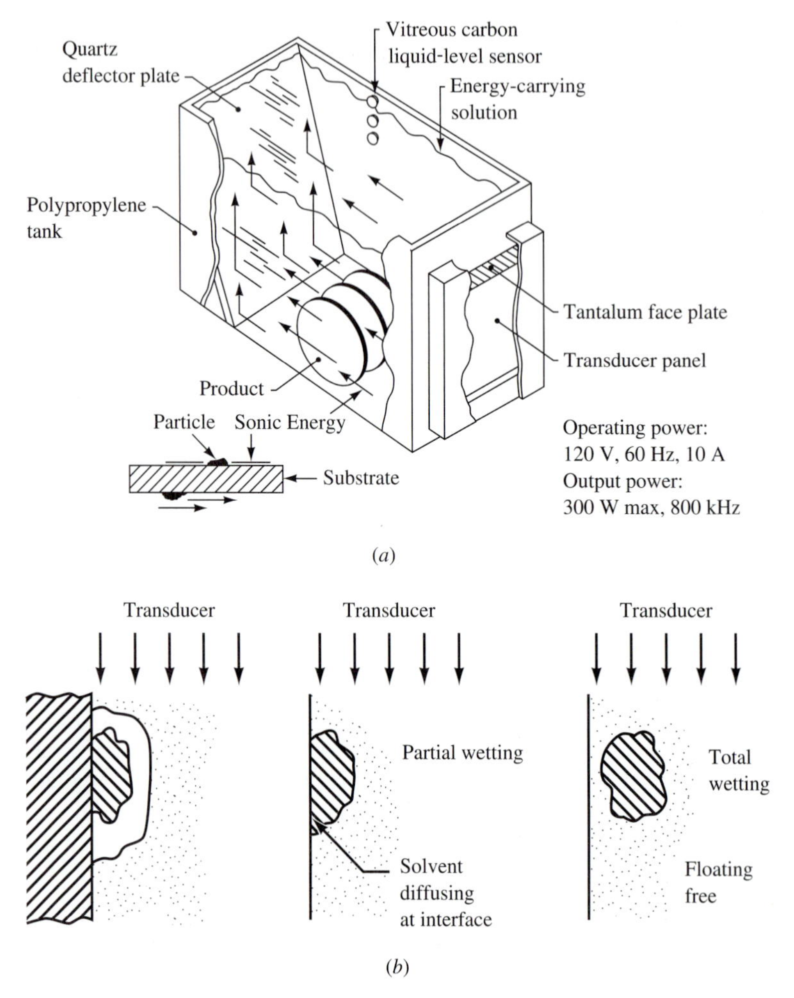

FIGURE 5
(*a*) Detailed schematic of static megasonic cleaning tank. The transducer plate incorporates eight transducers. The water holder is not shown, for clarity. The megasonic action for particle removal is indicated schematically. In an alternative system design the transducers are mounted in the shape of strips, eliminating the need for the reflector. (*b*) Model of megasonic action on a particle held on a silicon wafer surface: frequency 900 kHz; pressure $9.1 \times 10^5 N/m^2$; maximum instantaneous velocity 30 cm/s; wavelength 1.3 mm; acceleration $\sim 10^5$ g; motion of H_2O molecule ~ 0.1 μm. (*After Shwartzman et al., Ref. 7.*)

Metal contamination

The source of metal contamination can be the chemical solution, the ion implantation, or the reactive ion etching (RIE) process. Metal contaminations start at a level of 10^{10} atom/cm^2. Such a low level is below the lowest limit of measurement of the commonly used total-reflection x-ray fluorescence (TRXF).[8] The effects of metal contamination on devices are structural defects at the interface,[9] stacking faults during later oxidation or epitaxial process,[10] increased leakage current of p-n junctions, and reduced minority carrier lifetime.[11] Figure 6 shows the metallic impurities on a silicon wafer surface resulting from different processes.[12,13] Steam oxidation has the lowest level, less than 10^{10} atom/cm^2. Ion implantation exhibits the largest metal contamination, at a level of 10^{12} to 10^{13} atom/cm^2.

There are two mechanisms for precipitation of metal impurities onto the Si surface. The first is the direct binding to the Si surface by the charge exchange between a metallic ion and hydrogen atom that terminated on the Si substrate.[13] This type of impurity is not easy to remove in the wet cleaning process. Metals of this kind are usually noble metal ions, such as Au, whose electronegativity is higher than that of Si and that tend to be neutralized by taking an electron from Si and to precipitate on the Si surface. The electronegativity of various metals is shown in Table 4.

The second mechanism for precipitation of metal impurities occurs when oxide forms on the surface and metallic impurities are included simultaneously. Metals such as Al, Cr, and Fe tend to oxidize when the Si surface is oxidized and are included in the oxide film. These metallic impurities can be removed by etching the oxide with dilute HF. The enthalpy of metallic oxide formation is shown in Table 5.

Currently, the wet cleaning process is the most effective method for removing metallic contaminations. HF–H_2O_2(HF : 0.5%, H_2O_2 : 10%) cleaning is effective in removing metallic impurities that bind directly with Si surface. Some additives, such as hydrocarbon-type surfactants, which are added to BHF (buffered HF) to improve its wettability of the Si surface, are also found to be effective in removing Cu at room temperature.[14]

Both SC-1 and SC-2 have the capability of removing metallic impurities on the Si wafer. This is believed to result from the high oxidizing mechanism of the H_2O_2. Metallic contamination on the surface also induces microroughness,[15] which is shown in Fig. 7. Compared to the RCA-cleaned wafer, surfaces contaminated with Ca show a pronounced increase in haze after annealing in Ar at 950°C for 5 min.

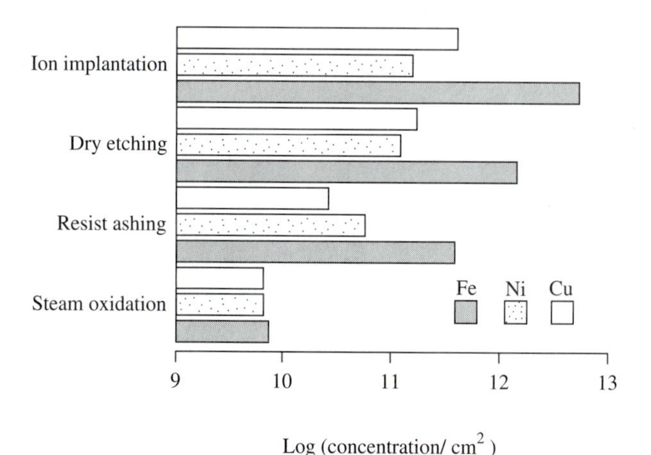

FIGURE 6
Contamination induced by various processes. (*After Anzai et al., Ref. 12, and Ohmi et al., Ref. 13.*)

TABLE 4
Electronegativity scale of metals:[13] $M^+ + e^- \rightarrow M$

Element	Electronegativity (Pauling)	Half-cell reduction potential (V)	
Au	2.4	1.68	↑
Pt	2.2	1.19	
Ag	1.9	0.80	
Hg	1.9	0.79	
Cu	1.9	0.34	
Si	1.8	0.10	Tendency to be precipitated on bare Si
Pb	1.8	−0.13	
Sn	1.8	−0.14	
Ni	1.8	−0.23	
Fe	1.8	−0.41	
Zn	1.6	−0.76	
Al	1.5	−1.66	
Mg	1.2	−2.34	
Ca	1.0	−2.87	
Na	0.9	−2.71	
K	0.8	−2.92	

The defect density for these contaminated wafers after oxidation is shown in Fig. 8. Fortunately, Cu contamination has little effect on the oxide quality and the surface roughness. The most detrimental effect is from the Ca contamination, which results in a rougher surface during the annealing process and also increases defect densities in the oxide. Figure 9 shows the yield as the function of Ca contamination.[16] As the Ca contamination increases to 10^{11} atom/cm^2, the yield drops abruptly from 80% to 20%. This suggests that the threshold value for Ca contamination is 10^{11} atom/cm^2 for a reasonable yield. Figure 10 shows a typical *I-V* curve of an oxide grown on a Ca-contaminated wafer.[17] In the low-field region the current is enhanced, because of the rough surface caused by Ca contamination, and results in early breakdown. Although

TABLE 5
Enthalpy of oxide formation:[13]
$M + O_2 \rightarrow MO_2$, $\Delta H < 0$ (heat-releasing process)

Oxide	ΔH_{25}^{293} (kJ/mol)	
Al_2O_3	−1675	↑
Cr_2O_3	−1130	
CrO_2	−583	
CrO_3	−580	Tendency to be included in the oxide film
Fe_3O_4	−1118	
Fe_2O_3	−822	
SiO_2	−909	
NiO	−241	
CuO	−155	

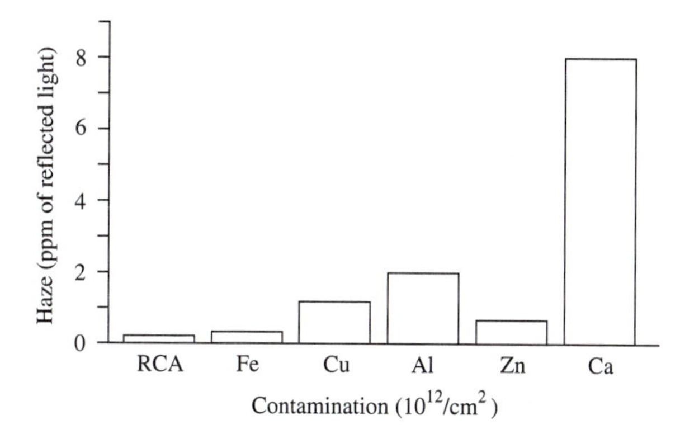

FIGURE 7
Haze number on contaminated Si wafers after annealing in Ar at 950°C for 5 min. (*After Verhaverbeke et al., Ref. 15.*)

Ca contamination will cause serious problems in the gate oxide, most of it can be removed by the HF-last cleaning steps and DI water rinsing. Table 6 shows typical metallic contamination after different last cleaning steps.[18] HF-dipping reduces the metal contamination most. The $HF–H_2O_2–H_2O$ cleaning process can remove more Ca than HF can, but larger amounts of the other contaminants remain.

Another heavy metal prevalent on the Si is Fe, due to the common use of stainless steel in the equipment and facility. Figure 11 shows the histogram of oxide failure for Fe contamination levels of 1×10^{13} to 5×10^{14} atom/cm^3.[9] When the Fe contamination is less than 1×10^{10} atom/cm^3, there is no effect on the failure rate, but when Fe contamination increases to 5×10^{13} atom/cm^3, all oxides exhibit an early breakdown. The contamination threshold decreases as the oxide thickness is decreased[9] as shown in Fig. 12. For an oxide only 100 Å thick, the threshold of Fe concentration is as low as 10^{11} atom/cm^3. Therefore, the Fe contamination level must be controlled after each process to achieve a high gate oxide yield.

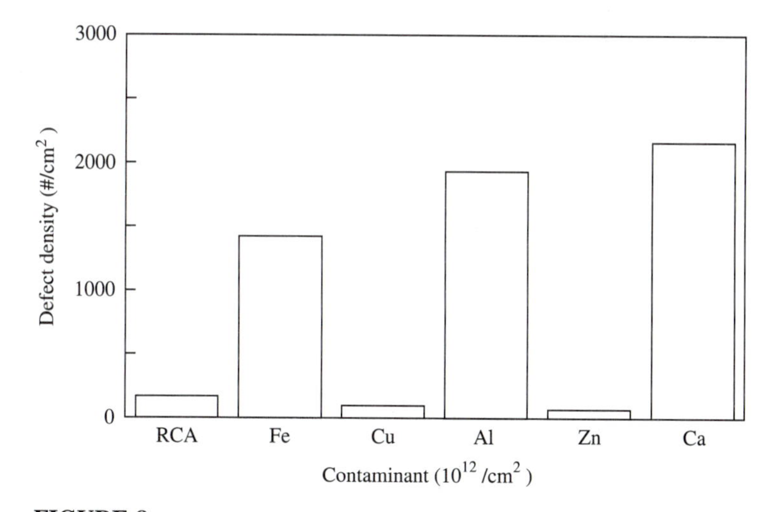

FIGURE 8
Oxide defect density on the RCA-cleaned and contaminated samples as obtained from extreme value statistics. The oxide thickness was 15 nm. (*After Verhaverbeke et al., Ref. 15.*)

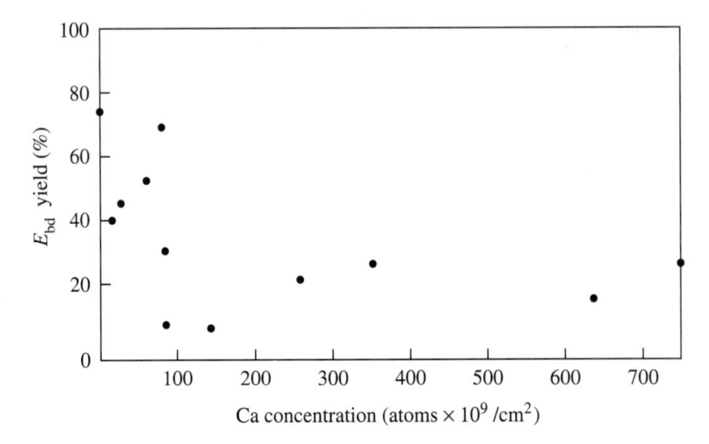

FIGURE 9
Relation between the Ca concentration and the yield. (*After Verhaverbeke et al., Ref. 16.*)

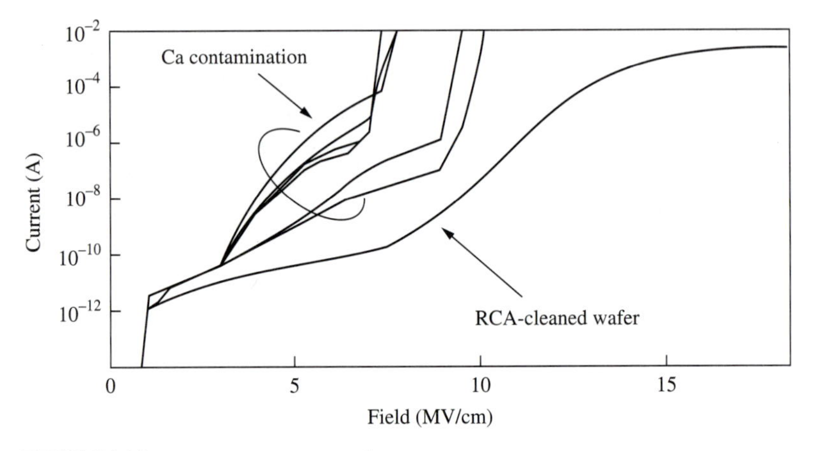

FIGURE 10
A typical *I-V* curve obtained from a Ca-contaminated wafer. (*After Verhaverbeke, Ref. 17.*)

TABLE 6
Typical metallic contamination after the last cleaning step followed by rinsing ($\times 10^{10}$ atom/cm^2), measured by VPD (vapor-phase decomposition)-TRXF[18]

	K	Ca	Cr	Fe	Ni	Cu	Zn
RCA	0.3	8.6	0.2	5.1	3.3	0.3	0.4
HF	0.1	3.8	0.05	0.3	0.1	0.06	0.1
HF–H_2O_2	0.6	1.6	0.3	2.2	0.2	0.09	1.2
BHF	0.2	1.4	0.4	2.6	0.3	3.7	0.7

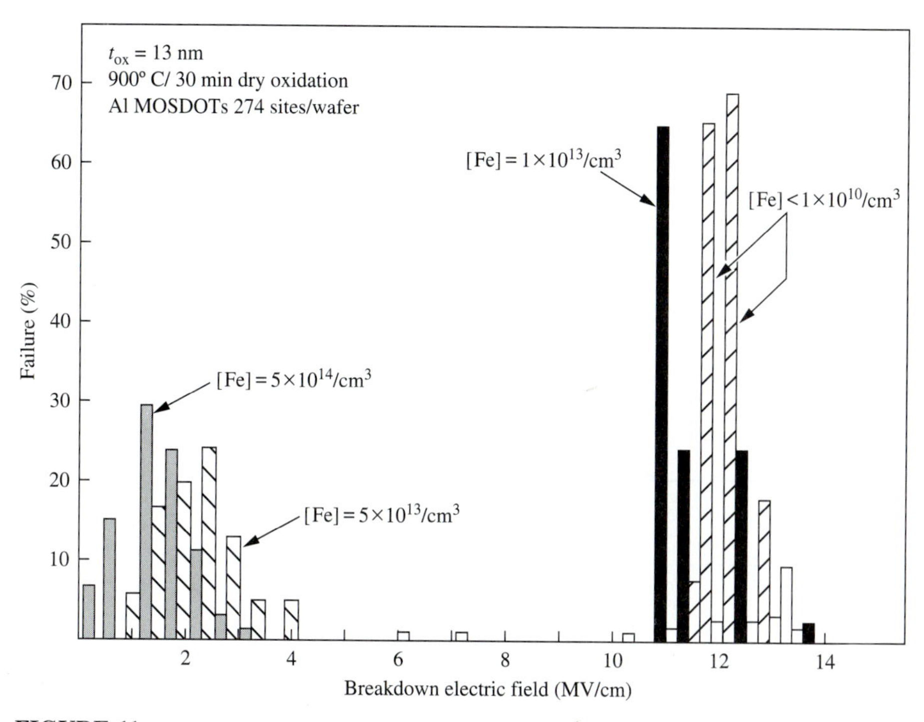

FIGURE 11
Combined summary breakdown field histograms for 13 nm oxides. All oxides were grown in dry O_2 at 900°C. The threshold contamination level for this oxide thickness is between 1×10^{13} and 5×10^{13} cm^{-3}. (*After Henley et al., Ref. 9.*)

Characterization of metal contamination level is important when one wants to control the metallic contamination. Characterization techniques are compiled in Table 7. Note that the sensitivity limits are quoted in cm^{-2} or cm^{-3}, corresponding to surface or bulk concentration. Atomic absorption spectroscopy (AAS), surface photovoltage (SPV), and total reflection x-ray fluorescence (TRXF) have the highest sensitivity. The SPV measurement can identify only Fe contamination. On the other hand, it is suitable for routine monitoring of Fe because it is nondestructive.

Organic contamination

Organic contaminations on the silicon surface are usually from organic vapor in the ambient, storage containers, and the residue of photoresist. The presence of organic impurities on the surface will cause incomplete cleaning of the surface, leaving contaminants such as the native oxide or metal impurities,[19,20] and will also cause a micromasking effect in the later RIE process. Residual photoresist is the main source of organic contamination in the IC processes. In the present cleaning processes, photoresist is usually stripped by a combination of dry ashing by ozone with subsequent wet cleaning in H_2SO_4–H_2O_2 = (3 : 1 to 4 : 1 at 120 to 130°C). Most of the organic residue is removed in the first dry ashing process; the wet process is employed to clean the wafer completely. However, the high cleaning temperature depletes the concentration of H_2O_2 and makes the process difficult to control. Recently, ozone-injected ultrapure water[21] or the sequence of UV and a filter system[22] to

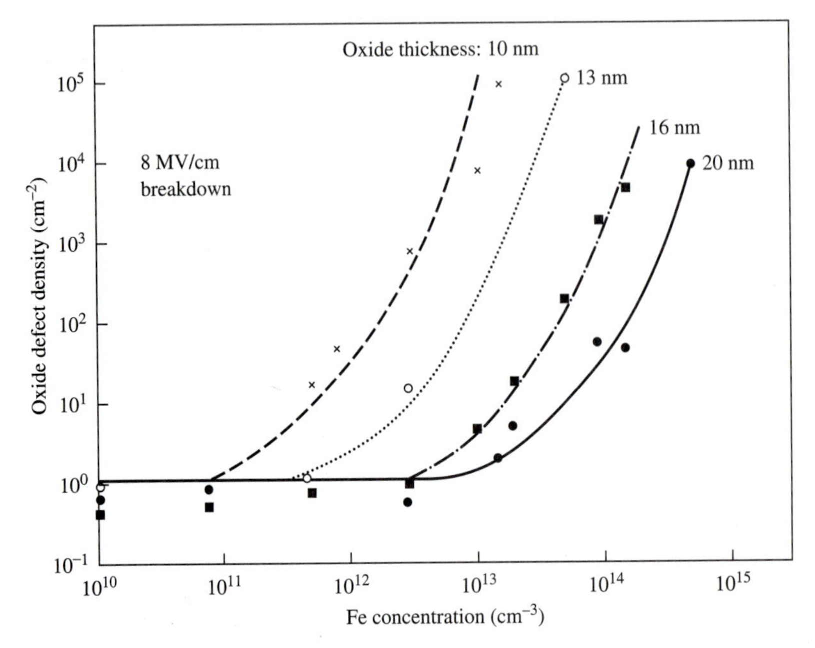

FIGURE 12
MOSDOT capacitor experimental data: oxide defect density versus iron contamination level for 900°C dry oxides. A defect is defined as breakdown occurring at less than 8 MV/cm during a ramp voltage *I-V* test. (*After Henley et al., Ref. 9.*)

remove organic impurities was proposed and designed. The mechanism is as follows: When ozone is dissolved in ultrapure water, it decomposes and becomes a strong oxidizing agent that decomposes organic impurities. This process has the advantages of lower temperature, simplicity, and reduced chemical consumption and chemical waste. However, this method causes native oxide to grow on the Si wafer because of

TABLE 7
Analytical techniques for the detection of metallic impurities in silicon[9]
Sensitivities are typical reported values for routine operation.

Techniques	Sensitivity	Chemical ID?	Destructive?	SPC†† line monitor?
SIMS†	10^{15} cm^{-3}	Yes	Yes	No
TRXF‡	10^{10} cm^{-2}	Yes	No	Maybe
AAS§	10^9 cm^{-2}	Yes	Yes	No
DLTS¶	10^{11} cm^{-3}	Yes	Yes	No
SPV††	10^9 cm^{-3}	Iron only	No	Yes

† Secondary-ion mass spectroscopy
‡ Total-reflection x-ray fluorescence
§ Atomic absorption spectroscopy
¶ Deep-level transient spectroscopy
†† Surface photovoltage
†† Standard process control

the ozone. The oxide thickness increases as the immersion time increases, and it also increases with the concentration of ozone. Figure 13 shows the native oxide growth caused by ozone-injected ultrapure water.[21] Rinsing in 2 ppm ozoned ultrapure water results in a very rapid growth of native oxide. The surface roughness using the ozone-injected cleaning process is the same as that from the H_2SO_4–H_2O_2 cleaning process.

Figure 14 shows the influence of residual surfactant molecule on the silicon wafer surface.[21] From the measurement of x-ray photoelectron (Si_{2p}) spectra, the wafer with the surfactant still has native oxide on the surface after the dilute HF dip (DHF) process. The other sample, with ozone treatment, exhibits an almost native-oxide-free surface after the DHF process. The organic contaminant–free surface resulting from ozone treatment helps the subsequent cleaning to function properly.

The contact angle of a droplet of DI water on the silicon surface is usually used to indicate the chemical condition of the surface. With poor passivation the surface free energy is high and results in a low contact angle (Fig. 15*a*), whereas a surface with good passivation exhibits a high contact angle (Fig. 15*b*). Figure 15*c* shows the contact angle as the function of the ozone concentration on a Si surface with absorbed surfactant.[21] The contact angle of the absorbed surfactant on Si surface is around 56.6°. After 10 min rinsing in ultrapure water, the contact angle is only 62°, which is lower than the initial contact angle, 78.9°. However, if the ozone concentration increases to as little as 0.28 ppm, the contact angle increases to the initial value at the same time. Increasing the concentration to 1.5 ppm reduces the clean time. Hence, the ozone-injected ultrapure water is an effective cleaning process to remove

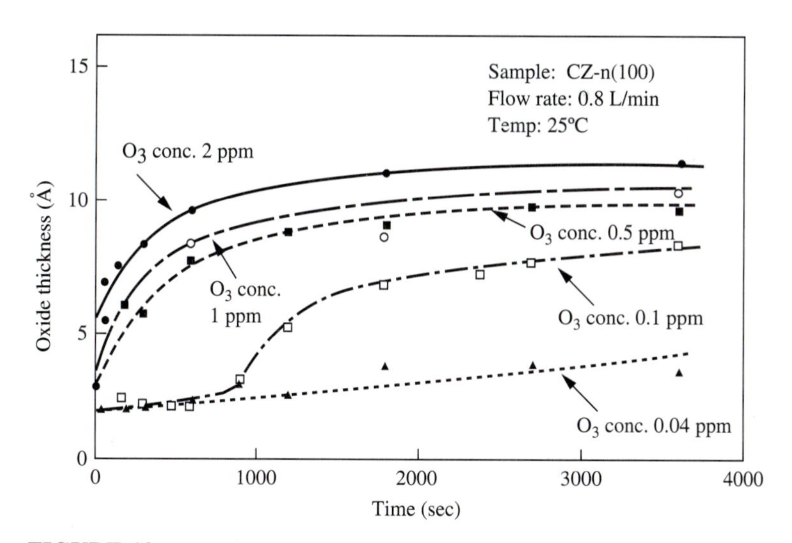

FIGURE 13
Native oxide growth on the wafer surface by ozone-injected ultrapure water as a function of dipping time and ozone concentrations. The horizontal axis shows immersion time, and the vertical axis shows the thickness of the native oxide formed in the ozone-injected ultrapure water. Ozone concentrations are 2, 1, 0.5, 0.1, and 0.04 ppm. Very rapid growth of native oxide is confirmed in the 2 ppm ozonized ultrapure water. (*After Ohmi et al., Ref. 21.*)

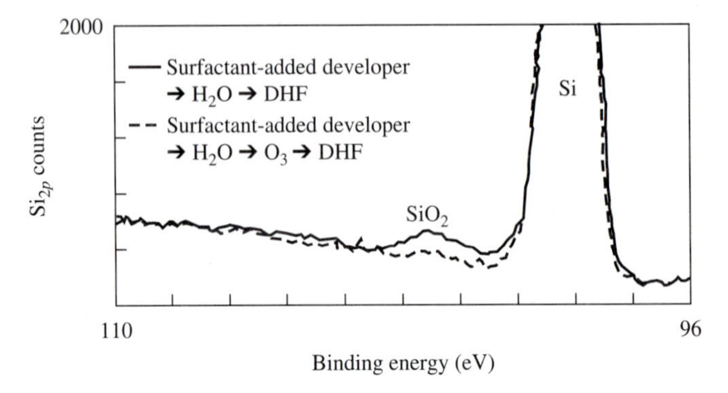

FIGURE 14
Influence of residual surfactant molecules on the silicon wafer surface. The solid line is the sample without the ozone-injected ultrapure water cleaning; the broken line is the cleaned sample. (*After Ohmi et al., Ref. 21.*)

the residual organic impurities on Si surface. It is a room-temperature process and can replace the conventional H_2SO_4–H_2O_2 process.

Surface microroughness

Surface microroughness is an important factor in the manufacture of high-performance and high-reliability submicrometer and deep submicrometer ULSIs.[23–26] This is because the thickness of the thin dielectric, which is grown after the conventional RCA cleaning process, is reduced to less than 100 Å for 0.35 μm technology. For the 0.1 μm technology of the year 2000, the thickness of the gate oxide will be reduced to only 40 Å. Then the surface requirement should be atomic flatness. In the conventional RCA cleaning processes, NH_4OH–H_2O_2–H_2O is used as the first step to remove particle, organic, and some metal impurities. It has been reported that this conventional process with NH_4OH–H_2O_2–H_2O = 1 : 1 : 5 at 70°C for 10 to 15 min results in a rough surface. The mechanism producing this rough surface is that the NH_4OH acts as the etchant of the oxide while H_2O_2 acts as the oxidant. The processes of etching and oxidizing progress simultaneously in the SC-1 solution; consequently, the surface becomes rough. Methods of reducing the microroughness can be summarized as

1. Reduce the proportion of NH_4OH (the etchant).
2. Reduce the temperature of the bath.
3. Reduce the cleaning time.

The scanning tunneling microscope (STM) is the most commonly used instrument to investigate the topography of the silicon wafer. Note that the native oxide on the silicon surface affects the results of an STM study. The existing native oxide reduces the tunneling current and influences the resolution. The atomic force microscope (AFM) is also used to measure the topography of an insulator, such as thin oxide. Figure 16 shows the surface microroughness of an n-type Czochralski (CZ) wafer as a function of various etching solutions (10 min immersion) such as HF (0.5%), conventional BHF (NH_4F: 35 to 38%), and advanced surface-active BHF

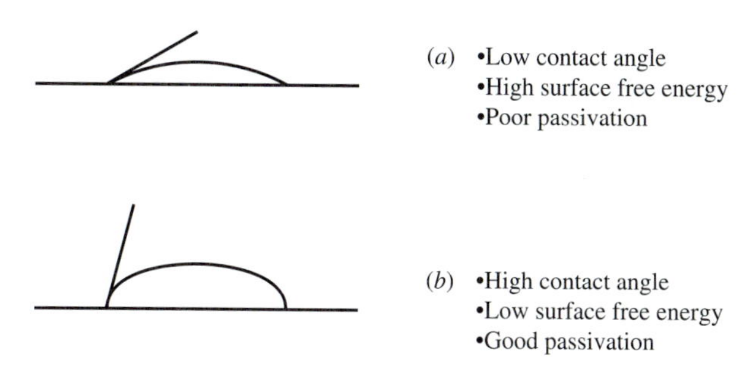

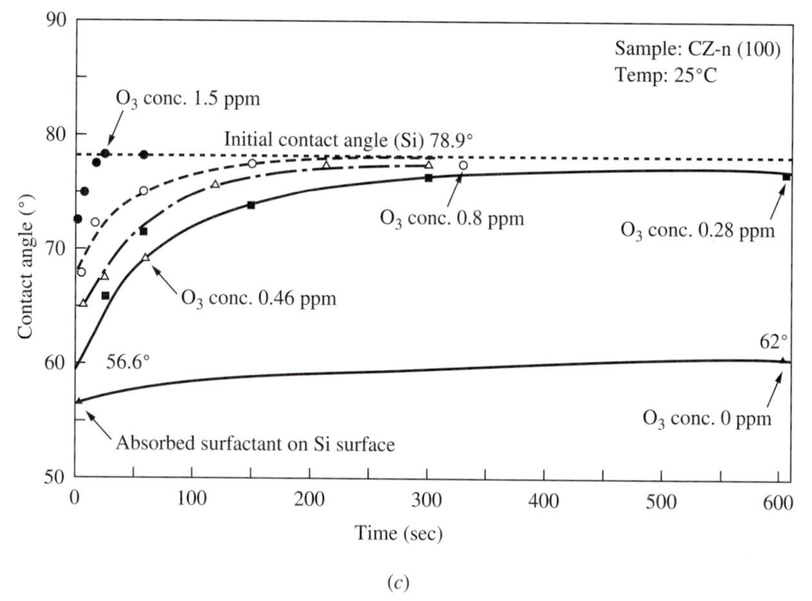

(c)

FIGURE 15
Schematic representation of the contact angle measurements for (*a*) poor passivation and (*b*) good passivation. (*c*) Time dependence for eliminating absorbed surfactant molecule from the silicon surface using ozonized ultrapure water, as measured by the contact angle. The horizontal axis shows immersion time; the vertical axis shows contact angle. Ozone concentration in the ozone-injected ultrapure water was varied. Very rapid recovery, up to the initial contact angle, has been observed under 1.5 ppm ozone concentration. (*After Ohmi et al., Ref. 21.*)

(NH_4F : 17%) with surfactant.[25] It is clearly seen from the figure that the average roughness, R_a, is largest with conventional BHF, followed by the HF. The advanced BHF has the same level as the control sample treated in advanced BHF for 1 min.

The R_a values of the other conventional cleaning processes are shown in Fig. 17:

H_2SO_4(98%)–H_2O_2(30%) (4 : 1) (SPM)
NH_4OH(28%)–H_2O_2(30%)–H_2O (1 : 1 : 5) (SC-1)
HCl(36%)–H_2O_2(30%)–H_2O (1 : 1 : 6) (SC-2)

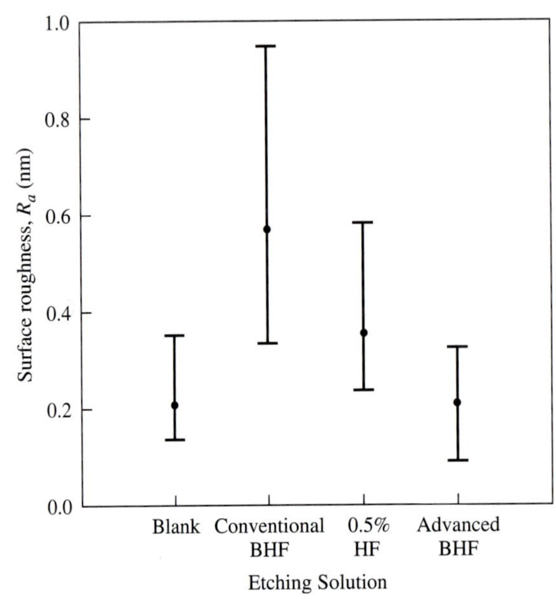

FIGURE 16
Relationship between the average surface microroughness (R_a) of an n-type CZ wafer and various etching solutions. A "blank" wafer is a wafer treated in advanced BHF for 1 min. (*After Miyashita et al., Ref. 25.*)

No significant increase of roughness is found for the SPM and SC-2 cleaning processes. However, the SC-1 cleaning process has the largest R_a and deviation.[24] The proportion of NH_4OH in the cleaning solution has a significant effect on microroughness. Figure 18*a* shows the dependence of the surface microroughness of CZ and floating-zone (FZ) wafers on the SC-1 cleaning process with various NH_4OH ratios.[25] The figure shows that the R_a is reduced to the level of SC-2 cleaning when

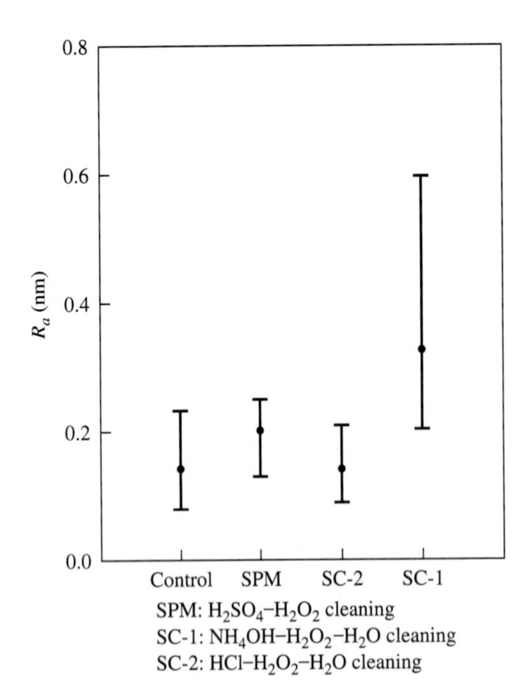

FIGURE 17
Surface microroughness on a silicon wafer treated with SC-2 cleaning, SC-1 cleaning, and fourth-time cyclic SPM cleaning. (*After Ohmi et al., Ref. 24.*)

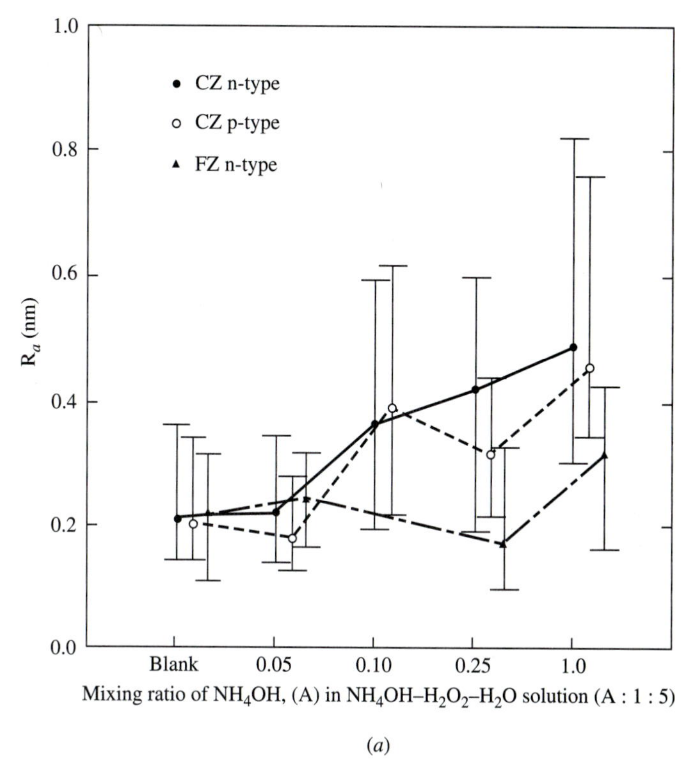

(*a*)

FIGURE 18
Relationship of surface microroughness of (*a*) CZ and FZ wafers having various NH_4OH mixing ratios in SC-1 solution. (*After Miyashita et al., Ref. 25.*)

the ratio of NH_4OH is reduced from 1 to 0.05. The FZ wafer exhibits weak dependence on the NH_4OH ratio, because of the lower Si vacancy concentration in an FZ wafer than in the CZ wafer; a 4-hr wet oxidation at 1000°C can overcome this problem.[25] The surface microroughness of a wafer surface treated by a complete RCA cleaning process with four different mixing ratios in SC-1 is shown in Fig. 18*b*. As the ratio is increased, the surface becomes rougher. The succeeding steps of SC-1 cleaning also influence the surface condition. Rinsing at high temperature results in a larger R_a.[24] Hence, for the advanced SC-1 cleaning process the ratio of NH_4OH must be reduced and the wafers rinsed in DI water at room temperature for an atomically flat surface.

One problem that must be taken into consideration is that when the concentration of NH_4OH is reduced, the efficiency of particle and metal impurity removal is reduced. The particle removal efficiencies as a function of NH_4OH mixing ratio are shown in Fig. 18*c*; in fact, the particle removal efficiency optimizes at a ratio from 0.05 to 0.10. Nor does the metal-impurities removal efficiency change when the ratio of NH_4OH is reduced from 1.0 to 0.05 (Fig. 18*d*).

The etched thermal oxide depth in various SC-1 mixtures as a function of NH_4OH concentration at 70°C for 10 min is shown in Fig. 19. The etching rate

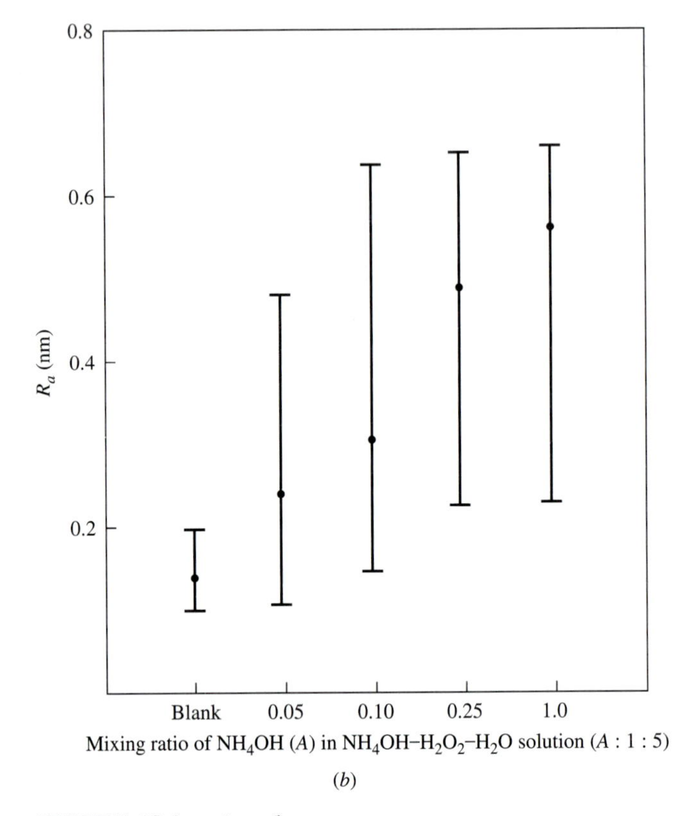

FIGURE 18 (*continued*)
(*b*) silicon wafers treated in an entire RCA cleaning process with four different NH_4OH mixing ratios in SC-1 solution. (*After Miyashita et al., Ref. 25.*)

increases as the concentration of NH_4OH increases. For the conventional concentration the etching rate is 3 nm for a 10-min immersion at 70°C. The higher the bath temperature, the higher the etching rate.

Hence, to avoid microroughness the ratio of NH_4OH can be reduced to between 0.05 and 0.25. This ratio can still maintain the efficiency of removing particles and metallic contaminations.

The effects on reliability of the gate oxide and the device performance resulting from the microroughness have been discussed.[26] The dependence of the breakdown field on microroughness R_a is shown in Fig. 20*a*. The larger the R_a, the lower the breakdown field. The trend is the same for oxide thicknesses of 9.8 and 8.2 nm. Figure 20*b* shows the charge to breakdown, Q_{bd}, versus the surface microroughness.[26] As R_a increases to 0.8 nm, the Q_{bd} value decreases almost an order of magnitude.

In addition to the influence on the quality of oxide, microroughness also has a significant influence on the performance of the devices. Typical current–voltage characteristics for two MOS transistors with average surface microroughness of 0.6 and 0.3 nm are shown in Fig. 21. The transistor with an R_a of 0.6 nm has lower current derivability than that with an R_a of 0.3 nm. The channel electron mobility,

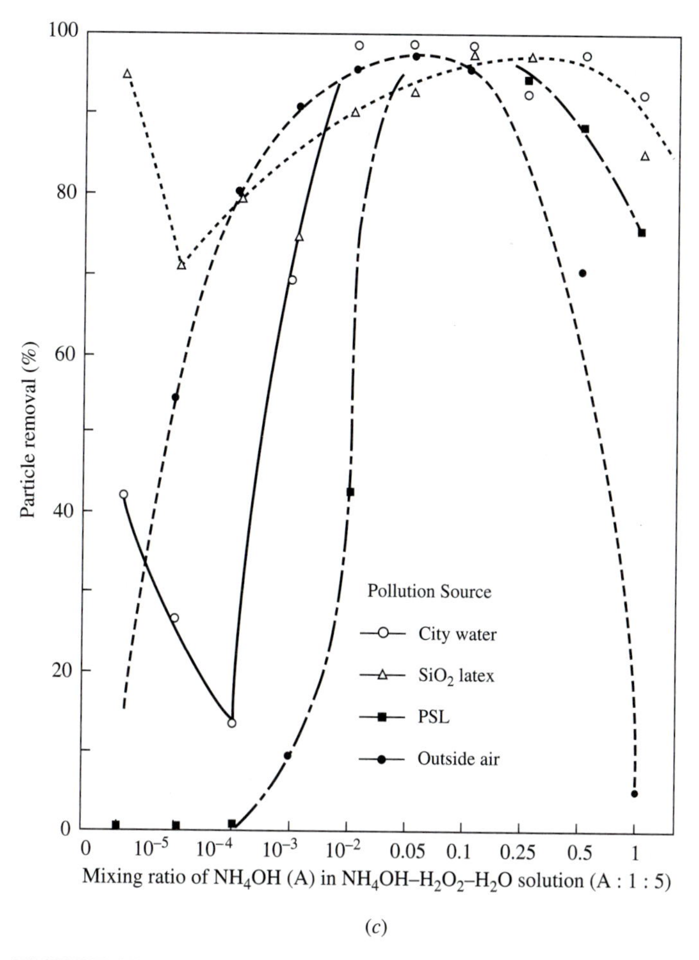

FIGURE 18 (*continued*)
(*c*) Particle removal efficiencies as a function of NH_4OH mixing ratio in SC-1 cleaning for four different particles: PSL (polystyrene latex spheres), silica latex, particles from city water, and particles from outside air. (*After Ohmi et al., Ref. 24.*)

as a function of SiO_2–Si interface microroughness, is shown in Fig. 22. The mobility of electrons in the sample with $R_a = 1.5$ nm reduces to 35% of the reference value, 0.2 nm. These results clearly indicate the importance of surface microroughness for MOS transistor performance, especially in speed and reliability.

Native oxide

Native oxide that is free on a silicon surface becomes very important in the advanced ULSI process.[27–29] A silicon surface with native oxide results in uncontrollable ultrathin oxide growth, high contact resistance, and the inhibition of selective chemical vapor deposition (CVD) or epitaxy. In addition, some metallic impurities are included in native oxide during the cleaning process. If this native oxide is not completely removed, it serves as the source of metallic impurities, which diffuse

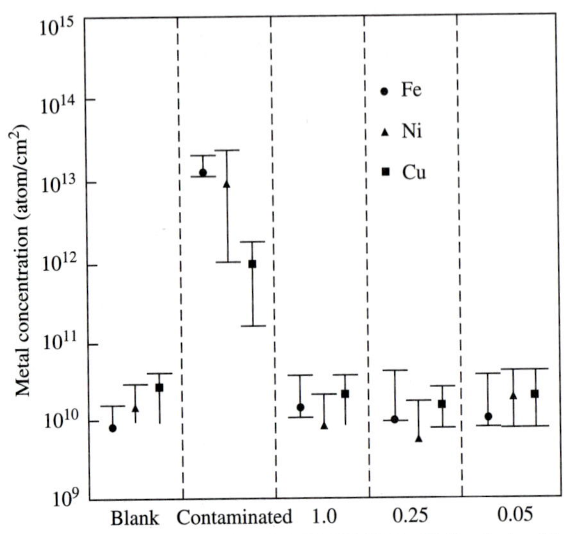

FIGURE 18 (*continued*)
(*d*) Metallic impurities Fe, Ni, and Cu removal efficiency by SC-1 cleaning with three different NH_4OH mixing ratios. (*After Ohmi et al., Ref. 24.*)

into the silicon or precipitate at the interface of the SiO_2–Si, resulting in defects in the succeeding high-temperature process. Therefore, a silicon surface that is native oxide–free is a key factor in obtaining high performance and reliability of semiconductor devices. Native oxide thickness is usually determined by x-ray photoelectron spectroscopy (XPS) and ellipsometry. It is difficult to determine the

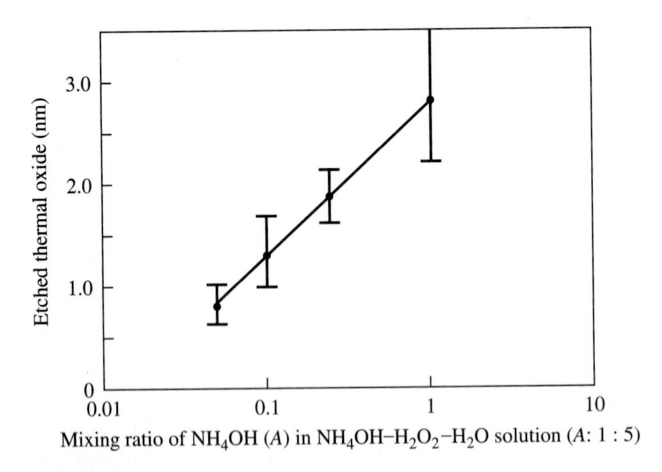

FIGURE 19
Etch depth of thermal oxide after 10 min in various SC-1 mixtures as a function of ammonia concentration of the mixture at 70°C. (*After Verhaverbeke, Ref. 2.*)

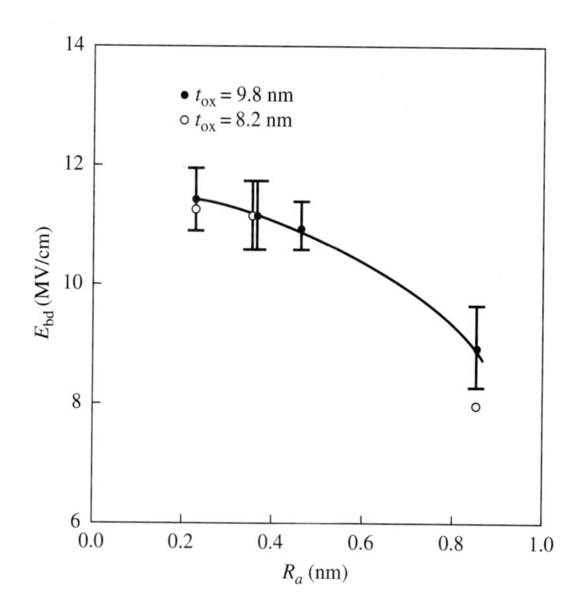

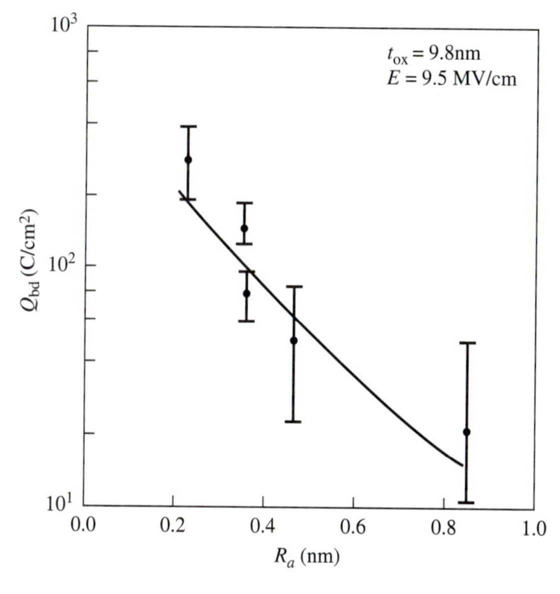

FIGURE 20
(*a*) Dielectric breakdown field intensity (E_{bd}) plotted as a function of average surface microroughness for two oxide films, having thicknesses of 9.8 and 8.2 nm formed on an n-type substrate. (*b*) Surface microroughness dependence of Q_{bd} (C/cm^2) under a constant field of 9.5 MV/cm. (*After Ohmi et al., Ref. 26.*)

thickness and refractive index simultaneously with an ellipsometric measurement when the thickness is less than 200 Å. Hence, the native oxide thickness is determined by using a fixed refractive index of 1.46. The Si_{2p} spectra of XPS is also used to measure the thickness of native oxide. However, this method does not obtain the result directly. The area ratio of the signal from thicker oxide to that of the silicon substrate must be calibrated first by ellipsometry. Then the thickness of the native oxide can be obtained from the area of the XPS.

The growth of native oxide in air at room temperature (23.7°C) with a humidity of 42% is covered in Ref. 30. The oxide thickness increases as the exposure time increases. The thickness of the native oxide shows a steplike increase of 2 to 3 Å. This

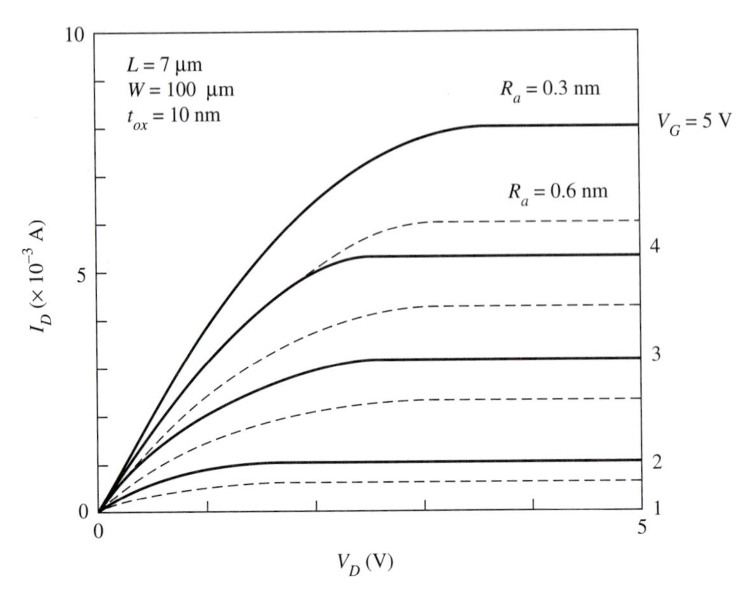

FIGURE 21
Typical current-voltage characteristics for two MOS transistors with average surface microroughness of 0.6 nm (dashed line) and 0.3 nm (solid line). (*After Ohmi et al., Ref. 26.*)

can be considered as a layer-by-layer growth of native oxide on a wafer at room temperature. N-type silicon with a high doping concentration (10^{20} atoms cm^{-3}) grows thicker native oxide than less-doped silicon (10^{15} cm^{-3}). After 7 days of exposure to cleanroom air containing 1.2% H_2O, the thickness of native oxide is 6.7 Å, which is larger than that of samples in an $N_2 + O_2$(1.7Å) or N_2(1.9Å) environment with moisture less than 0.1 ppm.[27]

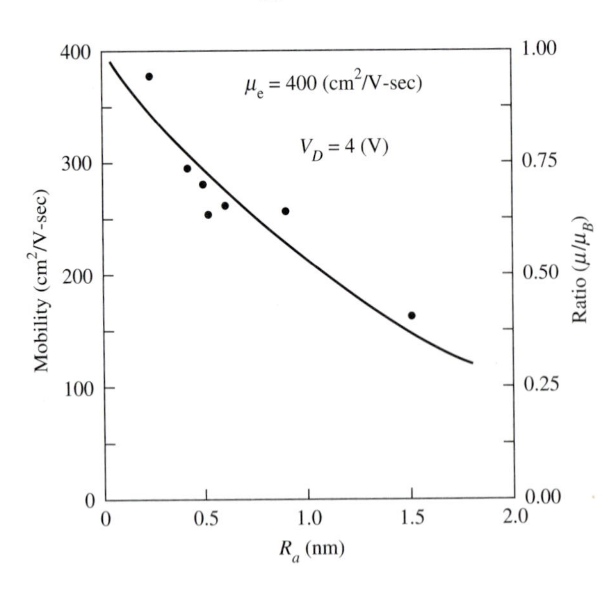

FIGURE 22
Relationship of channel electron mobility (μ_e) to the Si–SiO_2 interface microroughness obtained in the current saturation region. The channel electron mobility normalized by the bulk electron mobility is also plotted. (*After Ohmi et al., Ref. 26.*)

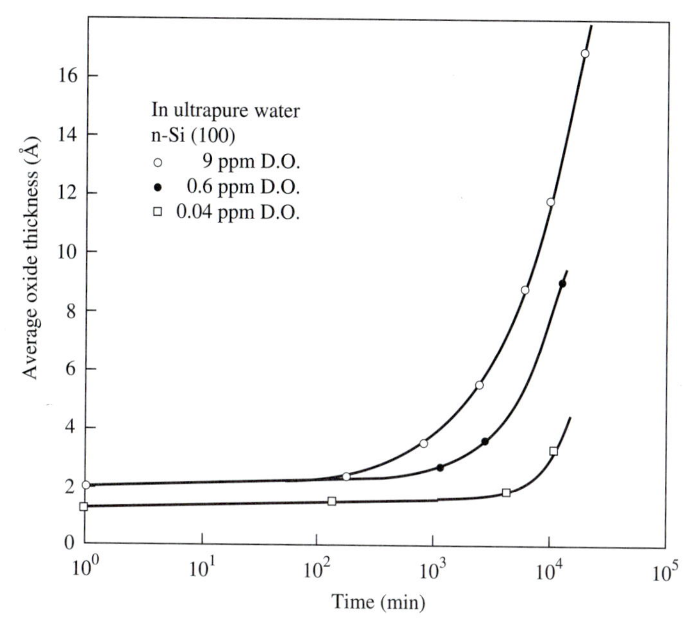

FIGURE 23
Oxide thicknesses of wafers as a function of immersion time in ultrapure water at room temperature for different dissolved oxygen (D.O.) concentrations. (*After Morita et al., Ref. 30.*)

The dissolved oxygen in DI water enhances the growth of native oxide.[30] Figure 23 shows the oxide thickness of wafers as a function of immersion time in ultrapure water for different dissolved oxygen concentrations. The thickness of native oxide increases as the dissolved oxygen concentration increases. The curves increase parabolically, unlike those of layer-by-layer growth in air. Si dissolves into the water simultaneously during native oxide growth. The number of dissolved Si atoms is larger than the number of Si atoms in native oxide by one order of magnitude. This will result in a rougher surface.

A model of how native oxide grows in air and in ultrapure water has been proposed.[30] Figure 24 shows the model of native oxide grown in air. The oxygen

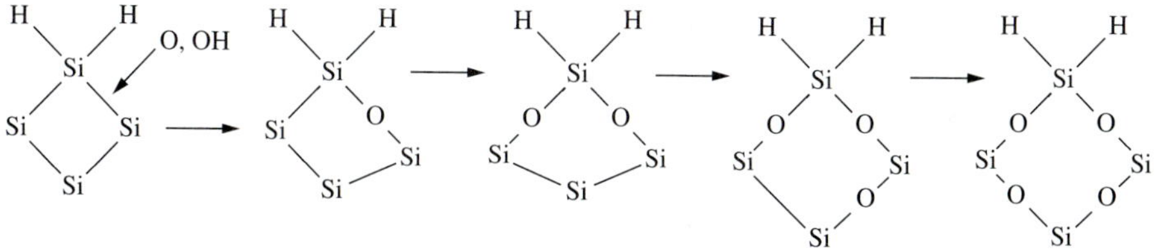

FIGURE 24
A model of native oxide growth in air. (*After Morita et al., Ref. 30.*)

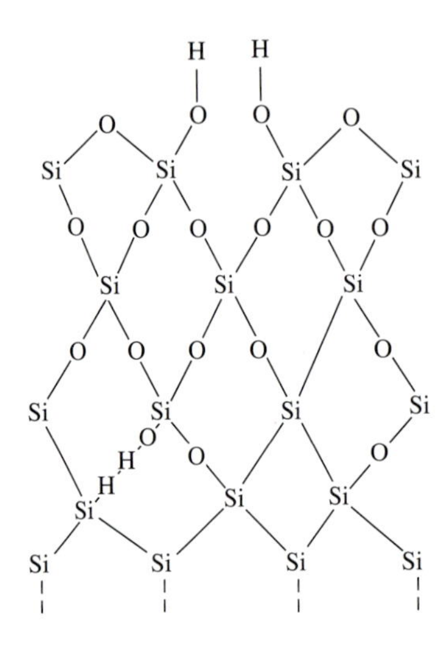

FIGURE 25
A model of native oxide growth in ultrapure water. *(After Morita et al., Ref. 30.)*

species break Si–Si bonds first. The surface still remains hydrogen-terminated and hydrophobic. For the surface grown in ultrapure water shown in Fig. 25, the surface is terminated by O or OH; the Si–H bond exists only at the SiO_2–Si interface.

Some methods have been proposed to reduce the dissolved oxygen.[31] The first stage is a degassing-membrane module. Oxygen in the water rapidly diffuses through the membrane, which is surrounded by a low-pressure vacuum. After this stage, the oxygen concentration is reduced to 300 to 400 ppb; then two methods can be employed. The first is reduction with catalytic resins by N_2H_4 or H_2 (Pd catalyst) according to the following reactions:

$$O_2 + N_2H_4 \rightarrow 2H_2O + N_2$$

or

$$O_2 + 2H_2 \rightarrow 2H_2O$$

The residual concentration of dissolved oxygen can be reduced to 5 ppb or less by using this method. The other method is N_2 gas bubbling, which reduces the partial pressure of oxygen in the water. Theoretically, the oxygen concentration can be reduced to less than 0.1 ppb with this method.

An Al–Si Schottky diode contact made with a native oxide–free process exhibits a great improvement, such as a low contact resistance of 0.4 $\mu\Omega\text{-cm}^2$, and an ideality factor, n, of 1.02 without any thermal treatment.[32] The native oxide plays a more important role when the gate oxide thickness is only 50 to 60 Å.[33,34] A *preoxide* process, to prevent the increase of surface microroughness during the high-temperature treatment of Si wafers, has been reported. The silicon surface is intensively oxidized at 300°C in ultraclean oxygen to form one molecular layer of oxide as a passivation layer after the desorption of hydrogen from the surface by the HF-clean. Dielectric breakdown histograms of this process are shown in Fig. 26*a* and, with the control samples, in Fig. 26*b*. Early breakdown of the dielectric is found for controlled

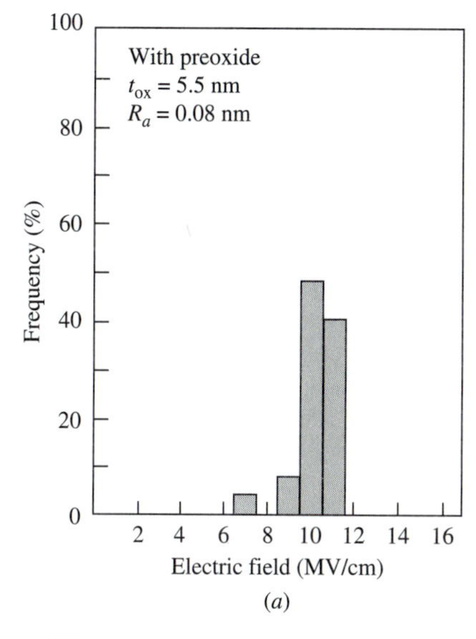

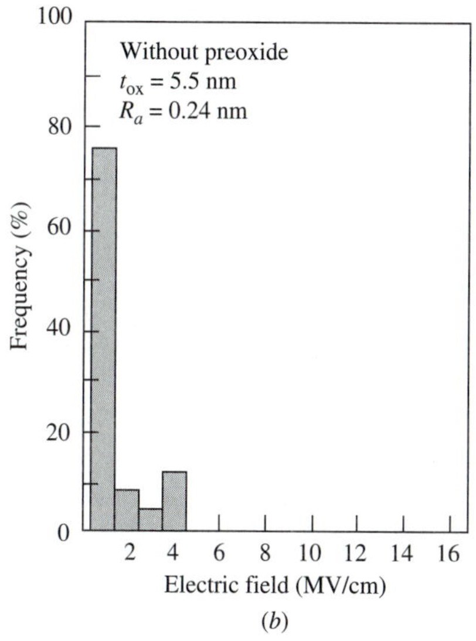

FIGURE 26
Dielectric breakdown histograms for Al/SiO_2/n-Si(100) diodes (*a*) with and (*b*) without preoxide. (*After Makihara et al., Ref. 34.*)

samples. Figure 27 shows the current density–voltage characteristics of 5.5 and 9.0 nm oxide with and without the preoxide processes.[34] The leakage current in the conventional dry oxide is much higher than that in the ultraclean preoxide. This indicates that the preoxide-controlled, ultraclean oxide has a higher electrical insulating capability, particularly for ultrathin oxide.

Last step of wafer cleaning for gate oxidation/epitaxy

Recently, epitaxy has become more important in high-speed devices, such as high-electron-mobility transistors (HEMTs) or SiGe devices.[35] Gate oxide yield with HF-last cleaning increases to 74%, comparing to the conventional RCA-last cleaning

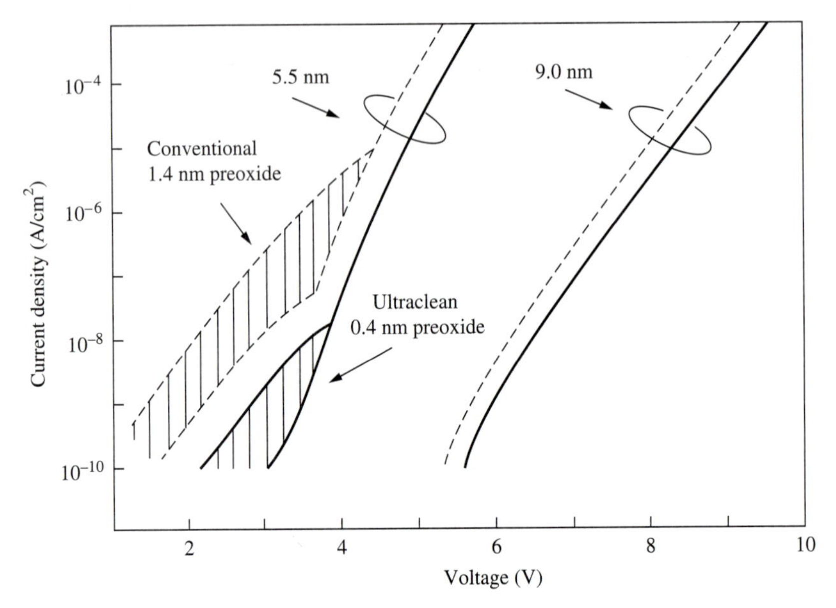

FIGURE 27
Current density-voltage characteristics of n^+-polycrystalline Si/SiO_2/p-Si(100) diodes under negative-biased metal electrodes with and without preoxide. (*After Makihara et al., Ref. 34.*)

(47%). This increase is due to the reduced metallic contamination as contaminants included in the oxide can be removed in the HF-last cleaning.[36] Native oxide–free silicon is the key requirement for producing a high-quality epitaxy film.[37,38] Dilute HF solution is usually used to remove the native oxide on the semiconductor surface. After the HF dipping, most of the silicon surface (80%) is covered by H-terminated bonds, and the rest are covered by F, O, and C terminations. However, if the hydrogen termination of the surface is incomplete, subsequent exposure to air causes absorption of carbonic impurities. The F coverage also depends on the orientation, HF concentration, and immersion time.[39] Figure 28 shows the ideal hydrogen termination on a Si(100) and a Si(111) surface. On a Si(100) surface, a surface Si atom has two dangling bonds that can be passivated with hydrogen; on a Si(111) surface, each surface Si atom has only one dangling bond for hydrogen passivation. The real surface is always terminated by both monohydrides and dihydrides, resulting from an imperfect, defective, or rough surface.

High-resolution electron energy loss spectroscopy (HREELS) is commonly used to analyze the chemical composition of a treated surface. Figure 29 shows the vibrational spectra of Si(100) and Si(111) surfaces after a 1 min dip in 40% HF.[39] The Si(100) is covered by dihydride at 2100 cm^{-1} (stretching mode), 900 cm^{-1} (scissor mode), and a wagging or bending mode at 640 cm^{-1}. On the other hand, the Si(111) surface is covered predominantly by monohydride (stretching mode at 2080 cm^{-1} and bending mode at 640 cm^{-1}). CH_x bonds can be found on both surfaces. They have wavenumbers of 2950 and 1350 cm^{-1} for stretching and deformation vibration. The spectra of the surfaces after they are rinsed in DI water are shown in

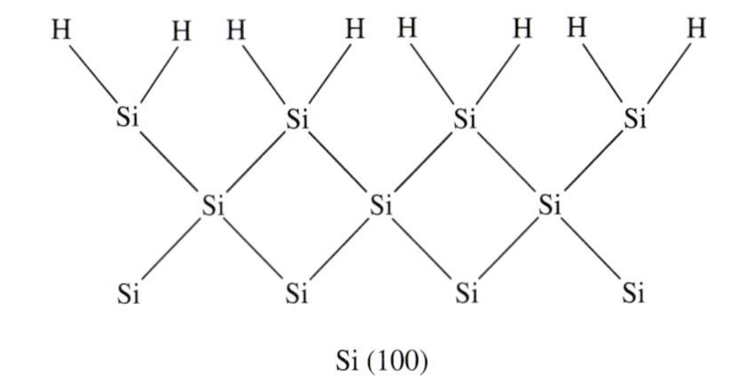

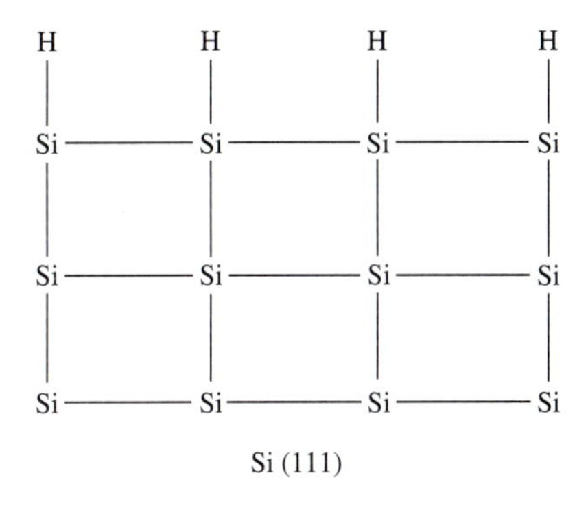

FIGURE 28
Schematic representation of the ideal hydrogen terminations on Si(100) and Si(111) surfaces.

Fig. 30. A Si–OH vibration is found after only a 30-sec water rinse. This presence of Si–OH is explained by the fast exchange of Si–F groups with water. The silicon surface, covered with hydrogen from the HF-last process, is hydrophobic. The HREELS spectrum of a hydrophilic surface after the SC-1 and SC-2 wet cleaning process has a strong asymmetric Si–O–Si vibration. Hence, the OH groups are responsible for the hydrophilic nature of the Si surface.[39]

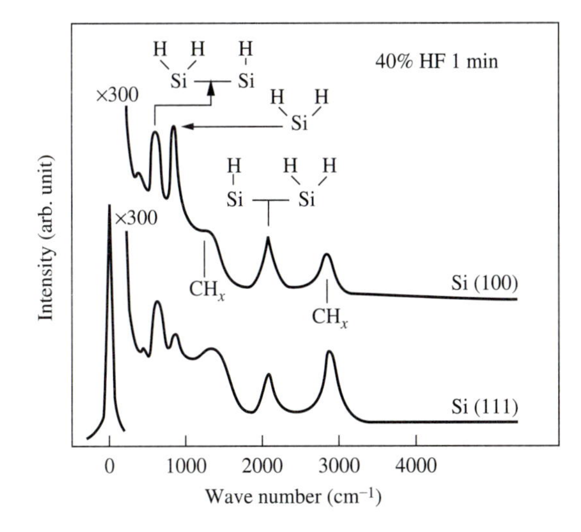

FIGURE 29
Vibrational spectra of Si(100) and (111) surfaces after a 1-min dip in 40% HF. Only Si–H (640 cm^{-1}, 900 cm^{-1}, and 2100 cm^{-1}) and hydrocarbon vibrations (2950 cm^{-1} and 1350 cm^{-1}) contribute to the spectra. (*After Grundner et al., Ref. 39.*)

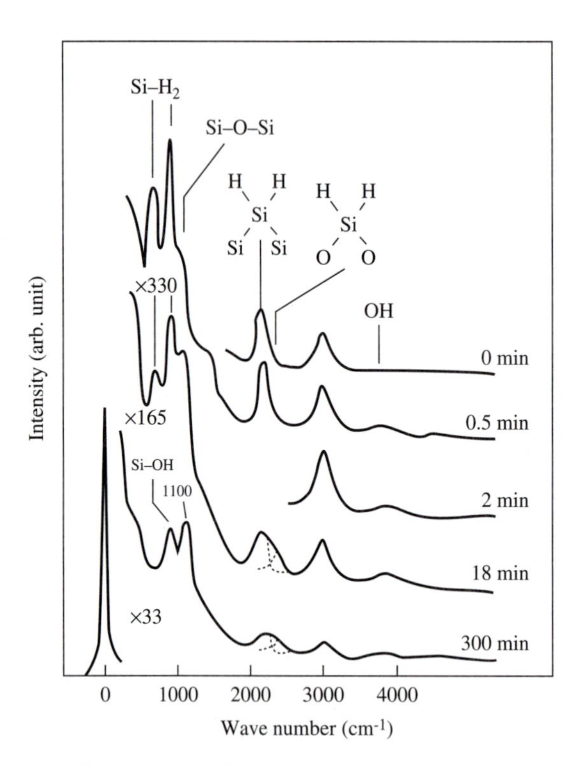

FIGURE 30
Vibrational spectra of a Si(100) sample (treated in 40% HF for 1 min) after different water rinsing times. Changes in the spectra occur around the OH, Si–O–Si, and Si–H stretching frequencies. The latter develops a schematically shifted component, indicated as O_2–Si–H_2. After 300 min there is still an unshifted contribution, suggesting that part of the surface has not yet oxidized. (*After Grundner et al., Ref. 39.*)

Contact angle is the simplest and most effective method to indicate the level of passivation. High passivation results in a high contact angle. Figure 31 shows the contact angle on a silicon surface after HF dipping (0.5%) with and without a DI water rinse.[36] A dip of about 30 seconds increases the contact angle to a maximum of 70.0°. The surface with the DI water rinse has a lower contact angle because of

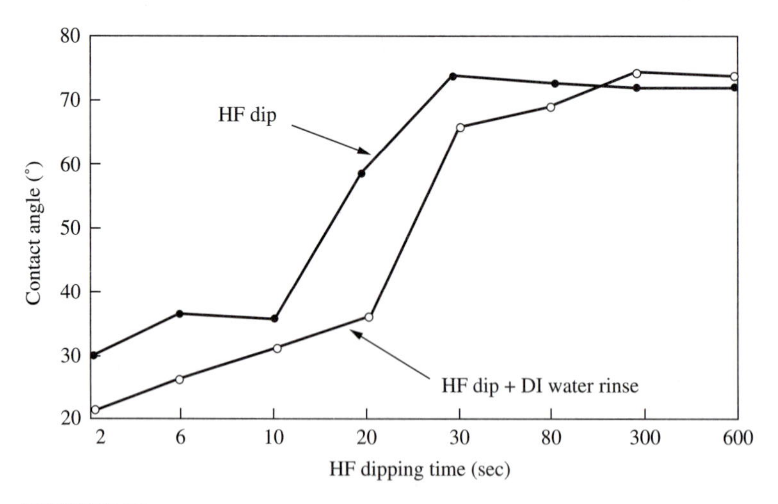

FIGURE 31
Contact angle on Si after HF dipping (0.5%) with and without a DI water rinse. (*After Verhaverbeke et al., Ref. 36.*)

partial reoxidation. The contact angle is also a function of the dipping time and the temperature of water. Hot HF (70.0°C) can be used effectively to get the surface completely oxide-free. Some modified HF-last processes have been proposed:

HF/IPA: A small amount of isopropyl alcohol (IPA) (< 1000 ppm) mixed with HF (0.5%) prevents the addition of particles to the Si surface in the traditional HF-last process and preserves the chemical state of the surface. Figure 32 shows the particle density after the HF/IPA cleaning process. When 1000 ppm of IPA is added to the HF mixture, the number of particles decreases to the level of the standard RCA-clean. After a 60-sec dip in HF/IPA, the contact angle of the silicon surface reaches the same maximum value as that achieved in the HF-last process. The correlation between XPS and contact angle measurement for an HF/IPA-dipped Si surface is shown in Fig. 33. The intensity of O_{1s} to SiO_2, as determined by XPS, decreases as the contact angle increases. Therefore, the contact angle is very sensitive to the degree of oxidation of a clean surface.[40]

HF/H_2O_2: H_2O_2 (10%) added to HF (0.5%) can remove metallic impurities to a level of 10^{10}atom/cm^2 in 10 min and can also reduce the number of particles. This has been discussed in connection with metal impurities.[14] This process has been proposed to replace the conventional HF-last process because it improves gate oxide integrity. However, a high H_2O_2 ratio results in a slightly oxidized silicon surface and, consequently, reduces the contact angle.

There are some other additives that can be used in the HF solution, such as HF/ethanol and HF/acetone. However, neither ethanol nor acetone can reduce the number of particles; hence, they are not suitable for the HF-last process.

Conventionally, prior to epitaxy in ultrahigh vacuum (UHV), a high-temperature (1200°C) thermal etching of native oxide is used. Figure 34 shows the in situ Auger

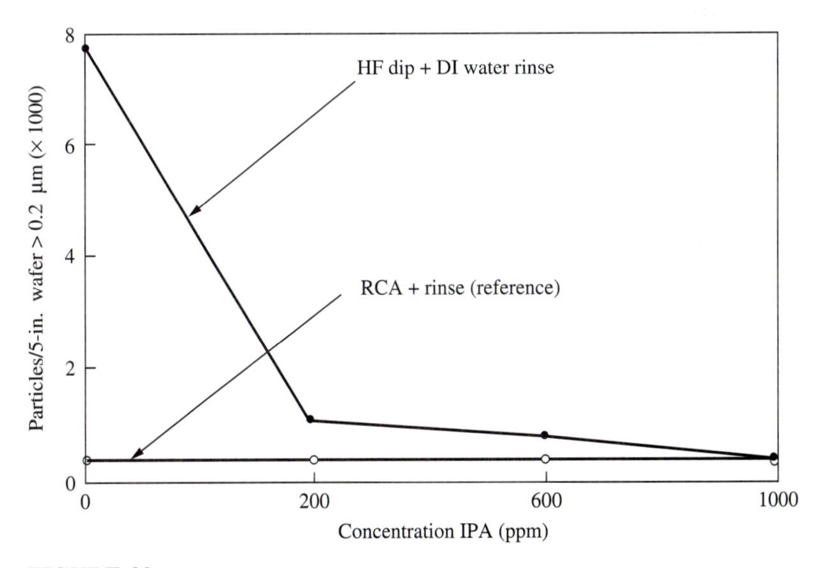

FIGURE 32
Particle density on a 5-inch wafer after dipping in HF with various amounts of IPA added and subsequent DI-water rinsing. (*After Verhaverbeke et al., Ref. 36.*)

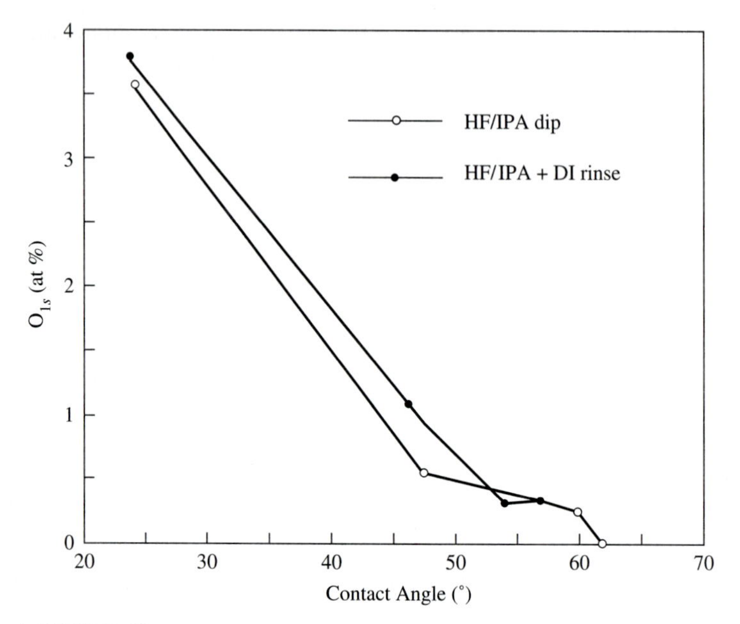

FIGURE 33
Correlation between XPS O_{1s} concentration and contact-angle measurement for HF/IPA-dipped Si surfaces. (*After Alay et al., Ref. 40.*)

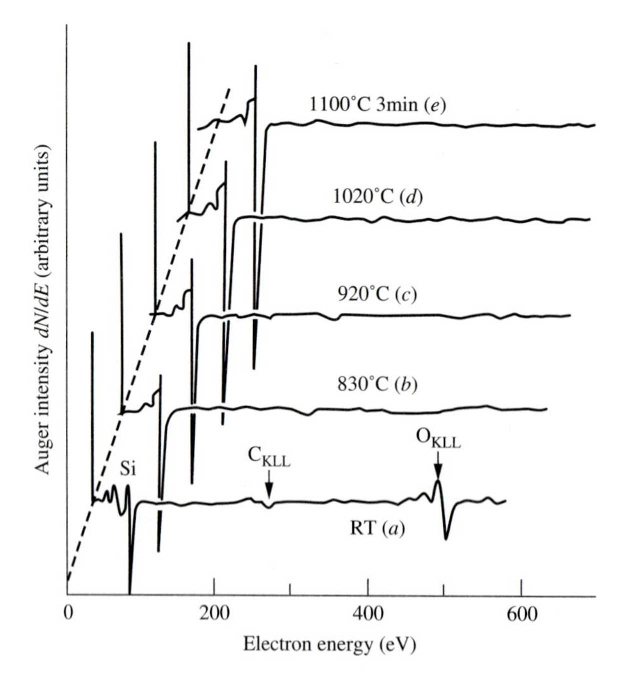

FIGURE 34
In-situ Auger spectra of a Si surface cleaned by the conventional high-temperature thermal etching method: (*a*) freshly oxidized, room temperature (RT), (*b*) after heating at 830°C, (*c*) 920°C, (*d*) 1020°C, and (*e*) 1100°C. (*After Ishizaka et al., Ref. 41.*)

spectra of the Si surface cleaned by the conventional high-temperature etching method.[41] Line *a* is the surface after conventional chemical treatment; O_{KLL} (515 eV) and C_{KLL} (272 eV) Auger signals, which originate from the surface contamination, are observed. The O_{KLL} signal disappears as the temperature rises to 850°C, but the C_{KLL} peak still remains. Lander et al.[42] first reported the removal of oxide at temperatures between 800 and 1000°C in a high vacuum by

$$Si + SiO_2 \rightarrow 2SiO$$

The C_{KLL} signal disappears after the temperature rises to 1100°C. That is why the conventional epitaxy process needs such a high-temperature treatment. Hence, if a carbon-free, oxidized Si surface can be prepared, low-temperature thermal etching of an oxide-free surface is possible (< 800°C). The high-temperature treatment causes impurity diffusion and changes the designed impurities concentration profile, crystal defects, such as dislocation and stacking faults, high wearout of the substrate-heating elements, and low throughput. Therefore, a low-temperature surface cleaning of silicon has been proposed (Shiraki clean process).[41] It includes

1. Degreasing
2. HNO_3 boiling + DHF dip (2.5%)
3. NH_4OH–H_2O_2–H_2O(1 : 1 : 3) + DHF dip (2.5%)
4. HCl–H_2O_2–H_2O(3 : 1 : 1)
5. Rinsing + spin dry to let the surface become uniformly hydrophilic

Figure 35 shows the in situ Auger spectra after heating at 785°C for 15 min. The surface becomes oxide- and carbon-free. This method results in an atomically clean surface, and the epitaxy can be achieved at a low temperature of 785°C. There is

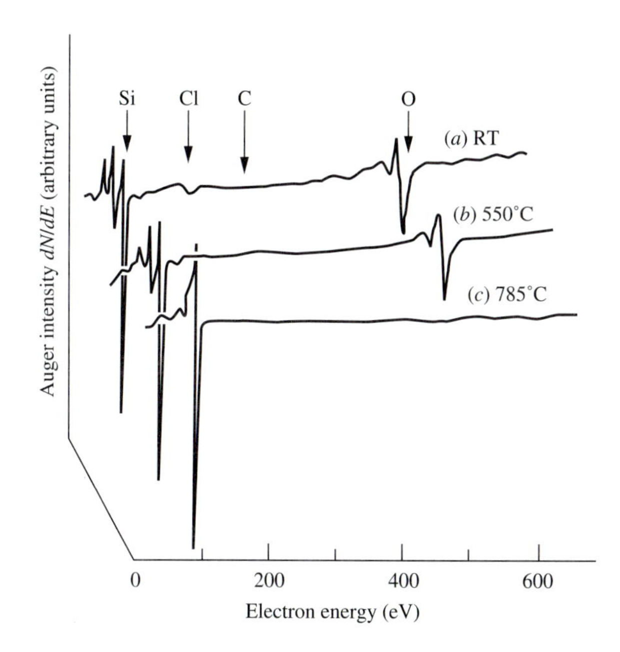

FIGURE 35
In-situ Auger spectra of Si surface cleaned by the low-temperature thermal etching method: (*a*) as freshly oxidized, (*b*) after heating at 550°C for 15 min, (*c*) after heating at 785°C for 15 min. (*After Ishizaka et al., Ref. 41.*)

also another method to reduce carbon. A modified Shiraki-clean process has been proposed that omits the last HCl boiling and uses 4% HF as the last step to produce a hydrogen-terminated surface without native oxide. The growth temperature using this process can be reduced to as low as 650°C and maintains the same high quality as that of the Shiraki cleaning process.[43]

2.3 WET CLEANING TECHNOLOGY

Some optional processes steps have been added into the conventional RCA cleaning process.[44] A preliminary cleanup treatment with a hot $H_2SO_4 : H_2O_2(2:1$ to $3:1$, at 120°C) mixture is used to remove the greasy contamination, which may be from the cassette or residues from the photoresist layers. After the SC-1 process, a 15-sec immersion in 1% HF–H_2O solution may be beneficial for removing any trace impurity. Some techniques have been proposed for advanced cleaning processes.[44]

Immersion technique: Vessels of fused silicon have been introduced to prevent leaching of Al, B, and alkalis, which can result if Pyrex glass is used. A DI water overflow-quenching system is also used for the automatic processing.

Megasonic cleaning: This cleaning process can remove particles and contamination with diluted SC-1 solution at a temperature of only 35 to 42°C.

Centrifugal spray cleaning: The chemical solutions, including SC-1, SC-2, and DI water, are pressure-fed directly onto the spinning wafer. This process can reduce the volume of fresh chemical and is faster than immersion. The spin-drying process requires antistatic protection to prevent static-induced particle deposition on the wafers.[45]

Closed-system chemical cleaning: In this process wafers are cleaned in a closed system during the entire cleaning, rinsing, and drying process to avoid contamination from the operator and environments.

Advanced wet cleaning processes. The steps of the advanced wet-chemical cleaning process proposed by Ohmi,[46] and the contaminants each removes, are

$H_2O + O_3$	Organic contamination
$NH_4OH + H_2O_2$	Particle, organic, and metallic impurities
$HF + H_2O_2$	Native oxide, metallic impurities
Ultrapure water	Rinsing

This process omits the conventional RCA SC-2 step when highly purified HF solution is used. In the clean process, most of the organic contamination is removed first in the ozoned, ultrapure water (low oxygen dissolved < 1 ppb) cleaning. Then $NH_4OH–H_2O_2–H_2O = 0.05 : 1 : 5$ is used to remove the particles, organic, and metallic impurities, maintaining an atomically flat surface. A HF (0.5%)–H_2O_2 (10%) cleaning is proposed as the final step to remove the native oxide and noble metals.

TABLE 8
Concept of IMEC cleaning[47]

Step 1† Oxide growth	Step 2‡ Oxide removal	Step 3§ Optional	Drying¶
H_2SO_4/H_2O_2	Wet DHF	Chemical oxide	IPA vapor
H_2SO_4/O_3	HF vapor + rinsing		Hot DI water
H_2O/O_3			Spin dry
UV-O_3			N_2 blow

† Step 1: growth of about 1.5 nm of chemical oxide.
‡ Step 2: removal of the oxide: drying procedure.
§ Step 3 (optional): growth of chemical oxide.
¶ Step 4: easier drying due to the growth of a hydrophilic Si surface.

Another new cleaning concept, IMEC (for Interuniversity Microelectronic Center) cleaning, is illustrated in Table 8. The basic concept of this new cleaning process is to separate the oxidizing and the etching action.[47] In step 1 the organic contamination is removed and a thin chemical oxide is grown; then this oxide is etched in step 2. By using this method, the total etch depth should be sufficient to remove the particles. The advantage of separating the cleaning into two steps is that the formation of the oxide layer is a self-limiting process of the chemical reaction. This layer is easy to control by oxide removal to obtain minimum surface roughness. Step 3, a chemical oxide growth, is optional. A cleaning process based on the IMEC cleaning concept is as follows:

$$\begin{array}{ll} H_2SO_4\text{–}H_2O_2 = 4:1 & \text{at } 90^\circ\text{C for 10 min} \\ \text{HF}(0.5\%)/\text{IPA}(0.1\%) & \text{at room temperature for 2 min} \end{array}$$

It has been shown that the particle removal efficiency using the IMEC clean is better than SC-1 on the native oxide, and slightly weaker on the thermal oxide. IMEC cleaning is also quite superior to the other cleaning process in removing metallic contamination.[48] The yields of the gate oxide for four different clean methods are

IMEC + RCA(spray) + HF(bath)	80%
IMEC + RCA(spray)	65%
RCA(bath)	60%
IMEC(bath)	83%

The IMEC (bath) clean has the highest yield. The yield is defined as the percentage of gate oxide with a breakdown field larger than 12 MV/cm.

2.4 DRY CLEANING TECHNOLOGY

Wet cleaning technology still remains the major method for wafer cleaning. However, the ULSI industry has begun to recognize problems associated with wet cleaning.

Among them are particle generation, drying difficulties, cost, chemical waste disposal, incompatibility with advanced concepts in integrated processing, and general inflexibility.[49] The dry cleaning process may provide a way to solve these problems, although currently it is not completely successful. The integrated dry cleaning processing system is shown in Fig. 36, in which the dry cleaning chamber is connected to the "cluster" processing system. The dry cleaning process, a kind of gas-phase chemistry, usually requires excitation energy to enhance the chemical reaction at low temperature. This added energy may be plasma, particle beam, short-wavelength radiation, or thermal heating, which enhances the surface cleaning result but must avoid the side effects of damage to the wafer.

Recently, many dry cleaning technologies have been developed, namely HF/H_2O vapor cleaning,[49–52] ultraviolet-ozone cleaning (UVOC),[53–56] H_2/Ar plasma cleaning,[29,57–59] and thermal cleaning.[37] These processes will be discussed in the following paragraphs. Usually, wafers receive a wet cleaning first. Although RCA clean[1,44] has been used for more than 30 years in semiconductor manufacturing because it is effective in removing organic, particulate, and metallic contaminations, the resulting Si wafer surface is covered with 4 to 6 nonuniform monolayers (ML) of silicon oxide resulting from liquid surface tension, bubbles, and generated particles.[49] The diluted HF dip cleaning (1 to 4%HF in H_2O) usually suffers some problems such as contamination by hydrocarbons on the surface.[56] It is desirable to have a hydrogen-passivated surface[3,38] for the subsequent processing steps, such as poly-Si or epitaxial silicon growth. However, in dry cleaning, after the HF/H_2O vapor process, the surface is fluorine (F)-terminated.[49–50]

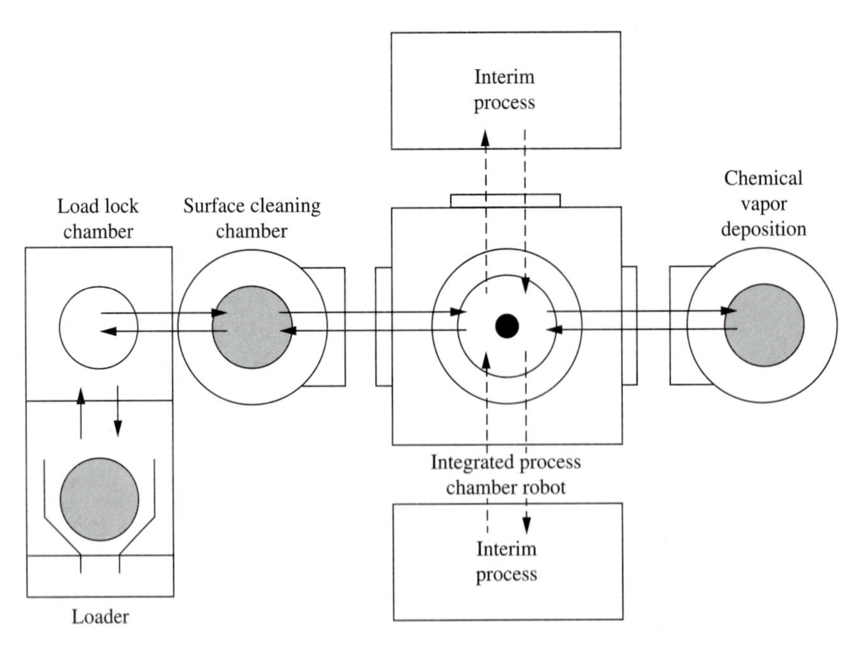

FIGURE 36
An integrated processing system concept incorporating a dry cleaning chamber.

Ultraviolet-ozone clean (UVOC)[53–56]

This is a very effective way to remove hydrocarbons, although the surface is oxide-passivated. The mechanism of UVOC can be explained by the following process equations.

$$\text{Absorbed impurity } + h\nu \xrightarrow{2000\text{–}3000\text{ Å UV}} \text{Excited impurity} \tag{2.1}$$

In addition, oxygen atoms are excited by UV light to form atomic oxygen. This is a surface excitation process.

$$O_2 + h\nu \xrightarrow{1849\text{ Å UV}} 2O \tag{2.2}$$

$$O + O_2 \rightarrow O_3 \tag{2.3}$$

$$O_3 + h\nu \xrightarrow{2537\text{ Å UV}} O + O_2 \tag{2.4}$$

$$\text{Then} \quad \text{Excited impurity} + (O, O_3) \rightarrow \text{volatile compound} \tag{2.5}$$

The reaction in Eq. (2.1) is a surface-excitation process, whereas the reactions in Eqs. (2.2) through (2.5) are gas-phase excitation processes that convert impurities, such as hydrocarbons, into volatile gases.

Figure 37 shows the XPS spectra of a HF-dipped surface and a UV/O_3 exposed surface, which indicates the formation of surface oxide.[29] The high resolution of surface energy loss spectroscopy (HREELS) is used to examine the decomposition of hydrocarbons. As shown in Fig. 38, after a 300 sec exposure in UV/O_2, peaks of δ(CH) and ν(CH) disappear, while ν(SiO) and δ(SiO) peaks are intensified. The

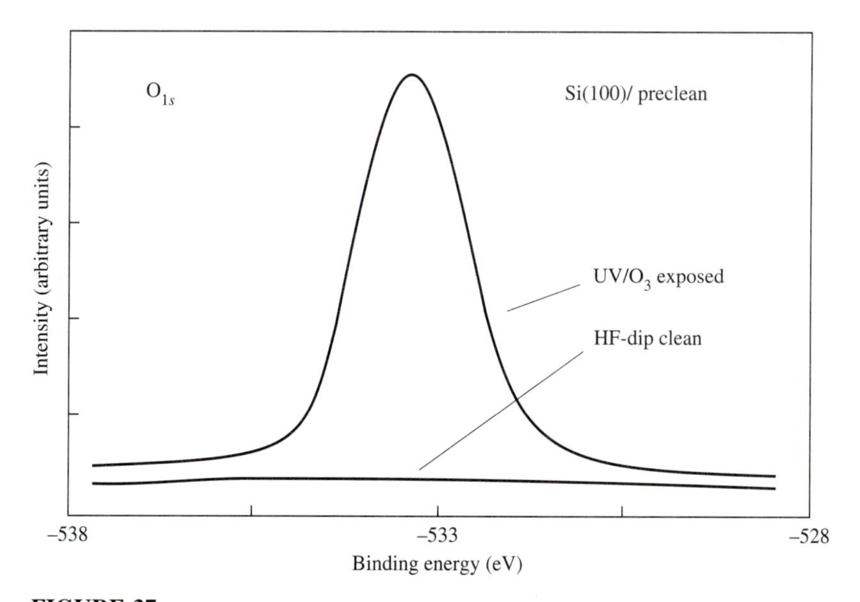

FIGURE 37
X-ray photoemission spectra displaying the O_{1s} regions of Si(100) wafers after 15 min UV-ozone and after 10 sec 10% HF dip clean. (*After Offenberg et al., Ref. 29.*)

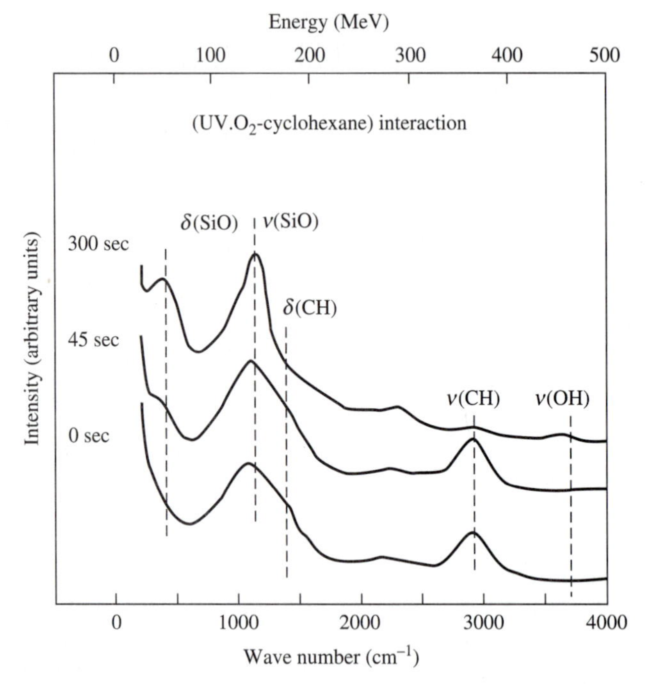

FIGURE 38
HREELS spectra for a Si(100) wafer that was HF-dip cleaned, subsequently contaminated with cyclohexane, then cleaned with UV/O_2 for 45 sec, and for 300 sec. (*After Kasi and Liehr, Ref. 50.*)

subsequent HF/H_2O vapor or Ar/H_2 plasma cleaning processes in vacuum would be helpful in removing the surface oxide.

HF/H_2O vapor clean

The diluted HF dip promotes a hydrogen-passivated surface, but the HF/H_2O vapor clean induces a fluorine-terminated surface.[51,52] Table 9 indicates that "clean 1" ($2H_2SO_4 + 1H_2O_2$) results in a 5.8-monolayer (ML) oxide on the surface, "clean 2" (HF dip, then H_2O rinse) results in an OH termination with 0.9 ML oxide cover on the surface, and "clean 3" (1% HF in H_2O) results in a hydrogen-terminated surface with only 0.3 ML oxide. The contact angle measured is 19° for clean 1, 138° for clean 2, and 155° for clean 3. This indicates a tendency to change from a hydrophilic surface to a hydrophobic surface. When wafers receive a HF/H_2O vapor clean, the content of water vapor varies with the composition ratios. Although the surfaces are F-terminated, the [C], [O], and [F] concentrations are different with different process conditions. Contact angles decrease from 140° for 0.4 ML of [O] to 20° for 5.9 ML of [O], indicating surface-state transformations.[51,52] Figure 39 indicates the concentration of fluorine, C[F], versus the concentration of oxygen C[O], where $F_1 = C$[Si–F bond], $F_2 = C$[O–Si–F bond]. Clean 3, with a 155° contact angle, is slightly more hydrophobic than process 1, with 140° of contact angle.[51]

TABLE 9
XPS surface residues and contact angle of Si wafers[51, 52]

	Termination	[C], ML	[O], ML	[F], ML	Contact angle (°)
Wet clean					
$2H_2SO_4 + 1H_2O_2$ (clean 1)	O	1.4 ML	5.8		19
HF, then H_2O (clean 2)	OH	1.6	0.9		138
HF/H_2O last (1:100) (clean 3)	H	0.9	0.3	0.08	155
HF/H_2O vapor clean					
Process 1	F	0.4	0.4	0.2	140
Process 2, water-rich	F	0.37	0.7	0.48	60
Process 3	F	0.5	1.4	0.4	—
Process 4	F	0.7	5.9	0.7 ML	20

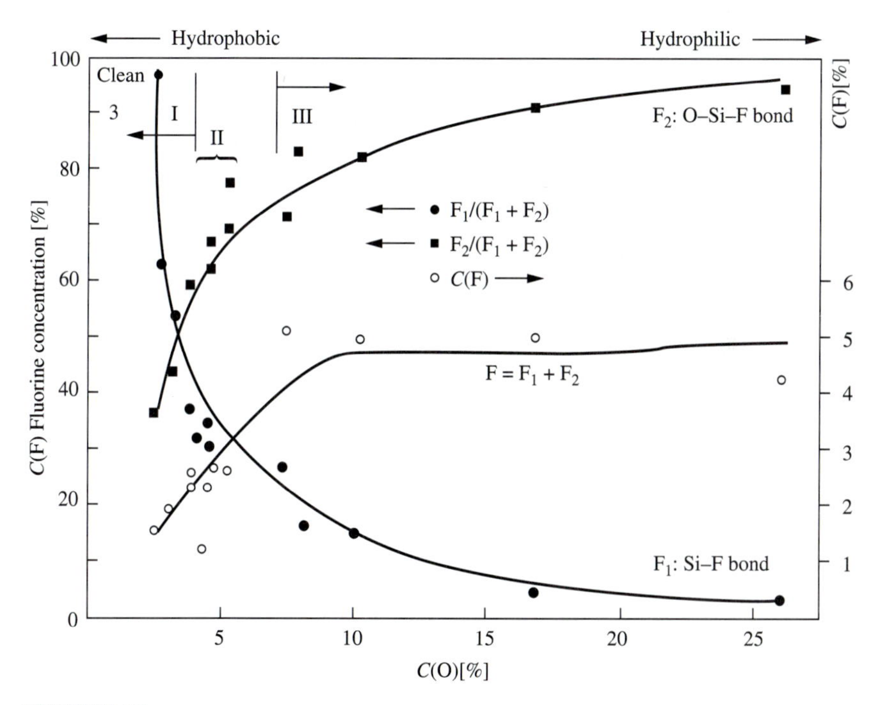

FIGURE 39
Fluorine concentration versus oxygen concentration and respective F–Si and O–Si–F concentrations versus oxygen concentration C[O]. (*After Ermolieff et al., Ref. 51.*)

AR/H_2 plasma cleaning

Usually a remote plasma is used to reduce the bombardment damage to the wafer. Gases, Ar or H_2 or both, are passed through an rf (13.6 MHz) inductive-coil or capacitive-coupled region where a plasma region is formed. At a pressure of 1 torr, the Ar or H_2 gases molecules are excited or ionized and generate a plasma. These gaseous molecules are carried to another region of the wafer, where a cleaning process begins. Excited Ar ions physically sputter the surface impurities away, while the excited H_2 ions chemically etch the surface. By proper adjustment of the physical/chemical etching (cleaning) ratio, an optimum cleaning condition can be obtained that produces minimum damage of the surface.

Instead of RF plasma excitation, electron cyclotron resonance (ECR) can do the same job at a higher plasma density and at a lower kinetic energy of the excited atom or molecules (~100 eV). ECR occurs at a frequency of $f_c = eB/2\pi m^*$. The f_c is 2.45 GHz, for a B of 875 gauss. The pressure is about 1 mtorr. The detailed construction of ECR will be discussed in Chapter 7. Native oxide and hydrocarbons are removed by H_2-plasma cleaning or by ECR Ar/H_2 plasma[60] at room temperature or at a temperature below 250°C. Figure 40 shows the effective carbon and oxygen removal by a remote hydrogen plasma[58] at several temperatures up to 305°C, monitored by Auger spectroscopy.

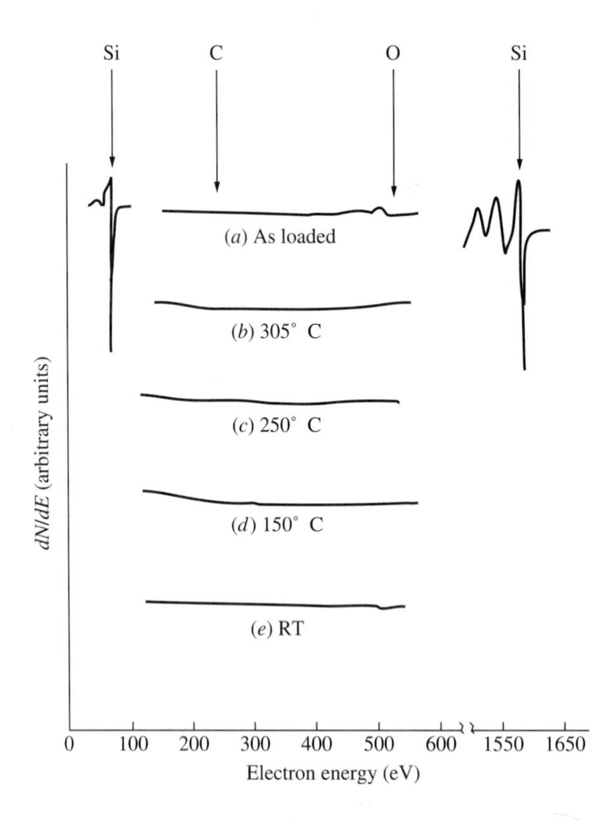

FIGURE 40
AES analysis of a Si(100) surface after (*a*) a 40 : 1 H_2O : HF (49%) dip and remote hydrogen plasma clean for 45 min at (*b*) 305°C, (*c*) 250°C, (*d*) 150°C, and (*e*) room temperature (RT). (*After Hsu et al., Ref. 58.*)

Thermal cleaning

The native oxide can be removed by heating the wafer to 800°C or above in ultrahigh vacuum ($< 10^{-10}$ torr) to vaporize the oxide. The native oxide is SiO_x, in which x depends on the previous cleaning process. In the Shiraki cleaning process,[41] an 850°C, 10 min thermal clean is necessary, whereas in the HF dip process (4% HF) a prebake of 200°C, or no bake process, is used for a 650°C epitaxial growth.

In all these processes, high-temperature cleaning should be carefully examined because at high temperature ($>$ 800°C), the following reaction process occurs at low oxygen partial pressure,[52]

$$\mathrm{Si} + \mathrm{SiO_2} \longrightarrow 2\mathrm{SiO} \tag{2.6}$$

The SiO is volatile at temperature above 750°C. When the SiO_2 film is removed, the silicon wafer starts to etch:

$$\mathrm{Si} + \mathrm{O_2} \longrightarrow 2\mathrm{SiO} \tag{2.7}$$

produces aggravated etching, which induces microroughness on the surface. The deposited gate oxide, therefore, has a low breakdown voltage (~1 MV/cm) due to microroughness. The low oxygen partial-pressure oxidation kinetics can be explained by the work of Smith and Ghidini,[61] as shown in Fig. 41. At low temperature and

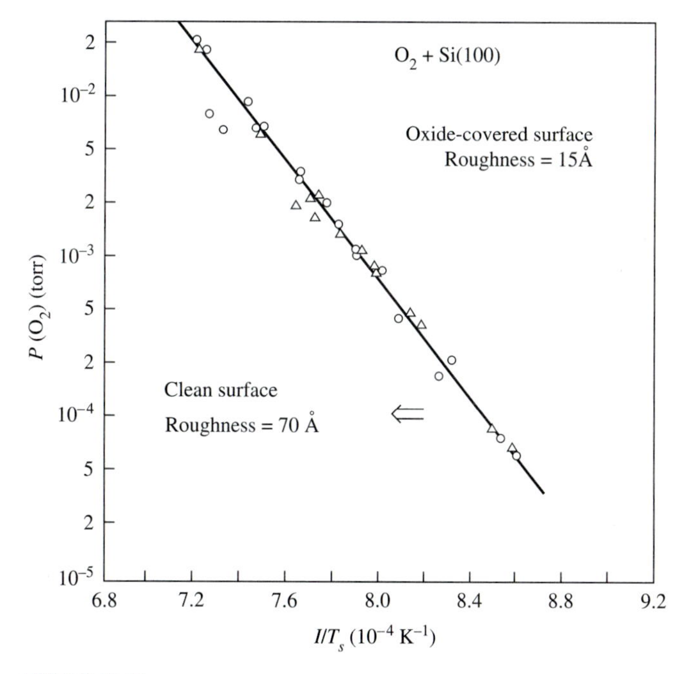

FIGURE 41
Boundary separating oxide-covered surface (upper right) from clean surfaces (lower left) for Si(100) exposed to a partial pressure of O_2 at elevated temperature. (*After Smith and Ghidini, Ref. 61.*)

TABLE 10
Comparison of wet and dry cleaning

Method	Surface termination	Main effects	Remarks	References
Wet				
RCA	4~6 ML SiO_2	Removal of organic, metallic, and other particulate residues	Nonuniformity due to bubble, particle, surface tension, etc.	1, 44
HF-last	H^-	Oxide removal		3, 40
Dry				
HF + H_2O (vapor)	F^-	Oxide removal		49–52
UVOC	OH^-, SiO^-	CH removal		53–56
H_2/Ar plasma	H^-	H_2 chemical clean Ar physical clean	1 torr	57–59
Remote plasma	H^-	C/O removal	1 torr	
ECR	H^-	C/O removal	1 mtorr	
Thermal clean		Microroughness	> 800°C, UHV	52

a high O_2 partial pressure a smooth surface with a roughness of 1.5 nm is obtained, whereas at high temperature and a low O_2 partial pressure a roughness of 7 nm indicates that a thermal etching process Eq.(2.6) is taking place.

Table 10 summarizes the effects of various cleaning processes. In surface termination, the RCA clean results in 4 to 6 monolayers of SiO_2 while removing organic and metallic residues on the surface. The HF-last cleaning process forms a hydrogen-passivated surface and removes the oxide on the surface. These processes suffer the result of incomplete cleaning due to forming gas bubbles, surface tension, and remaining particles. The dry cleaning processes summarized in Table 10 have therefore been developed; they produce different surface terminations and impurities removals. The HF + H_2O vapor cleaning forms a F-terminated surface. The UVOC results in an OH^-, SiO^- covered surface; it is an effective way to remove hydrocarbons. The H_2/Ar plasma processes result in a hydrogen-passivated surface; however, the surface residues are removed by H_2 chemical etch and by Ar physical sputtering processes. Good control of H_2/Ar ratio will give a good result of "soft" cleaning with minimum damage. Remote plasma and ECR plasma cleaning processes are effective in carbon and oxide removal. Finally, a thermal clean is presented that removes oxide, but sometimes a microrough surface is obtained if oxygen remains in the vacuum chamber.

2.5 SUMMARY AND FUTURE TREND

Removal of silicon surface contaminations such as heavy metals, organics, and particles can be achieved by wet cleaning with a modified RCA cleaning process. Recent studies of surface-state physics have increased the efficiency of cleaning methods. An oxygen-free DI water rinse promotes hydrogen surface termination by eliminating the Si–Si bond breakage resulting from oxygen atoms, thus making the surface more hydrophobic. IPA added to the HF solution enhances particle removal. An O_3-containing DI water rinse helps to remove the organic and hydrocarbon contaminants. Advanced wet cleaning processes have been proposed for the more sophisticated ULSI processes. Microroughness is crucial to ULSI device quality, and a detailed mechanism and method of obtaining a good microsmooth surface has been proposed.

To meet future requirements for an integrated processing system, dry etching processes have been introduced. A summary of these technologies were given, which should help readers make a judgment about using these processes in their individual process applications.

The present state of these technologies still does not meet all the future requirements for more advanced ULSI devices. More in-depth understanding of the surface-state physics and chemistry is required, not only to eliminate the surface contaminations but also to provide the atomically flat surface and interface required for future semiconductor manufacturing. In addition, the development of more efficient dry etching and wet cleaning processes is essential for future integrated processing systems.

REFERENCES

1. W. Kern and D. A. Puotinen, "Cleaning Solutions Based on Hydrogen Peroxide for Use in Silicon Semiconductor Technology," *RCA Rev.* **31,** 187 (1970).
2. S. Verhaverbeke, *Dielectric Breakdown in Thermally Grown Oxide Layers,* Katholieke Universiteit, Leuven, 1993.
3. C. Werkhoven, E. Granneman, R. De Blank, S. Verhaverbeke, P. Mertens, M. Meuris, W. Vandervorst, and M. Heyns, "Wet and Dry HF-Last Cleaning Process for High-Integrity Gate Oxides," *Tech. Dig., IEDM,* 633 (1992).
4. M. Liehr, "Science Issues Related to Wafer Cleaning in Silicon Technology," *Mat. Res. Soc., Symp. Pro.,* **259,** 3 (1992).
5. P. H. Singer, "Trends in Wafer Cleaning," *Semic. Intl.* 36 (Dec., 1992).
6. M. Itano, F. W. Kern, Jr., R. Reed, W. Rosenberg, M. Miyashita, I. Kawanabe, and T. Ohmi, "Particle Deposition and Removal in Wet Cleaning Processes for ULSI Manufacturing," *IEEE Trans. Semic. Manuf.* **5,** 114 (1992).
7. S. Shwartzman, A. Mayer, and W. Kern, "Megasonic Particle Removal from Solid-State Wafers," *RCA Rev.* **46,** 81 (1985).
8. S. Verhaverbeke, C. Werkhoven, M. Meuris, H. F. Schmidt, K. Dillenbeck, P. Mertens, M. Heyns, and A. Philipossian, "Investigation of Sources of Metallic Contamination by TXRF," *Institute of Environmental Science Annual Technical Meeting, Las Vegas, Nevada,* 1993, p.423.
9. W. B. Henley, L. Jastrzebski, and N. F. Haddad, "Monitoring Iron Contamination in Silicon by Surface Photovoltage and Correlation to Gate Oxide Integrity," *Mat. Res. Soc., Symp. Proc.,* **315,** 299 (1993).
10. H. Park, O. R. Helms, D. Ko, M. Tran, and B. B. Triplett, "Effect of Surface Iron on Gate Oxide Integrity and Its Removal From Silicon Surfaces," *Mat. Res. Soc., Symp. Proc.,* **315,** 353 (1993).

11. L. Jastrzebski, O. Milic, M. Dexter, J. Lagowski, D. Debusk, K. Nauka, R. Witowski, M. Gordon, and E. Persson, "Monitoring of Heavy Metal Contamination During Chemical Cleaning with Surface Photovoltage," *J. Electrochem. Soc.,* **140,** 1152 (1993).
12. N. Anzai, Y. Kureishi, S. Shimizu, and T. Nitta, *Proceeding of the 1st Workshop on ULSI Ultra Clean Technology,* 1989, p.75.
13. T. Ohmi, T. Imaoka, I. Sugiyama, and T. Kezaka, "Metallic Impurities Segregation at the Interface between Si Wafer and Liquid during Wet Cleaning," *J. Electrochem. Soc.,* **139,** 3317 (1992).
14. T. Ohmi, T. Imaoka, T. Kezuka, J. Takano, and M. Kogure, "Segregation and Removal of Metallic Impurity at Interface of Silicon and Fluorine Etchant," *J. Electrochem. Soc.,* **140,** 811 (1993).
15. S. Verhaverbeke, P. W. Mertens, M. Meuris, M. M. Heyns, A. Schnegg, and A. Philipossian, "The Influence of Metallic Impurities on the Dielectric Breakdown of Oxide and New Processes to Avoid Them," Technical Conference SEMICON, Europe, 1992.
16. S. Verhaverbeke, M. Meuris, P. W. Mertens, A. Kelleher, M. M. Heyns, R. F. Keersmaecker, M. Murrell, and C. J. Sofield, "The Effect of Metallic Contamination on Void Formation, Dielectric Breakdown and Hole Trapping in Thermal SiO_2 Layers," *Proceeding, in Cleaning Technology in Semiconductor Device Manufacturing,* Electrochemical Society, 1991, p.187.
17. Ref. 2, p. 76.
18. Ref. 2, p. 178.
19. K. Hashimoto, K. Egashira, M. Suzuki, D. Matsunaga, "Gate Oxide Deterioration Caused by Organic Contamination onto the Oxide," *1991 International Conference on Solid-State Devices and Materials, Yokohama,* 1991, p.143.
20. M. Kogure, T. Futatsuki, J. Takano, T. Isagawa, K. Kimura, Y. Ogato, F. Tanaka, and T. Ohmi, "Influence of Organic Impurities in Hydrogen Peroxide for Advanced Wet Chemical Processing," *J. Electrochem. Soc.,* **140,** 3321 (1993).
21. T. Ohmi, T. Isagawa, M. Kogure, and T. Imaoka, "Native Oxide Growth and Organic Impurity Removal on Si Surface with Ozone-Injected Ultrapure Water," *J. Electrochem. Soc.,* **140,** 805 (1993).
22. R. A. Governal, A. Bonner, and F. Shadman, "Effect of Component Interactions on the Removal of Organic Impurities in Ultra Pure Water Systems," *IEEE Trans. Semic. Manuf.,* **4,** 298 (1991).
23. Ref. 2, p. 112.
24. T. Ohmi, T. Tsuga, J. Takano, M. Kogure, K. Makihara, and T. Imaoka, "Influence of Vacancy in Silicon Wafer of Various Types on Surface Microroughness in Wet Chemical Process," *IEICE Trans. Electron.,* **7,** 800 (1992).
25. M. Miyashita, T. Tsuga, K. Makihara, and T. Ohmi, "Dependence of Surface Microroughness of CZ, FZ, and Epi Wafers on Wet Chemical Processing," *J. Electrochem. Soc.,* **139,** 526 (1992).
26. T. Ohmi, M. Miyashita, M. Itano, T. Imaoka, and I. Kawanabe, "Dependence of Thin Oxide Films Quality on Surface Microroughness," *IEEE Trans. Electron. Dev.,* **39,** 537 (1992).
27. M. Hirose, T. Yasaka, M. Takakura, S. Miyazaki, "Initial Oxidation of Chemically Cleaned Silicon Surfaces," *Sol. State Technol.,* 43 (Dec. 1991).
28. J. M. de Larios, D. B. Kao, C. R. Helms, B. E. Deal, "Effect of SiO_2 Surface Chemistry on the Oxidation of Silicon," *Appl. Phys. Lett.,* **54,** 715 (1989).
29. M. Offenberg, M. Liehr, and G. W. Rubloff, "Surface Etching and Roughening in Integrated Processing of Thermal Oxides," *J. Vac. Sci. Technol. A,* **9,** 1058 (1991).

30. M. Morita, T. Ohmi, E. Hasegawa, M. Kawakami, and M. Ohwada, "Growth of Native Oxide on a Silicon Surface," *J. Appl. Phys.* **68,** 1272 (1990).
31. Y. Yagi, T. Imaoka, Y. Kasama, and T. Ohmi, "Advanced Ultrapure Water Systems with Low Dissolved Oxygen for Native Oxide Free Wafer Processing," *IEEE Trans. Semic. Manuf.,* **5,** 121 (1992).
32. M. Miyawaki, S. Yoshitaka, and T. Ohmi, "Improvement of Al-Si contact Performance in Native Oxide Free Processing," *IEEE Electron Dev. Lett.,* **11,** 448 (1990).
33. T. Ohmi, M. Morita, A. Teramoto, K. Makihara, and K. S. Tseng, "Very Thin Oxide Film on a Silicon Surface by Ultraclean Oxidation," *Appl. Phys. Lett.,* **60,** 2126 (1992).
34. K. Makihara, A. Teramoto, K. Nakamura, M. Y. Kwon, M. Morita, and T. Ohmi, "Preoxide-Controlled Oxidation for Very Thin Oxide Films," *Jpn. J. Appl. Phys.,* **32,** 294 (1993).
35. S. M. Sze, Ed., *High-Speed Semiconductor Devices,* John Wiley and Sons, Inc., New York, 1990.
36. S. Verhaverbeke, M. Meuris, M. Schaekers, L. Haspeslagh, P. Mertens, M. M. Heyns, R. De Blank, and A. Philipossian, "A New Modified HF-Last Cleaning Process for High-Performance Gate Dielectrics," 1992 Symposium on VLSI Technology, Washington, June 22, 1992.
37. T. Yamazaki, N. Miyata, T. Aoyama, and T. Ito, "Investigation of Thermal Removal of Native Oxide from Si(100) Surface in Hydrogen for Low-Temperature Si CVD Epitaxy," *J. Electrochem. Soc.,* **139,** 1175 (1992).
38. T. C. Chang, C. Y. Chang, T. G. Jung, and W. C. Tsai, "Nanometer Thick Si/SiGe Strained-Layer Superlattice Grown by an UHVCVD Technique," *J. Appl. Phys.,* **75,** 3441 (1994).
39. M. Grundner, D. Graf, P.O. Hahn, and A. Schnegg, "Wet Chemical Treatments of Si Surface: Chemical Composition and Morphology," *Sol. State Technol.,* 69 (Feb. 1991).
40. J. L. Alay, S. Verhaverbeke, W. Vandervorst, and M. Heyns, "Critical Parameters for Obtaining Low Particle Densities in an HF-Last Process," *1992 International Conference on Solid-State Devices and Materials, Yokohama,* 1992, p.123.
41. A. Ishizaka and Y. Shiraki, "Low Temperature Surface Cleaning of Silicon and Its Application to Silicon MBE," *J. Electrochem. Soc.,* **133,** 666 (1986).
42. T. J. Lander and J. Morrison, "Low Voltage Electron Diffraction Study of Oxidation and Reduction of Si," *J. Appl. Phys.,* **33,** 2089 (1962).
43. P. E. Thompson, M. E. Twigg, D. J. Godbey, and K. D. Hobart, "Low Temperature Cleaning Processes for Si Molecular Beam Epitaxy," *J. Vac. Sci. Technol. B,* **11(3),** 1077 (1993).
44. W. Kern, "The Evolution of Silicon Wafer Cleaning Technology," *J. Electrochem. Soc.,* **137,** 1887 (1990).
45. H. Inada, S. Sakata, T. Yoshida, T. Okado, and T. Ohmi, "Antistatic Protection in Wafer Drying Process by Spin-Drying," *IEEE Trans. on Semic. Manuf.,* **5,** 541 (1992).
46. T. Ohmi, "Advanced Wet Chemical Cleaning for Future ULSI Fabrication," *Electrochem. Soc., 184th Meeting, New Orleans,* 1993, p.495.
47. M. M. Heyns, S. Verhaverbeke, M. Meuris, P. W. Mertens, H. Schmidt, M. Kubota, A. Philipossian, K. Dillenbeck, D. Graf, A. Schnegg, and R. De Blank, "New Wet Cleaning Strategies for Obtaining Highly Reliable Thin Oxides," *Mat. Res. Soc., Symp. Proc.,* **315,** 35 (1993).
48. S. Verhaverbeke, M. Meuris, P. Mertens, H. Schmidt, M. M. Heyns, A. Philipossian, D. Graf, and K. Dillenbeck, "Advanced Wet Cleaning Technology for High Reliable Thin Oxides," *Proceedings of the 4th International Symposium on ULSI Science and Technology,* Electrochemical Society, 1993, p.199.

49. B. E. Deal, M. A. McNeilly, D. B. Kao, J. M. de Larios, "Vapor Phase Wafer Cleaning Processing for the 1990s," *Sol. State Technol.,* 73 (July 1990).
50. S. R. Kasi and M. Liehr, "Vapor Phase Hydrocarbon Removal for Si Processing," *Appl. Phys. Lett.,* **57,** 2095 (1990).
51. A. Ermolieff, F. Martin, A. Amouroux, S. Marthon, and J. F. M. Westendorp, "Surface Composition Analysis of HF-Vapor-Cleaned Silicon by X-ray Photoelectron Spectroscopy," *Semic. Sci. Technol.,* **6,** 98 (1991).
52. X. Xu, R. T. Kuehn, M. C. Ozturk, J. J. Wortman, R. J. Nemanich, G. S. Harris, and D. M. Maher, "Influence of Dry and Wet Cleaning on the Properties of Rapid Thermal Grown and Deposited Gate Dielectrics," *J. Electron. Matls.,* **22,** 335 (1993).
53. J. R. Vig, "UV/Ozone Cleaning of Surfaces," *J. Vac. Sci. Technol. A,* **3,** 1027 (1985).
54. J. Ruzyllo, G. T. Duranko, and A. M. Hoff, "Preoxidation UV Treatment of Silicon Wafers," *J. Electrochem. Soc.,* **134,** 2052 (1987).
55. M. Suemitsu, T. Kaneko, and N. Miyamoto, "Low Temperature Silicon Surface Cleaning by HF Etching/Ultraviolet Ozone Cleaning (HF/UVOC) Method (I)" *Jpn. J. Appl. Phys.,* **12,** 2421 (1989).
56. T. Kaneko, M. Suemitsu, and N. Miyamoto, "Low Temperature Silicon Surface Cleaning by HF Etching/Ultraviolet Ozone Cleaning Method (II): In Situ UVOC," *Jpn. J. Appl. Phys.,* **12,** 2425 (1989).
57. T. R. Yew and R. Reif, "Low-Temperature In Situ Surface Cleaning of Oxide Patterned Wafers by Ar/H_2 Plasma Sputter," *J. Appl. Phys.,* **68,** 4681 (1990).
58. T. Hsu, B. Anthony, R. Qian, J. Irby, S. Banerjee, A. Tasch, S. Lin, H. Marcus, and C. Magee, "Cleaning and Passivation of the Si (100) Surface by Low Temperature Remote Hydrogen Plasma Treatment for Si Epitaxy," *J. Electron. Matls.,* **20,** 279 (1991).
59. J. Cho, T. P. Schneider, J. Vander Weide, H. Jeon, and R. J. Nemanich, "Surface Electronic States of Low-Temperature H-Plasma Cleaned Si (100)," *Appl. Phys. Lett.,* **59,** 1995 (1991).
60. S. J. Pearton, F. Ren, A. Katz, U. K. Chakrabarti, E. Lane, W. S. Hobson, R. F. Kopf, C. R. Abernathy, C. S. Wu, D. A. Bohling, and J. C. Ivankovits, "Dry Surface Cleaning of Plasma-Etched High Electron Mobility Transistor," *J. Vac. Sci. Technol. B,* **11,** 546 (1993).
61. F. W. Smith and G. Ghidini, "Reaction of Oxygen with Si(111) and (100)-Critical Conditions for the Growth of SiO_2," *J. Electrochem. Soc.,* **129,** 1300 (1982).

PROBLEMS

1. What is needed for the complete achievement of an ultraclean wafer surface?

2. What are the advantages of the HF-last step?

3. The RCA SC-1 process results in a rough silicon surface. What can be done to reduce the roughness of the silicon surface?

4. What are the advantages of the dry cleaning processes?

CHAPTER 3

Epitaxy

Pei-Jih Wang

3.1 INTRODUCTION

Epitaxy is one of the most fundamental processes for semiconductor device fabrication. It is the foundation on which most devices are built. The term *epitaxy* means "upon-ordered" in Greek. In reality, epitaxy is simply a process to grow a single-crystal layer on a single-crystal substrate. When the single-crystal layer and the single-crystal substrate are of exactly the same material, it is called *homoepitaxy;* if the grown single-crystal layer and its single-crystal substrate are different in any aspects, the process is called *heteroepitaxy.* Strictly speaking, almost all epitaxial processes are heteroepitaxial, because there are always some differences between the layer and the substrate, even if they are of the same material chemically. Such microscopic differences may result from variations in doping or defect density. However, it is widely accepted that homoepitaxy takes place when the two materials have the same crystalline and chemical structures, whereas heteroepitaxy involves different materials.

In principle, it is much easier to achieve high-quality epitaxial layers with homoepitaxy, because there are no problems associated with lattice mismatch. In contrast, it is much more difficult to grow a high-quality layer with heteroepitaxy, because two dissimilar crystals of different lattice constants or orientations are thermodynamically unfavorable to combine without generating defects, such as dislocations. This is not to say that perfect heteroepitaxial layers are not obtainable. Most of the theories, together with experimental results, agree that the strain due to lattice mismatch can be accommodated elastically if the layer is thinner than a certain critical thickness, h_c.[1] Figure 1 shows schematic drawings of a typical unstrained and strained heteroepitaxy. It has also been reported by many researchers that the value of h_c is inversely proportional to the misfit strain between two

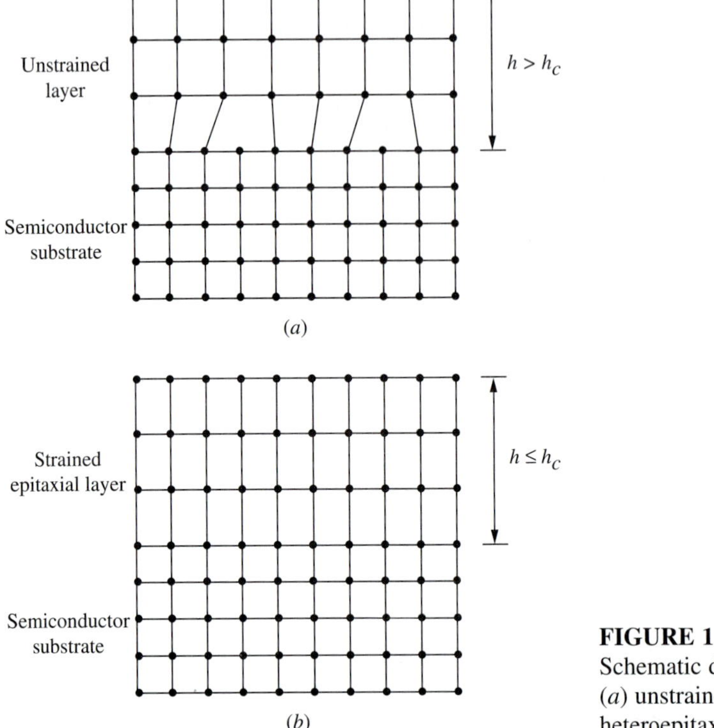

FIGURE 1
Schematic drawings of typical (*a*) unstrained and (*b*) strained heteroepitaxy.

materials and is greatly increased with reduced growth temperatures. At low growth temperatures the epitaxial growth process is far from equilibrium and is dominated by a kinetic reaction, resulting in a larger h_c.[2]

The term *epitaxy* is applied to all kinds of thin-film depositions as long as they are "arranged on in order." This process has been widely employed to prepare III–V and II–VI compound semiconductor materials, such as GaAs, InP, AlGaAs, InGaAsP, CdSe, and HgCdTe, and devices, such as the long-wavelength infrared detector, light-emitting diode (LED), semiconductor laser, heterojunction bipolar transistor (HBT), and modulation-doped field-effect transistor (MODFET). However, in this chapter we will focus only on silicon epitaxy. For readers who are interested in pursuing work in the field of III–V or II–VI compound semiconductors, please refer to other articles.[3]

Silicon epitaxy provides flexibility for a device designer to tailor or optimize the device performance. Unlike the diffusion or ion implantation process, epitaxy is a much easier method for controlling doping layer thickness, doping concentration, and doping profile. In addition, it can produce a buried layer structure or a lightly doped epitaxial layer on top of a heavily doped substrate, a combination that is desirable for tuning device performance. For example, a bipolar device built directly on a heavily doped substrate has a low collector resistance but suffers from a low collector/substrate junction breakdown voltage. On the other hand, if a lightly doped epitaxial layer is deposited on top of a heavily doped substrate, the collector/substrate junction breakdown voltage of a bipolar device can be greatly improved without sacrificing the advantage of low collector resistance, which is essential for high-speed operation. An epitaxial wafer with a lightly doped layer on top of a heavily doped substrate has also been employed for advanced complementary metal-oxide-silicon (CMOS) devices to minimize latch-up effects.[4]

An epitaxial process involves many complex physical and chemical steps. In this chapter, we will first introduce the fundamental aspects of epitaxy, including basic growth kinetics and various growth models, in Section 3.2. Once the background concepts are established, conventional Si epitaxy is introduced in Section 3.3. The term *conventional* is used to differentiate it from other novel approaches such as low-temperature epitaxy, described in Section 3.4, and selective Si epitaxy, described in Section 3.5. These novel growth techniques have received a great deal of attention recently, because of the demands for a low thermal budget and much more miniaturized devices for ULSI technology. Device performance is ultimately influenced by the quality of an epitaxial layer; in Section 3.6 we will introduce various characterization techniques that are commonly used to evaluate the quality of an epitaxial layer. Finally, the trends and future developments in the field of Si epitaxy will be summarized in Section 3.7.

3.2 FUNDAMENTAL ASPECTS OF EPITAXY

Epitaxial growth can be achieved from solid-phase, liquid-phase, vapor-phase, and molecular-beam deposition. For a Si epi-layer, vapor-phase epitaxy (VPE) has received the widest acceptance of all these techniques. Therefore, in this section, we will focus on the basics of the VPE process. However, some of the models and mechanisms, which are also applicable to other epitaxial growth processes, will be discussed as well.

A typical VPE process is illustrated schematically in Fig. 2. Prior to the layer deposition, the growth system is purged by nitrogen or hydrogen for a short period, and followed by a vapor HCl etching. The deposition process is then initiated by

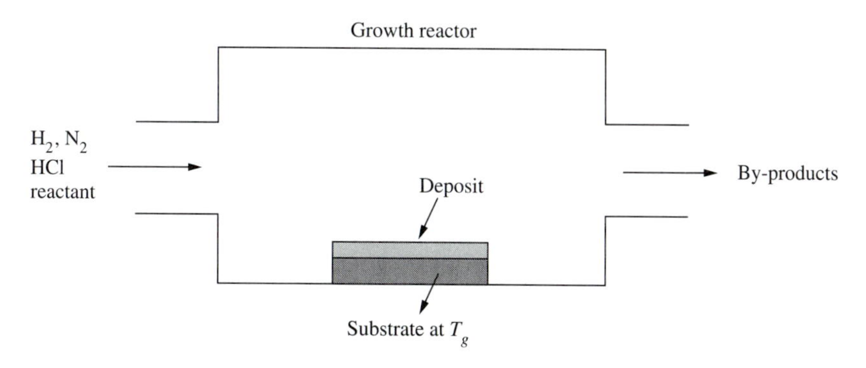

FIGURE 2
Schematic illustration of a typical VPE process.

directing the reactant gases into the growth reactor, where the substrate is located and kept at a desired growth temperature, T_g. Once the reactant gases are fed into the growth reactor, these chemical species undergo a series of physical and chemical reactions, which result in the layer deposition. In a steady state, the sequence of the growth process[5] is best illustrated by Fig. 3. According to this figure, a growth process can be broken down into the following steps:

1. Introduction of the reactant species to the substrate region
2. Transfer of the reactant species to the substrate surface
3. Adsorption of the reactant species on the substrate surface
4. Surface diffusion, site accommodation, chemical reaction, and layer deposition
5. Desorption of residual reactants and by-products
6. Transfer of residual reactants and by-products from the substrate surface
7. Removal of residual reactants and by-products from the substrate region

During growth these steps may occur consecutively or in series, depending on the growth conditions such as growth temperature and growth pressure. These sequential steps describe only the fundamental reactions involving the VPE growth process. As with all chemical reactions, the slowest step of these reactions determines the overall rate of the growth process. This slowest step is called the *rate-limiting step.* This concept is very important to describe a growth process. If a growth process is dominated by step 3, 4, or 5 in the list above, one may say it is a *surface-controlled* process. On the other hand, if a growth process is dominated by step 1 or 2, it is called a *mass-transport-controlled* process. In all cases a growth process can be either surface-controlled or mass-transport-controlled at different growth conditions. We will discuss this point in some detail in the next section.

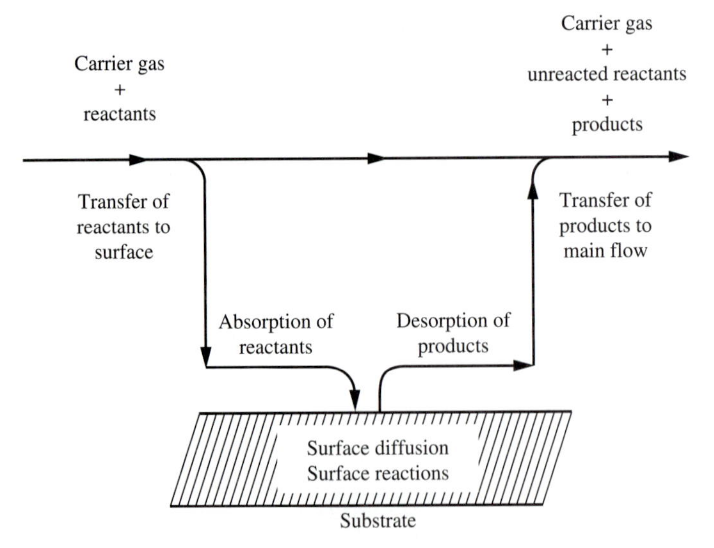

FIGURE 3
Schematic description of the sequential steps of a VPE process. (*After Shaw, Ref. 5.*)

This growth sequence is only applicable to cases where the input reactant gases are flowing in one direction and passing over the substrate to the exhaust. This kind of flow pattern is called *laminar flow* and is a characteristic of a low-growth-pressure process. Conventionally, the growth system is operated at atmospheric pressure. Under this condition, turbulent flow inevitably occurs, which complicates the growth process. In addition, desorption of the reactant species from the walls of the tubing and furnace and the back-diffusion of gases from the exhaust make the growth process even more complicated. Even though a VPE growth process is too complicated to understand fully, many models have been proposed to describe the basic phenomena observed in this process. Before introducing these models, we shall briefly review the basic growth kinetics.

3.2.1 Basic Growth Kinetics of Epitaxy

Deposition of a thin epitaxial layer on the substrate can be described by two processes: nucleation and growth. In the nucleation process a small region of a new phase is created on another phase, in this case the surface of the substrate. When the newly formed phase grows to approximately a critical size, it becomes stable and is usually called a *nucleus;* below the critical size it is called an *embryo* and is unstable. A nucleation process proceeds in two ways: homogeneously and heterogeneously. The former is characterized by the formation of an embryo without the aid of other phases, such as foreign materials or surfaces, whereas the heterogeneous nucleation forms in the presence of other objects.[6] For a VPE process the former represents the formation of an embryo via gas-phase reactions; the latter, via the substrate surface.

In homogeneous nucleation, for a VPE process, a number of embryos are formed in the vapor phase. Subsequently, these embryos coalesce and condense into the solid phase. Such a process involves the release of free energy (ΔG_v), resulting from the condensation process, and an increase of free energy (γ_s), due to the formation of surfaces in the coalescence process. The former is a function of reactant gas partial pressure and increases as the reactant gas partial pressure increases. The latter depends on the surface energy of the associated phases in the growth region, such as vapor, substrate surface, and nucleus. The total free-energy change (ΔG_t) for an embryo of a radius r with homogeneous nucleation can be expressed as

$$\Delta G_t = \left(\frac{4\pi r^3}{3}\right)\Delta G_v + (4\pi r^2)\gamma_s \tag{3.1}$$

From the thermodynamic viewpoint, the critical radius r_c of an embryo can be obtained at maximum total free-energy change, where

$$r_c = \frac{-2\gamma_s}{\Delta G_v} \tag{3.2}$$

After substituting r_c in Eq. (3.2) into r in Eq. (3.1), the critical formation free-energy change ΔG_{fc} of a stable embryo is then written as

$$\Delta G_{\text{fc}} = \frac{16\pi\gamma_s^3}{3(\Delta G_v)^2} \tag{3.3}$$

An embryo with a radius larger than r_c will tend to lower its free energy by growing. Under a quasi-equilibrium condition, the number of stable embryos, n_c, can be expressed by the following equation:

$$n_c = n_o \exp(-\Delta G_{\text{fc}}/kT) \tag{3.4}$$

where n_o is the total number of embryos, k is Boltzmann's constant, and T is the substrate temperature in kelvins. Finally, the nucleation rate of a homogeneous nucleation process, R_n, is given by

$$R_n = (n_o/t_e) \exp(-\Delta G_{\text{fc}}/kT) \tag{3.5}$$

where t_e is the average lifetime for stable embryos.

As already mentioned, most nucleation processes may occur heterogeneously on a substrate in the VPE process. Figure 4 illustrates the initial wetting characteristics on a substrate surface, where γ_{sv}, γ_{nv}, and γ_{ns} are surface energies associated with surface (s)/vapor (v), nucleus (n)/vapor (v), and nucleus (n)/surface (s), respectively, and θ is the contact angle. A mathematical derivation[6] gives the critical energy for heterogeneous nucleation:

$$\Delta G_{\text{hetero}} = \Delta G_{\text{homo}}[(2 + \cos\theta)(1 - \cos\theta)^2/4] \tag{3.6}$$

According to Eq. (3.6), with the aid of a substrate surface, ΔG_{hetero} is lower than ΔG_{homo}. Therefore, thermodynamically, heterogeneous nucleation is always preferable to homogeneous nucleation.

It can be noted from Eq. (3.6) that surface wetting (i.e., the contact angle θ) is a predominant factor for a heterogeneous nucleation process. Thus, surface preparation prior to the epitaxy is one of the most important steps that determine the quality of an epitaxial layer. In addition, since the value of ΔG_v decreases with decreasing reactant partial pressure and increasing temperature, performing the VPE process at reduced partial pressure and higher temperature is preferable.

An epitaxial process proceeds by impingement, adsorption, diffusion, reaction, and desorption of the reactant species. Many models have been proposed to describe this process. However, none of them can account for all aspects of the various growth

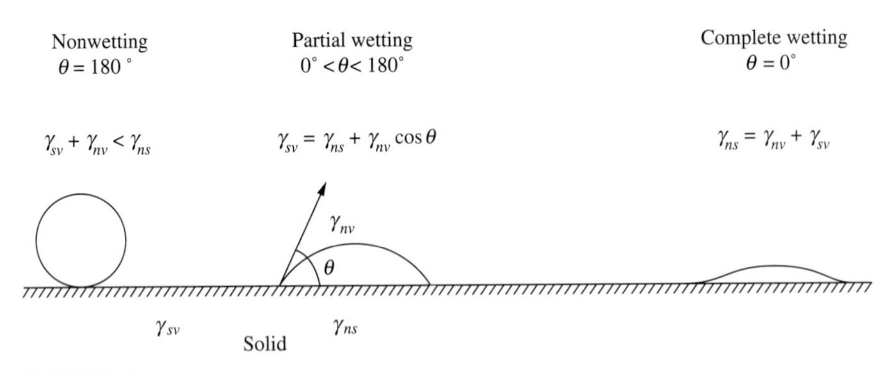

FIGURE 4
Wetting characteristics of a heterogeneous nucleation process. (*After Ghandhi, Ref. 3.*)

results obtained by different growth techniques. In the next section we will introduce a few generally accepted models to explain some characteristics of an epitaxial growth process.

3.2.2 Growth Models for Epitaxy

A model[7] based on the assumption that only transport of the reactant species and chemical reactions are operative is shown in Fig. 5. In the schematic drawing of the model, F_i is the flux difference between the reactant in the input gas stream, C_g, and that at the substrate surface, C_s, and F_o is the consumption of the reactant at the substrate surface. Assuming a linear change from C_g to C_s, F_i can be expressed by

$$F_i = h_g(C_g - C_s) \tag{3.7}$$

where h_g is the gas-phase mass transfer coefficient and is relatively independent of temperature. For simplicity, by assuming F_o proportional to C_s, F_o is expressed as

$$F_o = k_s C_s \tag{3.8}$$

where k_s is the chemical surface reaction constant. The value of k_s, on the other hand, is a strong function of temperature and is usually expressed in the form of $k_s = k_o \exp(-E_a/kT)$, where k_o is a temperature-independent constant and E_a is the activation energy of the reaction.

At steady-state conditions the two fluxes must be equal. By equating Eqs. (3.7) and (3.8), one obtains

$$C_s = \frac{C_g}{1 + (k_s/h_g)} \tag{3.9}$$

If $h_g \gg k_s$, then $C_s \approx C_g$. This condition is called a surface-controlled process. On the other hand, if $h_g \ll k_s$, $C_s \approx 0$, and the process is called mass-transport-controlled. Simple mathematical derivations lead to the growth rate G of a VPE process given by

$$G = C_t k_s M/N \qquad \text{(surface-controlled case)} \tag{3.10}$$

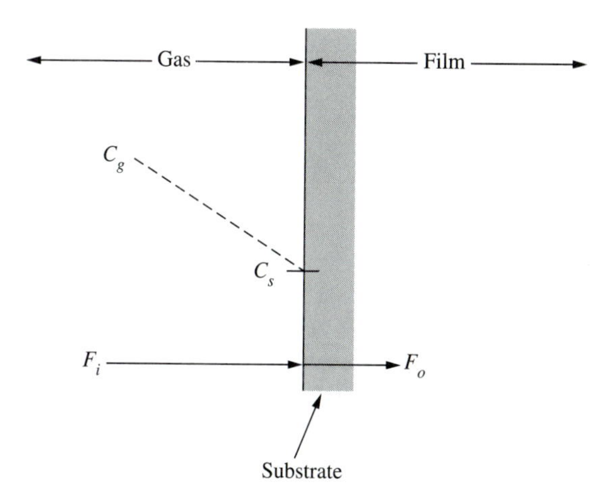

FIGURE 5
Schematic representation of the Grove (Ref. 7) model.

or by

$$G = C_t h_g M/N \qquad \text{(mass-transport-controlled case)} \tag{3.11}$$

where C_t is the total number of molecules per cm^3 in the gas stream, M is the mole fraction of the reactant species in the gas stream, and N is the number of atoms incorporated into a unit volume of the layer.

Another model assumes that the gas stream above the substrate surface is divided into two regions.[8] As shown in Fig. 6*a*, in the region away from the substrate surface, the gas stream moves with a constant speed V in parallel to the substrate surface, where in the region next to the substrate surface, a stagnant layer exists in which the gas velocity is zero. In this model, mass transfer of the reactant species from the upper region across the stagnant layer to the substrate surface is dominated by a diffusion process. With this assumption, F_i in Eq. (3.7) can be rewritten as

$$F_i = D_g[(C_g - C_s)/t_s] \tag{3.12}$$

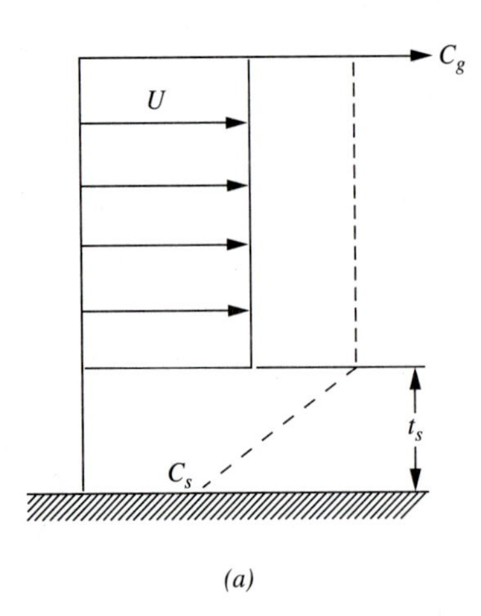

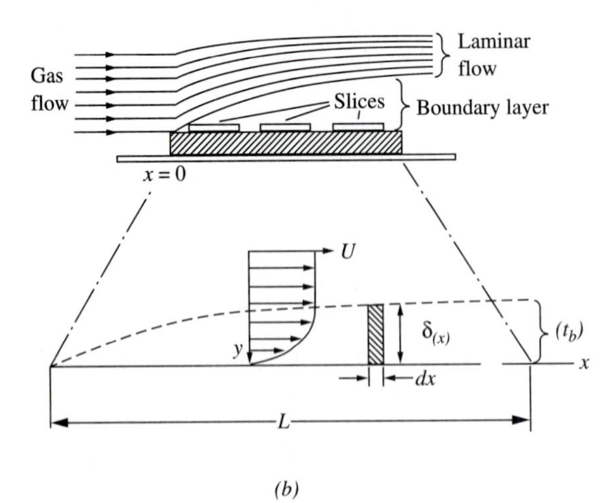

FIGURE 6
Schematic representations of (*a*) stagnant-layer model and (*b*) boundary-layer model. (*After Wolf and Tauber, Ref. 20.*)

where D_g is the diffusion coefficient of the reactant species and t_s is the thickness of the stagnant layer. From Eqs. (3.7) and (3.12) we find

$$h_g = D_g/t_s \tag{3.13}$$

A model similar to the stagnant-layer model was developed to provide a more realistic estimate of the mass transfer coefficient h_g.[8] As shown in Fig. 6*b*, a boundary layer is present between the substrate, where the gas velocity is zero, and a free gas stream, where the gas velocity is u. In this model, u is not a constant value but changes rapidly across the entire range of substrate or susceptor. The boundary layer thickness t_b is a function of distance along the flat plate, x. Therefore, it can be expressed as

$$t_b(x) = \left(\frac{\mu x}{\rho u}\right)^{1/2} \tag{3.14}$$

where μ is the gas viscosity and ρ is the gas density. Upon integration, the average boundary layer thickness over an entire flat plate with length L is expressed as

$$t_b = \frac{(2/L)}{3(\text{Re})^{1/2}} \tag{3.15}$$

where Re is the Reynolds number for the gas, given by

$$\text{Re} = \frac{\rho u L}{\mu} \tag{3.16}$$

For Re < 2000 the gas flow has a laminar pattern, whereas for larger Re values the gas flow is turbulent.

The average boundary layer thickness, t_b in Eq. (3.15), can be substituted for t_s in Eq. (3.12), resulting in an expression for h_g as

$$h_g = \frac{3D_g\text{Re}^{1/2}}{2L} \tag{3.17}$$

According to this equation, in the mass-transport-controlled regime the growth rate of a layer depends on the square root of the free gas stream velocity u. This predicted dependency is observed in a typical experimental result for epi-Si grown at low flow rates, as shown in Fig. 7. At higher flow rates the growth rate reaches a maximum and becomes independent of the flow rate. Figure 8 shows the formation of boundary layers in a horizontal reactor. The spatial dependencies of reactant concentration, gas velocity, and temperature across the entire susceptor are illustrated. The fluxes of the various reactant species that cross this layer to the substrate surface are therefore dependent on factors such as boundary layer thickness, reactant concentration, growth pressure, and growth temperature.

Once the reactant species are adsorbed on the substrate surface (in which condition they are often called *adatoms*), the growth of the epitaxial layer is believed to occur via the migration of adatoms (position A in Fig. 9) to steps on the substrate surface (positions B, C in Fig. 9). From the thermodynamic viewpoint, the kink position (position C in Fig. 9) along a step is the most favorable site for growth to occur. Such a layer growth is assumed to proceed by lateral movements of the steps. In

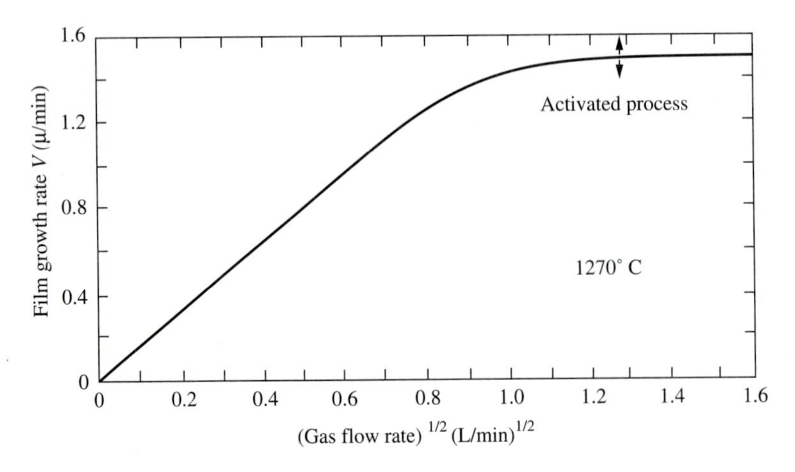

FIGURE 7
The growth rate of an epi-Si versus the free gas flow rate. (*After Wolf and Tauber, Ref. 20.*)

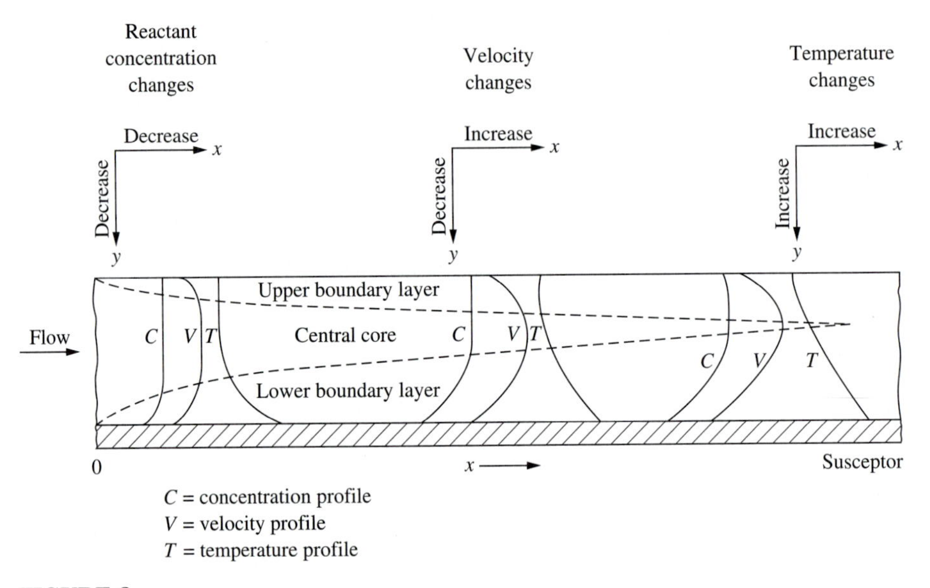

FIGURE 8
Schematic illustration of the formation of boundary layers in a horizontal reactor. (*After Ban, Ref. 8.*)

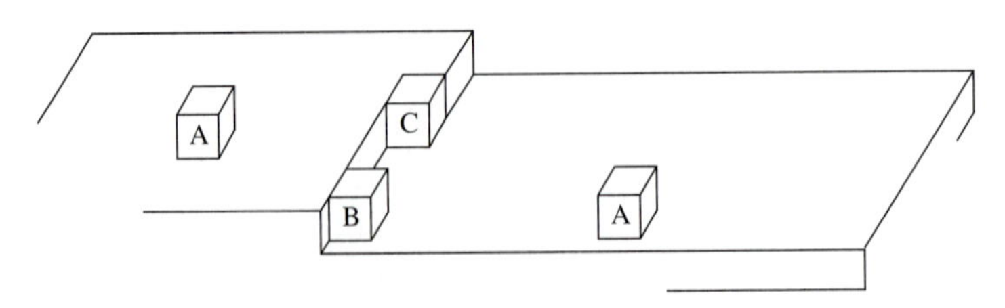

FIGURE 9
Schematic illustration of epitaxial growth with adatoms. (*After Wolf and Tauber, Ref. 20.*)

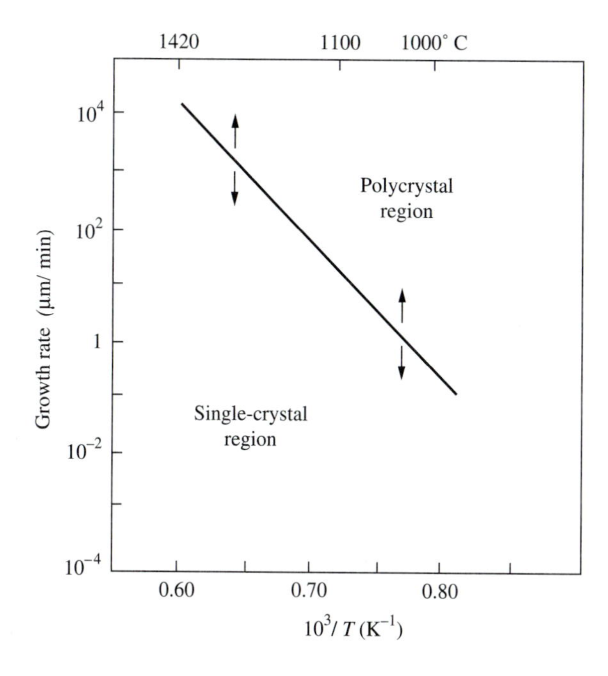

FIGURE 10
Crystallinity with respect to growth rate and growth temperature. (*After Bloem, Ref. 10.*)

this growth model, a "maximum growth rate" is imposed to ensure the crystallinity of the resulting layer. At high growth rates, not enough time is available for the adatoms to migrate, resulting in a polycrystal layer. However, as the growth temperature increases, the surface migration rate increases, facilitating single-crystal layer growth. Figure 10 shows the crystallinity of a layer with respect to temperature and growth rate. For a fixed temperature, there is a maximum growth rate for an epi-layer. Likewise, for a fixed growth rate, a minimum temperature is required to achieve an epi-layer. For Si epitaxy, an activation energy of $\approx$5 eV is observed. This value is comparable to the activation energy of self-diffusion in Si. Therefore, self-diffusion in Si is believed to be the dominant mechanism for the growth of Si epi-layers.[10]

3.3 CONVENTIONAL Si EPITAXY

The term *conventional Si epitaxy* is used to differentiate it from the other approaches, such as low-temperature Si epitaxy and selective Si epitaxy, which will be discussed in the following sections. Conventional Si epitaxy was originally developed to overcome the problems associated with high concentrations of impurities present in the substrate. This technique was then used extensively to optimize the performance of bipolar transistors and bipolar integrated circuits. It enables the growth of a lightly doped Si epi-layer over a heavily doped Si substrate. As a result, a higher collector–substrate breakdown voltage is achieved with minimal increase of collector resistance. Similarly, if CMOS devices or circuits are built in the lightly doped Si epi-layer over the heavily doped Si substrate, the latch-up effect is greatly reduced.[4]

The advantages of conventional Si epitaxy are its flexibility and accuracy in controlling the doping concentration and doping profile. However, since this technique is usually carried out at temperatures above 1100°C, it suffers from problems such as autodoping and pattern shift. In addition, because the epitaxy is an "arranged on" process, it is susceptible to defect generation due to strain, impurities, and particles on the substrate prior to or during the growth process. In this section, we present (*a*) growth chemistry; (*b*) doping and autodoping; (*c*) defect generation and minimization; (*d*) pattern shift, distortion, and washout; and (*e*) epitaxial deposition equipment.

3.3.1 Growth Chemistry

The success of an epitaxial growth relies heavily on the chemistry used in the process. The major source gases used to deposit Si epi-layer commercially contain hydrogen and chlorosilanes, such as silicon tetrachloride ($SiCl_4$), trichlorosilane ($SiHCl_3$), and dichlorosilane (SiH_2Cl_2). In addition, silane (SiH_4) has also been used when a lower growth temperature is desired. The choice of each source gas is based on the growth conditions and layer requirements. One of the most important parameters that affect the selection of the Si source gas is the growth rate. Figure 11 shows the growth rate of Si layers as a function of growth temperature for various silicon sources.[11] Two distinct growth regimes are noted in this figure. At low growth temperatures (region A) the growth rate of Si layer is exponentially dependent on temperature, which is evidence of a surface-controlled reaction. In the high-temperature regime (region B), the growth rate is nearly independent of the growth temperature, which is typical of a mass-transport or diffusion-controlled

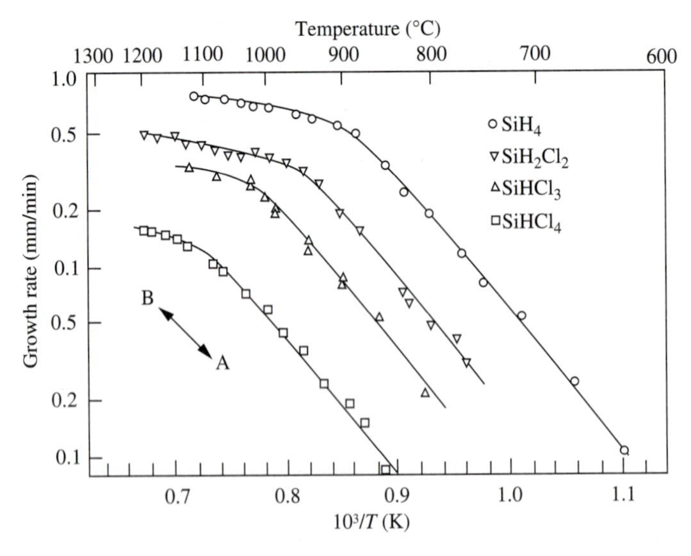

FIGURE 11
Growth rate of epi-Si as a function of growth temperature for a variety of Si sources. (*After Eversteyn, Ref. 11.*)

reaction. It is important to note that at the lower temperature, as-grown Si layers are in polycrystal form. Epi-Si layers are formed at temperatures above the transition point of each curve seen in Fig. 11. The position of the transition point varies with the mole fraction of reactant, gas flow rate, and reactor type. However, it can be inferred from this figure that epi-Si can be grown from the SiH_4 gas at temperatures above 900°C, whereas for $SiCl_4$ gas it is above 1100°C.

Reductions of $SiCl_4$ and SiH_4 to solid Si encompass two extremes of the chemical reactions involving conventional Si epitaxy. The overall reaction for the reduction of $SiCl_4$ in H_2 can be written as

$$SiCl_4 + 2H_2 \longrightarrow Si + 4HCl \tag{3.18}$$

In reality, a study of thermodynamics of the H–Cl–Si system showed that a number of chemical reactions were occurring either simultaneously or sequentially in the deposition process. It is postulated[12] that the net reaction of Eq. (3.18), may proceed as follows:

$$SiCl_4 + H_2 \longleftrightarrow SiHCl_3 + HCl \tag{3.19}$$

$$SiHCl_3 + H_2 \longleftrightarrow SiH_2Cl_2 + HCl \tag{3.20}$$

$$SiH_2Cl_2 \longleftrightarrow SiCl_2 + H_2 \tag{3.21}$$

$$SiHCl_3 \longleftrightarrow SiCl_2 + HCl \tag{3.22}$$

$$SiCl_2 + H_2 \longleftrightarrow Si + 2HCl \tag{3.23}$$

Note that all the above reactions, Eqs. (3.19) to (3.23), are reversible. Therefore, both reduction and etching processes compete with each other, depending on the mole fraction of reactants and growth temperature. Figure 12 shows the boundary between etching and deposition in a $SiCl_4$ and H_2 mixture at atmospheric pressure as a function of temperature and partial pressure of $SiCl_4$.[13] A separate study also showed that the growth rate of silicon varied with the growth temperature with $SiCl_4 + H_2$ chemistry. Figure 13 shows that an etching process is favored at both low and elevated

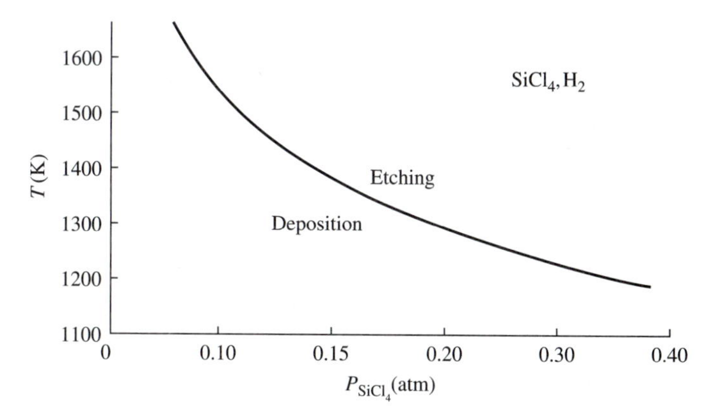

FIGURE 12
Growth temperature as a function of the partial pressure of $SiCl_4$ in a $SiCl_4 + H_2$ mixture at one atmosphere. (*After Ban and Gilbert, Ref. 13.*)

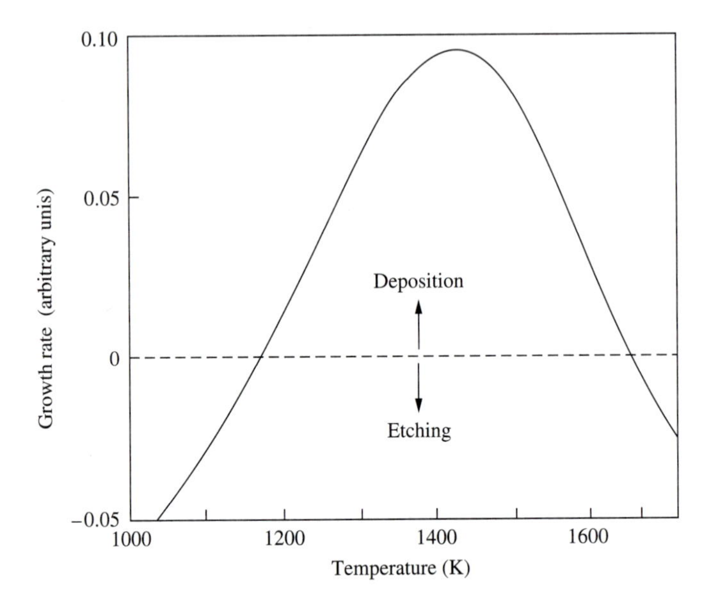

FIGURE 13
Growth rate of Si films as a function of growth temperature for $SiCl_4 + H_2$ chemistry. (*After Sirtl, Hunt, and Sawyer, Ref. 14.*)

temperatures.[14] Therefore, the deposition process is usually performed at temperatures between 1100 and 1300°C in this case. For thicker layer growth $SiHCl_3$ is often used because its growth rate is higher than that of $SiCl_4$.

In contrast to $SiCl_4$ chemistry, the reduction reaction of SiH_4 is not reversible and proceeds by the following overall reaction:

$$SiH_4 \longrightarrow Si + 2H_2 \tag{3.24}$$

As noted in Fig. 11, the major advantage of the SiH_4 is that epi-Si can be obtained at relatively lower temperatures than any of the chlorosilane chemistries. However, it is very difficult to avoid some degree of gas-phase nucleation of Si via the homogeneous reaction of SiH_4. As a result, Si particulate is formed, which results in poor surface morphology or even polycrystal Si growth. This problem can be minimized by raising the gas velocity through the growth region via a low-pressure operation or by controlling the deposition temperature. In constrast to the chlorosilane chemistry, SiH_4 is more susceptible to oxidation and is explosive. Therefore, it is not commonly used in the conventional Si epitaxial process. Because there is no HCl in the SiH_4 growth system, there is no etching process. This may result in higher metallic impurities in the film grown with SiH_4 unless a careful preclean process is used.

3.3.2 Doping and Autodoping

The conductivity type and electrical resistivity of an epitaxial layer are ultimately dependent on the doping characteristics. In an epitaxial process, dopants are incorporated into the grown layer simultaneously or intermittently during the layer deposition reaction. Typically, hydrides of various types are used to introduce these dopants. For Si epitaxy, diborane (B_2H_6) is used to incorporate the p-type dopant B, whereas phosphine (PH_3) and arsine (AsH_3) are used for the n-type dopants P and As, respectively. These gases are extremely toxic and are unstable above room temperature. Therefore, they are normally diluted with a large amount of H_2. Because of this, the hydride doping gases would not follow a simple rule for incorporation into the growing Si film from the gas phase. The dopant/Si ratio in solid phase corresponding to that in gas phase can only be obtained through experiments for each growth condition and each growth reactor. The many factors that affect the dopant incorporation include the growth temperature, growth rate, dopant/Si ratio in gas phase, and reactor geometry.

In addition, the interactions or competing reactions between the dopant gases and Si source gases make the doping process even more complicated. It has been found that B_2H_6 and PH_3 have opposite influences on the silicon deposition rate.[15] The former enhances the silicon deposition rate, whereas the latter suppresses it. It has been found that PH_3 chemisorbs dissociatively with the formation of Si–H bonds at high temperatures. Hydrogen can be desorbed at temperatures above 400°C to form a phosphorus layer, which has a maximum adsorption at 550°C. Once adsorbed, PH_3 is stable under a SiH_4 flux, effectively passivating the silicon surface and suppressing the growth rate in the phosphorus-doped Si epitaxy.

On the other hand, B_2H_6 has a very small sticking coefficient and is easy to decompose directly on the silicon surface to deposit boron. The adsorbed B_2H_6 facilitates the heterogeneous reaction of SiH_4, resulting in enhanced growth rate. The amount of dopant incorporation in epi-Si is influenced by the growth rate. Figure 14

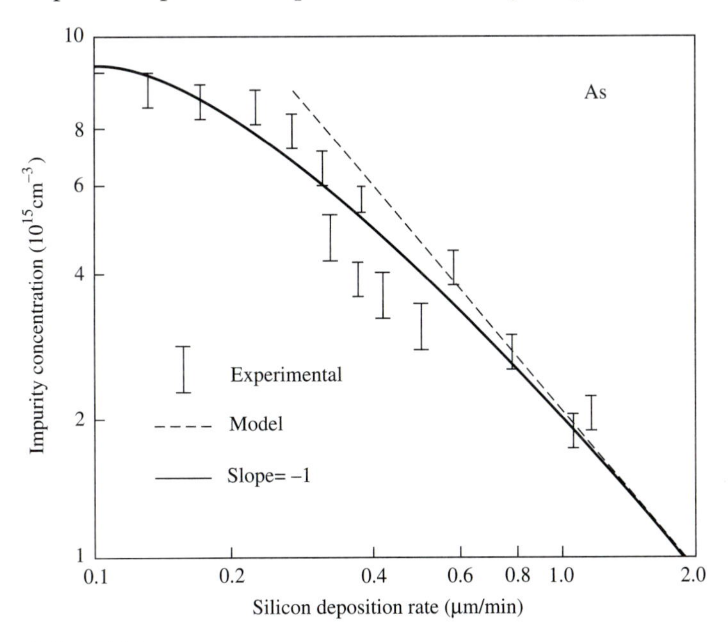

FIGURE 14
A typical plot of arsenic doping concentration as a function of silicon deposition rate. (*After Reif, Kamins, and Saraswat, Ref. 16.*)

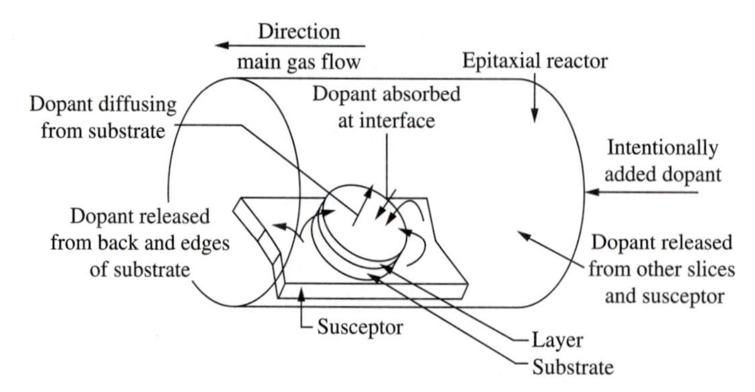

FIGURE 15
Schematic illustration of various sources of dopants for a VPE growth process. (*After Srinivasan, Ref. 17.*)

shows typical results of impurity incorporation as a function of Si deposition rate for arsenic doping.[16] At high growth rates the incorporation of As is dominated by surface kinetic processes such as surface adsorption and surface diffusion, while at low growth rates it is controlled by the mass transport process.

The foregoing discussion applies only to the doping process that is performed intentionally to obtain different conductivity types and various resistivities in the epi-Si layer. In practice, the doping process is complicated by an unintentional process named *autodoping*. Figure 15 shows a schematic drawing of various sources of dopants for an epitaxial growth.[17] Dopants can be released from the substrate or susceptor through outdiffusion or outgassing. The outdiffusion is in the solid state, whereas the outgassing is in gas phase and is often termed the *memory effect*. Figure 16 shows a plot of a typical doping profile of an epitaxial layer grown over a heavily

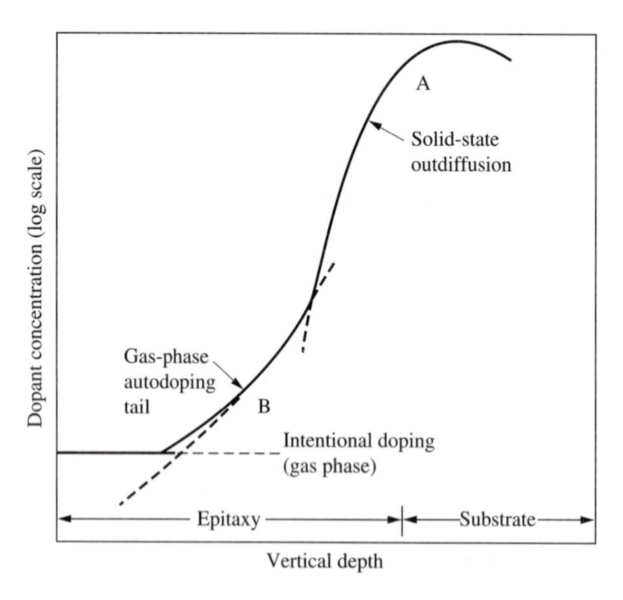

FIGURE 16
A typical doping profile of an epi-Si grown over a heavily doped Si substrate. (*After Srinivasan, Ref. 18.*)

doped substrate.[18] Several regions of autodoping are noted. Close to the substrate/epi interface, outdiffusion in solid phase from the substrate is predominant, resulting in a broad transition region. This effect is soon limited, because the growth rate is normally much higher than the speed of solid-state outdiffusion. Thereafter, the dopant incorporation is controlled by the sources in gas phase. If the outgassing dopant exceeds that intentionally introduced, an autodoping tail develops. However, this effect becomes insignificant as the growth process proceeds, because of the coverage of a low-doped layer over substrate and susceptor, which slows down the supply of the dopant to the surface (i.e., the outgassing speed). The extent of the autodoping tail is a function of substrate dopant species and its concentration, reactor geometry, growth temperature, growth rate, and growth pressure. Once the autodoping is minimized, the intentional doping process predominates.

For most ULSI applications lightly doped epi-Si layers over a heavily doped substrate are required. Autodoping imposes a constraint on the minimum layer thickness and minimum doping level for an epi-Si layer with controlled doping. Therefore, it is a subject of great technological importance. Various techniques are proposed to minimize this effect. Low growth temperature seems to reduce boron autodoping, whereas lower growth pressure works well for phosphorus and arsenic. In-situ vapor-phase etching, using HCl, or cap-seal, with a thin lightly doped Si over the substrate and susceptor, are also employed. With the advent of low-temperature epitaxial growth techniques, the autodoping effect has been greatly minimized.

3.3.3 Defect Generation and Minimization

In an epitaxial process various types of defects are often observed. These defects may result from substrate imperfection, contamination or inclusion in the reactor or substrate, or any mismatch between the substrate and epi-layer. In principle, the defect density of an epi-layer cannot be lower than that of its parent substrate. Defects due to the substrate are related to the surface conditions prior to the deposition process. Therefore, care must be taken to remove organic and metallic impurities and particulate on the front and rear substrate surfaces. In addition, a defect-free wafer is preferred for an epitaxial process.

There are four types of crystalline defects in an epitaxial layer: *point defects,* such as impurities and vacancies; *line defects,* such as dislocations; *area defects,* such as stacking faults; and *volume defects,* such as voids and precipitates. Figure 17 illustrates the defects that commonly occur in epitaxial layers.[19] In this figure, 1 represents a dislocation line propagated from the substrate through the growing layer; 2 is a stacking fault nucleated from the impurity or precipitate on the substrate surface; 3 is an impurity precipitate incorporated by contamination in the growth system; 4 is a *growth hillock* caused by the process or surface finish of the substrate; and 5 is a bulk stacking fault or void caused by an existing surface defect in the substrate. In addition to these, defects such as dislocations or slip lines can be generated from the thermal stresses set up by the temperature gradients between the front and rear substrate surfaces during the deposition process, as illustrated in Fig. 18.[20] Furthermore, the existence of lattice mismatch between the substrate and epi-layer leads to the formation of dislocations to accommodate the misfit strain.

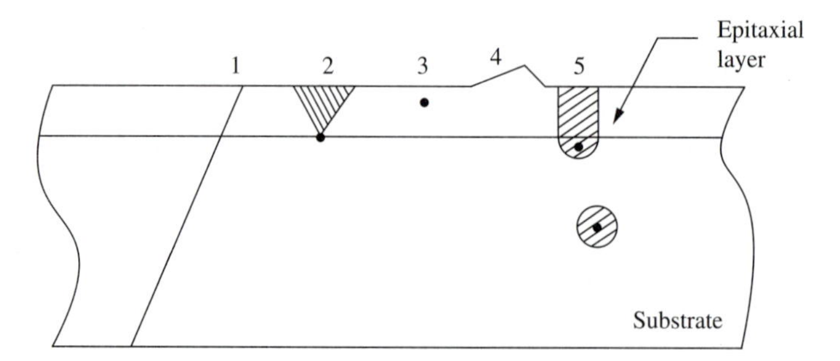

FIGURE 17
Schematic illustration of common defects in epitaxial layers. (*After Ravi, Ref. 19.*)

One of the most important measures for minimizing the defects in epi-Si is the substrate cleaning. The pre-epitaxy substrate cleaning (see Chapter 2) consists of a wet clean followed by a dilute HF dip and an in-situ HCl, HF or SF_6 vapor etch.[21] This preclean process is especially important for low-temperature Si epitaxy and will be discussed separately in Section 3.4.2. In addition to the wafer cleaning, defects in epitaxial layers can be reduced by employing "denuded" substrates.[22] Such a substrate is obtained by using an oxygen intrinsic gettering technique to minimize the defect and metallic impurity levels on the substrate surface and thus reduce the defect density in the growing layer. For heteroepitaxy, misfit dislocations due to lattice mismatch are inevitably formed. It has been found that a strain field near the misfit dislocation serves as a gettering site to trap impurities.[23] A strained-layer structure grown between the substrate and active device layers was also found to be effective in gettering heavy metals without introducing misfit dislocations into the device layer, leading to better device performances.

3.3.4 Pattern Shift, Distortion, and Washout

Pattern shift, distortion, and washout (the extreme case of pattern shift) occurs when an epitaxial layer is deposited over a buried layer in the fabrication of modern bipolar integrated circuits. Schematic representations of these phenomena[24] are shown

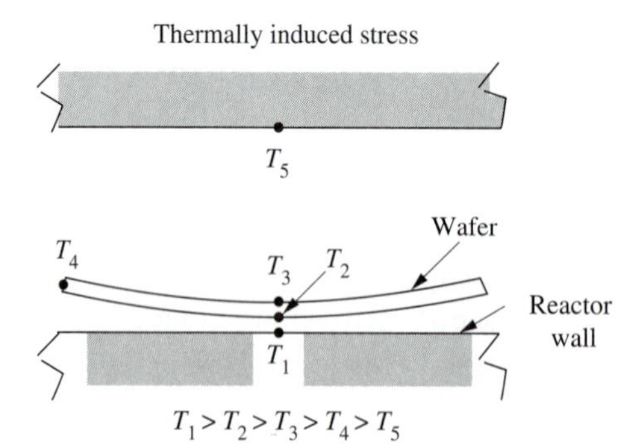

FIGURE 18
Schematic drawing of wafer bow, due to a temperature gradient between front and back of a wafer during a growth process. (*After Wolf and Tauber, Ref. 20.*)

in Fig. 19. The main cause of these effects is the different growth rates on various crystal orientations. However, the amount of shift or distortion also depends on deposition rate, growth temperature, growth pressure, and silicon source. Therefore, to compensate these effects the circuit designer must adjust the position of features on the subsequent mask levels.

It has been found that pattern shift is at a minimum when the pattern is oriented on-axis with respect to a (100)-oriented substrate. For the case of (111)-oriented substrates, a misorientation of 2 to 5 degrees from the [111] to the [110] direction is needed to minimize these effects. For both cases the amount of shift seems to decrease with increasing growth temperature and decreasing growth rate.[25] Pattern shift can be completely prevented if SiH_4 is used. In addition, reduced growth pressure seems to minimize the amount of shift. In contrast to pattern shift, the amount of pattern distortion increases with increasing temperature and decreasing growth rate,[26] and the use of SiH_4 results in more severe pattern distortion than use of chlorosilanes. Moreover, pattern distortion occurs in both (100) and (111) substrates. Reduced growth pressure may be effective for controlling such a distortion for the (111)-oriented substrate, but not for the (100)-oriented substrate. However, increased layer thickness seems to reduce the distortion for both cases. Therefore, an optimized epitaxial process is required to minimize both pattern shift and pattern distortion.

3.3.5 Epitaxial Deposition Equipment

A typical epitaxial growth equipment consists of a gas distribution system, a high-purity quartz bell jar or reactor tube, a susceptor, a substrate heating system, an electrical control system, a cooling system, and an exhaust system.

The gas-handling system is used to feed the reactant gases, such as silane or chlorosilanes, H_2, and HCl; doping gases, such as AsH_3, PH_3, and B_2H_6; and purging gas, H_2 or N_2. Note that except for N_2, all other gases are explosive, corrosive, or highly toxic. Therefore, built-in safety interlocks, together with hydrogen and

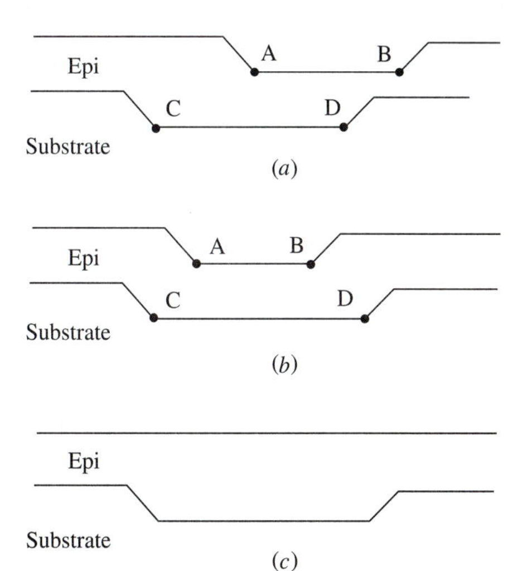

FIGURE 19
Schematic illustrations of (*a*) pattern shift, (*b*) pattern distortion, and (*c*) pattern washout. (*After Wolf and Tauber, Ref. 24.*)

toxic-gas monitors, are required for preventing accidents. In addition, the entire growth system must be operated under leaktight conditions to ensure safety. In order to meet the stringent requirements of an epitaxial process for ULSI applications, high-precision mass-flow controllers with a vent-run configuration are used to obtain tight control on gas flows. Electropolished stainless steel tubing is used to transport all gases fed into the growth reactor. All gases must be of high purity or passed through a gas purifier (e.g., Pd-diffused H_2) to ensure minimum impurity incorporation. An exhaust system, consisting of a burn box and a wet or dry scrubber, is necessary to safely remove or treat unreacted by-products from the growth system.

A high-purity quartz bell jar is normally used as the growth reactor. The susceptor in the reactor is used to support substrates and to supply thermal energy for the reaction. Therefore, it should be mechanically strong and not reactive to the reactants and their by-products. In addition, a susceptor must be noncontaminating to the growth system. Graphite is commonly used as the primary material for susceptors. Silicon nitride– or silicon carbide–coated polysilicon or quartz susceptors are often used as alternatives. Coating is also required for the graphite susceptor, because it is impure and soft. In addition to SiC and Si_3N_4, glassy carbon and porolytic graphite are often used to coat the susceptor. Care must be taken to avoid any crack or scratch on the susceptor and to eliminate any contaminating source from the susceptor.[27]

Commercially, Si epitaxy is carried out in reactors of the types shown in Fig. 20. Susceptors of horizontal, pancake, and barrel shapes are widely used.[27] The horizontal reactor is the most commonly used system. It offers high capacity and throughput. However, it suffers from an inability to achieve a uniform deposition over the entire susceptor. Tilting the susceptor a few degrees (1.5 to 3 degrees) seems to reduce this problem greatly. In contrast, the vertical pancake reactor is capable of very uniform growth with minimal autodoping problems. Disadvantages of this system include its mechanical complexity, its low throughput, and its susceptibility to particulate incorporation. The barrel reactor is an expanded version of the horizontal reactor

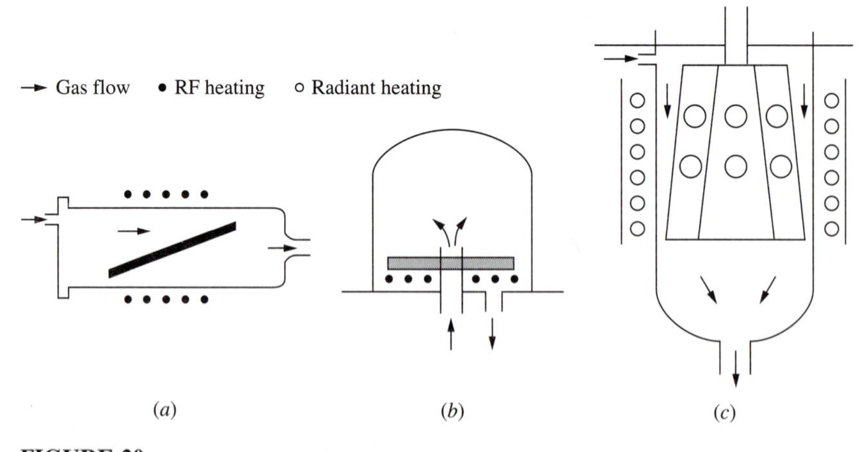

FIGURE 20
Types of conventional epitaxial reactors: (*a*) horizontal reactor; (*b*) vertical reactor; (*c*) barrel reactor. (*After Pearce, Ref. 27.*)

in a different configuration. When used with a tilted susceptor, this system allows high-volume production and uniform growth.

Two types of substrate-heating systems are used to provide the thermal energy to the reaction: inductive heating and radiant heating. In the former case the susceptor is heated inductively by the underlying water-cooled RF coils. These coils are usually set outside the reaction chamber but close enough to the susceptor to obtain RF coupling. Once the susceptor is heated, the thermal energy is transported to the substrate by conduction and radiation. In the latter case the energy is supplied by radiant heating using banks of quartz-halogen lamps via IR absorption. This method provides more uniform heating than inductive heating does.[28] Temperature measurement is usually done with an optical or IR pyrometer. To avoid contaminating the growth environment, thermocouples should not be used in the reactor.

3.4 LOW-TEMPERATURE EPITAXY OF Si

Low-temperature epitaxy (LTE)[29] of Si is a process that produces epitaxial growth of the Si layer at temperatures of 550°C or less, much lower than the conventional Si epitaxy process does. Because tight controls over interfacial quality, spatial dopant distribution, and layer thickness are required for many advanced Si devices such as heterojunction bipolar transistors (HBTs), low processing temperature becomes necessary to minimize thermal diffusion and/or mass-transport-controlled processes. This section will start with an introduction of the chemical aspects of the LTE, followed by a discussion of a key step for this technique: surface preparation. There are many growth techniques that can be classified as LTE; this section will introduce the techniques that attract the most research interest. Finally, device applications of the LTE process are reviewed.

3.4.1 Chemical Aspects

The reason a Si layer can be deposited epitaxially onto a substrate at low temperature is primarily the growth conditions, such as substrate surface material, purity of Si source gas, and growth pressure.

To grow an epitaxial layer, the seed substrate surface should be of single-crystal material. It is well known that SiO_2 is readily formed on Si substrate under equilibrium conditions and becomes unstable at a temperature higher than 1000°C at atmospheric growth pressure. It has been found that the stability of the SiO_2 film depends on the partial pressures of oxygen and water vapor in the growth environment. As shown in Fig. 21, the crossover between the oxidation and the oxide-free regions is a function of water-vapor partial pressure and temperature.[29] Similar results have been observed for oxygen. Therefore, one of the most important criteria to realize LTE is to control the oxygen and water-vapor partial pressures in the growth reactor. In order to grow the Si epi-layer at a temperature lower than 700°C, the partial pressure of water vapor should be lower than 10^{-8} torr, maintaining a bare single-crystal

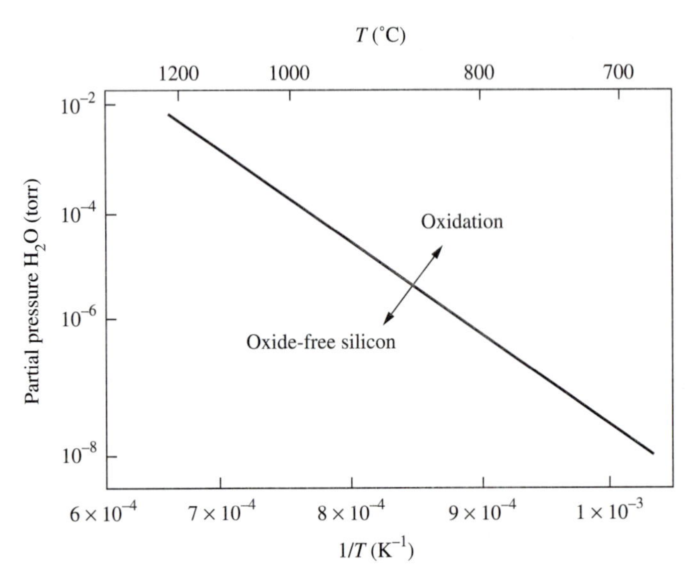

FIGURE 21
Water vapor pressure vs. reciprocal temperature for oxide-free and oxidized Si surfaces. (*After Meyerson, Ref. 29.*)

Si substrate surface. Presumably, in the oxidation region, the stable SiO_2 is formed on the silicon surface by the chemical reaction

$$\text{Si (solid)} + 2\text{H}_2\text{O (gas)} \rightarrow \text{SiO}_2\text{ (solid)} + 2\text{H}_2\text{ (gas)} \tag{3.25}$$

In the oxide-free region, on the other hand, the water vapor serves to etch the silicon surface by the following chemical reaction:

$$\text{Si(solid)} + \text{H}_2\text{O(gas)} \rightarrow \text{SiO(gas)} + \text{H}_2\text{(gas)} \tag{3.26}$$

There are many ways to achieve a bare Si single-crystal surface prior to the epitaxial growth process. As mentioned above, the straightforward approach is to keep the growth environment at an ultrahigh vacuum. Growth techniques employing this approach are ultrahigh vacuum/chemical vapor deposition (UHV/CVD) and molecular beam epitaxy (MBE). Another approach is to employ ultraclean gases with ultralow oxygen and water-vapor contents. Growth techniques using this approach are atmospheric-pressure chemical vapor deposition (APCVD) and ultraclean low-pressure CVD. In addition, plasma-enhanced CVD (PECVD) and photochemical CVD (photo-CVD) have been proposed to achieve LTE-grown epi-Si layers.

The surface preparation of the Si substrate prior to the layer deposition is a key step in a successful LTE process. This subject will be discussed in the next section. To date, most LTE studies use SiH_4 as the Si source gas; other gases, such as Si_2H_6 and $SiCl_2H_2$, have also been used, but to a lesser extent. The major advantage of the SiH_4 source is its low decomposition temperature, which facilitates the LTE process. In addition, H_2 plays an important role in the LTE process. It has been reported that the desorption rate of H_2 dominates the growth rate of the Si layer in the UHV/CVD LTE process.[30]

3.4.2 Surface Preparation

Surface preparation is a key step in the LTE process. The conditions of the initial growth interface determine the quality and the content of unintended impurities at addition to providing an ultraclean growth environment, all LTE growth techniques rely strongly on the preparation of an ultraclean substrate surface prior to growth. A wide variety of surface cleaning techniques have been proposed and implemented in the LTE process. Those techniques can be classified into two groups, namely ex-situ and in-situ. A combination of both ex-situ and in-situ methods is frequently used. The ex-situ techniques perform the surface cleaning process outside the growth reactor, whereas the in-situ techniques perform it inside the growth reactor.

The most widely used ex-situ cleaning technique involves the removal of organic impurities and particles using H_2SO_4: H_2O_2 and NH_4OH: H_2O_2 and the removal of metallic impurities by growth of a thick protective SiO_2 layer using a Si oxidizer, such as a HCl: H_2O_2 solution. This method is often called the *RCA clean* or *Shiraki clean* in the literature.[31] Following the wet chemical ex-situ cleaning process, two methods have been used to remove the oxide. One is to dip the substrate in the HF: H_2O (e.g., 1 : 10) solution ex-situ, and the other is to desorb the surface oxide in the growth reactor. In the case of the former, it has been observed that after a dip in HF solution, a hydrophobic (i.e., hydrogen-passivated) silicon surface is formed. This hydrogen-passivated silicon surface is air-stable and remains oxide-free for more than 10 minutes.[32] Using this method, device-quality Si and SiGe epi-layers were obtained at 550°C or lower without employing any in-situ surface-cleaning process. In the case of the latter, the thick oxide was removed by thermally desorbing it at a temperature above 950°C in ultrahigh vacuum or by etching it at a temperature above 800°C in H_2 inside the growth reactor.[33] The drawback to this approach is the use of a high-temperature process, which results in rapid impurity diffusion, extensive outgassing, and wafer warpage. In addition, the high-temperature process complicates integrated device fabrication; hence, it is not desirable from a manufacturing viewpoint.

Recently, a number of low-temperature Si-cleaning techniques have been developed. Among them *Si beam cleaning,* which employs a small flux of Si to desorb a thin Si oxide, has brought the temperature down to 700°C. Furthermore, with an in-situ exposure of a clean Si surface to H atoms, a successful Si epitaxy can be performed at 600°C. Etching in gaseous HF has also been reported to result in a H-passivated surface that allows the Si epitaxy at 550°C. However, a 200°C prebake is essential for low-temperature epitaxy on HF-dipped Si.[34] By incorporating this process, Si epitaxy has been performed at a temperature of 370°C.

3.4.3 Low-Temperature Si Epitaxial Growth Techniques

There are many ways that Si epi-layers can be grown at low temperatures. All of these techniques rely on both an ultraclean growth environment and a unique Si surface-cleaning process. CVD[29] and MBE[35] are the most popular methods used by researchers. However, solid-phase epitaxy (SPE) has also been reported.[36]

CVD has received the most research interest because of its simplicity and manufacturability. Many CVD approaches have been reported, such as ultrahigh vacuum/chemical vapor deposition (UHV/CVD),[29] ultraclean low-pressure CVD,[37] rapid-thermal-processing CVD (RTPCVD),[38] limited-reaction-processing CVD (LRP-CVD),[38] atmospheric-pressure CVD (APCVD),[39] plasma-enhanced CVD (PECVD),[40] photo-CVD,[41] and excimer laser–assisted CVD.[42] These techniques can be divided into three groups based on their characteristics, namely ultrahigh vacuum and/or ultraclean environment, rapid thermal processing, and enhanced-mode processing. In this section, we will only consider the two most promising techniques: UHV/CVD and MBE. For details of the other techniques, see the references herein.

A schematic drawing of a UHV/CVD system[29] is shown in Fig. 22. The wafers used in this technique are subjected to an RCA clean followed by a 10 : 1 H_2O/HF dip for 10 seconds. These precleaned wafers are placed coaxially into a quartz boat and loaded into the load-lock chamber. The load-lock chamber is kept at approximately 100°C and is pumped down to a pressure below 10^{-6} torr. Once this pressure is reached, the wafers are introduced into the growth chamber together with flowing hydrogen. The growth chamber is kept at a constant temperature of 550°C and is pumped down to a base pressure of below 10^{-9} torr. The source gases are subsequently fed into the growth chamber prior to the desorption of hydrogen atoms from the HF-passivated surface. Typical source gases used for this system are silane (SiH_4) and germane (GeH_4) for Si and SiGe epi-layer growths. The dopant gases are diborane (B_2H_6) for p-type and phosphine (PH_3) for n-type. Nominally, the operating growth pressure is maintained at 10^{-3} torr as the growth process proceeds. As a result of the low-pressure process, the reactant residence time within the growth zone is greatly reduced. This technique makes an abrupt control of the Ge composition and doping profile possible.

A typical stepwise boron profile in epi-Si achieved by a UHV/CVD technique is shown in Fig. 23. Abrupt transitions between each doping step are clearly evident. In addition, this growth technique yields a growth rate of about 3 Å/min for intrin-

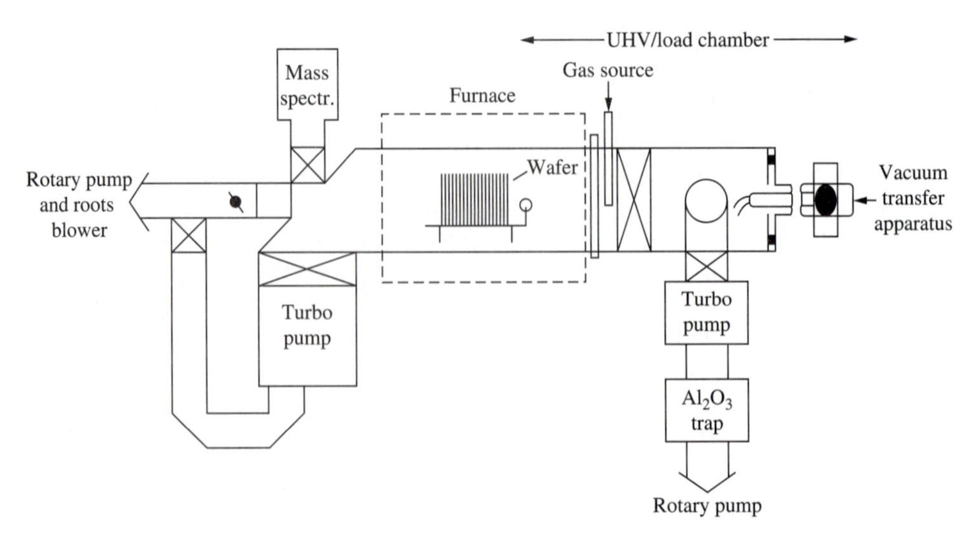

FIGURE 22
Schematic drawing of a UHV/CVD system. (*After Meyerson, Ref. 29.*)

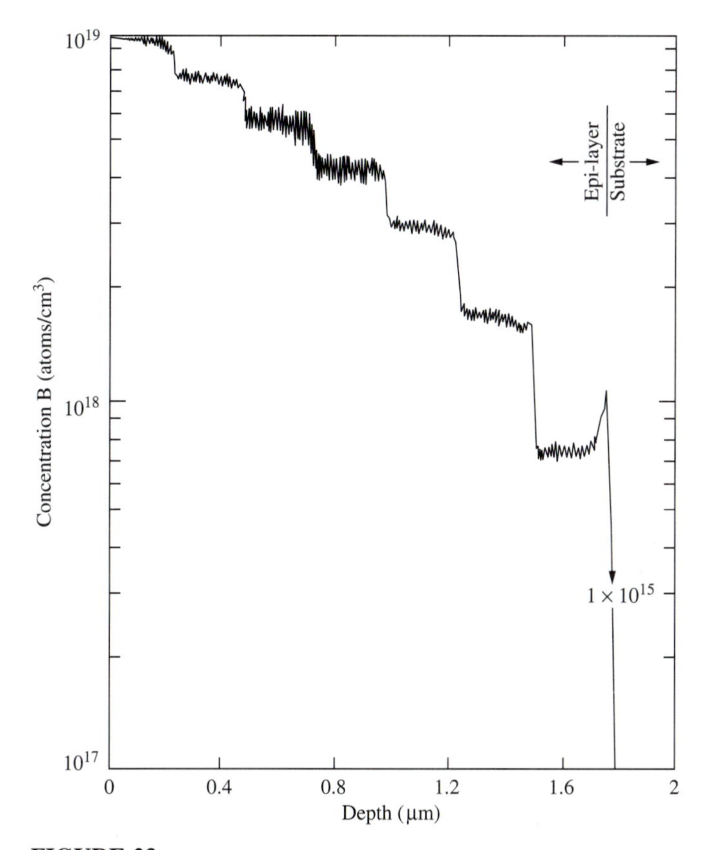

FIGURE 23
A typical stepwise boron profile in epi-Si achieved by a UHV/CVD technique. (*After Meyerson, Ref. 29.*)

sic or boron-doped silicon epi-layers. Therefore, it enables a precise control of layer thickness for advanced device applications. Since the growth of epi-layers of this technique is thermally activated, with an activation energy of about 1.6 eV, it can be increased rapidly at higher growth temperature. By using this growth technique, many novel heterostructures have been realized for Si-based materials.

MBE is a technique that was developed to fabricate devices that demand very tight controls over epi-layer thickness and doping uniformity. A conventional MBE technique, as shown in Fig. 24, is similar to that of an evaporation process.[43] For Si-MBE, the silicon and desired dopants are simply evaporated onto the Si substrate under ultrahigh-vacuum conditions. Typically, the base pressure of the MBE process is kept at 10^{-11} torr or higher. The evaporated atoms or molecules impinge on the heated substrate and then condense to form layers epitaxially. The Si-MBE process is usually performed in a temperature range of 500 to 900°C. Therefore, it has been, and remains, one of the most sophisticated low-temperature growth techniques for epi-Si. Because of its complexity and noncompatibility with device fabrication requirements, MBE has received little attention until very recently, when a low thermal budget has become more and more important for ULSI device fabrication.

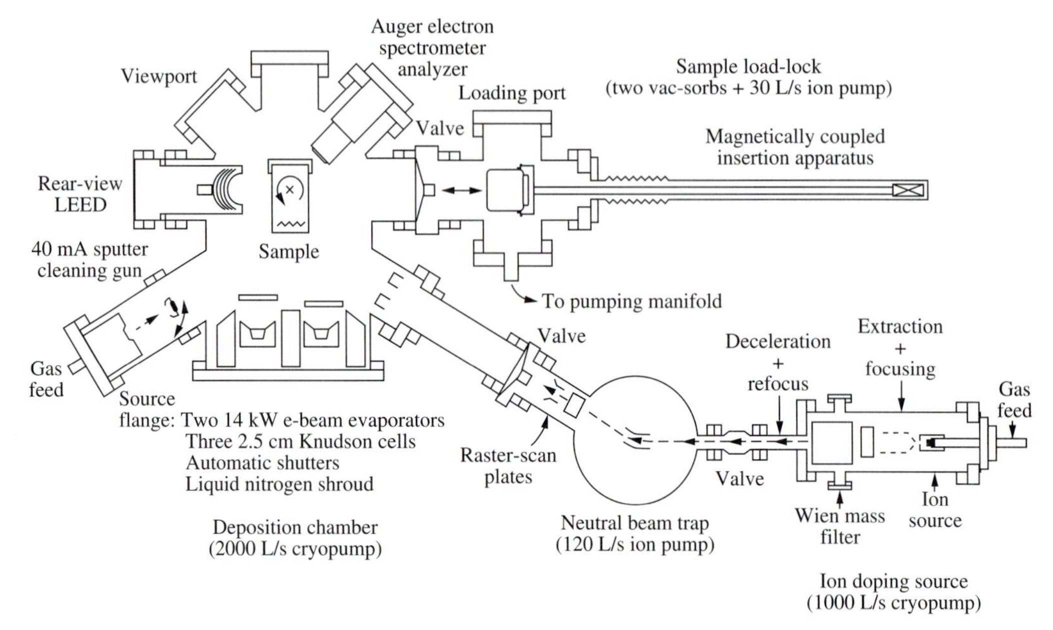

FIGURE 24
Schematic drawing of a typical MBE growth system. (*After Bean, Ref. 43.*)

Over the last two decades the MBE process has been improved significantly. However, its conventional electron-gun evaporator configuration still suffers from problems, such as clustering and spitting defects. To solve these problems, a modified MBE system, called gas-source MBE (GSMBE),[44] has been reported. This growth technique employs gaseous rather than solid sources to deposit epi-Si. As a result, spitting-type defects are eliminated. Conceptually, GSMBE is similar to UHV/CVD, except that the former is a cold-wall system whereas the latter is a hot-wall system. From the manufacturing point of view, MBE still suffers from low throughput. However, GSMBE opens a new approach toward the development of LTE for ULSI applications.

3.4.4 Device Applications

The LTE process has been employed to fabricate advanced bipolar transistors, MOSFETs, resonant tunneling diodes, infrared waveguide detectors, and other devices. Because the scope of this book is ULSI, only bipolar and MOSFET devices using an LTE process will be discussed. For readers who are interested in the other device applications of the LTE process, please refer to the references herein.

Both UHV/CVD and MBE techniques have been used to fabricate SiGe-based heterojunction bipolar transistors (HBTs). It is well known that the speed of a bipolar transistor can be improved by scaling the vertical profile and by reducing the parasitic resistance and capacitance of the extrinsic portion of the device. Theoretically, with a proper doping profile, a homojunction Si bipolar transistor can reach unity-current-gain cutoff frequencies (f_T) as high as 50 GHz. However, such a device would be practically of no use, because of the limitations imposed by base resistance, punch-through, and breakdown voltage. These limitations are mitigated with a heterojunction device. This concept has been implemented extensively for III–V compound semiconductor devices for more than a decade. However, only very recently, it was applied to the Si-based devices, because it is now possible to make device-quality Si epi-layers at low temperatures. In the case of the Si-based bipolar transistor, the heterojunction is formed by using SiGe as the base material.[45, 46]

Improved transport properties of electrons and holes have been demonstrated by Si/SiGe heterostructures grown with UHV/CVD and MBE techniques.[47, 48] N-channel and p-channel FETs of various types have been fabricated to exploit the advantages of Si/SiGe heterostructures for high-speed devices.[49, 50] Modulation-doped SiGe p-channel MOSFETs, utilizing the enhanced hole mobility characteristics of Si/SiGe heterostructures, have achieved a hole mobility of 220 cm^2/V-s at 300 K and 980 cm^2/V-s at 82 K in the SiGe channel, which is much higher than those of conventional p-channel MOSFETs. A similar study, which simply used high-quality UHV/CVD-grown SiGe layers to form p-channel MOSFETs, achieved a 70% higher transconductance at 300 K than conventional p-channel devices.[51] To avoid degradation of LTE-grown layers during device processing, a PECVD oxide was selected to form the gate at low temperatures. However, it was found that thermal and PECVD gate oxides gave comparable device characteristics. Therefore, the choice of a gate oxide formation process depends on the design of the MOSFET device. Even though enhanced electron mobility has been reported for Si/SiGe heterostructures grown by the UHV/CVD technique,[47] n-channel MOSFETs have been fabricated only by using MBE,[50] presumably because of the difficulties with n-type doping control using the UHV/CVD technique. Enhanced electron mobility has been reported by incorporating an n-type modulation-doped structure.

3.5 SELECTIVE EPITAXIAL GROWTH OF Si

Selective epitaxial growth (SEG) of Si is a process that allows the deposition of a Si epitaxial layer on a bare Si-substrate surface without the simultaneous growth of amorphous Si thin film on the silicon dioxide or silicon nitride surface. Therefore, the SEG process has attracted a great deal of interest for novel device applications.[52]

The SEG process was initially developed to achieve an advanced dielectric isolation structure. This approach is intended to replace the conventional local oxidation of Si (LOCOS) process, because of the scalability limits of LOCOS.[53] $SiCl_4$ has often been used for the SEG of Si. Both silicon dioxide and silicon nitride have been employed as the mask materials. The addition of HCl to the $SiCl_4$ is believed to prevent spurious nucleation and growth of silicon on the silicon dioxide mask surface; hence, it increases the selectivity of the growth process.[54] To a lesser extent, SiH_4 chemistry has also been used for the SEG process.[55]

SEG of Si was performed using a commerical cylindrical radiant-heating epitaxy reactor at a growth temperature below 1000°C using $SiCl_2H_2$ + HCl at a reduced pressure of 80 torr.[56] A lower growth temperature plus an increased HCl additive flow were found to lower the stacking fault density in the Si epi-layer. This work marked the major advance in SEG of Si. Since then, most of the SEG of Si has been performed at reduced temperatures and pressures. In 1985 a successful SEG-Si deposition down to 826°C, using SiH_2Cl_2 at 25 torr, was reported. Proper epitaxial precleaning and deposition techniques would prevent the oxide sidewall lifting (i.e., oxide undercutting) problem.[57] In 1989 the SEG-Si process was carried out with a growth temperature down to 650°C at an operating pressure less than 2 torr using SiH_2Cl_2. Over a temperature range of 650 to 1100°C no loading effect was observed.[58] This process was further performed at a growth temperature down to 600°C using the $SiH_4 + H_2$ chemistry at low deposition pressures. Even at a deposition pressure as low as 1 mtorr an ultraclean growth environment was required to achieve a successful SEG process.[37, 59] In 1991 a UHV/CVD system was employed to use the SEG process in novel device fabrication.[60]

A novel approach, called epitaxial lateral overgrowth (ELO), using the SEG-Si process, was invented to obtain silicon-on-insulator (SOI) substrates.[61] As IC technology advances, the quality of SEG-Si layers becomes crucial for device applications. The use of a sacrificial oxide was found effective for removing the contamination and damage on the Si substrate, greatly improving the SEG-Si surface quality. The defects at the interface between the silicon dioxide and the SEG-Si layer could be minimized when the mask edges were oriented in a [100] direction.[62]

3.5.1 Fundamental Aspects

Strictly speaking, SEG-Si symbolizes a process that deposits Si only on the bare Si substrate, with no Si in any form deposited on the silicon dioxide or silicon nitride mask. However, in practice, the SEG-Si process can be divided into two types. *Type 1* is the standard SEG-Si process, in which Si epitaxy occurs only on the bare Si substrate surface within the openings in the dielectric mask film, with no Si deposition whatsoever on the dielectric films. *Type 2* SEG-Si is a process that allows simultaneous deposition of a Si epitaxial layer on the single-crystal Si substrate within the openings in the dielectric film and deposition of polycrystalline Si (poly-Si) on the dielectric mask surface.[63]

Type 1 SEG-Si is achieved when Si atoms possessing high surface mobility are deposited from the silicon sources. Si atoms are able to migrate to favored sites on the single-crystal substrate, where nucleation occurs. It has been found that the surface mobility of Si atoms is affected by the presence of halides in the silicon sources. The degree of selectivity increases as the number of chlorine atoms in the silicon sources increases. Accordingly, $SiCl_4$ becomes the silicon source of choice for Type 1 selective deposition. However, as shown in Fig. 11, $SiCl_4$ requires the highest reduction temperature among all silicon sources; thus, the use of this source gas has been limited because of the degradation of the silicon dioxide mask at high temperatures. As an alternative, the silicon nitride mask, which can sustain higher temperature than the silicon dioxide mask, was employed for Type 1 SEG-Si process using $SiCl_4$ source gas. However, silicon nitride is a better nucleation source for silicon deposition than silicon dioxide is, and its use results in poor selectivity. As a result, $SiHCl_3$ becomes the optimum silicon source of choice to perform

Type 1 SEG-Si, simply because of its high chlorine content and lower reduction temperature compared to $SiCl_4$. SiH_2Cl_2 and SiH_4 have also been used frequently as the source gases for Type 1 SEG-Si process. In these cases, HCl or Cl_2 has been added to the growth system to increase selectivity. The presence of crystallographic defects, such as microtwins and stacking faults, is commonly observed at epi/dielectic mask interface. It has been reported that these defects are greatly reduced, because the wall of the silicon dioxide step is coated with poly-Si prior to the SEG deposition.

In contrast, the Type 2 SEG-Si process can be achieved more easily by using SiH_4, since selectivity is not strictly required and a low deposition temperature is needed. This process results in an interface between the epitaxial layer on the substrate surface and polysilicon on the dielectric mask film. The angle of this interface, relative to the film-growth direction, depends on the crystallographic orientation of the substrate. For example, this angle would be 90 degrees for (110), 72 degrees toward poly-Si for (100), and 70 degrees tapered toward the single-crystal silicon for (111)-oriented substrates.

Many factors affect the selective nature of Si deposition, including growth pressure, growth temperature, HCl concentration, silicon source, Si substrate surface condition, dielectric opening size, and the surface ratio of bare Si to oxide (dielectric film). Applications of the SEG-Si process are discussed in the next section.

3.5.2 Applications

One of the most important applications of the SEG process is for device isolation at submicron levels. The conventional LOCOS isolation process has a problem known as "bird's beak encroachment." Therefore, the scalability of the LOCOS is limited to about the μm range. To increase the device integration level, different isolation techniques are required. A typical SEG process for device isolation[57] is shown in Fig. 25. This approach offers many advantages, such as near-zero lateral encroachment, deep submicron-device separation, excellent planarity, and excellent latch-up immunity. In addition, it allows the formation of independent n and p wells, making retrograde wells possible by buried-layer epitaxy or graded epitaxy techniques.[57] For CMOS devices, the SEG process can be used to tailor the substrate and well resistance, increasing the latch-up immunity. Recently, low-temperature epitaxy was used in the SEG process, resulting in a sharp epi/substrate transition region due to minimal dopant outdiffusion and autodoping.

Another application of the SEG process is epitaxial lateral overgrowth (ELO), which is one way to form silicon-on-insulator (SOI) structures.[64] A schematic illustration of ELO process is shown in Fig. 26. In this process the lateral-to-vertical growth ratio depends on the step height and the substrate orientation.[65] It has been found that for the same substrate orientation, the lateral-to-vertical growth ratio decreases as the step height increases. For the same step height the lateral-to-vertical growth ratio decreases according to orientation, (100) > (110) > (111). To avoid facet formation, both lateral and vertical growth should be in the [100] directions.[52]

Device applications of the SEG process have been reported by many researchers. It has been used to develop an epitaxy-over-trench (EOT) process for dynamic

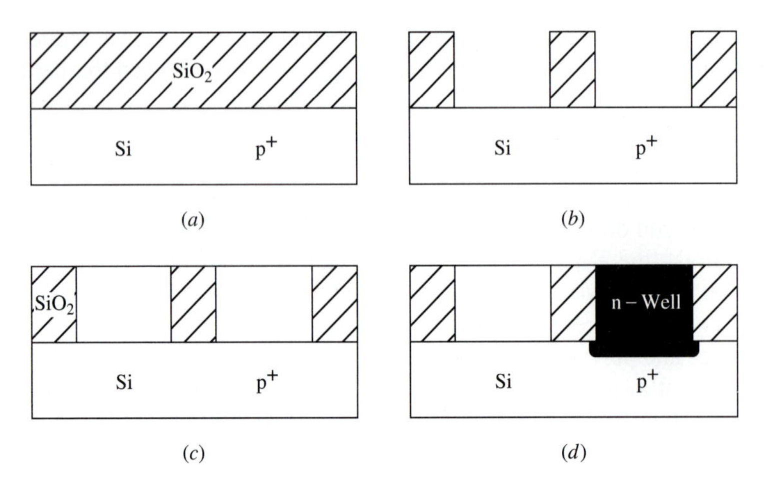

FIGURE 25
Schematic illustration of a typical SEG process for device isolation: (*a*) oxide deposition; (*b*) window formation; (*c*) epi growth; (*d*) n-well drive-in. (*After Borland and Drowley, Ref. 57.*)

random-access memory (DRAM) technology.[66] This approach permits the transfer transistor to be fabricated directly over the storage capacitor, resulting in a high-density DRAM. For bipolar technology a method using the SEG-Si process to form epitaxial bases has been proposed. Such a selective-epitaxy-base transistor (SEBT) allows the scaling of bipolar devices to less than that obtainable with the ion implantation technique to produce a narrow base width in the deep submicron regime.[67] In addition, a selective-epitaxy emitter-window (SEEW) process has been used. In this case, the SEG-Si is deposited with high boron concentration to define the geometry of the emitter window opening and to provide electrical contact to the extrinsic base of the transistor.[60]

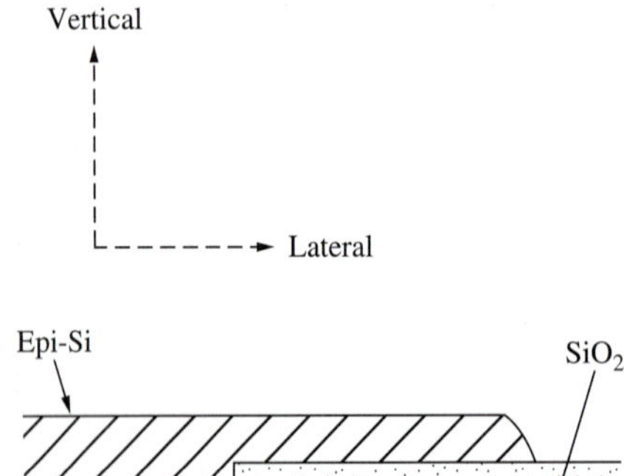

FIGURE 26
Schematic illustration of an ELO process.

3.6 CHARACTERIZATION OF EPITAXIAL FILMS

To develop and then implement an epitaxial process would not be successful without employing proper characterization techniques. In this section, we will review some of the measurements that are essential to evaluate epitaxial films. By nature, these techniques can be used to investigate the physical, chemical, and electrical properties of an epi-layer.

3.6.1 Physical Properties

The physical properties of an epitaxial layer include surface morphology, crystallographic defects, stress, and layer thickness.

Surface morphology is normally examined by eyes under high-intensity illumination or microscopically by conventional optical microscopy (in the brightfield or darkfield mode), Nomarski interference-contrast microscopy, and scanning electron microscopy (SEM). The selection of a proper technique depends solely on the need for each application. In an industrial environment, human eyes with high-intensity illumination are frequently used, because the process is simple and has a high screening speed. When a detailed surface morphology is required, microscopic methods of various modes are employed. In the brightfield mode of optical microscopy the image is formed by the reflection of the light received by the sample; in the darkfield mode only light that is reflected by the features on the wafer surface is collected. The former gives the best overall image and information about the surface and is, therefore, commonly used. In contrast, the latter is very powerful in resolving small structural features on the surface and is often used to scan a wide range of defects (including crystallographic defects after etching) on the surface. Using a splitting polarized light source, Nomarski interference-contrast microscopy is capable of resolving surface features of 3 to 5 nm in depth, far beyond what can be achieved by ordinary optical microscopy. SEM operates by scanning a focused electron beam over a surface and sensing the secondary electrons emitted from the surface. This technique has a large depth of field and high resolution and is capable of resolving submicron surface features at a magnification much higher than that obtainable with any optical microscopy.[68]

An epitaxial layer may contain a variety of crystallographic defects. In an as-deposited epi-layer, these defects are not always readily visible. Etchants are used to delineate defects preferentially prior to microscopic examination. For Si epitaxy, Wright etch, a mixture of HF, HNO_3, CrO_3, $Cu(NO_3)_2$, acetic acid, and deionized water, is the most frequently used etchant. Stacking faults in an epi-layer can be revealed with a 5-min Wright etch. Other solutions, such as the Secco or Sirtl etches, are also often used to delineate dislocations and pits.[69]

When a thin epi-layer is deposited on a substrate of another material, it is under either tensile or compressive stress, causing the substrate to deflect downward or upward, respectively. A variety of techniques may be used to measure the wafer center deflection and, hence, the stress in the epi-layer. The simplest method is to place the wafer on the stage of an optical microscope and calibrate the vertical adjustment for the in-focus vertical positions at the wafer edge and the wafer center.[70] A commercial wafer deflection gauge, using the degree of light reflection, is also used to evaluate the amount of wafer deflection.

Depending on the nature of a film, a measurement of the film's thickness can be fairly simple or quite complex. In general, a thickness measurement technique can be either destructive or nondestructive. Although there are many thickness measurement techniques, only a few, especially nondestructive ones, are preferable for Si epitaxial layers. Destructive methods include *angle-lap and stain* and *groove and stain.*[71] These techniques can be used only for an epi-layer that has a different doping type or doping concentration than the substrate. For example, the cross section of the p-n or n-n^+ junction is differentiated by chemical etching (staining) and then examined with optical microscopy or SEM to measure the epi-layer thickness.

The most widely used nondestructive epi-layer thickness measurement techniques are based on optical properties of the Si epi-layer. In the near infrared region (2.5 to 50 μm), lightly doped Si is relatively transparent, whereas heavily doped Si behaves as a reflective surface. Therefore, when an electromagnetic radiation with various wavelengths is applied to an epi-layer on a Si substrate, interference and optical patterns are formed because of the interaction between the reflections from the air/epi-layer surface and epi-layer/substrate interface. Based on these optical interference patterns, the epi-layer thickness can be obtained. The infrared (IR) interference spectrophotometer is, therefore, often used as a quick and nondestructive thickness measurement technique for Si epi-layers.[72] A Fourier-transform infrared (FTIR) spectroscopy procedure using a Michelson interferometer has been developed to measure Si epi-layer thickness. This technique is capable of a rapid ($\approx$30 sec), accurate ($\pm$0.05 μm), and precise ($\pm$0.005 μm) measurement and, therefore, is widely accepted.[73] More recently, a high-resolution double-crystal x-ray diffraction (HRXRD) method[74] has been employed to measure the layer thickness of Si/SiGe multilayer structures with an accuracy within $\pm 2\%$.

3.6.2 Chemical and Electrical Properties

The chemical and electrical properties of an epi-layer are usually closely related. The resistivity of an epi-layer is proportional to the activated dopant concentration. When a dopant is incorporated into an epi-layer, it can be either electrically active or nonactive, depending on the lattice site. Electrically active dopant concentration can be determined only by electrical measurement techniques, such as the four-point probe, capacitance-voltage (C-V), and spreading resistance. On the other hand, total dopant concentrations, including electrically active and nonactive species, are usually measured by surface spectroscopic techniques, such as secondary-ion mass spectroscopy (SIMS), Auger electron spectroscopy (AES), electron spectroscopy for chemical analysis (ESCA), and Rutherford backscattering spectroscopy (RBS).

The four-point probe is one of the fastest techniques to measure the dopant concentration.[75] However, it is applicable only to an epi-layer that is grown on a lightly doped substrate of an opposite conductivity type to the layer. In this technique, the resistivity ρ of the epi-layer is expressed by

$$\rho = R_s T F \tag{3.27}$$

where R_s is the sheet resistivity, T is the epi-layer thickness, and F is the geometric correction factor. Low current is preferred, to avoid ohmic heating or reaching punch-through voltage. When a thin layer is measured, care must be taken to avoid the tips penetrating the layer. Overall, this technique can achieve an accuracy of only $\pm 10\%$ and is not capable of measuring doping profiles.

The C-V technique is performed by using a mercury contact to form a Schottky barrier diode with the epi-layer.[76] This is a rapid, nondestructive technique to measure the doping profile in the layer. By applying a reverse bias, the capacitance, as a function of voltage, is recorded, and the doping profile is obtained according to the following equation:

$$N(x) = -C^3[q\epsilon A^2(dC/dV)]$$

and

$$x = \epsilon A/C(V) \tag{3.28}$$

where C is the capacitance, q is the electric charge, ϵ is the permittivity of silicon, A is the junction area, x is the distance from the junction, and V is the bias voltage. This technique works well with doping in the range of 10^{14} to $10^{17}/\text{cm}^3$.

For higher doping concentrations the spreading resistance technique is used. This technique employs a cleaved and beveled sample. A two-point probe is used to measure the resistivities sequentially at a number of points, starting from the epi-layer surface through the layer/substrate interface.[77] By correlating the change of resistivity between each point with carrier concentration, the doping concentration and its profile are deduced. This technique is capable of measuring a wide range of doping concentrations (10^{14} to $10^{20}/\text{cm}^3$). However, it suffers from poor accuracy and complex data reduction. Besides, special care must be taken to compensate for the effect of the wear of the probe tip, which significantly affects the accuracy of this measurement technique.

In contrast to these electrical measurements, surface spectroscopic techniques are used when a total doping profile is needed. All surface spectroscopic techniques are carried out under high-vacuum conditions. Therefore, they are not useful for routine process control. However, as Si technology becomes more and more miniaturized, these techniques become more and more important because of their excellent depth-profiling capability.

Secondary-ion mass spectroscopy (SIMS) uses ions, that are accelerated by high voltage (500 to 5000 eV), as the probe beam to sputter ions, called secondary ions, from the surface of interest.[78] These secondary ions are then analyzed with a mass spectrometer to characterize the elemental constitutes of the surface. Both oxygen and cesium ion beams are used for SIMS measurements. The former is more effective for electropositive elements such as B, Al, and Cr, and the latter for electronegative elements such as C, O, and As. Since this technique involves sputtering of the surface, depth profiling can be obtained readily. Depth resolution in the 2 to 5 nm range can be achieved routinely. Layers of ≈1 μm thickness can be depth-profiled with this technique. For Si with doping levels down to 10^{15} cm^{-3}, SIMS is one of the most powerful techniques for measuring the total doping profile.

Auger electron spectroscopy (AES) employs an energetic electron beam (up to 10 keV) to probe the surface.[79] The energetic electron can ionize an atom by dislodging an inner-core electron. When an outer-core electron falls back to the inner core to replace the ejected electron, the atom gives up its excess energy by emitting an x-ray. Alternatively, it can eject a second, "Auger" electron. The energy of this Auger electron is characterized by the atom from which it comes, allowing the elemental composition to be measured. Although the electron beam can penetrate many hundreds of angstroms into the surface, only Auger electrons that escape from the top 1 to 3 nm of the surface can be collected. When combined with a sputtering process, AES is also capable of depth profiling.

Electron spectroscopy for chemical analysis (ESCA) is also known as x-ray photoelectron spectroscopy (XPS). This technique uses a beam of low-energy x-rays to bombard a sample.[80] The x-ray photon energy, when it is absorbed, removes an outer-shell electron from an atom. This electron is then emitted with a kinetic energy of a magnitude equal to the difference between the x-ray and the bonding energy of the electron. The energy of the emitted electron is analyzed to determine the type of atom, and the number of electrons at this energy is used to obtain the density of such atoms in the layer. Like AES, this technique is truly a surface analysis technique (1 to 3 nm from the surface). Since an x-ray beam is used, ESCA allows rapid depth profiling. In addition, it is often used to obtain information about chemical bonding around an atom, which is not achievable with other surface analysis techniques.

Rutherford backscattering spectroscopy (RBS) employs high-energy He^{2+} ions (≥1 MeV) to penetrate the sample surface to obtain the depth profile in the layer.[81] This technique is nondestructive and is based on the elastic collision theory to obtain the chemical information in a layer. When He^{2+} ions impinge on a surface, penetrate to the subsurface, and collide with atoms, some ions are backscattered with an energy loss that is characteristic of the atom struck and the distance from the sample surface where the collision occurs. Therefore, both the elemental composition and its depth profile can be obtained simultaneously. RBS is a surface-analysis technique that gives quantitative chemical information without using standards. However, it has relatively poor depth resolution (≈2 nm) compared to other techniques.

3.7 SUMMARY AND FUTURE TRENDS

Epitaxy will remain as a unique process that provides doping capability and flexibility for IC technology. As the industry moves into the ULSI era, LTE will play an important role in preparing the active parts of IC devices, such as the intrinsic base for a bipolar transistor. In addition, low temperature is required to minimize autodoping effects, which become more and more critical for ULSI circuits. In the near future, LTE will undoubtedly receive most attention in the field of epitaxy. Among all LTE techniques, UHV/CVD seems to be most promising for being incorporated into ULSI manufacturing. However, APCVD also shows a high potential for this application. In any case, an ultraclean environment is inevitably required for any epitaxial processes in the future. Because of the limitation imposed by the Si material and Si homojunction, the development of Si/SiGe heteroepitaxy will receive a great deal of attention in the foreseeable future. Defect minimization remains as one of the major topics in the pursuit of heteroepitaxy. In terms of growth chemistry, SiH_4 will be the preferred chemical in the future, because LTE will be a dominant technology. Finally, more sophisticated instrumentation, such as the FTIR and SIMS apparatus, will be widely used for characterizing physical, chemical, and electrical properties of an epi-layer to meet the stringent requirements of ULSI circuits.

REFERENCES

1. J. W. Matthews, "Defects Associated with the Accommodation of Misfit Between Crystals," *J. Vac. Sci. Technol.,* **12,** 126 (1975).
2. R. People and J. C. Bean, "Calculation of Critical Layer Thickness versus Lattice Mismatch for GeSi/Si Strained-Layer Heterostructures," *Appl. Phys. Lett.,* **47,** 322 (1985).
3. S. K. Ghandhi, *VLSI Fabrication Principles: Silicon and Gallium Arsenide,* Wiley, New York, 1983.
4. S. M. Sze, *Physics of Semiconductor Devices,* 2nd ed., Wiley, New York, 1981; C. J. Hu et al., "A Self-Aligned 1 μm CMOS Technology for VLSI," *Tech. Dig. IEDM,* 731 (1983).
5. D. W. Shaw, "Fundamental Aspects of Epitaxy," in *Crystal Growth, Vol.1,* C. H. L. Goodman, Ed., Plenum, New York (1974).
6. R. E. Reed-Hill, *Physical Metallurgy Principles,* Van Nostrand, New York, 1973.
7. A. S. Grove, "Mass Transfer in Semiconductor Technology," *Ind. Eng. Chem.,* **58,** 48 (1966).
8. V. S. Ban, "Mass Spectrometric Studies of Chemical Reactions and Transport Phenomena in Silicon Epitaxy," *Proceedings of the 6th International Conference on Chemical Vapor Deposition,* The Electrochemical Society 1977, p. 66.
9. J. Bloem and L. J. Gilling, "Epitaxial Growth by Chemical Vapor Deposition," in *VLSI Electronics,* N. G. Einspruch and H. Huff, Eds., *Vol. 12,* 89, Academic Press, Orlando, Florida, 1985.
10. J. Bloem, "Nucleation and Growth of Silicon by CVD," *J. Cryst. Growth,* **50,** 581 (1980).
11. F. C. Eversteyn, "Chemical-Reaction Engineering in the Semiconductor Industry," *Philips Res. Rept.,* **29,** 45 (1974).
12. J. Nishizawa and M. Saito, "Growth Mechanism of Chemical Vapor Deposition of Silicon," *Proceedings of the 8th International Conference on Chemical Vapor Deposition,* The Electrochemical Society 1981, p. 317.
13. V. S. Ban and S. L. Gilbert, "Chemical Processes in Vapor Deposition of Silicon," *J. Electrochem. Soc.,* **122,** 1382 (1975).
14. E. Sirtl, L. P. Hunt, and D. H. Sawyer, "High Temperature Reactions in Silicon-Hydrogen-Chlorine System," *J. Electrochem. Soc.,* **121,** 919 (1974).
15. M. L. Yu, D. J. Vitkavage, and B. S. Meyerson, "Doping Reaction of PH_3 and B_2H_6 with Si(100)," *J. Appl. Phys.,* **59,** 4032 (1986).
16. R. Reif, T. I. Kamins, and K. C. Saraswat, "A Model for Dopant Incorporation into Growing Silicon Epitaxial Films," *J. Electrochem. Soc.,* **126,** 644 (1979).
17. G. R. Srinivasan, "Autodoping Effects in Silicon Epitaxy," J. Electrochem. Soc., **127,** 1334 (1980).

18. G. R. Srinivasan, "Kinetics of Lateral Autodoping in Silicon Epitaxy," *J. Electrochem. Soc.,* **125,** 146 (1978).
19. K. V. Ravi, *Imperfection and Impurities in Semiconductor Silicon,* Wiley, New York, 1981.
20. S. Wolf and R. N. Tauber, *Silicon Processing for the VLSI Era, Vol.1,* Lattice Press, California, 1986, p. 140.
21. B. J. Baliga, "Defect Control During Silicon Epitaxial Growth Using Dichlorosilane," *J. Electrochem. Soc.,* **129,** 1078 (1982).
22. W. Dyson, L. G. Hellwig, J. M. Moody, and J. A. Rossi, "n^+ and p^+ Substrate Effects on Epitaxial Silicon Properties," in *VLSI Science and Technology,* K. E. Bean and G. Rozgonyi, Eds., The Electrochemical Society, 1984, p. 107.
23. Nalih, A. S., Rozgonyi, R. F. Davis, and H. J. Kim, "Extrinsic Gettering with Epitaxial Misfit Dislocations," *Semiconductor Processing,* ASTM, STP 850, Gupta (1984).
24. S. Wolf and R. N. Tauber, *Silicon Processing for the VLSI Era, Vol.1,* Lattice Press, California, 1986, p. 144.
25. P. H. Lee, M. T. Wauk, R. S. Rosler, and W. C. Benzing, "Epitaxial Pattern Shift Comparisons in Vertical, Horizontal, and Cylindrical Reactor Geometries," *J. Electrochem. Soc.,* **124,** 1824 (1977).
26. S. P. Weeks, "Pattern Shift and Pattern Distortion During CVD Epitaxy on <111> and <100> Silicon," *Sol. State Technol.,* **24,** 111 (1981).
27. C. W. Pearce, "Epitaxy," *VLSI Technology,* 2nd edn., S. M. Sze, Ed., McGraw-Hill, 1988, p. 51.
28. M. L. Hammond, "Silicon Epitaxy," *Sol. State Technol.* **21,** 68 (1978).
29. B. S. Meyerson, "Low Temperature Si and Si:Ge Epitaxy by Ultrahigh Vacuum/Chemical Vapor Deposition: Process Fundamentals," *IBM J. Res. Devel.,* **34,** 806 (1990); T. G. Jung, C. Y. Chang, T. C. Chang, H. C. Lin, T. Wang, W. C. Tsai, G. W. Huang, and P. J. Wang, "Low-temperature Epitaxial Growth of Silicon and Silicon-Germanium Alloy by Ultrahigh-Vacuum Chemical Vapor Deposition," *Jpn. J. Appl. Phys.,* **33,** 224 (1994).
30. B. S. Meyerson, K. J. Uram, and F. K. LeGoues, "Cooperative Growth Phenomena in Silicon/Germanium Low Temperature Epitaxy," *Appl. Phys. Lett.,* **53,** 2555 (1988).
31. W. Kern and D. A. Puotinen, "Cleaning Solutions Based on Hydrogen Peroxide for Use in Silicon Semiconductor Technology," *RCA Rev.,* **31,** 187 (1970).
32. B. S. Meyerson, F. J. Himpsel, and K. J. Uram, "Bistable Conditions for Low-temperature Silicon Epitaxy," *Appl. Phys. Lett.,* **57,** 1034 (1990).
33. T. O. Sedgwick and P. D. Agnello, "Atmospheric Pressure Chemical Vapor Deposition of Si and SiGe at Low Temperatures," *J. Vac. Sci. Technol.,* **A10,** 1913 (1992).
34. D. J. Eaglesham, G. S. Higashi, and M. Cerullo, "370°C Clean for Si Molecular Beam Epitaxy Using a HF Dip," *Appl. Phys. Lett.,* **59,** 685 (1991).
35. J. C. Bean, "Silicon Molecular Beam Epitaxy: 1984–1986," *J. Cryst. Growth,* **81,** 411 (1987).
36. E. D. Ahlers, R. M. Ostrom, P. Datta, and F. G. Allen, "Crystal Quality of Solid Phase Epitaxially Grown Silicon," *J. Vac. Sci. Technol. B,* **6,** 723 (1988).
37. J. Murota, N. Nakamura, M. Kato, N. Mikoshiba, and T. Ohmi, "Low-Temperature Silicon Selective Deposition and Epitaxy on Silicon Using the Thermal Decomposition of Silane Under Ultraclean Environment," *Appl. Phys. Lett.,* **54,** 1007 (1989).
38. T. Y. Hsieh, K. H. Jung, D. L. Kwong, and S. K. Lee, "Silicon Homoepitaxy by Rapid Thermal Processing Chemical Vapor Deposition (RPTCVD)—A Review," *J. Electrochem. Soc.,* **138,** 1188 (1991).
39. C. M. Gronet, et al., "Thin, Highly Doped Layers of Epitaxial Silicon Deposited by Limited Reaction Processing," *Appl. Phys. Lett.,* **48,** 1012 (1986).

40. K. Baert, P. Deschepper, J. Poortmans, J. Nijs, and R. Mertens, "Selective Si Epitaxial Growth by Plasma-Enhanced Chemical Vapor Deposition at Very Low Temperature," *Appl. Phys. Lett.,* **60,** 442 (1992).
41. A. Yamada, Y. Jia, M. Kohagai, and K. Takahashi, "Heavily P-doped Si and SiGe Films Grown by Photo-CVD at 250°C," *J. Electron. Matls.,* **19,** 1083 (1990).
42. A. Yamada, A. Satoh, M. Konagai, and K. Takahashi, "Low Temperature (600–650°C) Silicon Epitaxy by Excimer Laser-Assisted Chemical Vapor Deposition," *J. Appl. Phys.,* **65,** 4268 (1989).
43. J. C. Bean, "Silicon Molecular Beam Epitaxy as a VLSI Processing Technique," *IEEE Proc. IEDM,* 1981, p. 6.
44. H. Hirayama, T. Tatsumi, and N. Aizaki, "Gas Source Molecular Beam Epitaxy Using Disilane," *Appl. Phys. Lett.,* **52,** 1484 (1988).
45. G. L. Patton, J. H. Comfort, B. S. Meyerson, E. F. Crabbé, G. J. Scilla, Z. De Frésart, J. M. C. Stork, J. Y. C. Sun, D. L. Harame, and J. N. Burghartz, "75-GHz f_T SiGe-Base Heterojunction Bipolar Transistors," *IEEE Electron Dev. Lett.,* **11,** 171 (1990); G. L. Patton, S. S. Iyer, S. L. Delage, S. Tiwari, and J. M. C. Stork, "Silicon-Germanium-Base Heterojunction Bipolar Transistors by Molecular Beam Epitaxy," *IEEE Electron Dev. Lett.,* **9,** 165 (1988).
46. E. F. Crabbé, B. S. Meyerson, J. M. C. Stork, and D. L. Harame, "113-GHz f_T Graded Base SiGe HBT's" 1993 Device Research Conference, Paper #IIA-3, June, 1993.
47. K. Ismail, B. S. Meyerson, and P. J. Wang, "High Electron Mobility in Modulation-doped Si/SiGe," *Appl. Phys. Lett.,* **58,** 2117 (1991).
48. P. J. Wang, B. S. Meyerson, F. F. Fang, J. Nocera, and B. Parker, "High Hole Mobility in Si/SiGe/Si P-type Modulation-Doped Double Heterostructures," *Appl. Phys. Lett.,* **55,** 2333 (1989).
49. S. Verdonckt-Vandebroek, E. F. Crabbé, B. S. Meyerson, D. L. Harame, P. J. Restle, J. M. Stork, A. C. Megdanis, C. L. Stanis, A. A. Bright, G. M. Kroesen, and A. C. Warren, "High-Mobility Modulation-Doped Graded SiGe-Channel P-MOSFET's," *IEEE Electron Dev. Lett.,* **12,** 447 (1991).
50. H. Daembkes, H. J. Herzog, H. Jorke, H. Kibbel, and E. Kaspar, "The N-Channel SiGe/Si Modulation-Doped Field-Effect Transistor," *IEEE Trans. Electron. Dev.,* **ED-33,** 633 (1986).
51. S. Subbanna, V. P. Kesan, M. J. Tejwani, P. J. Restle, D. J. Mis, and S. S. Iyer, "Si/SiGe P-channel MOSFET's," *Symposium on VLSI Technology, Japan,* 1991, p. 103.
52. B. J. Ginsberg, J. Burghartz, G. B. Bronner, and S. R. Moder, "Selective Epitaxial Growth of Silicon and Some Potential Applications," *IBM J. Res. Devel.,* **34,** 816 (1990).
53. A. Ishitani, H. Kitajima, K. Tanno, and H. Tsuya, "Selective Silicon Epitaxial Growth for Device-Isolation Technology," *Microelectron. Eng.,* **4,** 3 (1986).
54. D. M. Jackson, "Advanced Epitaxial Process for Monolithic Integrated-Circuit Application," *Trans. Met. Soc. AIME,* **233,** 596 (1965).
55. P. Rai-Choudhury and D. K. Schroder, "Selective Silicon Epitaxy and Orientation Dependence of Growth," *J. Electrochem. Soc.,* **120,** 664 (1973).
56. K. Tanno, N. Endo, H. Kitajima, Y. Kurogi, and H. Tsuya, "Selective Silicon Epitaxy Using Reduced Pressure Technique," *Jpn. J. Appl. Phys.,* **21,** L564 (1982).
57. J. O. Borland and C. I. Drowley, "Advanced Dielectric Isolation through Selective Epitaxial Growth Techniques," *Sol. State Technol.,* **28,** 141 (1985).
58. J. L. Regolini, D. Bensahd, E. Scheid, and J. Mercier, "Selective Epitaxial Silicon Growth in the 650–1100°C Range in a Reduced Pressure Chemical Vapor Deposition Reactor Using Dichlorosilane," *Appl. Phys. Lett.,* **54,** 658 (1989).

59. T. R. Yew and R. Reif, "Selective Silicon Epitaxial Growth at 800°C by Ultralow-Pressure Chemical Vapor Deposition Using SiH_4 and SiH_4/H_2," *J. Appl. Phys.*, **15,** 2509 (1989).
60. J. N. Burghartz, S. R. Mader, B. S. Meyerson, B. J. Ginsberg, J. M. Stork, C. Stanis, and Y. C. Sun, "Self-Aligned Bipolar NPN Transistor with 60nm Epitaxial Base," *Technical Digest, IEEE International Electron Device Meeting,* 229 (1989).
61. L. Jastrzebski, J. F. Corboy, and R. Pagliaro, Jr., "Growth of Electronic Quality Silicon Over SiO_2 by Epitaxial Lateral Overgrowth Technique," *J. Electrochem. Soc.,* **129,** 2645 (1982).
62. J. T. McGinn, L. Jastrzebski, and F. Corboy, "Defect Characterization in Monocrystalline Silicon Grown over SiO_2," *J. Electrochem. Soc.,* **131,** 398 (1984).
63. S. Wolf and R. N. Tauber, *Silicon Processing for the VLSI Era, Vol.1,* 155, Lattice Press, California (1986).
64. J. Borland, M. Gangani, R. Wise, S. Fong, Y. Oka, and Y. Matsumoto, "Silicon Epitaxial Growth for Advanced Device Structures," *Sol. State Technol.,* **31,** 111 (1988).
65. D. D. Rathman, D. J. Silversmith, and J. A. Burns, "Lateral Epitaxial Overgrowth of Silicon on SiO_2," *J. Electrochem. Soc.,* **129,** 2303 (1982).
66. G. B. Bronner, N. C. C. Lu, T. V. Rajeevakumar, B. Ginsberg, and B. Machesney, "Epitaxy over Trench for ULSI DRAMs," *Technical Digest, IEEE Symposium on VLSI Technology,* 1988, p. 21.
67. J. N. Burghartz, B. J. Ginsberg, S. R. Mader, T. C. Chen, and D. L. Harame, "Selective Epitaxy Base Transistor (SEBT)," *IEEE Electron Dev. Lett.,* **9,** 259 (1988).
68. P. R. Thronton, *Scanning Electron Microscopy,* Chapman and Hall, London, 1968.
69. D. G. Schimmel, "A Comparison of Chemical Etches for Revealing ⟨100⟩ Silicon Crystal Defects," *J. Electrochem. Soc.,* **123,** 734 (1976).
70. R. Glang, R. Holmwood, and R. Rosenfeld, "Determination of Stress in Films on Single Crystalline Silicon Substrates," *Rev. Sci. Inst.,* **36,** 7 (1965).
71. "Standard Test Method for Thickness of Epitaxial and Diffused Layers in Silicon by the Angle Lapping and Staining Technique," *1984 Annual Book of ASTM Standards,* F110-84, Vol.10.05, 1984, p. 230.
72. "Thickness of Epitaxial Layers of Silicon on Substrates of Same Type by Infrared Reflectance," *1984 Annual Book of ASTM Standards,* F95, Vol.10.05, 1984, p. 213.
73. A. L. Smith, *Applied Infrared Spectroscopy,* Wiley, New York, 1979.
74. P. J. Wang, M. S. Goorsky, B. S. Meyerson, F. K. Le Goues, and M. J. Tejwani, "Characterization of Si/SiGe Strained-Layer Superlattices Grown by Ultrahigh Vacuum/Chemical Vapor Deposition Technique," *Appl. Phys. Lett.,* **59,** 814 (1991); T. C. Chang, C. Y. Chang, T. G. Jung, W. C. Tsai, P. J. Wang, T. L. Lee, and L. J. Chen, "Nanometer Thick Si/SiGe Strained-Layer Superlattices Grown by an Ultrahigh Vacuum/Chemical Vapor Deposition Technique," *J. Appl. Phys.,* **75,** 1 (1994).
75. Am. Soc. Test. Mater., ASTM Standard, F374, Part 43.
76. D. L. Rehrig and C. W. Pearce, "Production Mercury Probe Capacitance-Voltage Testing," *Semic. Intl.,* **3,** 151 (1972).
77. Y. Isda, H. Abe, and M. Knodo, "Impurity Profile Measurement of Thin Epitaxial Wafers by Multilayer Spreading Resistance Analysis," *J. Electrochem. Soc.,* **124,** 1118 (1977).
78. J. A. Mchugh, "Secondary Ion Mass Spectroscopy," in *Methods of Surface Analysis,* A. W. Czanderna, Ed., Elsevier, New York, 1975, p. 223.
79. A. Joshi, I. E. Davis, and P. W. Palmberg, "Auger Electron Spectroscopy," in *Methods of Surface Analysis,* A. W. Czanderna, Ed., Elsevier, New York, 1975, p. 164.
80. K. F. J. Heinrich, *Electron-Beam X-Ray Microanalysis,* Van Nostrand Reinhold Co., New York, 1981.

81. W. K. Chu, J. Mayer, and M. Nicolet, *Backscattering Spectroscopy,* Academic, New York, 1978.

PROBLEMS

1. A heteroepitaxial layer with a lattice constant of a_1 is deposited on a (001) substrate with a lattice constant of a_s. Assuming that the layer is in a perfectly strained condition, what are its lattice constants along the directions normal and parallel to the substrate? Is the layer in tensile or compressive stress along these two directions if $a_1 > a_s$?

2. Following Problem 1, assume that the lattice constants of the layer normal and parallel to the substrate are c and a, respectively. How are a, c, and a_1 correlated?

3. Derive mathematically the relationship between the critical layer thickness h_c and the lattice-misfit strain ϵ under equilibrium conditions. Explain why a larger h_c value is obtained at lower growth temperatures.

4. Derive Eq. (3.6) in the text. Explain mathematically why reduced pressure and high temperature are preferable for a VPE process from a thermodynamic viewpoint.

5. Intrinsic diffusivity of B in Si can be expressed by $D = 0.76e^{-3.46(\mathrm{eV})/kT}$. Calculate the minimum growth rate that is required when a Si-epi is grown on a heavily B-doped Si substrate for 20 min at 1200°C. Discuss why a low growth temperature is essential to achieve a submicron Si-epi layer.

6. The UHV/CVD technique is capable of achieving a boron doping concentration two orders higher than the solubility of boron in Si at a growth temperature of 550°C. Why? Discuss why a conventional MBE technique is not able to achieve such a high boron doping concentration.

7. Use a schematic drawing to illustrate how facets can be avoided when a Si-epi is grown with the SEG process on a (100)-oriented Si substrate.

8. Develop a method to determine a dislocation density of $10^2/\mathrm{cm}^2$ in a Si-epi routinely using an optical microscopy.

CHAPTER 4

Conventional and Rapid Thermal Processes

Richard B. Fair

4.1 INTRODUCTION

The chemical and physical processes applied to silicon (Si) wafers are, in general, thermally activated; therefore, the wafers require heating at elevated temperatures so that these processes proceed at reasonable rates. Typical silicon-based semiconductor factories use *batch furnaces* for thermal fabrication steps, where a batch consists of 20 to 100 wafers that are simultaneously processed together in a single system. Batch-furnace thermal processes include thermal oxidation, dopant diffusion, annealing, glass reflow, and various chemical vapor deposition (CVD) processes. Such conventional thermal processing involves heating by convection and conduction where the wafers are in thermal equilibrium with the furnace surroundings. For example, batch wafer processing may typically occur in a clean, high-purity fused-quartz tube that is inserted into the core of a furnace that is heated resistively. Wafers are placed inside the open tube and pushed into the hot zone of the furnace where a uniform temperature is maintained. Thus, the heating elements surrounding the quartz tube heat the tube, the ambient inside the tube, and the wafers with long-wavelength thermal radiation (spectral distribution peak at 2 μm). In this approach, many wafers can be processed simultaneously, and the cost of the operation is divided among the number of wafers in each batch. However, in this "hot wall" process, the wafers are susceptible to contamination from the hot processing chamber on which deposited films may accumulate and flake off or through which impurities may diffuse and contaminate the wafers inside. For additional details on conventional, batch-furnace thermal processes, the reader is referred to the appropriate chapters on oxidation, diffusion, epitaxy, ion implantation, and dielectric and polysilicon film deposition in the previous edition of this book, *VLSI Technology*, (2nd ed.).[1]

Processing of ULSI wafers requires tight control of contamination, process parameters, and reduced manufacturing costs. Toward this end, producers are beginning to use microprocessing environments for single wafers rather than batches of wafers. In single-wafer machines, it is necessary to extract a wafer out of a carrier and present it to the process chamber. If multiple process chambers are connected together, wafer transport among the different modules can be done best in a modest vacuum. Thus, a compatible heating technology is required that allows thermal processing to occur in closed microenvironments and that matches the wafer throughput of the conventional open-tube, batch-furnace systems. Rapid thermal processing (RTP) using transient lamp heating can meet this requirement nicely. With RTP, a single wafer is heated quickly under atmospheric conditions or at low pressure under isothermal conditions. The processing chamber is made of either quartz, silicon carbide, stainless steel, or aluminum and has quartz windows through which the optical radiation passes to illuminate the wafer. The wafer holder is often made of quartz and contacts the wafer in a minimum number of places. A measurement system is placed in a control loop to set wafer temperature. The RTP system is interfaced with a gas-handling system and a computer that controls system operation.[2] A schematic diagram of a basic RTP system is shown in Fig. 1*a* and is compared with a batch-furnace system that is resistively heated, shown in Fig. 1*b*. Typically, wafer temperature in an RTP system is measured with a noncontact optical pyrometer that determines temperature from radiated infrared energy.

The small thermal mass inherent in the RTP heating system stems from the fact that the wafer heats up but the surroundings remain cool. This results from

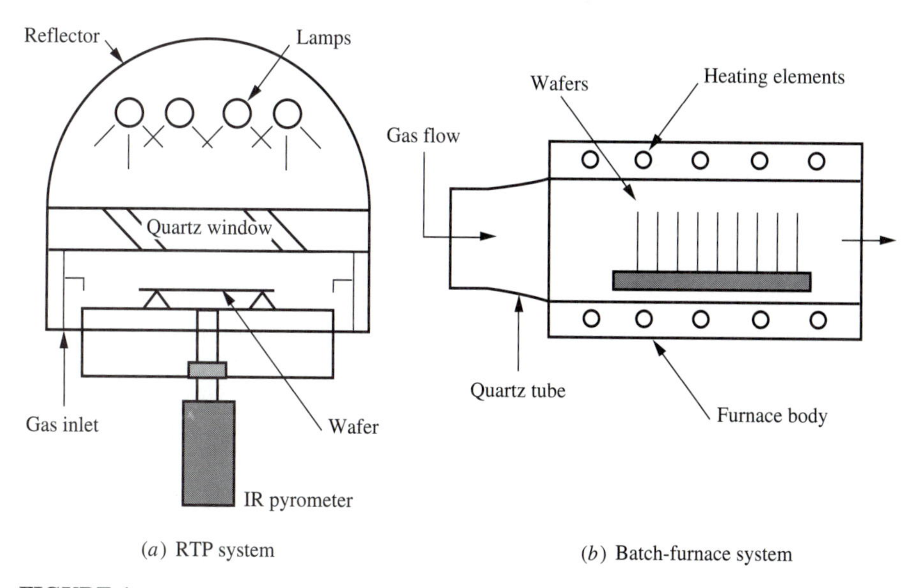

(*a*) RTP system

(*b*) Batch-furnace system

FIGURE 1
(*a*) Rapid thermal processing (RTP) system that is optically heated compared with (*b*) batch-furnace system that is resistively heated.

the Si wafers selectively absorbing radiation from the lamps, which produce short-wavelength radiation (spectral distribution peaks in the range of 0.5 to 1.0 μm). Thus, RTP is based on the energy transfer between a radiant heat source and a wafer, and the wafer is not, typically, in thermal equilibrium with the reactor wall. The response of a processed Si wafer to near-surface radiant energy deposition is a highly complex process in which the absorption of the energy depends on its wavelength, reflections of energy in the RTP chamber, as well as light diffusion, refraction, and transmission. Current knowledge is limited for the purposes of accurate modeling of wafer heating dynamics, so equipment manufacturers must rely on empirical methods of measurement and design. Some of the basic concepts are presented here.

4.1.1 Thermophysics in Rapid Thermal Processing (RTP)

The basic physical laws that describe radiative heat flow begin with the total radiant exitance, M, expressed in units of watts emitted per square meter of surface by an object with absolute surface temperature T, and given by the Stefan–Boltzmann law[3]

$$M(T) = \varepsilon \sigma T^4 \tag{4.1}$$

where ε is the emissivity and σ is the Stefan–Boltzmann radiation constant with a value of $5.6697 \times 10^{-8} \mathrm{W/m^2K^4}$. In an RTP system with lamps and reflectors removed some distance from the wafer and with an intervening quartz optical window in place, it is necessary to produce a high level of radiative heat flow. From Eq. (4.1), this requires that the temperature of the source be very high. In addition, each type of radiation source has the wavelength of the maximum intensity related to its temperature according to Wien's displacement law[3]

$$\lambda_{\max} \cdot T = 2.89783 \times 10^{-3} \tag{4.2}$$

where $\lambda_{\max}$ is in units of meters. Typical lamps in practical RTP systems have theoretical blackbody color temperatures from Eq (4.2) between 6000 K (arc lamp) and 2900 K (tungsten filament).[4] However, the color temperature of the heated wafer is much lower, usually 900 to 1400 K. Thus, in the optical-heating reactor, the absorption and emission spectra of the Si wafer never overlap. The result is that temperature nonuniformity is created in the wafers, and this is the greatest challenge to the acceptance of RTP as a manufacturing technology. By contrast, conventional furnaces that are resistively heated produce a blackbody color temperature of 1450 K, which allows for the thermal-uniformity characteristic of wafers that have seen furnace heating.

The total emissivity of Si in an RTP system depends on the optical properties of the starting wafer (*intrinsic emissivity*), the dielectric or conducting layers that may exist on top of the wafer surface or on the wafer bottom, buried layers within the wafer (*extrinsic emissivity*), and the specific optical properties of the reflective chamber and the components contained therein (*effective emissivity*).[4] The intrinsic emissivity of a perfect blackbody is $\varepsilon = 1$, causing it to emit the maximum possible amount of radiation [see Eq. (4.1)]. Also, under thermal equilibrium conditions, a diffuse radiator will have an emissivity that is equal to its absorptivity.

The emissivity, ε, of a 1.8-mm-thick, 15-ohm-cm, phosphorus-doped silicon wafer is shown in Fig. 2 as a function of wafer temperature and radiation wavelength.[5] It can be seen that for $\lambda > 1.5$ μm, ε varies from near zero at low temperatures up to 0.7 as the temperature increases. The value of 0.7 is characteristic of an ideal gray body. The deviation from ideal gray body behavior is caused by electronic valence-to-conduction band absorption, which is efficient for wavelengths below the silicon bandgap of 1.2 μm. Above 6 μm, absorption due to lattice vibrations becomes efficient. Between these two wavelengths (1.2 to 6 μm) and below 600°C, the main Si wafer heating process is by intrinsic free-carrier absorption. As a

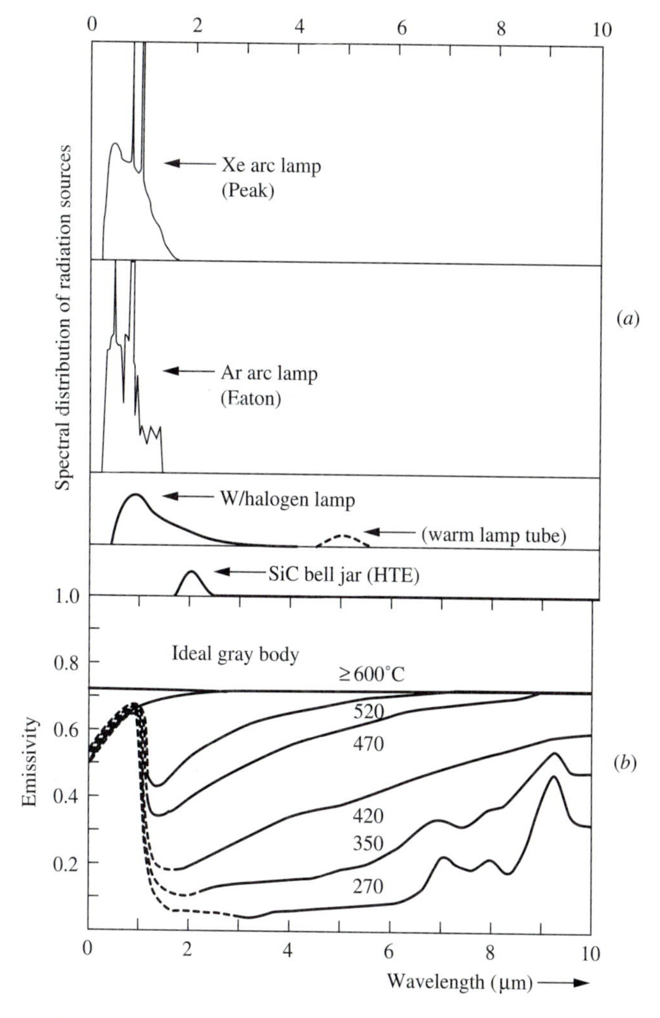

FIGURE 2
Spectral features in RTP systems: (*a*) heat sources (*After Roozeboom, Ref. 6.*); (*b*) temperature dependence of emissivity of a 1.8-mm-thick, 15 ohm-cm, P-doped Si wafer. (*After Sato, Ref. 5.*)

result, the emissivity in this range depends strongly on the free-carrier density and, therefore, dopant concentration, wafer thickness, and temperature.

From this discussion it is evident that proper selection of the spectral output of the heating lamps in an RTP system is important, since this determines the response of the processed semiconductor to the rapid near-surface energy deposition. For comparison, the spectral distributions of several popular radiation sources used in commercial RTP systems are shown[6] in Fig. 2*a*. Since a portion of these spectra occur at wavelengths above 1.2 μm, the initial heating rate of Si below 600°C depends on free-carrier absorption and, thus, on doping concentration. And the fraction of radiation that falls in this range is different for the various lamps. For example, the fraction is smallest for arc lamps since they emit their spectra with a peak wavelength of 0.5 μm. Tungsten-halogen lamps emit radiation with a peak at 1.0 μm, producing a larger fraction of energy above the silicon bandgap. Finally, resistively heated silicon carbide bell jars exhibit a peak at 2.0 μm. Thus, the influence of lamp spectral characteristics on wafer heating is only important up to 600°C, at which point the wafer is no longer transparent to radiation but rather becomes an opaque gray body with $\varepsilon = 0.7$.

Device wafers contain layers of deposited films and patterns in those films, which further complicates the concept of emissivity. Parameters that affect the extrinsic emissivity include film thicknesses (front and backside), material optical properties, and etched features. Optical interference will produce changes in emissivity, and thus absorptivity, during the deposition of films. For example, in Fig. 3, emissivity changes are shown for Si with 0.5 μm of SiO_2 and variable thicknesses of deposited polycrystalline Si.[7] This graph reveals that the spectral emissivity oscillates as a function of film thickness. The implications for such variations are serious, since the surface temperature varies in a similar fashion. It has been reported that temperature variations can exceed 100°C during poly-Si deposition in an RTP system, causing alternating layers of amorphous and polycrystalline Si if the temperature swings below the polycrystalline-to-amorphous transition value.[8]

The walls of an RTP chamber are made highly reflective to overcome large variations in extrinsic emissivity as described above. For example, Fig. 3[7] shows two emissivity curves for poly-Si thickness changes; one in a chamber with 0% chamber reflectivity and one with 70% reflectivity. The incidence of direct and reflected radiation on a wafer is shown schematically for a typical RTP reactor in Fig. 4.[9] It can be seen that reflections occur off of all surfaces onto both the front and the backside of the Si. This phenomenon complicates the determination of the emissivity. Thus, the extrinsic wafer parameters together with the RTP chamber and its internal components make up the overall effective emissivity.[4] Effective emissivity is, therefore, dependent on the specific reactor and must be measured at the wafer surface in conjunction with an infrared pyrometer to determine the temperature at any given time during an RTP process.

If a wafer is patterned, an additional complication occurs during lamp heating, since the various surfaces absorb energy differently, creating a lateral temperature distribution across the wafer. This nonisothermal nature is due to larger differences in the time constants of the physical processes involved. For instance, the process of photon absorption has a very short reaction time (10^{-12} to 10^{-14} sec), whereas the

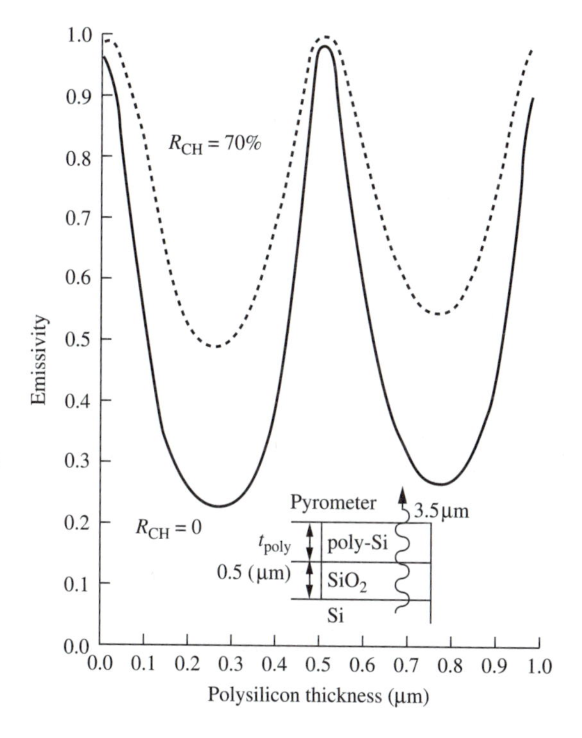

FIGURE 3
Emissivity changes with varying polycrystalline Si thickness deposited on 0.5 μm SiO_2 as measured with a pyrometer at 3.5 μm in a black chamber (reflectivity, R_{CH} = 0) and in a 70% reflecting chamber. (*After Hill, Ref. 7.*)

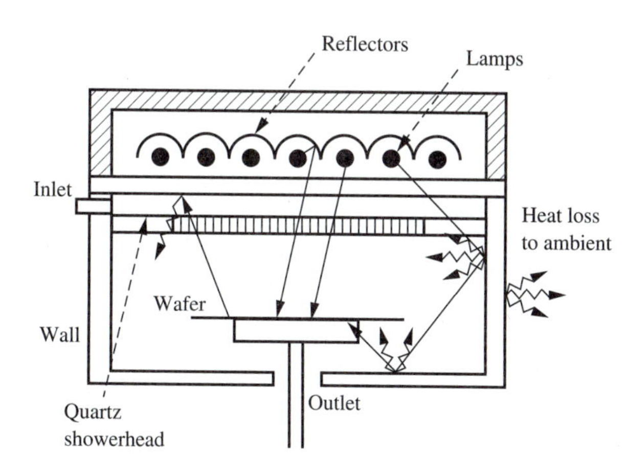

FIGURE 4
Modes of radiation heat transfer in the system and to the ambient in a typical RTP reactor. (*After Merchant et al., Ref. 9.*)

time constant for thermal conductivity is a function of the square of the distance. Consequently, there is a great difference between the vertical response time (temperature changes in 5 to 50 msec) and the lateral thermal response time (temperature changes in 1 to 10 sec) of the wafer among different features on the surface.[10] Thus, RTP is rarely an isothermal process in practice.

4.1.2 Other RTP Approaches

While the challenges inherent in optical heating are being tackled by many equipment companies, alternative approaches have been implemented that use wafer susceptors with lamps to improve wafer temperature nonuniformity[11] or that use continuous heat sources. Rapid wafer heating can be accomplished with continuous, resistively heated sources when the wafer is moved quickly in and out of the vicinity of the heat source. The difference between optical heating and this approach is that thermal equilibrium is maintained within the hot wall–ambient system while the wafer is heating or cooling. A diagram of the system is shown in Fig. 5.[12] The wafer is supported on three quartz pins on an elevator that is rapidly inserted into a silicon carbide bell jar. The top of the bell jar is resistively heated and the bottom of the bell jar is water-cooled, establishing a smooth temperature gradient from top to bottom. Annealing temperatures are dependent on the position of the elevator in the jar, and control in the range of 2 to 4°C is possible across a 200 mm wafer. Temperature ramp rates are 10 to 100°C/sec, whereas optical systems can operate at 50 to 300°C/sec.

4.1.3 Comparisons between RTP and Furnaces

A comparison between conventional furnace and RTP technologies is shown in Table 1. In order to achieve short processing times, tradeoffs must be made in temperature

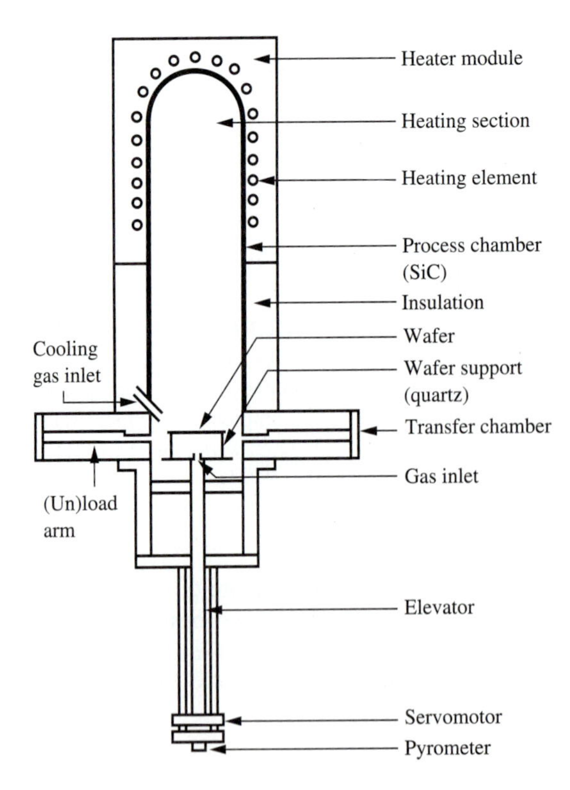

FIGURE 5
A schematic view of a continuous heat source, vertical furnace system for rapid thermal heating. (*After Lee and Chizinsky, Ref. 12.*)

TABLE 1
Technology comparisons

Furnace	RTP
Batch	Single-wafer
Hot wall	Cold wall
Long time	Short time
Small dT/dt	Large dT/dt
High cycle time	Low cycle time
Environment temperature measurement	Wafer temperature measurement
Issues	Issues
Large thermal budget	Uniformity
Particles	Repeatability
Atmosphere control	Throughput
	Wafer stress
	Absolute temperature measurement

and process uniformity, temperature measurement and control, and wafer stress and throughput. In addition, there are concerns over the introduction of electrically active wafer defects during the very fast (100 to 300°C/sec) thermal transients.[13] Rapid heating with temperature gradients in the wafers can cause wafer damage in the form of slip dislocations induced by thermal stress. On the other hand, conventional furnace processing brings with it significant problems such as particle generation from the hot walls, limited ambient control in an open system, and a large thermal mass that restricts controlled heating times to tens of minutes. In fact, it is these latter issues with the added concern of manufacturing costs that have driven semiconductor producers to consider RTP over conventional furnace technologies. Requirements on contamination, process control, and cost of manufacturing floor space are driving a paradigm shift to a single-wafer, microprocessing methodology.

To date, RTP has been integrated into some noncritical ULSI manufacturing applications where the advantages of a reduced thermal cycle are obtained. For RTP to replace existing manufacturing technologies it must demonstrate technical and cost advantages such as higher process yield, higher throughput, and no lost processing potential. The major limitation on RTP has been uncertainty in temperature uniformity and repeatability. And high process yield can only be obtained by improving the repeatability and uniformity intrinsic to the equipment.[14] We now investigate some additional manufacturing requirements for thermal processes.

4.2 REQUIREMENTS FOR THERMAL PROCESSES

ULSI processing equipment must produce chips with high reliability, tight tolerances, and high yield. The specific requirements on thermal processes can be stated in terms of a thermal budget, which is the allowed time-at-temperature for a given step, as well as a manufacturing budget that deals with temperature control, reproducibility, process step yield, and uniformity across a wafer. For example, typical requirements on equipment to meet the thermal and manufacturing budget needs of junction formation and oxidation include the following:

Mechanical control budget for processing and positioning large-diameter wafers
Temperature uniformity and reproducibility budget within each wafer and from run to run
Atmospheric budget for controlling gases used in the equipment
Time control budget
Absolute, repeatable temperature measurements on wafers

4.2.1 Thermal Budget

Dopants are introduced into Si by processes involving ion implantation, gaseous sources, solid sources, or crystal growth. Each method creates n- or p-type conductivity layers with a desired spatial distribution depending on the application. The desired time-at-temperature (thermal budget) to produce pn junctions at the desired depth from the Si surface for p-channel metal-oxide-silicon (MOS) transistor sources and drains is shown in Fig. 6. Assuming a gaseous dopant source is used that gives a boron (B) surface concentration of 1×10^{20} atoms/cm^3, then the thermal budget as a function of temperature can be determined. In Fig. 6, the requirements for four MOS technologies specified by the minimum patterned feature sizes (2, 1, 0.5, and 0.25 μm) are plotted.[15] Thus, for each generation of technology the time allowed at each temperature to form a junction at the necessary depth can be determined from the calculated lines. For example, for 0.25-μm devices, the time allowed to form the required 700 Å deep source-drain junction at 1000°C is only 24 seconds! This compares to junction-formation times of over 1000 sec at 1000°C for 2-μm technology.

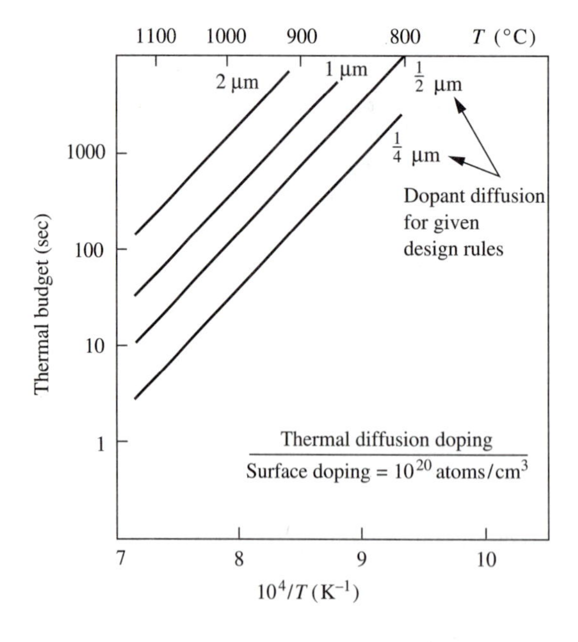

FIGURE 6
Available time-at-temperature (thermal budget) to produce scaled p^+n junction depths for four MOS technologies. Thermal diffusion sources are assumed with a fixed surface concentration of 1×10^{20} atoms/cm^3. (*After Fair, Ref. 15.* © 1990 IEEE)

This example shows us that the thermal budget requirements for MOS technology are driving a change in the type of furnaces needed to form pn junctions by diffusion. The natural question is, Why not keep the conventional batch furnaces and reduce the temperature to get more thermal budget time? If this were done for the last example, we could operate at 800°C and have 1000 seconds of processing time available. Although this certainly is feasible, gaseous source diffusion is generally not used, because it is difficult to get rid of the source after the junctions are formed. Traditional diffusion sources form doped glasses in the open windows that are defined in SiO_2. These glasses etch marginally faster than SiO_2, making them difficult to remove without losing the definition of the window itself. In addition, doping control is difficult with gaseous sources. The preferred process for introducing dopants into Si is ion implantation, since it is controllable and there is no residual source material to remove. However, implanting high-energy ions into a Si substrate produces extensive damage to the crystal through collisions between the incoming ions and the lattice atoms.[16] Once the temperature of the implanted wafer is raised sufficiently, the implant damage begins to anneal or change form, producing large quantities of point defects, i.e., Si vacancies and Si self-interstitial atoms. If the implant dose exceeds approximately 2×10^{14} ions/cm^2, more stable, extended defects may form, such as dislocation lines and loops. As we will see, an 800°C anneal is not sufficient to remove these extended forms of damage, which may cause incomplete dopant activation or leaky junctions. Thus, the appropriate annealing temperatures for ion-implanted junctions are above 900°C, which constrains the thermal budgets for shallow junctions for ULSI devices to tens of seconds.

Perhaps the most significant factor associated with the use of ion implantation in junction formation for ULSI is the production of large quantities of point defects during the damage anneal. The basic process of thermal diffusion of dopants requires the presence of such point defects, usually found in the Si crystal at equilibrium concentrations. However, damage annealing can produce point defects in concentrations that may reach many orders of magnitude above equilibrium levels at the annealing temperature. As a result, dopant diffusion can be significantly enhanced during ion-implantation damage annealing when excess point defects are present.[17]

The time required to remove the simplest forms of damage produced by ion implantation (point defect clusters) is measured in seconds at temperatures above 1000°C. This time is considered insignificant for older MOS technologies with deep source/drain junctions that require considerable diffusion times. However, this is not the case for ULSI technologies, where junctions must be established in seconds. As a result, thermal budgets can be dominated by the effects of implantation damage annealing. The thermal budgets for ion-implanted p^+ junctions for the same four generations of MOS technologies previously discussed in Fig. 6 are now shown in Fig. 7. Implants consisted of 6- to 10-keV BF_2 ions to form B-doped layers in Si with surface concentrations of 1×10^{20} atoms/cm^3.[5] The thermal-budget curves were calculated using the PREDICT simulator,[18] which contains models for damage effects on diffusion. For 0.25-μm technology, only 2.5 seconds are available for a diffusion temperature of 1000°C, which is a factor of 10 less than the thermal budget for a comparable gaseous-source process in Fig. 6. And, as we have seen, reducing the diffusion temperature to buy more process time is a limited option. It is usually

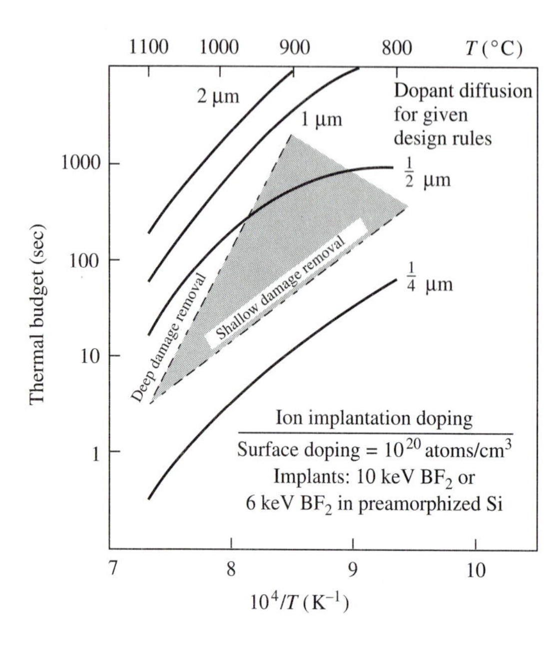

FIGURE 7
Available time-at-temperature (thermal budget) to produce p^+ scaled junction depths for four MOS technologies. Ion-implantation doping using BF_2 ions is assumed with a final surface concentration of 1×10^{20} atoms /cm^3. Shown for reference is the range of times required to remove shallow and deep extended dislocations caused by implantation damage. (*After Fair, Ref. 15.* © 1990 IEEE)

important to remove the implantation damage before the desired junction depth has been achieved. Otherwise, the annealing of the damage with the resulting generation of excess point defects in the Si will dominate the junction formation process rather than normal, thermally assisted diffusion. This results in poor control, because implantation damage is not very reproducible from one ion implanter to the next. The variables in damage production include wafer temperature during implantation, ion beam currents (dose rates), and wafer orientation with respect to the beam. Additional uncertainties follow when the wafer is heated. The production of excess point defects depends on temperature ramp rates as well as the final annealing temperature. It follows, then, that the preferred junction formation process should be controlled by thermally assisted diffusion (no damage) rather than by damage-assisted diffusion.

Times to remove shallow and deep dislocations produced by ion implantation are superimposed on the diffusion-time curves in Fig. 7.[15] Such dislocations result when heavy ion implants are performed at doses sufficient to turn the implanted crystal amorphous. The shallow dislocation boundary is based on data for damage 400 Å deep, and the deep dislocation boundary is for damage 1000 Å deep. The time to remove damage continues to decrease as the proximity to the surface increases. For example, the time required to form 700-Å-deep junctions will be less than the minimum time needed to remove the 400-Å-deep dislocations. Thus, the damage will remain after annealing, unless the dislocations are within 200 Å of the Si surface. To do this requires ultralow-energy ion implants.

Experience has shown that rapid thermal annealing will be required to achieve defect-free junctions for ULSI. The technology allows rapid heating to high temperatures for short times, which is best for removing damage and limiting diffusion of dopants. In addition, the technology road map for shallow junctions shown in Fig. 8 requires the use of either very-low-energy ion implantation ($E < 1$ keV for B—see Section 4.3.4, "RTA of ion-implanted junctions")[19] or solid diffusion sources that

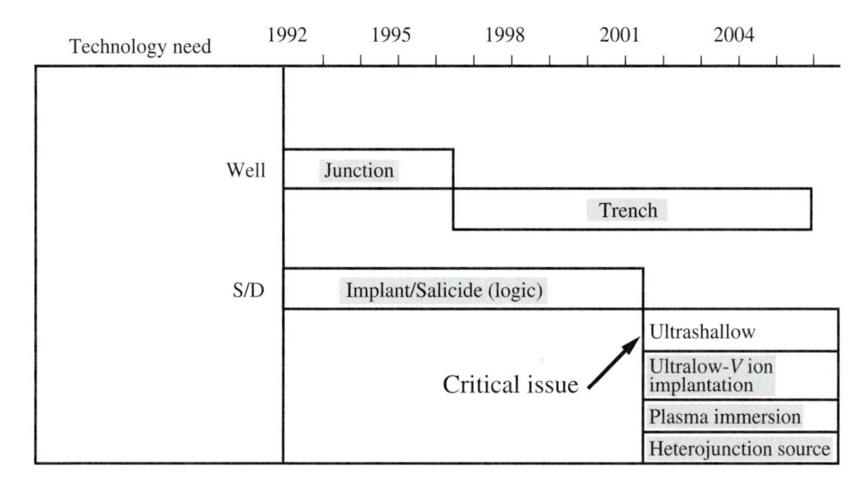

FIGURE 8
Semiconductor Industry Association front-end-of-the-line road map for isolation wells and source /drain junction formation showing current and future technologies.

become part of the device structure. Both approaches will yield diffusion-controlled junction processing, which is necessary for manufacturability.

4.2.2 Ambient Control Budget

The smallest fabricated dimension in an MOS transistor is the gate-oxide thickness, which is produced by thermal oxidation. For 0.25-μm devices, a 70-Å gate oxide controlled to within 7 Å is needed; this meets stringent reliability requirements with defect densities on the order of $0.5/cm^2$. To manufacture devices with ultrathin oxides requires careful process control and oxidation-furnace optimization. For example, the use of 150 to 200-mm diameter Si wafers requires large-diameter furnace tubes in which careful ambient control is difficult. The backstreaming of air from the large open ends of the furnace tubes is a major problem. In addition, the partial pressures of the oxidants, dry O_2 or wet O_2, would have to be maintained within 6% of nominal.[15] For dry O_2 oxidation, there is the additional requirement that trace amounts of water be controlled. A relatively small amount of water (125 ppm) in O_2 can increase the oxidation rate at 980°C, enough to throw the oxidation process out of spec for a sub-0.5 μm technology.[20]

Silicon surface control prior to the growth of thin oxide layers is also important because a native oxide grows on bare Si at room temperature. Thus, 70-Å film growth can be controlled if the Si surface is cleaned with HF, which leaves the Si surface bonds terminated with H.[21] Desorption of the H at 300°C in a highly pure Ar gas ambient followed by the formation of one monolayer of oxide passivates the surface for subsequent gate-oxide growth.

For ambient control, single-wafer RTP chambers offer a microenvironment approach that can satisfy the stringent requirements for ultrathin growth using rapid thermal oxidation (RTO).

4.2.3 Contamination Budget

Scaling transistors to dimensions suitable for ULSI has a profound effect on the manufacturing yield and reliability of integrated circuits. Processing complexity naturally increases with each generation of chips, caused in part by the need for additional mask levels to interconnect the increased number of subcircuits on the chip. This trend makes devices more susceptible to contamination introduced by particulate and chemical impurities at each process step. In addition, smaller devices are vulnerable to smaller defects and smaller amounts of chemical impurities that may result in chip loss.[22]

Particles can cause yield loss through the presence of random defects in the patterning of film levels. Chip yield is expressed in terms of the defect density through various statistical models such as the Poisson distribution,[23]

$$\text{Yield} = e^{-A\rho} \tag{4.3}$$

where A is the chip area and ρ is the density of defects per unit area. As devices are scaled down in their dimensions, the density of potentially fatal defects increases rapidly. This result follows because killer defects are at least one-third the size of a lithographic feature or one-half of a film thickness. With a gate oxide 70 Å thick, a 35-Å particle could be fatal.

Measured particle densities are plotted in Fig. 9 versus particle size for bulk gases, semiconductor chemicals, deionized water, and cleanroom air at MCNC (Microelectronics Center of North Carolina).[15, 22] All of these distributions show increasing particle densities with decreasing particle size. And this is the environment in which smaller devices will be made. Device scaling by a factor of two occurs in a processing environment in which the number of potentially fatal particles increases by four to eight times.[22] The impact on yield may be devastating. Under these conditions, a process that yields 25% and is limited by particle contamination would yield nothing after scaling it down by a factor of two. Consequently, improved means must be found for controlling particle size and numbers in the processing environment.

Improved contamination control is possible in single wafer, in-situ processing where process modules are clustered together and interconnected by vacuum transport means. An example of such an arrangement is shown in Fig. 10,[24] where the connected processing modules include wafer cleaning, silicon oxidation, and low-pressure chemical vapor deposition (LPCVD). Rapid thermal processing technology allows for the use of cold-wall systems, which can better meet the stringent requirements for low-level contamination. It should be noted that recent developments of vertical batch furnaces can help reduce the contamination problem inherent in hot-wall, conventional furnaces. However, there is no methodology to integrate such furnaces with an ultralow-contamination environment that protects the wafers during pre- and post-processing.

4.2.4 Device Electrical Budget

The electrical budget refers to the control of electrical device parameters that are determined by device-scaling rules. Included in this budget are specifications

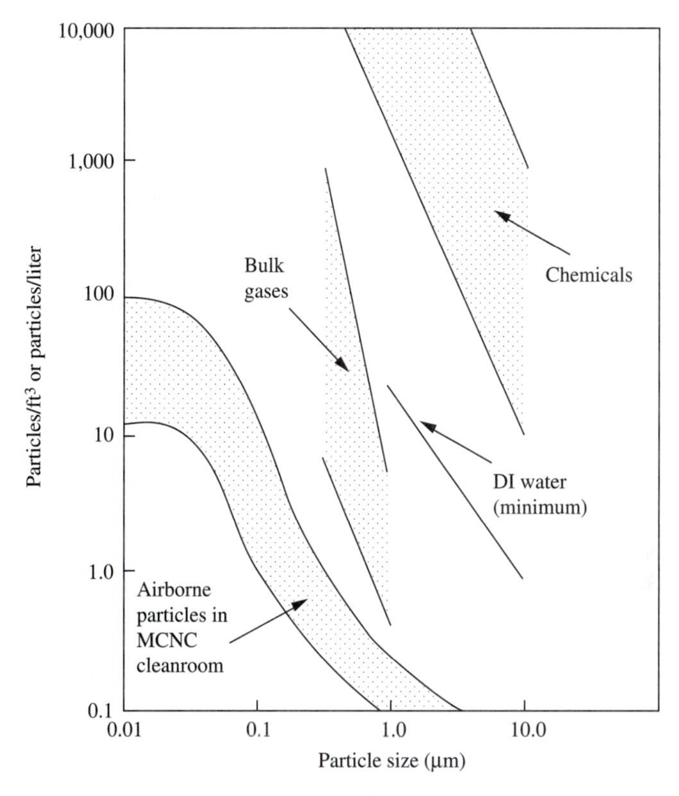

FIGURE 9
Comparison of particle size distributions in bulk gases, chemicals, deionized water, and processed air in the MCNC semiconductor cleanroom. The particle densities increase rapidly with decreasing particle size; this is the environment in which smaller silicon devices are being scaled. (*After Osburn et al., Ref. 22.*)

for contacts, interconnections, electric-field levels, and process-induced electric charge.

For ULSI, it is desirable that contacts with low specific resistance (1×10^{-8} ohm-cm^2) to semiconductor junctions be made with high yield. These contacts must also be reliable, serving as barriers to unwanted metal reactions with Si. Barrier layers of titanium/tungsten and titanium nitride have proven to be good choices.

Low-resistance contacts are imperative. If contact dimensions are halved, contact resistance increases by a factor of four. Thus, specific contact resistances must be decreased by factors of 10 or better for ULSI.

Refractory silicides of transition metals have been used to improve contact resistances. Recent work with Al–TiW–$TiSi_2$ contacts to shallow n^+ junctions has been reported.[25] By performing sputter etching of the $TiSi_2$ surface to remove any oxides prior to TiW deposition in the same vacuum environment, specific contact resistances below 2×10^{-8}ohm-cm^2 can be achieved. Thus, the contact-resistance budget is driving the use of in-situ vacuum processing in an RTP, single-wafer module.

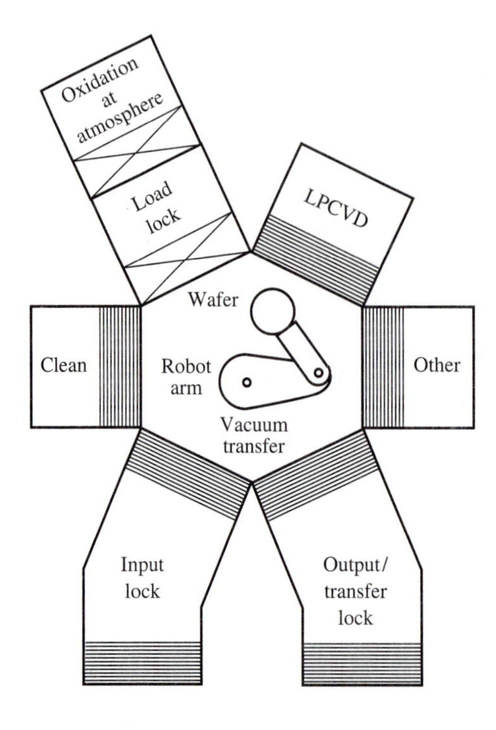

FIGURE 10
Example of rapid thermal processing modules in a radial, multichamber cluster designed for better particle exclusion, interprocess ambient control, and flexible operation. (*After Rosser, Moynagh, and Affolter, Ref. 24.*)

The intrinsic electrical-parameter budget of submicrometer semiconductor devices becomes increasingly bounded as device dimensions decrease. For example, submicrometer MOS transistors must be designed and manufactured without exceeding physical limits imposed by drain-junction avalanche breakdown, bulk punch-through, short-channel effects, and hot-electron effects.[26–29] In addition, device design for one-transistor DRAM cells is constrained by noise-margin requirements that are dominated by the threshold-voltage mismatch at the input of the on-chip sense amplifier.[27] Although there are several sources of threshold-voltage mismatch, one that will become much more important as device dimensions approach 0.1-μm minimum feature size is the variation resulting from channel doping distribution statistics. Indeed, channel doping by ion implantation causes scattering in the location of the impurity atoms in the Si, which causes a one-sigma variation in V_t of 15 mV for a 0.1-μm channel length.[30] However, this is a control problem that can be solved by rapid thermal epitaxy and atomic-layer doping in a single-wafer processing module (see Section 4.3.1). Such new doping technologies will also affect control of the subthreshold conduction current between source and drain of a MOSFET. This current can vary in a 0.5-μm device by a factor of ten with a 500-Å change in channel profile depth. In addition, such process parameter variations also change the subthreshold slope (the change in gate voltage produces a change in drain current) by 25 mV/decade.[31]

4.2.5 Cost Issues and the Manufacturing Budget Crisis

The cost of manufacturing submicrometer ULSI chips is increasing in inverse proportion to the decrease in device feature sizes. These expenses are driven by a manufacturing budget crisis associated with the technology and complexity limits of integrated circuit (IC) design, costs of research and development to address manufacturing issues, and facility capitalization.[6] Thus, examination of the rate of progress in microelectronics over the past 30 years suggests that the primary challenge in reaching the gigachip age in the year 2001 will be the semiconductor industry's ability to change the cost trend lines; i.e., to change the economics of how ICs are developed and manufactured.[2]

Several approaches have been suggested for reducing the costs associated with developing and manufacturing ULSI chips. Requirements on contamination, process control (manufacturing-parameter budgets), and cost of manufacturing floor space are driving a paradigm shift to a microprocessing methodology. Thus, single-wafer processing environments with highly controlled, ultraclean ambients clustered together in specialty process modules are being considered. Today, in-situ vacuum processing equipment accounts for 40% of the total equipment, and it is projected that by the year 2001, 80% of ULSI equipment could be in-situ vacuum-based modules clustered together and controlled by a factory information system.[2]

4.3 RAPID THERMAL PROCESSING

Perhaps the most important feature of a rapid thermal processing system is its generation and delivery of radiant energy to the wafer from tungsten-halogen lamps in a wavelength band of 0.3 to 4.0 μm. Because of the optical character and wavelength of the energy transfer, the quartz walls do not absorb light efficiently, whereas the Si wafer does. Thus, the wafer is not in thermal equilibrium with the cold walls of the system. This allows for short processing times (seconds to minutes) compared to minutes to hours for conventional furnaces.

With the addition of reactive ambients to RTP reactors, other applications have opened up besides simply heating wafers. The merger of single-wafer RTP and single-wafer chemical vapor deposition (CVD) has brought continuous improvements to film deposition technology. Thus, rapid thermal CVD (RTCVD) technology provides a means of controlling chemical vapor deposition processes with temperature ramps rather than by turning gas flows on and off. This methodology is called *limited reaction processing* (LRP).[32] The use of LRP in silicon epitaxy is demonstrated in Fig. 11 where temperature-versus-time profiles are drawn comparing conventional furnace epitaxy to the LRP growth process.[33] The reduction in temperature-time exposure afforded by the RTP-based method is dramatic.

4.3.1 Epitaxy

Processing demands on the growth of high-purity epitaxial Si include ambient purity (oxygen and water levels in the parts per billion range), optimization of gas flow patterns, minimum wall deposition, and vacuum compatibility. Thus, an RTP system suitable for epitaxy will look different from other RTP systems. A schematic

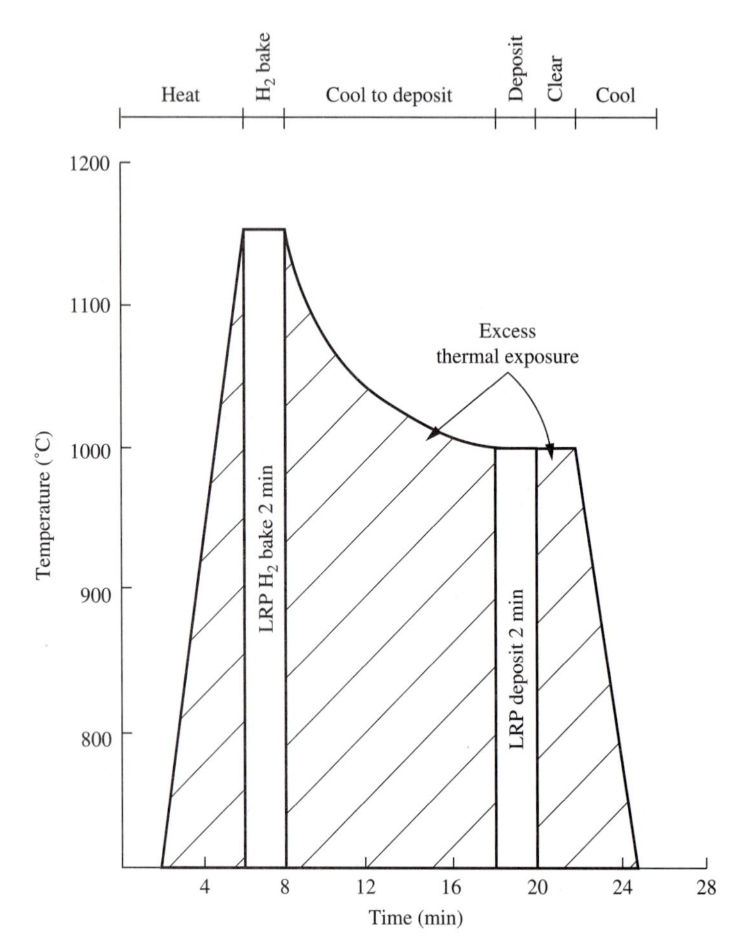

FIGURE 11
Schematic temperature versus time profile comparing conventional furnace, silicon epitaxy (hatched regions) to the LRP growth process. (*After Hoyt, Ref. 33.*)

diagram of an advanced, low-pressure, epitaxial Si reactor based on RTP technology is shown in Fig. 12.[33] This technology is an alternative to other growth methods such as plasma-enhanced CVD, ultrahigh-vacuum (UHV) CVD, and molecular beam epitaxy (MBE).

Conventional epitaxial growth of Si is usually performed in a quartz reaction chamber with a batch of wafers placed flat against a graphite susceptor. Deposition occurs with the wafers held at an elevated temperature while the reactant gas stream flows across the wafers' surfaces. Typical Si source gases include silane and dichlorosilane, and a hydrogen carrier gas provides the required gas velocity. Overall, the deposition process comprises a mass-transport process with a weak temperature dependence and a sequential surface-reaction process that is exponentially dependent on wafer temperature (see Chapter 3). Since the properties of epitaxial

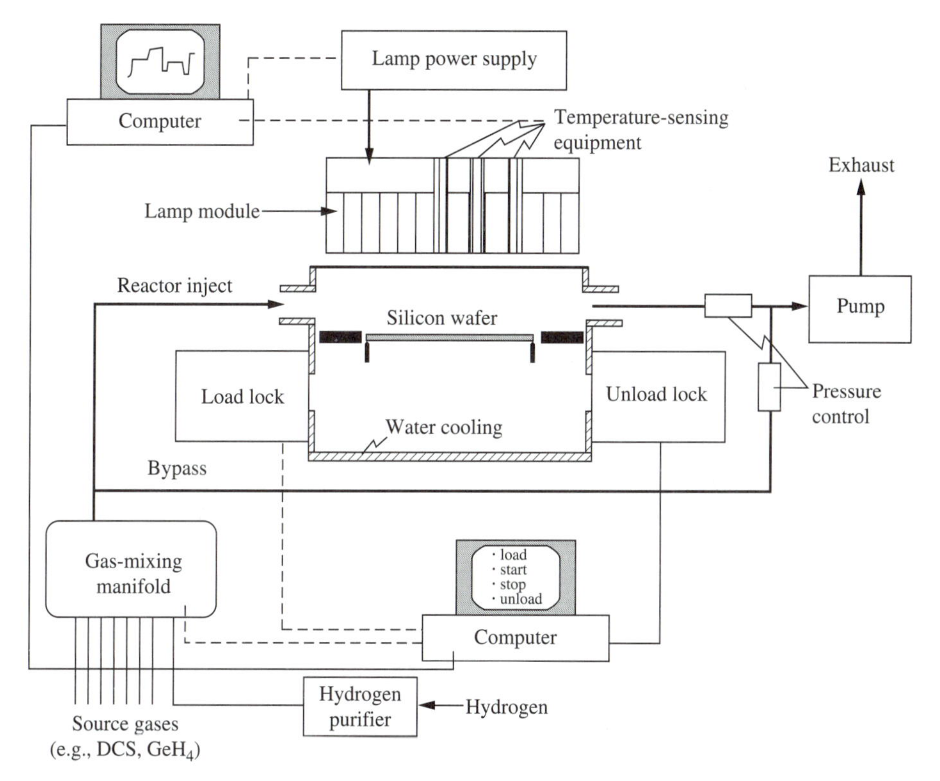

FIGURE 12
Diagram of an advanced, low-pressure, RTP, epitaxial, Si reactor. (*After Hoyt, Ref. 33.*)

layers improve as the growth pressure decreases, most conventional systems operate at either atmosphere (760 torr) or reduced pressures (20 to 100 torr). Typically, the graphite susceptor is heated inductively or by infrared lamps to around 1150°C, and the quartz chamber is water- or air-cooled or both.[33]

The requirements for ULSI devices include thinner epitaxial layers, better dopant profile control, and selective epitaxial growth.[34] In addition, lower cost-per-wafer requirements translate into factors such as equipment costs, energy consumption, and raw materials. If such costs remain constant, then the wafer cost varies inversely with the reactor throughput.[35] But throughput in conventional reactors is achieved primarily by the number of wafers that can be accommodated on the susceptor. As wafer sizes increase, fewer wafers can be processed in a single run. Thus, the next-generation machine must be purchased with a larger susceptor and a higher price. On the other hand, for RTP, throughput depends minimally on wafer size and mostly on processing time, which can be reduced with improved automatic wafer-handling tools and high-rate processes.

Selective growth

To achieve full selective growth of silicon epitaxy from dichlorosilane in a conventional tool requires the addition of HCl. The presence of HCl is important for etching small Si nuclei and clusters that will try to grow over the masking layer where growth is undesirable. The widely accepted vapor phase reaction for dichlorosilane is

$$SiH_2Cl_2 \rightarrow SiCl_2 + H_2 \tag{4.4}$$

Under conditions of limited reaction processing, full, selective, epitaxial growth of Si has been observed over the temperature range from 650 to 1100°C without the addition of HCl.[33] It has been proposed that the reaction kinetics are different under LRP conditions and that HCl is produced as a by-product of the dichlorosilane reaction.[36] Thus, LRP deposition kinetics are controlled by this reaction:

$$SiH_2Cl_2 \rightarrow SiHCl + HCl \tag{4.5}$$

The dissociation energy of Eq. (4.5) is 2.7 eV, whereas Eq. (4.4) yields 1.4 eV. Measurements of growth rate versus temperature for three deposition conditions are shown in Fig. 13.[36] Unlike conventional epitaxy, where surface reaction energies are independent of gas-phase species,[37] LRP conditions produce variable reaction energies that depend on Cl concentration in the gas stream. The activation energies for the three curves are (*a*) 1.7 eV, (*b*) 2.6 eV, and (*c*) 3.2 eV. The temperature dependence of the surface reaction-limited deposition rate is exponential and follows the Arrhenius relation

$$R = Ae^{-E_a/kT} \tag{4.6}$$

where E_a is the activation energy in eV, T is the absolute temperature in K, and A is the frequency factor. One conclusion from this result is that good temperature control is even more important for low-temperature LRP reactions because of the higher activation energies and, consequently, the larger dependence of epitaxial growth on

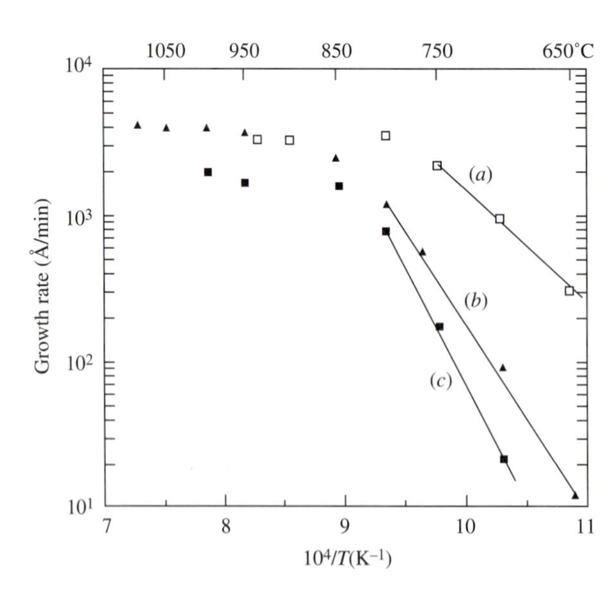

FIGURE 13
Silicon growth rate in an RTP reactor operating at 2 torr with 2 slm H_2 and (*a*) 40 sccm (standard cm^3/min) of SiH_4, (*b*) 80 sccm of SiH_2Cl_2, and (*c*) 20 sccm of SiH_4 and 2 sccm of HCl (*After Regolini et al., Ref. 36.*)

temperature. However, temperature control and uniformity of today's RTP systems do not approach the levels of conventional systems, which is a major drawback for production control.

Doping profiles

The growth of thin epitaxial layers on heavily doped substrates requires that autodoping and outdiffusion from the substrate be minimized. Otherwise the required doping in the epi-layer can be lost by diffusion of impurities from the epi/substrate interface or from vapor-phase doping of impurities that have evaporated from the substrate. Diffusion decreases with temperature and time, and autodoping decreases similarly through lower gas-residence time, lower dopant surface segregation, and dopant evaporation rates. RTCVD has been shown to produce intrinsic Si layers grown on n^+ Si with results comparable to molecular beam epitaxy,[32] i.e., abrupt doping-transition regions. An example of boron delta-doping spikes obtained by epitaxial Si layer growth at 900°C in a UHV-RTCVD reactor is shown in Fig. 14.[38] The use of B_2H_6 with Si_2H_6 was preferred over SiH_4 in this application because higher growth rates could be obtained at lower growth temperatures, allowing higher throughput and better control of gas impurities. Precision temperature and time profile control allows the growth of such films with dimensional control within tens of angstroms.

The ability of RTCVD to produce abrupt n^+ doping layers has met with mixed success. Whereas abrupt i-n-i layers have been demonstrated using dichlorosilane

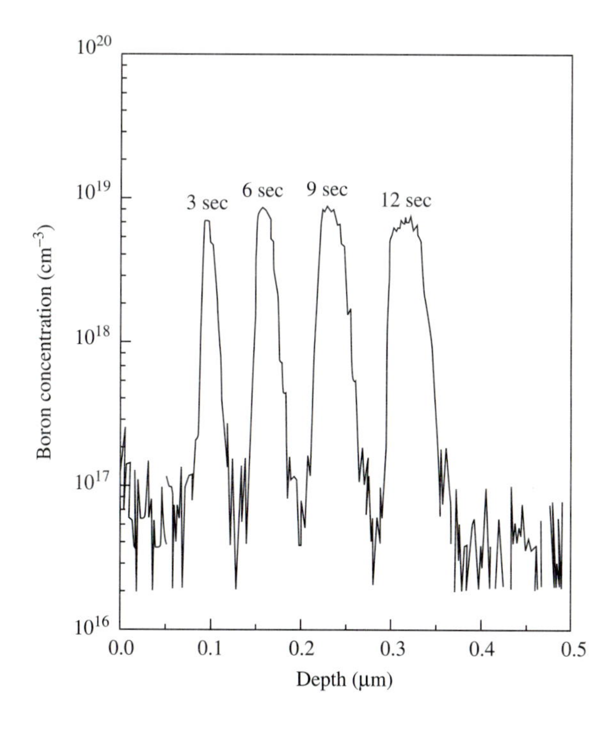

FIGURE 14
Boron delta-doping spikes obtained by epitaxial growth using Si_2H_6, B_2H_6, and H_2 in a UHV-RTCVD reactor. The growth times in seconds for each spike are shown. (*After Ozturk and Wortman, Ref. 38.*)

and phosphine in a single-wafer reactor,[39] similar arsenic-doped layers have not been possible, because of the accumulation of arsenic (As) in the growing Si surface.[40] As the layer is grown, the As atoms available for doping the layer are slowly released, causing a gradual buildup of doping to its steady-state level.

4.3.2 Thin-Film Deposition

Chemical vapor deposition can be performed over a wide pressure range from atmospheric to ultrahigh vacuum. Low-pressure CVD (LPCVD), atmospheric CVD (APCVD), and ultrahigh-vacuum CVD (UHVCVD) all rely on the pyrolysis of source gases at elevated temperatures. At present, LPCVD is the preferred conventional deposition technology for many thin films because of high deposition rates and excellent film-thickness uniformity.

The application of RTP to CVD processes is particularly well-suited to single-wafer cluster tool technology. A typical design of an RTCVD system appears in Fig. 15, which shows a cross-sectional view of a RAPRO™ RTCVD chamber.[41] The chamber walls are water-cooled, thus eliminating deposition on the chamber walls. In LPCVD, the chamber walls are hot and in equilibrium with the rest of the system. Another difference between conventional CVD and RTCVD is the sequence of required tasks. The process sequences for LPCVD and RTCVD are shown in Fig. 16.[41] It can be seen that RTCVD requires higher temperatures but shorter times to deposit the same amount of material as LPCVD. In addition, temperature is the switch that turns the RTP deposition process on or off, avoiding the long ramp-up and ramp-down times required in the conventional method.

Oxide deposition processes

To be commercially viable, RTCVD systems must be able to process a single wafer in 1 to 2 min. The corresponding throughput is 30 to 60 wafers per hour. For relatively thick deposited films of 1000 to 2000 Å, this requires deposition rates greater than 1000 Å/min. However, conventional batch LPCVD processes are much slower (100 Å/min). Therefore, if one plans on transferring a recipe for an LPCVD deposition step to an RTCVD system, it will be necessary to raise the temperature in the RTCVD system to increase the system throughput. However, there are limits on the deposition rate for a given gas flow once the temperature is raised sufficiently, so the deposition kinetics become mass-transport-limited. For example, in Fig. 17 the mass transport region occurs when the deposition temperature exceeds 850°C, and further increases in temperature do not yield further increases in the rate.

As an example, consider the deposition of thick SiO_2 layers as either an inter-level dielectric between metallization layers or as a sidewall spacer on a polysilicon gate. These applications require oxides that are several thousands of angstroms in thickness. An important requirement for both applications is good oxide step coverage. The physical parameter indicative of step coverage is the sticking coefficient of the CVD process.[42] A small sticking coefficient of absorbed atoms means that they will tend to diffuse on the surface of the deposited layer or be reemitted. A CVD process with a small sticking coefficient provides better step coverage

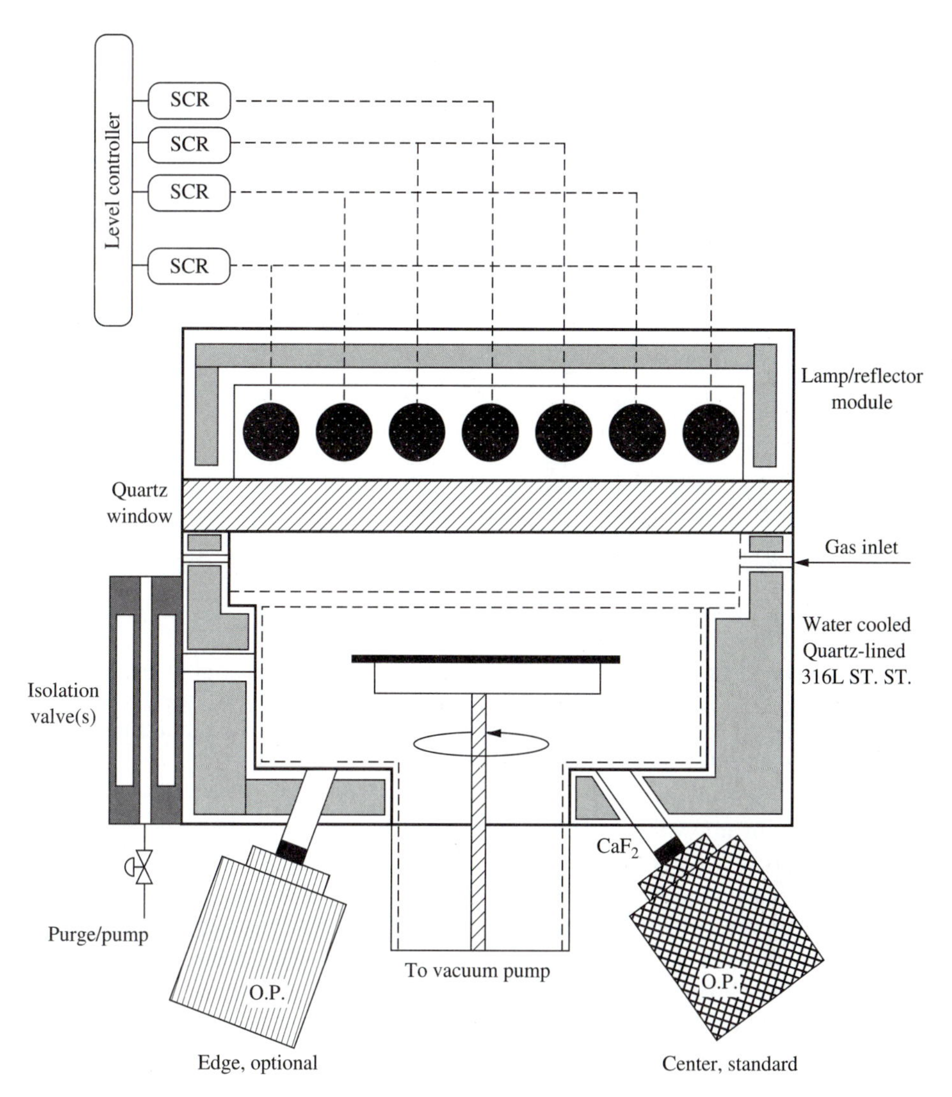

FIGURE 15
Cross-sectional view of the RAPRO™ RTCVD chamber (*After Ozturk, Ref. 41.*)

and conformality.[41] This requirement favors the use of TEOS (tetraethoxysilane or tetraethylorthosilicate) and TMCTS (tetramethylcyclotrasiloxane) in the CVD process. Thus the RTCVD of SiO_2 by pyrolysis of TEOS is believed to occur by the reaction[43]

$$Si(OC_2H_5)_4 \rightarrow SiO_2 + 2H_2O + 4C_2H_4 \tag{4.7}$$

The temperature dependence of the oxide deposition rate obtained in an RTCVD system according to Eq. (4.7) is shown in Fig. 18.[44] Below 800°C, the deposition rate is controlled by surface reaction processes with a 3.3 eV activation energy. This large sensitivity to temperature would require tight temperature control for thin

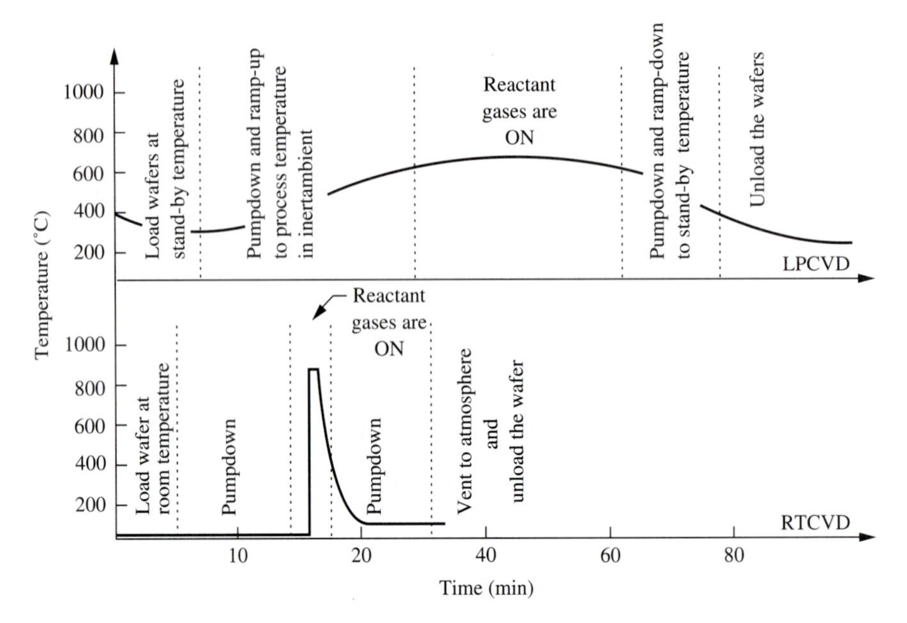

FIGURE 16
Process sequences for conventional LPCVD and RTCVD. The major difference in the two technologies lies in the way the deposition is initiated; in LPCVD, the gas flow is turned on and off for control, whereas in RTCVD, temperature is used for control with the gas flowing. (*After Ozturk, Ref. 41.*)

oxide deposition at low temperatures. Above 800°C, the deposition rate of SiO_2 approaches 1000 Å/min with lower activation energy, which meets the throughput and deposition control requirements of the RTCVD system. Such a high operating temperature, however, restricts the process to oxide deposition steps at the front end of the line. The process is unsuited for depositing interlevel SiO_2 at the back end of the line in multilevel-metallization technologies that use aluminum. For this purpose, a process is needed that can provide acceptable throughput at temperatures below approximately 450°C. Such an RTCVD process has not yet been found.[41]

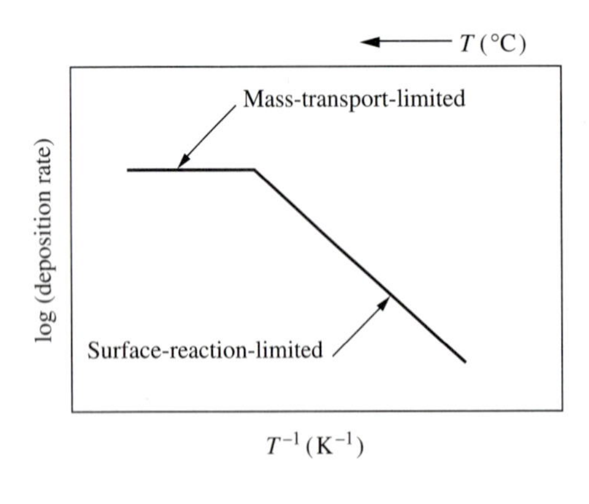

FIGURE 17
Temperature dependence of the deposition rate for a typical CVD process. (*After Ozturk, Ref. 41.*)

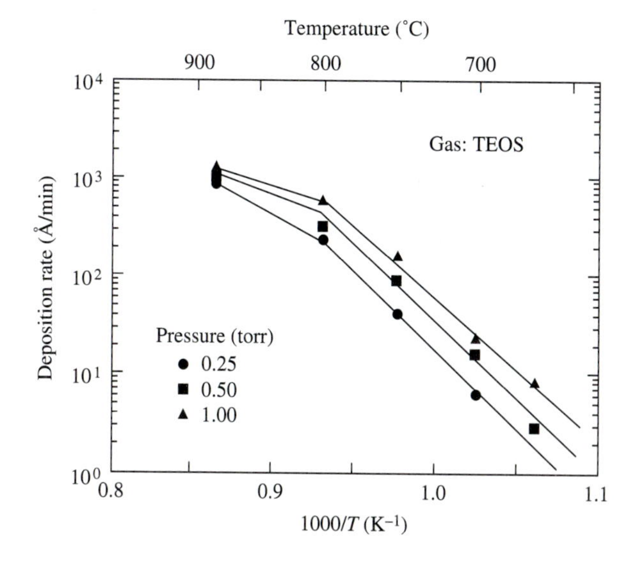

FIGURE 18
Temperature dependence of the SiO_2 deposition rate at different pressures using TEOS in an RTCVD reactor. The activation energy below 800°C is 3.3 eV. (*After Miller et al., Ref. 44.*)

Thin oxide applications include in-situ deposition of MOS gate structures where Si surface cleaning, gate oxide formation, and polysilicon gate electrode deposition would all occur in a low-pressure, multichamber cluster tool. This process would require the low-temperature deposition of SiO_2 as a substitute for the preferred, thermally grown oxide. However, oxidation or rapid thermal oxidation (RTO) is performed at atmospheric pressure, which would not be consistent with the low-pressure cluster tool approach. In addition, RTO is performed at temperatures exceeding 950°C, which requires process times up to 2 min. Therefore, both throughput and thermal-budget issues could become important for RTO. As a result, work is being done to bring the quality of low-temperature CVD oxides up to that attainable with thermal oxides.

RTCVD offers a promising technology for high-quality oxide deposition through the use of ultraclean gases in a chamber with a highly controlled ambient. Most conventional LPCVD furnaces use silane and oxygen in the oxide deposition reaction described in Eq. (4.8).[45]

$$SiH_4 + O_2 \rightarrow SiO_2 + 2H_2 \tag{4.8}$$

Typical deposition temperatures are in the 300 to 450°C range. However, Eq. (4.8) results in excessive particulate generation, which may contaminate the wafers. Thus, alternate gas combinations are being explored for RTCVD. One potential gate-quality RTCVD oxide-deposition process being studied utilizes SiH_4 and N_2O:[41]

$$SiH_4 + 2N_2O \rightarrow SiO_2 + 2N_2 + 2H_2O \tag{4.9}$$

At 800°C and with $SiH_4 : N_2O$ ratios less than 4%, stoichiometric films are obtained[46] with activation energies in the range of 1.5 to 1.75 eV and deposition rates around 50 Å/min, which are nearly 100 times larger than the thermal oxidation rate at atmospheric pressure. The films also have acceptable electrical properties.

Silicon nitride deposition

Silicon nitride has found important uses in integrated circuit processing: as a mask against O_2 diffusion during local oxidation, as a passivation layer because it is a barrier to contaminants, as a gate dielectric in MNOS memory transistors, and a new use as an interlevel dielectric in oxide-nitride-oxide stacked-gate structures.[47]

Deposition of Si_3N_4 in conventional systems involves the reaction of ammonia with either silane in APCVD or dichlorosilane in LPCVD. This latter reaction has become the preferred approach and is described as follows:

$$3SiH_2Cl_2 + 10NH_3 \rightarrow Si_3N_4 + 6NH_4Cl + 6H_2 \tag{4.10}$$

This reaction is unsuitable in a cold-wall RTCVD system. The reaction by-product, ammonium chloride (NH_4Cl), deposits in the form of a fine powder on cool surfaces in the chamber. Although this is a minor concern in hot-wall LPCVD systems, it would be a major problem in an RTCVD system, in which deposits would occur essentially everywhere, including the quartz windows and chamber walls.[41]

Alternative reactions have been studied for application to RTCVD systems, including the use of SiH_4 and NH_3.[48] This reaction can be described as

$$3SiH_4 + 4NH_3 \rightarrow Si_3N_4 + 12H_2 \tag{4.11}$$

It is interesting that Eq. (4.11) has been tried in LPCVD systems with little success. Wafers with deposited nitride films showed radial nonuniformities in which less material was deposited in the wafer center than on the perimeter. This "bullseye" effect was believed to be caused by mass-transfer deficiencies.[41] However, the reaction in Eq. (4.11) works well in a single-wafer RTCVD system, where reactant-depletion effects are not observed.[48]

It has been observed[48] that stoichiometric nitride films are obtained for depositions using $NH_3 : SiH_4$ ratios greater than 120 : 1. The Si_3N_4 deposition rate versus $NH_3 : SiH_4$ reactant gas ratio at 785°C is shown in Fig. 19. The useful deposition rates at high flow ratios are very small (100 Å/min), and therefore this process is

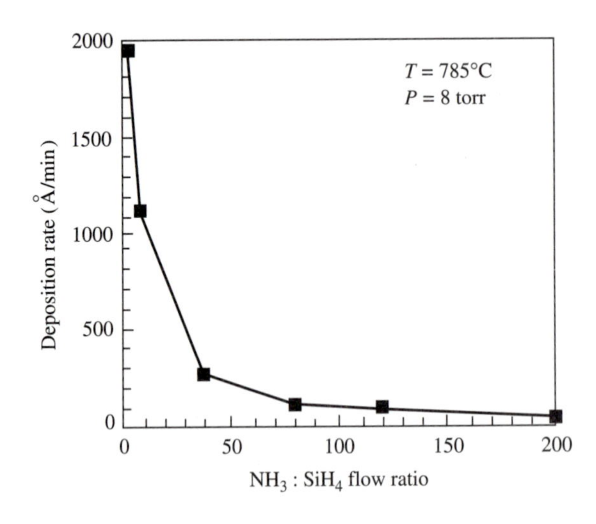

FIGURE 19
Si_3N_4 deposition rate versus reactant gas ratio. Deposition conditions were NH_3 and 10% SiH_4 in Ar at 785°C and a total pressure of 8 torr. *(After Johnson et al., Ref. 48.)*

suited only for thin-film applications such as stacked-gate structures. However, no suitable single-wafer RTCVD processes exist yet for typical thick-film uses of silicon nitride.

Polycrystalline silicon deposition

Polysilicon is the preferred gate electrode material in ULSI MOS devices. It is also used as an interconnect layer with a low-resistivity, silicide-forming metal such as titanium or cobalt. In addition, polysilicon resistors are used in analog ICs as well as in static random-access memory (SRAM) chips. In conventional LPCVD systems, polysilicon is deposited by the pyrolysis of silane:[49]

$$SiH_4(gas) \rightarrow SiH_4(absorbed) \rightarrow Si + 2H_2 \tag{4.12}$$

Deposition usually occurs around 600°C at 0.2 to 1 torr with rates of 100 to 200 Å/min. Since these rates are too low for single-wafer processing, it becomes necessary to raise the temperature of the reaction above 700°C with the goal of attaining rates in the 2000 Å/min range. This approach, however, has been tried in LPCVD systems with the result that higher deposition temperatures resulted in polysilicon films with rough surfaces.[50] Conversely, very smooth polysilicon films have been deposited in RTCVD systems at elevated temperatures and pressures. The main factors that control polysilicon deposition rates are temperature and the partial pressure of the reactants. For example, the temperature dependence of deposition using 10% SiH_4 in a carrier gas of Ar is shown in Fig. 20.[49a] The surface reaction activation energy of 1.7 eV is similar to that achieved in LPCVD systems, and Fig. 20 shows that 1000 Å/min rates are possible at 700°C. The figure also

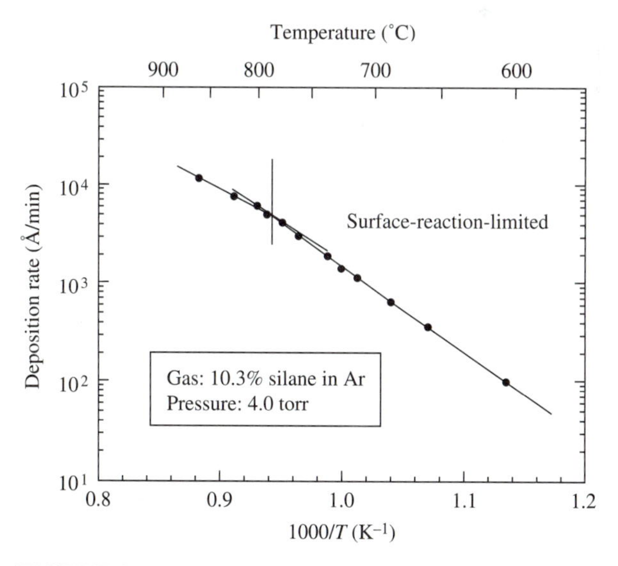

FIGURE 20
The deposition rate of polysilicon from SiH_4 in an RTCVD reactor. Silane was diluted with Ar to 10.3%. (*After Ozturk et al., Ref. 49a.*)

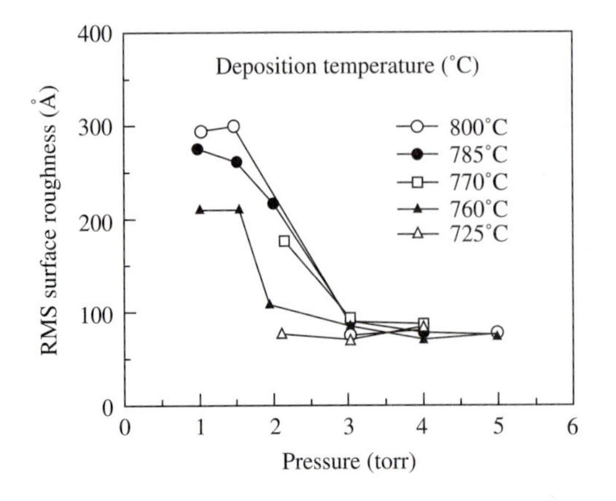

FIGURE 21
Root-mean-square surface roughness of 2000-Å-thick polysilicon deposited on SiO_2 at several temperatures and pressures. Data were obtained using either cross-section TEM or UV surface-reflectance measurements. (*After Ren et al., Ref. 51.*)

shows that there is no true mass-transport-limited deposition region above 780°C. With regard to surface roughness of the polysilicon films, it appears that very smooth films can be achieved in RTCVD systems at deposition temperatures above 725°C and at elevated pressures (above 3 to 4 torr).[51] Surface roughness measurements on 2000-Å-thick films are shown in Fig. 21, where root-mean-square roughness data measured on 2000-Å-thick polysilicon layers are illustrated. These data were obtained from either cross-section TEM or UV surface-reflectance measurements.

Perhaps the biggest challenge for RTCVD systems designed to deposit polysilicon is film-thickness uniformity. Deposition requirements for ULSI manufacturers range from $\pm 1\%$ to $\pm 3\%$ film uniformity across a 200-mm wafer. The influence of the deposition system on uniformity depends on the process chemistry through the activation energy. If a process has zero activation energy, then there is no temperature dependence and perfect uniformity is possible. Assuming a 1.6-eV process for polysilicon deposition, the deposition uniformity as a function of the percentage of temperature uniformity across a wafer is plotted in Fig. 22 for three temperatures.[51a] An assumed $\pm 3\%$ film uniformity extrapolates to $\pm 0.2\%$ temperature uniformity.

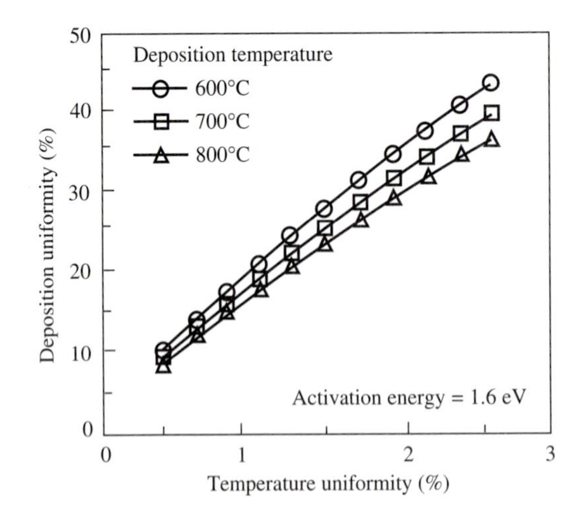

FIGURE 22
Polysilicon deposition-thickness uniformity versus the percentage of temperature uniformity across the wafer for three deposition temperatures and an activation energy of 1.6 eV. (*After Ozturk et al., Ref. 51a.* © 1991 IEEE)

This means that at 700°C the temperature would have to be held to within ±1.4°C across the wafer. Only a very few RTP tools have this level of control today, whereas this level of control is routine in conventional furnaces.

4.3.3 Oxidation

Rapid thermal oxidation (RTO) technology is well-suited to the growth of thin, high-quality dielectric layers in ULSI devices. An ideal RTO system is also an ideal RTP system, but with some additional requirements.[52] The system must be designed to handle reactive gases without introducing contamination or particles onto the wafers, and the gas-handling system must be vacuum-compatible and capable of switching quickly from one process ambient to another. In addition, the heating system must have a radiation spectrum that is compatible with the absorption properties of the wafer to allow uniform heating across the wafer.

The primary issues that differentiate RTO from conventional thermal oxidation are the more complex chamber design in RTO compared to the quartz tube in a furnace, the design of the radiation source, and the temperature measurement and calibration. From the point of view of oxide-growth kinetics, it is believed that oxide growth during RTO may be influenced by both thermally activated processes and a nonthermal, photon-induced component observed in the ultraviolet and visible range of the optical spectrum.[53] The basic photochemistry is described by the following reactions:[54]

$$O_2 + h\nu \rightarrow O + O \tag{4.13a}$$

$$O + O_2 + O_2 \rightarrow O_3 + O_2 \tag{4.13b}$$

$$O + O_3 \rightarrow 2O_2 \tag{4.13c}$$

$$O_3 + h\nu \rightarrow O + O_2 \tag{4.13d}$$

Enhanced Si oxidation has been observed at 450 to 550°C resulting from the presence of UV radiation. The cause of this is speculated to be the absorption of UV light by the oxygen gas, leading to photodissociation and the production of the species in Eq. (4.13). The highly reactive, monatomic O atoms produced are believed to create a parallel oxidation reaction that dominates at low temperatures.[54] However, it is not known whether this reaction is important at normal RTO temperatures above 900°C.

Thermally induced stresses in wafers processed in RTO systems have also been shown to affect thin-oxide growth[55] by up to 10%. It has been found that regions of the Si wafer in tensile stress yield thicker oxides than compressive regions. A tensile stress in Si stretches the Si atom-to-atom spacing, making oxidation easier. In conventional oxidation furnaces, thermal stresses are controlled by uniform heating and slow-temperature ramping.

Oxide growth comparisons

An example of a two-step RTO process is shown in Fig. 23.[56] The wafer is ramped up in O_2 to a preset temperature below the oxidation temperature, and then ramped again to the oxide growth temperature. A second, high-temperature, rapid thermal annealing (RTA) step is performed in N_2 to anneal and improve the electrical properties of the Si-SiO_2 interface. Figure 23 compares operation of the system under two temperature control methods: pyrometer temperature control and preset control of the heating source power. In conventional furnaces, the temperature is fixed and the wafers are inserted under N_2 ambient conditions. Switching the O_2 gas flow on and off determines the duration of oxidation.

RTO growth kinetics exhibit activation energies that differ from those measured in conventionally grown oxides. In the initial 20 seconds, the RTO growth is observed to be linear followed by nonlinear growth.[57] The duration of the linear region is found to depend on the type of heating lamp used (tungsten halogen vs. water-cooled arc), and the linear region showed activation energies of 1.44 eV and 1.71 eV, respectively, for the two lamps. This result compares to a linear-rate activation energy of 1.76 eV reproduced in conventional furnaces. Numerous other studies of linear-growth kinetics in RTO systems have been conducted, each producing different activation energies. The origin of these differences has not been modeled, and work in this direction would not be warranted now because of uncertainties about wafer temperatures in different RTO systems. As an example, discrepancies in RTO growth rates on (100) Si in dry O_2 at 950°C are shown in Fig. 24.[52] The data were taken from a number of published reports in which these specific oxidation conditions were used. For comparison, the analytical growth relationship of Massoud and Plummer,[58] using rate constants for conventional oxidation, is also plotted at 950°C and at 1000°C. Figure 24 shows that only one RTO growth experiment was consistent with the calculated 950°C curve. This result implies an uncertainty of up to 50°C in the RTO systems used.

Temperature control in closed-loop RTO systems is particularly difficult. Such systems use pyrometry as a temperature-sensing method, but this method is viable only if the effective emissivity of the wafer is known. Section 4.1.1 stated that many

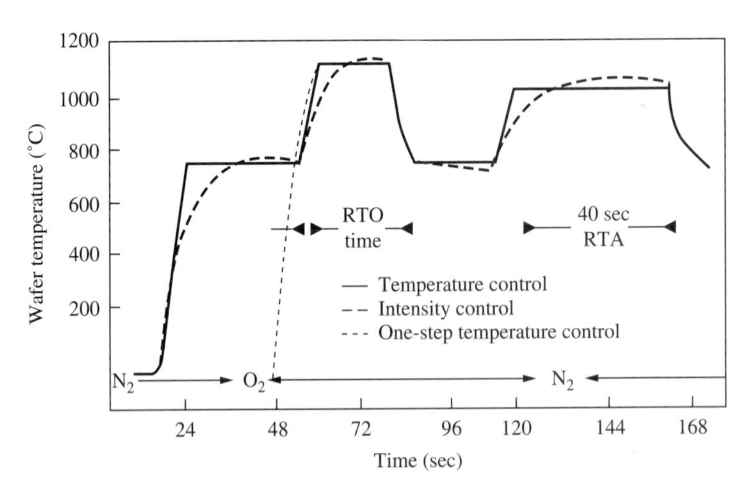

FIGURE 23
Example of a two-step RTO cycle showing wafer temperatures and ambients for RTO followed by RTA. (*After Nulman, Krusius, and Renteln, Ref. 56.*)

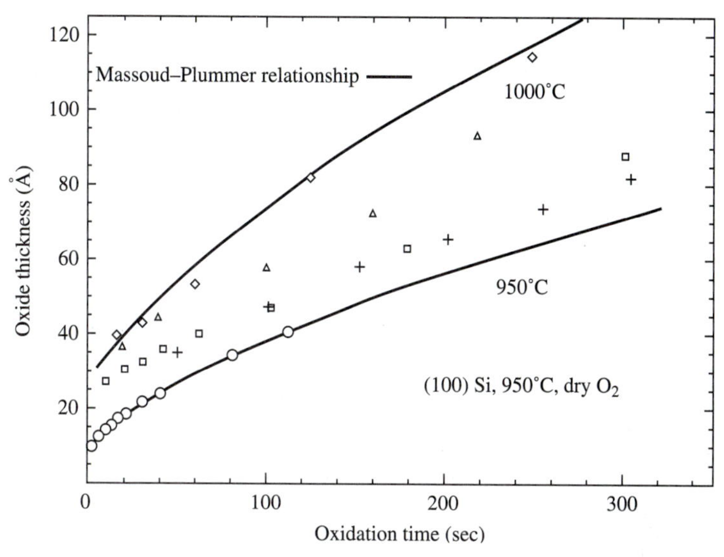

FIGURE 24
Discrepancy in RTO growth kinetics on (100) Si in dry O_2 at 950°C. Data symbols represent oxide growth in a number of different RTO systems. Solid lines are the analytical thermal-oxidation-growth curves calculated from the model of Massoud and Plummer.[58] *(After Massoud, Ref. 52.)*

factors influence emissivity, such as absorption in the ambient gas and surface film thicknesses. Wafer emissivity also depends on the presence of dielectric films which may have been deposited or grown on the reverse of the wafer during processing. The layers act as interference filters, leading to constructive or destructive interference of the emitted radiation in the wavelength range of the pyrometer. This causes a critical emissivity dependence on thickness, stacking sequence, and optical constants of these layers.[14] An example of this effect is shown in Fig. 25. RTO was performed on low-doped Si wafers using closed-loop pyrometric feedback. A six-layer stack of deposited films was put on the reverse side of the wafer in the following sequence:[14]

1. Oxide—4900 Å
2. Nitride—1600 Å
3. Oxide—variable
4. Polysilicon—2100 Å
5. Oxide—4900 Å
6. Nitride—250 Å

RTO was performed on the bare, front side for layer-three oxide thicknesses of 4000 to 8300 Å, and the resulting RTO oxid thicknesses are shown in Fig. 25. The calculated temperatures corresponding to the oxide thickness grown vary by 1081 ± 85°C because of the changes in layer-three oxide thickness. Significant improvements in wafer temperature control are possible when the RTO system is operated in a power-control mode in which lamp power, cooling air flow, and optical characteristics of

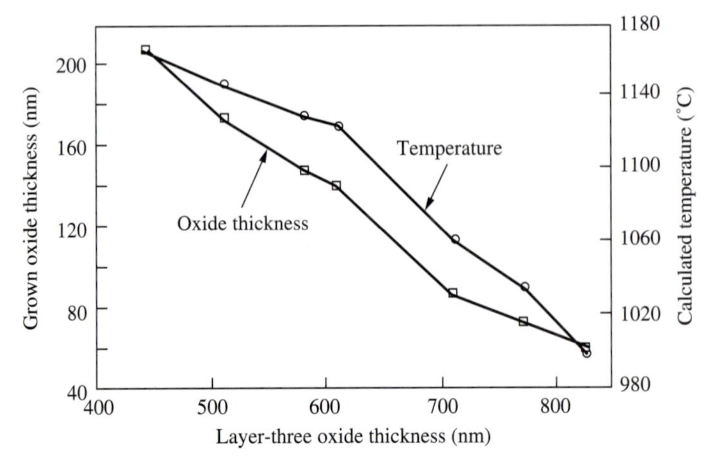

FIGURE 25
Rapid thermal oxidation results obtained with closed-loop pyrometric feedback on the front side of a wafer with a six-layer stack of deposited films on the reverse. The third deposited oxide layer thickness is the variable. Emissivity variations result in significant wafer-temperature changes and RTO film thickness. (*After Nakos, Ref. 14.*)

the system are controlled. Indeed, ±3.5°C temperature control was achieved in the same experiment described above under a power-control mode.

Properties of rapid, thermally processed oxides

Rapid thermal processing is used to grow dielectric films as well as to postprocess the films to impart the desired chemical or electrical properties. For example, in Fig. 23 we saw a two-step RTO process followed by an RTA process to improve the interface electrical properties. Other RTP processes used in thin-gate dielectric formation for MOS devices include rapid thermal nitridation (RTN), rapid thermal oxynitridation, and all the RTCVD processes previously discussed.

Oxides annealed in an RTA system generally possess electrical and dielectric properties that are comparable to or better than those of furnace oxides.[52] Parameters included in this statement are oxide fixed charge, interface traps, and charge-to-breakdown levels, Q_{BD}. Indeed, it has been shown that Q_{BD} increases from 20 C/cm^2 for conventional furnace oxides to 80 C/cm^2 for RTO-RTA oxides of 100 Å thickness.[59] The charge-to-breakdown is directly related to the trapping of electrons that are injected into the oxide. A higher Q_{BD} implies that RTO-RTA oxides have fewer trapping centers, resulting in better reliability. In studies of the annealing of interface traps by RTA, it was shown that the RTA temperature was dominant in reducing traps. Since RTA can be performed in the RTO chamber at higher temperatures than postfurnace annealing, the expected improvement in interface traps is observed.[60]

High-frequency, quasi-static capacitance-voltage measurements on oxides yield interface trap distributions and oxide-fixed charges at the Si-SiO_2 interface. A comparison of interface trap distributions in furnace oxides (FO) and RTO oxides is

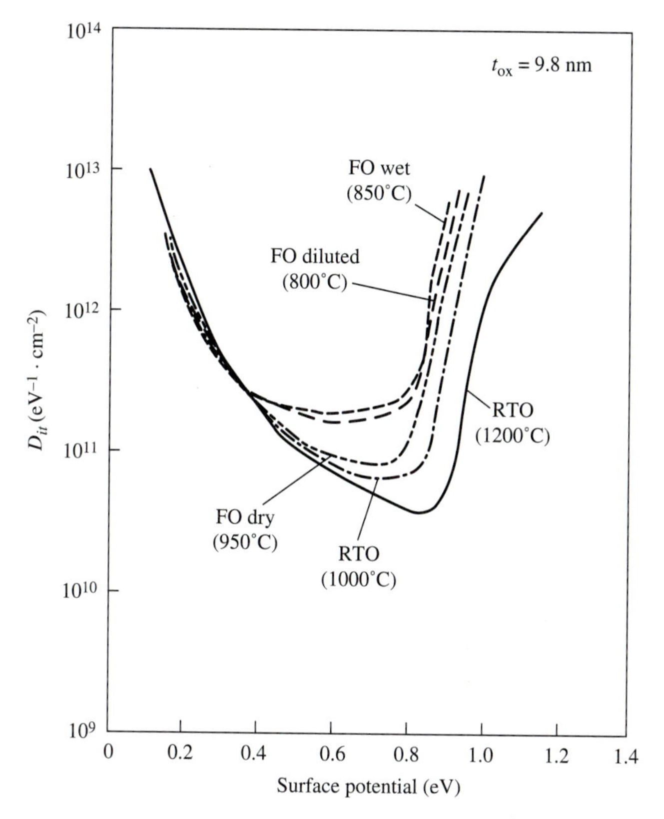

FIGURE 26
Comparison of interface-trap distributions in furnace oxides (FO) and RTO oxides. (*After Fukuda, Ref. 61.*)

shown in Fig. 26.[61] It should be pointed out that the results shown in Fig. 26 are not state-of-the-art, but do illustrate the possible advantages of RTO over furnace annealing. The growth temperatures and ambient conditions for FO growth are noted in the figure. All RTO processes were performed in dry O_2. The higher RTO temperature results in lower mid-gap trap levels, but the trap densities for both FO and RTO are similar at lower temperatures. Thus, the apparent advantage of RTO over FO may occur only because of the higher oxidation temperatures used in RTO processing, which cause a reduction in the amount of unoxidized, excess Si at the Si-SiO_2 interface and, thus, the fixed charge.

RTO oxides seem to exhibit improved dielectric breakdown properties over FO oxides. Time-zero dielectric breakdown (TZDB) testing involves applying a negative gate voltage to an MOS capacitor and recording the breakdown voltage when the measured current is 1 μA. The TZDB distributions of FO and RTO oxides of the same thickness are compared in Fig. 27. The FO samples were grown at 950°C in dry O_2 and the RTO samples were grown at 1000°C in dry O_2.[61] The TZDB data for the RTO samples are better in two respects. First, no low- or medium-field breakdowns (1 to 8 MV/cm) were seen; second, the maximum breakdown frequency occurs at higher fields than in the FO samples.[52, 61] Both of these observations are related to

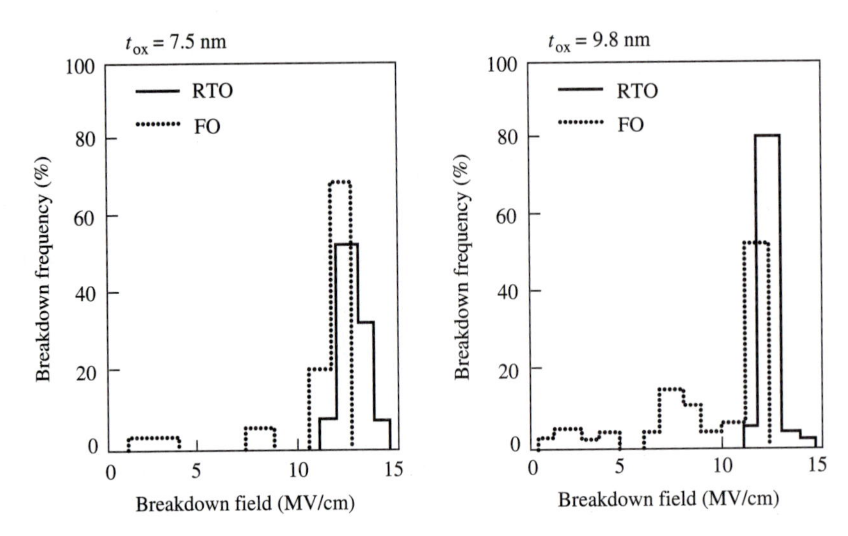

FIGURE 27
Time-zero dielectric breakdown (TZDB) histograms of FO and RTO MOS capacitors with 7.5- and 9.8-nm oxide thickness. (*After Fukuda, Ref. 61.*)

the lower charge-trapping properties of RTO oxides. It is believed that there is a reduction in strained Si–O–Si and Si dangling bonds at the Si-SiO_2 interface when the RTO oxide is grown at a higher temperature.

4.3.4 Rapid Thermal Annealing for Junction Formation and Defect Removal

It has been shown that RTA differs from furnace annealing in the way the wafers are heated (radiative versus convective and conductive) and in the rates at which wafers are heated and cooled (10 to 200°C/sec versus 5 to 50°C/min, respectively). Even though RTA rates are rapid and optical excitation of the semiconductor produces large numbers of electron-hole pairs, the relaxation of the excess free carriers is much shorter than the temperature ramp rates. Indeed, it has been shown theoretically that, under high-intensity light irradiation, nonradiative processes limit electron-hole recombination times to the 10^{-9} to 10^{-12}/sec range. For these values, it has been found that an excess carrier density will be too small to change the position of the Fermi level in Si at 1000°C.[62] Thus, carrier levels are considered to be at equilibrium, and processes such as impurity diffusion and defect annealing in the semiconductor are believed to be in thermal equilibrium during the RTA temperature cycle.

The application of RTA to producing pn junctions by annealing ion-implanted layers has been studied since 1983. However, the semiconductor industry has put RTA into limited production because only a few vendors of equipment have provided systems that meet the production requirements for reproducibility and uniformity, and these systems were not available until 1992. Noncritical processing steps with wide processing latitude, such as silicide formation, have found their place in production. However, there have been more fundamental reasons why RTA has not gained widespread acceptance in processes such as junction formation. Some of these are listed below:

High RTA temperatures produce slip lines in wafers caused by thermal stresses.
RTA-annealed, ion-implanted junctions show excess diffusion for a given thermal budget that is dominated by defect annealing rather than by thermally assisted diffusion.
The excess diffusion of dopants during RTA has not been modeled adequately to allow engineers to design devices and processes with confidence.
Localized heating on patterned wafers causes higher wafer temperature than on bare control wafers, making process control difficult.

Nevertheless, RTA has several important advantages over furnace systems:

Low thermal budget control for shallow junctions
High level of dopant activation and defect annealing

In the following two sections, the application of RTA to the annealing of ion implantation defects and to the diffusion of dopants for junction formation is discussed.

Defect annealing

Nuclear-stopping processes of ions implanted into Si are responsible for producing displacement damage. The types and amount of damage produced depend on implant species, energy, dose, wafer temperature and orientation, dose rate, and materials covering the substrate. The as-implanted defect morphologies then change during wafer annealing, depending on temperature, time, furnace ambient, and ramp-up/ramp-down rates. Many of the point-defects (vacancy and self-interstitial defects) are generated during the changes from as-implanted damage to stable or dissolved-damage structures as a result of annealing. The impact of the damage evolution during annealing increases as the lateral and vertical dimensions of Si devices decrease, because interactions between point defects and extended defects that form during annealing dominate dopant diffusion. Thus, it is important to have doping and annealing technologies that deal with these effects and produce defect-free junctions of the desired depth and controllability.

Depending on the dose and species, ion implantation can produce different types of damage, as shown schematically in Fig. 28. For example, when the implant dose is below a critical level required for the formation of extended defects such as dislocations (at a critical dose level of approximately 2×10^{14}ions/cm^2), the dominant damage species are isolated point defects or point-defect clusters. Figure 28*a* shows an example in which I-type (interstitial) defects are distributed throughout the depth of the implanted region and beyond.

Defects also arise whenever an amorphous layer is generated by a high-dose implant, as shown in Fig. 28*b*. V-type (vacancy) defects are observed forming near the Si surface. I-type defects occur outside the amorphous region and are probably interstitial clusters. Type II dislocation loops are located just beyond the amorphous/crystal interface in the heavily damaged, but still crystalline, substrate. They are also called end-of-range (EOR) damage.[63] The EOR defects may be significant to successful junction formation if they are present in the resulting space-charge

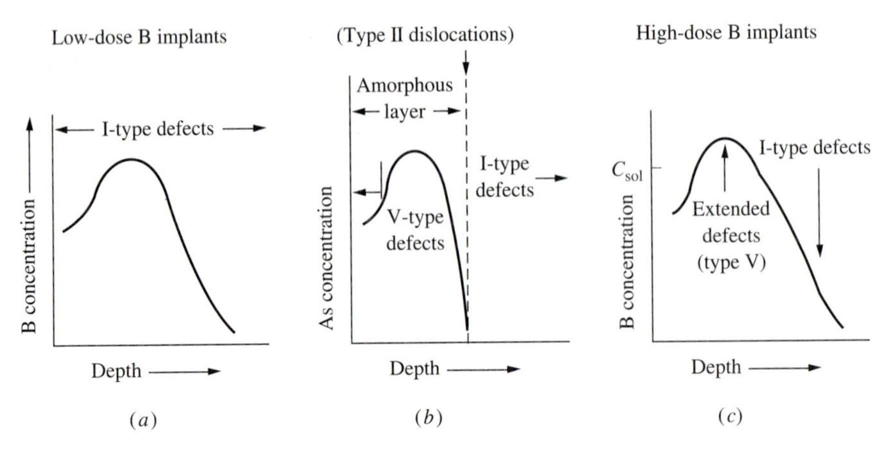

FIGURE 28
Ion-implantation defect production for three classes of implants, showing dopant concentrations versus depth. I-type defects are of the self-interstitial type and V-type defects are of the vacancy type. (*a*) I-type defects are distributed through a lightly implanted B layer. (*b*) An amorphous layer forms after high-dose implantation—V-type defects form in the Si surface, I-type defects form in the crystalline region beyond the amorphous layer, and dislocation loops form at the amorphous-crystalline interface. (*c*) Projected range dislocations are shown at the peak of a high-dose B implant, and I-type defects are shown in the Si bulk when the peak B concentration exceeds the B solubility limit, C_{sol}. (*After Fair, Ref. 66.*)

region, and their presence may affect the diffusion process occurring during RTA. Other types of defects related to the amorphous-layer regrowth may be avoided by choosing implantation and annealing conditions carefully, but the EOR defect is unavoidable once an amorphous layer is formed. Therefore, when the dose exceeds the amorphization threshold (1×10^{14} to 1×10^{15}ions/cm^2), implantation at room temperature of most Group III and V elements would result in extrinsic dislocation loops upon annealing.

The third type of implantation defect, shown in Fig. 28*c*, is the projected-range defect, which is caused by the formation of precipitates of the implanted species distributed near the projected range when solid solubility at the annealing temperature is exceeded. These defects are most noticeable in high-dose B implants where no amorphous layer is formed. The defects are in the form of extrinsic dislocation loops extending to the Si surface. It is difficult to remove these defects by annealing since they are very stable at temperatures up to 1050°C. During annealing by RTA or conventional furnace, self-interstitial defects are generated by the dissolving projected-range dislocations, causing enhanced diffusion for extended periods.

Annealing characteristics of EOR defects may be the most important process to understand for ULSI, since the preamorphization of the Si surface is performed before shallow junctions form, to prevent dopant channeling during low-energy implantation. In addition, high-dose implantations to reduce sheet resistance are likely to form surface amorphous layers when the implantation is done in a crystalline substrate. The location and annealing properties of the EOR defects are important factors in determining the implanted dopant diffusion and the quality of implanted junctions. The annealing rate and the inverse of the time required to completely remove the loops were found to have approximately 5 eV activation energy, indicating that the self-diffusion of Si atoms limits the annealing process.[63–65] This high activation energy is very significant and is greater than the activation energy for dopant diffusion, which is 3.5 to 4.2 eV. Thus, effective removal of the defects with minimum dopant diffusion would occur at high temperatures for short times, favoring RTA[66] over furnace annealing.

RTA of ion-implanted junctions

Any modern junction formation process must be capable of creating "good quality" pn junctions without excessive process complexity. In a "good quality" junction the final junction depth must be controllable and reproducible. These characteristics are important for manufacturing. Additionally, the pn junctions must have low reverse-bias leakage current and good (near ideal) forward characteristics. This requirement translates into a need for well-controlled, junction-region defect densities. Both the number and the location of any process-induced defects (i.e., EOR damage from implantation) are important. Diffused or implanted regions must have low sheet resistance. Thus, high carrier mobility and a high level of dopant activation are required. Finally, there are compatibility issues. An advanced junction process should be as compatible as possible with existing, well-established processing steps. A suitable contact technology must be available, either independent of the junction formation process or as a part of it. Thus, the designer is faced with numerous tradeoffs in what is described in Fig. 29 as the "junction processing circus."[65] Implant-damage annealing may increase dopant diffusivity and decrease dopant electrical activation, whereas increased activation increases diffusivity in additional ways, with solid solubility levels determining the rates of damage dissolution.

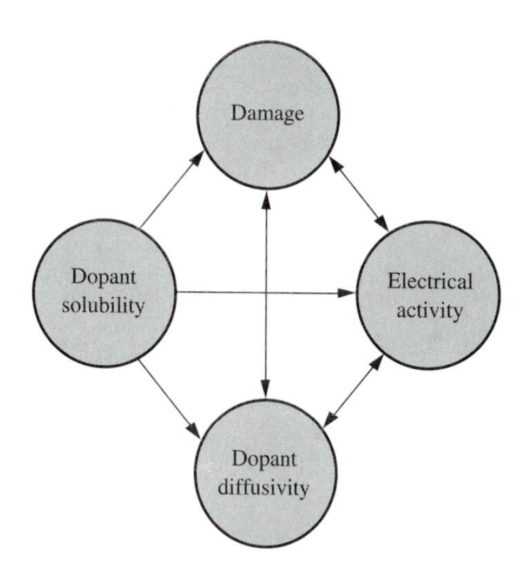

FIGURE 29
The junction processing circus, otherwise known as the implantation-diffusion interaction matrix, showing the interactions between damage, dopant solubility limits, dopant diffusivity, and electrical activity of the implanted dopants. (*After Jones, Ref. 65.*)

It has been widely observed that the diffusivity of dopants in Si may be greatly enhanced during implant-damage annealing because of the generation of point defects. The extent of the observed transient diffusion enhancement is believed to be related to which point defects are required by a dopant for diffusion (vacancies or self-interstitials) and which of the two types of point defects is generated. For example, B requires Si self-interstitials for diffusion, whereas As diffusion is predominantly vacancy-based.

The nuclear collisions between incoming, energetic ions and the Si lattice atoms produce atomic displacements and, subsequently, vacant lattice sites (vacancies). The displaced Si atoms become energetic self-interstitials that displace additional lattice atoms and produce cascades of Si self-interstitials. The distributions of vacancies and self-interstitials throughout the implanted region follow the general shape of the ion-implanted impurity distribution. An example of vacancy and interstitial distributions following an ion implantation is shown in Fig. 30. The distributions were determined using Boltzmann transport damage calculations and also show the net defect distribution, which is obtained by subtracting the vacancy concentrations from the self-interstitial concentrations.[67] It can be seen that vacancy-rich and interstitial-rich regions are predicted, and this result has been verified experimentally. However, the self-interstitials can move in Si at low temperatures and tend to clump together to form clusters during the implantation step, causing a distortion in the interstitial profile shown in Fig. 30. As soon as the Si wafer is heated, these interstitial clusters

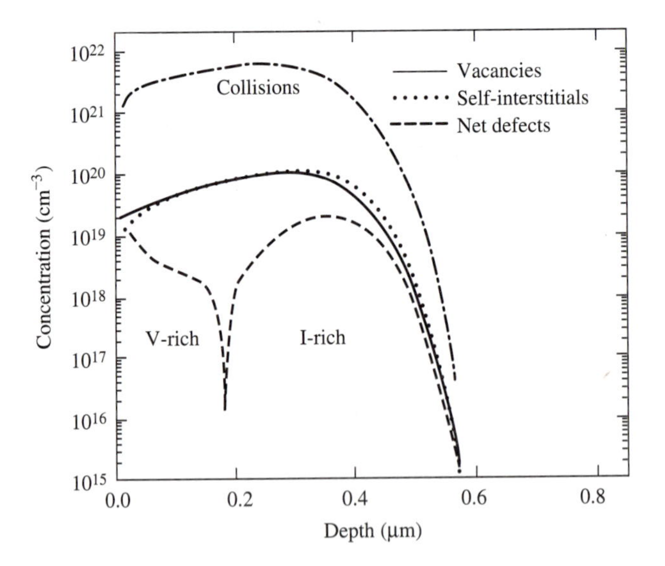

FIGURE 30
Ion implantation damage calculations, calculated with the Boltzmann transport equation, to determine production of vacancy and self-interstitial distributions. The total distribution of collisions, the resulting distributions of vacancies (V) and self-interstitials (I), and the net defects are shown. (*After Giles, Ref. 67.*)

begin to dissolve, producing an excess interstitial transient that can enhance the diffusion of dopants such as B that are transported in Si via an interstitialcy mechanism.

From Fig. 28*a,* low-dose implants introduce isolated interstitial-type cluster defects (the origin of which has been described). *Low-dose,* in this case, refers to implants below 2×10^{14}ions/cm^2, which is the threshold dose for creating extended dislocations in Si. Such low-dose implants might be used in setting the base doping in a bipolar transistor process or in doping the channel of an MOS transistor. The number of such isolated defects, $N(E)$, depends on the energy, E, and dose of the implant through the production of displaced atoms for each incident ion:

$$N(E) = \frac{E_n}{2E_{oo}} \tag{4.14}$$

where E_n is the energy that goes into nuclear collisions and E_{oo} is the energy required to displace a Si-lattice atom. Thus, if 25 eV is required to displace a Si-lattice atom, a single 50 keV ion would produce 1000 displaced atoms along its trajectory in the Si. Some of these displaced atoms coalesce in stable clusters of various sizes and remain until the Si is heated, whereupon they dissolve.

The energy effect on the time-averaged pn-junction displacement can be seen for low-dose phosphorus (P) implants in Fig. 31.[66] The Δx_j data represent the difference between the initial, as-implanted junction depth at a level of 1×10^{17}atoms/cm^3, and the final depth of the profile after annealing. Data are shown for both RTA and furnace anneals. The diffusion coefficient of P, D_p, is directly proportional to the number of point defects produced, $N(E)$, and the junction displacement is proportional to $D_p^{1/2}$. This accounts for the $E^{1/2}$ dependence of the curves in Fig. 31. The magnitude of the displacement is defined at a P concentration of 1×10^{17} atoms/cm^3. Similar results are observed for B implants. Both RTA- and furnace-annealed samples show damage-assisted diffusion with the same $E^{1/2}$ dependence. The difference in slopes between the RTA curve and the furnace annealing curve is due to the differences in temperatures of the anneals and associated damage annealing times,

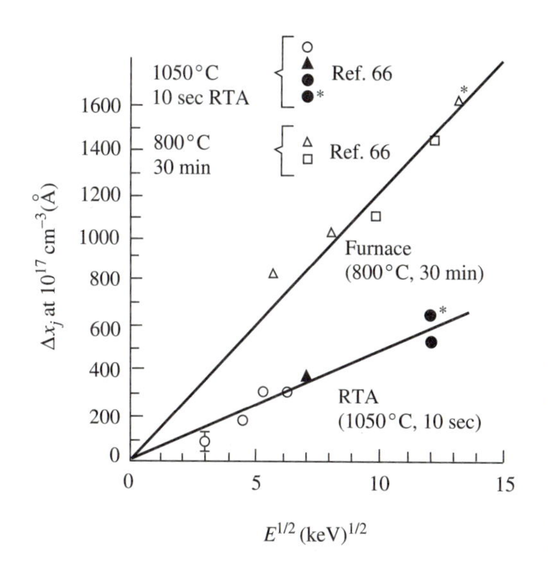

FIGURE 31
Transient junction displacement during RTA and furnace annealing for low-dose, phosphorus implants versus ion-implantation energy. Both annealing methods show an $E^{1/2}$ dependence on junction displacement. (*After Fair, Ref. 66.*)

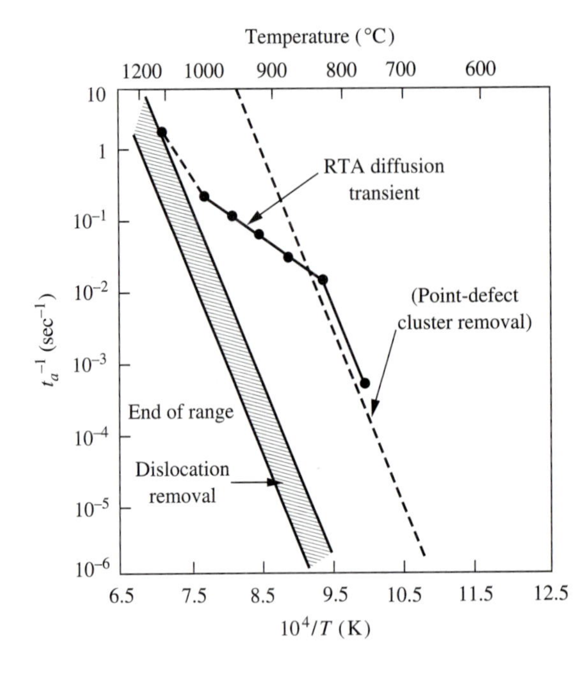

FIGURE 32
Measured boron diffusion transients during RTA and damage annealing times in ion-implanted Si. The annealing of both extended dislocations and clusters shows a 5-eV activation energy, and the diffusion transient data make a transition from domination by point-defect generation from extended dislocations at high temperatures, to domination by point-defect cluster dissolution at low temperatures. (*After Fair, Ref. 71.*)

t_a. This implies that the $D_p \cdot t_a$ product is lower for RTA even though the RTA was performed 250°C higher in temperature.

To understand this anomalous result, note that during normal, thermally assisted diffusion of low-concentration P, D_p is given by the following expression that has been obtained from experimental data:[68]

$$D_p = 3.85 \exp(-3.66 \text{ eV}/kT) \text{ (cm}^2\text{/sec)} \tag{4.15}$$

During damage-assisted P diffusion, the activation energy is reduced from 3.66 eV to approximately 2.2 eV as measured in transient-diffusion experiments:[69]

$$D_p(\text{enhanced}) = 0.007 \exp(-2.2 \text{ eV}/kT) \tag{4.16}$$

It is believed that the 2.2 eV may be the migration energy of a phosphorus self-interstitial pair. In the presence of excess self-interstitials caused by ion-implantation damage annealing, the diffusion-activation energy is reduced by approximately the amount of energy needed for the Si crystal to thermally generate equilibrium levels of self-interstitials. Estimates of this energy vary from 1.5 to 2.5 eV.

The times for damage annealing, t_a, can be measured by studying the dissolution of damage for various annealing time increments with an electron microscope. It is also believed that t_a is related to the time constant for transient dopant diffusion. The time constants for transient diffusion of low-dose B implants (1×10^{14} ions/cm^2, 150 keV) were measured and fit to an exponential function associated with the decay of excess point defects with annealing time:[70]

$$D(t) = D_i + D_o \exp(-t/\tau_o) \tag{4.17}$$

where D_i is the intrinsic B diffusivity, τ_o is the decay time of generated point defects, and D_o is an empirical prefactor for the transient diffusion coefficient.

Decay-time data are plotted in Fig. 32 along with the dissolution rate curves for end-of-range dislocations and point-defect clusters.[71] Other diffusion transient data are also shown for low-dose B implants annealed at temperatures as low as 750°C and as high as 1100°C. These results show that short-time anneals below 1050°C start to be dominated by relatively rapid dissolution of point-defect clusters, with good agreement between diffusion-coefficient time constants and damage-annealing times. At higher temperatures, enhanced diffusion is assisted by the annealing of extended dislocations. The t_o data show how the transition occurs between the dominance in dopant diffusion of the two different sources of point defects as a function of temperature.

Other data for the enhanced diffusion time of low-dose B implants versus annealing temperature are shown in Fig. 33.[66] If we arbitrarily take the data from Ref. 72 as the most representative, then t_a is proportional to exp (3.7 eV/kT), and now the $D_p \cdot t_a$ product becomes

$$D_p(\text{enhanced}) \cdot t_a \propto \exp\,(1.5\ \text{eV}/kT) \tag{4.18}$$

The result in Eq. (4.18) can be used to explain the differences in the slopes of the RTA and furnace diffusion junction-displacement curves in Fig. 31. Higher anneal

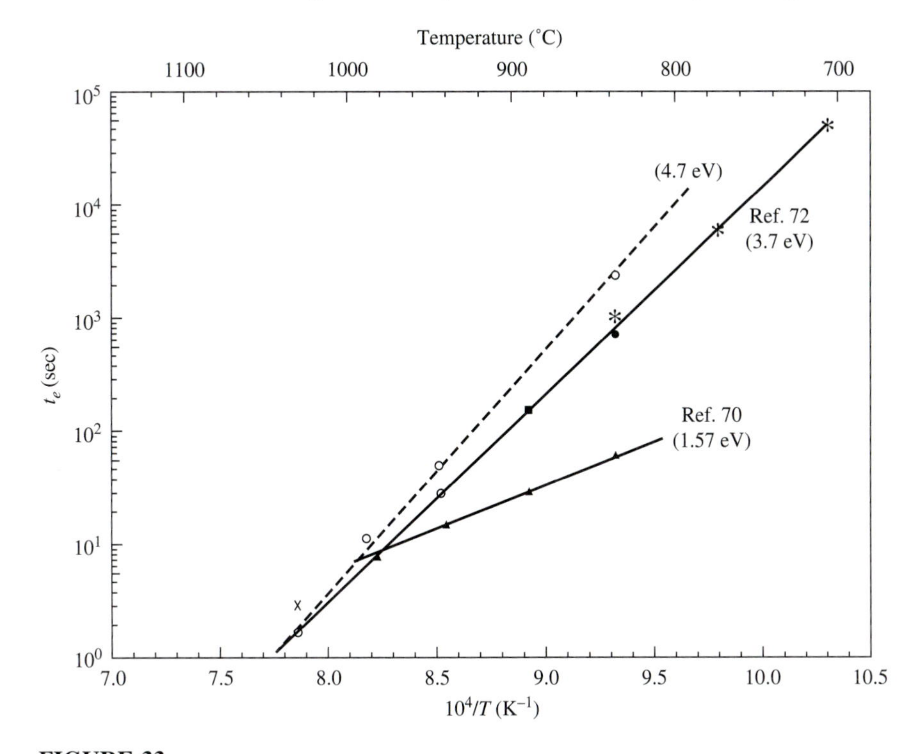

FIGURE 33

Transient-enhanced diffusion time, t_e, for low-dose boron implants versus the inverse annealing temperature showing the measured results of different researchers. (*After Miyake, Aoyama, and Kurchi, Ref. 70, and Solmi, Baruffaldi, and Canteri, Ref. 72.*) The data symbols are identified in Ref. 66; their disparity indicates uncertainties in the annealing temperatures and detection of the transient durations. (*After Fair, Ref. 66.*)

temperatures result in higher dopant diffusion (Eq. 4.16), but they also resultin correspondingly smaller transient time constants, with the latter dominating the product of the terms in Eq. (4.18). The result is less diffusion for the damage annealing times associated with high-temperature RTA than for the low-temperature furnace anneals.

The formation of shallow emitters in bipolar transistors or source-drains in MOS transistors requires high-dose implants in the range of 5×10^{14} to 1×10^{16}atoms/cm^2, which are sufficient to create extended defects. Annealing conditions for meeting the requirements of shallow, highly doped junctions must, therefore, include removing extended defects from junction space-charge regions as well as controlling junction depths. The implantation damage produced for amorphizing implants and high-dose B implants is shown in Fig. 28*b* and 28*c*, respectively. Therefore, as the implant dose increases, more defects are created that can contribute to additional damage-assisted diffusion for longer times and, subsequently, diminished control of junction depths. The onset of the longer diffusion transient time with increasing B dose is illustrated in Fig. 34.[73] At 1×10^{14} ions/cm^2, the transient saturates in less than 10 seconds at 1000°C. As the dose reaches 5×10^{14} ions/cm^2, lattice damage increases and the transient becomes 20 seconds long. Finally, above a dose of 2×10^{15} ions/cm^2, the diffusion transient is over 40 seconds. In Fig. 34*d*, the B concentration in the peak of the implant has exceeded the solid solubility level of B in Si at 1000°C, and projected range dislocations (type V) have formed, as illustrated in Fig. 28*c*. High-temperature RTA is better suited to remove these defects than low-temperature furnace anneals, since at 950°C these types of defects can pump out Si self-interstitials for hours, causing continued enhanced B diffusion.[73, 74, 75]

If the implant dose is sufficient to transform the crystalline Si into the amorphous phase, then different point-defect generation processes occur when the wafer is annealed. During heating, the amorphous region will rapidly regrow to single-crystal Si at temperatures above 500°C by epitaxially aligning with the underlying crystalline substrate. The recrystallized Si is shown schematically in Region A of Fig. 35. However, damage produced near the original amorphous/crystalline Si interface evolves into end-of-range (EOR) dislocation loops by absorbing self-interstitials that are produced by the dissolution of interstitial clusters in Region B of Fig. 35. The clusters are formed during ion implantation. The annealing of the clusters produces an excess of self-interstitials in Region B, but not in Region A, since the growing EOR loops absorb the interstitials, thus screening them out of A. Once the clusters have dissolved (times of one sec at 1000°C), then the loops begin to dissolve into the Si bulk, producing a second self-interstitial transient that will remain as long as the loops are present (times of seconds to hours at 1000°C, depending on the distance from the Si surface). The resulting distribution of self-interstitials is shown in Fig. 35.

An example of implant-damage-assisted diffusion resulting from BF_2 implants that have formed amorphous layers in the first 500 Å of the surface is shown in Fig. 36.[76] RTA was performed first to show that diffusion saturated within 15 seconds at 900°C. However, when this same sample was subsequently placed in a furnace and annealed at 850°C for 1 hr, substantial additional diffusion occurred. This result highlights the compounding effect of damage-assisted diffusion from the annealing of point-defect clusters (short transient) and EOR dislocations produced near the amorphous/crystalline interface (long transient). If the EOR dislocations are more than a few hundred angstroms from the Si surface, then it becomes difficult to remove

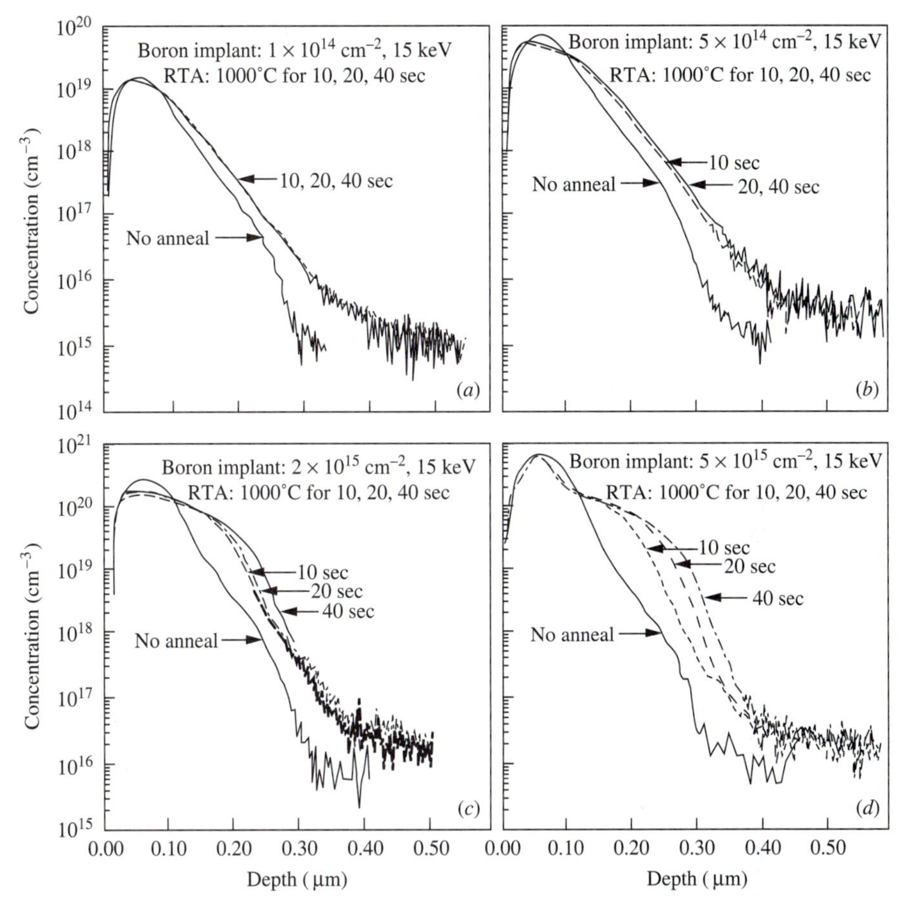

FIGURE 34
Boron secondary-ion mass spectroscopy (SIMS) profiles before and after RTA at 1000°C for 10, 20, and 40 sec at 15 keV. (*a*) 1×10^{14} cm^{-2}; (*b*) 5×10^{14} cm^{-2}; (*c*) 2×10^{15} cm^{-2}; (*d*) 5×10^{15} cm^{-2}. (*After Kim, Konoshita, and Kwong, Ref. 73.*)

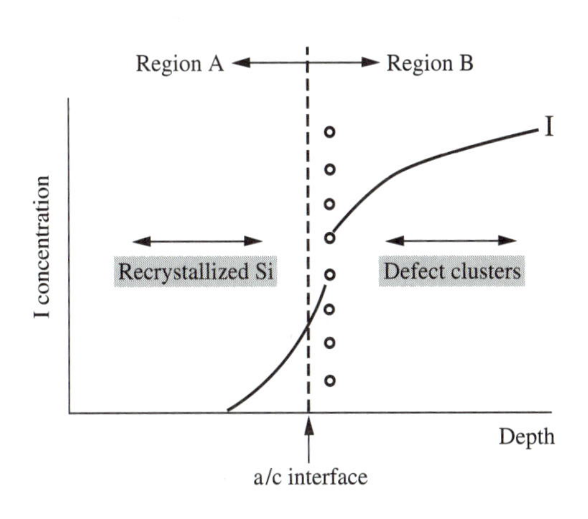

FIGURE 35
Schematic drawing of the annealing of an amorphizing implant with the resulting production of loop defects and self-interstitials. The amorphous layer in Region A rapidly regrows to single-crystal Si upon heating, and interstitial clusters in Region B dissolve, producing excess interstitials. Loops at the amorphous/crystalline interface grow in response, but screen the interstitials out of Region A by absorption. Finally, the loops begin to dissolve and create excess interstitials in Region B.

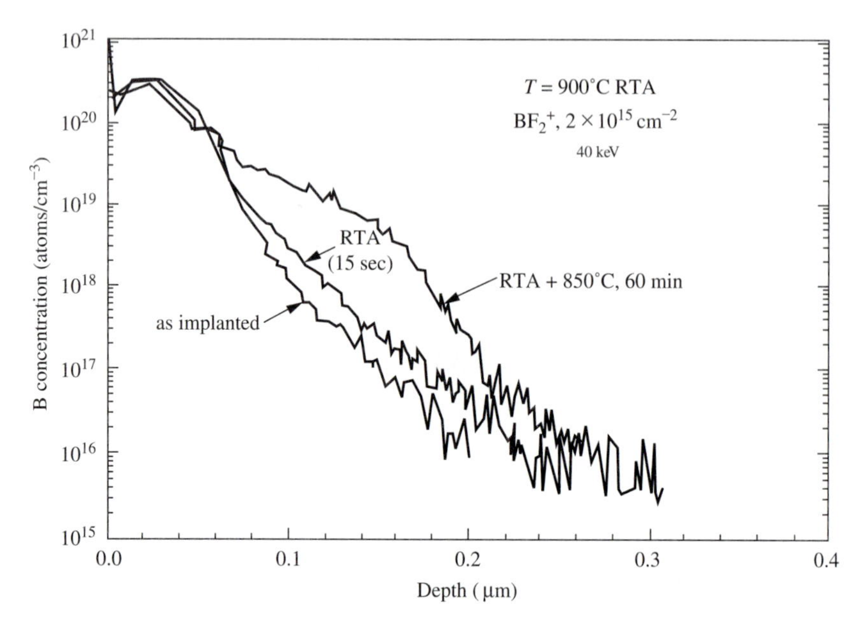

FIGURE 36
SIMS boron profiles for 2×10^{15} cm^{-2} BF_2 implants in (100) Si at 40 keV, annealed by RTA and RTA plus furnace annealing. (*After Kim, Massoud, and Fair, Ref. 76.*)

them within the time period allowed for junction anneal. For illustration, the first and second diffusion transients are identified in Fig. 37[77] for P with deep EOR damage. Annealing was performed at 900°C in a furnace. The first transient-enhanced diffusion (TED) is completed within 30 sec, whereas the second TED continues for hours while the EOR damage anneals. At higher temperatures and with shallower implants, the first and second TEDs can be made very short so that they become a minor part of the total junction diffusion time. This requires RTA and very low-energy implants.

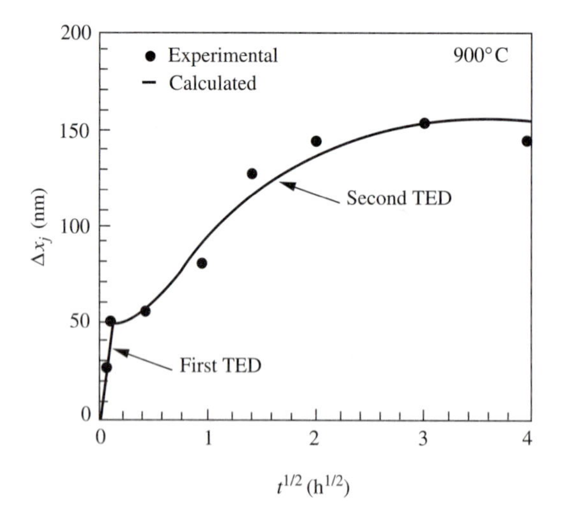

FIGURE 37
Junction depth displacement as a function of the square root of annealing time at 900°C in P-doped specimens, showing the time ranges for the first and second transient-enhanced diffusions (TED). The continuous line is calculated. (*After Solmi, Ref. 77.*)

The conditions for creating defect-free, minimum depth p^+ junctions using RTA for 10 seconds are shown in Fig. 38.[19] Simulations were performed to determine the annealing temperatures necessary to remove implant damage at each energy and for each implant species (B or BF_2 in crystalline and Ge-preamorphized Si). For all combinations of implants, energies below 2 keV and RTA are required to form the 500- to 800-Å-deep junctions for ULSI. These results also show that, although preamorphization is important for reducing junction depths of B and BF_2 implants above 5 keV, little advantage exists at lower energies except that higher dopant activation is achieved for preamorphized implants.

The advantages of very shallow implants combined with RTA are

Lowered excess point-defect levels because of less implant damage and the proximity of the Si surface, which *sinks* point defects
Lowered point-defect transient time because of rapid damage annihilation
Junction depth control by thermal diffusion, not by damage annealing

Dopant activation

The electrical activation of implanted impurities by RTA or furnace heating is an important issue, especially for shallow junctions created by low-thermal-budget processing. Activation is related to the fraction of impurity atoms that are on substi-

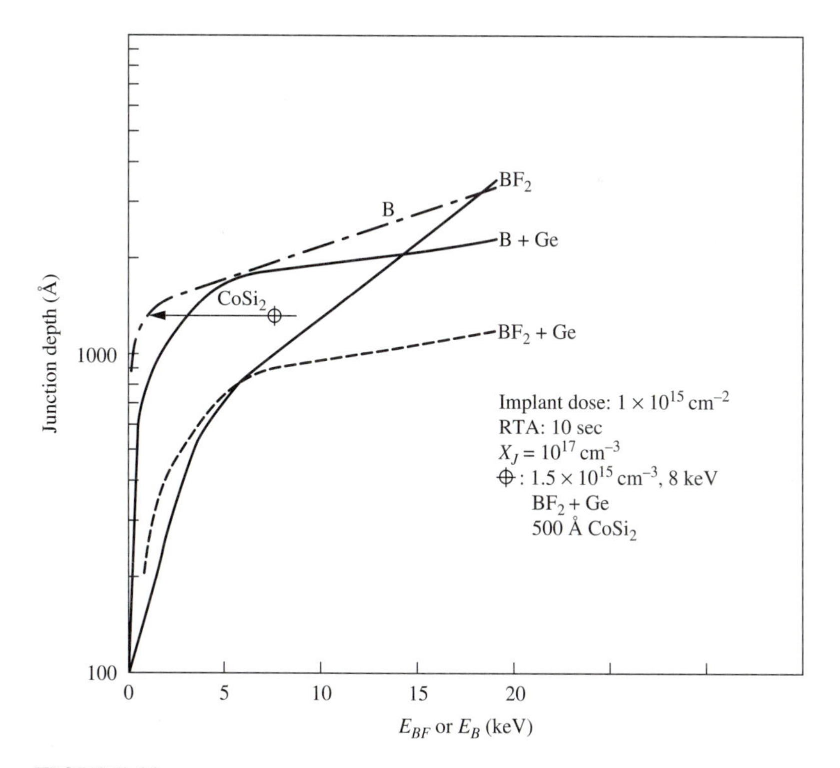

FIGURE 38
Minimum depth, damage-free p^+ junctions versus implant energy. The curves are simulated with the annealing temperatures necessary to remove damage from the junction region in 10 sec. (*After Fair, Ref. 19.* © 1990 IEEE)

tutional lattice sites acting as donors or acceptors. Thus, free carrier concentrations as well as carrier mobility are indicators of activation as manifested in the measured sheet resistance of a junction layer.

A key question is, How can a high activation be achieved without loss of junction depth control? It is desirable to achieve 100% electrical activation up to the solubility limit of the dopant. High electrical activity is needed to ensure low contact resistances to junctions, since contact resistance is sensitive to active surface concentrations. In addition, high activation allows for the lowest possible sheet resistances of the implanted and diffused layers. The attainment of high carrier mobility is also important and requires that residual defects be minimized in the doped layer, i.e., the junction processing circus shown in Fig. 29.

The activation of B in high-dose implants into Si proceeds slowly at low annealing temperatures of 800 to 850°C. An example of activation at 800°C is shown in Fig. 39.[77] The total as-implanted B profile and the free-carrier profiles result from annealing in a furnace for 5 min and 2 hr. The carrier profiles have spread into the Si by diffusion. Although the hole concentration increases with time, it largely remains below the solid solubility level of 3.5×10^{19} atoms/cm^3 at 800°C. The inactive B above this level is either in the form of B precipitates or interstitial B.

By contrast, improved activation of B is possible by performing RTA at or above 900°C. An example is shown in Fig. 40.[77] After 5-sec and 15-sec RTA at 1000°C, a maximum electrical activity is achieved for a 2×10^{15} cm^{-2} implant that is 2 to 3 times greater than the equilibrium solubility. Similar results are seen at 900°C.[78] These results demonstrate that RTA, with its high-temperature and short-time capability, is a preferred technology in the trade-off between activation and junction-depth.

The basic mechanism of activation involves the transition of B atoms from interstitial positions in the Si lattice to substitutional positions. Since this process involves point defects, the activation energy is approximately 5 eV (diffusion energy for Si atoms), which makes it highly temperature-dependent. In addition, the damage

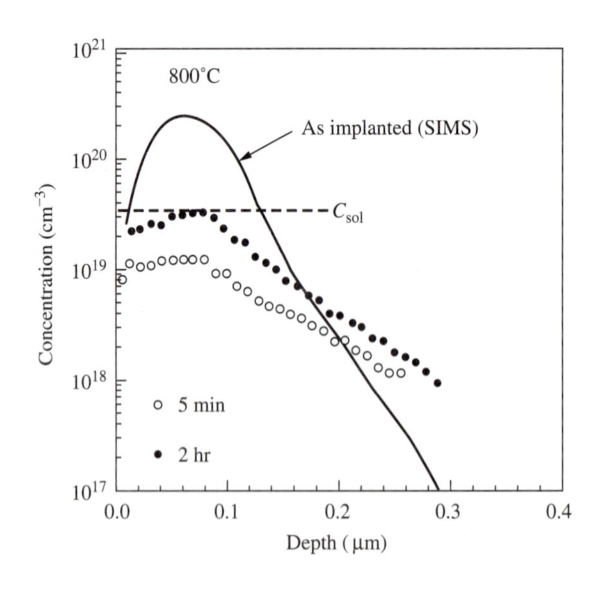

FIGURE 39
Carrier concentration profiles obtained in Si that was B-implanted at 20 keV with a dose of 2×10^{15} cm^{-2} and then annealed at 800°C twice. The as-implanted profile is total B obtained by SIMS. Hole concentration increases with time but remains below the solubility limit. (*After Solmi, Ref. 77.*)

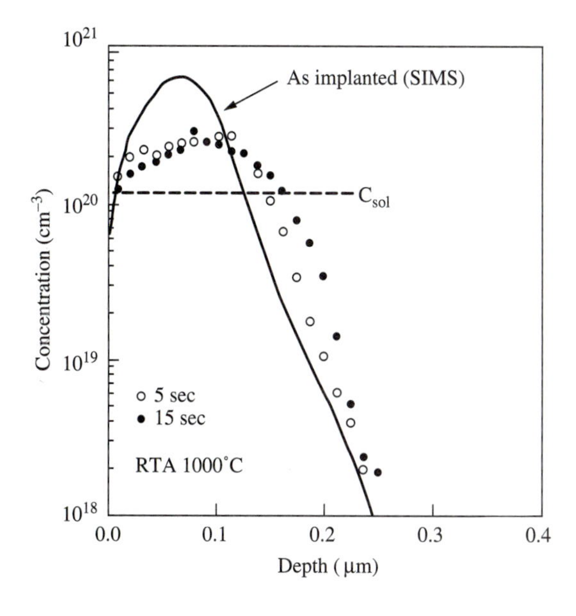

FIGURE 40
Carrier concentration profiles obtained in samples implanted at 30 keV with a dose of 1×10^{16} cm^{-2} after RTA for 5 and 15 sec. The as-implanted B profile was measured by SIMS. Hole concentrations over twice the equilibrium solubility limit are obtained. (*After Solmi, Ref. 77.*)

produced by the implantation is known to retard B activation,[78] and damage removal also proceeds with a 5-eV energy (see Fig. 32). Consequently, high temperature favors damage removal and activation. On the other hand, B precipitation is a 2.1-eV process[77] and will prevail over activation at lower temperatures, whereas the opposite is true at high temperatures.

The competition between activation and precipitation is diagrammed in Fig. 41, where peak free-carrier concentrations are plotted against annealing time.[77] The dashed curve shows the decrease in free carriers resulting from precipitation, and the solid curve shows the increase in carriers resulting from activation. Figure 41*a* shows that, at high temperature, activation can occur quickly up to the limit allowed by precipitation, which exceeds the solubility limit for short times. Figure 41*b* shows that, at low temperature, by the time the activated carrier concentration reaches its highest level, all the B above solid solubility is precipitated.

A special case exists for low-energy B or BF_2 implanted into preamorphized Si. The fraction of activated B as a function of annealing temperature depends on the depth of the amorphous layer relative to the implant depth. Two effects have been observed if the amorphous layer is shallower than the B implant:[79]

Reduced B activation occurs if EOR loops exist in the implanted region. A decrease in electrically active B occurs as the dislocation loops coalesce from defect clusters at moderate annealing temperatures (600 to 800°C).

Decreasing the preamorphization depth decreases the fraction of activated dopant, because fewer B atoms reside within the regrown layer.

Over the annealing range of 500 to 900°C, preamorphization produces higher levels of activation for B implants than implants into crystalline Si.[80] This is caused by the incorporation of B atoms onto substitutional sites during the process of solid phase epitaxial regrowth of the amorphous layer, and this process begins at temperatures as low as 550°C.

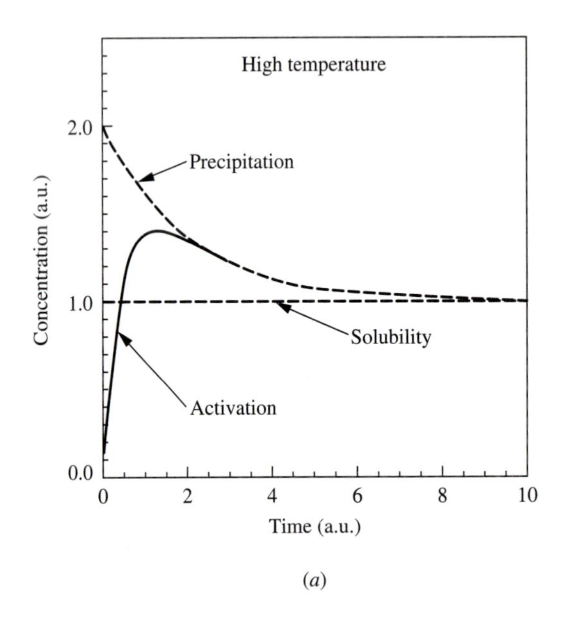

(*a*)

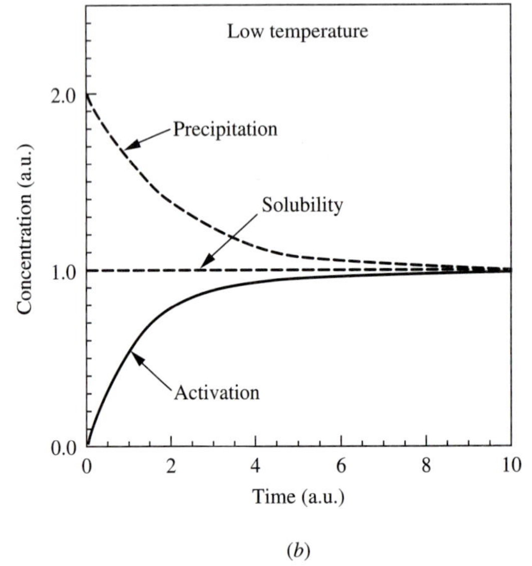

(*b*)

FIGURE 41
Diagrams of the peak free-carrier concentrations allowed by precipitation of B and by activation are plotted versus time in arbitrary units (a.u.) for (*a*) high-temperature annealing and (*b*) low-temperature annealing in samples implanted with concentrations above solid solubility. (*After Solmi, Ref. 77.*)

Activation limits of P-doped Si have been defined in terms of a solubility of electrically active P. Studies have shown that the electrically inactive P is in the form of phosphide precipitates.[81] However, it has also been shown that 100% activation of P implants is possible using laser annealing.[82] Free-carrier concentrations of up to 5×10^{21} cm^{-3} were obtained in laser-annealed layers free of structural imperfections. All of the P was on substitutional lattice sites, since the high wafer temperature produced by the laser caused rapid activation. Subsequent thermal anneals reduced the carrier densities to equilibrium values dependent only on temperature.

For the annealing time associated with high-temperature RTA, the equilibrium, active P concentrations are achieved[83] rapidly for diffusions from spin-on phosphosilicate glasses. Activation of P-implanted layers at lower temperatures is complicated by the presence of lattice defects. In general, the following observations have been made:

Implanted P in very-low-damage regions can be activated by annealing at temperatures as low as 400°C.

For damage densities of about 20%, carrier trapping centers may be formed to compensate the activation. Temperatures above 500°C are needed to achieve electrical activity.

For damage densities up to 75%, amorphous clusters are formed by ion-implantation overlap, producing larger damage regions. Decomposition of these regions during annealing may produce vacancies that enhance activation of P atoms onto substitutional sites.

For fully amorphized regions (100% damage density), movement of the amorphous/crystalline (a/c) interface during epitaxial regrowth incorporates P atoms into lattice sites. However, the activation of P atoms deeper than the a/c interface depends on the annealing temperature.

The activation of As is similar to that of P. However, controversy still remains regarding the form of electrically inactive As. It has been proposed that complex point defects exist in thermal equilibrium that result in electrically inactive As species. The effect of low-thermal-budget anneals and furnace ramps on the electrical activation of As was recently studied.[84] The equilibrium between active and inactive As can generally be described by a rate equation such as

$$\frac{\partial C_{\text{active}}}{\partial t} = K_D F(C_{\text{inactive}}, C_{\text{chemical}}) - K_c R(C_{\text{active}}, C_{\text{chemical}}) \tag{4.19}$$

where K_D is a declustering or deprecipitation coefficient, K_c is a clustering or precipitation coefficient, and F and R are functions of the concentrations indicated and depend on the model chosen. Computer simulation of activation or deactivation requires that Eq. (4.19) be solved simultaneously with the diffusion equation for As,[84] given as follows:

$$\frac{\partial C_{\text{chemical}}}{\partial t} = \frac{\partial}{\partial x} D \frac{\partial C_{\text{active}}}{\partial x} + Z\mu C_{\text{active}} \frac{\partial \Psi}{\partial x} \tag{4.20}$$

where D and μ are the mobile As diffusivity and mobility, respectively, Z is the charge state, and Ψ is the crystal potential. The second term on the right-hand side of Eq. (4.20) is the electric field dependence.

Depending on the initial conditions chosen, the solution to Eq. (4.20) will yield different solutions for active As concentration versus anneal time. The activation model of Tsai et al.[85] shows that, for high As doping, less than 5 minutes are required to achieve equilibrium at 1000°C. However, at 900°C, the time approaches 60 minutes. Thus, high-temperature RTA is favored for As as well as for B and P. During most common RTA cycles, the activated As concentration for high-dose

As implants is in transition from the initial value after epitaxial regrowth to a final equilibrium value. This also means that the activation depends on the ramp-up and ramp-down rates of the RTA cycle.

4.3.5 Metallization

The application of RTCVD as an alternative to LPCVD or PECVD of metal films has received very limited attention. On the other hand, rapid thermal processing has been widely applied to forming metal silicides for contacts to junctions and gate structures. RTA has been found to be superior to conventional furnace techniques for silicides because of the limited thermal stability of many silicide materials and because of their extreme sensitivity to trace impurities in the annealing ambient.[86] Thus, RTA is used more for control in heating and ambient than for deposition. Current applications include Ti, Co, W, Ta, and Pt silicides and TiN formation and annealing for local interconnections on the chip.

RTA of silicides

The early development of self-aligned silicide technology was performed in furnace systems that required a significant amount of attention to avoiding oxygen contamination. The mechanics of oxygen control basically rendered the approach impractical. Two features of RTP enabled silicide technology to reach production:

Atmosphere control
Short, high-temperature annealing, which allows the nucleation and growth of the desired high-conductivity silicide phase

In addition, the reaction of thin metal films with Si to form silicides proceeds rapidly at moderate temperatures and thus provides a self-limiting, uniform, thin-film process that is relatively independent of temperature variations inherent in RTP technology. Indeed, silicide processes are tolerant of thermal gradients and temperature uncertainties that occur in the wafers. The temperature dependencies of the resulting sheet resistances of silicide films are independent of the annealing system.[86]

One-and two-step processes used to form self-aligned silicides in RTP systems are shown in Fig. 42.[86] The metal reacts only where it comes in contact with Si, so after the first annealing step the unreacted metal is etched away selectively. One early difficulty with using silicides in patterned MOS devices was the lateral migration of Si from the substrate through the formed silicide to the unreacted metal, where it then formed additional silicide outside the patterned window. When the selective metal etch was applied, there would remain a silicide layer that stretched beyond the contact window in the form of an electrical short to the surrounding features of the device, i.e., the gate. This lateral overgrowth is minimized by the anneal-etch-anneal process shown in Fig. 42*b*. RTP enables the use of the first low-temperature, short-time step to form only a modest amount of Si/metal interdiffusion. The etch step then removes the excess metal outside the patterned contact windows, and a subsequent high-temperature step forms the high-conductivity silicide phase. This approach is unique to RTA, since it requires short annealing times.

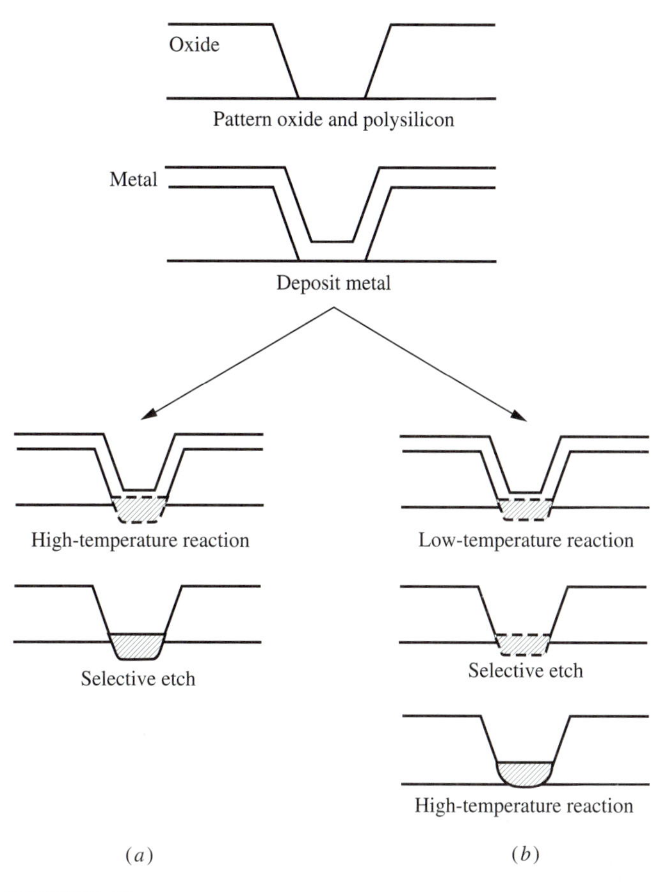

FIGURE 42
Thermal reaction of a metal thin film with Si to form a self-aligned silicide: (*a*) single-step anneal and (*b*) two-step anneal. (*After Osburn, Ref. 86.*)

The process windows giving the acceptable range of initial RTA temperatures and times needed to avoid lateral bridging in MOSFETs are shown in Fig. 43*a* for $TiSi_2$[87] and Fig. 43*b* for $CoSi_2$.[88] Those processing conditions within the acceptable window avoid lateral bridging of the junction and the gate in MOSFETs that have oxide spacers separating the poly-Si gate and the junction. Figure 43*b* shows the data on spacer-bridging yield for $CoSi_2$ formed from 38 nm of cobalt as a function of the initial 30-sec RTA temperature. The graphs demonstrate that improved yield is obtained for lower initial RTA temperatures, with 600°C being an upper limit.

Silicide growth kinetics—RTA versus furnace

The reactions of metals with Si are either diffusion-controlled or limited by nucleation of the phase that is undergoing growth. Diffusion-controlled kinetics are observed in most silicides, and the growth rate increases with the square root of time.[86] For example, the silicide growth-rate parameter of $TiSi_2$ measured in terms of the

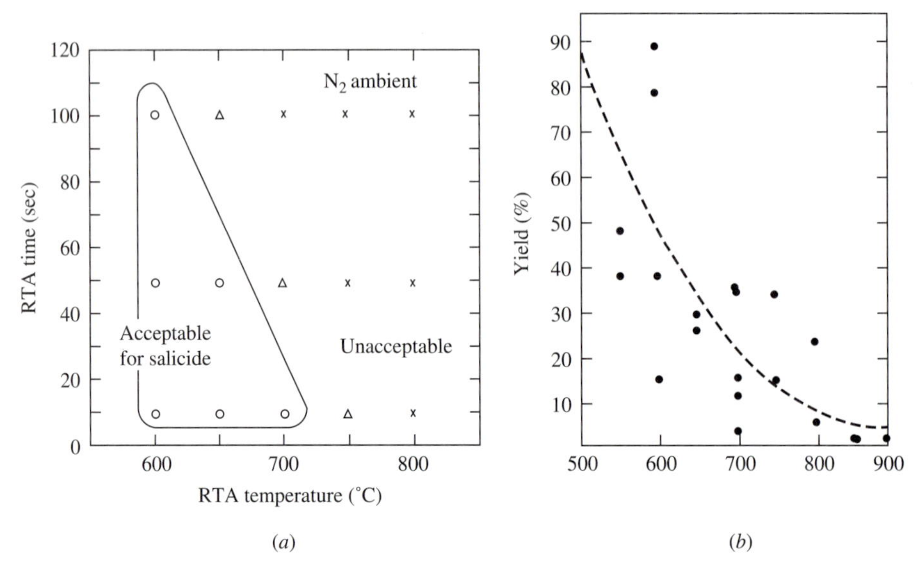

FIGURE 43
Process windows for (*a*) $TiSi_2$[87] and (*b*) $CoSi_2$[88] showing the acceptable range of RTA temperatures and times required to avoid lateral bridging (contact-to-gate shorts). The devices tested were MOSFETs with oxide spacers separating the polysilicon gate from the contact windows.

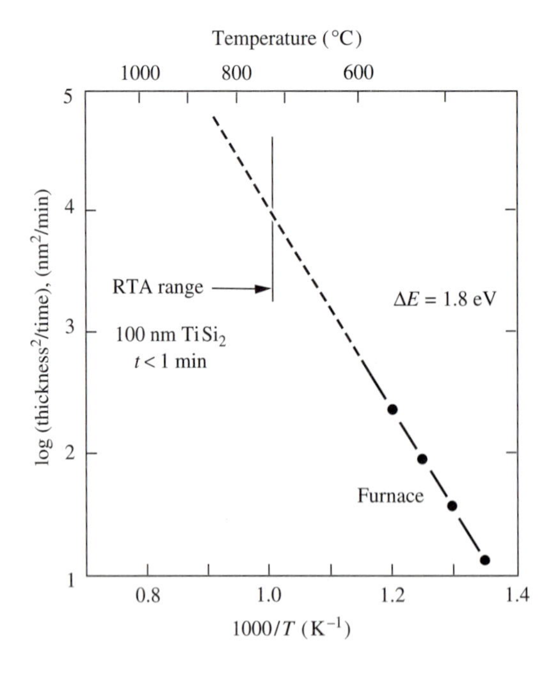

FIGURE 44
Kinetics of silicide formation in the furnace and RTA regimes. (*After d'Heurle, Ref. 89.*)

log of thickness^2/time is plotted in Fig. 44 over a wide range of times, from seconds to hours, and a wide range of temperatures, including the low-temperature furnace range and the higher-temperature RTA range.[89] Since a single activation energy of 1.8 eV characterizes the reaction, the use of RTA at moderate temperatures (500 to 800°C) can fully react the metal film in a few seconds. Extended annealing times are undesirable after silicide formation, since agglomeration of the silicide may occur, destroying film uniformity.

Factors that can influence silicide kinetics are the presence of a thin native-oxide film on the Si surface, dopant impurities, and oxidation of the surface of the deposited metal film by oxygen or by moisture that competes with the silicide reaction at the metal/Si interface. The sensitivity of the sheet resistance of titanium silicide to trace amounts of oxygen is shown in Table 2.[90] Keeping oxygen contamination below these levels is difficult in large-diameter furnace tubes with air backstreaming from the outlet end. However, the microenvironment achievable within an RTA system along with the low impurity levels in high-purity ULSI gases, N_2 or Ar, are suitable for silicides.

Finally, the application of RTA to silicide film growth is not without some issues. Among the most serious are the following:[86]

- Nonuniform silicide/Si interface, which limits the depth of the silicide relative to the pn junction depth
- Silicide agglomeration and instability
- Forces at the silicide/Si interface that can lead to dislocation generation at the higher RTA temperatures
- Depletion of dopants on the Si side of the interface via grain-boundary diffusion through the silicide, which leads to irregular junctions

The effects of surface and junction roughening and dopant depletion are illustrated in Fig. 45.[91] During silicide growth, silicon is consumed in the reaction with metal. Dopants in the Si can diffuse out through the silicide by grain-boundary transport toward the surface where they either accumulate or evaporate. Dopants may be introduced by ion implanting through the layer once the silicide is formed. However, the resulting damage produced by the process can increase the sheet resistance of the silicide. Last, the annealing step causes top and bottom surface roughening as a result of silicide agglomeration. The angles that occur between the grain boundaries present in the layer change in response to surface-tension forces, producing thermal grooving. Continued heating can cause increased roughness and the formation of silicide islands, and this result drives annealing requirements in the direction of controlled, low-thermal-budget processing, achievable with RTA.

TABLE 2
Effect of oxygen impurities on $TiSi_2$ sheet resistance[90]

Gas	Impurities (ppm of oxygen)	Silicide sheet resistance (Ω / □)
Nitrogen	0.1	2.6
Nitrogen	1.0	3.4
Nitrogen	10.0	100.0
Argon	<0.1	1.8

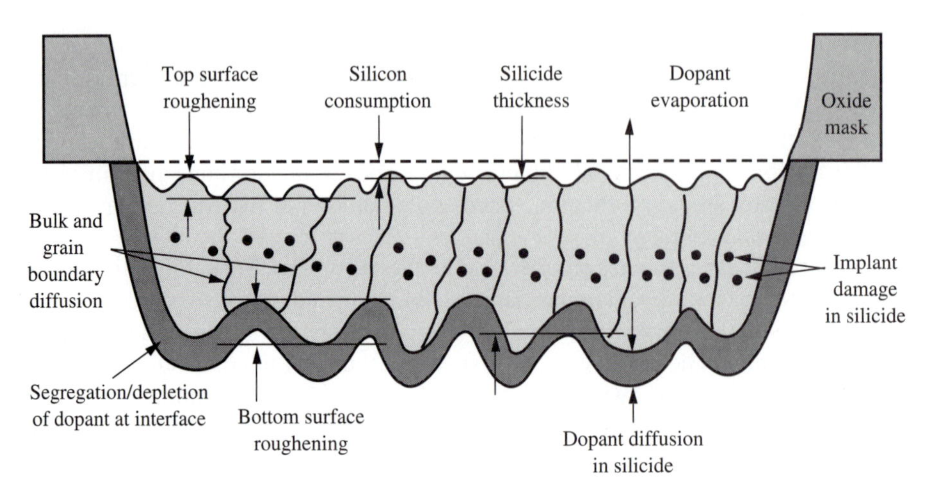

FIGURE 45
Key phenomena in silicide processes. (*After Osburn et al., Ref. 91.*)

4.4 SUMMARY AND FUTURE TRENDS

Rapid thermal processing (RTP) has the advantages over conventional furnace technology of operating in a controlled microenvironment with low thermal mass and rapid heating rates. Because high temperatures are attainable in a short time, single-wafer thermal processing can occur in multiple-use chambers or in clusters of single-use chambers, thermal budgets can be achieved for defect-free junctions with high dopant activation, and sharp interfaces can be created in layered structures.

A road map showing the introduction of RTP into manufacturing DRAMs is depicted in Table 3. Each of the key processing areas that will be affected by RTP is listed along with estimated time for the transition from batch to single-wafer technology. The road map shows that the potential of RTP increases the further the technology moves into the sub-half-micrometer design rule regime. For example, the "processing" row shows that batch processing will give way to single-wafer processing to meet the requirements of 0.5-μm manufacturing technologies. Once this choice is made, there are no alternatives left except to implement RTP for wafer throughput. As more and more processes are converted to single-wafer tools, it eventually becomes economically feasible and necessary to begin integrating the tools into clusters, as is expected by the end of this decade for 0.15-μm complementary MOS.

Within the next few years, all batch processes will have been demonstrated in RTP. However, the introduction of these RTP processes into manufacturing will occur only when the process is absolutely needed because of technical and cost requirements.[4] For example, existing CVD processes in conventional, batch CVD furnaces do not always work well in RTP systems. New chemistries are needed, especially for thick-film deposition, to obtain the deposition rates that are required for wafer throughput. Otherwise, the processes will be too costly. However, there may

TABLE 3

RTP technology road map

	1985		1990		1995		2000	
Design rule (μm)	2	1.3	0.8		0.5	0.3	0.2	0.15
DRAM equivalent	256k	1M	4M		16M	64M	256M	1G
Wafer size (mm)	125	125	150–200		200	200–250		300
t_{ox}(nm)	35	25	20		15	12	10	8
L_{eff}(μm)	2	1.3	0.8		0.5	0.3	0.2	0.15
x_j(μm)	0.3	0.25	0.2		0.15	0.10	0.08	0.06
Repeatability (°C)	±15		±5		±2		±1	
Uniformity (°C)	±7–10		±3–5		±2–3		< ±2	
Accuracy (°C)	±20		±5–10		±3–5		< ±3	
Thermal budget			Furnace				RTA	
t_{ox} control			Furnace				RTO	
Thin dielectrics			LPCVD				RTCVD-ONO	
Silicides			Furnace		RTA			
Epitaxy			LPCVD, MBE,[†] APCVD, etc.		RTCVD			
Polysilicon			LPCVD		RTCVD-SiGe			
Metals			LPCVD sputtered, evaporated		RTCVD			
Cleaning			Wet chemistry		RTC			
Processing	Batch				Single wafer		Clusters	
Atmospheric budget	0.1/cm^2	> 0.5 μm	0.01/cm^2	> 0.3 μm	0.005/cm^2	> 0.2 μm	0.001/cm^2	> 0.1 μm

[†] Molecular beam epitaxy

be good justification for introducing thin-film CVD processing in the near future. As single-wafer technology evolves, CVD processes may first become integrated into single-process, multichamber tools with some level of in-situ etch cleaning. For fabricating MOSFETs in a single chamber, it is desirable to have a high-quality CVD process for depositing the gate dielectric rather than using RTO with its attendant problems of temperature control and uniformity. Thus, the use of RTCVD-ONO (oxide-nitride-oxide) deposition precedes the advent of RTO in Table 3.

Much of the time-to-introduction of the various technologies listed in Table 3 depends on the ability of equipment manufacturers to design RTP tools with suitable process chambers, temperature measurement and control systems, sensors, and reproducible processes. Technology advancement will probably focus on RTP, single-wafer chambers with multizone heating reflectors, and on real-time, independent control of the zones in the multilamp heating systems. End point detection techniques for process control should be introduced, rather than relying on temperature measurement. End point detection involves a measurement of some property associated with a process, such as a film thickness or the chemical composition of the chamber ambient. Typical end point detection methods include ellipsometry, reflectometry, interferometry, and possibly emissivity measurements.[4] Although end point detection will not solve the problems of temperature control in RTP, it may improve the repeatability of a process.

Finally, it is evident that RTP technologies are evolving without any major breakthroughs, but rather with constant and diligent attention to engineering details. Indeed, radiant heating had a deceptively simple veneer that covered all the underestimated problems. And no producer of RTP tools has solved the fundamental issues of depositing radiant energy into the materials common to semiconductor devices.[92] As a result, progress is made by fixing and patching. The preferred methodology would be one in which the basic issues and science were well understood so that problems could be avoided.

REFERENCES

1. S. M. Sze, Ed., *VLSI Technology,* 2nd ed., McGraw Hill, New York, 1988.
2. R. B. Fair, "Rapid Thermal Processing—A Justification," in R. B. Fair, Ed., *Rapid Thermal Processing: Science and Technology,* Academic Press, Boston, 1993, p. 1.
3. J. K. Roberts, *Heat and Thermodynamics,* 4th ed., Blackie and Son, Ltd., Glasgow, 1955.
4. F. Roozeboom, "Manufacturing Equipment Issues in Rapid Thermal Processing," in R. B. Fair, Ed., *Rapid Thermal Processing: Science and Technology,* Academic Press, Boston, 1993, p. 349.
5. T. Sato, "Spectral Emissivity of Silicon," *Jpn. J. Appl. Phys.* **6,** 339 (1967).
6. F. Roozeboom, "Temperature Control and System Design Aspects in RTP," in *Mater. Res. Soc. Symp. Proc.,* **224,** 9 (1991).
7. C. Hill, S. Jones, and D. Boys, "Rapid Thermal Annealing—Theory and Practice," in R. A. Levy, Ed., *Reduced Thermal Processing for ULSI,* Plenum Press, New York, 1989, p. 143.

8. J. C. Liao and T. I. Kamins, "Power Absorption during Deposition of Polycrystalline Silicon in a Lamp-Heated CVD Reactor," *J. Appl. Phys.,* **67,** 3848 (1990).
9. T. P. Merchant, K.-H. Lee, J. V. Cole, and K. F. Jensen, "Strategies for Modeling of Rapid Thermal Processing Systems," in R. B. Fair and B. Lojek, Eds., *1st International Rapid Thermal Processing Conference, RTP'93, Scottsdale, AZ* 1993, p. 376.
10. Z. Nenyei and A. Tillmann, "Reaction Time Analysis in Rapid Thermal Annealing," in R. B. Fair and B. Lojek, Eds., *1st International Rapid Thermal Processing Conference, RTP'93, Scottsdale, AZ,* 1993, p. 429.
11. I. Beinglass and S. S. Schwartz, "Single Wafer Integrated Processing System for High Temperature Thin Films Formation," in R.B. Fair and B. Lojek, Eds., *1st International Rapid Thermal Processing Conference, RTP'93, Scottsdale, AZ,* 1993, p. 416.
12. C. Lee and G. Chizinsky, "Rapid Thermal Processing Using a Continuous Heat Source," *Solid State Tech.,* **32 (1),** (1989), p. 43.
13. A. Kamgar and S. Hillenius, "Rapid Thermal Anneal Induced Effects in Polycrystalline Silicon Gate Structures," *Appl. Phys. Lett.,* **51,** 1251 (1987).
14. J.S. Nakos, "Application of Rapid Thermal Processing in Manufacturing: The Effect of Emissivity and Coupling," in R. B. Fair and B. Lojek, Eds., *1st International Rapid Thermal Processing Conference, RTP'93, Scottsdale, AZ,* 1993, p. 421.
15. R. B. Fair, "Challenges in Manufacturing Submicron, Ultra-Large Scale Integrated Circuits," *Proc. IEEE,* **78,** 1990, p. 1687.
16. R. S. Nelson, "The Physical State of Ion Implanted Solids," in G. Dearnaley, J. H. Freeman, R. S. Nelson, and J. Stephen, Eds., *Ion Implantation,* North Holland Pub. Co., Amsterdam, 1973, p. 154.
17. A. E. Michel, "Diffusion Modeling of the Redistribution of Ion Implanted Impurities," in T. O. Sedgwick, T. E. Seidel, and B. Y. Tsaur, Eds., *Rapid Thermal Processing,* Mat. Res. Soc., Pittsburgh, **52,** 3 (1986).
18. *PREDICT®—PRocess Estimator for the Design of IC Technologies,* MCNC, Research Triangle Park, NC, 1987.
19. R. B. Fair, "Damage Removal/Dopant Diffusion Trade-offs in Ultra-Shallow Implanted p^+n Junctions," *IEEE Trans. Electron Dev.,* **37,** 2237 (1990).
20. E. A. Irene and R. Ghez, "Silicon Oxidation Studies: The Role of H_2O," *J. Electrochem. Soc.,* **124,** 1757 (1977).
21. N. Yabumoto, K. Saito, M. Morita, and T. Ohmi, "Oxidation Process of Hydrogen-Terminated Silicon Substrate Studied by Thermal Desorption Spectroscopy," *Jpn. J. Appl. Phys.,* **30,** 1419 (1991).
22. C. M. Osburn, H. Berger, R. Donovan, and G. Jones, "The Effect of Contamination on Semiconductor Manufacturing Yield," *Proc. of Institute of Environmental Science,* **31,** 45 (1988).
23. B. T. Murphy, "Cost-Size Optima of Monolithic Integrated Circuits," *Proc. IEEE,* **52,** 1537 (1963).
24. P. J. Rosser, P. B. Moynagh, and K. B. Affolter, "Multi-Chamber Rapid Thermal Processing," *Soc. Photo-Opt. Instrum. Eng. Symp. Proc.* **1393,** 49 (1990).
25. K. Shenai, P. A. Piacente, and B. J. Baliga, "Reliable Low Contact Resistance Al-TiW-$TiSi_2$-n^+ Si Contacts for Silicon VLSI and Smart Power Applications," in M. Scott, Y. Arkasaka, and R. Rief, Eds., *Advanced Materials for VLSI,* Electrochem. Soc., Pennington, NJ, 155 (1988).
26. Y. El-Mansey, "MOS Device and Technology Constraints in VLSI," *IEEE Trans. Electron. Dev.,* **ED-29,** 567 (1982).
27. E. Takeda, G. A. C. Jones, and H. Ahmed, "Constraints of the Application of 0.5 μm MOSFETs to VLSI Systems," *IEEE Trans. Electron. Dev.,* **ED-32,** 322 (1985).

28. E. Sun, J. Moll, J. Berger and B. Alders, "Breakdown Mechanism in Short-Channel MOS Transistors," in *Tech. Dig. IEDM,* 478 (1983).
29. W. H. Lee, T. Osakama, K. Asada, and T. Sugano, "Design Methodology and Size Limitations of Submicrometer MOSFETs for DRAM Application," *IEEE Trans. Electron. Dev.,* **35,** 1876 (1988).
30. S. Asai, "The Material Matters in Future ULSIs," in *Symposium on Adv. Sci. and Tech. of Si Materials,* Japan Society for Promotion of Science, Kona, 1 (1991), (unpublished).
31. K. M. Cham and S. Y. Chiang, "Device Design for the Sub-Micrometer p-Channel FET with n^+ Polysilicon Gate," *IEEE Trans. Electron. Dev.,* **ED-31,** 964 (1984).
32. J. F. Gibbons, C. M. Gronet, and K. E. Williams, "Limited Reaction Processing: Silicon Epitaxy," *Appl. Phys. Lett.* **47,** 721 (1986).
33. J. L. Hoyt, "Rapid Thermal Processing-Based Epitaxy," in R. B. Fair, Ed., *Rapid Thermal Processing: Science and Technology,* Academic Press, Boston, 1993.
34. J. O. Borland, "Historical Review of SEG and Future Trends in Silicon Epi Technology," in G.W. Cullen, Ed., *Proceedings of the 10th International Conference on Chemical Vapor Deposition, Pennington, NJ,* Electrochemical Society, 87 (1987).
35. M. Ogirima and R. Takahashi, "Some Problems and Future Trends in Silicon Epitaxial Technology," in G.W. Cullen, Ed., *Proceedings of the 10th International Conference on Chemical Vapor Deposition, Pennington, NJ,* Electrochemical Society, 87 (1987).
36. J. L. Regolini, D. Bensahel, J. Mercier, and E. Scheid, "Selective Epitaxial Silicon Growth in the 650–1100°C Range in a Reduced Pressure Chemical Vapor Deposition Reactor Using Dichlorosilane," *Appl. Phys. Lett.* **54,** 658 (1989).
37. F. C. Eversteyn, "Chemical-Reaction Engineering in the Semiconductor Industry," *Philips Res. Rept.* **29,** 45 (1974).
38. M. C. Ozturk and J. J. Wortman, "Rapid Thermal Chemical Vapor Deposition for Silicon Based Microelectronics Manufacturing," in R.B. Fair and B. Lojek, Eds., *1st International Rapid Thermal Processing Conference, RTP'93, Scottsdale, AZ,* 1993, p. 297.
39. T. O. Sedgwick, P. D. Agnello, D. Nguyen Ngoc, T. S. Kuan, and G. Scilla, "High Phosphorus Doping of Epitaxial Silicon at Low Temperature and Atmospheric Pressure," *Appl. Phys. Lett.* **58,** 1896 (1991).
40. J.-E. Sundgren, J. Knall, W.-X. Ni, M.-A. Hasan, L. C. Markert, and J. E. Green, "Dopant Incorporation Kinetics and Abrupt Profiles during Silicon Molecular Beam Epitaxy," *Thin Solid Films,* **183,** 281 (1989).
41. M. C. Ozturk, "Thin-Film Deposition," in R. B. Fair, Ed., *Rapid Thermal Processing: Science and Technology,* Academic Press, Boston, 1993, p. 79.
42. L. Y. Cheng, J. C. Rey, J. P. McVitte, and K. Saraswat, *IEEE VMIC Conference, June 12–14, Extended Abstracts,* 1990, p. 404.
43. F. S. Becker, D. Pawlik, H. Anzinger, and A. Spitzer, "Low Pressure Deposition of High Quality SiO_2 films by Pyrolysis of Tetraethylorthosilicate," *J. Vac. Sci. Technol.* **B5,** 1555 (1987).
44. R. Miller, M. C. Ozturk, J. J. Wortman, F. S. Johnson, and D. T. Grider, "LPCVD of Silicon Dioxide by Pyrolysis of TEOS in a Rapid Thermal Processor," *Mat. Lett.* **8,** 353 (1989).
45. A. C. Adams, "Dielectric and Polysilicon Film Deposition," in S. M. Sze, Ed., *VLSI Technology,* 2nd ed., McGraw-Hill, New York, 1988, p. 233.
46. X. Xu, R. Kuehn, J. J. Wortman, and M. C. Ozturk, unpublished (1992).
47. W. Ting, S. N. Lin, and D. L. Kwong, "Thin Stacked Oxide/Nitride Dielectrics Formation by in situ Multiple Reactive Rapid Thermal Processing," *Appl. Phys. Lett.* **55,** 2312 (1989).
48. F. S. Johnson, R. M. Miller, M. C. Ozturk, and J. J. Wortman, "Characterization of LPCVD of Silicon Nitride in a Rapid Thermal Processor," in *Rapid Thermal Annealing/*

Chemical Vapor Deposition and Integrated Processing Symposium, MRS Symposia Proceedings, **146** 1989, p. 345.
49. R. F. C. Farrow, "The Kinetics of Silicon Deposition by Pyrolysis of Silane," *J. Electrochem. Soc.,* **121,** 899 (1974).
49a. M. C. Ozturk, J. J. Wortman, Y. Zhong, X. Ren, R. M. Miller, F. S. Johnson, D. T. Grider, and D. A. Abercrombie, in *Rapid Thermal Annealing/Chemical Vapor Deposition and Integrated Processing Symposium,* MRS Symposia Proceedings, **146** 1989, p. 109.
50. K. L. Chiang, C. J. Dell'Oca and F. N. Schwettmann, "Optical Evaluation of Polycrystalline Silicon Surface Roughness," *J. Electrochem. Soc.,* **126,** 2267 (1979).
51. X. Ren, M. C. Ozturk, J. J. Wortman, B. Zhang, and D. Maher, "Deposition and Characterization of Polysilicon Films Deposited by Rapid Thermal Processing," *J. Vac. Sci. Technol. B,* **10,** 1081 (1992).
51a. M. C. Ozturk, F. Y. Sorrell, J. J. Wortman, F. S. Johnson, and D. T. Grider, "Manufacturability Issues in Rapid Thermal Chemical Vapor Deposition," *IEEE Trans. Semicond. Manuf.* **4,** 155 (1991).
52. H. Z. Massoud, "Rapid Thermal Growth and Processing of Dielectrics," in R. B. Fair, Ed., *Rapid Thermal Processing: Science and Technology,* Academic Press, Boston, 1993.
53. A. Kazor and I. W. Boyd, "UV-assisted Growth of 100 Å Thick SiO_2 at 550°C," *Electron. Lett.,* **27,** 909 (1991).
54. V. Craciun and I. W. Boyd, "Low Temperature UV Induced Oxidation of Silicon and Silicon-Germanium Strained Layers," in R. B. Fair and B. Lojek, Eds., *1st International Rapid Thermal Processing Conference, RTP'93, Scottsdale, AZ,* 1993, p. 363.
55. R. Deaton and H. Z. Massoud, "Effect of Thermally-Induced Stresses on the Rapid Thermal Oxidation of Silicon," *J. Appl. Phys.* **70,** 3588 (1991).
56. J. Nulman, J. P. Krusius, and P. Renteln, "Material and Electrical Properties of Gate Dielectrics Grown by Rapid Thermal Processing," *Mat. Res. Soc., Symp. Proc.,* **52,** 341 (1985).
57. C. A. Paz de Araujo, J. C. Gelpey, Y. P. Huang, and R. Kwor, "Comparison of the Growth Kinetics of Oxides Grown in Tungsten-Halogen and a Water-cooled Arc Lamp System," *Mat. Res. Soc., Symp. Proc.,* **92,** 133 (1987).
58. H. Z. Massoud and J. D. Plummer, "Analytical Relationship for the Oxidation of Silicon in Dry Oxygen in the Thin-Film Regime," *J. Appl. Phys.,* **62,** 3416 (1987).
59. J. Nulman, J. Scarpulla, T. Mele, and J. P. Krusius, "Electrical Characteristics of Thin Gate Implanted MOS Channels Grown by Rapid Thermal Processing," in *Tech. Dig. IEDM,* Cat. No. 85CH2252-5, 376, IEEE, New York (1985), p. 376.
60. M. L. Reed and J. D. Plummer, "Kinetic Studies of Silicon-Silicon Dioxide Interface Trap Annealing Using Rapid Thermal Annealing," in *Mat. Res. Soc., Symp. Proc.* **52,** 333 (1986).
61. H. Fukuda, "Thin-Gate SiO_2 Films Formed by in situ Multiple Rapid Thermal Processing," T. Arakawa and S. Ohno, Eds., *IEEE Trans. Electron. Dev.* **39,** 127 (1992).
62. B. Lojek, "Point Defect Behavior During Rapid Thermal Annealing," unpublished (1992).
63. K. S. Jones, S. Prussin, and E. R. Weber, "Enhanced Elimination of Implantation Damage Upon Exceeding the Solid Solubility," *J. Appl. Phys.* **62,** 4114 (1987).
64. T. E. Seidel, D. J. Lischner, C. S. Pai, R. V. Knoell, D. M. Maher, and D. C. Jacobson, " A Review of Rapid Thermal Annealing (RTA) of B, BF_2 and As Implanted into Silicon," *Nucl. Inst. Methods Phys. Res. B,* **7/8,** 251 (1985).
65. K. S. Jones, "Extended Defects from Ion Implantation and Annealing," in R. B. Fair, Ed., *Rapid Thermal Processing: Science and Technology,* Academic Press, Boston, 1993.

66. R. B. Fair, "Junction Formation in Silicon by Rapid Thermal Annealing," in R. B. Fair, Ed., *Rapid Thermal Processing: Science and Technology,* Academic Press, Boston, 1993.
67. M. D. Giles, unpublished (1990).
68. R. B. Fair, "Concentration Profiles of Diffused Dopants in Silicon," in F. F. Y. Wang, Ed., *Impurity Doping Processes in Silicon,* North Holland Press, Amsterdam, 1981.
69. F. Morehead and R. Hodgson, "A Simple Model for the Transient Enhanced Diffusion of Ion-Implanted Phosphorus in Silicon," in D. K. Biegelsen, G. A. Rozgonyi, and C. V. Shanks, Eds., *Energy Beam-Solid Interactions and Transient Thermal Processing,* **35,** Mat. Res. Soc., Pittsburgh, 1985.
70. M. Miyake, S. Aoyama, and K. Kurchi, "Transient Enhanced Diffusion of Ion Implanted B in Si during Rapid Thermal Annealing," in *Extended Abstracts,* Meeting of the Electrochem. Soc., Abs. #691, Honolulu (1987).
71. R. B. Fair, "The Role of Transient Damage Annealing in Shallow Junction Formation," *Nucl. Instr. Meth. Phys. Res. B,* **37/38,** 371 (1989).
72. S. Solmi, F. Baruffaldi, and R. Canteri, "Diffusion of Boron in Silicon During Post-implantation Annealing," *J. Appl. Phys.,* **69,** 2135 (1991).
73. Y. M. Kim, G. Q. Lo, H. Konoshita, and D. L. Kwong, "Roles of Extended Defect Evolution on the Anomalous Diffusion of Boron in Si during Rapid Thermal Annealing," *J. Electrochem. Soc.* **138,** 1122 (1991).
74. D. K. Sadana, S. C. Shatas and A. Gat, in *Microscopy of Semiconducting Materials,* Institute of Physics Conference Series, Institute of Physics, London (1983).
75. R. B. Fair, "Low-Thermal-Budget Process Modeling with the PREDICT™ Computer Program," *IEEE Trans. Electron. Dev.,* **35,** 285 (1988).
76. Y. Kim, H. Z. Massoud, amd R. B. Fair, "The Effect of Ion-Implantation Damage on Dopant Diffusion in Silicon During Shallow-Junction Formation," *J. Electron. Matls.,* **18,** 143 (1989).
77. S. Solmi, "Dopant Diffusion and Activation in Implanted Silicon," in R. B. Fair and B. Lojek, Eds., *1st International Rapid Thermal Processing Conference, RTP'93, Scottsdale, AZ,* 1993, p. 179.
78. S. Solmi, F. Baruffaldi, and R. Canteri, "Diffusion of Boron in Silicon During Post-Implantation Annealing," *J. Appl. Phys.*, **69,** 2135 (1991).
79. S. N. Hong, G. A. Ruggles, J. J. Wortman, and M. C. Ozturk, "Material and Electrical Properties of Ultra-Shallow p^+-n Junctions Formed by Low-Energy Ion Implantation and Rapid Thermal Annealing," *IEEE Trans. Electron. Dev.,* **38,** 476 (1991).
80. J. J. Wortman, unpublished (1991).
81. M. Finetti, P. Negrini, S. Solmi, and D. Nobili, "Electrical Properties and Stability of Supersaturated Phosphorus-Doped Silicon Layers," *J. Electrochem. Soc.,* **128,** 1313 (1981).
82. D. Nobili, A. Armigliato, M. Finetti, and S. Solmi, "Precipitation as the Phenomenon Responsible for the Electrically Inactive Phosphorus in Silicon," *J. Appl. Phys.,* **53,** 1484 (1982).
83. J. Kato and Y. Ono, "Phosphorus Diffusion Using Spin-On Phosphosilicate-Glass Source and Halogen Lamps," *J. Electrochem. Soc.*, **132,** 1730 (1985).
84. M. Orlowski, R. Subrahmanyan, and G. Huffman, "The Effect of Low Thermal Budget Anneals and Furnace Ramps on the Electrical Activation of Arsenic," *J. Appl. Phys.*, **71,** 164 (1992).
85. M. Y. Tsai, F. F. Morehead, J. E. E. Baglin, and A. E. Michel, "Shallow Junctions by High-Dose As Implants in Si: Experiments and Modeling," *J. Appl. Phys.,* **51,** 3230 (1980).

86. C. M. Osburn, "Silicides," in R. B. Fair, Ed., *Rapid Thermal Processing: Science and Technology,* Academic Press, Boston, 1993, p. 227.
87. V. Q. Ho and D. Poulin, "Formation of Self-aligned $TiSi_2$ for VLSI Contacts and Interconnects," *J. Vac. Sci. Technol.,* **A5,** 1396 (1987).
88. R. D. J. Verhaar, A. A. Bos, H. Kraaij, R. A. M. Wolters, K. Maex, and L. Van denhove, "Self-aligned $CoSi_2$ in a submicron CMOS Process," in A. Heuberger, H. Ryssel, and P. Lange, Eds., *Proc. ESSDERC '89,* Springer-Verlag, Heidelberg, 229 (1989).
89. F. M. d' Heurle, "Material Properties of Silicides for VLSI Technology," in P. Balk and O. G. Folberth, Eds., *Solid State Devices,* Elsevier, Amsterdam, 1986, p. 213.
90. C. M. Osburn, H. Berger, R. P. Donovan, and G. W. Jones, "The Effects of Contamination on Semiconductor Manufacturing Yield," *J. Environ. Sci.,* **31,** 45 (1988).
91. C. M. Osburn, J. Y. Tsai, Q. F. Wang, J. Rose, and A. Cowen, "PREDICT 1.6: Modeling of Metal Silicide Processes," *Electrochem. Soc. Meeting, Honolulu,* May, 1993, p. 352.
92. B. Lojek, "Rapid Thermal Processing of Semiconductors 1963–1993. Where To from Here?" in R.B. Fair and B. Lojek, Eds., *1st International Rapid Thermal Processing Conference, RTP'93, Scottsdale, AZ,* 1993, p. 2.

PROBLEMS

1. Define emissivity and give examples of factors that influence intrinsic, extrinsic, and effective emissivity in an RTP chamber.

(*a*) Why does the intrinsic emissivity of a Si wafer become relatively independent of the wavelength of incident radiation at temperatures greater than 600°C ?

(*b*) Explain why the temperature of a deposited overlayer on a wafer undergoing RTA might depend on the overlayer's thickness.

(*c*) In (*b*), what would be the difficulty associated with trying to control wafer temperature during film growth by using an optical pyrometer in the temperature-control feedback loop?

(*d*) In (*c*), what factor does the reflectivity of the RTP chamber play in temperature measurement?

2. In a 0.25-μm CMOS process, what is the thermal budget for forming ion-implanted p^+ source/drain layers with surface doping concentrations of 1×10^{20} B/cm^3 if the RTA temperature is 900°C ?

(*a*) What kinds of device performance issues might a process engineer encounter with this thermal treatment?

(*b*) What must be done to create defect-free junctions for the 0.25-μm ion-implantation process without increasing the junction depth? (Assume 700 Å junction depth.)

(*c*) Why is it that doping processes such as epitaxy and CVD of solid dopant sources onto Si have larger thermal budgets than ion-implantation processes?

3. Discuss the pros and cons of batch CVD epitaxy versus LRP, single-wafer epitaxy from the standpoint of throughput, control, Si film uniformity, and cost.

4. RTCVD of dielectric films requires high deposition rates for single-wafer processing to achieve high wafer throughput.

(*a*) What are the issues for applying RTCVD to thin-gate dielectric film deposition?

(*b*) What are the issues for applying RTCVD to interlevel dielectric formation at the back-end of the process?

(*c*) If SiH_4 and N_2O react to deposit SiO_2, what temperature uniformity across the wafer is required to deposit films with a 2% uniformity, assuming a 1.8-eV activation energy and a maximum wafer temperature of 800°C ?

5. Conventional batch-furnace oxidation has been used by manufacturers for over thirty years to grow high-quality, MOS-grade gate oxides, which are highly uniform across the wafer. Is there a compelling case to be made for replacing such a successful technology with single-wafer RTO for ULSI? Describe your analysis.

6. Explain why high-temperature RTA is preferable to low-temperature RTA for defect-free shallow junction formation.

(*a*) Why do you obtain shallower junctions with a high-temperature RTA (1000°C, 20 sec) than with a low-temperature RTA (800°C, 20 sec)? Assume a P implant.

(*b*) What is the basis for the activation of high-dose, ion-implanted B or P by high-temperature RTA when the resulting free-carrier concentrations exceed the solid solubility limit at the RTA temperature?

7. Assuming BF_2 ion implantation in Si with a dose of 1×10^{15} cm^{-2}, what is the maximum implant energy allowed to produce a shallow, defect-free p^+ junction less than 1000 Å deep using a 10-sec RTA?

(*a*) Does preamorphization using Ge implantation provide any benefits?

(*b*) What type of ion-implantation defects are produced by the BF_2 implant?

(*c*) Describe the two steps involved in the transient diffusion of the BF_2-implanted Si during RTA.

8. Why has silicide formation using RTA been easily incorporated into the manufacturing of ICs?

(*a*) Why is it necessary to use a two-step RTA process in silicide formation?

(*b*) What are the limitations of silicides for ultrashallow junction contacts?

(*c*) Using the concept of surface tension, explain top silicide surface roughening. *Hint:* The process of roughening is referred to as thermal grooving.

9. What are the primary design considerations in building an RTA system that yields uniform wafer temperature distributions (not necessarily uniform radiation)?

CHAPTER 5

Dielectric and Polysilicon Film Deposition

H. C. Cheng

5.1 INTRODUCTION

Dielectric and polysilicon film deposition have been used extensively in integrated circuits (ICs) since the device planarization technique was developed. Dielectric films, including silicon dioxide and silicon nitride, are used as the isolation, mask, and passivation layers. Silicon dioxide deposited by chemical vapor deposition (CVD) has been shown to fit the requirements of gate dielectrics in ultralarge-scale integration (ULSI) circuits. On the other hand, polysilicon film can be used as the conducting layer, semiconductor, or resistor by proper doping with different impurities. Various deposition processes, such as CVD and physical vapor deposition (PVD), are described in Section 5.2. Different deposition schemes and the film qualities resulting from these deposition methods are described in Sections 5.3, 5.4, and 5.5 for silicon dioxide, silicon nitride, and polysilicon films. Other deposition techniques, e.g., plasma-assisted deposition, photo CVD, laser CVD, rapid-thermal-processing CVD (RTPCVD), and electron-cyclotron-resonance (ECR) CVD, are demonstrated. Finally, the applications of these deposited films and their future trends are presented in Sections 5.8 and 5.9, respectively.

5.2 DEPOSITION PROCESSES

5.2.1 Reaction Theories of Chemical Vapor Deposition (CVD) and Physical Vapor Deposition (PVD)

There are many deposition techniques for the material formations in ULSI technology. These methods can be classified into two main reaction mechanisms:

chemical vapor deposition and physical vapor deposition. These two technologies are described as follows.

Chemical vapor deposition (CVD)

Chemical vapor deposition is defined as the formation of a nonvolatile solid film on a substrate by the reaction of vapor-phase chemicals (reactants) that contain the required constituents. It is most often used for semiconductor processing. It is a material synthesis process whereby the constituents of the vapor phases react chemically near or on a substrate surface to form a solid product.

Several steps must occur in every CVD reaction:[1,2]

1. Transport of reacting gaseous species to the substrate surface
2. Absorption, or chemisorption, of the species on the substrate surface
3. Heterogeneous surface reaction catalyzed by the substrate surface
4. Desorption of gaseous reaction products
5. Transport of reaction products away from the substrate surface

The sequence of reaction steps in a CVD process is illustrated in Fig. 1. In practice, the chemical reactions of the reactant gases leading to the formation of a solid material may take place not only on (or very close to) the wafer surface (heterogeneous reaction) but also in the gas phase (homogeneous reaction). Heterogeneous reactions are much more desirable, because such reactions occur selectively only on the heated surfaces and produce good-quality films. Homogeneous reactions, on the other hand, are undesirable, because they form gas-phase clusters of the depositing material, resulting in poorly adhering, low-density films with defects. In addition, such reactions also consume the reactants and can cause a decrease in deposition rates. Thus, one important issue of a chemical reaction for CVD application is the degree to which heterogeneous reactions are favored over gas-phase reactions.

The most common deposition methods are atmospheric-pressure CVD (APCVD), low-pressure CVD (LPCVD), and plasma-enhanced CVD (PECVD).[3,4,5] A comparison of APCVD and LPCVD shows that the benefits of the low-pressure deposition processes are uniform step coverage, precise control of composition and structure, low-temperature processing, high enough deposition rates and throughput, and low

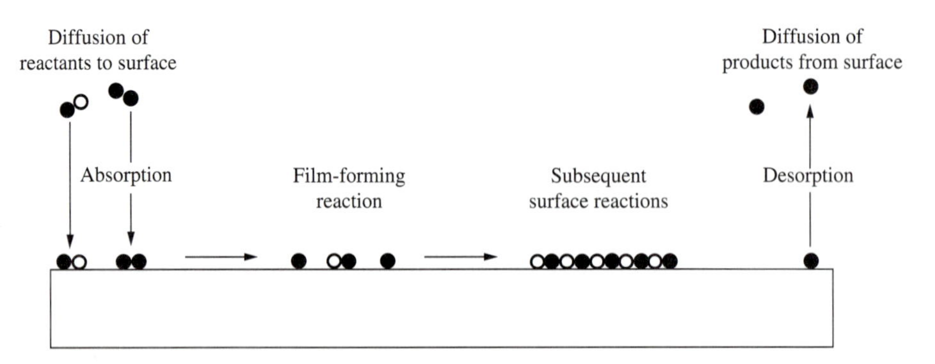

FIGURE 1
The sequence of reaction steps in a CVD reaction.

processing costs.[6] Furthermore, no carrier gases are required in LPCVD, which reduces particle contamination. Hence, LPCVD is used widely in the highly cost-competitive semiconductor industry for film depositions. Epitaxial growth of silicon at a reduced pressure minimizes autodoping (contamination of the substrate by the dopant), a major problem in atmospheric-pressure epitaxy. The most serious disadvantage of LPCVD and APCVD is that their operating temperatures are high. PECVD is an appropriate method to solve this problem, as illustrated later.

Because the deposition process includes force convection, boundary-layer diffusion, surface absorption, decomposition, surface diffusion, and incorporation, there are several variables to be controlled. Temperature, pressure, flow rate, position, and reactant ratio all are important factors for high-quality films. The industry has optimized these conditions to improve the film properties.

Since the aforementioned steps for a CVD process are sequential, the one that occurs at the slowest rate will determine the deposition rate. The rate-determining steps can be grouped into gas-phase processes and surface processes. For the gas-phase process, the concern is the rate at which gases impinge on the substrate. This model considers the rate at which gases cross the boundary layer that separates the bulk regions of flowing gas and substrate surface. Such transport processes occur by gas-phase diffusion, which is proportional to the diffusivity of the gas and the concentration gradient across the boundary layer. The rate of mass transport is only relatively weakly influenced by the deposition temperature.

On the other hand, at low temperatures the surface reaction rate is reduced, and eventually the arrival rate of reactants exceeds the rate at which they are consumed by the surface reaction process. Under such conditions, the deposition rate is surface-reaction-rate-limited. Thus, at high temperatures, the deposition is usually *mass-transport-limited,* while at low temperatures it is *surface-reaction-rate-limited,* as shown in Fig. 2. In actual processes the temperature at which the deposition condition moves from one of these growth regimes to the other depends on the activation energy of the reaction and the gas flow conditions in the reactor. In processes that are under surface-reaction-rate-limited conditions, the deposition temperature is an important parameter. That is, uniform deposition rates throughout a reactor require conditions that maintain a constant reaction rate. This, in turn, implies that a constant temperature must also exist everywhere at all wafer surfaces. On the other hand, under such conditions the rate at which reactant species arrive at the surface is not so important, because their concentrations do not limit the growth rate. Thus,

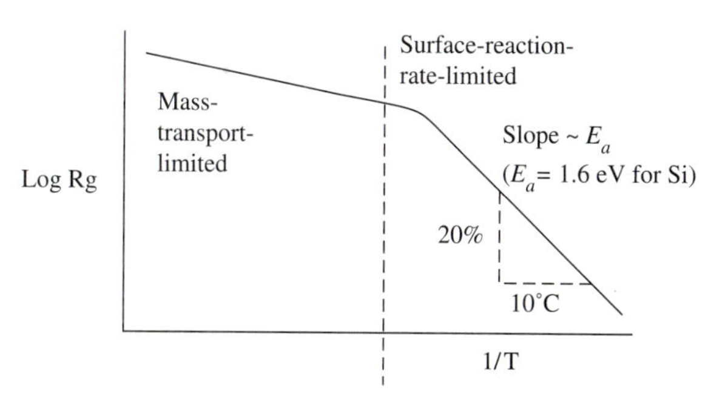

FIGURE 2
The deposition rate R_g is a rapid varying function of temperature T in the surface-reaction-limited regime of operation (low temperatures), whereas it changes only slowly with temperature in the mass-transport-limited regime (higher temperatures).

it is not so critical that a reactor be designed to supply an equal flux of reactants to all locations of a wafer surface. It will be seen that in low-pressure CVD (LPCVD) reactors wafers can be stacked vertically and at very close spacing, because such systems operate in a surface-reaction-rate-limited mode. Under the low pressure of an LPCVD reactor ($\sim$1 torr), the diffusivity of the gas species is increased by a factor of 1000 over that at atmospheric pressure, and this is only partially offset by the fact that the boundary layer (the distance across which the reactants must diffuse) increases by less than the square root of the pressure. The net effect is that there is more than an order-of-magnitude increase in the transport of reactants to (and by-products away from) the substrate surface, and the rate-limiting step is thus the surface reaction. Since the surface reaction rate also generally increases with increasing surface concentration, nonuniform gas-phase concentrations produced by local depletion of reactants within the reactor can result in depletion nonuniformities. An example of such an effect is the depletion of reactants by their deposition on wafers located at the inlet of an end-feed reactor tube. Wafers near the outlet end are consequently exposed to lower concentrations of reactants than those at the inlet end of the tube. For film uniformity, the deposition temperatures at these two ends will thus have to be properly adjusted.

In deposition processes that are mass-transport-limited, the temperature control is not so critical. The mass transport process, which limits the growth rate, is only weakly dependent on the temperature. On the other hand, it is very important that the same concentration of reactants be present in the bulk gas regions adjacent to all locations of a wafer, because the arrival rates of reactants are directly proportional to the concentration gradient in the bulk gas. Thus, to ensure that films are uniform across a wafer, reactors operated in the mass-transport-limited regime must be designed so that all locations of the wafer surface and all wafers in a run are supplied with an equal flux of reactant species. Atmospheric-pressure reactors that deposit SiO_2 at $\sim$400°C are operated in the mass-transport-limited regime. The most widely used APCVD reactor designs provide a uniform supply of reactants by horizontally positioning the wafers and moving them under a gas stream.

Physical vapor deposition (PVD)

Physical vapor depositions are different from CVD in the deposition mechanism. The deposition rate for CVD will be proportional to the deposition temperature. However, the deposition rate for PVD will generally decrease with increasing deposition temperature. The PVD technologies usually used are described as follows.

Evaporation technology. This is one of the oldest techniques used for depositing thin films. A vapor is first generated, by evaporating a source material in a vacuum chamber, then transported from the source to the substrate and condensed to a solid film on the substrate surface.

Molecular beam epitaxy (MBE). This is a sophisticated, finely controlled method for growing epitaxial films in an ultrahigh vacuum at 10^{-8} to 10^{-10} torr. The most widely studied materials are epitaxial layers of III–V semiconductor compounds and SiGe compounds. Figure 3 shows the schematic diagram of a molecular

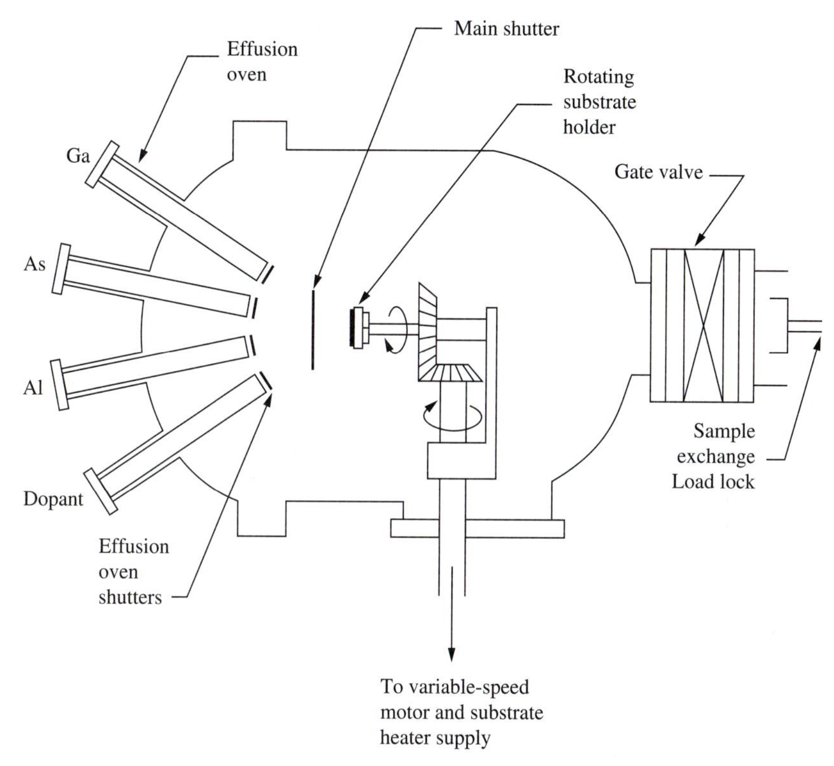

FIGURE 3
Schematic of a molecular beam epitaxy growth chamber.

beam epitaxy growth chamber. The evaporated species are transported at a very high velocity to the substrate. The relatively low vapor pressures of dopants and silicon ensures the condensation of deposited films on a low-temperature substrate.

Sputtering. This process involves the ejection of surface atoms from an electrode surface by momentum transfer from the bombarding ions to the electrode surface atoms. The generated vapor of electrode material is then deposited on the substrate. Sputtering processes, unlike evaporation, are very well controlled and are generally applicable to all materials, such as metals, insulators, semiconductors, and alloys. The schematic diagram of a sputtering system is shown in Fig. 4.

5.2.2 Equipment

The design and operation of CVD reactors depend on a variety of factors; hence, these reactors can be categorized in several ways. One way of grouping CVD reactors depends on the method used to heat the wafers. Another criterion used to distinguish reactor types is the pressure regime of operation (atmospheric-pressure reactors or reduced-pressure reactors). The reduced-pressure group can be further split into

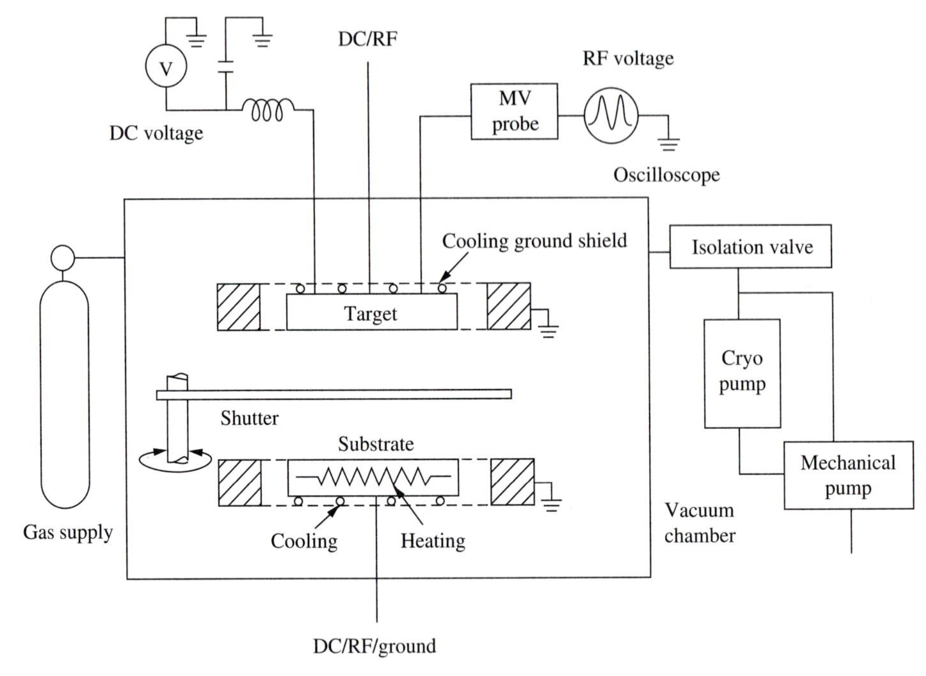

FIGURE 4
Schematic drawing showing some of the components of a sputtering system.

1. Low-pressure reactors, called low pressure-CVD (LPCVD) reactors, in which the energy input is entirely thermal
2. Those in which energy is partially supplied by a plasma as well, known as plasma-enhanced CVD or PECVD reactors

Each of the reactor types in these two pressure regimes is further divided into subgroups defined by reactor configuration and method of heating.

Four methods of heating the wafers have been adopted:

1. Resistance heating
2. RF induction heating
3. Heating by energy from a glow discharge (plasma)
4. Heating by photon energy

Energy may be transferred either to the reactant gases themselves or to the substrate. When radiant heating, from resistance-heated coils surrounding the reaction tube, is utilized, not only the wafer but also the reaction chamber walls become hot, and hence such designs are known as hot-wall reactors. In these systems, film-forming reactions occur on reaction chamber walls as well as on the substrate. This implies that such systems will require frequent cleaning to avoid particle contamination. On the other hand, energy input via RF induction or infrared lamps mounted within the reactor will heat the wafers as well as susceptors and does not cause appreciable heating of the reaction chamber walls. Consequently, systems heated by these methods are called cold-wall reactors. In some cold-wall systems, however, significant chamber wall heating can still take place, so means for cooling the walls must be implemented (e.g., water cooling) to prevent reactions or depositions on the wall.

Reactor geometry is constrained by the pressure regime and energy source and is an important factor in throughput. Since atmospheric-pressure reactors operate in the mass-transport-limited regime, they must be designed so that an equal flux of reactants is delivered to each wafer. As a result, wafers are never stacked vertically at close spacing but rather are laid flat on a horizontal surface. An undesirable consequence of this design is the high susceptibility of the wafers to incorporated falling particles. LPCVD reactors are not constrained by the mass-transfer-rate limitation, allowing designs that accommodate a large number of wafers per run. Wafers can be stacked side by side, only a few mm apart in a quartz reaction tube. Quartz wafer holders (*boats*) can hold up to 200 wafers. However, LPCVD reactors must be capable of precise temperature control, because they operate in the surface-reaction-limited mode. Table 1 summarizes the characteristics and applications of various CVD reactor designs.

For high-volume production of integrated circuits, the number of wafers processed simultaneously should be as large as possible. The requirement for high wafer capacity led to the development of the commercially important hot-wall, low-pressure CVD reactor, shown in Fig. 5. For maximum wafer capacity, the wafers are held vertically and separated from each other by a narrow space. As in a diffusion or oxidation furnace, the wafers are placed perpendicular to the gas flow in a quartz tube with a circular cross section. The gas flows by forced convection through the annular space between the wafers and the walls of the deposition chamber and then must travel along the narrow space between the walls. The pressures in the reaction chamber are typically 0.1 to 5.0 torr; the temperatures are between 300 and 900°C; and the gas flow rates range between 100 and 1000 standard cm^3/min (sccm). Special inserts that alter the gas flow dynamics are sometimes used. Large load space, good uniformity, and the ability to feed large-diameter wafers are the major advantages of this reactor. However, low deposition rates and the frequently used special gases are still problems.

TABLE 1
Characteristics and applications of CVD processes[58]

Process	Advantages	Disadvantages	Applications
APCVD (low-temperature)	Simple reactor, fast deposition, low temperature	Poor step coverage, particle contamination, low throughput	Low-temperature oxides (both doped and undoped)
LPCVD	Excellent purity and uniformity, conformal step coverage, large wafer capacity, high throughput	High temperature, low deposition rate	High-temperature oxides (both doped and undoped), silicon nitride, poly-Si, W, WSi_2
PECVD	Low temperature, fast deposition, good step coverage	Chemical (e.g., H_2) and particle contamination	Low-temperature insulators over metals, passivation (nitride)

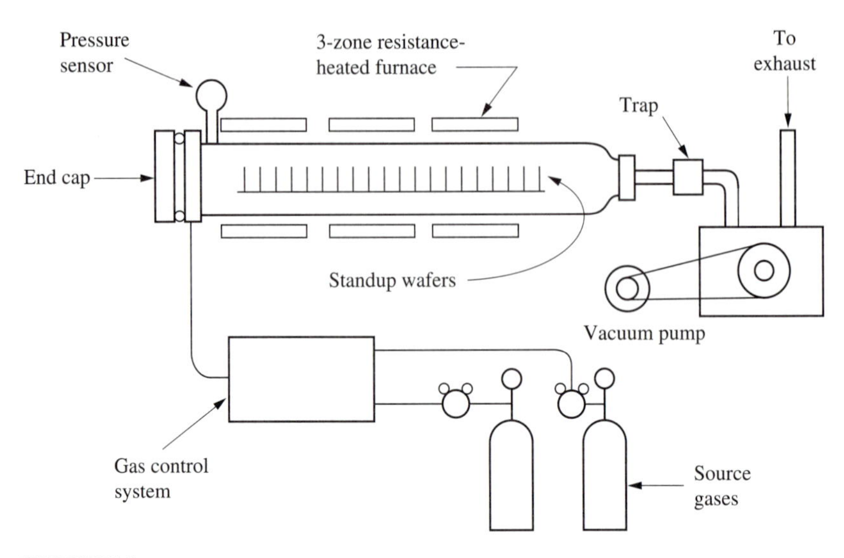

FIGURE 5
The hot-wall, low-pressure reactor is used for routine deposition of polysilicon because of its high wafer capacity and simplicity.

Atmospheric-pressure CVD (APCVD) reactors were the first to be used in the microelectronics industry.[7,8] Operation at atmospheric pressures keeps reactor design simple and allows high film deposition rates. APCVD, however, is susceptible to gas-phase reactions, and the films typically exhibit poor step coverage. Since APCVD is generally conducted in the mass-transport-limited regime, the reactant flux to all parts of every substrate in the reactor must be precisely controlled. Figure 6*a* shows the horizontal tube type. Such systems consist of a horizontal quartz tube, with the wafers lying flat on a fixed horizontal plate, while gas flows parallel to the wafer surface. Figures 6*b* and *c* show continuous-processing APCVD reactors. These configurations are the most widely used designs for depositing low-temperature CVD oxide films in production applications.[8] The samples are carried through the reactor on a conveyor belt. Reactant gases flowing through the center of the reactor are contained by gas curtains formed by a fast flow of nitrogen. These continuous reactors possess the benefits of high throughput, good uniformity, and the capability to process large-diameter wafers. However, they have the problems of high gas consumption and frequent need of reactor cleaning.

Figure 7 shows the diagram of a typical commercial PECVD system. Rather than relying solely on thermal energy to sustain chemical reactions, PECVD systems use an rf-induced glow discharge to transfer energy into the reactant gases, allowing the substrate to remain at a lower temperature than those in APCVD or LPCVD processes.[9,10,11] Low deposition temperature is the major advantage of PECVD, and in fact, PECVD provides a method of depositing films on substrates that do not have the thermal stability to accept coatings by other methods. In addition, PECVD can enhance the deposition rate, as compared with thermal reactions alone, and can produce films of unique compositions and properties. However, the limited capacity

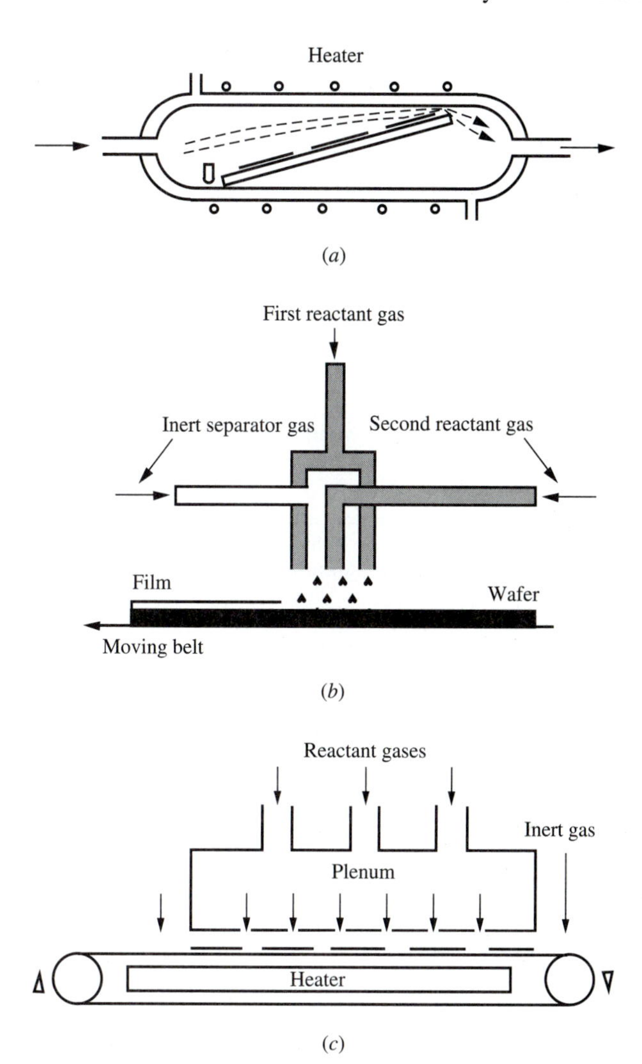

FIGURE 6
(*a*) Horizontal tube APCVD reactor. (*b*) Gas injection–type continuous-processing APCVD reactors. (*c*) Plenum-type continuous-processing APCVD reactor.

(especially for large-diameter wafers) and the possibility of particle contamination by loosely adhering deposits are still major concerns.

5.2.3 Safety Issues

Most of the gases used for film deposition are toxic. Those used for LPCVD are more unsafe than those used for atmospheric-pressure CVD, because pure silane is used

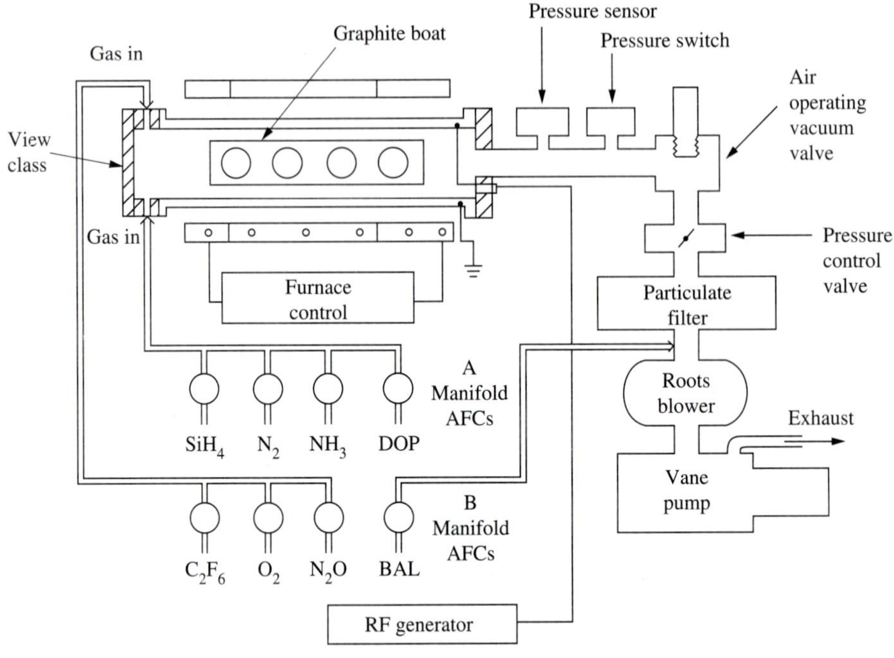

FIGURE 7
Diagram of a typical commercial PECVD system. (*Courtesy of Pacific Western Systems.*)

in LPCVD systems. In addition, hazardous gases can also cause a reaction with the vacuum pump oil.

These hazardous gases can be divided into four general classes: pyrophoric (flammable or explosive); poisonous; corrosive; and dangerous combinations of gases. Gases commonly used in CVD are listed in Table 2. Gas combinations such

TABLE 2
Gases commonly used in CVD

Gas	Formula	Hazard	Flammable limits in air (vol%)	Exposure limit (ppm)
Ammonia	NH_3	Toxic, corrosive	16–25	25
Argon	Ar	Inert	—	—
Arsine	AsH_3	Toxic	—	0.05
Diborane	B_2H_6	Toxic, flammable	1–98	0.1
Dichlorosilane	SiH_2Cl_2	Flammable, toxic	4–99	5
Hydrogen	H_2	Flammable	4–74	—
Hydorgen chloride	HCl	Corrosive, toxic	—	5
Nitrogen	N_2	Inert	—	—
Nitrogen oxide	N_2O	Oxidizer	—	—
Oxygen	O_2	Oxidizer	—	—
Phosphine	PH_3	Toxic, flammable	Pyrophoric	0.3
Silane	SiH_4	Flammable, toxic	Pyrophoric	0.5

as silane with halogens, silane with hydrogen, and oxygen with hydrogen will cause safety problems.[12]

In addition, silane will react with air to form solid products, causing particle contaminations within the gas lines. These particles can plug the pipes and perhaps create combustion. Hence, the piping of the gas supply system must be carefully installed and maintained. The gas piping should be monitored and inspected before each operation to prevent the leakage of residual gases. To avoid the plugging of the gas products, burn-box systems for the hydride gases have been widely used in ULSI operations.

5.3 ATMOSPHERIC-PRESSURE CHEMICAL-VAPOR-DEPOSITED (APCVD) AND LOW-PRESSURE CHEMICAL-VAPOR-DEPOSITED (LPCVD) SILICON OXIDES

Dielectric layers have been inevitably used as the gate materials in transistors and capacitors. Dielectric layers can also be used to electrically isolate one level of conductor from another in multilevel-interconnect systems. The requirements of dielectric layers in multilevel interconnects are stringent, including

1. No residual constituents that will outgas during later processing
2. Good step coverage
3. Ease of etching
4. Stability up to temperatures of 500°C
5. Good adhesion
6. Low dielectric constant for frequencies up to ~20 MHz (to keep capacitance low between metal lines)
7. High breakdown field strength (>5 MV/cm)
8. No moisture absorption or permeability to moisture

Furthermore, there is a significant difference between the dielectric films used as poly–metal interlevel dielectrics (e.g., between poly, or other local interconnect–level materials, and metal 1) and the dielectric films employed between the metal layers (intermetal dielectrics). Poly–metal interlevel dielectric (PMD) films can be deposited (and densified if necessary) at a higher temperature than is possible for the intermetal dielectric layers. Furthermore, PMD films can be flowed and reflowed at temperatures in excess of 800°C. On the other hand, the maximum temperature of the intermetal dielectric layers is limited to 450°C or below, because Al is present on the wafer surface.

In general, silicon dioxides can be formed by a thermal annealing in oxygen ambient or by a chemical reaction (SiH_4 with O_2, $SiCl_2H_2$ with N_2O, or decomposition of $Si(OC_2H_5)_4$). The former can attain excellent electrical properties and is commonly used to fabricate the gate dielectrics in ULSI devices. The latter, called chemical vapor deposition (CVD), is often used for PMD, intermetal isolation, dopant masks, capping or passivation layers, diffusion sources, and even gate dielectrics in ULSI processes. This is because CVD can achieve more flexible characteristics, including large thickness and low-temperature processing.[13,14] Doped or undoped silicon dioxide films are widely used in various applications. Undoped silicon dioxide is used as an ion-implantation or diffusion mask, an insulating layer between multilevel metallization, a capping layer over doped regions to prevent outdiffusion, and an oxide layer to increase the thickness of thermally grown oxides. Undoped CVD oxides used as gate dielectrics and insulators between conducting layers are deposited and then densified by annealing. A solution containing fluoride or a CHF_3 plasma can etch the oxides to open the contact windows and via holes. Phosphorus-doped silicon dioxide is used as an insulator between metal layers, a final passivation over devices, and a gettering source.[15] Oxides doped with phosphorus, arsenic, or boron are used occasionally as diffusion sources.

Silicon nitride films have generally not been used as stand-alone PMD layers because they possess a much higher dielectric constant than SiO_2 films and cannot be flowed or reflowed. Hence, doped CVD SiO_2 films have found the widest application as PMD layers in MOS ICs. Although high-temperature CVD oxide films (deposited at 900°C by the reaction of dichlorosilane and nitrous oxide) possess properties almost identical to those of thermally grown SiO_2 films, these films cannot, unfortunately, be doped under the high temperatures at which they are deposited.[16] As a result, they cannot be flowed or reflowed at temperatures lower than 1100°C. Generally, phosphorus-doped CVD oxides are heated to a temperature between 950 and 1100°C so that the oxides soften and flow and form a smooth surface to obtain a good step coverage for the next metallization.[15,17] Boron-doped phosphosilicate oxides will further lower the reflow temperatures.

5.3.1 Deposition Methods

The methods used to deposit the silicon dioxide are classified by heating source, chemical reactants, reactor, and deposition temperature.

We first consider a method in which oxide films are deposited at temperatures below 500°C by reacting silane with oxygen. If phosphorus doping is included, the chemical reactions are expressed as

$$SiH_4(g) + O_2(g) \rightarrow SiO_2(s) + 2H_2(g) \tag{5.1}$$

$$4PH_3(g) + 5O_2(g) \rightarrow 2P_2O_5(s) + 6H_2(g) \tag{5.2}$$

This process can be conducted in an atmospheric-pressure chamber (atmospheric-pressure chemical vapor deposition, APCVD) or in a reduced-pressure chamber (low-pressure chemical vapor deposition, LPCVD). Owing to the advantages of low deposition temperatures this method can be used to deposit silicon dioxides on the metal interconnect. However, the poor step coverage and high particle concentration characteristic of this technique require careful control.

Another method uses tetraethyl orthosilicate, $Si(OC_2H_5)_4$, abbreviated TEOS.[18] By decomposing the vaporized liquid TEOS, silicon dioxide can also be deposited in an LPCVD reactor at temperatures between 650 and 750°C. The reaction is written as

$$Si(OC_2H_5)_4(l) \rightarrow SiO_2(s) + \text{by-products}(g) \tag{5.3}$$

where the by-products are organic and organosilicon compounds. LPCVD TEOS is often used to deposit the spacers beside the polysilicon gates. TEOS depositions have very good uniformity and step coverage, but the high-temperature process limits the application of LPCVD TEOS on aluminum interconnects.

Silicon dioxide can also be deposited by LPCVD at about 900°C. The reaction is expressed as

$$SiCl_2H_2(g) + 2N_2O(g) \rightarrow SiO_2(s) + 2N_2(g) + 2HCl(g) \tag{5.4}$$

The advantage of this process is film uniformity. Like LPCVD TEOS, this process can deposit silicon dioxide over polysilicon, but its disadvantage is the chlorine-containing damage on the polysilicon.

Silicon dioxides can be doped by feeding dopant hydrides such as phosphine, arsine, and diborane into the CVD reactors. Such dopant compounds as halides or organic compounds can also be utilized.[20–23]

Recently, atmospheric-pressure and low-temperature CVD processes using TEOS and ozone (O_3) have been proposed,[24] as shown in Fig. 8. This CVD technology enables the formation of oxide films with high conformality and low viscosity under low deposition temperatures. The step angle depends on the ozone concentration, as shown in Fig. 9. In addition, the shrinkage of the oxide film during annealing is also a function of the ozone concentration, as shown in Fig. 10. Furthermore, the deposition temperature will determine the selectivity ratio of the O_3–TEOS films deposited on the SiO_2 to the Si substrate. This selectivity ratio will decrease with increasing deposition temperature, as shown in Fig. 11. Because of their porosity, O_3–TEOS CVD oxides are often accompanied by plasma-assisted oxides to execute the planarization technology in ULSI processing.

High-pressure techniques have been studied by many groups. The major advantages of high-pressure techniques stem mainly from their ability to use lower processing temperatures, which lead to reduced impurity diffusion and fewer substrate

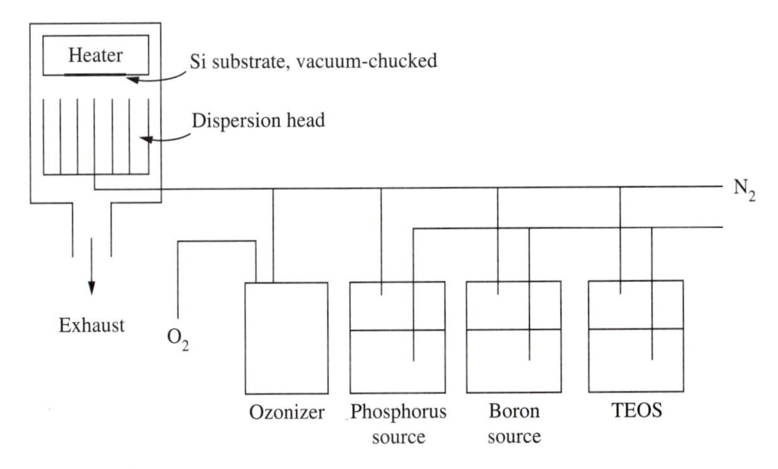

FIGURE 8
Experimental apparatus for the O_3–TEOS CVD system.

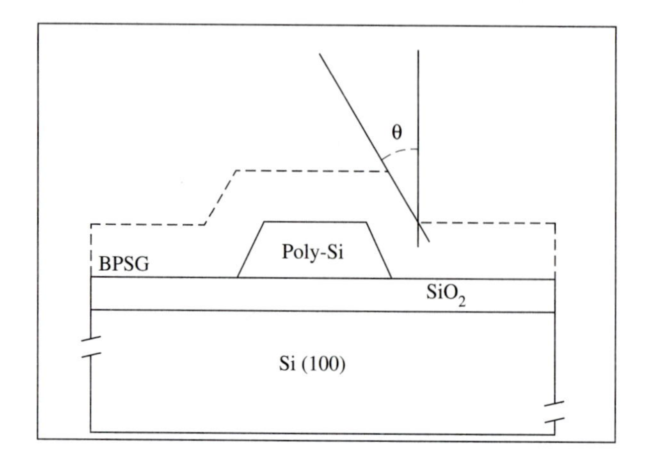

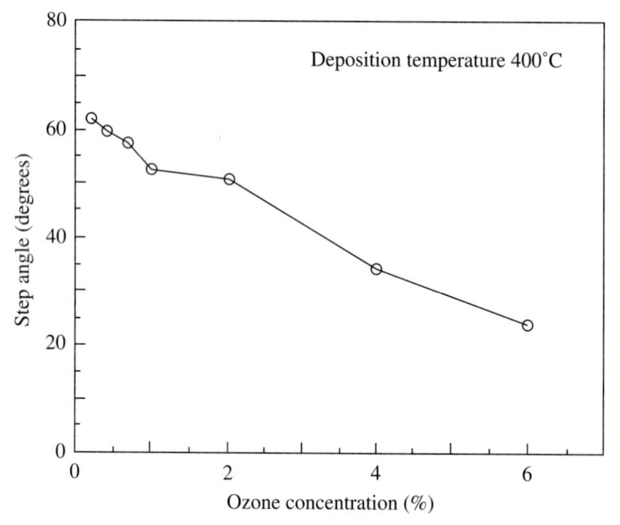

FIGURE 9
(*a*) Definition of step angle. (*b*) The step angle of the O_3–TEOS film as a function of the ozone concentration. (*Courtesy of SAMCO Company, Japan.*)

defects. However, the cost of facilities and the risk of danger discourage interest in high-pressure techniques.

5.3.2 Deposition Variables

Pressure, deposition temperature, reactant concentration, and dopants will affect the quality of silicon dioxide. For CVD the deposition rate of the silicon dioxide increases with deposition temperature. The silane–oxygen reaction needs less decomposition energy and, therefore, lower activation energy than TEOS decomposition. As the phosphorus dopants are added, the activation energies for these two cases

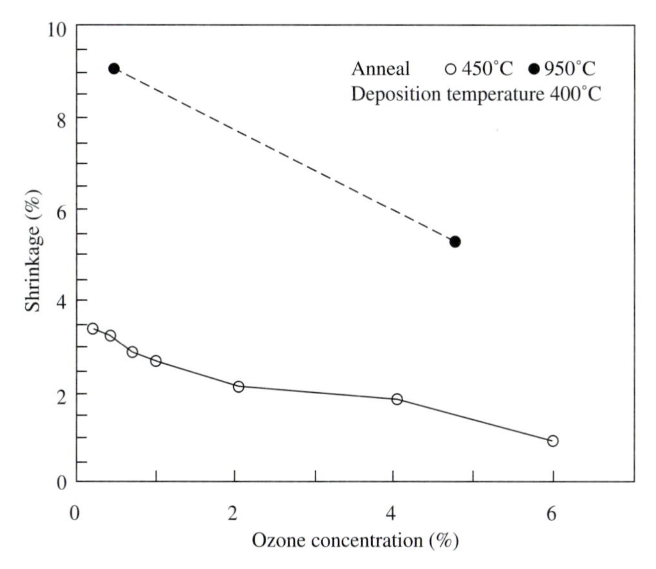

FIGURE 10
Dependence of the shrinkage of the O_3–TEOS film on ozone concentration during annealing. (*Courtesy of SAMCO Company, Japan.*)

will decrease. By introducing trimethylphosphite (TMOP) into the silane–oxygen or TEOS, phosphorus-doped oxides can also be formed. Figure 12 shows the dependence of deposition rate and the phosphorus content (mole % P_2O_5) on TMOP flow rate with a total TEOS/N_2 flow rate of 3.5 standard liters per minute (slm). Ozone concentration is another parameter that influences film characteristics, as shown in Fig. 13. The surface-reaction-rate-limited mode and the surface catalysis by phosphorus and ozone can be used to explain the differences of the activation energies.

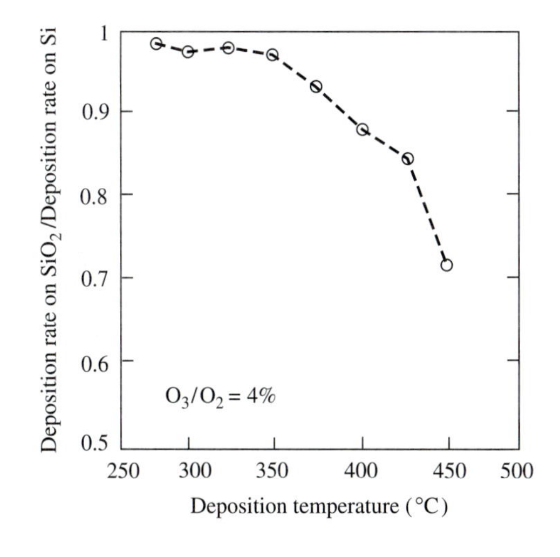

FIGURE 11
The selectivity ratio of the O_3-TEOS films deposited on the SiO_2 to the Si substrate as a function of the deposition temperature. (*Courtesy of SAMCO Company, Japan.*)

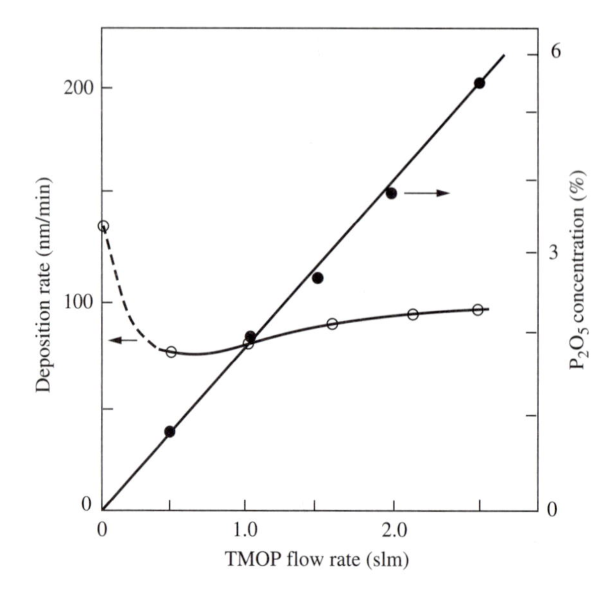

FIGURE 12
Dependence of deposition rate and P_2O_5 concentration of PSG films on TMOP/N_2 flow rate, deposited at 2.4% ozone and 400°C. (*After Fujino et al., Ref. 24.*)

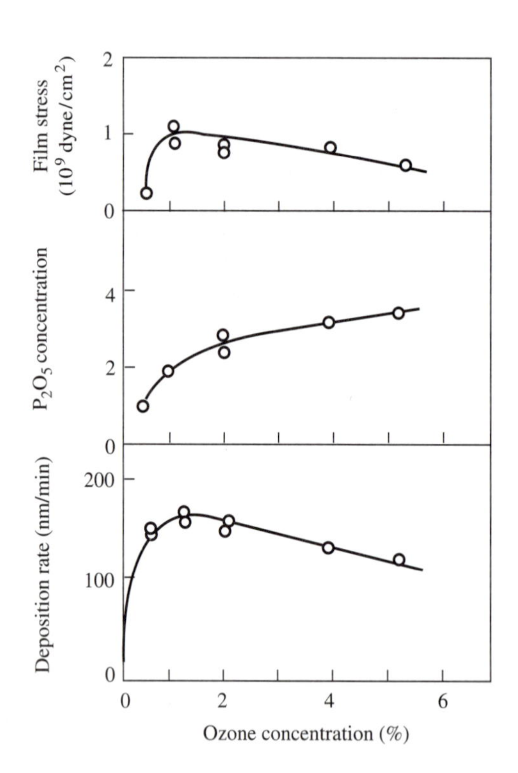

FIGURE 13
Dependence of deposition rate, P_2O_5 concentration, and film stress of PSG films on ozone concentration, deposited at 400°C. (*After Fujino et al., Ref. 24.*)

As for the case of silicon dioxide deposited with dichlorosilane and nitrous oxide, a high deposition temperature (900°C) results in a mass-transport-limited reaction, exhibiting a strong pressure dependence. For the dichlorosilane–nitrous oxide reaction this doping is impossible, because the phosphorus oxides are volatile at the deposition temperature of 900°C.[25] Inert gases often make the gas flow over the wafer surfaces uniform and improve film uniformity. On the other hand, by introducing diborane, boron tricholoride, or trimethylborate into the silane–oxygen or TEOS reactions, boron-doped oxides can be obtained.

5.3.3 Step Coverage and Reflow

Figure 14 shows three types of step coverage often found for the deposited dioxides in a trench. A completely uniform or *conformal* step coverage means that the film thickness along the walls and at the bottom of the step is constant, as shown in Fig.

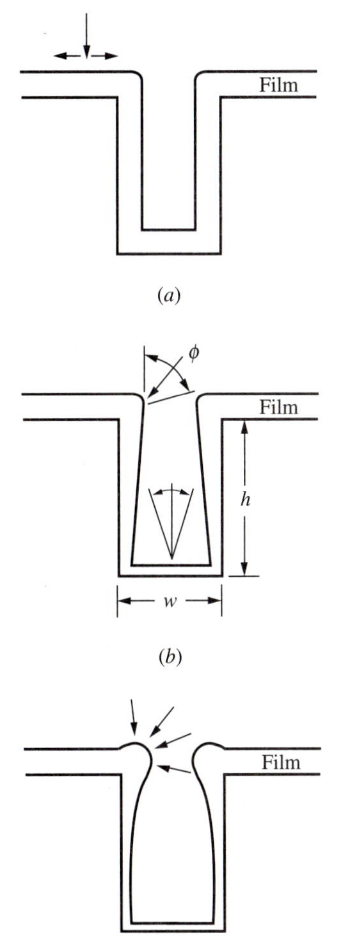

FIGURE 14
Step coverage of deposited films. (*a*) Uniform coverage resulting from rapid surface migration. (*b*) Nonconformal step coverage for long mean free path and no surface migration. (*c*) Nonconformal step coverage for short mean free path and no surface migration.

14*a*. As the reactants or reactive intermediates adsorb on the surface and then rapidly migrate along the surface before reaction, the resulting films will have a uniform surface concentration on the substrate and a constant thickness.

On the other hand, the deposition rate is proportional to the arrival angle of the gas molecules if the absorbed reactants predominantly do not migrate. The arrival angle ϕ is expressed as

$$\phi = \arctan\left(\frac{w}{h}\right) \tag{5.5}$$

where w is the width of the via hole and h is the height from the top surface. Thus, the film thickness will gradually decrease along the step wall and even form an opening at the bottom, as shown in Fig. 14*b*. Furthermore, if the gas atoms have only a short mean free path, the arrival angle at the top of the step is 270°. Consequently, a thick cusp at the top and a thin tail at the bottom of the step will be formed, as shown in Fig. 14*c*; it will be unfavorable for subsequent metal deposition.

In general, doped oxides used for the reflow process contain 6 to 9 wt % phosphorus.[20] Silicon dioxide (glass) with higher phosphorus concentration will facilitate the reflow smoothing but is detrimental to the metallization, because of aluminum corrosion. After the doped silicon dioxide is deposited, a subsequent heating is necessary until the oxide softens and flows, as shown in Fig. 15. Therefore, the procedure is named *reflow*. In addition to the phosphorus concentration, the reflowing morphology of the doped silicon dioxide will also be determined by the heating temperature, heating time, heating rate, and heating ambient. In gen-

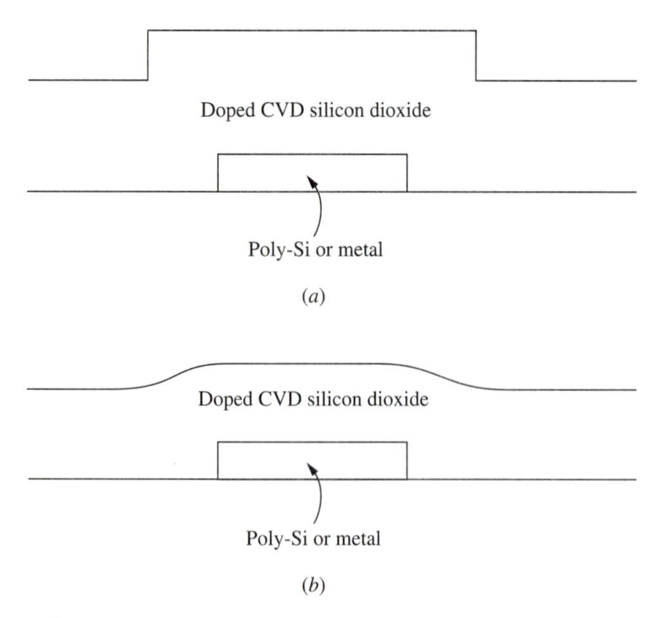

FIGURE 15
Schematic diagram of the reflow process. Polysilicon or metal step covered with doped CVD silicon dioxide (*a*) before and (*b*) after reflow process.

eral, P-glass reflow processes need a heating temperature as high as 950 to 1100°C with a 15 to 30-min cycle. The temperatures will be determined by the phosphorus concentration.

Sometimes, boron dopants are added to the phosphorus-doped silicon dioxide to further reduce the softening temperature by decreasing the glass viscosity.[21] Silane-based BPSG films are deposited at low temperatures (400 to 450°C) and are then immediately densified at ~800°C for one hour.[22,23] The purpose of this step is to completely stabilize the BPSG films, which would otherwise be prone to blistering during subsequent processing. The higher the boron concentration in such films, the lower the reflow temperature; however, BPSG films with high boron concentrations (i.e., greater than 5%) are not stable. Typical dopant concentrations of boron are 1 to 4 wt % for the oxide films with 4 to 6 wt % phosphorus. On a weight percent basis, boron is 1.5 times more effective in the reflow process than phosphorus.

Furnace annealing times coupled with high temperatures often result in significant dopant diffusion, leading to remarkable junction movement, autodoping, and outdiffusion. In addition, high-temperature reflow for a long time also possibly causes further oxidation of the contact hole or substrate. These phenomena must be avoided in the fine-geometry ULSI devices. In order to prevent these problems, and also to prevent leach-out of phosphorus caused by steam reflow, a two-layer film is sometimes used for the PMD; that is, a thin (120-nm-thick) silicon nitride layer is deposited before the BPSG film. The nitride prevents dopants in the BPSG from diffusing into the polysilicon or substrate device regions during the flow and reflow thermal cycles. In advanced ULSI technology, an alternative method, utilizing rapid thermal annealing (RTA) at 1000 to 1150°C for 10 to 60 seconds, has been performed to reflow the BPSG without attacking the underlying devices, as shown in Fig. 16. Obviously, the step angle, reflecting the smoothing topography, will increase with rapid thermal annealing temperature and time.

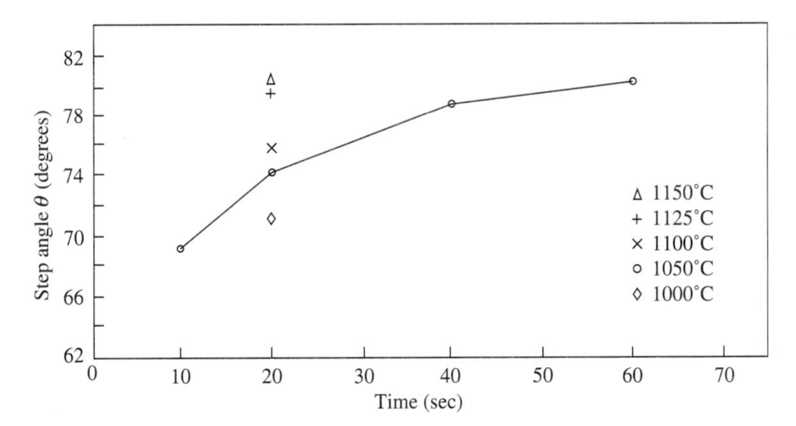

FIGURE 16
Step angles of LPCVD BPSG with concentrations of 3.6 wt % B and 3.9 wt % P, reflowed in the RTA system at different temperatures for various times. (*Courtesy of Peak Systems, Inc., California.*)

The step coverage of deposited oxides can be further improved by *planarization* or *etch-back* techniques. An abrupt step in phosphorus-doped silicon dioxide (P-glass) is often covered with a sacrificial organic coating, called SOG (spin-on-glass) because the organic material usually has a smooth surface. Then the sample is plasma-etched to remove the organic material and the P-glass at equal rates. Since plasma etching preserves the original smooth surface of the coating, the P-glass is left with the same smooth contour. The planarization process reduces step heights independent of the phosphorus concentration in P-glass and requires temperatures less than 200°C. The disadvantages are the increased thickness of the original doped oxide film and the need for additional processing steps. Another planarization process uses PSG films accompanied by chemical/mechanical polishing. In this planarization technique, stacked O_3–TEOS/plasma oxide films are used to cover the polysilicon or metal layers[24] and are subsequently polished by chemical/mechanical methods to obtain a smooth surface. These planarizations will be further discussed in Section 5.8.

5.3.4 Properties of Silicon Oxide, Phosphosilicate Glass (PSG), Borophosphosilicate Glass (BPSG), and Tetraethyl Orthosilicate (TEOS) Films

In order to improve the device characteristics and to use the deposited dielectric and polysilicon films in various applications, the properties of these films are important. As an example, low-dielectric-constant films will be used for intermetal dielectric, whereas high-dielectric-constant films will be used for MOS capacitors. Hence, dielectric constant is one important property of the deposited films. Deposited oxide films commonly have an amorphous structure. By varying deposition temperatures or conditions, it is possible to achieve oxide films with different composition and density. In general, the thermal CVD schemes produce films with a thermodynamic-equilibrium composition. A subsequent densification heating of the CVD oxides at 600 to 900°C is necessary, because higher density of the deposited films is desirable. On the other hand, higher deposition temperatures improve the step coverage. Silicon dioxides deposited from TEOS and dichlorosilane will attain better conformal characteristics. To avoid deterioration during subsequent processing, the deposited films used in integrated circuits generally need high thermal stability. Low deposition temperature and impurities will be harmful to thermal stability. Other parameters, such as refractive index, film thickness, film stress, dielectric strength, and leakage current, all significantly determine the film properties and even the applications of these deposited oxide films. Refractive index and film thickness are often measured by the ellipsometer and the surface profiler. Silicon dioxide has a refractive index of about 1.46, depending on the process. Dielectric strength and leakage current are tested with the semiconductor parameter analyzer. A silicon-rich oxide film will have higher leakage current and is often used as the tunneling gate oxide in flash memory and EEPROM (electrically erasable programmable read-only memory) devices. Stoichiometric SiO_2 is used as the gate oxide because of its lower leakage current. For use in ULSI devices the reliability of the silicon dioxide is also very important. Hence, the time to failure and charge to breakdown under constant-voltage or constant-current stress are analyzed to determine the oxide quality. The stress in silicon dioxide can change the film quality. Tensile stress in silicon dioxide may induce cracks in the films. Stress in silicon dioxide depends on the deposition rate, deposition temperature, post-annealing cycle, dopant concentration, film porosity, and water content. It can be measured with a Michelson interferometer or optical reflectance meter. Hence, the processing of silicon dioxide should be carefully controlled.

5.4 LPCVD SILICON NITRIDES

Silicon nitride (Si_3N_4) is commonly used in passivation layers on integrated devices because of its ability to protect against the diffusion of impurities and water, which would make the devices unstable. The local oxidation of silicon (LOCOS) process also uses the silicon nitride as the mask. The patterned silicon nitride will prevent the underlying silicon from oxidation and leave the exposed silicon to be oxidized. Silicon nitride is also used as the dielectric for DRAM MOS capacitors when it combines with silicon dioxide. Silicon nitride can be chemical-vapor-deposited by reacting silane and ammonia at atmospheric pressure and temperatures ranging from 700 to 800°C:

$$3SiH_4(g) + 4NH_3(g) \longrightarrow Si_3N_4(s) + 12H_2(g) \qquad \text{(APCVD)} \qquad (5.6)$$

or by reacting dichlorosilane and ammonia at reduced pressure and temperature between 700 and 800°C:

$$3SiCl_2H_2(g) + 4NH_3(g) \longrightarrow Si_3N_4(s) + 6H_2(g) \qquad \text{(LPCVD)} \qquad (5.7)$$

Better uniformity and higher wafer throughput can be achieved by the low-pressure process.

5.4.1 Deposition Variables

Total pressure, reactant concentrations, deposition temperature, and temperature gradients in the furnace will affect the properties of silicon nitride. For example, increasing the total pressure and the partial pressure of dichlorosilane will increase the deposition rate, but increasing the ammonia will reduce the deposition rate. The deposition rates will increase with deposition temperature.

5.4.2 Properties of Silicon Nitride Films

As deposited, silicon nitride is in an amorphous phase and often contains a large amount of hydrogen, depending on the deposition temperature and the ratio of ammonia to dichlorosilane.[26,29] High ammonia ratio and low deposition temperature will increase the hydrogen content. These hydrogen atoms can passivate the dangling bonds of silicon and can sometimes be used for passivation of polysilicon devices. The refractive index of the deposited silicon nitride is about 2.01, but it changes with

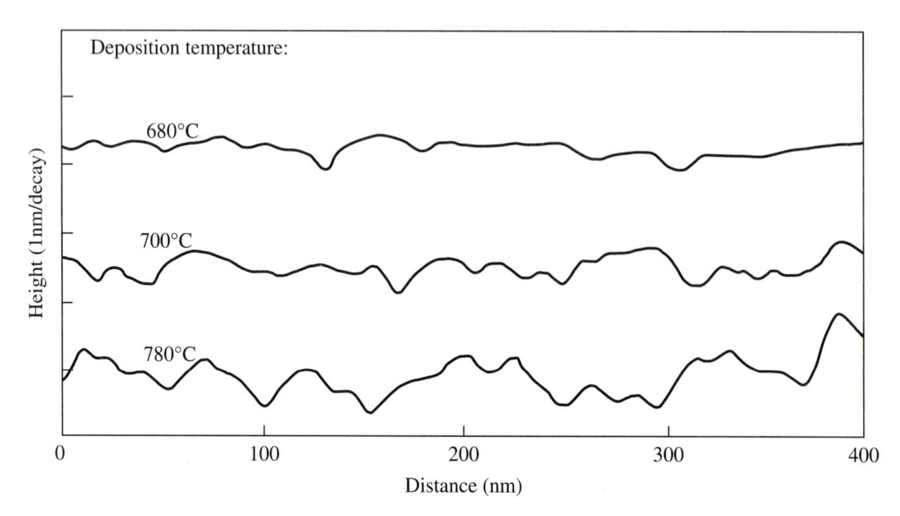

FIGURE 17
AFM surface profiles of Si_3N_4 films deposited at various deposition temperatures. (*After Tanaka et al., Ref. 28.*)

nitride composition. Silicon nitride films that contain oxygen are *silicon oxynitride* and have lower refractive indices. The composition of silicon oxynitride can vary from silicon dioxide, with a refractive index of 1.46, to silicon nitride, with an index of 2.01. On the other hand, silicon nitride has a refractive index that increases with increasing amounts of excess silicon. Because of high tensile stress, silicon nitride thicker than 200 nm may crack. The resistivity of silicon nitride at room temperature depends on film stoichiometry, deposition temperature, and the amount of impurities in the film.

In recent ULSI devices, thin CVD nitrides are extensively used in the dielectrics of the storage capacitors for dynamic random-access memories (DRAMs) and as interpoly dielectrics for nonvolatile memory devices. Because of their high dielectric constant and bulk-limited current, thin nitrides, oxidized in a pyrogenic ambient, have been used for high-density DRAMs. In hopes of shrinking the dimension of the dielectric, thin nitrides grown at different temperatures (680 to 780°C) were analyzed by atomic force microscopy (AFM) as shown in Fig. 17. Obviously, higher deposition temperatures result in rougher nitride surfaces. Hence, the ability of resisting oxidation increases with decreasing deposition temperature.[28]

5.5 LPCVD POLYSILICON FILMS

The low-pressure chemical-vapor deposition (LPCVD) method is also widely used in integrated-circuit manufacturing for the deposition of polycrystalline silicon films. Polysilicon films are still used as the gate electrodes in modern ULSI MOS devices. Furthermore, a metal or metal silicide, such as tungsten or tantalum silicide, is often deposited over the polysilicon gate to form a *polycide* or *salicide* structure and reduce the electrical resistivity.[29,30] Polysilicon can also be used for resistors, conductors, and diffusion sources. Other applications of poly-Si films, such as thin-film transistors in ULSI SRAM, stacked MOS capacitors in ULSI DRAM,[31,32] photovoltaic conversion, thermal and mechanical sensors, and large-area LCDs,[33,34] have been demonstrated. The polysilicon is deposited by dissociating silane at a temperature between 575 and 650°C. Either pure silane or 20 to 30% silane, diluted in nitrogen, can be fed into the LPCVD system at a pressure of 0.2 to 1.0 torr to fabricate the polysilicon films. The chemical reaction is expressed as

$$SiH_4(g) \rightarrow Si(s) + 2H_2(g) \tag{5.8}$$

For practical use a deposition rate of about 10 to 20 nm/min is required.

5.5.1 Deposition Variables

The properties of LPCVD polysilicon films are determined by deposition pressure, silane concentration, deposition temperature, and dopant content. In a low-pressure CVD reactor, deposition pressure can be changed by varying the pumping speed at a constant inlet gas flow or by varying the gas flow but keeping the pumping speed constant. To obtain better deposition reproducibility, the former is used. Figure 18 shows the results of the polysilicon depositions for different silane partial pressures.[35] At first, silane is adsorbed on the surface. The adsorbed silane then decomposes to form SiH_2 and adsorbed hydrogen. The SiH_2 further decomposes to form polysilicon and more adsorbed hydrogen. Finally, the adsorbed hydrogen

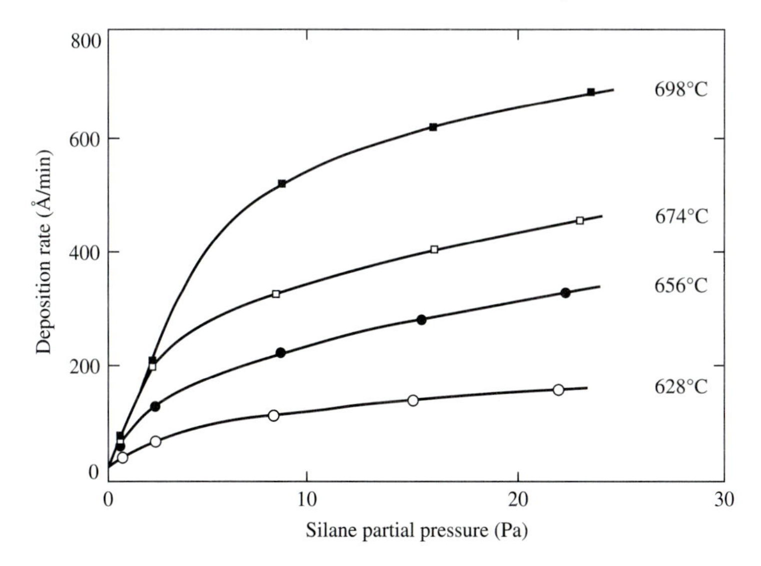

FIGURE 18
Deposition rate as a function of silane partial pressure for the polysilicon films deposited at different deposition temperatures. (*After Claassen et al., Ref. 35.*)

leaves, and additional silane is adsorbed to continue this reaction. The reaction sequence is written as

$$SiH_4(g) \longleftrightarrow SiH_4(ad) \tag{5.9}$$

$$SiH_4(ad) \longleftrightarrow SiH_2(ad) + H_2(ad) \tag{5.10}$$

$$SiH_2(ad) \longleftrightarrow Si + H_2(ad) \tag{5.11}$$

$$2H_2(ad) \longleftrightarrow 2H_2(g) \tag{5.12}$$

If the last step, desorption of the hydrogen, is slow as compared to the other steps, the overall reaction rate is given by

$$\text{Rate} = K_1 P_S^{1/2}(1 + K_2 P_S^{1/2}) \tag{5.13}$$

where P_S is the partial pressure of silane and K_1 and K_2 are derived from the equilibrium constants for the individual reaction steps of Eqs. (5.10) and (5.11). Hence, the deposition rate is correlated to the square root of the silane partial pressure.

At high temperatures, the surface reaction rate becomes higher than the arrival rate of the unreacted silane to the surface. When this occurs, the deposition rate no longer increases with deposition temperature, and the reaction is mass-transport-limited. Deposition occurring in the mass-transport-limited regime depends on the reactant concentration, reactor geometry, and gas flow. When the surface reaction rate is smaller than the arrival rate of the reactant, as with low deposition temperature, the deposition is surface-reaction-limited and the critical variables are reactant concentration and deposition temperature. Depositions that are surface-reaction-limited have excellent thickness uniformity and step coverage. The deposition rate increases with deposition temperature. The temperature dependence is usually exponential and follows the Arrhenius equation

$$R = A\exp(-E_a/kT) \tag{5.14}$$

where R is the deposition rate, A is the frequency factor, E_a is the activation energy in eV, T is the absolute temperature in K, and k is the Boltzmann constant. The activation energies for polysilicon deposition are found to be about 1.7 eV. The Arrhenius equation is useful for finding the deposition rate at different deposition temperatures. For example, the deposition rates of 10 nm/min and 2.5 nm/min can be obtained directly from Eq. (5.14) for the deposition temperatures of 600°C and 550°C, respectively, if the deposition rate is 18.75 nm/min at 625°C and the activation energy is 1.7 eV. For ULSI technology, polysilicon films are generally deposited at the temperatures ranging from 500 to 650°C. Higher temperatures result in a rough, loosely adhering deposit and poor uniformity. For temperatures much lower than 550°C, the deposition rate is too low. However, an amorphous structure will be achieved for deposition temperatures below 575°C. A subsequent annealing can be used to transform the amorphous silicon films into polycrystalline silicon with even larger grain sizes. These large-grain polysilicon films have been used for many applications.

Polysilicon can be doped in situ by adding phosphine, arsine, or diborane to the reaction gases. Here an inert gas is often accompanied to improve the film uniformity. The addition of diborane will significantly enhance the deposition rates of polysilicon films. However, the addition of phosphine or arsine seriously decreases the deposition rates, because the phosphine or arsine is strongly adsorbed on the silicon substrate surface, inhibiting the dissociative chemisorption of SiH_4. However, diborane forms borane radicals, BH_3, that catalyze gas-phase reactions and increase the deposition rate.

5.5.2 Structures

The structure of deposited polysilicon will be determined by the deposition temperature, deposition pressure, postdeposition annealing, dopant, and impurity contents. For deposition temperatures below 600°C the deposited silicon films will exhibit a microcrystalline or amorphous structure. For the microcrystalline phase, only grains with a size of tens to hundreds of angstroms (Å) will be detected using transmission electron microscopy (TEM). For the amorphous silicon, only the diffraction pattern of diffuse rings will be found in the TEM. Silicon films deposited above 600°C will possess the polycrystalline structure. The typical polysilicon deposited at 625°C generally has a grain size of about 0.1 μm with columnar grain structure. Note that the transition temperature, here 600°C, from the amorphous or microcrystalline to polycrystalline silicon films is not a constant but depends on the deposition pressure, deposition rate, and impurity or dopant incorporation. For example, the transition temperature will be lowered from 625°C to 575°C as the total pressure decreases from 50 Pa to 15 Pa. In general, the transition temperatures range between 550°C and 625°C. Polysilicon deposited at 600 to 650°C will have a {110} preferred orientation. For higher deposition temperatures, {100} orientation will increase. The texture will be affected by the deposition temperature, dopant, substrate material, and impurities.

After the silicon films are deposited, a subsequent annealing is required. The subsequent heating cycle can be used to crystallize the amorphous film and to increase the polysilicon grain size. The annealing processes can also be used for oxidation or dopant activation of the deposited silicon films. The polysilicon film as deposited has columnar grains, with grain size between 0.03 and 0.1 μm exhibiting a larger size on the top surface and a smaller size on the bottom interface. The deposited polysilicon grains do not significantly grow after a subsequent high-temperature annealing, as shown in Fig. 19*a*. On the other hand, the deposited amorphous silicon films or microcrystalline films will grow into the polycrystalline structure with a grain size even larger than those grown from a polycrystalline phase.[36] With a longer annealing time, the amorphous silicon can achieve a grain size larger than 0.2 μm, as shown in Fig. 19*b*. The technique has been popularly used in the thin-film transistors (TFTs) of SRAMs or LCDs addressed by TFTs to improve the electrical characteristics of the poly-Si TFTs. Other annealing schemes such as laser heating[37,38] can increase the polygrains from the amorphous type to a grain size as large as 10 μm, as shown in Fig. 19*c*. The electrical properties of TFTs fabricated with these polysilicon films are shown in Fig. 20. Higher I_{on}/I_{off} current ratio and field effect mobility have been achieved for the TFTs with larger grain sizes obtained by laser annealing (LA) as compared with those by furnace annealing (FA).[37,38] Recrystallization temperatures of the amorphous silicon films will also be affected by the dopants, impurities, and the heating methods. Oxygen, nitrogen,

(a)

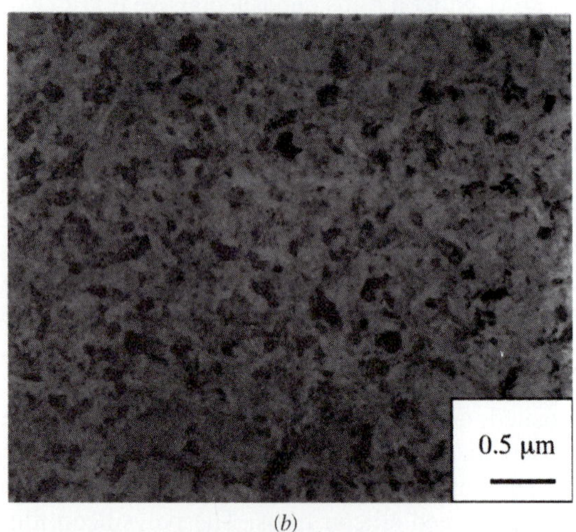

(b)

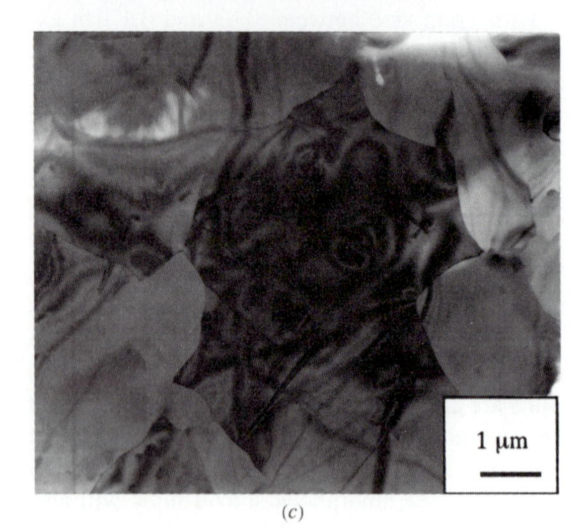

(c)

FIGURE 19
(*a*) Brightfield (BF) TEM showing the as-deposited poly-Si films furnace-annealed at 600°C for 48 h. (*b*) BF showing the amorphous-Si films furnace-annealed at 600°C for 48 h. (*c*) BF showing the amorphous-Si specimens CO_2-laser-annealed at 22.5 W. (*After Tsai and Cheng, Ref. 36.*)

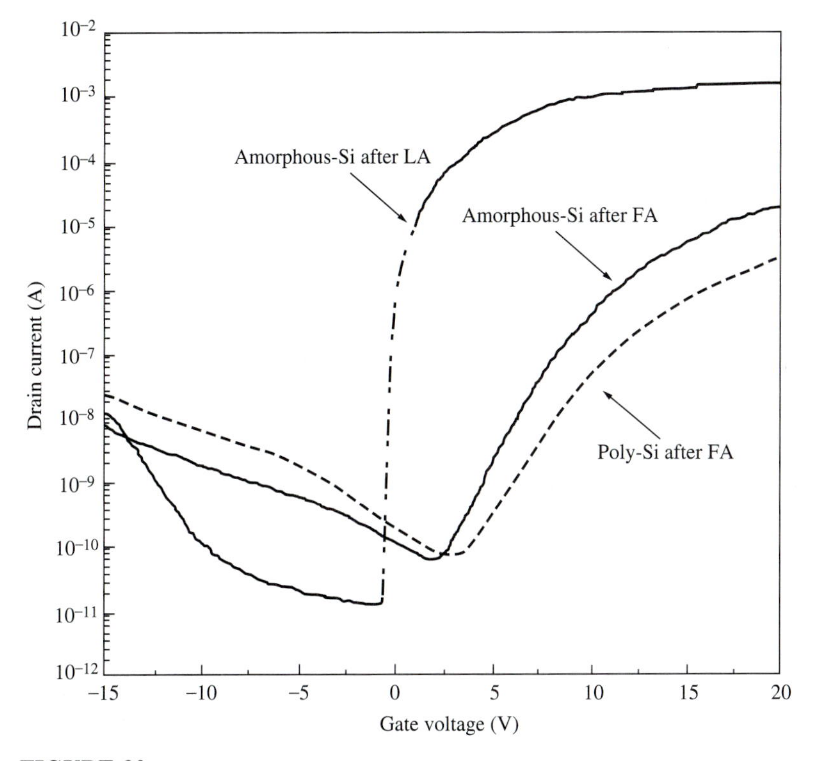

FIGURE 20
Transfer characteristics of TFT with the various recrystallized films shown in Fig. 19. $W/L = 25\ \mu m/10\ \mu m$ and $V_d = 5$ V. (*After Tsai and Cheng, Ref. 36.*)

and carbon will stabilize the amorphous films and raise the crystallization temperatures. Furthermore, the dopants, impurities, and heating methods will influence the preferred orientations. Table 3 lists the preferred orientations of the amorphous and polysilicon films after furnace annealing and laser annealing.

5.5.3 Doping Polysilicon

In general, implantation, diffusion, and in-situ doping are used to dope polysilicon films in ULSI technology.[39–41]

For dopant-implanted polysilicon films the resistivities are influenced by implant energy, implant species, implant dose, annealing temperature, and annealing time. Polysilicon films implanted with a low dose will lead to n-type grains alternating with intrinsic ones, because of the dopant traps in the grain boundaries. Polysilicon resistors can therefore be fabricated by this technique. Once these traps are saturated with dopants, the resistivity rapidly decreases. Diffusion of dopants results in low resistivities of polysilicon films. Because the dopants become segregated in the grain boundaries, the dopant concentrations in polysilicon films often exceed

TABLE 3
Orientation of amorphous and polysilicon films

Samples	X-ray texture (%)				
	(111)	(110)	(113)	(100)	(133)
As-deposited poly-Si	1.9	92.6	5.5	0	0
As-deposited amorphous-Si	—	—	—	—	—
Furnace-annealed poly-Si	3.4	85.5	6.8	2.6	1.7
Furnace-annealed amorphous-Si	30.2	25.8	27.3	9.1	7.6
Laser-annealed poly-Si (in solid-phase regime)	2.1	73.2	8.1	15.4	1.2
Laser-annealed amorphous-Si (in solid-phase regime)	11.5	14.2	26.3	38.2	9.8
Laser-annealed poly-Si (in liquid-phase regime)	14.7	13.6	8.2	58.8	4.7
Laser-annealed amorphous-Si (in liquid-phase regime)	4	2.4	4	79.4	10.2

the solid solubility limit. Polysilicon films doped in situ with phosphine, arsine, or diborane have resistivities that depend on the deposition temperature, dopant concentration, and annealing temperature. Increasing the deposition temperature will facilitate the decrease in the resistivities of polysilicon films; however, the resistivity still remains high. For in-situ boron doping, polysilicon can obtain a lower resistivity at a much lower deposition temperature than can be obtained for in-situ phosphorus doping.

For applications in the active region of TFTs, polysilicon films need to be grown to have larger grain size. Hence, an amorphous initial phase should be deposited and subsequently annealed for a long time or with a laser source. The device characteristics of the TFTs with polysilicon active layers can be improved by passivating the dangling bonds at the grain boundaries with dopants or hydrogen impurities, as mentioned in Section 5.5.4. For polysilicon films to be used as resistors, phosphorus implantation with appropriate low dose or double implantations of boron and phosphorus ions with an almost equal dose are utilized to increase the barriers at the grain boundaries. For polysilicon films to be used as conductors, such as gate electrodes or interconnections, high-dose doping will be necessary. In the future, low-temperature processes such as implantation and in-situ doping look promising. Implantation will offer precise control, and in-situ doping will have the benefit of a low-temperature process.

5.5.4 Properties of Polysilicon

Because polysilicon films have various electrical characteristics that depend on the doping processes, deposition conditions, post-annealing conditions, and passivation effects, they have been widely used in ULSI devices and other thin-film microelectronics. As mentioned above, the resistivity of polysilicon used in conductors is a main consideration. For such applications the doping concentration must be raised as

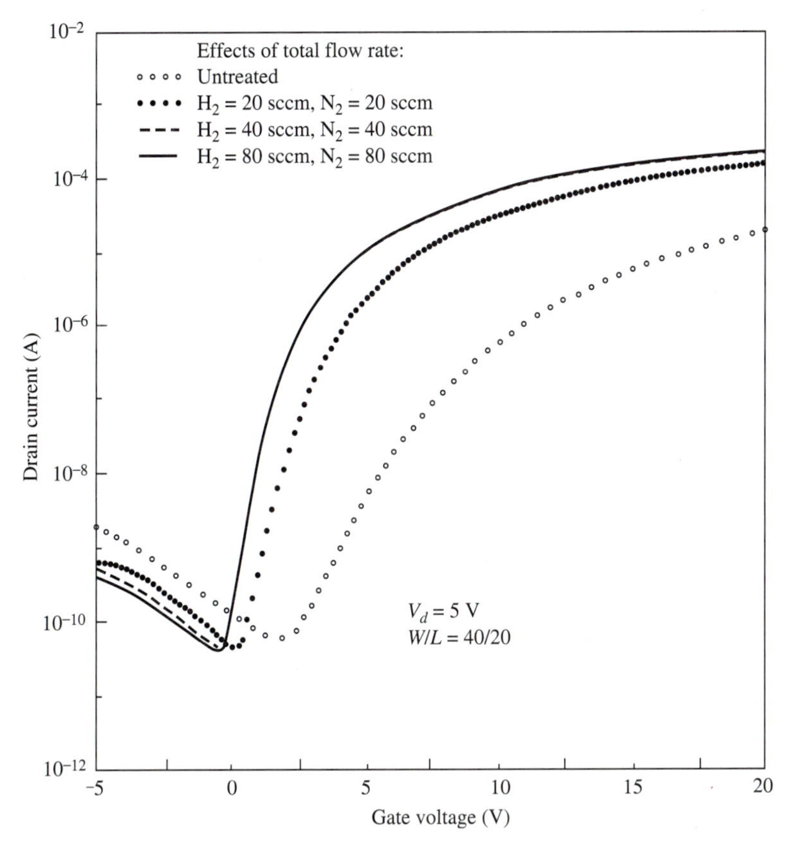

FIGURE 21
Characteristics of a TFT passivated with N_2–H_2 plasma for different flow rates. (*After Tsai and Cheng, Ref. 36.*)

high as possible. If polysilicon films are used as resistors, the doping concentration and the grain structure and silicon phase must be carefully controlled to increase the sheet resistance significantly. As for the use of polysilicon as a semiconductor, the deposition film phase, grain size, and passivation effects by hydrogen or nitrogen, etc., become very important. The previous section has described the electrical characteristics of TFTs fabricated with various polysilicon films. The electrical properties of TFTs are mainly determined by the grain sizes of these polysilicon films.

The passivation effects of polysilicon films on the electrical characteristics of TFTs are shown in Fig. 21. The on currents exhibit an increase for all samples passivated with hydrogen, nitrogen-diluted hydrogen, and nitrogen. Hydrogen has been reported to be capable of passivating the dangling bonds of polysilicon and improving the electrical properties of polysilicon.[42] The role of nitrogen in passivation is attributed to three effects. First, the added nitrogen would cause the formation of hydrogen radicals, which contain more hydrogen atoms to passivate the dangling bonds. Secondly, the added nitrogen suppresses the accumulation of hydrogen on the field oxide, which could induce surface leakage current. Thirdly, the nitrogen plasma

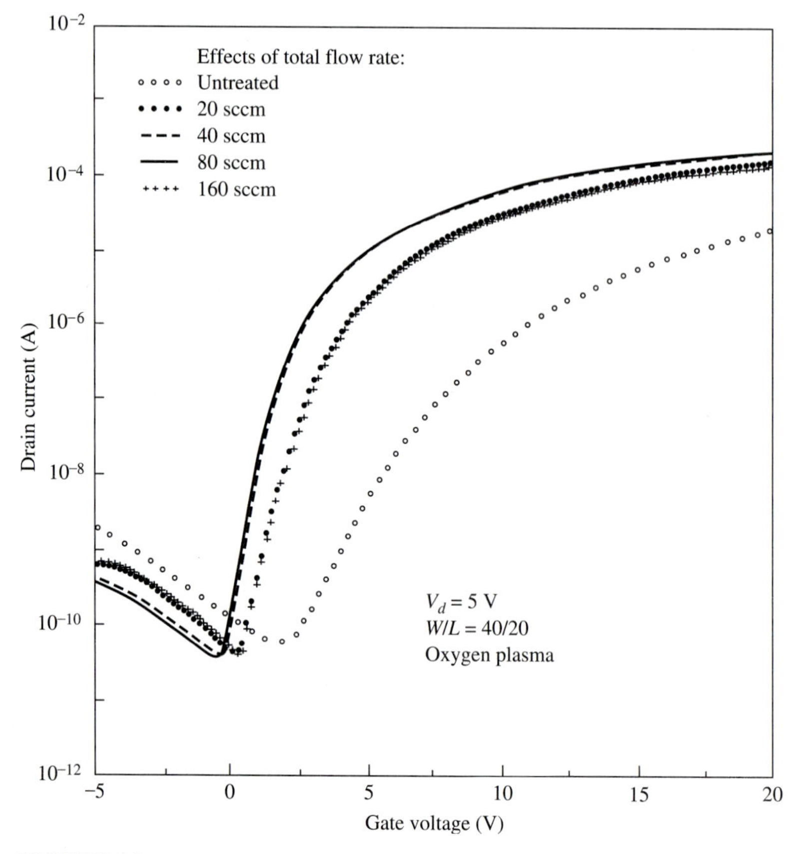

FIGURE 22
Characteristics of a TFT passivated with O_2 plasma for different flow rates. (*After Tsai and Cheng, Ref. 36.*)

could effectively passivate the poly-Si TFT by means of the three valence electrons of nitrogen. Hence, the TFT using nitrogen passivation also exhibits the lowest leakage current. Oxygen passivation effects on the polysilicon have also been found, as shown in Fig. 22. The passivation effects of various grain structures on polysilicon films are listed in Table 4. After hydrogen passivation, the laser-annealed samples still possessed better performance than the furnace-annealed specimens, because of the lower density of grain-boundary traps.

Polysilicon films are also utilized as dopant sources. As ULSI devices are scaled down, junction depths also shrink. With conventional implantation and activation, as well as the silicidation process, the control of shallow junction depth is difficult. A novel scheme, *implantation through polysilicon* (ITP), has been proposed to improve this phenomenon by using polysilicon as the diffusion source. By implanting dopant into thin poly-Si films, good junctions can be formed without these problems.[43,44] Figure 23 shows the leakage current J_r as a function of annealing temperature for the 1500Å-thick polysilicon films on silicon substrate with BF_2^+ implanted at various energies to a dose of 5×10^{15} cm^{-2}. Poly-Si serves as the implantation buffer layer

TABLE 4

Effects of passivation on performance of TFTs with various grain structures

	Before hydrogen passivation						After hydrogen passivation					
Samples	$I_{d,\min}$(pA)	S(V/decade)	V_t(volt)	μ_{eff}(cm²/v · cm)	I_{on}/I_{off}	N_t(1/cm²)	$I_{d,\min}$(pA)	S(V/decade)	V_t(volt)	μ_{eff}(cm²/v · cm)	I_{on}/I_{off}	N_t(1/cm²)
AP†	124	2.76	18.5	3.36	3.1E4	1.85E12	119	0.85	4.0	16.2	1.2E6	7.23E11
HA‡	58	1.94	13.2	7.23	2.9E5	1.67E12	52	0.76	3.2	22.7	3.9E6	6.78E11
LA§	65	1.76	11.3	8.7	4.1E5	1.48E12	47	0.62	2.0	37	5.6E6	5.57E11
LL¶	10	0.198	0.49	162	1.0E8	5.08E11	8	0.178	-0.42	188	1.2E8	4.9E11
LS††	66	2.14	14.9	5.4	1.3E5	1.65E12	65	0.74	2.82	23.2	2.7E6	6.33E11

The channel width and channel length were 25μm and 10μm, respectively.

† AP: As-deposited poly-Si film.

‡ HA: Amorphous-Si film after high-temperature grain growth.

§ LA: Amorphous-Si film after low-temperature annealing.

¶ LL: Amorphous-Si film after laser annealing under liquid-phase scheme.

†† LS: Amorphous-Si film after laser annealing under solid-phase regime.

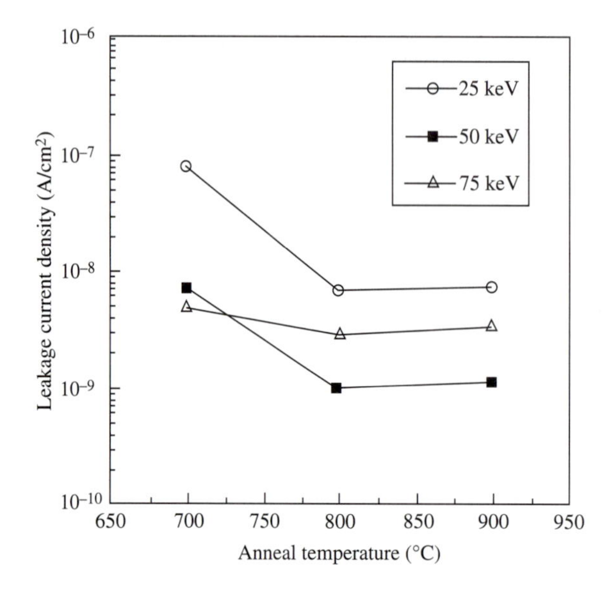

FIGURE 23
Dependence of leakage current density on annealing temperature for diodes fabricated with polysilicon films on silicon substrate samples BF_2^+-implanted at various energies to a dose of 5×10^{15} cm^{-2}. (*After Juang et al., Ref. 43.*)

as well as the doping source here. Excellent shallow p^+n junctions, which show a leakage of 1 nA/cm^2 and a junction depth of about 0.05 μm, have been formed by implanting BF_2^+ ions into thin poly-Si films and subsequent annealing.[43]

5.5.5 Oxygen-Doped Polysilicon

Oxygen-doped polysilicon, which has very high resistivity, is known as *semi-insulating polysilicon* (SIPOS).[45] SIPOS is deposited at temperatures ranging from 600 to 700°C. Silane mixed with a small quantity of nitrous oxide is fed into the reactor; the reactions are expressed as

$$SiH_4(g) \rightarrow Si(s) + 2H_2(g) \tag{5.15}$$

$$SiH_4(g) + xN_2O(g) \rightarrow SiO_x(s) + 2H_2(g) + xN_2(g) \tag{5.16}$$

SIPOS will have multiple phases consisting of crystalline silicon, amorphous silicon, silicon dioxide, and silicon monoxide. The resulting properties of SIPOS will be influenced by the amount of nitrous oxide, the deposition temperature, and the post-annealing condition. For practical use, SIPOS films generally have oxygen concentration of about 30 at%.

5.5.6 Oxidation of Polysilicon

Recently, polycrystalline silicon has become an important material in silicon technology. In most integrated circuits, *polyoxides* grown on polysilicon films have been widely applied in SRAMs, EPROMs,[46] EEPROMs, flash memories, and thin-film transistors. Polyoxides are often used to isolate the devices and are generally fabricated by annealing the polysilicon films in dry or wet oxygen ambient at 800 to 1000°C. The polyoxide thickness depends on the ambient, annealing temperature, and annealing time. Furthermore, polysilicon oxidation must be achieved in a short duration if the polyoxides are to be used as gate dielectrics.

For polysilicon films by LPCVD at around 625°C, the oxidation mechanism is predominantly surface-reaction-limited. The oxidation rates depend on the orientations of silicon grains, which cause the roughness of the interface between the polyoxide and the polysilicon layer. At higher oxidation temperatures the oxidation mechanism is predominantly diffusion-controlled, resulting in a smoother polyoxide/polysilicon interface.[47,48] The high electrical conductivity found for the polyoxides is ascribed to the enhanced local electric field. On the other hand, the polyoxides of amorphous silicon films deposited at 560°C and subsequently annealed at 600°C to form the polycrystalline structures are reported to have breakdown fields higher than those of the polysilicon films as deposited at 620°C. Both polysilicon films deposited at 625°C and amorphous films deposited at 550°C with subsequent annealing at 600°C for 24 hours to form polysilicon films are used to grow polyoxides at 850°C in wet or dry oxygen. For the recrystallized amorphous silicon specimens, the polyoxide/polysilicon interface is very flat, as shown in Fig. 24. On the other hand, the polyoxide/polysilicon interface is rough for the recrystallized as-deposited polysilicon samples, as shown in Fig. 25. Cross-sectional HRTEM (high-resolution transmission electron microscope) observation shows that intergranular oxidation occurs for the polysilicon specimens after wet oxidation at 850°C for 60

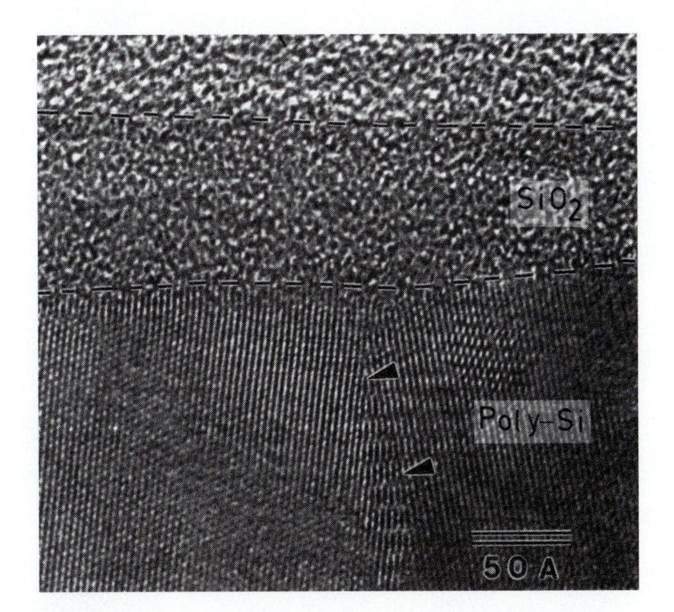

FIGURE 24
Cross-sectional high resolution TEM (HRTEM) for wet polyoxide grown on the amorphous specimen at 850°C for 150 s in wet oxygen ambient, showing the smooth polyoxide/polysilicon interface on the low-angle grain boundary. (*After Wang et al., Ref. 48.*)

FIGURE 25
Cross-sectional HRTEM image of polysilicon samples after dry oxidation at 850°C for 18 min, indicating the annealed as-deposited polyoxide thickness fluctuation on the high-angle grain boundary, marked by the arrows. (*After Wang et al., Ref. 48.*)

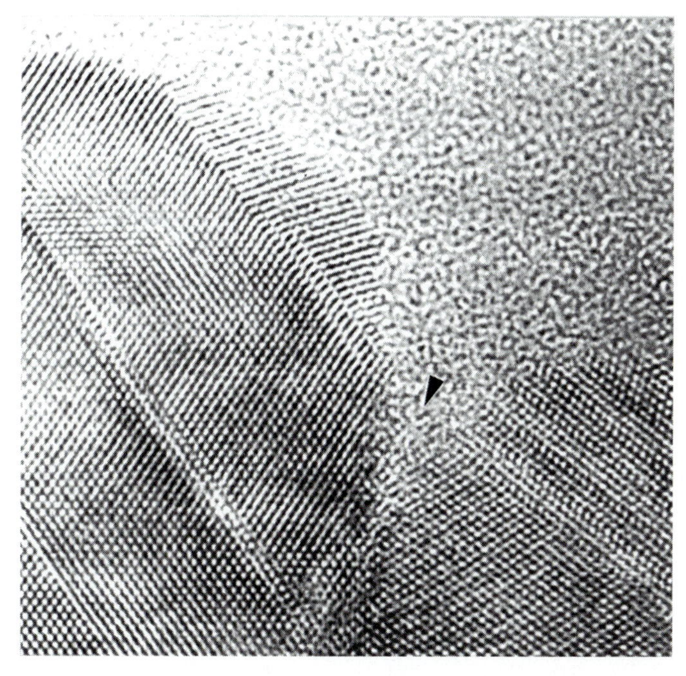

FIGURE 26
Cross-sectional HRTEM image showing the intergranular oxidation phenomenon, marked by arrows, for the polysilicon recrystallized as-deposited specimens wet-oxidized at 850°C for 60 min. (*After Wang et al., Ref. 48.*)

min, as shown in Fig. 26. The oxidation phenomenon of recrystallized amorphous samples is illustrated in the schematic diagram in Fig. 27*a*. Due to the poor crystallinity and low-angle grain boundaries of the recrystallized amorphous films, defect density is not much higher at grain boundaries than in the other parts of polysilicon films. Hence, the oxidation rates are similar for regions of highly damaged large grains, grain boundaries, and small or microcrystalline grains. On the other hand, the recrystallized as-deposited polycrystalline specimens possess good crystallinity and high-angle boundaries. The oxidation rates at the highly damaged region and grain boundaries are very different from those at the grains with good crystallinity. This results in the polyoxide/polysilicon interface roughness. The oxidation phenomenon of the as-deposited polysilicon specimens is illustrated in Fig. 27*b*. Figure 28 shows the Weibull plots of the time-zero-dielectric-breakdown (TZDB) characteristics for the polyoxides grown on the annealed as-deposited polysilicon films and the annealed amorphous silicon films. A wide range of breakdown events, which distribute from 1 to 7 V for the as-deposited polysilicon specimens and a narrow distribution for the recrystallized amorphous samples, is observed. Much better electrical characteristics have been achieved by using polyoxides on the recrystallized amorphous silicon films.[48]

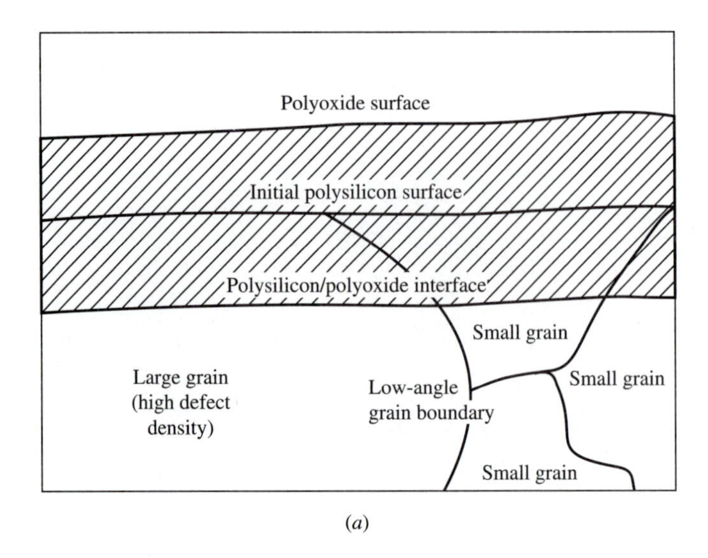

(*a*)

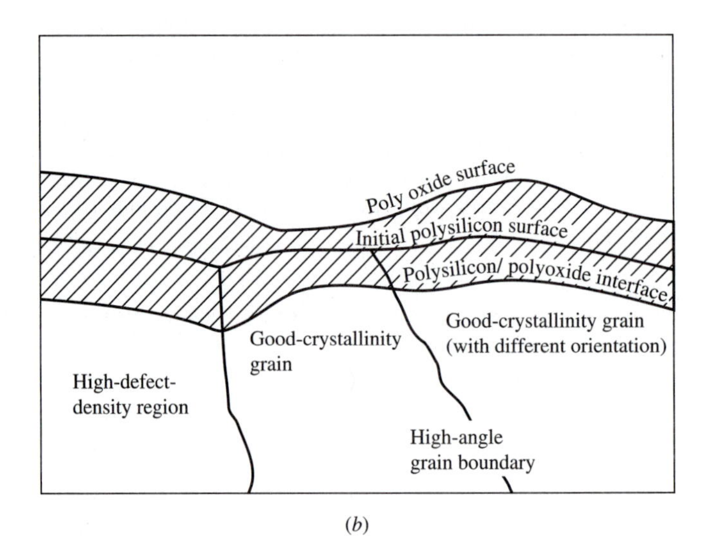

(*b*)

FIGURE 27
(*a*) Schematic diagram illustrating the oxidation phenomenon of the recrystallized amorphous samples. (*b*) Schematic diagram depicting the oxidation phenomenon of the as-deposited polysilicon samples.

5.6 PLASMA-ASSISTED DEPOSITIONS

A CVD method is categorized not only by the pressure regime but also by its energy input. As mentioned in Section 5.2.2, PECVD uses a radio frequency (rf) power

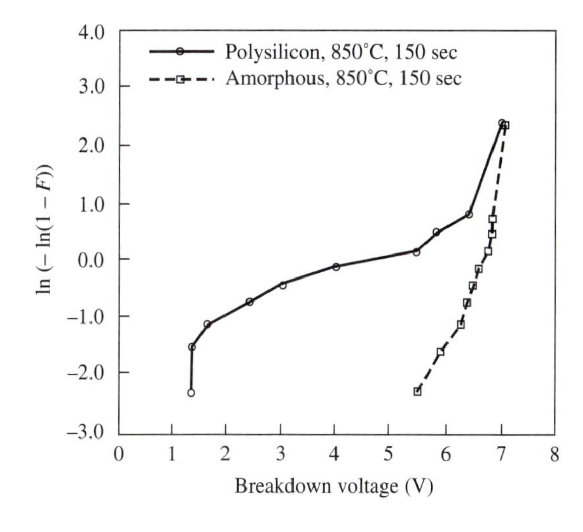

FIGURE 28
Weibull plots of the TZDB characteristics for the polyoxides grown on both recrystallized amorphous-Si and poly-Si samples. (*After Wang et al., Ref. 48.*)

to generate glow discharge to transfer the energy into the reactant gases, allowing the deposition on the substrate at a lower temperature than in the APCVD or the LPCVD process.[9,49–55] Hence, low deposition temperatures are a major advantage of PECVD. Desirable properties such as good adhesion, low pinhole density, good step coverage, adequate electrical properties, and compatibility with fine-line pattern transfer processes have made these PECVD films useful in ULSI circuits. Owing to good protection properties of scratch resistance, sodium diffusion, and moisture protection, plasma-deposited silicon nitride is commonly used as the final passivation layers of devices. Plasma-deposited nitride and oxide can be used as insulators between the metal layers, especially since the underlying metals have low melting points.[56] PECVD also provides deposition of amorphous silicon films, although these films have not yet been used in the ULSI circuits. The PECVD amorphous silicon has been widely applied in TFT LCDs.

5.6.1 Deposition Methods

The plasma is generated by the application of an rf field to a low-pressure gas, thereby creating free electrons within the discharge region. The electrons gain energy from the electric field. When they collide with gas molecules, the reactant gases (e.g., silane and nitrogen- or oxygen-containing species) dissociate and ionize. The energetic species are then adsorbed on the film surface. These radicals tend to have high sticking coefficients and also appear to migrate easily along the surface after adsorption. These two factors can lead to excellent film conformality. Upon being adsorbed on the substrate, the radicals are subjected to ion and electron bombardment and rearrangements and react with other adsorbed species to form new bonds and the resulting film. Atom rearrangement includes diffusion of the adsorbed atoms onto stable sites and concurrent desorption of reaction product. The desorption rates depend on the substrate temperatures; higher temperatures produce films with fewer entrapped by-products. Note that homogeneous gas-phase nucleation should be avoided to reduce particulate contamination.

Three types of PECVD reactors are available:

1. The parallel plate type
2. The horizontal tube type
3. The single wafer type

As compared with rf sputtering systems, the key difference in designing the reactor and electrode configuration in PECVD systems is that the potentials of both the powered and the grounded electrodes, relative to the plasma, must be approximately equal in PECVD systems. The walls of parallel plate reactors are made of either quartz-, ceramic-, or aluminum oxide–coated steel in order to place them at a floating potential with respect to the plasma. This can minimize bombardment and sputtering of the wall and reduce contamination on the growing films.

Parallel-plate PECVD reactors

The radial parallel plate reactor was developed by Reinberg in the early 1970s, and the first commercial model was offered in 1976. A schematic diagram of this reactor type is shown in Fig. 29*a*. The reaction chamber is a short, vertically oriented cylinder, typically constructed of aluminum-coated stainless steel. The rf power (at frequencies of 380 kHz to 13.56 MHz), which establishes the plasma, is applied to the upper electrode, and the wafers reside on the bottom, grounded electrode, which can be rotated, for improved uniformity, at substrate temperatures up to 400°C. The electrode spacing is typically 5 to 10 cm, and such systems operate in the pressure range of 0.1 to 5 torr. The parallel plate system, however, suffers from low throughput for large-diameter wafers. In addition, particulates flaking off from the walls or the upper electrode can fall onto the horizontally positioned wafers.

Horizontal-tube PECVD reactors

A new PECVD reactor design, the horizontal tube reactor, is shown in Fig. 29*a*. This design allows the throughput of PECVD reactors to be greatly increased. The reactor resembles a "hot-wall" LPCVD system. It consists of a long horizontal quartz tube that is radiantly heated. Gas is fed into one end and flows linearly to the other. Special long rectangular graphite plates serve both as the electrodes needed to establish a plasma and as the holders of the wafers. The electrode configuration is also designed to provide a uniform plasma environment for each wafer to ensure film uniformity. These vertically oriented graphite electrodes are stacked parallel to one another, side by side, with alternating plates serving as power and ground for the rf voltage. A plasma is formed in the space between each pair of plates.

The use of several long slabs allows an increased number of wafers (up to 120 pieces of 100 mm in diameter) to be loaded into the reactor at one time. Since the wafers are held vertically, most particles do not fall on the wafer surfaces. The entire graphite assembly is withdrawn from the reactor and then reinserted for processing. Care must be taken, however, to prevent particulate contamination when the wafers are loaded and unloaded.

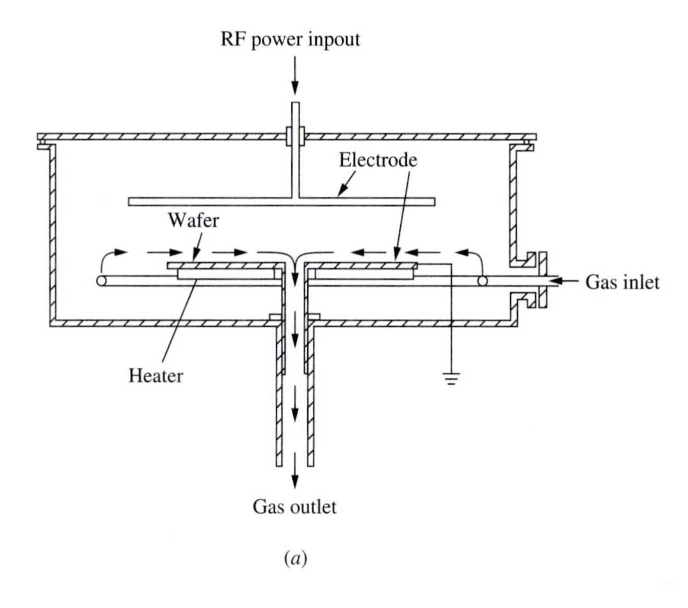

(*a*)

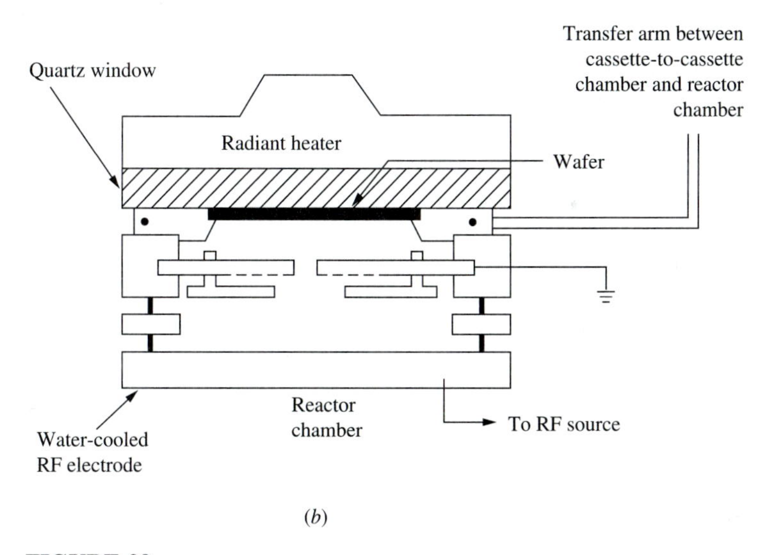

(*b*)

FIGURE 29
Schematic diagrams of plasma deposition reactors: (*a*) parallel plate type, and (*b*) single-wafer type.

Single-wafer PECVD reactors

The most recently introduced PECVD reactor is the single-wafer design shown in Fig.29*b*. The reactor, which is load-locked, offers cassette-to-cassette operation and provides rapid radiant heating of each wafer as well as allowing in-situ monitoring of the film deposition. Wafers larger than 200 mm in diameter can be adopted. Such reactors are being designed to allow the fabrication of large wafers with low particulate contamination. Multichamber PECVD, with or without a reactive ion etching (RIE) system, has also been developed and used to deposit the oxide, nitride, and silicon film free of air-exposure contamination.[57]

5.6.2 Deposition Variables

Plasma-assisted CVD requires the control and optimization of rf power density, frequency, and duty cycle in addition to the conditions similar to those of an LPCVD process, which possesses the parameters of gas composition, flow rate, deposition temperature, and total pressure. Like the LPCVD process at low temperatures, the PECVD process is surface-reaction-limited, and adequate substrate temperature control is thus necessary to ensure film thickness uniformity.

By reacting silane and oxygen or nitrous oxide in plasma, silicon dioxide films can be deposited. The reactions are expressed as

$$SiH_4(g) + O_2(g) \rightarrow SiO_2(s) + 2H_2(g) \tag{5.17}$$

$$SiH_4(g) + 4N_2O(g) \rightarrow SiO_2(s) + 4N_2(g) + 2H_2O(g) \tag{5.18}$$

Furthermore, silicon nitride films are deposited by reacting silane and ammonia or nitrogen in plasma, with the reactions

$$SiH_4(g) + NH_3(g) \rightarrow SiN : H(s) + 3H_2(g) \tag{5.19}$$

$$SiH_4(g) + N_2(g) \rightarrow 2SiN : H(s) + 3H_2(g) \tag{5.20}$$

Amorphous silicon is prepared mainly by the glow discharge decomposition of silane, with the reaction expressed as

$$SiH_4(g) \rightarrow Si(s) + 2H_2(g) \tag{5.21}$$

Processing parameters such as deposition rate will be affected by such deposition variables as total pressure, reactant partial pressure, discharge frequency and power, electrode materials, gas species, reactor geometry, pumping speed, electrode spacing, and deposition temperature. The higher the deposition temperature and the rf power, the higher the deposition rate is. Furthermore, the properties of all these PECVD films will be influenced by these deposition conditions. These properties will be discussed in the following section.

5.6.3 Properties of Plasma-Deposited Films

The fact that the radicals formed in the plasma discharge are highly reactive causes some stoichiometric problems, because the reactions are so varied and complicated. Moreover, the by-products and incidental species are incorporated into the reactant films, especially hydrogen, nitrogen, and oxygen. Excessive incorporation of these contaminants may lead to threshold voltage shifts in MOS circuits that use these PECVD films.

The refractive indices of plasma-deposited silicon nitride are distributed in a large range (1.8 to 2.6). The resistivity and breakdown field of the PECVD silicon

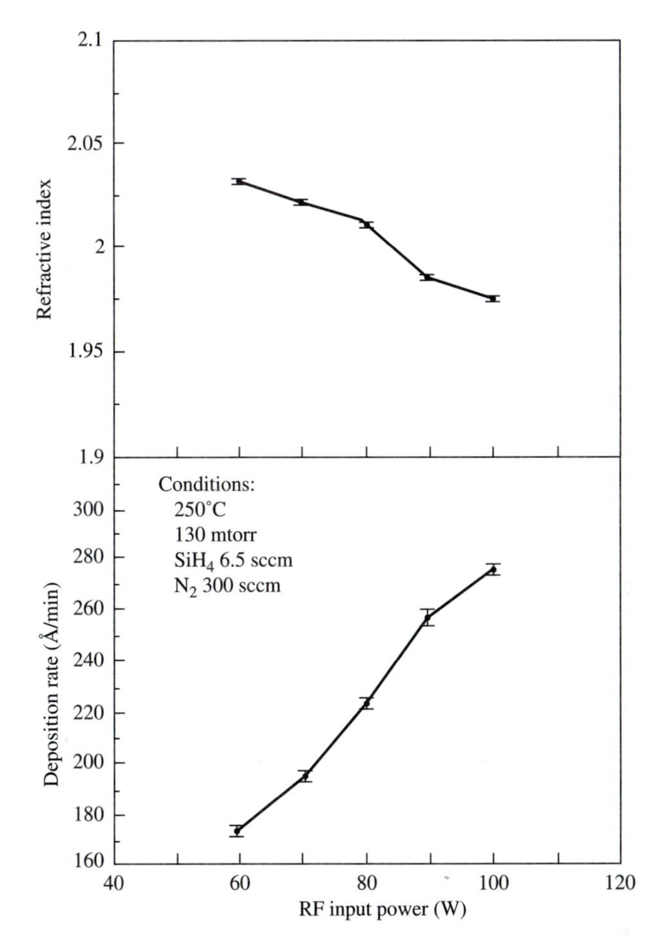

FIGURE 30
Refractive index and deposition rate as functions of rf power for plasma nitride films.

nitride also vary with different deposition conditions (10^5 to 10^{21} ohm-cm and 3 to 9 MV/cm, respectively). Figure 30 shows the refractive index and deposition rate as functions of rf power for plasma nitride films. For silicon-rich nitride films, which lower rf power tends to form, refractive indices will be high and resistivities will be low. Low rf powers also exhibit lower deposition rates. In addition, the dielectric breakdown field, dielectric constant, and band gap of plasma nitrides also vary with the compositions.

Plasma silicon nitrides often contain high hydrogen concentrations of between 10 and 35 at% hydrogen. The relative concentrations of Si–H and N–H vary widely and cause a large variation in film properties. Because high stress will cause cracking of the nitride film, minimizing stress in PECVD silicon nitride is most important. Crack-free films with low tensile stress can be deposited by precisely controlling the process parameters. Stress depends on the local bonding around individual atoms and is affected by deposition variables such as rf frequency and deposition temper-

TABLE 5
Properties of silicon nitride films

	Deposition	
	Plasma	LPCVD
Temperature(°C)	250–350	700–800
Composition	SiN_xH_y	$Si_3N_4(H)$
Stress (10^9 dyne/cm)	2C-5T	10T
Dielectric constant	6–9	6–7
Dielectric strength (10^6 V/cm)	5	10
Resistivity (Ω-cm)	10^6–10^{15}	10^{16}
Refractive index	1.8–2.5	2.01
Density (g/cm^3)	2.4–2.8	2.9–3.1
Energy gap (eV)	4–5	5
Si/N ratio	0.8–1.2	0.75
Atom % H	20–25	4–8

ature. Low rf frequency facilitates the plasma nitride films with compressive stress. The general properties of plasma-deposited silicon nitride are listed in Table 5. Since the film composition is a variable, the properties of plasma nitride vary over large ranges.

In contrast, the properties of plasma oxide are nearly constant, because the oxide composition is nearly stoichiometric. The refractive index and deposition rate of plasma silicon oxide as functions of deposition temperature are shown in Fig. 31. High deposition temperatures attain higher deposition rates. However, the refractive indices of these plasma oxide films are almost a constant, of about 1.45. Table 6 lists the properties of PECVD silicon oxide films. Note that the plasma oxides become conformal if the reaction of SiH_4 with O_2 or N_2O is replaced by that of TEOS with O_2 or N_2O. This feature makes PECVD TEOS very useful in ULSI technology. Combining the isotropic flow characteristics of the APCVD or slightly-reduced-pressure CVD O_3–TEOS films with the PECVD TEOS provides the planarization scheme for ULSI technology, especially for refilling the aluminum interconnects, which require low-temperature processing. Here, PECVD TEOS films can provide moisture protection for the APCVD O_3–TEOS in addition to the conformality feature. The processing conditions of the PECVD TEOS are a TEOS flow rate of 7 sccm, an O_3 to TEOS ratio of 13 : 1, a total pressure of 0.6 torr, an rf power of 0.5 W/cm, and substrate temperature of 350°C. The resulting deposition rate of PECVD TEOS is 250 Å/min.[58]

Some new dielectric films prepared by PECVD have been applied in ULSI memory devices (e.g., MOS capacitors of DRAMs), including Ta_2O_5,[59] TiO_2, $LiNbO_3$, and PLZT (lead lanthanum zirconate titanate, (Pb, La)(Zr, Ti)O_3).[57] Note that the use of Ta_2O_5 films has been announced by some microelectronics companies to increase the storage capacitance in 256-Mbit DRAM devices by a factor of about 4.[60] Conventional MOS capacitors are fabricated with the stacked oxide/nitride/oxide (ONO) structure prepared by LPCVD. In the ONO structure the tensile stress in nitride film is offset by the compressive stress of the oxide. However, all of the *oxynitrides,* or stacked oxide/nitride/oxide films, will not provide enough

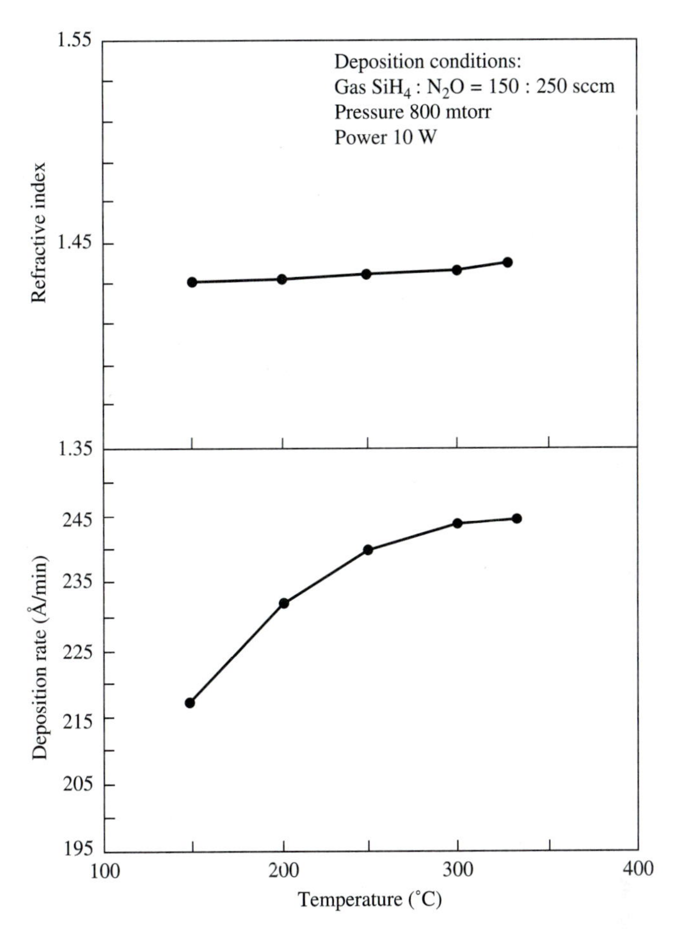

FIGURE 31
Refractive index and deposition rate of plasma silicon oxide as functions of deposition temperature.

capacitances for future ULSI memory devices. Hence, Ta_2O_5 films are fabricated in CVD from solid sources such as $TaCl_5$, or related halogens and liquid materials, such as $Ta(OCH_3)_5$, $Ta(OC_2H_5)_5$, and related alkoxides. In addition to PECVD, Ta_2O_5 can be prepared by LPCVD, metal-organic CVD (MOCVD), and photon-induced CVD. Photon-induced CVD and MOCVD will be described in Section 5.7.

5.7 OTHER DEPOSITION METHODS

5.7.1 Photon-Induced Chemical Vapor Deposition

Photon-induced chemical vapor deposition (PHCVD, photo CVD) is a technique that may fulfill the need for an extremely-low-temperature deposition process without the

TABLE 6
Properties of silicon oxide films

Deposition	**Plasma** $SiH_4 + O_2$(or N_2O)	**Plasma** TEOS + O_2	**APCVD** TEOS + O_3	**LPCVD** $SiH_4 + O_2$	**LPCVD** TEOS + O_2	**LPCVD** $SiCl_2H_2 + N_2O$	**Thermal**
Temperature(°C)	250	400	400	450	700	900	1000
Step coverage	Nonconformal	Conformal	Isotropic Flow	Nonconformal	Conformal	Conformal	Conformal
Composition	$SiO_{1.9}$(H)	SiO_x	$SiO_2(-OH)$	SiO_2(H)	$SiO_2(-OH)$	$SiO_2(-Cl)$	SiO_2
Stress (10^9 dyne/cm^2)	3C-3T	C-T	T	3T	1C	3C	3C
Dielectric constant	4.9	—	—	4.3	4.0	—	3.9
Dielectric strength (10^6 V/cm)	3–6	6	3–6	8	10	10	11
Refractive index	1.45	1.45	—	1.44	1.46	1.46	1.46
Density (g/cm^3)	2.3	—	1–2	2.1	2.2	2.2	2.2
Etch rate, nm/min (100 : 1 H_2O : HF)	40	164 20 : 1 H_2O : HF	—	6	3	3	2.5
Comments	High density Cusp problem and microparticle contamination	High density Limited surface coverage (combinations with SOG or O_3-TEOS)	Excellent surface coverage Self-planarization Low density High moisture affinity Stress-induced crack over 2mm thick	High density Low temperature	High density High temperature High moisture affinity Backing-induced crack	High density High temperature No doping capability	Excellent quality High temperature

problems of PECVD.[61] PHCVD uses high-energy, high-intensity photons to heat the substrate surface or to dissociate and excite reactant species in the gas phase. In the case of substrate surface heating the reactant gases are transparent to the photons, and the energy for gas-phase reactions is completely eliminated. On the other hand, in the case of reactant gas excitation the energy of photons can be chosen for energy-efficient transfer either to the reactant molecules themselves or to a catalytic intermediary such as mercury vapor. This technique enables deposition at extremely low substrate temperatures (e.g., room temperature). Unlike PECVD, PHCVD possesses the advantage of deposited films with a low defect density. Photo CVD has been used to deposit SiN_x film with the underlying substrate free of ion-bombardment damage that can be induced by PECVD. In addition, PHCVD films also show good step coverage. However, they may suffer from low density and molecular contamination as a result of the low deposition temperature.

There are three classes of PHCVD reactors, depending on the energy source: (1) UV lamp, (2) laser,[62] and (3) IR lamp. UV lamp reactors generally use mercury vapor for the energy transfer between photons and the reactant gases. UV radiation at the wavelength of 253.7 nm is efficiently absorbed by mercury atoms, which then transfer energy to the reactant species. Deposition rate for UV PHCVD reactors are typically much slower than those for other low-temperature techniques. The process data and film quality of Si_3N_4, amorphous silicon, and SiO_2 are listed in Table 7. The reactor chamber and the PHCVD system are schematically shown in Figs. 32*a* and *b*. The chemical reactions of PHCVD of a silicon film are expressed as

$$\mathrm{Hg(g)} + h\nu\,(253.7\ \mathrm{nm}) \rightarrow \mathrm{Hg^*(g)} \tag{5.22}$$

$$\mathrm{Hg^*(g)} + \mathrm{SiH_4(g)} \rightarrow \mathrm{Hg(g)} + \mathrm{SiH_4^*(g)} \tag{5.23}$$

$$\mathrm{SiH_4^*(g)} \rightarrow \mathrm{Si(s)} + 2\mathrm{H_2(g)}, \tag{5.24}$$

where Hg^* is the excitation state of Hg and $h\nu$ (253.7 nm) indicates the UV light. As for the deposition of the oxide film, its chemical reactions are written as

TABLE 7

Properties of films deposited by PHCVD

	Film type		
	Si_3N_4	α-Si	SiO_2
Thickness uniformity (%)	±5	±5	±5
Deposition rate (nm/min)	7–10	6	12
Deposition temperature (°C)	50–250	50–250	50–200
Deposition pressure (torr)	0.3–1.5	0.2–1.0	0.3–1.0
Dielectric constant	5.5	—	4.0
Refractive index	1.8–2.0	3.3–3.4	1.45–1.50
Density (g/cm^3)	1.8–2.5	2.2–2.3	2.1
Hydrogen content (at %)	10–15	8–20	—
Etch rate (nm/min) 20 : 1 H_2O : HF	5–10	—	10–15

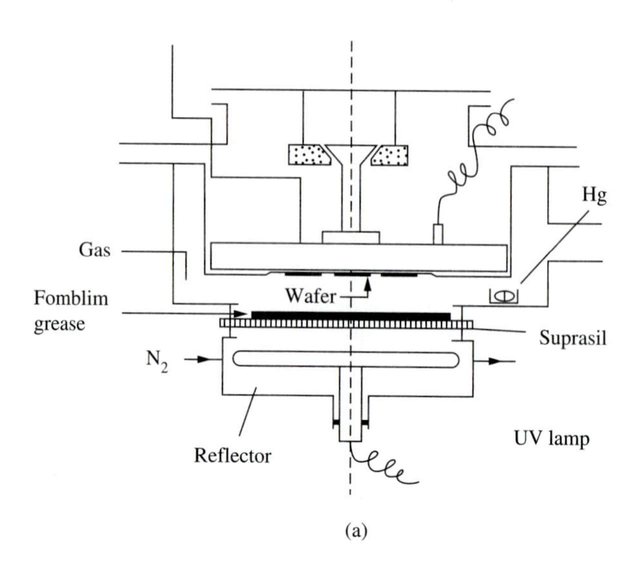

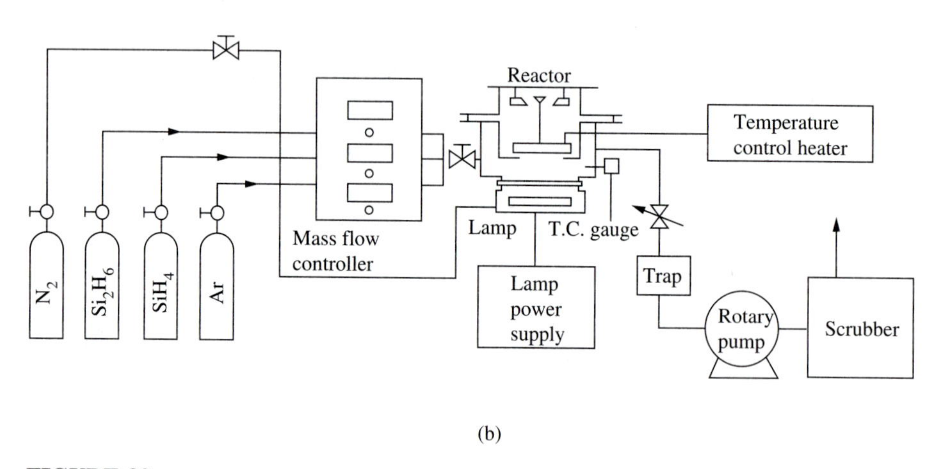

FIGURE 32
Schematic configuration of (*a*) the photo CVD reactor chamber, (*b*) the photo CVD system.

$$Hg(g) + h\nu\,(253.7\text{ nm}) \rightarrow Hg^*(g) \tag{5.25}$$

$$Hg^*(g) + N_2O(g) \rightarrow Hg(g) + N_2(g) + O(g) \tag{5.26}$$

$$SiH_4(g) + 2O(g) \rightarrow SiO_2(s) + 2H_2(g) \tag{5.27}$$

Similarly, Si_3N_4 films can be deposited by reacting SiH_4 with NH_3. Figure 33 shows the deposition rate of silicon dioxide deposited by PHCVD as a function of substrate temperature. Higher substrate temperatures will facilitate higher deposition rates, and the activation energy of films deposited by PHCVD is lower than that in the thermal CVD process.

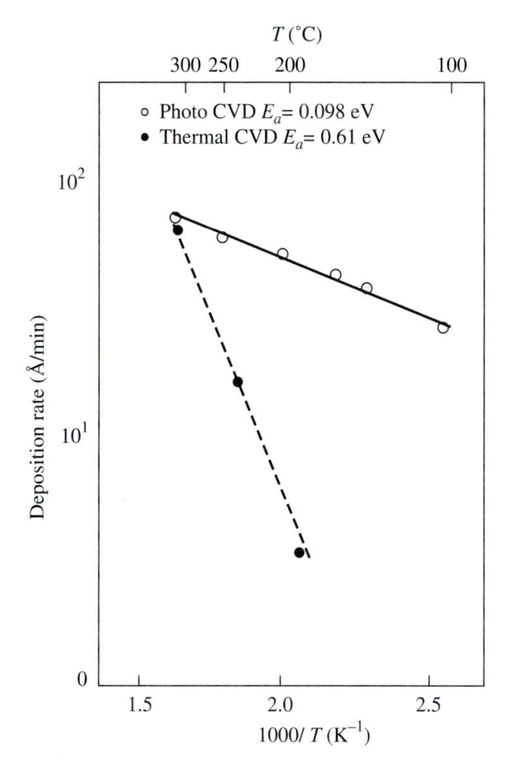

FIGURE 33
The deposition rates of silicon dioxide by thermal CVD (dashed line) and photo CVD (solid line) as a function of substrate temperature. (*After Peters et al., Ref. 61.*)

Laser PHCVD reactors offer the advantages of frequency tunability and a high-intensity light source. Tunability is useful to target the specific photon energies for particular dissociation reactions, enabling more flexible control of the depositions. The energy from the high-intensity laser also increases the reaction rate. Laser PHCVD opens the possibility of CVD writing, i.e., selective deposition by spatial control of the laser. Nevertheless, the deposition rates of the current PHCVD processes are still too low to allow them to be adopted for production applications.

On the other hand, an IR lamp has often been used in deposition of silicon dioxide, silicon film, and silicon nitride. Due to the rapid heating rate of the IR lamp, the reactors with an IR lamp heating source are also called rapid thermal processing (RTP) systems. If the reactions for formation of SiO_2, Si_3N_4, and polysilicon by chemical vapor deposition (CVD) are performed using the IR lamp, this technology is named as RTPCVD. The resulting film properties are all similar to those for the LPCVD or APCVD processes. However, the short processing duration can reduce autodoping of CVD epitaxial silicon or polysilicon films.

5.7.2 Electron Cyclotron Resonance Chemical Vapor Deposition

The microwave electron cyclotron resonance (ECR) CVD method is expected to realize low-temperature, low-ion-damage, and high-deposition-rate processes.[63–65] Silicon nitride films formed by ECR CVD without intentional heating of substrates can show good electrical insulating and protective properties as passivation films.

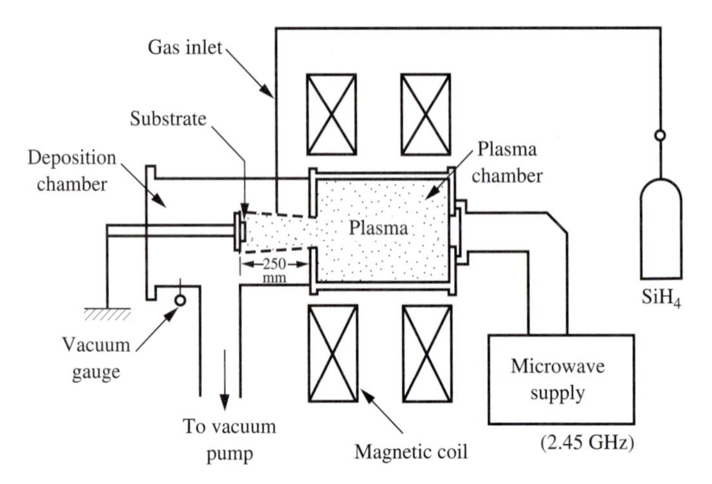

FIGURE 34
Schematic diagram of the microwave ECR CVD system.

Figure 34 illustrates a schematic configuration of the microwave ECR plasma CVD system. Microwave power is supplied to a plasma chamber through a rectangular waveguide. The magnetic field for the ECR plasma excitation is applied perpendicularly to the substrate surface. The plasma stream is introduced into the deposition chamber in the direction of the substrate by the magnetic field. The distance from the substrate to the plasma chamber is 250 mm. The gases are introduced into the deposition chamber. For ECR CVD the deposition variables include the microwave power, total pressure, gas species, and flow rate. These conditions are very similar to those for PECVD, except that the rf power is replaced by the microwave power. However, the ECR source generally creates a large number of electrons to generate a high density of plasma. Hence, the total pressure for the ECR CVD is much less than that for PECVD. The required power is also much lower than that for PECVD. Consequently, the induced plasma possess a lower ion energy, and the resulting film will achieve a higher quality due to low defect density. Typical deposition conditions are listed in Table 8. Figure 35 shows the effect of silane gas flow and the total pressure on the deposition rate of silicon oxide. Higher total pressures and flow rates result in higher deposition rates. The chemical reactions of SiO_2,

TABLE 8
ECR deposition conditions

Back pressure (torr)	2×10^{-6}
Gas flow rate (sccm)	4–30
Pressure (torr)	1×10^{-4} - 2×10^{-3}
Microwave frequency (GHz)	2.45
Microwave power (W)	200
Magnetic field (gauss)	875
Substratte temperature	Without intentional heating

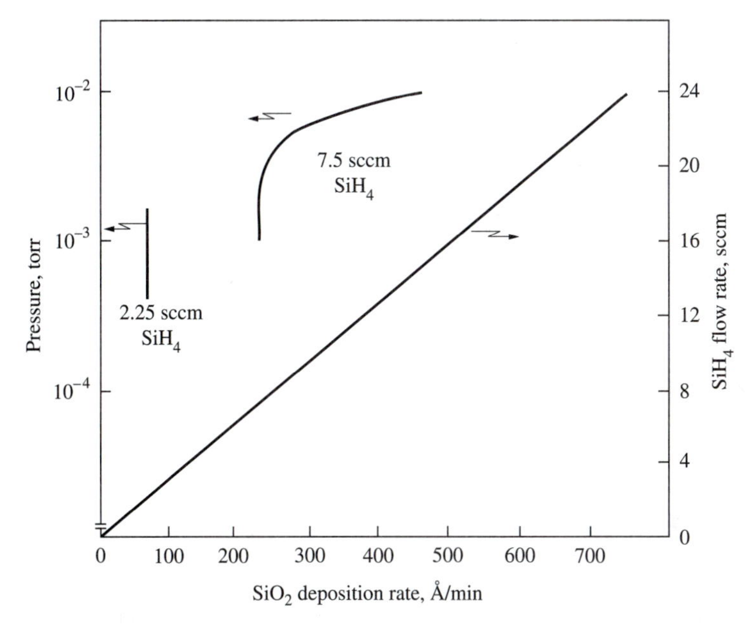

FIGURE 35
The effect of silane gas flow and the total system pressure on the deposition rate of ECR CVD silicon oxide film. (*After Popov and Waldron, Ref. 65.*)

Si_3N_4, and amorphous silicon (α-Si) films in ECR CVD are all the same as those in PECVD.

5.7.3 Ultrahigh-Vacuum Chemical Vapor Deposition

The ultrahigh-vacuum chemical vapor deposition (UHVCVD) apparatus is shown schematically in Fig. 36.[66] Prior to its initial use, the entire UHV section of the apparatus was subjected to a thorough bakeout as well as hydrogen plasma scouring until an ultimate vacuum in the 10^{-9} torr range was achieved. The UHVCVD system features a load-lock and growth chamber with a base pressure maintained at about 10^{-8} torr. Such an ultraclean environment ensures an undoped film of minimum impurity incorporation. In-situ mass spectrometry shows hydrogen to be in the greatest abundance, with water vapor and oxygen present at less than 10^{-10} torr partial pressure. For as-grown films, the oxygen and carbon concentrations were below the detection limit of secondary ion mass spectroscopy, which are about 10^{17} and 10^{18} cm^{-3} for O and C, respectively.[67] The low pressure also prevents the formation of a porous Si structure,[68] which will dramatically degrade the device performance.

Recently, many researchers have reported the growth of poly-Si and poly-$Si_{1-x}Ge_x$ at reduced pressure (e.g., below 10 mtorr) in UHVCVD systems. It was found that lower pressures led to larger grain size, lower transition temperature, or both. Moreover, the use of poly-$Si_{1-x}Ge_x$ instead of poly-Si would lower the process

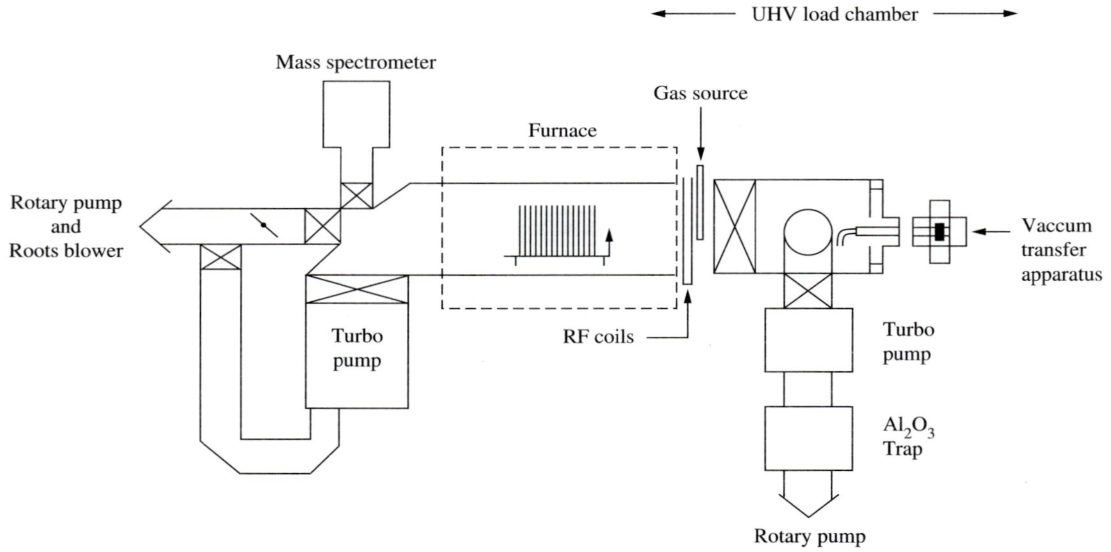

FIGURE 36
UHVCVD system schematic. (*After Meyerson, Ref. 66.*)

temperature as well as the process thermal budget. At a reduced-pressure condition, the gas flow is in a molecular mode, and the surface reaction plays a dominant role during growth. Liehr et al.[69] have shown by the desorption rate of surface-bonded hydrogen that the growth of poly-Si films was dominant. This is very similar to the growth mechanism observed for low-temperature (below 600°C) Si epitaxy from a hydride source (e.g., SiH_4).[70] Figure 37 shows the deposition rates of poly-Si films at four different growth temperatures, namely 600, 575, 550, and 525°C, respectively. The SiH_4 flow rate was kept at 20 sccm. It seems that the deposition rates are essentially the same for epitaxial and poly growth at temperatures below 600°C. These results indicate an identical deposition mode between these two cases. Figure 38 shows that the effect of GeH_4 flow rate on the deposition rate for poly films was smaller than epi-layer growth at a higher GeH_4 flow. This indicates that the addition of GeH_4 would change the deposition mode remarkably.[71] Poly-$Si_{1-x}Ge_x$ is potentially more advantageous than poly-Si as a gate electrode for the metal-oxide-semiconductor field-effect transistor (MOSFET).[72] With the former, the threshold voltage can be adjusted simply by varying the Ge content inside the film, which is much easier than the conventional channel implantation method.

The UHVCVD system could deposit in-situ-doped poly-Si films by using SiH_4 and 1% B_2H_6 in H_2. Figure 39 shows the resistivity as a function of the B_2H_6 flow rate for poly-Si films; a very low resistivity (about 4 mΩ-cm) was obtained.

Hence, the use of poly films deposited by a UHVCVD system can significantly lower the deposition temperature and the process thermal budget, which is required for deep submicron Si technology and some other applications demanding a low

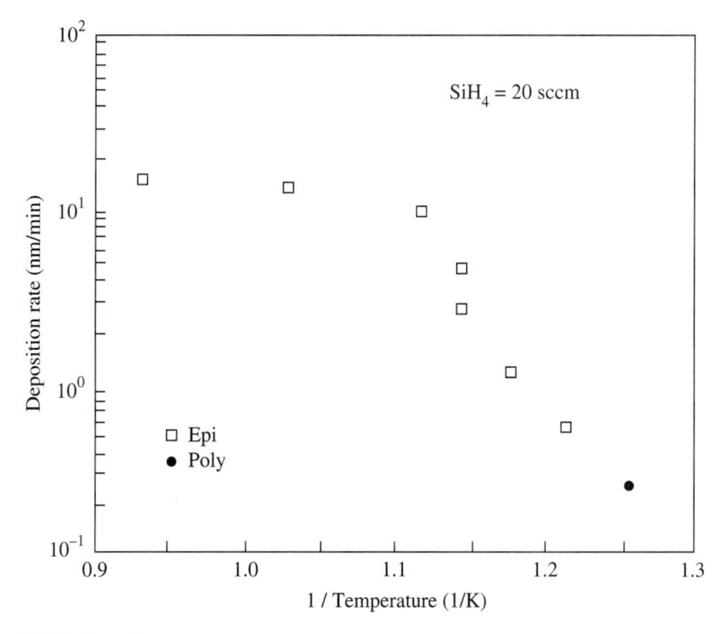

FIGURE 37
An Arrhenius plot of deposition rate for the poly-Si and epitaxial Si films. (*After Lin, Chang, and Lin, Ref. 71.*)

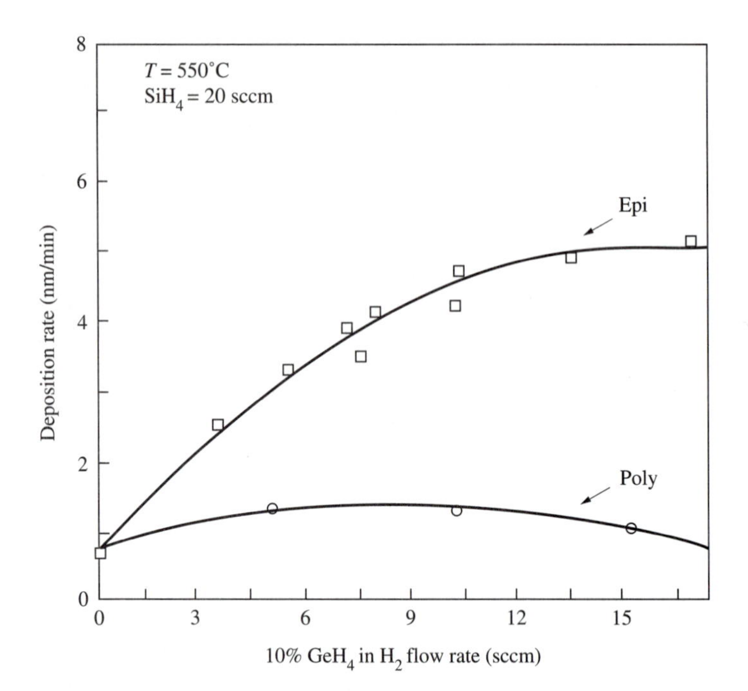

FIGURE 38
Deposition rate for poly and epitaxial films as a function of GeH_4 flow. (*After Lin, Chang, and Lin, Ref. 71.*)

processing temperature, such as thin-film transistor (TFT) fabrication for active-matrix liquid crystal displays (AMLCDs)[73,74] and photoconductors.[75]

5.7.4 Other CVD Techniques

Other CVD techniques, such as metal-organic CVD (MOCVD), laser ablation, and hybrid-excitation CVD, are used to deposit dielectric and silicon films.[76] Like the ferroelectric dielectrics, $LiNbO_3$ and PLZT, they can be prepared by MOCVD or laser ablation. MOCVD uses a metal-organic compound vapor that is decomposed or reacts with other reactants to form the desired film. The excitation is generally thermal but can also be plasma-assisted. Like the PECVD of Ta_2O_5 mentioned in Section 5.6.3, MOCVD of Ta_2O_5 adopts the liquid metal-organic compounds $Ta(OCH_3)_5$ and $Ta(OC_2H_5)_5$. However, MOCVD uses thermal energy to dissociate these organic compounds, rather than the rf power used in PECVD. The chemical reactions can be expressed as

$$Ta(OC_2H_5)_5(l) \rightarrow Ta(s) + OC_2H_5(g) \tag{5.28}$$

$$2O_2(g) \rightarrow 2O(g) + O_2(g) \tag{5.29}$$

$$Ta(s) + O(g) \rightarrow Ta_2O_5(s) \tag{5.30}$$

$$OC_2H_5(g) + O_2(g) \rightarrow CO_2(g), CO(g), H_2O(g), C_2H_4(g), CH_4(g) \tag{5.31}$$

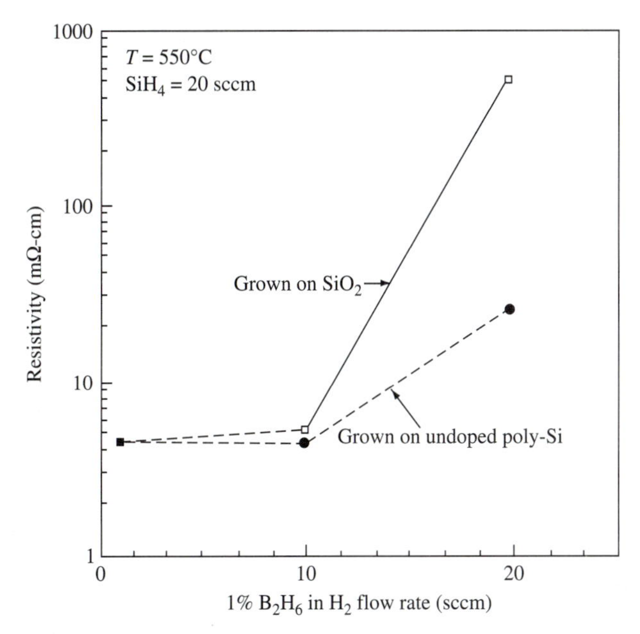

FIGURE 39
Effect of B_2H_6 (1% B_2H_6 in hydrogen) flow rate on the resistivity of poly-Si films grown on either SiO_2 or undoped poly-Si substrate. (*After Lin et al., Ref. 73.*)

Laser ablation deposits the dielectric films by directly scanning the laser beam on the dielectric target. The advantage is the low-temperature process, but the disadvantage is the uniformity control and lack of large-wafer capability.

The hybrid-excitation idea has been used in depositing some dielectric films. The hybrid-excitation techniques consist of plasma-assisted direct photolysis and photo-enhanced CVD methods. The environment contamination caused by the sensitizer encountered in photo CVD will be prevented. A schematic diagram of the hybrid-excitation CVD apparatus is shown in Fig. 40. The plasma of reactive gas is generated in a glow discharge tube and then diffuses into a process chamber. Therefore, the samples will also be free from plasma damages.

5.8 APPLICATIONS OF DEPOSITED POLYSILICON, SILICON OXIDE, AND SILICON NITRIDE FILMS

Polycrystalline silicon, also called polysilicon or poly-Si, thin films have many important applications in integrated circuit (IC) technology. The most widespread application of poly-Si is in metal-oxide-semiconductor (MOS) ICs. Heavily doped poly-Si films have been widely used as the gate electrodes and interconnections in MOS circuits. Poly-Si is utilized because of its compatibility with subsequent high-temperature processing, its excellent interface with thermal oxide (low interface

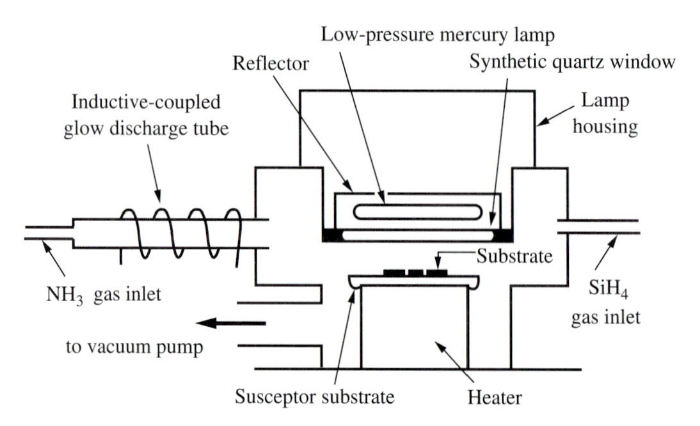

FIGURE 40
Schematic configuration of the hybrid-excitation CVD system.

state density), its higher stability than Al gate material, its ability to be deposited conformally over steep topography, and its capability of reacting with the overlaying metal to form *silicide* structures. Silicide films formed onto the poly-Si, known as *polycides,* on the gate dielectrics have been fabricated to lower the sheet resistance of poly-Si gates. This is especially important for ULSI devices. Moreover, self-aligned silicides (*salicides*) have also been used in ULSI circuits.[77] Through the reaction of metal with the underlying polysilicon and silicon substrate, the self-aligned silicide contacts on the source, drain, and gate will be simultaneously formed, avoiding any misalignment due to photolithography. Emitter structures with heavily doped poly-Si films are also utilized in advanced bipolar devices. Selective growth of poly-Si is studied as a planarization technique for filling the contact hole. Lightly doped poly-Si films are used as high-resistance resistors in static random access memories (SRAMs), and oxygen-doped poly-Si (semi-insulating poly-Si or SIPOS) can be used to form a barrier to the carrier injection. In addition, poly-Si films have been extensively used in ULSI DRAMs, SRAMs, and flash memory devices.[78] For 1-Mbit (and even higher packing density) DRAMs, high capacitances of the MOS capacitors are still needed. However, the decreasing area of the ULSI devices will require a three-dimensional capacitor structure. A layer of *rugged* polysilicon (hemispherical grains of polysilicon) is developed to increase the surface area. The hemispherical grains of polysilicon have been grown by selection of the deposition temperature and the subsequent annealing of the LPCVD system.[79]

Silicon dioxide has several uses: to serve as a mask against implant or diffusion of dopants into silicon, as a gate oxide and capacitor dielectric in MOS devices, and as a tunneling oxide in (EEPROMs). It can also be used to provide surface passivation, to isolate one device from another, to act as a component in MOS structure, and to provide electrical isolation of multilevel metallization systems. Several techniques for forming the oxide layers have been developed, such as wet or dry thermal oxidation (including rapid thermal techniques), wet anodization, chemical vapor deposition (CVD), PECVD, PHCVD, ECR CVD, and plasma anodization or oxidation. When the interface between the oxide and the silicon requires a low-trap

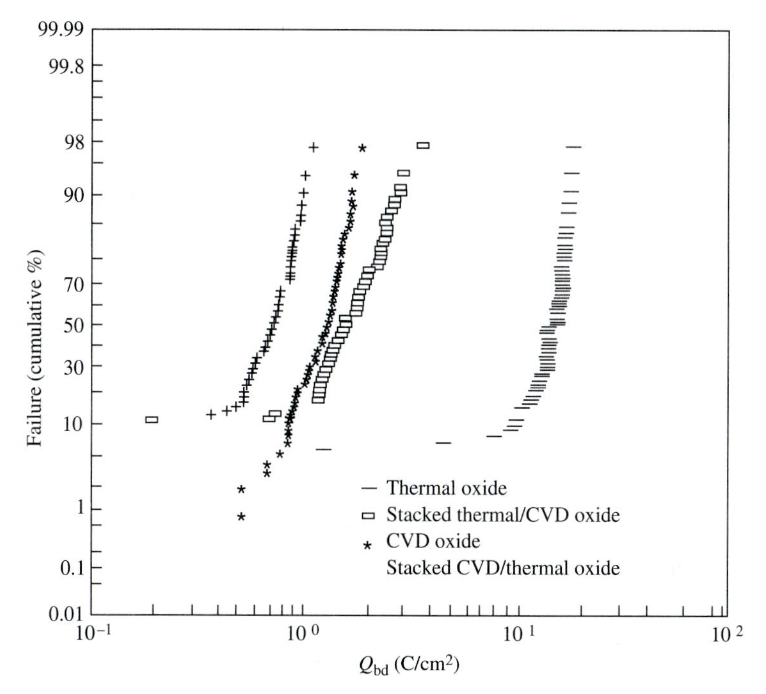

FIGURE 41
The time-dependent dielectric-breakdown charge Q_{bd} in coulomb/cm^2 by step current stress: thermal oxide (−); CVD oxide (*); stacked CVD/thermal oxide (+); stacked thermal/CVD oxide (o). *(After Tseng et al., Ref. 80.)*

density, thermal oxidation has been the preferred technique. In the past, CVD oxides have seldom been used in ULSI technology as the gate dielectrics, because an extrusion phenomenon of the CVD oxides forms at the n$^+$ polysilicon/CVD oxide interface. The resulting electrical properties of CVD and thermal oxides are compared in Fig. 41.[80] However, CVD oxides are particularly useful as the gate dielectrics on the polysilicon active channel. For example, CVD oxides can offer better characteristics than thermal polyoxides for the poly-Si TFTs in 4-Mbit and higher-density SRAMs because polyoxides generally have the problem of interface roughness. Owing to the improvement of the characteristics by the deposition schemes, CVD oxides will become more important in future ULSI technology, which requires multilayered structures. CVD oxides with low and high dielectric constants will be used in interlayer isolations and MOS capacitors, respectively.[81]

Silicon nitride films are amorphous insulating materials and have three main applications in ULSI technology:

1. As final passivation and mechanical protective layers for integrated circuits, especially for parts encapsulated in plastic packages
2. As a mask for selective oxidation of silicon
3. As a gate dielectric material in MOS devices

Silicon nitride also has a high dielectric constant, about 6–9 versus about 4.2 for CVD oxide, making it less attractive for the interlevel insulation because of the resultant higher parasitic capacitances between conductive layers. However, silicon nitrides are especially valuable in applications of ULSI circuits as the oxide/nitride/oxide (ONO) dielectrics. In ULSI technology, nitride films deposited by LPCVD are often accompanied by bottom thermal oxides and oxide overlayers in the MOS capacitors of 1-Mbit DRAMs, because the higher dielectric constant—7.6 for silicon nitride relative to 3.9 for silicon dioxide—can offer a higher capacitance for nitride under the constant film thickness. Especially for the DRAMs above 4 Mbit, the required oxide film thickness has approached the fabrication limit of thermal oxide. The use of silicon nitride resolves the problem, although numerous other techniques, such as rugged polysilicon, hemispherical-grained polysilicon, Ta_2O_5 film, and other ferroelectric films, have been developed and reported. Silicon nitride is highly suitable as a passivation layer because of its following properties:

1. It can be deposited by PECVD under low temperatures, which will not degrade the underlying aluminum interconnects.
2. It behaves as a nearly impervious barrier to the diffusion of moisture and sodium.
3. It can be prepared by PECVD to have a low compressive stress, which allows it to be subjected to severe environmental stress with less likelihood of delimitation or cracking.
4. Its coverage on underlying metals is conformal.
5. It can be deposited with acceptably low pinhole densities.

A notable application of deposited dielectric films is in planarization for ULSI technology. Multilevel metallization becomes an essential technology for ULSI devices because it determines the packing density as well as device performance and affects yield and reliability. For deep submicron applications the metallization schemes are becoming more complex and dense. Therefore, the intermetal planarization technology faces serious challenge.

The intermetal dielectric application has the following requirements:

1. Good step coverage properties and void-free filling of high-aspect-ratio gaps on the underlying metal lines
2. Good planarization capability at low temperature ($\leq$ 400°C), so as not to damage the underlying interconnect

The sandwich scheme of the silane-based or TEOS-based PECVD combined with SOG (spin-on-glass) is widely utilized for the intermetal dielectric. Figure 42 indicates the simple structure. Prior to SOG coating, PE-oxide or PE-TEOS is deposited as the underlayer to prevent SOG from attaching directly to the metal. After SOG annealing, one subsequent PECVD capping layer is also performed. The main concerns are high hydrogen concentration and moisture absorption of SOG films. Furthermore, because of concern about *poisoned* vias[82] (i.e., significant increase of via contact resistance caused by SOG outgassing) and related reliability issues, the SOG etchback process was well developed to remove SOG on the metal.

However, for subsequent gap filling of sub-half-micron geometry and for higher-temperature metallization processes such as tungsten plug (W-plug), the gap-filling

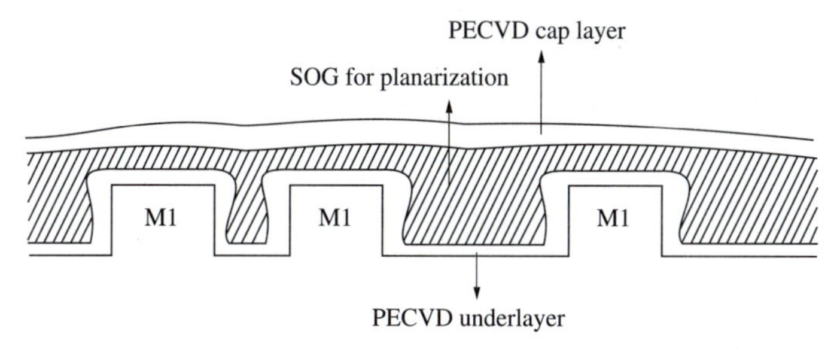

FIGURE 42
Schematic diagram of sandwiched intermetal dielectric scheme with SOG (spin-on-glass).

capability and thermal stability of SOG film suffers severe challenge. O_3–TEOS deposition process becomes another alternative. The chemistry is thermal reaction of O_2/O_3/TEOS. Basically, higher O_3/TEOS gas ratios achieve better film quality and gap-filling capability. However, there are still some problems in the O_3–TEOS process. The most important concerns are the serious moisture absorption of O_3–TEOS film and the associated impacts on reliability and device performance. So far, much effort has been expended on ways to eliminate the moisture absorption or strengthen the film characteristics by various treatments[83] and to integrate O_3–TEOS related modules.

Recently, electron cyclotron resonance (ECR) CVD technique has become more mature and has been shown to fill high-aspect-ratio gaps without voids and seams. Moreover, ECR CVD oxide possesses excellent film characteristics, including low dielectric constant ($\sim$4.07), high breakdown voltage ($>$ 7 MV/cm), and low moisture absorption. Therefore, it meets the requirements of devices in which capacitive coupling between metal lines causes propagation delays and crosstalk. The deposition chemistry includes the reaction of SiH_4 with O_2 as well as in-situ Ar sputtering, which can achieve good gap filling and oxide quality.[84]

For planarization, the etchback process for SOG or photo resist is already widely used for production. The structure is shown in Fig. 43. One new approach, CMP (chemical-mechanical polishing) technology, expresses very good planarization performance and has become a very hot research field. The factors that affect chemical-mechanical polishing are classified into two groups. Chemistry-related parameters include slurry type, pH value of slurry, solid content in slurry, slurry flow, as well as process temperature. Mechanical-related factors include polish pressure, back pressure, platen speed, and pad type. Currently, this process is widely studied for metal and dielectric polishing. The material and pattern sensitivities are also reported. With this planarization process, the step height of wafer topography can be reduced to within 1000 Å. The schematic diagram is shown in Fig. 44.

In summary, there are two major topics for deep submicron planarization. One is the filling of gaps with high aspect ratios. The other is to achieve actual global planarization, at distances exceeding one micron. Various approaches are being developed in parallel, and extensive room remains for investigation.

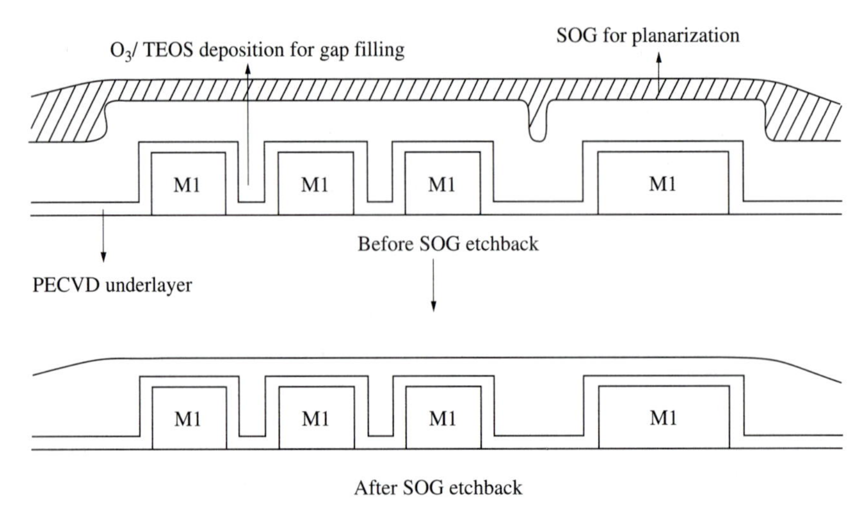

FIGURE 43
Planarization scheme with SOG etchback.

5.9 SUMMARY AND FUTURE TRENDS

To fabricate ULSI circuits, various dielectric and polysilicon film depositions are necessary. The major issues include small-geometry fabrication, low-temperature processing, particles or contamination, step coverage, planarization, selective deposition, throughput, uniformity, film properties, and large-wafer capability. ULSI devices with small dimensions require small channel length, shallow junctions, and small contacts. All of them will need precise lithography, fine pattern transfer with etching, and minimal diffusion under low-temperature processing. For precise lithography and anisotropic etching, conformal step coverage becomes more and

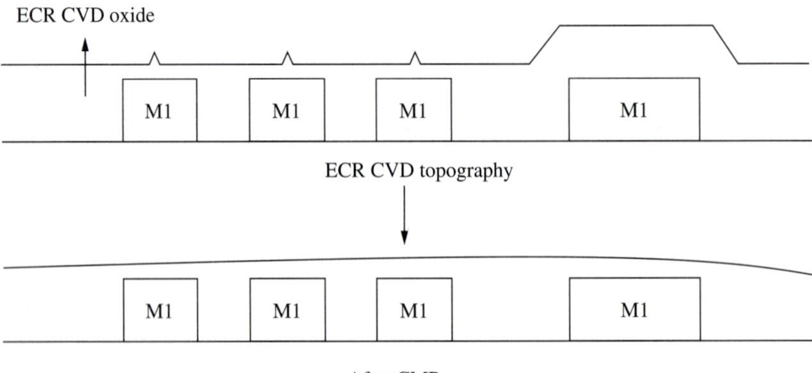

FIGURE 44
Planarization scheme with CMP.

more important. Low-pressure deposition of dielectric and polysilicon films will offer this advantage. In contrast, physical vapor deposition and atmospheric pressure deposition will have nonconformal step coverage. In addition, throughput and uniformity will also be challenged with physical vapor deposition, especially for future large wafers in ULSI technology (200-mm wafers have been introduced in newly established ULSI factories).

Moreover, reducing particles and contaminations also requires a low temperature and cold-wall deposition. Defects formed during the processing will seriously lower the ULSI device yield, because these devices are scaled down and more sensitive to particles and impurity contamination. Plasma-assisted and photo-assisted depositions have supplied the benefits of lowering the deposition temperatures. However, further study on these deposition schemes will be needed to improve the properties of the deposited films. Liquid-phase deposition of silicon dioxide via the decomposition of H_2SiF_6 supersaturated with silica at reaction temperatures below 100°C has been reported. Because the conventional holder will produce particles through the motion of the holder's rollers, the cantilever-type holder and vertical-type CVD furnace have been used to decrease particle contamination, as shown in Fig. 45.

High throughput is needed to reduce cost. High deposition rates are therefore required, since the single-wafer process will likely be adopted in future ULSI technology. Deposition through generation of high-density reacting radicals is therefore emphasized. Magnetic fields have been utilized to create the helicon source or to form the electron cyclotron resonance (ECR) for the radiofrequency (rf) or microwave systems, correspondingly. Cassette-to-cassette and multichamber concepts are also favorable for increasing throughput and decreasing contamination, as shown in Fig. 46.

Various deposited films with specific properties are required for ULSI devices. As an example, dielectric films with high dielectric constants have been investigated for the MOS capacitors in ULSI DRAMs above 1 Mbit. Oxide/nitride/oxide, ferroelectric films, and stacked or trench MOS capacitors need new deposition techniques for the dielectric and polysilicon films to satisfy various requirements for the film properties. In order to increase the capacitor area, polysilicon layers can also be deposited as semispherical-grained structures by tuning the deposition temperatures. Furthermore, flash memories or SRAMs above 1 Mbit need better characteristics of polysilicon films and dielectrics on the polysilicon layers. Polysilicon with large and small grain sizes and CVD dielectrics are utilized in these ULSI devices.

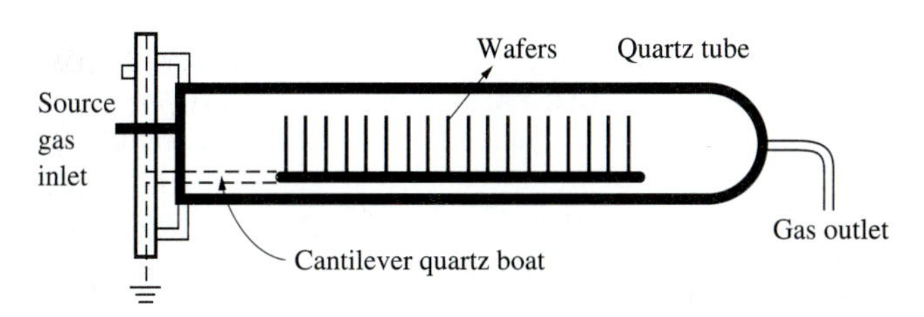

FIGURE 45
A cantilever-type LPCVD system.

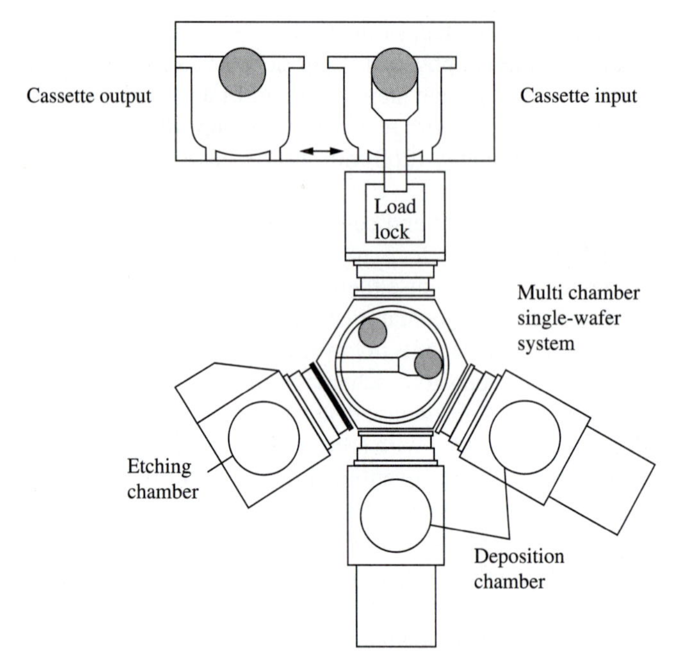

FIGURE 46
A cassette-to-cassette multichamber deposition system.

The research and development of these dielectric and polysilicon films in ULSI applications can have important impacts in other fields, such as optoelectronics, displays, microsensors, and bioelectronics. In the future these applications will need new and novel deposition techniques for these films with better properties.

REFERENCES

1. W. Kern and G. L. Schnable, "Low Pressure Chemical Vapor Deposition for VLSI Processing—A Review," *IEEE Trans. Electron Dev.* **ED-26,** 647 (1979).
2. P. Singer, "Techniques of Low Pressure CVD," *Semic. Intl.,* 72 (May, 1984).
3. M. de Fraiteur and J. Goldman, "Pressure Control in LPCVD System," *Semic. Intl.,* 250 (May, 1984).
4. A. Learn, "Modeling the Reaction of Low Pressure Chemical Vapor Deposition of Silicon Dioxide," *J. Electrochem. Soc.* **132,** 390 (1985).
5. B. Gorowitz, T. B. Gorczyca, R. J. Saia, "Application of PECVD in VLSI," *Sol. State Technol.,* 197 (June, 1985).
6. A. A. Chernov, "Growth Kinetics and Capture of Impurities during Gas Phase Crystallization," *J. Cryst. Growth,* **42,** 55 (Dec., 1977).
7. W. Kern and V. Ban, "Chemical Vapor Deposition of Inorganic Thin Films," *Thin Film Processes,* J. L. Vossen and W. Kern, Eds., Academic, New York, 1978, pp. 257–331.
8. M. Hammond, "Introduction to Chemical Vapor Deposition," *Sol. State Technol.,* 61 (Dec., 1979).

9. A. C. Adams, "Plasma-Assisted Deposition of Dielectric Films," in R. Reif and G. R. Srinivasan, Eds., *Reduced Temperature Processing for VLSI,* The Electrochemical Society, Pennington, NJ, 1986, p. 111.
10. A. Weiss, "PECVD: Silicon Nitride and Beyond," *Semic. Intel.* **6(7),** 88 (July, 1983).
11. W. L. Johnson, "Design of Plasma Deposition Reactors," *Sol. State Technol.,* 191 (April, 1983).
12. M. L. Hammond, "Safety in Chemical Vapor Deposition," *Sol. State Technol.,* **23,** 104 (1980).
13. C. Cobianu and C. Pavelescu, "A Theoretical Study of the Low-Temperature Chemical Vapor Deposition of SiO_2 Films," *J. Electrochem. Soc.,* **127,** 2254 (1980).
14. C. Cobianu and C. Pavelescu, "Silane Oxidation Study: Analysis of Data for SiO_2 Films Deposited by Low Temperature Chemical Vapor Deposition," *Thin Sol. Films,* **117,** 211 (1984).
15. A. J. Learn, "Phosphorus Incorporation Effects in Silicon Dioxide Films at Reduced Pressure," *J. Electrochem. Soc.,* **132,** 405 (1985).
16. A. C. Adams and C. D. Capio, "The Deposition of Silicon Dioxide Films at Reduced Pressure," *J. Electrochem. Soc.,* **126,** 1402 (1979).
17. R. M. Levin and C. D. Adams, "Low Pressure Deposition of Phosphosilicate Glass Films," *J. Electrochem. Soc.,* **129,** 1588 (1982).
18. H. Huppertz and W. L. Engl, "Modeling of Low Pressure Deposition of SiO_2 by Decomposition of TEOS," *IEEE Trans. Electron Dev.,* **ED-26,** 658 (1979).
19. K. Watanabe, T. Tanigaki, and S. Wakayama, "The Properties of LPCVD SiO_2 Film Deposited by SiH_2Cl_2 and N_2O Mixtures," *J. Electrochem. Soc.,* **128,** 2630 (1981).
20. A. C. Adams and C. D. Capio, "Planarization of Phosphorus-Doped Silicon Dioxide," *J. Electrochem. Soc.,* **128,** 423 (1981).
21. R. A. Levy and K. Nassau, "Reflow Mechanisms of Contact Bias in VLSI Processing," *J. Electrochem. Soc.,* **133,** 1417 (1986).
22. W. Kern and R. K. Smeltzer, "Borophosphosilicate Glasses for Integrated Circuits," *Sol. State Technol.,* **28,** 171 (1985).
23. T. Foster, G. Hoeye, and J. Goldman, "A Low Pressure BPSG Deposition Process," *J. Electrochem. Soc.,* **132,** 505 (1985).
24. K. Fujino, Y. Nishimoto, N. Tokumasu, and K. Meada, "Doped Silicon Oxide Deposition by Atmospheric Pressure and Low Temperature Chemical Vapor Deposition Using Tetraethoxysilane and Ozone," *J. Electrochem. Soc.,* **138,** 3019 (1991).
25. J. D. C-Sokol, C. J. Giunta and R. Gordon, "A Kinetic Study of the Atmospheric Pressure CVD Reaction of Silane and Nitrous Oxide," *J. Electrochem. Soc.,* **136,** 2993 (1989).
26. T. Makno, "Composition and Structure Control by Source Gas Ratio in LPCVD Silicon Nitride Deposited with Different NH_3/SiH_2Cl_2 Gas Ratios," *J. Electrochem. Soc.,* **132,** 3001 (1985).
27. J. V. Dalton and J. Drobek, "Structure and Sodium Migration in Silicon Nitride Films," *J. Electrochem. Soc.,* **115,** 865 (1968).
28. H. Tanaka, H. Uchida, N. Hirashida, and T. Ajioka, "The Effect of Surface Roughness of Si_3N_4 Films on TDDB Characteristics of ONO Films," *International Reliability Physics Symposium,* IEEE, New York, 1992, p. 31.
29. T. Kamins, *Polycrystalline Silicon for Integrated Circuit Application,* Kluwer Academic Publishers, Boston, 1988.
30. S. Wolf, R. N. Taauber, *Silicon Processing for the VLSI Era, vol. 1,* Chap. 6, Lattice Press, 1986, p. 161.
31. M. Ohkura, K. Kusukawa, and H. Sunami, "Beam-Induced Seeded Lateral Epitaxy with Suppressed Impurity Diffusion for a Three-Dimensional DRAM Cell Fabrication," *IEEE Trans. Electron Dev.,* **ED-36,** 333 (1989).

32. T. Yamanaka, T. Hashimoto, N. Hashimoto, T. Nishida, A. Shimizu, K. Ishibashi, Y. Sakai, K. Shimohigashi, and E. Takeda, "A 25 m^2, New Poly-Si PMOS Load (PPL) SRAM Cell Having Excellent Soft Error Immunity," *Tech. Dig. IEDM,* 48 (1988).
33. A. Mimura, N. Konishi, K. Ono, J.-I. Ohwada, Y. Hosokawa, Y. A. Ono, T. Suzuki, K. Niyata, and H. Kawakami, "High Performance Low-Temperature Poly-Si n-Channel TFT's for LCD," *IEEE Trans. Electron Dev.,* **ED-36,** 351 (1989).
34. M. Yuki, K. Masuino, and M. Kunigita, "A Full-Color LCD Addressed by Poly-Si TFT's Fabricated below 450°C," *IEEE Trans. Electron Dev.,* **ED-36,** 1934 (1989).
35. W. A. P. Claassen, J. Bloem, W. G. J. N. Valkenburg, and C. H. J. Van den Brekel, "The Deposition of Silicon from Silane in a Low-Pressure Hot-Wall System," *J. Cryst. Growth,* **57,** 259 (1982).
36. M. J. Tsai and H. C. Cheng, "Hydrogen Passivation of Thin Film Transistors with Different Film Structures," *Tech. Rep. IEICE,* **93,** 25 (1993).
37. A. Nakamura, F. Emoto, E. Fujii, A. Yamamoto, Y. Uemoto, "Analysis of Solid Phase Crystallization in Amorphized Polycrystalline Si Films," *J. Appl. Phys.,* **66,** 4248 (1989).
38. S. D. Brotherton, D. J. McCulloch, J. B. Clegg, and J. P. Gowers, "Excimer-Laser-Annealed Poly-Si Thin-Film Transistors," *IEEE Trans. Electron Dev.,* **ED-40,** 407 (1993).
39. R. E. Jones, Jr. and S. P. Wesolowski, "Electrical, Thermoelectric, and Optical Properties of Strongly Degenerate Polycrystalline Silicon Films," *J. Appl. Phys.,* **56,** 1701 (1984).
40. H. Kurokawa, "P-Doped Polysilicon Film Growth Technology," *J. Electrochem. Soc.,* **129,** 2620 (1982).
41. B. S. Meyerson and W. Olbricht, "Phosphorus-Doped Polycrystalline Silicon via LPCVD. 1. Process Characterization," *J. Electrochem. Soc.,* **131,** 2361 (1984).
42. I. W. Wu, T. Y. Huang, W. B. Jackson, A. G. Lewis, and A. Chiang, "Passivation Kinetics of Two Types of Defects in Polysilicon TFT by Plasma Hydrogenation," *IEEE Electron Dev. Lett.,* **EDL-12,** 181 (1991).
43. M. H. Juang, C. T. Lin, S. T. Jan, and H. C. Cheng, "The Process Limitation for Forming Ti Silicided Shallow Junction by BF_2^+ Implantation into Thin Polycrystalline Si Films and Subsequent Ti Silicidation," *Appl. Phys. Lett.,* **63,** 1267 (1993).
44. C. Y. Lu, J. J. Sung, R. Liu, N. S. Tsai, R. Singh, S. J. Hellenius, and H. C. Kirsch, "Process Limitation and Device Design Trade-offs of Self-Aligned $TiSi_2$ Junction Formation in Submicrometer CMOS Devices," *IEEE Trans. Electron Dev.,* **ED-38,** 246 (1991).
45. W. R. Knolle and H. R. Maxwell, Jr., "A Model of SIPOS Deposition Based on Infrared Spectroscopic Analysis," *J. Electrochem. Soc.,* **127,** 2254 (1980).
46. S. D. Chu and A. J. Steckl, "The Effect of Trench-Gate-Oxide Structure EPROM Device Operation," *IEEE Electron Dev. Lett.,* **EDL-9,** 284 (1988).
47. B. E. Deal and A. S. Grove, "General Relationship for the Thermal Oxidation of Silicon," *J. Appl. Phys.,* **36,** 3770 (1965).
48. P. W. Wang, H. P. Su, M. J. Tsai, G. Hong, M. S. Feng, and H. C. Cheng, "A New Portrayal of Oxidation of Undoped Polycrystalline Silicon Films for Short Duration," *Jpn. J. Appl. Phys.,* **33,** 429 (1994).
49. H. Dun, P. Pan, F. R. White, and R. W. Douse, "Mechanisms of Plasma-Enhanced Silicon Nitride Deposition Using SiH_4/N_2 Mixture," *J. Electrochem. Soc.,* **128,** 1555 (1981).
50. A. C. Adams, F. B. Alexander, C. D. Capio, and T. E. Smith, "Characterization of Plasma-Deposited Silicon Dioxide," *J. Electrochem. Soc.,* **128,** 1545 (1981).
51. S. Veprek, "Plasma-Induced and Plasma-Assisted Chemical Vapor Deposition," *Thin Sol. Films,* **130,** 135 (1985).
52. M. J. Rand and J. F. Roberts, "Silicon Oxynitride Films from the NO-NH_3-SiH_4 Reaction," *J. Electrochem. Soc.,* **120,** 446 (1973).

53. W. A. P. Claassen, H. A. J. Th. V. d. Pol, A. H. Goemans, and A. E. T. Kuiper, "Characterization of Silicon-Oxynitride Films Deposited by Plasma-Enhanced CVD," *J. Electrochem. Soc.,* **133,** 1458 (1986).
54. A. C. Carlson, T. H. Tom Wu, and H. B. K. Liang, "Deposition Method to Control Plasma-Enhanced Chemical Vapor Deposition Tetraethylortho Silicate Oxide Charge," *J. Electrochem. Soc.,* **140,** 774 (1993).
55. R. S. Rosler, "The Evolution of Commercial Plasma Enhanced CVD Systems," *Sol. State Technol.,* 67 (June, 1991).
56. I. Avigal, "Inter-metal Dielectric and Passivation-Related Properties of Plasma BPSG," *Sol. State Technol.,* **26,** 217 (1983).
57. A. Madan, P. Rava, R. E. I. Schropp, and B. v. Roedern, "A New Modular Multichamber Plasma Enhanced Chemical Vapor Deposition System," *Appl. Surf. Sci.,* **70/71,** 716 (1993).
58. A. C. Adams, "Dielectric and Polysilicon Film Deposition," in S. M. Sze, Ed., *VLSI Technology,* McGraw-Hill Book Company, New York, 1983.
59. P. A. Murawala, M. Sawai, T. Tatsuta, O. Tsuji, "Structural and Electrical Properties of Ta_2O_5 Grown by the Plasma-Enhanced Liquid Source CVD Using Penta Ethoxy Tantalum Source," *Jpn. J. Appl. Phys.,* **32,** pt. 1, 368 (1993).
60. G. Q. Lo, D. L. Kwong, P. C. Fazan, V. K. Mathews, and N. Sandler, "Highly Reliable, High-C DRAM Storage Capacitors with CVD Ta_2O_5 Film on Rugged Polysilicon," *IEEE Electron Dev. Lett.,* **EDL-14,** 216 (1993).
61. J. W. Peters, F. L. Gebhart, and T. C. Hall, "Low-Temperature Photo-CVD Silicon Nitride—Properties and Applications," *Sol. State Tech.,* **23,** 121 (1980).
62. R. Solaki, C. Moore, and G. Collins, "Laser Induced CVD," *Sol. State Technol.,* 220 (June, 1985).
63. M. Kitagawa, S. I. Ishihara, K. Setsune, Y. Manabe, and T. Hirao, "Low Temperature Preparation of Hydrogenated Amorphous Silicon by Microwave Electron-Cyclotron-Resonance Plasma CVD," *Jpn. J. Appl. Phys.,* **26,** L231 (1987).
64. M. Kitagawa, K. Setsune, Y. Manabe, and T. Hirao, "Preparation of Doped Hydrogenated Amorphous Silicon by Microwave Electron-Cyclotron-Resonance Discharge Deposition," *J. Appl. Phys.,* **61,** 2048 (1987).
65. O. A. Popov and H. Waldron, "Electron Cyclotron Resonance Plasma Stream Source for Plasma Enhanced Chemical Vapor Deposition," *J. Vac. Sci. Technol. A,* **7,** 914 (1989).
66. B. S. Meyerson, "Low-Temperature Silicon Epitaxy by Ultrahigh Vacuum/Chemical Vapor Deposition," *Appl. Phys. Lett.,* **48,** 797 (1986).
67. H. C. Lin, H. Y. Lin, C. Y. Chang, T. F. Lei, P. J. Wang, and C. Y. Chao, "Growth of Undoped Polycrystalline Si by an Ultrahigh Vacuum Chemical Vapor Deposition System," *Appl. Phys. Lett.,* **63,** 1351 (1993).
68. M. Miyasaka, T. Nakazawa, I. Yudasaka, and H. Ohshima, "TFT and Physical Properties of Polycrystalline Silicon Prepared by Very Low Pressure Chemical Vapor Deposition (VLPCVD)," *Jpn. J. Appl. Phys.,* **30,** 3733 (1991).
69. M. Liehr, S. S. Dana, and M. Anderle, "Nucleation and Growth of Silicon on SiO_2 during SiH_4 Low-Pressure Chemical Vapor Deposition as Studied by Hydrogen Desorption Titration," *J. Vac. Sci. Tech. A,* **10,** 869 (1992).
70. S. M. Gates and S. K. Kalkarn, "Kinetics of Surface-Reactions in Very Low Pressure Chemical Vapor Deposition of Si from SiH_4," *Appl. Phys. Lett.,* **58,** 2963 (1991).
71. H. C. Lin, C. Y. Chang, W. H. Chen, W. C. Tsai, T. C. Chang, T. G. Jung, and H. Y. Lin, "Effects of SiH_4, GeH_4, and B_2H_6 on the Nucleation and Deposition of Polycrystalline $Si_{1-x}Ge_x$ Films," *J. Electrochem. Soc.,* **141,** 2559 (1994).
72. T. J. King and K. C. Saraswat, "Low-Temperature (Less Than or Equal to 550°C) Fabrication of Poly-Si Thin-Film Transistors," *IEEE Electron Dev. Lett.,* **EDL-13,** 309 (1992).

73. H. C. Lin, H. Y. Lin, C. Y. Chang, T. F. Lei, P. J. Wang, R. C. Deng, J. Lin, and C. Y. Chao, "Deposition and Device Application of In-Situ Boron-Doped Polycrystalline SiGe Films Grown at Low Temperature," *J. Appl. Phys.*, **74,** 5395 (1993).
74. T. J. King, J. R. Pfiester, J. D. Shott, J. P. McVittie, and K. C. Saraswat, "A Polycrystalline-$Si_{1-x}Ge_x$-Gate CMOS Technology," *IEEE Proc. IEDM,* 253 (1990).
75. A. Hai, J. D. Morse, and R. W. Dutton, "1-GHz Integrated Poly-Si-and-SiGe Photoconductors with BiCMOS Compatibility," *IEEE Proc. IEDM,* 41 (1991).
76. S. Yamamoto and M. Migitaka, "Hybrid-Excitation Chemical Vapor Deposition of Silicon Nitride on (100)Si—The Film and Interface Properties," *Jpn. J. Appl. Phys.*, **31,** Pt. 1, 348 (1992).
77. D. Peters, "Implanted-Silicide Polysilicon Gate for VLSI Transistors," *IEEE Trans. Electron Dev.*, **ED-33,** 1391 (1986).
78. H. Sunami, "Cell Structure for Future DRAMs," *IEEE Proc. IEDM,* 694 (1985).
79. N. Matsuo and H. Ogawa, "Nucleation and Growth Mechanism of Hemispherical Grain Polycrystalline Silicon," *Appl. Phys. Lett.*, **60(21),** 2607 (1992).
80. K. S. Tseng, H. C. Cheng, C. C. Chang, C. G. Lou, and F. H. Yang, "Effects of n^+-Polysilicon/SiO_2 Interface on the Electrical Characteristics of MOS Capacitors," Unpublished.
81. H. H. Tseng, P. J. Tobin, J. D. Hayden, K. M. Chang, "Advantages of CVD Stacked Gate Oxide for Robust 0.5 μm Transistors," *IEEE Proc. IEDM,* 75 (1991).
82. N. Rutherford, M. Camenzind, and A. Belic, "Outgassing and Oxidative Damage in Non-Etchback Siloxane SOG Process," *Proc. 10th VMIC,* 141 (1993).
83. K. Fujino, Y. Nishimoto, N. Tokumasu, S. Fisher, and K. Maeda, "Plasma Post-Treatment for Tetraethoxysilane/O_3 Atmospheric Pressure CVD," *Proc. 10th VMIC,* 96 (1993).
84. C. Y. Chang, J. P. McVittie, J. Li, K. C. Saraswat, S. E. Lassig, and J. Dong, "Profile Simulation of Plasma Enhanced and ECR Oxide Deposition with Sputtering," *IEEE Proc. IEDM,* 853 (1993).

PROBLEMS

1. In order to increase the capacitance of MOS capacitors in ULSI DRAMs, Ta_2O_5 has been plasma-deposited to replace the oxide/nitride/oxide dielectric. The dielectric constants of SiO_2, nitride, and Ta_2O_5 are about 3.9, 7.6, and 25, respectively. What is the capacitance ratio for the capacitors with the Ta_2O_5 and oxide/nitride/oxide dielectrics for the same dielectric thickness, provided that the oxide/nitride/oxide has thickness ratio $1:1$ for the oxide to the nitride?

2. For the CVD of phosphosilicate glass, 20% silane is mixed in 80% nitrogen at a reaction temperature of 500°C. The flow rate is 600 sccm (standard cubic centimeter per minute). The reactor is loaded with 10 wafers, 10 cm in diameter. The growth rate and film density are 600 Å/min and $2 \times 10^{22}/cm^3$, respectively. With what efficiency is the silane transformed into silicon dioxide film? (*Hint:* A flow of one sccm is defined as a flux of one cm^3 of gas, as measured at 273 K and 760 torr, per minute.)

3. Phosphosilicate glass consists of SiO_2 and P_2O_5. Derive the relationships of weight percent (wt %), atom percent (at%), and mole percent (mole %), for the dopant concentration.

4. (*a*) What factors will affect the step coverage of the deposited oxide films?

   (*b*) What techniques have been used to improve the step coverage of the deposited oxide films?

5. The LOCOS process often causes a "bird's beak" phenomenon, i.e., field-oxide encroachment under the silicon nitride mask. A decrease of the stress between CVD nitride and Si substrate can reduce this effect. What methods can you suggest to attain this purpose?

6. If LPCVD polysilicon deposition has an activation energy of 1.65 eV and a deposition rate of 8 nm/min at 600°C, what is the deposition rate at 620°C?

7. In a PECVD system, the chamber pressure is 1 torr, the total gas flow rate is 100 sccm, the temperature is 27°C, and the volume of the chamber is 1000 cm^3. What is the average residence time for a gas molecule in the PECVD reactor?

8. For an ECR CVD system the magnetic field is 875 gauss. What is the frequency of the microwave power to generate electron cyclotron resonance?

CHAPTER 6

Lithography

K. Nakamura

6.1 INTRODUCTION

Lithography is a kind of art made by impressing, in turn, several flat embossed slabs, each covered with greasy ink of a particular color, onto a piece of paper. The various colors or levels must be accurately aligned with respect to one another within some registration tolerance. Many "originals" can be made from the same slabs as long as the quality remains adequately high.

Several methods can be used to make ULSI circuit patterns on wafers, as shown in Fig. 1*a*. The most common process is to make the master photomask using an electron beam exposure system and replicating its image by optical printers, as shown in Fig. 1*b*. The exposing radiation is transmitted through the "clear" part of a mask. The opaque part of the circuit pattern blocks some of the radiation. The resist, which is sensitive to the radiation and has resistance to the etching, is coated on the wafer surface. The mask is aligned within the required tolerance on the wafer; then radiation is applied through the mask, the resist image is developed, and the layer underneath the resist is etched.

Therefore, lithography for integrated circuit manufacturing is analogous to the lithography of the art world. The slabs correspond to masks for the various circuit levels. The press corresponds to the exposure system, which not only exposes each level but also aligns it to a completed level. The ink may be compared to either the exposing radiation or the radiation-sensitive resist; the paper can represent the wafer into which the pattern will be etched, using the resist as a stencil.

Lithography is the key technology in semiconductor manufacturing, because it is used repeatedly in a process sequence that depends on the device design. It determines the device dimensions, which affect not only the device's quality but also its product amount and manufacturing cost.

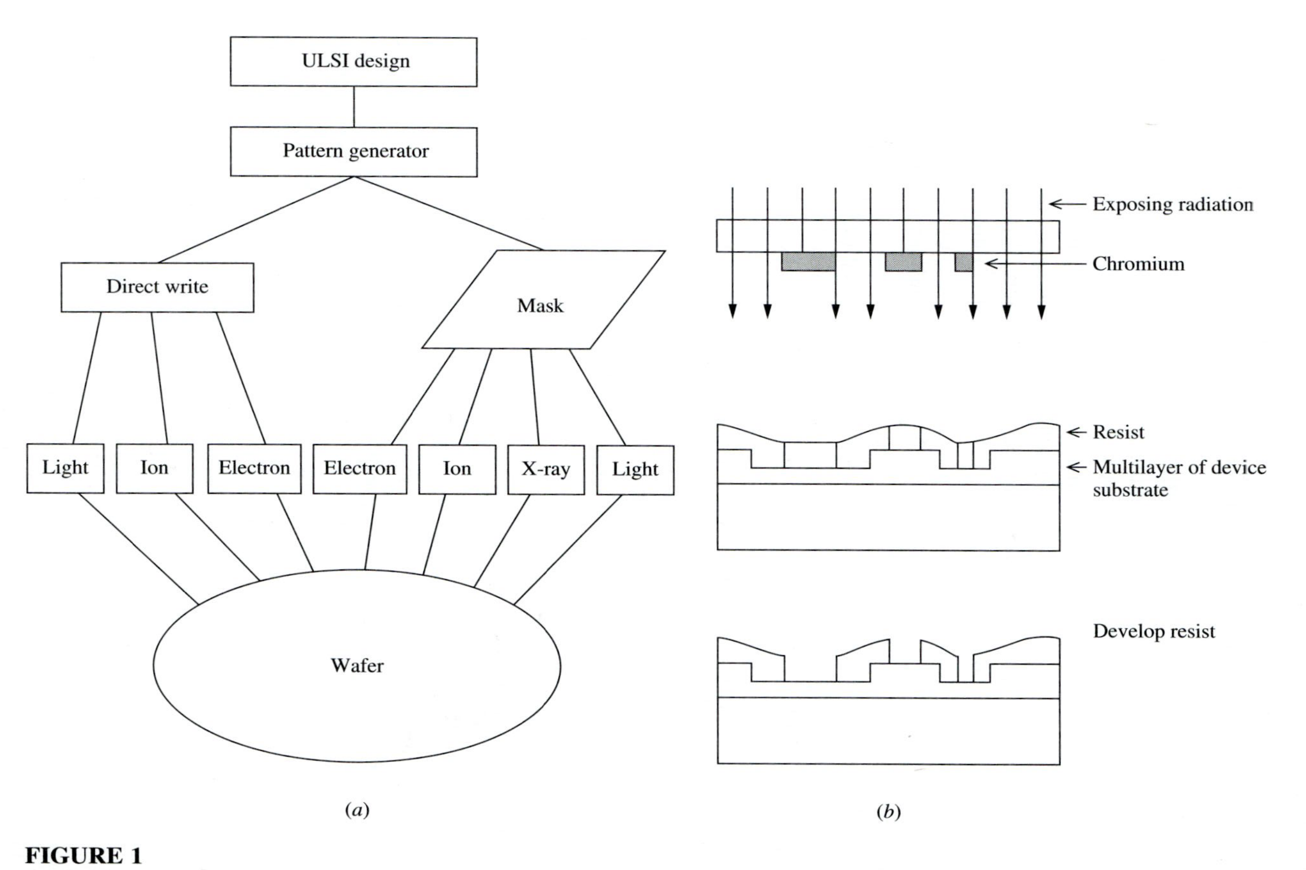

FIGURE 1
Device lithography generalization. (*a*) Lithographic process for ULSI. (*b*) Optical replication process.

6.2 OPTICAL LITHOGRAPHY

Optical lithography comprises the formation of images with visible or ultraviolet radiation in a photoresist using proximity or projection printing. Two methods are available to make masks for optical lithography: electron beam exposure and laser beam scanning. These are described in the next section. In the 1970s, the major technology was the combination of a negative resist and proximity printing. At present the most common method is the combination of a positive resist and a stepper. However, proximity printing is still used because of its convenience and low cost.

Table 1 lists examples of commercially available optical printers[1] designed to manufacture ULSI circuits. These machines are classified into three groups: proximity, reflective projection, and refractive projection. The key parameters such as numerical aperture (NA), depth of focus (DOF), resolution (usable linewidth), overlay accuracy, and throughput are also listed.

The most advanced ULSI has a minimum feature size of 0.3 μm to 0.4 μm, which has been reported by several organizations in 64-Mbit dynamic random access memories (DRAMs). It is believed that the linewidth limit of optical lithography lies near 0.2 μm using phase-shifting masks combined with a high numerical aperture projector and a short-wavelength light source. The performances of the machines listed in Table 1 are very close to the resolution limit of optical lithography.

6.2.1 Contact and Proximity Printing

Contact and proximity printings are relatively simple because they do not have any means of image formation between masks and wafers. A typical contact or proximity mask aligner consists of a light source, a condenser, a filter, a mirror, a shutter, the wafer stage, and the alignment microscope. In contact printing, a photomask is pressed against the resist-covered wafer, with pressures typically in the range of 0.05 atm to 0.3 atm, and is exposed by light with a wavelength near 400 nm. Very high resolution of less than 0.5 μm linewidth is possible, but because of spatial nonuniformity of the contact, resolution may vary considerably across the wafer. To provide better contact over the whole wafer, a thin (0.2 mm) flexible mask has been used; 0.2-μm space patterns have been formed by using 3-μm-thick PMMA (poly-methyl methacrylate) resist and 200 to 260 nm radiation.[2] Quartz or Al_2O_3 mask substrates must be used to pass these shorter wavelengths, since the usual borosilicate glass strongly absorbs wavelengths less than 300 nm.

Contact printing produces defects in both the mask and the wafer while the two are in contact and as they are separated from each other, so that the mask, whether thick or thin, may have to be replaced after a short period of use. Nevertheless, contact printing is still widely used. It is the most convenient way to get high resolution. Strictly speaking, *contact* printing is never actually practiced, because of the difficulty of uniform contact.[3] When a typical mask and wafer are brought into hard contact, the nonuniform gap can be as large as 15 μm because of unevenness of the surfaces. Besides that, the thickness of the resist between the mask and the wafer cannot be neglected for features of less than 1 μm.

TABLE 1

Commercially available optical printers[1]

Category	Model	Maker	NA†	DOF† (μm)	Field size (mm)	Usable linewidth (μm)	Overlay accuracy (μm)	Throughput/6″-hr
Contact/proximity	PLA600	Canon			160	1	0.5	120
	3HRP	JBA				0.25	0.5	45/10″
	Q-6000	Quintel				0.25		60
	MA150M	Karl Suss				0.5	0.5	60
Catadioptric/4×	Micrascan IIi	SVG	0.5	1.2	22 × 32.5	0.5	0.07	51/8″
/1×	Micralign 700	SVG	0.167	6	150 × 135	1.0–1.2	0.25	100
/1×	2000gh	Ultratech	0.24–0.40	2–3	18 × 18–39 × 11	0.8–1.0		78
I-line stepper	PAS5500/100	ASM	0.6	1	22 × 27.6	0.4	0.07	88
	XLS7500/29	GCA	0.55	1.1	21 × 21	0.5	0.09	70
	NSR2005i9C	Nikon			21.2 × 22.8/18 × 25.2	0.45	0.1	47/8″
	FPA2500i3	Canon	0.6	1	20 × 20	0.4	0.12	62
KrF stepper	PAS5500/90	ASM	0.5	1	21 × 27.7	0.35	0.07	80
	XLS7800/31	GCA	0.53	1.2	22 × 22	0.35	0.07	72
	ISR2005EXBA	Nikon			21.2 × 21.2/16 × 25.2	0.4	0.1	34/8″
	FPA3000EX1	Canon	0.45	1	20 × 20	0.3	0.1	66

†NA - numerical aperture
DOF - depth of focus

Masks used for proximity printing have the advantage of longer life because there is no contact between the mask and the wafer. Typical separations between mask and wafer are in the range of 20 to 50 μm. Resolution is not so good as in contact printing or projection printing. Commercial machines can select the mode automatically and switch to either contact or proximity printing.

Figure 2*a* shows proximity printing schematically with a slit of width W, illuminated by a monochromatic source, separated from a parallel image plane (wafer) by a gap g. We assume that g and W are larger than the wavelength λ of the imaging light and that $\lambda \ll g < W^2/\lambda$—the region of Fresnel diffraction. In this case, the diffraction that forms the image of the slit is a function only of the particular combination of λ, W, and g, which we call the parameter Q, where[4]

$$Q = W\sqrt{2/g\lambda} \tag{6.1}$$

This parameter can be considered as a normalized slit width. The lower limit of the nearfield diffraction region is at $Q = \sqrt{2}$ from Eq. (6.1), or $g = (W^2/\lambda)(2/Q^2) < W^2/\lambda$. Thus, the resolution, W in Eq. (6.1), becomes better at smaller gaps and shorter wavelengths. The resist image heights, which for practical purposes are converted to resist thickness and mask-wafer gap, have been measured as a function of linewidth, wavelength, and intensity tolerance.[3] The optimum image heights of 0.25-μm and 0.5-μm patterns are 0.9 μm and 3.2 μm, respectively, in the case of a deep UV (wavelength less than about 200 nm) light source with ±5% intensity tolerance.

Another parameter is the divergence of the illuminating beam as a result of the size of the light source. One effect is to smooth out the undulations in the image intensity profile, and the other is to produce a greater linewidth variation as the mask-to-wafer distance varies. The apparent source size of the illumination system must

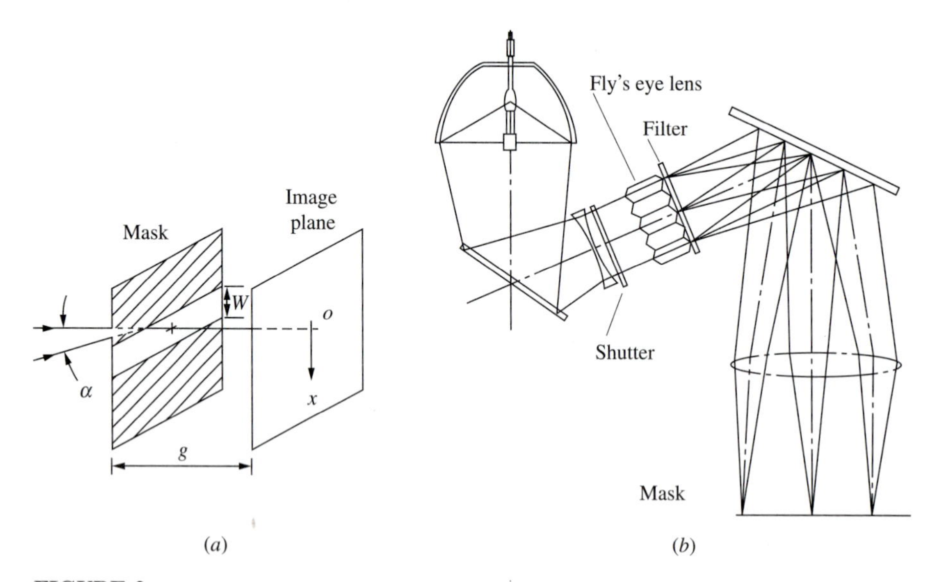

FIGURE 2
Schematic illustration of a proximity printer and illumination system. (*a*) Proximity printing schematic. (*b*) Illumination system of the Canon PLA600.

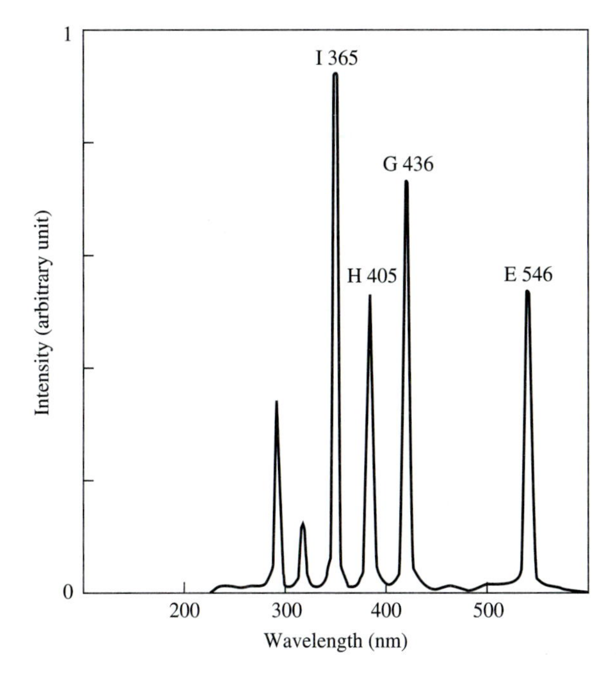

FIGURE 3
Spectrum of a high-pressure mercury lamp.

be large enough to give a value of α (Fig. 2) that allows the smallest features to be printed, typically, a few degrees. The mercury arc lamp used as the source[5] is too small to yield the required α. The illumination is telecentric or normally incident at the mask to prevent runout (magnification) error similar to that in x-ray lithography. The optical system must also minimize nonuniformity of the intensity across the field. The illumination system illustrated in Fig. 2*b* attains $\pm 3\%$ uniformity across a 6-inch wafer, as well as a large enough α and normal incidence. With a Hg arc source, the strong lines at 436 nm, 405 nm, and 365 nm provide the exposure flux shown in Fig. 3. The same printer is available with a Xe-Hg source for enhanced output in the 200–300 nm spectral region.

Commercially available machines are listed in the first group in Table 1.

6.2.2 Projection Printing

Projection printing offers higher resolution than proximity printing. It has a larger separation between the mask and the wafer because of its image formation system. According to Rayleigh's criterion, the resolution W of the optical systems and the depth of focus (DOF) are given in the following equations, where the numerical aperture NA $= n \sin \alpha$, n is the refractive index $=$ unity for air, 2α is the solid angle of the ray reaching the imaging point from the objective lens, and λ is the wavelength:[6]

$$W = 0.6\lambda/\mathrm{NA} \tag{6.2}$$

$$\mathrm{DOF} = \pm\lambda/2(\mathrm{NA})^2 \tag{6.3}$$

Modern projection printers employ diffraction-limited optics, which means that the design and fabrication of the optical elements are not a source of optical problems. Accordingly, the image characteristics are dominated by diffraction effects associated with the finite apertures in the condenser and with projection optics rather than by aberrations. When monochromatic light from a very small point is imaged by a diffraction-limited lens, the image consists of diffraction rings of light surrounding a central bright spot called the *Airy disc*.[6] The diameter of the pattern is $1.2\lambda/\mathrm{NA}$.

It is very useful to apply a modulation transfer function (MTF) to characterize the resolution capability of projection printers and proximity printers as well.[7] We consider the mask pattern as consisting of a periodic grid (line and space) with equal linewidths b. Light intensities in the center of dark lines and of bright lines are assigned as $I_{\min}$ and $I_{\max}$, respectively. Modulation is defined as

$$M = \frac{I_{\max} - I_{\min}}{I_{\max} + I_{\min}} \tag{6.4}$$

The MTF is the ratio of the modulation in the image plane (wafer) to that in the object plane (mask). It is a function of the spatial frequency ($\nu = 1/2b$) of the mask and the NA. The shape of the MTF, related to the spatial frequency curve, depends on the coherence of the illumination system.[7] The degree of coherence is defined by $\sigma = (\mathrm{NA})_c/(\mathrm{NA})_o$, where $(\mathrm{NA})_c$ is the numerical aperture of the condenser and $(\mathrm{NA})_o$ is the numerical aperture of the objective as shown in Fig. 4. In the coherent system, Fig. 4*a*, an object at point A is illuminated by only a narrow angle. Hence, all the light diffracted by A is coherent in amplitude at the image plane. In Fig. 4*b*, A is illuminated by rays from all portions of the extended incoherent source. Each ray is diffracted by the object A and forms an image in the image plane.

In Fig. 5, the solid lines show a plane wave front [$\sigma = (\mathrm{NA})_c/(\mathrm{NA})_o \to 0$] incident normal to the mask, which contains a grating pattern with b width lines and spaces (spatial frequency $\nu = 1/2b$).[8] The undiffracted component of light emerging from the mask contains no information about ν. This information is contained only in the diffracted light. The direction of the first diffraction peak is given by the grating formula $2b\sin\theta = \lambda$, so that $\nu = \sin\theta/\lambda$. If the light diffracted to direction θ is to reach the image plane, $\theta \approx \alpha$ where $\mathrm{NA} = \sin\alpha$. Therefore, the highest grating frequency that can be imaged by an optical system with coherent illumination becomes the result of Eq. (6.5).

For incoherent illumination (shown by dotted lines in Fig. 5), incident at a general angle i, the direction of the first maximum is given by $2b(\sin i + \sin\theta) = \lambda$. For both the undiffracted beam and the first diffraction peak to reach the image plane, i and θ both $\leq \alpha$. Therefore, $2b \geq \lambda/2\sin\alpha$ and $\nu_{\max} = 2b = 2\mathrm{NA}/\lambda$. Consequently, $\nu_{\max}$ (incoherent) $= 2\nu_{\max}$ (coherent). In modern printers the illumination is intermediate between the coherent and the incoherent limit.

The curve for MTF ($H(\nu)$) vs. spatial frequency[9] is shown in Fig. 6 and is described as[5]

$$\nu_{\max} = \mathrm{NA}/\lambda \qquad \text{(coherent)} \tag{6.5}$$

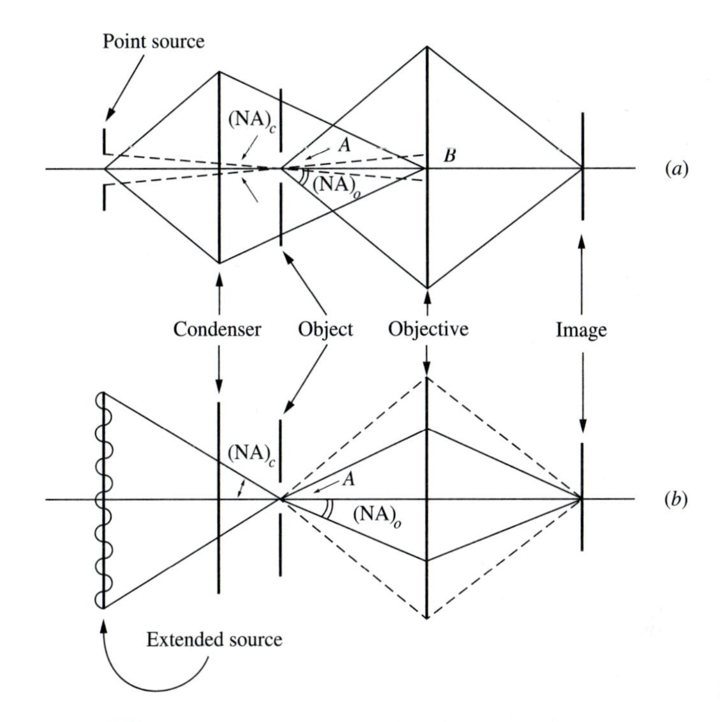

FIGURE 4
Coherent and incoherent illumination of projection printers. (*After King and Goldrick, Ref. 7.*) (*a*) Coherent system. (*b*) Incoherent system.

$$H(\nu) = \frac{2}{\pi}[\cos^{-1}(\nu/\nu_{\max}) - \nu/\nu_{\max}\sqrt{1-(\nu/\nu_{\max})^2}] \qquad \text{(incoherent)} \quad (6.6)$$

where $\nu_{\max} = 2\text{NA}/\lambda$

Figure 6 shows the modulation of the image intensity versus the spatial frequency for different values of coherence σ with a 0.3 NA lens. Note that the MTF of a fully coherent optical system has a spatial frequency cutoff that is half that of an incoherent system as explained in Eq. (6.5) and Eq. (6.6). Partially coherent illumination has a higher MTF in the region of ν larger than $\frac{1}{2}\nu_0$. The useful range, MTF > 0.6, is extended to higher spatial frequencies; edge gradients in the image become steeper, and the image is somewhat less sensitive to focusing.

The focus error is a simple but important aberration. The error is a displacement of best focus away from its intended position. In Fig. 6, the dashed curve shows the effect of a displacement of the image plane or wafer from the focal plane by one Rayleigh unit $w = \lambda/2(\text{NA})^2$, corresponding to a phase error of $\pi/2$ at the edge of the pupil.

It was once believed that a high NA is always better. However, in the submicrometer region there is an optimum NA if a resolution requirement and the imaging wavelength are given.[10] If the NA is too low, the resolution cannot be achieved, but if the NA is too high, the DOF, which is inversely proportional to $(\text{NA})^2$, becomes

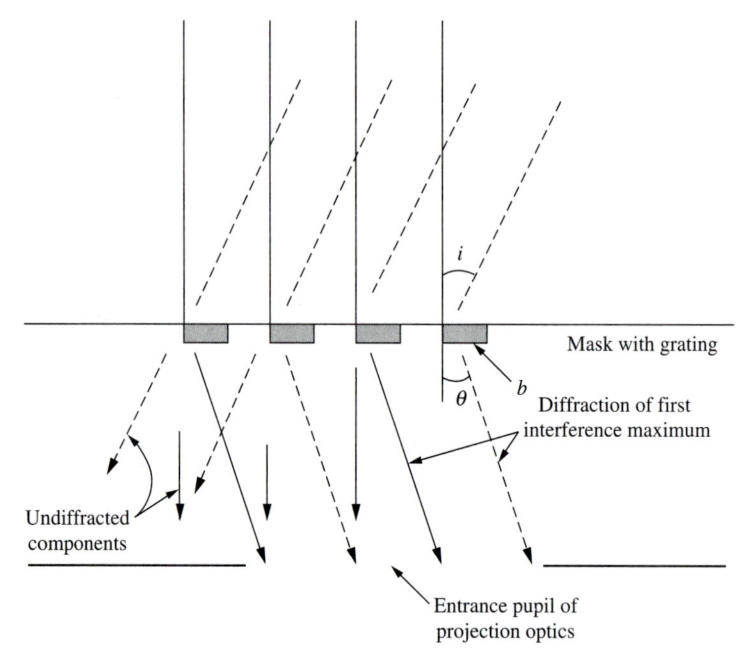

FIGURE 5
Diffraction of coherent and incoherent light by a grating pattern. (*After Cuthbert, Ref. 8.*)

unacceptable. There is an optimum NA at which the DOF is maximum. The normalization of the resolution W is given by

$$k_1 = W\frac{\text{NA}}{\lambda} \tag{6.7}$$

where k_1 is usually known as Rayleigh's coefficient for resolution but is now redefined as the normalized resolution. The normalized DOF is given by

$$k_2 = \Delta Z\frac{(\text{NA})^2}{\lambda} \tag{6.8}$$

where ΔZ is the physical axial displacement from the focal plane, and k_2, which is usually known as Rayleigh's coefficient for DOF, is now redefined as the normalized DOF.

With normalized resolution and defocus, the lithographic imaging behavior of a given feature and illumination can be universally plotted in the exposure-defocus (E-D) space in the form of constant linewidth contours. The exposure dosage required to keep the image linewidth constant at each defocal plane is evaluated to form contours. This set of contours, which is the E-D tree of the particular feature and illumination condition, is universally applicable for all W-λ-NA combinations leading to an identical k_1. The DOF is defined as the total amount of defocus allowed without violating a given linewidth tolerance. Typically, a $\pm 10\%$ tolerance is

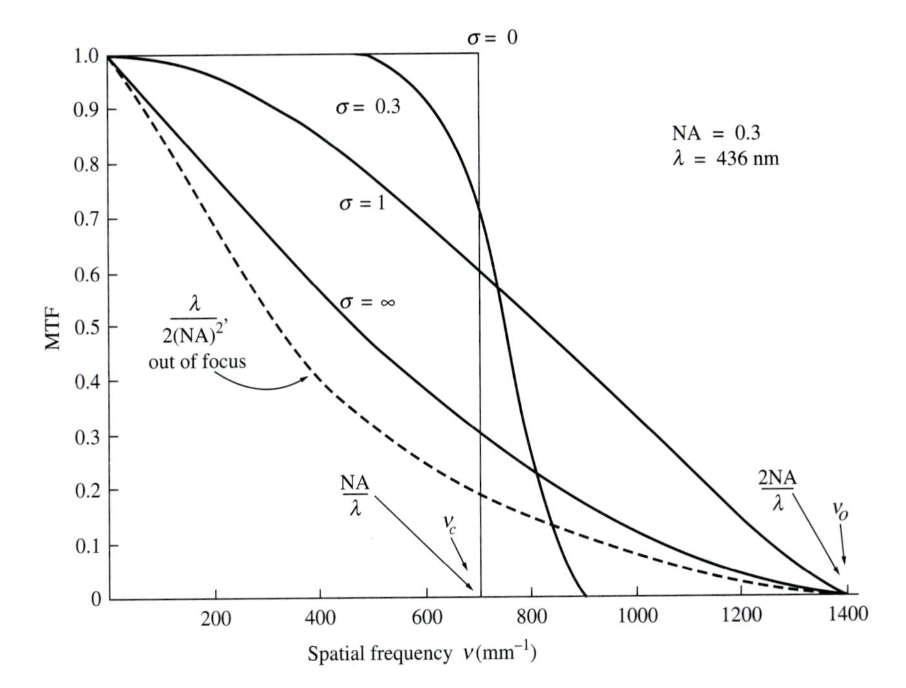

FIGURE 6
Comparison of MTFs for different optical conditions. (*After Lacombat et al., Ref. 9.*)

adopted. The DOF is the height of the largest window whose width is the exposure budget (process margin) that accommodates the tolerances that can affect the actual exposure dosage. Figure 7*a* shows the DOF determined by 10% and 30% exposure budgets, respectively. Figure 7*b* shows two E-D windows. One is bound by the common zone of three kinds of feature (line and spaces, isolated line openings, isolated spaces), and the other by five kinds of feature (including holes and islands).

The DOF equation is produced from Eq. (6.7) and Eq. (6.8) as

$$\text{DOF} = \frac{k_2}{k_1^2}\frac{W^2}{\lambda} \tag{6.9}$$

where λ/W = NA coefficient

W^2/λ = DOF coefficient

k_2/k_1^2 = DOF, normalized to W and λ

Figure 8*a* and 8*b* shows k_2 and k_2/k_1^2, respectively, as a function of k_1, and Fig. 8*c* shows W^2/λ and λ/W as a function of W. The normalized DOF and k_1 can be converted to physical DOF and NA by multiplying by the DOF coefficient and the NA coefficient, respectively, which are plotted in Fig. 8*c*.

The optimum NA for individual feature shapes is much lower than expected, as a result of Fig. 8*c*. The results lead to the problems that occur in reducing the usable k_1 factor for manufacturing. The optimum k_1 for the single-layer resist process (10% linewidth budget estimated) ranges from 0.57 to 0.87 depending on the feature shape.

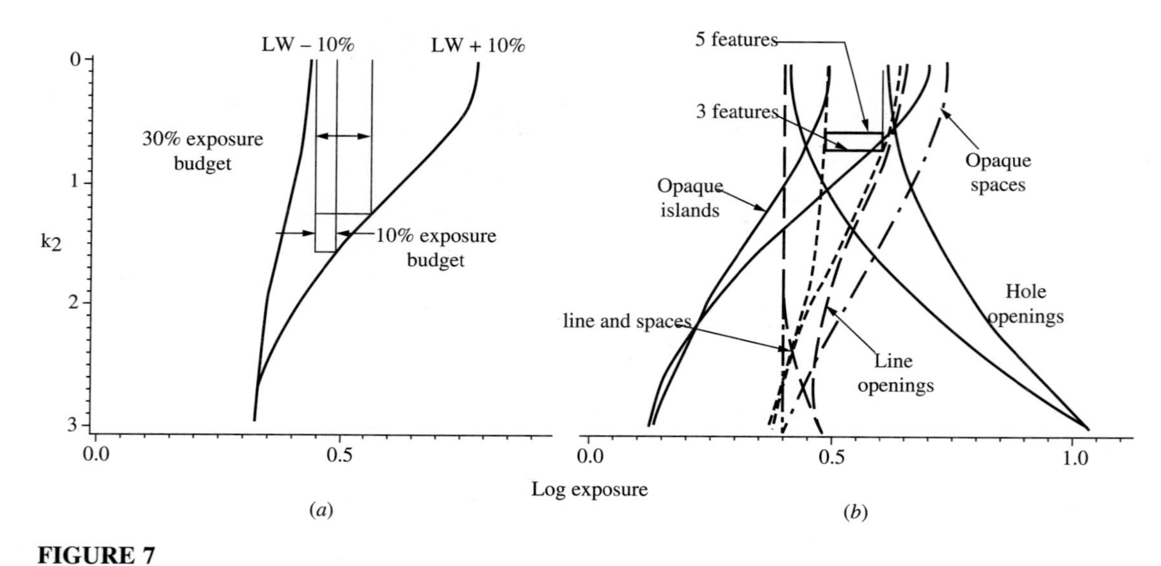

FIGURE 7
E-D windows. (*After Lin, Ref. 10.*) (*a*) The 10% and 30% E-D window bounded by ±10% E-D branches of line and space pairs at $k_1 = 0.8$, $\sigma = 0.4$. (*b*) The 3-feature and 5-feature E-D windows at $k_1 = 0.7$, $\sigma = 0.52$.

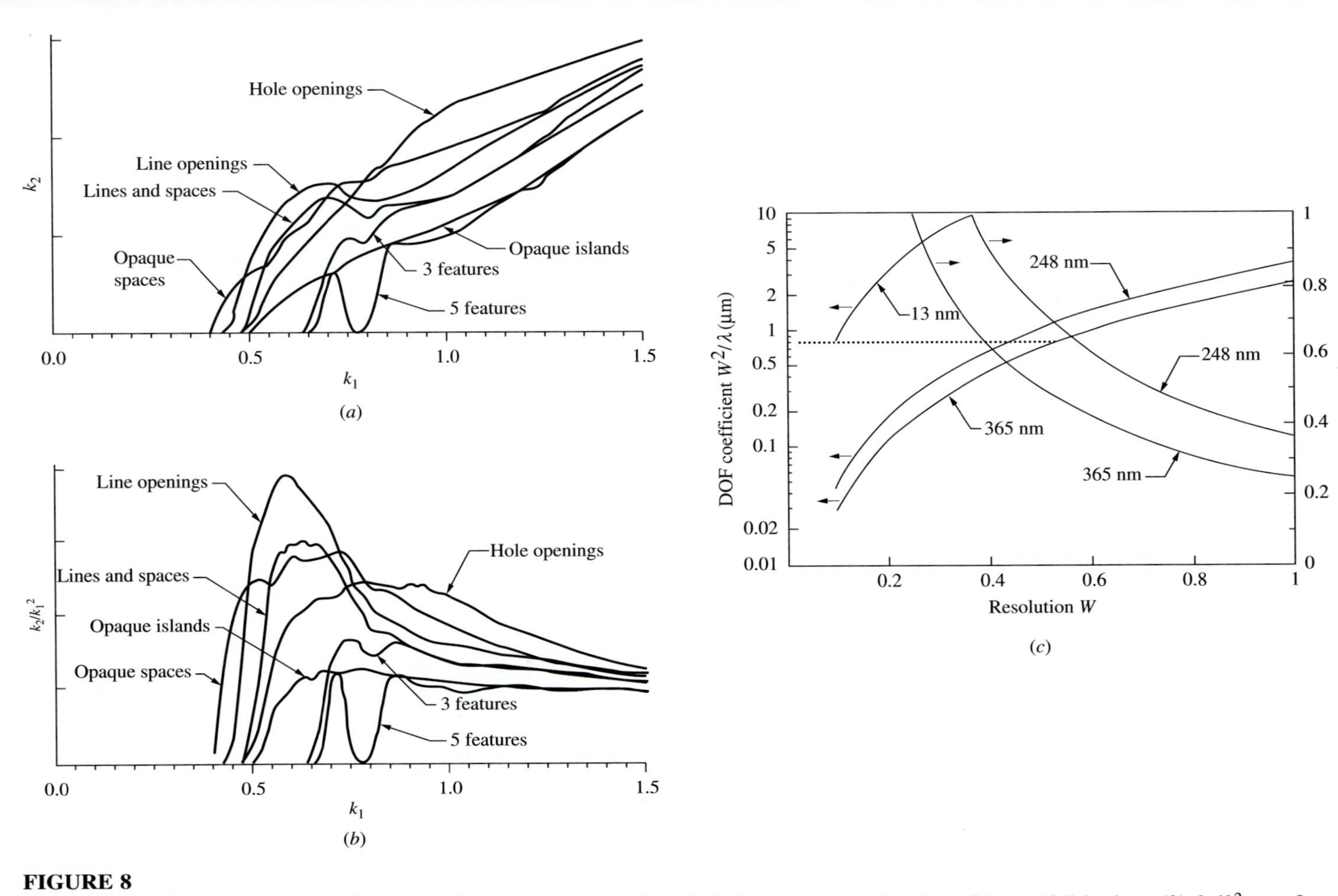

FIGURE 8
NA optimization parameters for optical projection printers. (*After Lin, Ref. 10.*) (*a*) k_2 as a function of k_1 at 10% budget. (*b*) k_2/k_1^2 as a function of k_1 at 10% budget. (*c*) The DOF coefficient and NA coefficient as a function of W. The dotted line shows the similar difficulty of achieving 0.1μm by X-ray and 0.5μm by I-line.

For a multilayer resist process (30% budget), k_1 ranges from 0.42 to 0.7, although it is generally defined as 0.6 as shown in Eq. (6.2).

Reflection and catadioptric projection

A significant advantage of the reflective or the catadioptric system[11] (combining reflecting and refracting components) is the larger spectral bandwidth. Reflection projection systems generally have higher throughput with less standing-wave interference than refraction systems because of their polychromatic characteristics. Catadioptric systems usually require more than just one optical axis and thus can lead to great difficulty in aligning the optical elements. Several systems of this type[12] have been developed as shown in the second category of Table 1 and in Fig. 9.

The Micralign®, made by Perkin Elmer, consists of only three optical components, as shown in Fig. 9*a*. It has a bandwidth of 400 nm. The aberrations vary as a function of the distance from the center and can be zero for a narrow zone about the center. Only a small zone of the spherical mirror is used, limited by the slit, providing nearly diffraction-limited imaging. The mask and the wafer are swept in unison through an arc to form an image of the whole mask. Figure 9*b* is the catadioptric system, also by Perkin Elmer, which is still widely used as a 1× full-scan system. The throughput is very high, as shown in Table 1, but the disadvantages of these systems are the low resolution and the difficulty of alignment during scanning; hence, the overlay accuracy is rather poor, as listed in Table 1.

Figure 9*c* is a catadioptric 1× stepper made by Ultratech (Table 1). Only the wafer has a step-and-repeat movement, and a die-by-die alignment is done. This system also has a very simple optical configuration, featuring a variable NA that depends on object field size—a low NA corresponds to a large field size. The practical resolution is 0.7 μm.

Figure 9*d* shows a 4× catadioptric reduction projector (Table 1) developed by Perkin Elmer that has 0.5-μm resolution. This machine also has a *ring field* similar to Fig. 9*a* and 9*b* with field-by-field alignment and a focus adjustment. The image field is larger than in I-line steppers.

Refraction projection

In refraction projection, the image of the mask is projected on the wafer through a high-resolution lens. A refraction projection system consists of a light source, a condenser, a heat-removing filter, beam-orienting mirrors, a shutter, and the mask and wafer stages. The refraction projection system is very similar to the proximity printers except for the imaging lens between the mask and the wafer. The optical stepper has become the most common lithography tool for ULSI production. The resolution is 0.4 μm using the I-line of the Hg lamp with the Kohler method, which images the exposure source through a condenser lens in an entrance pupil of the projection lens. However, the DOF is reduced to 1 μm because of the stepper's large NA and short wavelength. The features of the stepper are high resolution, and overlay accuracy as shown in Table 1.

The last group in Table 1 is an excimer laser stepper[13] whose wavelength is 248 nm. It can obtain the highest resolution of the equipment listed in Table 1 based on the excimer's shortest wavelength. The system configuration is almost the same as an I-line stepper, but the light source and the lens system are different. In the deep

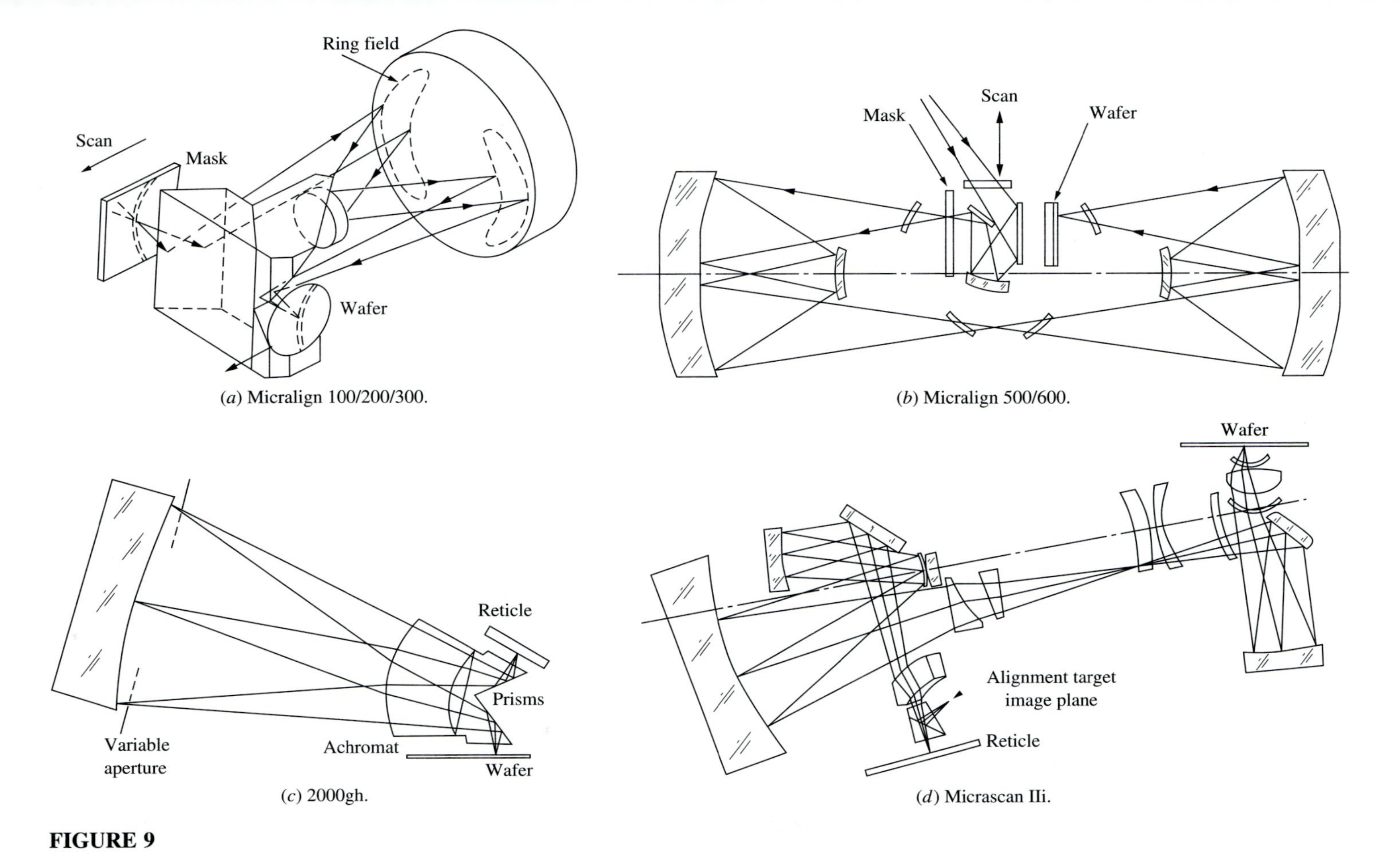

FIGURE 9
Examples of reflection projection printers. (*After Lin, Ref. 12.*)

UV region, the materials of the optical components are limited to fused silica and several crystalline fluorides because of UV transmittance characteristics. The lenses are made entirely of fused silica and therefore are chromatic. The source bandwidth must be narrow to achieve sufficient resolution. The focus and the magnification differences due to wavelength difference are 0.15 μm/pm and 0.3 ppm/pm, respectively. The current performance of the KrF laser source, such as a 2.2-pm bandwidth and a ±0.25-pm wavelength stability, is considered adequate to meet the requirement for practical use. A 0.35-μm resolution within the 0.03-μm critical dimension spread is reported.[13] The use of the excimer laser is another possible way to fabricate ULSI patterns without resists by using the damaging effect of a short-wavelength light.

6.2.3 Enhancement

Many efforts have been made to solve optical lithography problems arising from wavelength or DOF. Among these, the phase-shifting mask and the off-axis illumination technique seem to be useful. The basic concept of the phase-shifting mask[14] is shown in Fig. 10. At the transmission mask, the electric field $\mathscr{E}$ has the same phase at every aperture (clear area) in Fig. 10*a*. Diffraction and the limited resolution of the optical system spread the electric field $\mathscr{E}$ at the wafer, as shown by the dotted

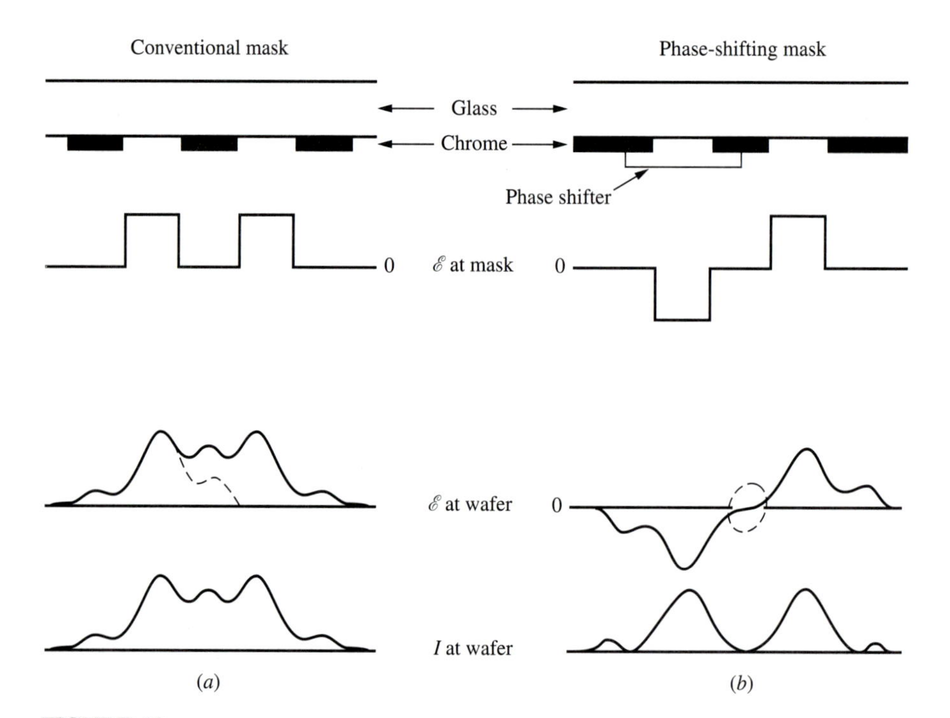

FIGURE 10
The principle of phase-shift technology. (*After Levenson et al., Ref. 14.* © 1992 IEEE.) (*a*) Conventional technology. (*b*) Phase-shifting technology.

line. Interference between waves diffracted by the adjacent apertures enhances the field between them. The intensity I is proportional to the square of the electric field.

The phase-shifting layer that covers adjacent apertures reverses the sign of the electric field as shown in Fig. 10*b*. The intensity at the mask is unchanged. The electric field of these images at the wafer, shown by the dotted line, can be canceled. Consequently, images that are projected close to one another can be separated completely. A 180° phase change occurs when a transparent layer of thickness $d = \lambda/2(n - 1)$, where n is the refraction index and λ is the wavelength, covers one aperture as shown in Fig. 10*b*.

The Fourier analysis[14] of phase-shifting and conventional transmission masks is explained in Fig. 11. The conventional resolution power of a projection system for transmission objects is the critical frequency $\nu_c = \text{NA}/\lambda$, which is related to the wavelength λ and numerical aperture NA and assumes coherent illumination. With incoherent illumination some spatial modulation is transmitted through such a system up to a frequency of $2\nu_c$; but for coherent illumination the modulation transmission function is unity up to ν_c and zero for $\nu > \nu_c$, as shown in Fig. 6. For the purpose of calculation it is convenient to define for the mask a transmission function that has sinusoidal intensity, as illustrated at the top of Fig. 11. In each case,

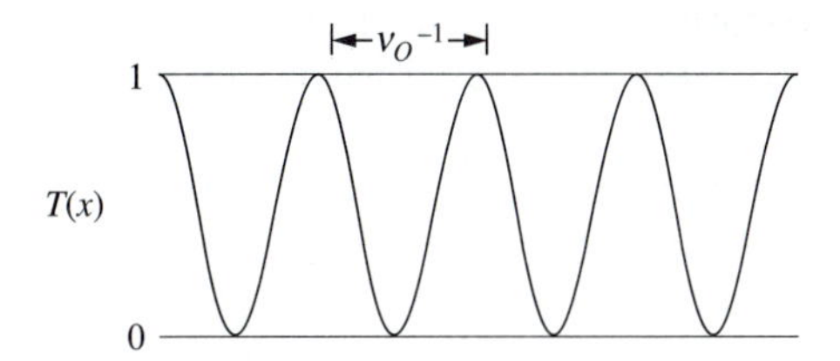

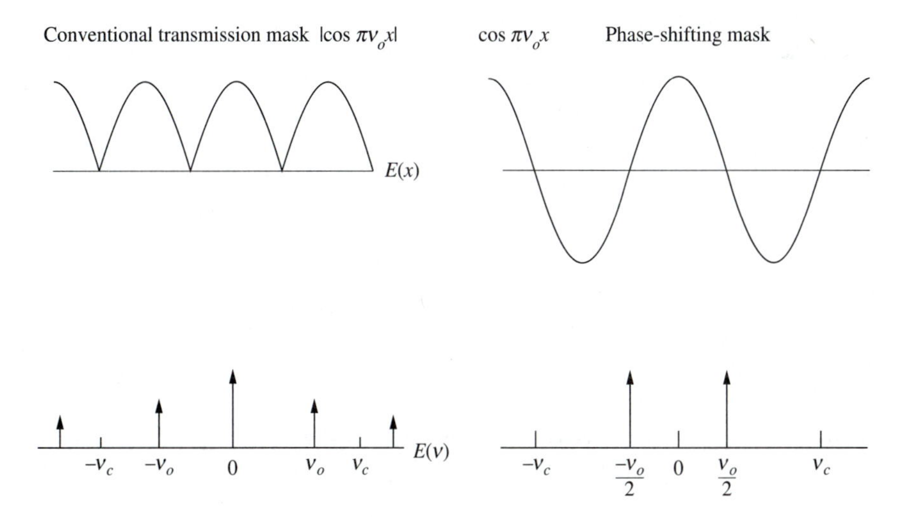

FIGURE 11
Fourier analysis of the optics of transmission and phase-shifting masks. (*After Levenson et al., Ref. 14.* © 1992 IEEE.)

the intensity transmitted by the mask is assumed to be

$$T(x) = \tfrac{1}{2}(1 + \cos 2\pi\nu_0 x) \tag{6.10}$$

The electric field at the mask plane can have two possible forms when the illumination is normally incident and coherent. A transmission mask would produce an electric field profile $\mathscr{E}$ that is proportional to $|\cos \pi\nu_0 x|$, as in the left side of Fig. 11, whereas a phase-shifting mask would produce the electric field on the right, or $\mathscr{E} \propto \cos \pi\nu_0 x$. Both profiles can be Fourier-analyzed as shown at the bottom of Fig. 11. The transmission mask yields a Fourier spectrum of the form

$$\begin{aligned}\mathscr{E}_T(\nu) = \frac{4}{\pi}\Bigg\{ &\frac{1}{2}\delta(\nu) + \frac{1}{6}[\delta(\nu - \nu_0) + \delta(\nu + \nu_0)] \\ &- \frac{1}{30}[\delta(\nu - 2\nu_0) + \delta(\nu + 2\nu_0)] + \cdots \Bigg\}\end{aligned} \tag{6.11}$$

where $\delta(\nu)$ is the Dirac delta function and higher harmonics have been ignored. The phase-shifting mask projects only two Fourier components:

$$\mathscr{E}_\phi(\nu) = \tfrac{1}{2}[\delta(\nu - \nu_0/2) + \delta(\nu + \nu_0/2)] \tag{6.12}$$

The same information, in the sense of an optical intensity, is transmitted in each case, but the phase-shifting mask requires less spatial bandwidth than the transmission mask.

The optical system through which these Fourier components must propagate can be modeled as having a response function equal to unity for $\nu \leq \nu_c$ and zero for $\nu > \nu_c$, where ν_c is the cutoff frequency. Thus, the Fourier components of the electric field at the image plane are (except for a scale factor that can be set to unity) identical to those at the mask plane for $\nu < \nu_c$ and zero otherwise. From this treatment, the critical frequency for the transmission of modulation produced by a phase-shifting mask is $\nu_0 = 2\nu_c$, twice that of a transmission mask.

It is expected that an I-line stepper can resolve features down to 0.2 μm using a phase-shifting mask without reducing the DOF. The problem with this technology is the restriction on the pattern layout. It is very difficult to apply this technology to random patterns, although it is very useful in a periodic layout. Several ideas have been proposed[15] to solve the problem, as shown in Fig. 12. The halftone method, shown in Fig. 12*a*, has advantages such as not requiring extra data and ease of processing, but the resolution is not so good as the alternative method of Fig. 12*c*.

Phase-shifting masks are made by using an electron-beam (e-beam) direct-writing method. After the first layer is etched, the second layer is exposed on a newly coated resist, referenced to the alignment marks made in the first layer. In Fig. 12*b* or 12*c*, two kinds of patterns are necessary for each wafer layer. The problems of phase-shift technology are the design of the shifter pattern and the complexity of the mask process. Other key technologies required to make phase-shifting masks are improved inspection and repair technologies for the shifter patterns and the opaque patterns.

Off-axis illumination[16] is another useful method to enhance the resolution and DOF. The basic principle is to tilt the illumination to an angle that passes through both the undiffracted light and the first-order diffracted light symmetrically as shown

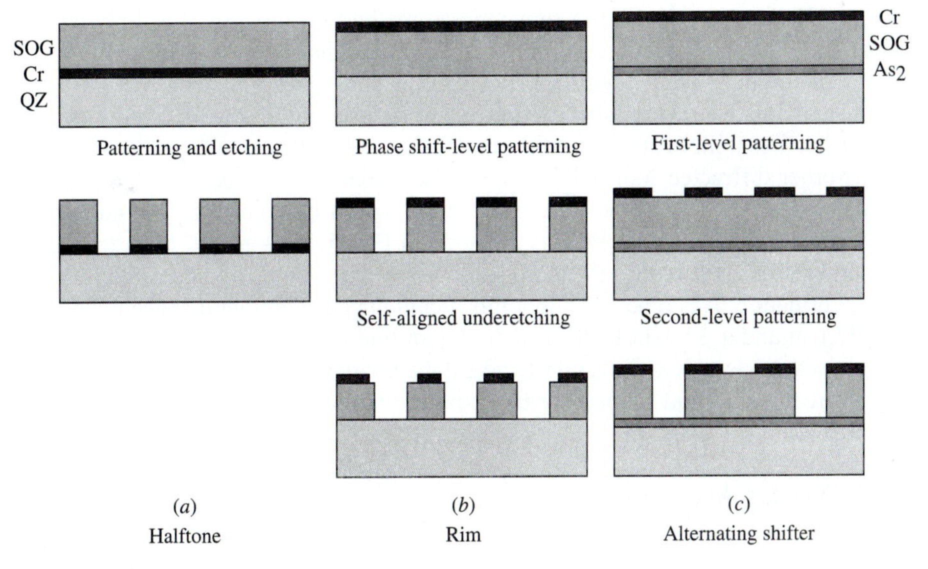

FIGURE 12
Examples of phase-shifting masks. (*After Ronse et al., Ref. 15.*) (*a*) Halftone. (*b*) Rim. (*c*) Alternating shifter.

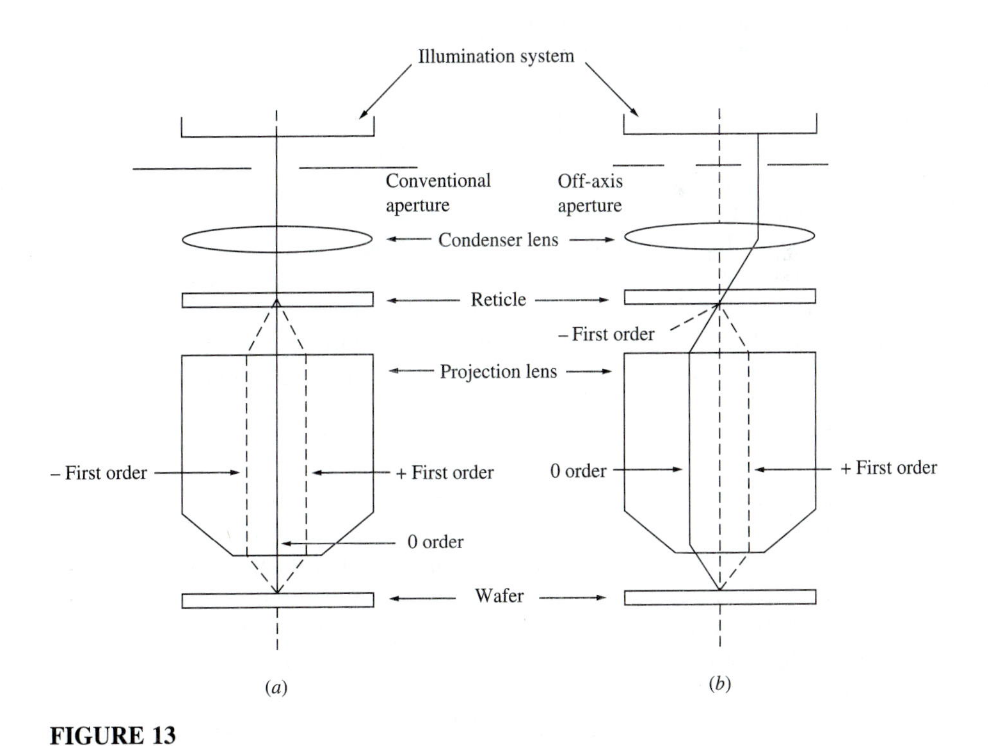

FIGURE 13
The off-axis illumination principle. (*After Shiraishi et al., Ref. 16.*) (*a*) Conventional illumination. (*b*) Off-axis illumination.

in Fig. 13*b*, compared with the conventional method in Fig. 13*a*. The light beams transmitted through the off-axis aperture illuminate the reticle pattern with a specific incident angle ϕ, defined as $f \sin\phi = x$, where x is the distance between the optical axis and the transmissive portion of the aperture and f is the focal length of the condenser. The light beams are diffracted by the reticle pattern. The −first-order or higher-order diffracted light will not enter the projection lens because the pitch of the pattern is so fine and the diffraction angle ($\sin\theta$) is larger than the numerical aperture of the projection lens, as explained in Fig. 5. As a result, only zero- and first-order diffracted light will interfere on the wafer surface and contribute to the image formation. It is reported[16] that this technique has a line-and-space resolution of 0.27 μm and a 2.8-μm DOF for a 0.35-μm line-and-space pattern with a 1.18-μm resist thickness. It is necessary to adjust the incident angle depending on pattern spatial frequencies and to increase the exposure time to compensate for the loss of illumination due to tilting.

6.2.4 Overlay Accuracy

To this point we have emphasized resolution, but overlay accuracy is another fundamental technology component.[12] Overlay error is generated mainly in two process steps—the alignment and the exposure. Many types of errors are included in the exposure process, such as the relative placement error between the mask and the wafer, distortion or magnification error of the lens, and mechanical instability of the exposure system caused by vibration, temperature drift, or atmospheric pressure drift.

Alignment methods are classified into two types: off-axis and TTL (through the lens). Off-axis offers the possibility of using nonactinic rays and of broad-band and high-NA viewing, with the flexibility of brightfield, darkfield, or even phase-contrast viewing. It is also easy to upgrade the off-axis alignment system because of its total separation from the imaging optics. However, off-axis requires high mechanical stability and precision as well as accurate means to refer the positions of the mask, the wafer, and the alignment microscopes to each other.

TTL offers the inherent advantage of a direct mask-to-wafer reference and thus relaxes the requirement for high mechanical stability and precision. However, some accuracy of the direct reference is lost if nonactinic alignment is used or if the wafer has to be moved into the exposure position after alignment. Both of these process steps require compensation.

Brightfield alignment suffers from multiple interferences of the wafer film stacks in case of TTL narrow-bandwidth illumination. The alignment signal is sensitive to variation in film thickness, on the order of 10 nm. This problem is reduced by using broad-band illumination. Alternatively, darkfield alignment removes the dependence on film thickness but does not completely nullify the effect of uneven resist coverage. Also, darkfield alignment has the potential problem of signal degradation on a grainy substrate. The solution to this problem is using the slit beam and the periodic alignment marks. One way to solve the aforementioned problems is to remove the resist over the alignment marks in the positive resist process, but this is impossible in the case of multilayer resist and negative resist processes.

An alignment system[17] is illustrated in Fig. 14. Off-axis global alignment measures the relative position of the wafer with a He-Ne laser. Die-by-die alignment is

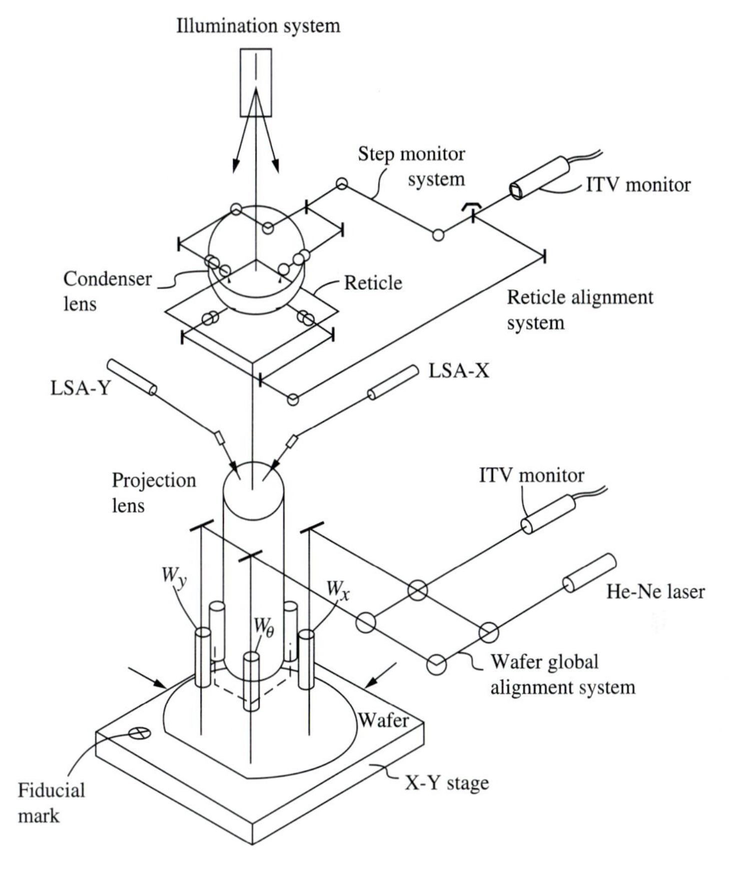

FIGURE 14
Wafer alignment system of an optical stepper. (*After Shiotake and Yoshida, Ref. 17.*)

more accurate because it uses an extra laser source (LSA-X, LSA-Y) through the projection optical path, i.e., the TTL method. Step-and-repeat exposure offers level-to-level registration, which is independent of wafer size, by separate alignment of each exposure field. In the reduction projection system, the mask pattern dimensions are larger than in systems imaged with unity magnification, and therefore it is more convenient to adjust the overlay position.

The basic sources of overlay error are listed in Table 2. The problem of overlay is common to both optical and x-ray lithography. X-ray lithography and e-beam direct write are considered distortionless. The registration error, defined as the ability to locate the alignment marks, is assumed to be identical for all optical alignment systems. Resist-induced alignment error is considered to be systematically removable. Wafer inplane distortion is caused by various hot processes. The major part of the magnification error of reduction systems is eliminated by fine tuning, but not easily, owing to the inherent symmetry of 1× optical systems. Coupled with the linear part of wafer inplane distortion, the magnification error of 1× optical systems can be as large as 0.15 μm. Similarly, x-ray systems have an error of 0.1 μm as a result

TABLE 2
Sources of overlay errors[12]

Items	5× optics (μm)	1× optics (μm)	1× x-ray (μm)	E-beam direct write (μm)
1. Lens distortion	0.07	0.06	0	0
2. Registration error	0.05	0.05	0.05	0.05
3. Wafer inplane distortion	0.05	0.05	0.05	0.05
4. Magnification error	0.01	0.1	0.1	0.01
5. Mask-to-mask placement error	0.05	0.05	0.05	0.05
6. Total contributions (RSS)	0.1	0.15	0.13	0.09
7. Total without lens distortion	0.07	0.13	0.13	0.09
8. Total without magnification error	0.1	0.11	0.09	0.09
9. Total without distortion or magnification error	0.07	0.09	0.09	0.09

of wafer inplane distortion and mask heating during exposure. The mask-to-mask placement error reflects the accuracy of e-beam mask making. For x-ray systems, the placement error of the membrane mask can be worse than for the optical mask, but it is not included here. In the case of e-beam direct write, the mask error is simply the composite placement error of the e-beam between two masking levels. In the case of 5× lithography, the mask-to-mask placement error is divided by 5 before being included in the RSS (square root of the sum of the squares) total of the contributions.

6.2.5 Optical Resists

Photoresist is the general term for polymers that can create patterns by the use of solvents after irradiation. The development of the resist is based on a chemical reaction and depends on the solubility difference between irradiated and unirradiated areas. Photoresists are of two types: negative, which on exposure to light become less soluble in a developer solution, and positive, which become more soluble.

Negative resists generally consist of a chemically inert polyisoprene rubber, which is the film-forming component, and a photoactive agent. The photoactive agent releases nitrogen gas on exposure to light, and the radicals generated react with the double bonds to form cross-links between rubber molecules, making the rubber less soluble in an organic developer solvent. The reactive species formed during the exposure can react with oxygen and be rendered ineffective for cross-linking, so an inert atmosphere is used. The developer solvent dissolves the unexposed resist. The exposed resist has low molecular weight, so it swells as the uncrosslinked molecules are dissolved. The swelling distorts the pattern features and limits resolution to 2 to 3 times the initial film thickness. For VLSI/ULSI applications, the use of negative resists has been supplanted by positive resists because negative resists have a resolution limit of about 2 μm, although they have many advantages such as resistance to etching and good adhesion to the substrate.

Positive resists have two components: a resin and a photoactive compound dissolved in a solvent. The photoactive compound is a dissolution inhibitor. When the photoactive compound is destroyed by exposure to light, the resin becomes more soluble in an aqueous developer solution. The unexposed regions have high molecular weight and so do not swell much in the developer solution; therefore higher resolution, suitable for ULSI, is possible with positive resists. The development process of projection-printed images in positive resists has been modeled theoretically;[18] it is an isotropic etching or removal process.

One of the fundamental properties of a resist, which determines resolution, is the contrast.[19] Contrast, in the case of negative resist, is related to the rate of formation of a cross-linked network and may be measured by exposing pads of known area to varying radiation doses. D_g^i is the minimum dosage required to form the first insoluble film. Thereafter, the film thickness, and consequently the thickness remaining after development, increases with increasing dose until, ultimately, the thickness of the developed pad is not detectably different from the original film thickness (in practice it is slightly less owing to volume contraction during cross-linking). This process is shown schematically in Fig. 15*a*. The resist contrast γ_n is then defined as

$$\gamma_n = 1/(\log D_g^0 - \log D_g^i) \tag{6.13}$$

For a positive resist, the film thickness of the irradiated region that remains after development decreases with increasing dose until eventually a dose D_c is reached, which results in complete removal of the film on development, as shown in Fig. 15*b*. The contrast of positive resist γ_p is defined as

$$\gamma_p = 1/(\log D_c - \log D_0) \tag{6.14}$$

When photoresist films are exposed using monochromatic radiation, standing waves are formed in the resist.[20] These are caused by coherent interference effects due to a reflecting substrate. Coherent interference results in periodic intensity distributions in the direction perpendicular to the plane of the resist with a period $\lambda_r/2$, where λ_r is the wavelength in the resist. The refractive index of most resists is about 1.6, leading to an optical mismatch at the air/resist interface. Figure 16 shows

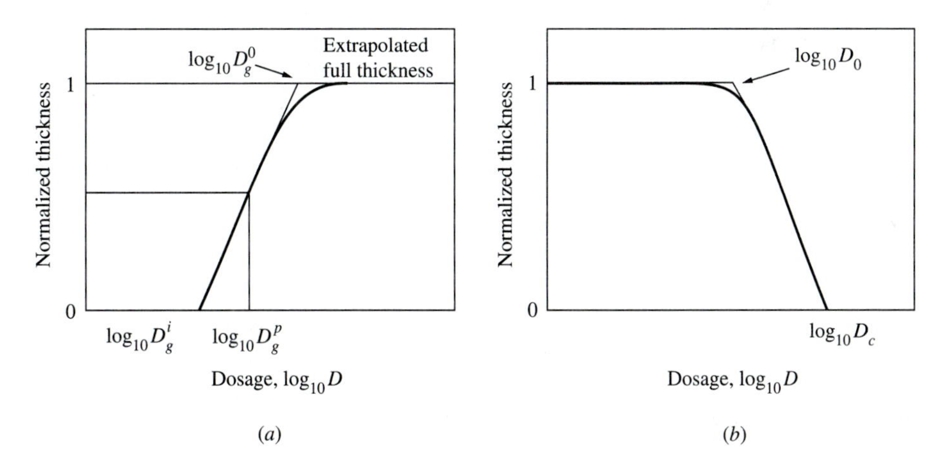

FIGURE 15
Sensitivity curves of negative and positive resists. (*After Thompson, Ref. 19.*) (*a*) Negative resist. (*b*) Positive resist.

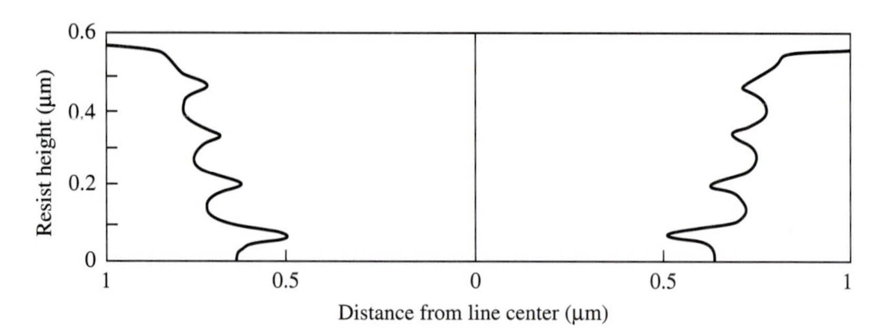

FIGURE 16
Simulated standing wave edge profile for a nominal 1-μm line in AZ1350 photoresist developed for 85 sec in 1 : 1 AZ developer : water. (*After Dill et al., Ref. 20.* © 1975 IEEE.) Substrate: Si + SiO_2, 60 nm; resist thickness: 583.6 nm; exposure wavelength: 435.8 nm (G-line); exposure energy: 57 mJ/cm^2

the standing wave. This variation in peak intensity with resist thickness becomes less with decreasing substrate reflectivity and increasing absorption by the resist. Since resist thickness varies at a step in the substrate topography, the resulting difference in effective exposure leads to size variations in the resist image.

Commercially available resists are listed in Table 3. The negative RD2000N-type resists, which have a diazide compound as a reactor with polyvinylphenol resin, have a resolution similar to a positive resist and do not swell on development. Photoresists are being developed for exposure at shorter wavelengths where higher resolution is possible. A few such deep-UV resists are PMMA, sensitive for $\lambda < 250$ nm; polybutene sulfone, sensitive for $\lambda < 200$ nm; and Microposit MP-2400, sensitive for $\lambda = 250$ nm. At these shorter wavelengths, the radiation quantum is large enough to produce scission of the molecular chain.

The other very important resists for deep UV are the chemically amplified resists, which exhibit high photo-speed, excellent resolution, and process tolerance.[21] The chemically amplified resist consists of a polymer host and a generator of acid in the presence of light. In the postexposure bake the photogenerated acid catalyzes thermal reactions that alter the solubility of the exposed region. Generally, resolution and sensitivity are affected by postexposure bake conditions. Some chemically amplified resists are applicable to electron lithography.

Other important properties of resists are adhesion to the substrate and resistance to wet and dry etch processing. In general, the commercially available optical resists are compatible with such processes.

Some inorganic materials are also effective as resists. Fundamental functions are based on the Ag-doping effect, which is induced by photoirradiation or electron-beam irradiation on a stacked-layer system of Ag and Se-Ge chalcogenide-glass film. When doped with Ag, the chalcogenide-glass films become almost insoluble in an alkaline solution. The light source, such as a Hg lamp, should have a wavelength that is shorter than the absorption edge of Se-Ge. This method is expected to have high resolution, with applications in dry processing.[22,23]

TABLE 3
Commercial optical resists

Name	Maker	Type	Chemical structure	Remarks
OMR83	Tokyo-oka	Negative	Isoprene rubber, azide compound	1× aligner
CIR	Nippon Synthetic Rubber	”	”	”
RD-2000N	Hitachi Chemical	”	Phenol resin, azide compound	Deep UV
MES-U	Nippon Synthetic Rubber	”	Chloropolystyrene	”
THMR-iN200	Tokyo-oka	”	Novolac resin, azide	I-line
TDUR-N7	”	”	Phenol resin	Deep UV
Microposit 1350J	Shipley	Positive	Novolac resin, O-kyniadiazide	G-line
AZ1350J	Hoechst	”	”	”
OFPR800	Tokyo-oka	”	”	”
Microposit2400	Shipley	”	”	Deep UV
ODUR-120	Tokyo-oka	”	PMMA	”
ODUR1000	”	”	PMIPK	I-line
THMR-IP3000	”	”	Novolac resin, azide	”
TSCR-80I	”	”	”	”
PFI28	Sumitomo Chemical	”	”	”
PFRIX060	Nippon Synthetic Rubber	”	”	”
PFRIX061	”	”	”	”
FHI-3950	Fuji Hunt	”	”	”
AZ7500	Hoechst	”	”	”
NPR-A18SH7	Nagase	”	”	”

6.2.6 Process Technologies

Increasing the NA and decreasing the wavelength are logical approaches to improving resolution. However, if these techniques are used, DOF decreases, based on Eq. (6.3). In the actual production process, many possible causes of interference to focusing arise. Some occur during the wafer process, such as surface topography and warping. Others are due to the exposure process, such as vibration, illumination nonuniformity, and aberrations. One solution to these problems is the combination of planarization and surface imaging. The thick bottom planarization layer reduces optical reflection and pattern-width degradation due to the surface topography beneath. The thin top layer enhances resolution, pattern-width control, and process latitude.

Surface-imaging resist processes are summarized in Fig. 17. The first item is treatment through a development cycle to enhance the contrast of the resist.[24] In this cycle the wafer is dipped in the developer, rinsed, and dried. This sequence is repeated until the resist film of the exposed area is completely dissolved. This interrupted development process promotes the formation of a passivation layer on the

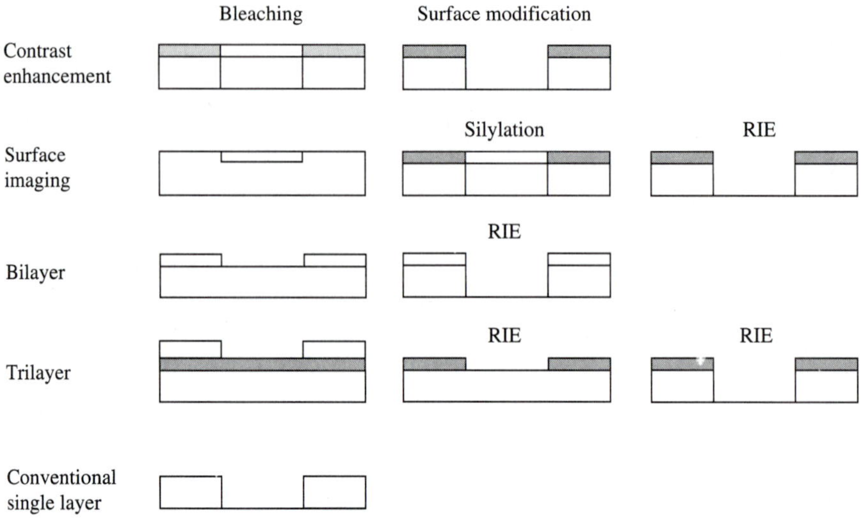

FIGURE 17
Advanced resist processes.

unexposed surface. As a result, the side walls become vertical and sensitivity is improved 30 to 50% in AZ resist. This technology is also used in electron lithography.

The second process in Fig. 17 is silylation.[25] A single layer of PLASMASK® (UCB from Belgium) resist is spincoated onto the substrate, with a thickness from 1.5 to 2.5 μm, and prebaked to a self-planarizing layer. After exposure, the wafers are treated with a gas-phase silylating agent (hexamethyldisilazane) at elevated temperature to reduce the silylation time. The resist material of the exposed areas selectively bonds chemically with silicon to a depth of 100–200 nm, and remains stable for a long time. The wafers are developed in oxygen plasma. During this treatment, the silicon is converted into silicon dioxide, which forms a thin protective mask that stops the etching of these exposed areas. It is reported that 0.35-μm features can be achieved using a 248-nm source.[26]

In the bilayer method, the image of the top layer formed by exposure and development is transferred to the bottom planarization layer by reactive ion etching (RIE). Resolution better than 0.5 μm is achieved by using a resist containing tungsten, which has a high resistance to etching, as the top layer.[26] There are some disadvantages in the bilayer method, such as intermixing between the resist layers and the formation of cracks while different materials are coated and baked.

The trilayer resist structure typically consists of an inorganic interlayer that is 0.1 to 0.2 μm thick, sandwiched between a top organic imaging layer that is 0.3 to 0.4 μm thick and a bottom organic planarization layer that is 1 to 2 μm thick. The trilayer method is more complex than the bilayer but is more flexible because of the intermediate layer.[27,28] Because of the intermediate barrier, the processes for the top layer and for the bottom layer are independently selectable, so many combinations of processes can be achieved by the proper choice of the intermediate layer materials and the top layer resists. The thickness of the top layer can be minimized to get high resolution with enough etching duration for the intermediate layer (usually oxide). This method is also used in electron lithography.

6.3 ELECTRON LITHOGRAPHY

Electron lithography has the possibility of higher resolution than optical lithography because of the small wavelength (less than 1 angstrom) of the 10–50 keV electrons. Resolution in electron lithography systems is not limited by diffraction but by electron scattering in the target materials including the resist and by the various aberrations of the electron optics. Scanning electron beam pattern generators have been under development for more than 20 years and were derived from the scanning electron microscope.[29,30] Because of the serial nature of the pattern writing, throughput is much lower than for optical systems. However, a wide variety of applications is available in the pattern-generating function for electron lithography, such as mask fabrication for optical or x-ray lithography, direct writing on the wafers, and direct reaction with some materials on the substrate. Electron lithography is classified into two types—scanning and projection—and the scanning type can be either raster or vector scanning.

The exposure scheme is illustrated in Fig. 18. The ULSI pattern is composed by a computer-aided design (CAD) system. The output format from the CAD system is converted into the internal format of the individual exposure systems. The electron

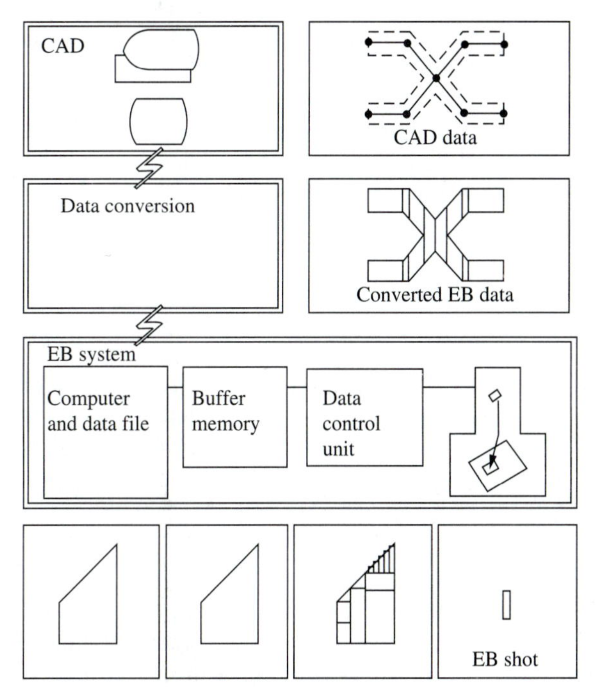

FIGURE 18
Flow of data in electron beam exposure systems.

exposure machine decomposes the data into simple elements (trapezoids or rectangles, depending on the machines) to control the electron beam irradiation. Electron beam exposure machines are bigger and more complicated than the optical printers because of their data-handling function.

Although electron projection and proximity-printing methods have been developed for full image exposure and are expected to be one of the alternatives to optical technology after an optical resolution limit is reached, these techniques have not become practical yet. Among many kinds of electron beam exposure machines, the MEBES® from Perkin Elmer has proven to be the best photomask pattern generator. Some special products, such as microwave transistors, have for many years been manufactured by direct writing. In some low-volume production devices, direct writing has also been applied in the development stages of ULSIs and in personalization of application-specific integrated circuits (ASICs) to minimize the mask-related cost and development time. Commercially available electron-beam machines are listed in Table 4. Generally, Gaussian spot vector machines are used for research and development of small-geometry devices, and the variable-shaped or raster spot machines for production. About 450 machines have been installed and used in the world. The last entry of Table 4 is included because it functions much like an electron-beam exposure machine but uses an optical system.

6.3.1 Electron Optics

Figure 19 shows the basic electron optical system. Magnetic lenses form a demagnified image of the source d_o on the image plane d_i. The position of an electron beam, irradiated on the substrate, is controlled by deflectors. The on-off control of an electron beam is carried out by a blanking plate with an aperture underneath.

The characteristics of some cathode materials are shown in Table 5. The one used most commonly is the LaB_6 single crystal, which has a long life, high stability, and high current capability compared to the conventional tungsten hairpin cathode. The thermal field emission (TFE) source consists of a tungsten tip with a radius of 0.5 to 1 μm heated sufficiently (about 1900 K) to emit high current with high stability.

Emission current density from the cathode J_c is given by

$$J_c = AT^2 \exp(-E_w/kT) \tag{6.15}$$

where A is the Richardson constant, k the Boltzmann constant, and E_w the work function of the cathode material. The electrons are accelerated by the voltage V (10 to 50 kV) and focused by the gun at a point near the anode called the *crossover,* which has a diameter d_o of 10 to 100 μm. The configuration of the electron gun is illustrated schematically in Fig. 20*a*. The maximum value of the brightness is

$$\beta = J_c eV/\pi kT \tag{6.16}$$

The lenses in the electron optical column (Fig. 19) are magnetic, and their structure is illustrated in Fig. 20*b*. If a parallel beam of radius r_o enters the field **B** of the lens, the electrons experience a force that causes those not on the axis to revolve

TABLE 4

Commercially available electron-beam exposure systems and a laser-beam pattern generator

Model	Maker	Type	Optics	Acceleration voltage (kV)	Beam size (μm)	Resolution (μm)	Overlay accuracy (μm)
MEBES4	ETEC	Raster/continuous	TFE, Gaussian	10	0.05	0.2	0.4
AEBLE150	”	Vector/continuous	LaB_6, variable shape	20	~2	0.5	0.15
EBPG4	Leica	Vector/step-repeat	LaB_6, Gaussian	50	0.01~	0.03	0.08
EBML300	”	”	”	50	0.025	0.1	”
JBX7000MV	JEOL	”	LaB_6, variable shape	20	~4	0.5	0.05
JBX8000DX	”	”	”	20	”	0.2	
JBX-5DII	”	”	LaB_6, Gaussian	25/50	0.008~	<0.1	0.06
HL700M/D	Hitachi	”	LaB_6, variable shape	30	~4	0.2	0.06
HL700F	”	”	TFE, Gaussian	30	0.03	0.03	0.1
CORE2564	ETEC	Raster/continuous	364 nm laser	Optical system		0.6	0.05

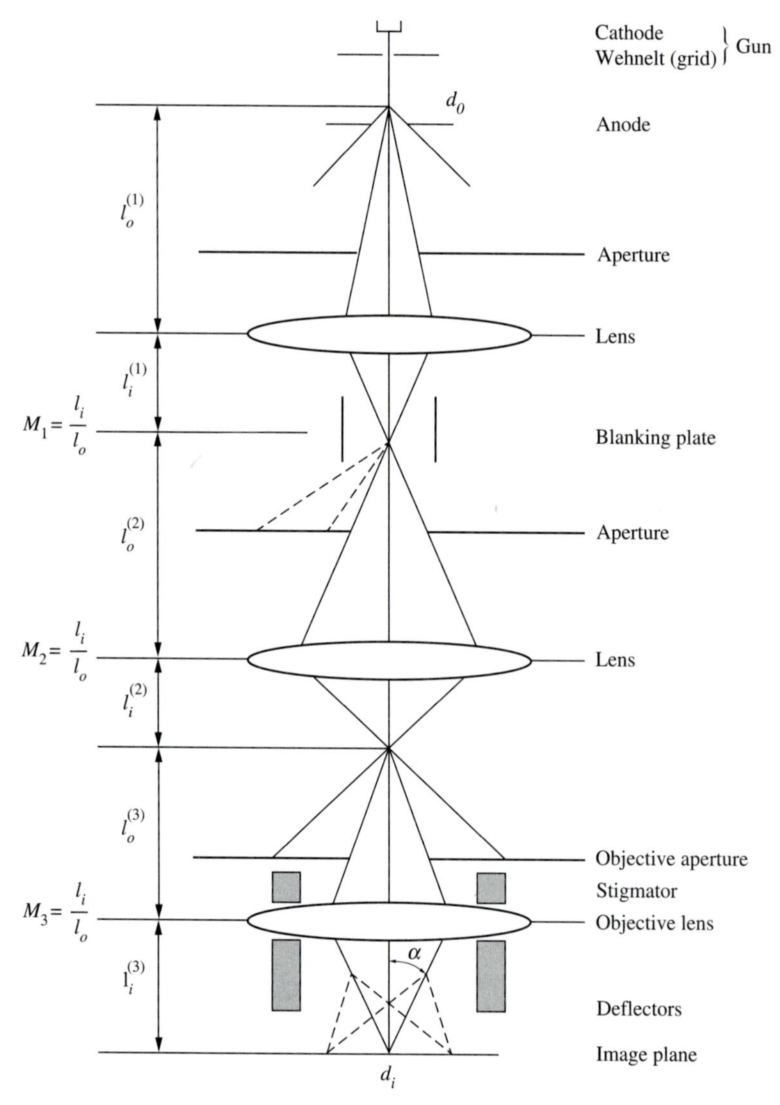

FIGURE 19
Basic electron optical column.

about the axis and turn toward it. For a thin lens, the electron path beyond the lens is given by

$$\frac{d_r}{d_z} \approx -\frac{r_o e}{8mV}\int_{-\infty}^{+\infty} B_z^2 dz \qquad \frac{d_r}{d_z} = \frac{-r_o}{f} \tag{6.17}$$

TABLE 5
Characteristics of cathodes used in electron scanning systems

Materials	Operating temperature (K)	Energy dispersion (eV)	Current density (μA/sr)	Source size (Å)	Brightness A/cm²sr	Drift (%/hr)	Life (hr)
W hairpin	2700	4.5	10^4	50×10^4	10^5	< 1	100
LaB_6	1800	2.7	10^5	10×10^4	10^6	< 1	~5000
W(310)-FE	Room temp.	0.3	50	30	10^8	30	~5000
ZrW-TFE	1900	3.1	10^3	100	10^9	< 1	~5000

where r_o = beam radius
e = electronic charge
m = mass of electron
B_z = axial component of **B**
f = focal length
V = acceleration voltage

Many rules of light optics apply here also, such as

$$1/l_o + 1/l_i = 1/f \qquad M = l_i/l_o \tag{6.18}$$

where l_o = object distance
l_i = image distance
M = magnification

The current density J and current I on the image plane are

$$J \leq \pi\beta\alpha^2 \qquad \text{and} \qquad I = J(\pi d_i^2/4) \tag{6.19}$$

where β = gun brightness
d_i = diameter of the electron beam on the image plane

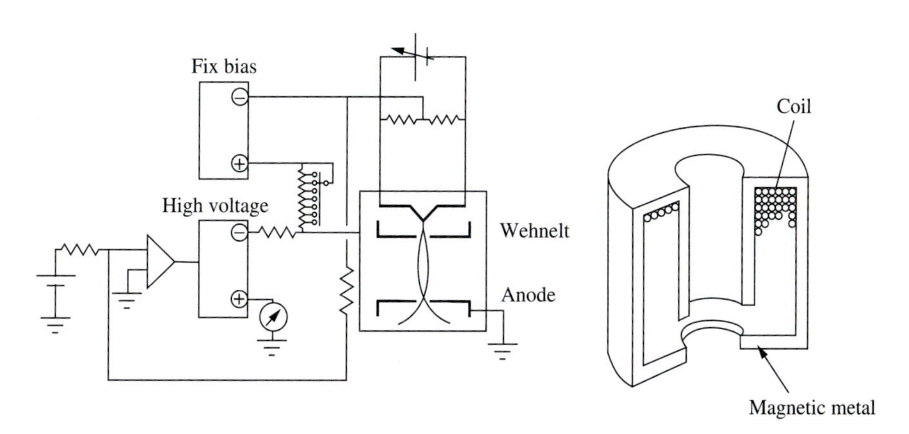

FIGURE 20
Configuration of the electron gun and magnetic lens. (*a*) Configuration of electron gun and its power supply. (*b*) Magnetic lens structure.

Spot sizes of d_i that are of interest range from 0.01 to 0.1 μm. From Section 6.2.2 the diameter d of the central spot of the Airy pattern is $1.2\lambda/\alpha$. NA($\approx \sin\alpha$, where 2α is the solid angle of the ray reaching the imaging point from the objective lens) is the numerical aperture. For 15-keV electrons, the wavelength $\lambda = 0.1$Å. Taking $\alpha = 10^{-2}$ radians, we have a diffraction spot width

$$d_{\text{diff}} = 1.2\lambda/\alpha = 10^{-3}\mu\text{m} \ll d \tag{6.20}$$

Thus, diffraction can be ignored. However, aberrations in the final lens and in the deflection system will increase the spot size and can change its shape as well. The actual spot size in the presence of aberrations is the square root of the sum of the squares of the contributions of the independent aberrations. The aberrations are of two types: undeflected beam aberrations and aberrations that are functions of the distance r, in the image plane from the axis, to the position of the deflected beam. The aberrations of the first type are

$$\begin{aligned} \text{Spherical aberration: } d_s &= \tfrac{1}{2}C_s\alpha^3 \\ \text{Chromatic aberration: } d_c &= C_c(\Delta V/V)\alpha \end{aligned} \tag{6.21}$$

Spherical aberration is a result of focusing rays passing through the lens at different distances from the axis. The spread of electron energies, $e\Delta V$, is usually a few eV. C_s and C_c are constants determined by the lens parameters. Astigmatism, a third type of aberration, results from the breaking of cylindrical symmetry in the column and can be removed by introducing compensating fields with *stigmator* coils as shown in Fig. 19. Deflection aberrations, such as coma and field curvature, are proportional to $r^2\alpha$ or $r\alpha^2$. Another source of spot broadening is the mutual Coulomb repulsion of the electrons as they traverse the column. For total column length L (cm) and beam current I (A) this contribution is given approximately by

$$d_{\text{ee}} \approx [LI/(\alpha V^{3/2})] \times 10^8 \mu\text{m} \tag{6.22}$$

This spreading becomes important only at large currents when small spot sizes or sharp edges of large spots are required. For example, if $I = 500$ nA, $L = 70$ cm, $V = 20$ kV, and $\alpha = 10^{-2}$ rad, then Eq. (6.22) gives $d_{\text{ee}} = 0.12$ μm.

Note that the distortion, which causes the positioning error of the electron beam landing on the substrate, depends on the deflection distance from the center. The actual amount of error depends on the condition of the electron optics, such as the manufacturing accuracy of the electron optical parts or deflection range, but an approximate value is on the order of 10^{-3} on the deflection edge. An electron lithography machine has many correction functions that differ from those in optical systems because of its sequential exposure method and the real-time response characteristic of the electron beam. These functions include corrections for distortion, beam drift, stage position, beam size related to deflection distance or substrate height, and beam unblanking time due to beam current drift.

6.3.2 Raster Scan and Vector Scan

The EBES® machine, developed by AT&T Bell Laboratories, and the commercially available MEBES® system from ETEC (Hayward, CA) use beam deflection in one

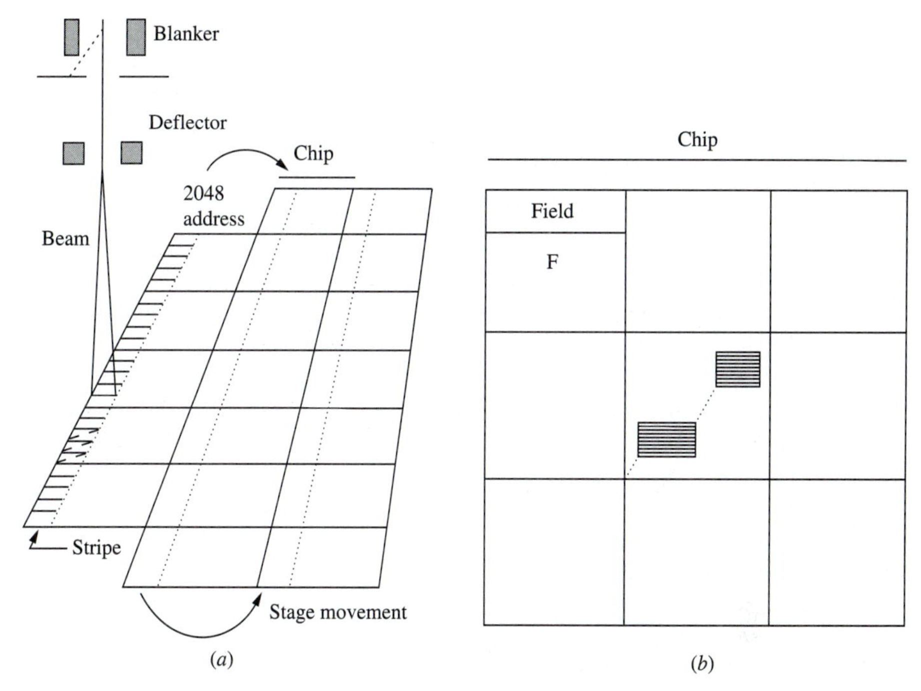

FIGURE 21
The principles of raster and vector scan. (*a*) Raster scan. (*b*) Vector scan.

dimension (mainly).[31] The writing scheme is shown in Fig. 21*a*. The stage moves continuously in a direction perpendicular to the beam scanning direction. The electron beam is moved repetitively in one direction while it is slowly deflected in the perpendicular direction, corresponding to the stage movement. The pattern data are decomposed into a number of stripes parallel to the stage movement, and one stripe is written on all chips of the same type before the next stripe is begun. The stripe is decomposed into many scan lines whose widths are basically similar to the feature addressing unit. The electron beam is irradiated on the substrate by controlling the blanker that turns the beam on at the exposure starting points and turns it off at the end point during rastering. The performance of MEBES4 is given in Table 4.

In vector scan the beam is directed sequentially to the parts of the chip pattern to be exposed.[32] The pattern is decomposed into a number of elements (rectangles, triangles, etc.) and each is exposed by the writing beam. Many vector scan machines expose in a step-and-repeat fashion. Figure 21*b* shows an exposure field of dimensions $F \times F$, which is the electron beam deflection area. The dotted lines indicate where the beam is turned off (deflected by the blanking plates), similar to blanking in a raster scan. After all elements in one field are exposed, the stage is stepped to the next field and the exposure process is repeated. The stage need not provide highly precise positioning. Stage position is monitored by a laser interferometer, and small differences from the desired stage location are compensated by small offsets of the beam. If a chip is larger than an exposure field, as in Fig. 21*b*, several fields can be used to expose the chip. The scan field must be as large as allowed by deflection

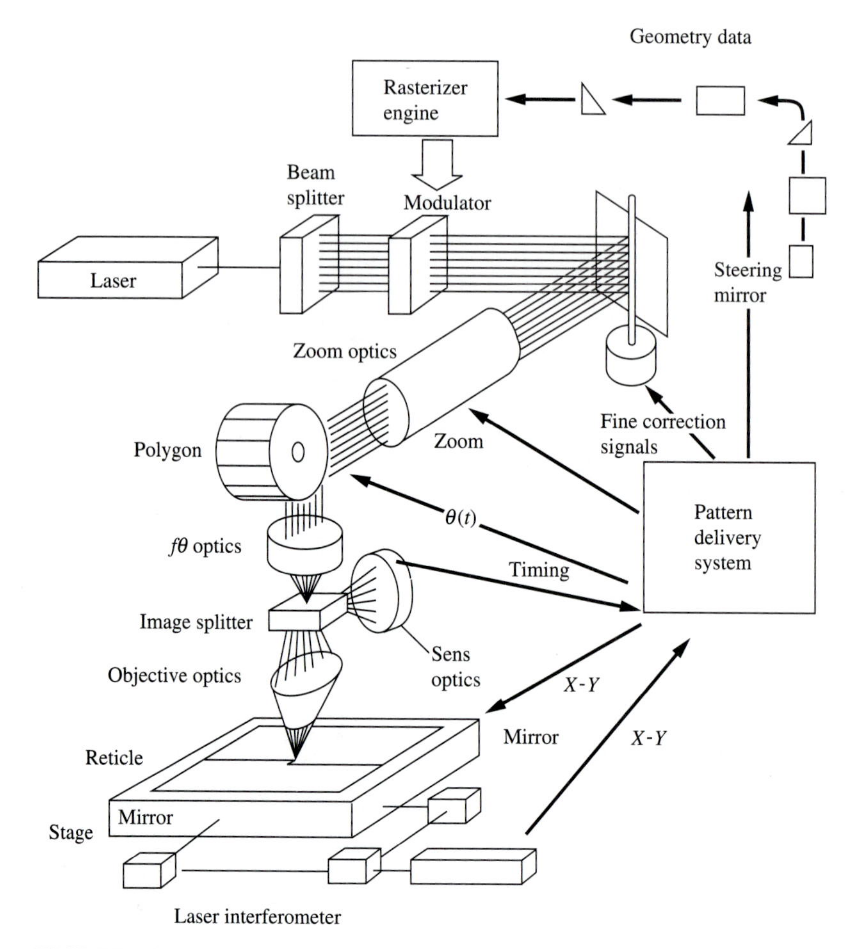

FIGURE 22
System configuration of a laser reticle generator. (*After Burns and Schoeffel, Ref. 34.*)

positioning accuracy because stage-stepping time affects throughput. There are many vector-type machines commercially available, as shown in Table 4.

The EBES4 machine developed by AT&T Bell Laboratories has a continuously moving stage, three hierarchical deflection systems, and a high-brightness gun that uses a thermal field emission cathode in order to manufacture deep-submicron-geometry (less than about 0.5 μm) masks or wafers within an acceptable exposure time.[33]

The CORE® laser reticle writer, the last entry in Table 4, is an optical pattern generator.[34] The writing concept is very similar to the MEBES® electron beam machine. A narrow laser beam is scanned very accurately by the combination of a polygon mirror, a piezo-controlled steering mirror, and a high-frequency beam modulator (on-off controller) on the continuous-movement mask stage. The system configuration is illustrated in Fig. 22. This machine can use the same process as with wafer fabrication for mask making because it is an optical method. It also has the capability of direct writing and phase-shifting mask making.

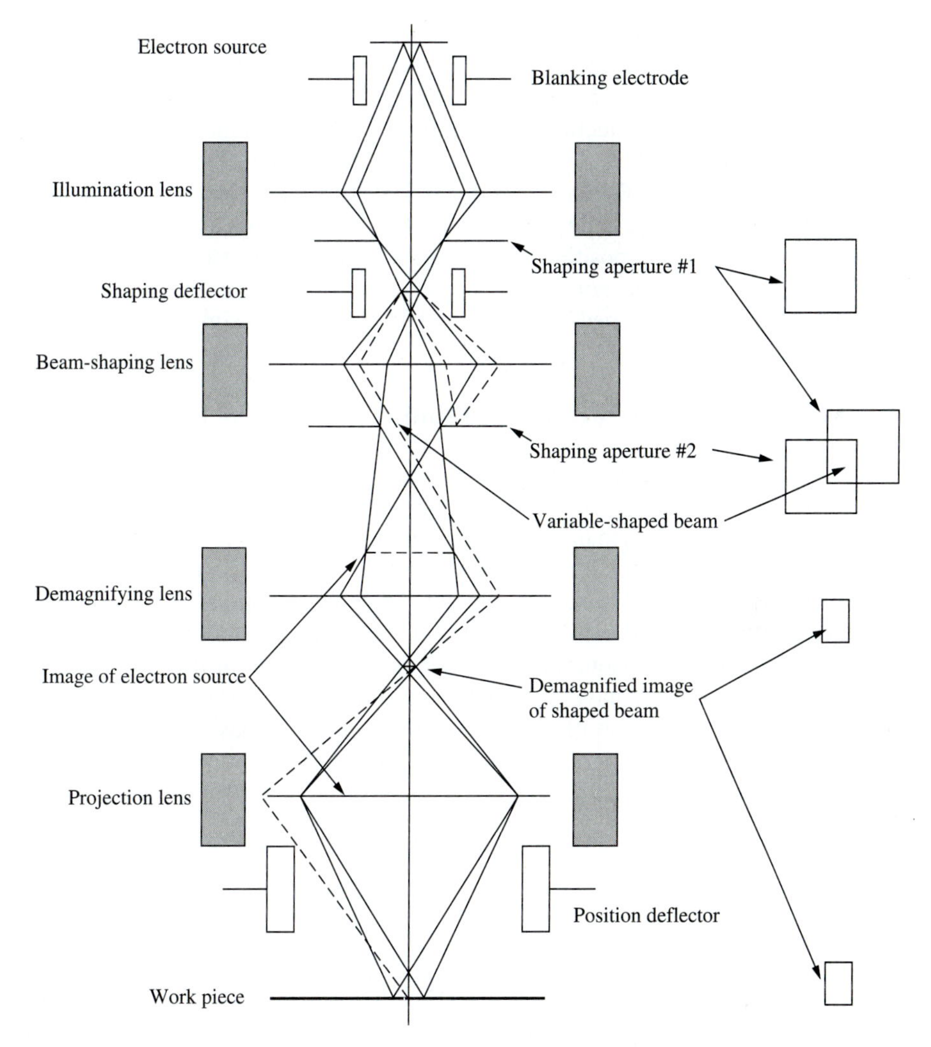

FIGURE 23
Variable-shaped-beam column of JEOL model JBX-6A3.

6.3.3 Variable Shaping

For vector scan machines the exposure time of a specified area is proportional to the number of electron beam irradiations if the current density and the resist sensitivity are fixed. The *shaped-beam* method offers a way to project many image points in parallel and to achieve faster exposure by reducing the irradiation numbers.[35,36] In the JEOL JBX-6A3 machine shown schematically in Fig. 23, the shape of the electron beam is rectangular, and the size is variable. The image of the first square aperture is shifted in two dimensions to cover various portions of the second aperture, which is the actual object of the image projected on the mask or wafer. The minimum and maximum rectangle widths are 0.2 μm and 6 μm, respectively. Maximum current density is 2A/cm^2 with the LaB_6 cathode. Electrostatic beam deflection is employed.

Another interesting approach is the combination of a variable shaped beam with continuous stage movement technology to minimize stage movement overhead. The first commercially available machine is AEBLE-150 from ETEC.[37] It has static and magnetic two-stage deflectors. The magnetic deflector decides the static deflection positions, and the static deflections control the electron-beam exposure positions corresponding to the stage positions. The stage is continuously moved like a raster machine, and the position, monitored by laser interferometer, is fed back to both deflectors. Machines developed by NTT[38] and Toshiba[39] use a similar concept.

6.3.4 Electron Proximity/Projection Printing

An electron proximity printing system[40] is illustrated in Fig. 24*a*. It needs step-and-repeat stage movement because of its small mask size and thus differs from an optical proximity printer. The distance between the mask and the wafer is 0.6 mm. The beam, with a diameter of approximately 1 mm, is scanned over the full chip-size mask, which is made on a silicon membrane. To permit electrons to pass through the transparent parts of the mask physical holes must be present, unlike an x-ray mask. Creating isolated opaque patterns such as the interiors of doughnut shapes is a problem because they have no attachment to the rest of the mask. A solution to this problem is to use a set of two complementary (half) masks. Registration is by reference to alignment marks on each chip. Resolution is reported as 0.3 μm.

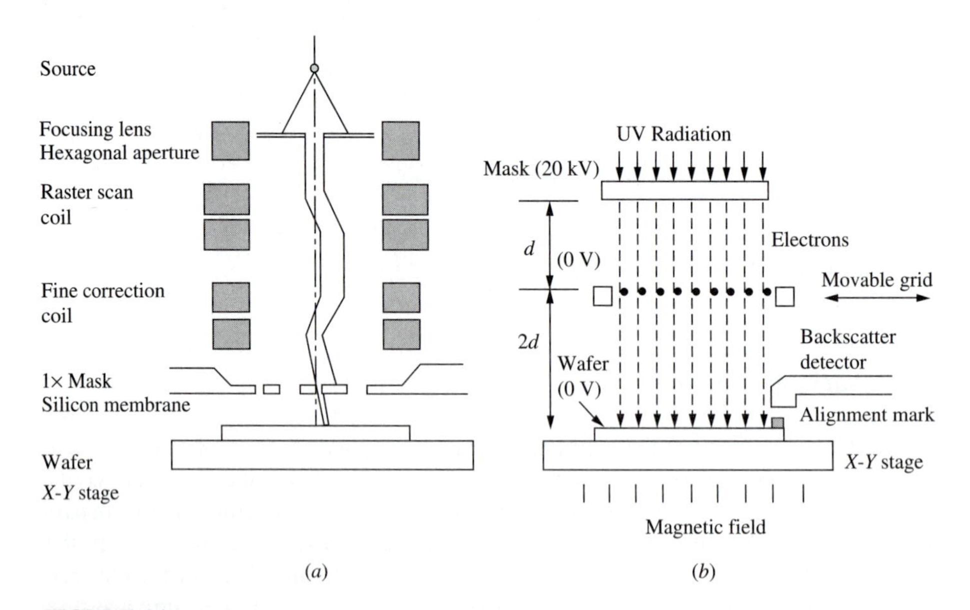

FIGURE 24
The principles of electron proximity and 1× electron projection. (*a*) Electron proximity. (*After Meissner et al., Ref. 40.*) (*b*) The 1× electron projection. (*After Ward et al., Ref. 41.*)

Overlay error is less than 0.05 μm (1 σ). Estimated exposure time is around 30 sec per 5-inch wafer with 10 $\mu C/cm^2$ sensitivity. The major disadvantages of this system are the need for two masks for each pattern and the heating problem associated with the stencil mask.

Electron projection systems[41] are another method of achieving high resolution over a large field with high throughput as shown in Fig. 24*b*. In a 1 : 1 projection system, parallel electric and magnetic fields image electrons onto the wafer. The "mask" is quartz, patterned with chrome and covered with CsI on the side facing the wafer. Photoelectrons are generated on the mask/cathode by backside UV illumination. The advantages of the system include a stable mask, good resolution, fast-repeat exposure with low-sensitivity resists, large field, and fast alignment. Proximity effects can be compensated for by undersizing or oversizing the features on the mask or by increasing the acceleration energy to 50 keV or more. Apparently neither method is entirely satisfactory for production use. Another problem is that the cathode has an unacceptably short life, only 50 exposures, before the CsI must be replaced by fresh material.

Recently, another projection method, a modification of the variable-shaping method, has been proposed by several organizations.[42] Several different unit patterns, such as the cell patterns of a memory chip, are located in the second shaping aperture of the variable-shaping column as shown in Fig. 25. The deflection range of the electron-beam shaping is larger than with the variable-shaping method in order to select the different types of cells. Hence, redeflection is necessary to eliminate the beam shape deformation passing through the field far from the center of the lenses. The reduction ratio of the cell patterns is the same as that of the variable rectangle. Repetitive patterns are exposed by using the cell aperture, which is designed and manufactured in advance, that is specific to the device layout. Random patterns or areas that connect the repeated patterns are exposed by a conventional variable rectangular beam. Cell apertures are manufactured by a method similar to the one used to make the electron proximity mask, but the pattern size of the cell aperture is some ten times as large as the actual wafer geometry. This technique is very flexible because it can use either a conventional variable-shaping function or preprogrammed shaping. The problems of this method are the formulation of the flexible algorithms that combine the data preparation of the ULSI patterns with exposure control, the adjustment of electron-beam optics, and the life of the cell aperture.

6.3.5 Electron Resists

Electron exposure of resists occurs through bond breaking (positive resist) or the formation of bonds between polymer chains (negative resist). The incident electrons have energies far greater than the bond energies in the resist molecules, so the exposure energies can cause direct molecular reactions. Both bond scission and bond formation occur simultaneously. Which one predominates determines whether the resist is positive or negative.

In a negative resist, electron beam–induced crosslinks between molecules make the polymer less soluble in the developer solution. One crosslink per molecule is sufficient to make the polymer insoluble. Resist sensitivity increases with increasing

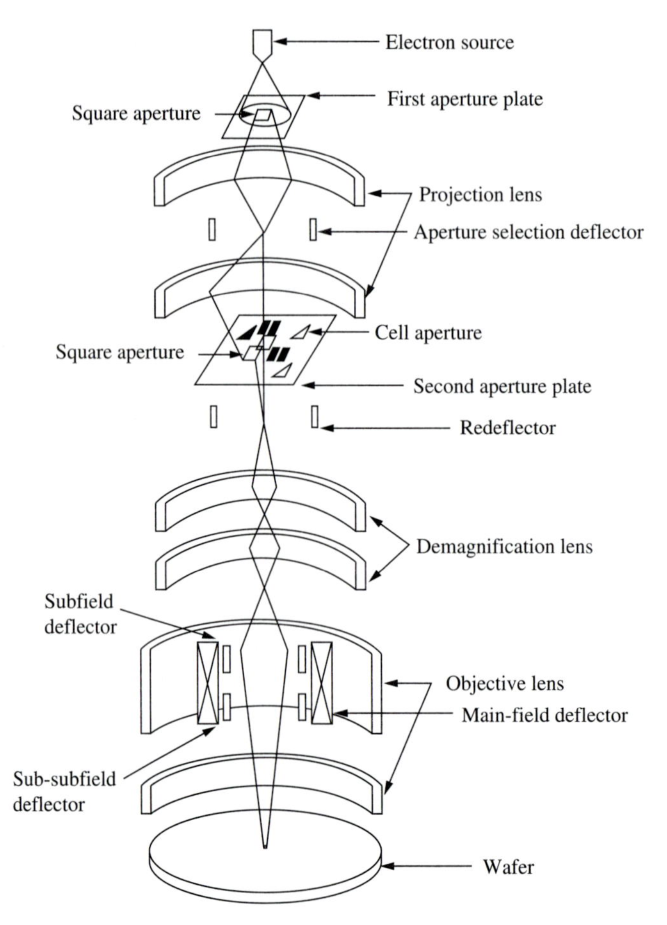

FIGURE 25
The electron optics of the cell projection method. (*After Sakitani et al., Ref. 42.*)

molecular weight. If the molecules are larger, then fewer crosslinks are required per unit volume for insolubility. The polymer molecules in the unexposed resist will have a distribution of lengths or molecular weights, and thus a distribution of sensitivities to radiation. The narrower the distribution, the higher the contrast γ—similar to an optical resist, as shown in Fig. 15. The exposure dose has units of charge deposited by the beam per unit area, or C/cm^2.

In a positive resist the scission process predominates; the exposure leads to lower molecular weights and greater solubility. Again, high molecular weight and narrow distribution are advantageous.

Two major factors limit resist resolution: swelling of the resist in the developer (more severe for negative resists) and electron scattering. Swelling of a negative resist, whether optical, electron, or x-ray, has two deleterious effects. First, two adjacent lines of resist may swell enough that they touch. Second, this expansion and contraction weakens the adhesion of very small resist features to the substrate and can cause small undulations in narrow lines. Both problems become less severe as resist thickness is reduced.

Recently developed chemically amplified resists have good resolution, sensitivity, and dry-etching resistance capabilities without swelling.[43] The basic process is explained in Section 6.2.5. It is reported to have a 0.2-μm resolution with a 4.0-μC/cm^2 sensitivity at 20 kV. Dry-etching performance is as good as that of conventional diazoquinone-based positive resists.

A few of the readily available electron resists are listed in Table 6. As the wavelength of optical lithography becomes shorter, several optical resists also have sensitivity to an electron beam. The resist PMMA has the highest resolution known. Note that the values for sensitivity and resolution are approximate. Because faster electrons penetrate more deeply, more current is required at higher voltages. A resist is about one-half as sensitive for 20-keV electrons as it is for 10-keV electrons.

When electrons are incident on a resist or other material, they enter the material and lose energy by a collision process known as scattering, producing secondary electrons, x-rays, and, finally, heat. This fundamental process limits the resolution of electron resists to an extent that depends on resist thickness, beam energy, and substrate composition. The envelope of the electron cloud in the material is shaped like an onion bulb; the cloud is pulled closer to the surface as beam voltage decreases. At higher beam voltage the electrons penetrate farther before being scattered over larger lateral distances. These phenomena are shown in Fig. 26, from the results of Monte Carlo simulations of the penetrating paths of the individual electrons in the target materials.[44] The scattering range of electrons in the Si substrate is about 1 μm for both distance and depth at 10 keV, and 2 μm for distance and 3.5 μm for depth at 20 keV.

For an electron beam of zero width incident at position r on the resist-covered substrate, the distribution of energy deposited in the resist at depth z is closely approximated by

$$f(r, z) = a_1 \exp[-r^2/\beta_f^2(z)] + a_2 \exp[-r^2/\beta_b^2(z)] \qquad (6.23)$$

where a_1, a_2 = ratios of forward- and backscattering of the total electrons, respectively

$\beta_f(z)$ = width for the forward-scattered beam in the resist

$\beta_b(z)$ = backscattering from the substrate

and $\beta_b(z) \gg \beta_f(z)$, $f(r, z)$ is the point spread function. Generally, the value of z of interest is that corresponding to the resist/substrate interface z_i. For a 25-keV electron beam penetrating a 0.5-μm thick resist substrate, for example, $\beta_f(z_i) = 0.06$ μm, $\beta_b(z_i) = 2.6$ μm, and $a_2/a_1 = 2.7 \times 10^{-4}$. As separation between lines decreases, the backscattered electrons contribute a greater dose between the lines, where the dose should be zero. This feature is somewhat similar to the reduced modulation in an optical image at higher spatial frequencies.

An exposed pattern element adjacent to another element receives exposure not only from the incident electron beam but also from scattered electrons from the adjacent element. This is called the *proximity effect* and is, of course, more pronounced the smaller the space between pattern elements. For example, an isolated 0.5-μm line requires 20–30% more exposure than 0.5-μm lines separated from each other

TABLE 6
Electron-beam resists

Name	Maker	Type	Polymer	Sensitivity ($\mu C/cm^2$) at 20 keV	Contrast	Resolution (μm)
CMS-EXS	Toyo Soda	Negative	Chloromethyl polystyrene	2	1.5	0.75
CMS-EXR	”	”	”	20	3	0.3
SEL-N	Somar	”	PGMA[†]–maleic methylate	0.4	0.9	1
RE4000N	Hitachi Chemical	”	Iodinated polystyrene	4	1.3	1
GMCIA	ATT/Mead	”	GMA[‡]–3-chlorostyrene	7	1.7	0.5
SAL 601	Shipley	”	Novolac	5	5	0.2
PBS	Mead Tech	Positive	Polybutene SO_2	2	1.7	0.5
PMMA	KTI Chemical	”	PMMA	100	2	<0.1
EBR-9	Toray	”	Poly-2,2,2-ethyl a-acrylate	1.5	3	0.5
FBM-120	Daikin Ind.	”	Fluoroallyl methacrylates	2	5	1.5
AZ 2400	Shipley	”	Novolac	200	2	0.5
RE 5000P	Hitachi Chemical	”	PMPS[§]-Novolac	4	1	0.8
ZEP 520	NihonZeon	”	a-Chloromethacrylate, a-Methylstyrene	5–20	2.5	<0.1

[†] Polyglycyl methacrylate
[‡] Glycyl methacrylate
[§] Polymethylpentane sulfone

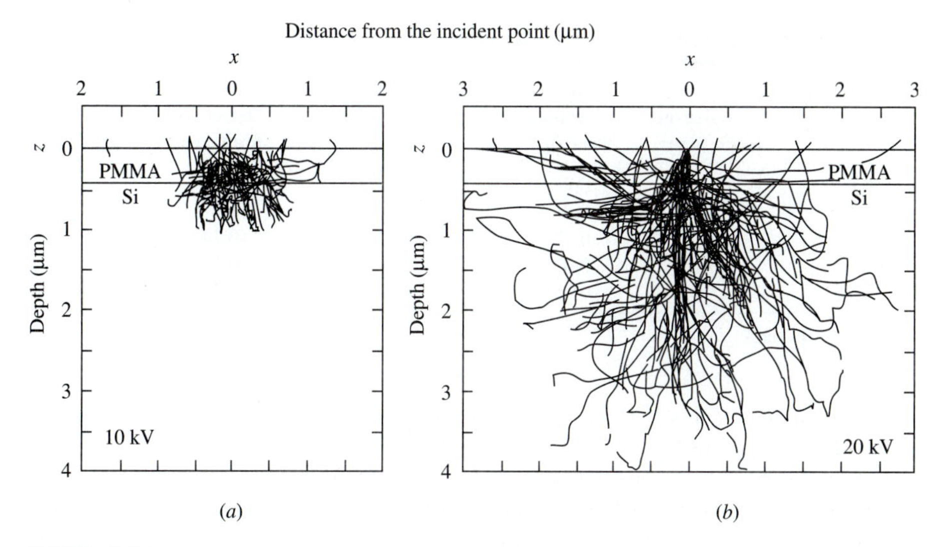

FIGURE 26
Simulated electron trajectories for a beam incident at the origin. (*After Kyser and Viswanathan, Ref. 44.*) The trajectories of 100 primary electrons are shown projected onto the *x*-*z* plane. (*a*) A 10-kV point source. (*b*) A 20-kV point source.

by 0.5 μm. Thus, as pattern density increases, it becomes necessary to adjust the exposure for various classes of elements or, in the extreme case, for different parts of an element.[45] This process is carried out through data preparation from the CAD output into the machine-oriented format in Fig. 18. In general, it takes a very long time, even for advanced computers, to apply perfect correction in ULSI patterns.

6.3.6 Electron-Beam Applications

Mask fabrication

Electron beam exposure machines have a wide variety of applications because of their pattern-generating function. Applications include direct writing and mask writing for the other lithographies such as the optical stepper, x-ray, ion, or electron projection. By the beginning of the 1970s the 10× reticle pattern was exposed by optical pattern generators. The rectangles are generated by a pair of L-shaped mechanical masks, and then the reduced images of these rectangles are projected on the blanks. A large, dense chip requires 20 hours or more of optical pattern generator time, but only two hours or less of electron beam pattern generator time. The first widespread use of electron beam pattern generations was in photomask making.

Table 7 lists specifications for the advanced masks required for optical steppers. In this table pattern placement errors are displacements of patterns from the desired locations on the mask. One of the types of error components that degrade these mask specifications is the relative-position error between the stage and the exposure position, which occurs because of mechanical slippage of the mask due to stage movement or thermal expansion. Fused-silica mask substrates, with their low

TABLE 7
Specifications of masks for optical steppers

Minimum feature size (μm)	Line size on mask (μm)	Pattern placement error (μm)	Critical dimension uniformity (μm)	Defect size limit (μm)	Applied DRAMs
0.8	4	0.2	0.15	1.5	1 M
0.65	3.25	0.2	0.15	1	4 M
0.5	2.5	0.12	0.1	0.74	16 M
0.35	1.75	0.09	0.07	0.5	64 M
0.25	1.0*	0.05*	0.04*	0.25*	256 M

*Values on 4× mask.

thermal expansion (coefficient of linear expansion is 0.4×10^{-6}/°C), can reduce thermal contributions that result in relative-placement errors among several masks. Another type of placement error is caused by the bending of the mask. To minimize this error, the use of 0.25-inch thick blanks is beneficial compared to the conventional 0.09-inch ones. Commercially available machines have metrology capabilities to measure exposure performance of the masks.

Several factors also contribute to decrease the critical dimension (CD) uniformity, that is, the exposed pattern width uniformity. If the chemical process for the resist development is perfect, the factors that affect the CD are the dose and beam profile. Usually dosage is controlled by beam irradiation time (beam-on time with unblanking) and depends on the beam current. Beam blanking and unblanking control has a very quick response—within several nanoseconds. The problem is the heating effect of electron bombardment, which is more important in direct writing because of rather thick resists.[46,47]

Another important parameter of mask performance is the defect density. Usually, the etched pattern is inspected by an optical method and compared with the original data base used for the electron exposure. Mask defects are either opaque spots in areas that should be transparent or pinholes where the chromium layer should be continuous. The allowable defect size, which depends on the layout rule of the devices, is shown in Table 7. The mask must be perfect within that size. Consequently, inspection and repair technology is very important not only to ensure the device performance but also to increase the mask production yield. The minimum thickness of the resist is set by the need to avoid pinholes and by its resistance to etching. The resist thickness of the photomask is approximately 0.2 to 0.4 μm. In the case of thin resists, it is easy to control the CD by using uniform process conditions, such as minimizing swelling, proximity effect, heating effect, and charging of electrons.

Mask making can be more difficult than direct writing in spite of its 5× larger geometry. Absolute position accuracy and pattern-integrity perfection on the completely clean blanks, without any references such as alignment marks on wafers, are required during a rather long exposure time. The reduction ratio of the optical stepper will become 4× in the case of 256-Mbit DRAM production instead of the conventional 5× because of the large chip size and the limit on field size of the stepper as shown in Table 1. The required specifications of masks for 256-Mbit DRAM become more severe than the conventional ones because of changing the reduction ratio.

Direct writing

Direct writing means making patterns directly on the resist-coated substrate. The advantages of direct writing are that no mask is needed and that it produces a high-resolution pattern. Because it is maskless, direct writing can be used in many applications. For example, quick design modification; shortened manufacturing time from design to device test, usually called QTAT for quick turnaround time; and the capacity to put different types of pattern layouts on one wafer (one wafer module) are advantages of maskless applications. In ASIC manufacturing, the basic design and process are common from chip to chip; only the wiring and through-hole layers are different based on the logical functions of the devices.[48] Hence, the common layers can be manufactured using a stepper, and the personalization layers can be exposed by electron beam. Direct writing is very efficient not only because it is economical but also because of its higher production speed. If the mask-and-stepper process is used, it needs almost one full day from receiving the exposure data to making a mask without defect. In the direct writing process, the exposed pattern data are transferred directly to an electron beam machine, without the use of external components such as magnetic tape, through a network from the CAD system connected to the electron exposure machines. The trilayer resist process has been applied to actual production to get higher throughput, to achieve better resolution, and to minimize radiation damage.[48]

The first 64-Mbit DRAM was made by a mix-and-match method using the electron beam and the stepper.[49] Depending on pattern density, several newly developed technologies were used, such as the positive resist OEBR200 (Tokyo-oka) and the negative resist SAL601 (Shipley). To control CD uniformity in 0.3-μm geometry, a two-step development process (contrast enhancement), a conductive material coating to eliminate charge accumulation, and a new proximity-correction algorithm (to classify the pattern density area instead of only the pattern width and gap) are employed.

Another direct writing application, with long experience, is GaAs field-effect transistor (FET) production. A short gate length and low resistance are very important for this device to have high performance. A T-shaped gate is one way to accomplish this. A double exposure with different exposure conditions is another.[50] Two types of PMMA, with different sensitivities (molecular weights), are coated on the wafer. The top resist has a higher sensitivity and lower molecular weight compared to the bottom one. After the top layer has been opened, the second layer is exposed by referring to the alignment mark as shown in Fig. 27. It is reported[50] that the gate contact width is 0.15 μm.

As the pattern size becomes smaller and smaller, it becomes more difficult to make the pattern on the resist and transfer it to the wafers. There have been many approaches to making patterns directly on the wafer without a resist, especially in applications that need a geometry of less than 0.1 μm.[51,52] For direct electron beam deposition, an electron beam reacts with the source gas directly. The gas is introduced to the specimen chamber, and a Au pattern is deposited. Fifty-nm lines and spaces are reported.[51]

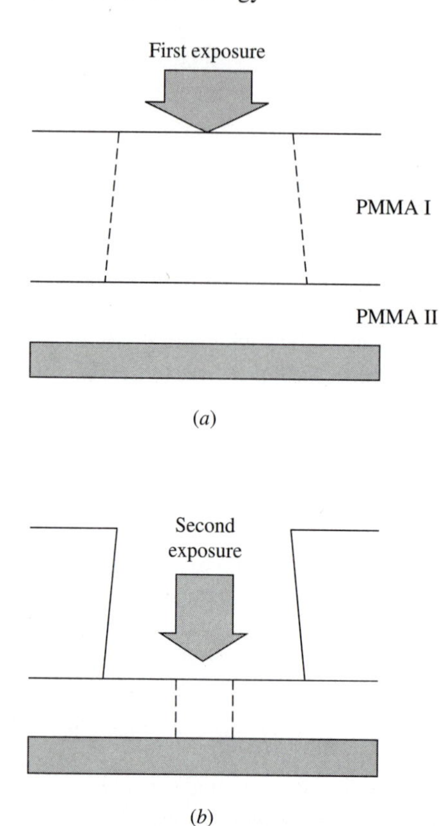

FIGURE 27
Two-step exposure principle. (*After Samoto et al., Ref. 50.*) (*a*) First exposure on high-sensitivity resist. (*b*) Second exposure on low-sensitivity resist.

6.4 X-RAY LITHOGRAPHY

Since x-ray lithography was proposed in 1972, development of the technology has been pursued in many laboratories.[53,54] In the previous section we noted that diffraction effects and resolution are improved by reducing the wavelength. If the wavelength is reduced further than deep UV, all optical materials become opaque because of fundamental absorption, but transmission increases again in the x-ray region. In x-ray lithography, an x-ray source illuminates a mask that casts shadows on a resist-covered wafer. There are several advantages in x-ray lithography in addition to short wavelength. Some contaminants, such as light organic materials, do not print as a defect; and the depth of focus is larger than that of optical printers. The essential technology components of this process are (1) a mask consisting of a device pattern made of x-ray-absorbing materials on transmitting material, (2) an x-ray source, and (3) an x-ray resist.

The x-ray absorption of several materials is shown in Fig. 28. The absorption coefficient of an elemental material of density ρ and atomic number Z is proportional to $\rho Z^4 \lambda^3$ over a wide range of wavelengths. The proportionality constant decreases in a step function fashion at the *absorption edge,* a wavelength that corresponds to

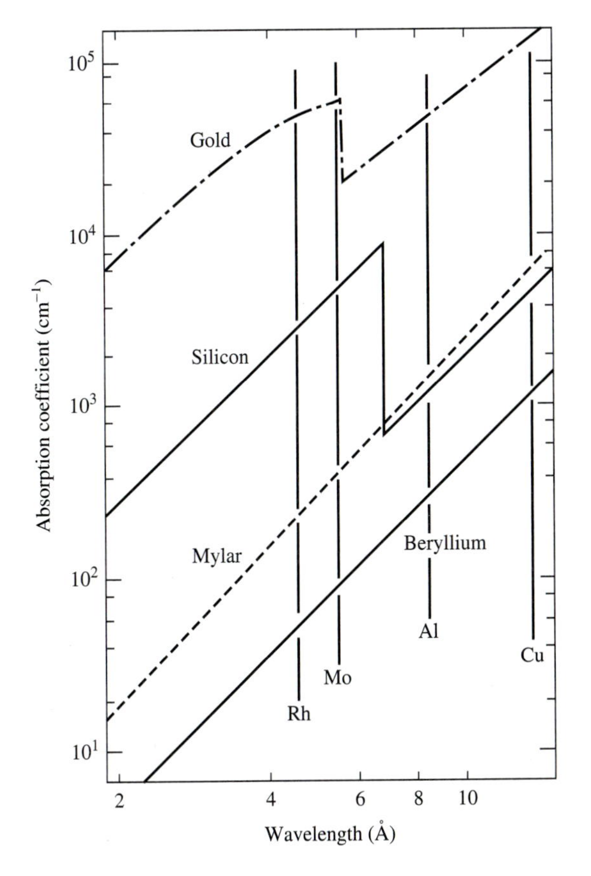

FIGURE 28
Typical x-ray absorption coefficients with some lines characteristic of commonly used materials.

the ionization energies of inner electrons of the K, L, and other shells. Note in Fig. 28 that the big differences in absorption coefficients for different materials are observed at the same wavelengths that are utilized for absorbing materials to make patterns and the transmitting substrate of the x-ray mask.

The performance required of x-ray lithography, as the alternative to optical lithography, has become more exacting as optical technology has continued to improve. The resolution and placement accuracy should soon surpass 0.2 μm and 0.03 μm, respectively.

6.4.1 Proximity Printing

Because the x-ray wavelength is short enough, simple geometrical considerations can be used to relate the image formation and the wafer without having to consider diffraction, as shown in Fig. 29*a*. The opaque parts of the mask cast shadows onto the wafer below. The edge of the shadow is not absolutely sharp because of the finite extent of the x-ray source at distance D from the mask. If the gap between mask and wafer is g, this blur δ is given by

$$\delta = Sg/D \tag{6.24}$$

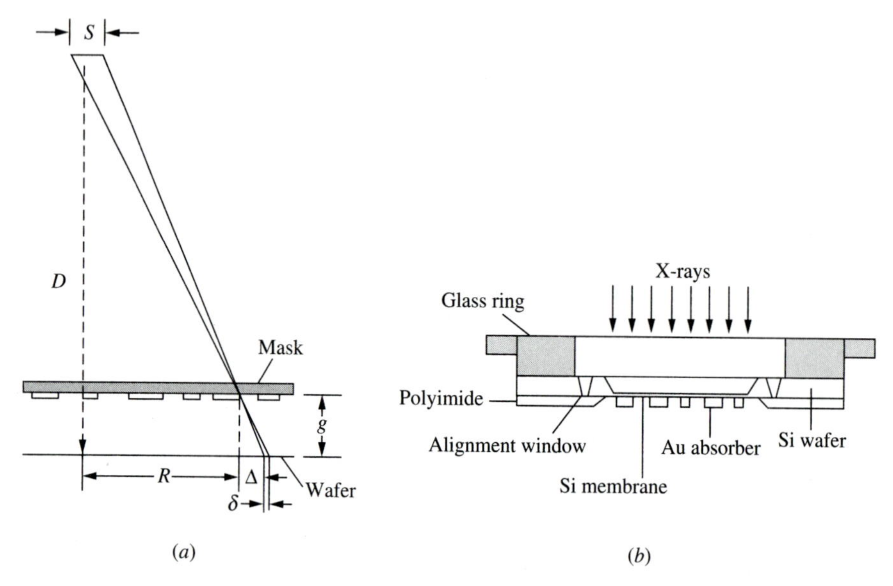

FIGURE 29
X-ray shadowing errors and x-ray mask structure. (*a*) X-ray proximity printing consideration. (*b*) X-ray mask structure example. (*After Fleming et al., Ref. 54.*)

The angle of incidence of the x-ray on the wafer varies from 90° at the center of the wafer to $\tan^{-1}(D/R)$ at the edge of the exposure field of radius R. The shadows are slightly longer at the edge by the amount

$$\Delta = g(R/D) \tag{6.25}$$

This small magnification is generally of no concern. In the special case where it may be undesirable, it can be compensated for when the mask is patterned. For multilevel devices the magnification must have the same value for each level, or at least its variation must be within the registration tolerance. This implies stringent control of the gap g. Wafer warping in processing can be nearly eliminated with a proper vacuum chuck. It is not necessary that the gap have the same value at all points on the wafer, only that the spatial variations be the same, within close tolerance, for all levels. The step-and-repeat motion is indispensable to the narrow beam line of x-ray lithography.

6.4.2 X-Ray Mask

An example of a typical mask structure is shown in Fig. 29*b*. An x-ray mask consists of a transmissive membrane substrate and an absorber patterned on it. The ratio of metal thickness to substrate thickness is greater than for a photomask because no materials are available that are fully transparent or fully absorbent, unlike the combination of glass and chromium in optical lithography. These thicknesses are determined by the transmission of the materials for the x-ray wavelength of interest. X-ray masks are made by electron lithography and use the technologies of mask making and direct writing. The pattern on the x-ray mask must be as perfect as a photomask, but the dimension is the same as on the wafer because of proximity printing. For features down to a 0.1 to 0.2-μm geometry it is harder to check the pattern integrity and to repair defects in x-ray than in optical lithography.[54]

Of the heavy metals with large ρZ^4 values, gold was once widely used, because it is relatively easily patterned by liftoff or electroplating. Tungsten and tantalum are used now, because they are easily etched by dry etching. The thicknesses of gold necessary for absorption of 90% of the incident x-ray flux are 0.7 μm, 0.5 μm, 0.2 μm, and 0.08 μm for x-ray wavelengths 4.4 Å (Pd_L), 8.3 Å (Al_K), 13.3 Å (Cu_L), or 44.8 Å (C_K), respectively. In general, the metal is considerably thicker than the chromium layer (0.1 μm) on a photomask. Methods for high-resolution patterning of the gold include electroplating and ion milling. Electroplating produces excellent definition with vertical walls but requires a vertical wall primary pattern in a resist that has a thickness equal to that of the metal to be plated.[55] More often, a subtractive process has been employed in which a thinner resist layer is used to pattern a thin layer of a refractive metal; the refractive metal serves as a mask for ion-milling the underlying gold. With this method, walls that depart from the vertical by 20° or less can be formed. The minimum linewidth attainable by ion milling 0.5-μm-thick gold is ~ 0.4 μm. For higher resolution, and where gold thickness can be reduced, longer wavelengths such as the 12 Å Cu_L radiation may be used. Lines as small as 0.16 μm have been replicated with this type of radiation.

The membrane forming the mask substrate should be as transparent as possible to the x-rays, smooth, flat, dimensionally stable, reasonably rugged, and transparent to visible light if an optical registration scheme is used. Materials that have been used include polymers such as polyimide and polyethylene terephthalate, silicon, SiC, Si_3N_4, Al_2O_3, and a Si_3N_4–SiO_2–Si_3N_4 sandwich structure. Although different mask substrates are appropriate for different portions of the soft x-ray spectrum, there is not yet general agreement on the best material for any particular wavelength.

The major questions remaining about x-ray masks concern their dimensional stability, minimum attainable defect densities, and ease of handling. The stress applied on the thin membrane during processing and formation of the absorber structure causes distortions. Absorber-induced distortion is not noticeable because it is evaluated by comparing the measured fiducial marks before and after electroplating. Resist films on the membrane produce tensile stress on the membrane. In the case of a multilayer process, RIE heating causes membrane distortion with a maximum error of 0.1 μm.[55] Dimensional stability can be degraded by radiation damage produced by x-ray flux, which also makes the mask substrate optically opaque. Pattern placement and CD accuracies are reported as 0.06 μm (3 σ) and are repeatable.[54]

6.4.3 X-Ray Sources

X-rays are produced by the interaction of incident electrons and a target material. The maximum x-ray energy is the energy E of the incident electrons. If E is greater than the excitation energy E_c of the characteristic lines of the atoms of the material, the x-ray spectrum will contain these lines.

X-ray generation by electron bombardment is a very inefficient process; most of the input power is converted into heat in the target. The x-ray flux is generally limited by the heat dissipation in the target. With electrons focused to a spot 1 mm in diameter on an aluminum target on a water-cooled stem, 400 to 500 W is a typical upper limit for the input power. The x-ray power produced is only about 10 mW, and this power is distributed over a hemisphere. The x-ray power is proportional to $(E - E_c)^{1.63}$.

Another x-ray source, which is capable of an order of magnitude greater flux, is plasma discharge.[56] There are several versions, but all function by heating a plasma to a temperature high enough to produce x-ray radiation. The radiation consists of strong lines superimposed on a weak continuum. The source is pulsed at a low rate. In one embodiment, the source size is 2 mm. The repetition rate is 3 Hz. Special problems with such a source are reliability and contamination produced in the plasma chamber.

Typical x-ray sources[57] are listed in Table 8. Although the electron bombardment and plasma sources are included,[58,59] their resolution and registration are limited because of their x-ray trajectory divergences and δ values shown in Table 8.

6.4.4 Synchrotron Radiation

We have considered many kinds of lithography equipment in this chapter. The synchrotron is the biggest and the most expensive; including its support facilities, it is more than two orders higher than any other lithography equipment. Some experiments have been reported in which the radiation from electron synchrotrons and storage rings was used for x-ray lithography. The small angular divergence of the radiation simplifies mask–wafer registration, and the high intensity of the radiation leads to short exposure times. A single storage ring could provide radiation to a large number of exposure stations.

In synchrotrons and storage rings, high-energy electrons are forced into closed curved paths, or orbits, by magnetic fields. An electron moving through a perpendicular magnetic field has an acceleration directed toward the center of the orbit and emits radiation converted from the kinetic energy. For the high-energy electrons of interest, which have velocities very nearly equal to that of light, the radiation is emitted in a narrow cone in the forward direction of motion of the electron. An observer looking along a tangent to the orbit sees a bright spot. The radiation is very different from that of a point source because of the narrow beam from each electron. The radiation from a circular divergence of the radiation in the vertical direction is $\Psi \approx (1957\,E)^{-1}$, where E is the electron energy in GeV and Ψ is in radians. Thus, for a 1 GeV machine the vertical divergence is only 0.5 mradians. The synchrotron is the only equipment that has a small divergence angle and high brightness.

High-energy electrons are provided to the storage ring by a small synchrotron or a small linear accelerator. The ring is operated, briefly, as a synchrotron to boost the electron energy to the final value. Then the electrons may circulate for several hours in a stable orbit. The loss due to the power radiated as synchrotron radiation is compensated by one or more acceleration cavities around the ring. Nevertheless,

TABLE 8
The characteristics of x-ray sources[57]

Source	Divergence (mrad)	Distance source/mask, D (m)	Source diameter, S (mm)	Proximity gap, g (μm)	Runout, Δ (nm)	Blurring, δ (nm)	Spectrum (nm)	Total power on mask (mW/cm^2)
Synchotron	5	4	0.5	50	250	< 10	Broadband, 0.4–2	> 100
Electron bombardment	50	0.4	3	50	250	400	Line, 0.44–1.3	> 1
Plasma focus	50	0.4	0.2	50	250	< 35	Line, 0.7–1.5	> 10
Laser plasma	50	0.4	0.1	50	250	15	Line, 0.9–1.7	> 1

the current slowly decays because electrons collide with residual gas molecules or the walls.

The peak of the power spectrum of the synchrotron radiation occurs at wavelength λ_p. This is related to the electron energy E (in GeV) and magnet bending radius R (in meters) by

$$\lambda_p = 2.35R/E^3 \tag{6.26}$$

A 0.83 GeV machine with $R = 2.1$ m would have a power spectrum with a peak at $\lambda_p = 8.4$ Å.

Oxford Instrument's compact synchrotron Helios® was installed at IBM, East Fishkill, NY. Figure 30 shows the total facility arrangement.[60] The size of Helios is about 2 m × 4 m. It needs many other components—the linear accelerator, beam lines, RF controllers, magnet controllers, pumping units, and a radiation protection building. It has a 0.9-nm wavelength, 0.4-mm source size, and a 45-hr beam life. The x-ray stepper made by Karl-suss has a 0.25-μm resolution and 0.1 μm overlay accuracy.[61] Synchrotron radiation actually has been used in 0.35-μm static RAM (SRAM) manufacturing. The critical layers (isolation, gate, contact hole, and metal 1) were fabricated using the synchrotron on the completed layers made by optical *i*-line or KrF steppers. The CD control of eight lots is reported[62] as 30–50 nm (3σ).

6.4.5 X-Ray Projection

The use of soft x-ray projection lithography has been demonstrated.[63] The key components of this system are x-ray mirrors, which consist of alternating layers of high-index and low-index materials deposited on a smooth substrate. The choice of materials is determined by the x-ray wavelength, and the individual layer thickness is determined by the wavelength and the incidence angle. The normal incidence reflectivity is better than 50% at $\lambda = 4.5$ nm. A laser-produced plasma source is adopted for the x-ray source. The system configuration is illustrated schematically in Fig. 31. The demagnification ratio is 1 : 5. The advantages, compared to proximity methods, are greater latitude with respect to placement of the mask pattern and less radiation-induced damage of the mask. Theoretical resolution is 8.9×10^3 line-pairs per millimeter, or a linewidth of 56 nm, based on $1/\lambda f^{\#}$, where $\lambda = 4.5$ nm and $f^{\#}$ is 125. The image field size is as small as 5 mm in diameter. The exposure time is estimated roughly as 15 minutes for a 6-inch wafer.

6.4.6 X-Ray Resists

An electron resist is also an x-ray resist, since an x-ray resist is exposed largely by photoelectrons produced during x-ray absorption. The energies of these photoelectrons are much smaller (0.3 keV to 3 keV) than the 10 keV to 50 keV energies used in electron lithography, making proximity effects negligible with x-rays and promising higher ultimate resolution.

On traversing a path of length z in a resist or any other material, an x-ray flux is attenuated by the factor $\exp(-\alpha z)$. The 1-μm thick resist film absorbs about 10%

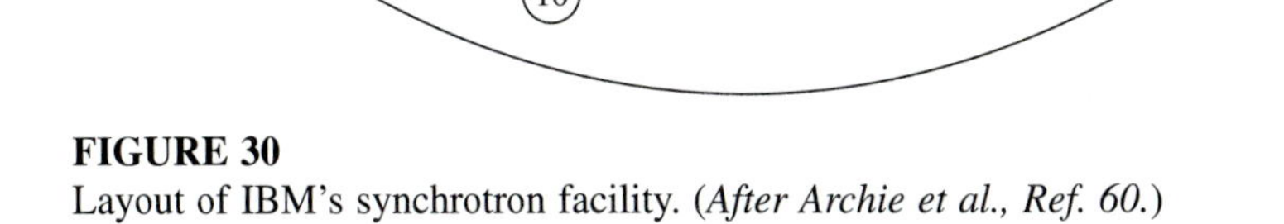

FIGURE 30
Layout of IBM's synchrotron facility. (*After Archie et al., Ref. 60.*)

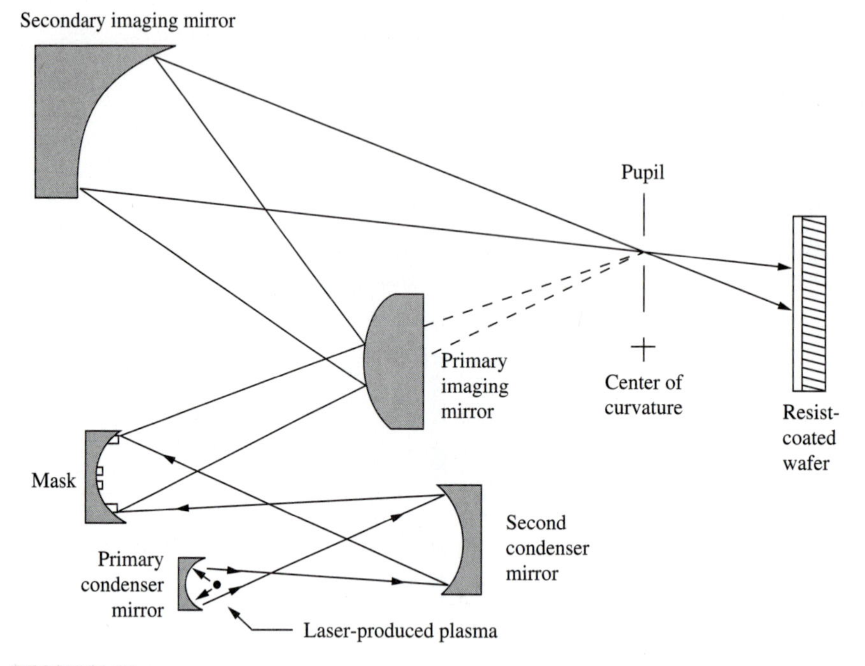

FIGURE 31
Schematic illustration of an x-ray projection system. (*After Hawryluk and Seppala, Ref. 63.*)

of the incident x-ray flux at the characteristic x-ray $Al_K\alpha$ (λ = 8.3Å), because most resists consist of H, C, and O only with a density $\rho \approx 1$ g/cm^3. This small absorption has the advantage of providing uniform exposure throughout the resist thickness z and the disadvantage of reduced sensitivity. As in the optical case, x-ray resist sensitivity is generally quoted in terms of incident dose Q (J/cm^2) required for exposure; sometimes absorbed dose αQ (J/cm^3) is used.

For the shorter x-ray wavelengths the λ^3 dependence of the absorption coefficient leads to low sensitivity. This can be offset by incorporating heavier elements in resist components to increase absorption (sensitivity). In x-ray lithography, we expect a special sensitivity enhancement due to the relationships between resist materials and x-ray absorption edges. For the characteristic x-ray $Pd_L\alpha$ radiation, λ = 4.37 Å, the DCOPA resist (a mixture of poly (3.3-dichloro-1-propyl acrylate) and PGMA/ZA) incorporates chlorine. The Cl_K absorption edge at 4.40 Å provides higher absorption, and chlorine is a chemically reactive species. Thus, two effects contribute to higher sensitivity.

6.5 ION LITHOGRAPHY

When an ion beam is used to expose a resist, higher resolution is possible than with an electron beam because of less scattering. In addition, resists are more sensitive to ions than to electrons. There is also the possibility of a resistless wafer process.

However, the most important application is the repair of masks for optical or x-ray lithography, a task for which commercial systems are available.

The sputtering yield increases with beam energy if the beam energy is larger than some small threshold value. There is an energy limit beyond which the yield decreases because the ions penetrate more deeply and fewer surface atoms receive enough energy to leave the surface. For example, the peak in the sputtering curve for Ar^+ ions incident on Cu occurs at 23 keV. For ion implantation, energies from 30 keV to 500 keV are used, and the dose ranges up to 10^{15} ions/cm^2 (or 1.6×10^{-4}C/cm^2 for monovalent ions). This represents a much larger dose than that used for resist exposure.

The sensitivity of PMMA resist has been measured for 30 keV, 60 keV, and 200 keV He^+ ions and for 100 keV and 150 keV Ar^+ ions. The required dose is nearly two orders of magnitude less than with 20 keV electrons.[64] The perpendicular straggle of the penetrating ion path and the range of low-energy secondary electrons produced are less than the range of backscattered electrons in electron lithography. The ion energies for exposing a resist depend on the ion species. If the ion must penetrate 2500 Å of resist, then a proton would need 14 keV; and a Au ion, 600 keV for the projected range to be 3000 Å.

Ion lithography systems are of two types: a scanning focused-beam system and a mask-beam system. The problems of ion optics for scanning ion systems are more serious than for electron optics. One problem is the ion source, and another is the beam-forming system. In the two practical types of field ionization source, ions are produced in the strong electric field near a pointed tungsten tip; the source of the ionized material is a gas or a liquefied metal surrounding the tip. The largest current densities obtained in the focused image of such a source are 1.5 A/cm^2 for Ga^+ in a 0.1 μm spot and 15 mA/cm^2 for H^+ in a 0.65 μm spot.[65] This value is two orders less than the conventional electron beam system. There are no bright sources for such useful implant species as B and P. Electrostatic lenses rather than magnetic lenses must be used to focus ion beams. If a magnetic lens were used to focus an ion beam, the field would have to be huge since the required field is proportional to $(mV/f)^{1/2}$ from Eq. (6.17), where m is the particle mass. Different isotopes would be focused at different points. Electrostatic optical systems generally have higher aberrations, necessitating a small aperture α (beam solid angle similar to optical or electron beams) and small scan fields.

A prototype scanning system[65] in which a beam of 57 keV Ga^+ ions is focused to a 0.1-μm diameter spot with current density 1.5 A/cm^2, is illustrated in Fig. 32. Spot size is limited by chromatic aberration of the electrostatic lens and the large 14-eV energy spread of the source.

Mask-based systems are of two types. One type is the 5×-reduction projection step-and-repeat system, which projects 60–100 keV light ions such as H_2^+ through a stencil mask.[66] The resolution and overlay accuracy are 0.15 μm and ±0.05 μm. The major disadvantage is the fragile foil mask, which can treat re-entrant patterns only by double exposure with two masks, similar to the technique used in the electron proximity method.

The second type is a step-and-repeat proximity printer in which 300-keV protons are projected through the 0.5-μm thick portion of an all-silicon mask.[67] The mask is aligned so that the ions travel along the channels in the [100] direction. Projected

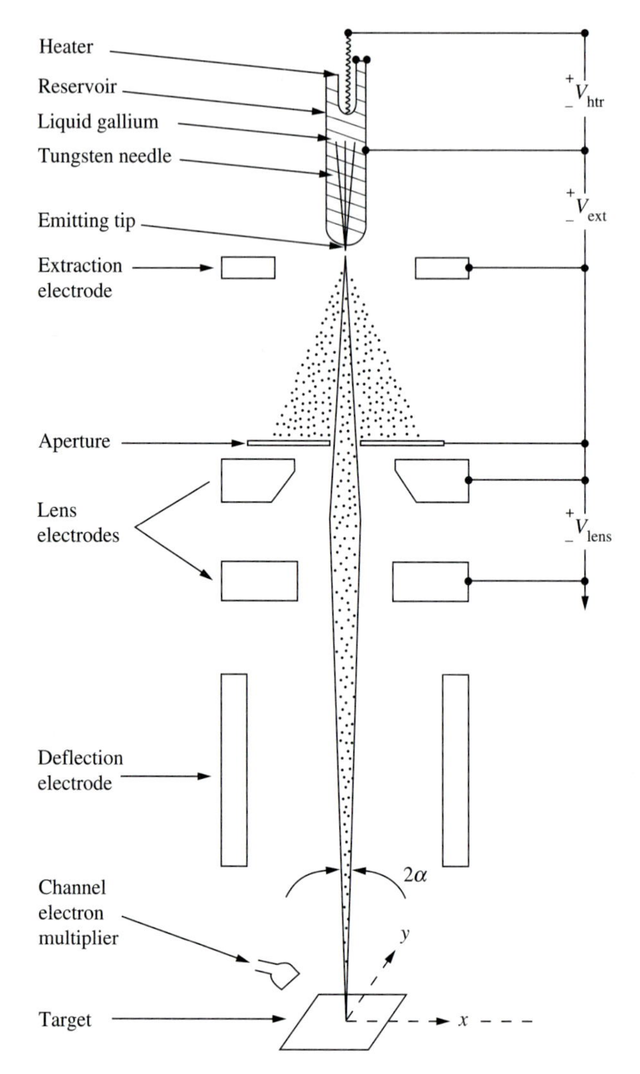

FIGURE 32
The optics of an ion lithography system. (*After Seliger et al., Ref. 65.*)

throughput is sixty 100-mm wafers/hr, with an overlay error of $\sim 0.1\mu m$—barely adequate for 0.5-μm lithography. Resolution is set by ions scattering as they emerge from the channels. Edge resolution is 0.1 μm for the beam-scattering angle of 0.3°. The main disadvantage is the fragile mask, with a 0.5 μm membrane. However, if a large fraction of the mask consists of the thicker silicon absorber, then the mask may be sufficiently rugged. The highest resolution in resist is obtained with a stencil mask and a beam of protons with an energy in the range of 40 to 80 keV. Lines as fine as 400 Å have been printed with this proximity printer.

6.6 SUMMARY AND FUTURE TRENDS

In general, DRAMs have been used as the indicator of progress in ULSI technology. The most advanced DRAM currently in mass production is the 4-Mbit type with 0.8-μm geometry. The 64 Mbit will start production at the end of this century, with a 0.35-μm feature size. If we consider the resolution, optical lithography systems have a capability very close to the required performance as listed in Table 1. Optical lithography is also better than the other methods from the overlay consideration as described in Table 2. Hence, it is likely that conventional optical lithography will be applied to 64-Mbit DRAM production with the combination of several sophisticated technologies such as the phase-shifting mask and off-axis illumination. Other reasons for the continued use of optical lithography are the wide variety of tools and technologies that already exist, its convenience, and its reasonable cost. Some commercially available resists can resolve down to 0.2 μm, so optical lithography still has a margin over next-generation device manufacturing.[68]

In addition, the DOF has become very tight because the wavelength has been reduced and the NA increased, as discussed in Section 6.2.2. One solution to the problem of tight DOF is the multilayer resist process, which has other problems because of its complex process steps.

The performance of various lithography methods based on throughput and resolution are summarized in Fig. 33. The borderline of each technology is very fuzzy. It will be difficult for another technology to replace optical lithography until 0.2-μm resolution is required. This chapter mainly discussed resolution. However, other key

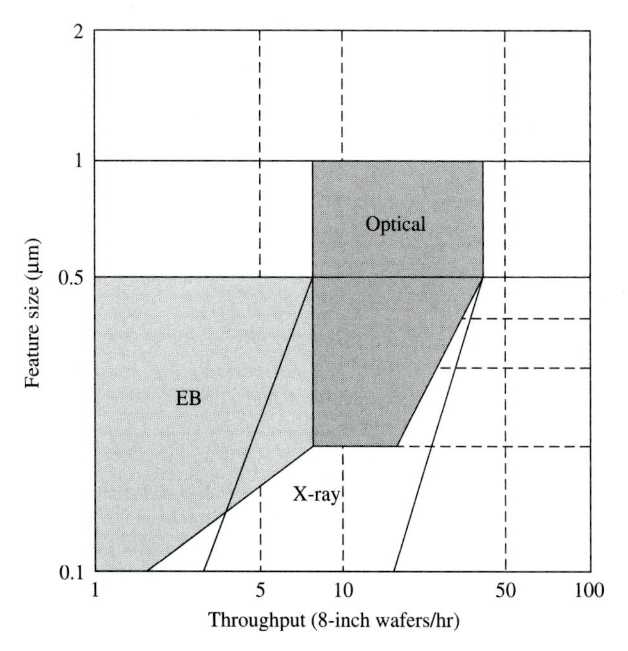

FIGURE 33
Resolution and throughput in the submicron region.

issues are the pattern-width uniformity and the overlay accuracy. If the tolerances of these items are defined as 10% of the feature size, they should be less than 0.02 μm. It is very difficult to guarantee these accuracies because the accuracy of present measurement tools cannot meet this requirement.

Pattern-geometry accuracy is degraded in each step of the device manufacturing process. In general, the patterning sequences are (1) electron beam irradiation, (2) resist patterning on the mask, (3) chromium patterning on the mask, (4) stepper exposure, (5) resist patterning on the wafer, and (6) pattern transfer to a specified layer on the wafer. To minimize error, two methods are available. One is to skip some manufacturing steps, for example, using a direct reaction on the wafer by electron beam with gas molecules. The second is a statistical approach, which is used because of the difficulty in analyzing the cause of errors in the intermediate steps. This method has recently been applied to mask writing to improve the pattern-placement accuracy by multiple exposures. The more active approach is the evaluation and repair of function units of the final product, such as memory or logic cells, instead of inspecting and repairing in the intermediate fabrication steps such as masks.

It is very difficult to predict the trend in lithography beyond 0.2 μm or the architecture or structure of future ULSI. Optical technology is considered almost impossible to use for devices that have less than 0.2-μm geometry because of its resolution limit. We have only two choices: electron beam direct writing or an x-ray technology. However, it is very difficult to make perfect x-ray masks, either 1× transmission or 5× reflection, because of the complicated process steps including inspection and repair. In conclusion, the only available choice is electron beam direct writing. The throughput of this method is just a few 8-inch wafers per hour, as shown in Fig. 33, if the cell projection method is applied to production. For the mass production application, we must minimize the cost and footprint (required floor area) of the machine and develop high-sensitivity resists that are easy to process. It is considered possible that all of these requirements can be fulfilled.

REFERENCES

1. E. G. Cromer, Jr., "Mask Aligners and Steppers for Precision Microlithography," *Sol. State Technol.*, **36,** 23 (Apr. 1993).
2. B. J. Lin, "Deep UV Lithography," *J. Vac. Sci. Technol.*, **12,** 1317 (1975).
3. B. J. Lin, "Optical Methods for Fine Line Lithography," *Fine Line Lithography,* Roger Newman, Ed., North-Holland, Amsterdam, Ch. 2, 1980.
4. J. G. Skinner, "Some Relative Merits of Contact, Near-Contact, and Projection-Printing," *Proceedings of Microelectronics Seminar, Kodak Interface,* **73,** 53 (1974).
5. R. K. Watts, "Advanced Lithography," *Very Large Scale Integration, Fundamentals and Applications,* D. F. Barbe, Ed., Springer, New York, Ch. 2, 1982.
6. M. Born and E. Wolf, *Principles of Optics,* 6th ed., Pergamon Press, New York, 1980.
7. M. C. King and M. R. Goldrick, "Optical MTF Evaluation Techniques for Microelectronic Printers," *Sol. State Technol.*, **20,** 37 (Mar. 1977).
8. J. D. Cuthbert, "Optical Projection Printing," *Sol. State Technol.*, **20,** 59 (Aug. 1977).
9. M. Lacombat, J. Massin, G. M. Dubroeucq, and M. Brevignon, "Laser Projection Printing," *Sol. State Technol.*, **23,** 115 (Aug. 1980).

10. B. J. Lin, "The Optimum Numerical Aperture for Optical Projection Microlithography," *Proc. SPIE,* **1463,** 42 (1991).
11. W. J. Smith, "Modern Optical Engineering," *The Design of Optical Systems,* 2nd ed., McGraw-Hill, New York, Ch. 13, 1991.
12. B. J. Lin, "The Path to Subhalf-Micrometer Optical Lithography," *Proc. SPIE,* **922,** 256 (1988).
13. R. Unger, C. Sparkes, P. A. DiSessa, and D. J. Elliott, "Design and Performance of a Production-Worthy Excimer-Laser-Based Stepper," *Proc. SPIE,* **1674,** 708 (1992).
14. M. D. Levenson, N. S. Viswanathan, and R. A. Simpson, "Improving Resolution in Photolithography with a Phase-Shifting Mask," *IEEE Trans. Electron. Dev.,* **ED-29,** 1828 (1982).
15. K. Ronse, R. Jonckheere, C. Juffermans, and L. Van den Hove, "Comparison of Various Phase Shift Strategies and Application to 0.35 μm ASIC Designs," *Proc. SPIE,* **1927,** 2 (1993).
16. N. Shiraishi, S. Hirukawa, Y. Takeuchi, and N. Magome, "New Imaging Technique for 64M-DRAM," *Proc. SPIE,* **1674,** 741 (1992).
17. N. Shiotake and S. Yoshida, "Recent Advances of Optical Step-and-Repeat System," *Proc. SPIE,* **537,** 168 (1985).
18. F. H. Dill, W. P. Hornberger, P. S. Hauge, and J. M. Shaw, "Characterization of Positive Photoresist," *IEEE Trans. Electron. Dev.,* **ED-22,** 445 (1975).
19. L. F. Thompson, "Design of Polymer Resist for Electron Lithography," *Solid State Technol.,* **17,** 27 (1974).
20. F. H. Dill, A. R. Neureuther, J. A. Tuttle, and E. J. Walker, "Modeling Projection Printing of Positive Photoresists," *IEEE Trans. Electron. Dev.,* **ED-22,** 456 (1975).
21. H. Ito and C. G. Willson, "Applications of Photoinitiators to the Design of Resists for Semiconductor Manufacturing," *Polymer in Electronics,* ACS Symposium Series **242,** T. Davidson, ed., American Chemical Society, 11, 1984.
22. A. Yoshikawa, O. Ochi, and Y. Mizushima, "Dry Development of Se-Ge Inorganic Photoresist," *Appl. Phys. Lett.,* **36,** 107 (1980).
23. K. L. Tai, R. G. Vadimisky, C. T. Kemmerer, J. S. Wagner, V. E. Lamberti, and A. G. Timko, "Submicron Optical Lithography Using an Inorganic Resist/Polymer Bilevel Scheme," *J. Vac. Sci. Technol.,* **17,** 1169 (1980).
24. K. G. Chiong, K. Petrillo, F. J. Hohn, and A. D. Wilson, "Contrast and Sensitivity Enhancement of Resists for High-Resolution Lithography," *J. Vac. Sci. Technol.,* **B6,** 2238 (1988).
25. B. Roland, R. Lombaerts, C. Jakus, and F. Coopmans, "The Mechanism of the DESIRE Process," *Proc. SPIE,* **771,** 69 (1987).
26. G. R. Misium, M. Tipton, and C. M. Garza, "Surface Imaging Lithography at 248 nm," *J. Vac. Sci. Technol.,* **B8,** 1749 (1990).
27. J. M. Moran and D. Maydan, "High Resolution, Steep Profile Resist Patterns," *J. Vac. Sci. Technol.,* **16,** 1620 (1979).
28. L. E. Stillwagon, A. Kornblit, and G. N. Taylor, "Thin Titanium Dioxide Films as Interlayers in Trilayer-Resist Structures," *J. Vac. Sci. Technol.,* **B6,** 2229 (1988).
29. R. F. M. Thorney and T. Sun, "Electron Beam Exposure of Photoresists," *J. Electrochem. Soc.,* **112,** 1151 (1965).
30. H. C. Pfeiffer, "Recent Advances in Electron-Beam Lithography for the High-Volume Production of VLSI Devices," *IEEE Trans. Electron. Dev.,* **ED-26,** 663 (1979).
31. D. R. Herriott, R. J. Collier, D. S. Alles, and J. W. Stafford, "EBES: A Practical Electron Lithographic System," *IEEE Trans. Electron. Dev.,* **ED-22,** 385 (1975).
32. A. J. Speth, A. D. Wilson, A. Kern, and T. H. P. Chang, "Electron-Beam Lithography Using Vector-Scan Techniques," *J. Vac. Sci. Technol.,* **12,** 1235 (1975).

33. D. S. Alles, C. J. Biddick, J. H. Bruning, J. T. Clemens, R. J. Collier, E. A. Gere, L. R. Harriott, F. Leone, R. Liu, T. J. Mulrooney, R. J. Nielsen, N. Paras, R. M. Richman, C. M. Rose, D. P. Rosenfeld, D. E. A. Smith, and M. G. R. Thomson, "EBES4: A New Electron-Beam Exposure System," *J. Vac. Sci. Technol.,* **B5,** 47 (1987).
34. G. A. Burns and J. A. Schoeffel, "Performance Evaluation of the ATEQ CORE-2000 Scanning Laser Reticle Writer," *Proc. SPIE,* **772,** 269 (1987).
35. E. Goto, T. Soma, and M. Idesawa, "Design of a Variable-Aperture Projection and Scanning System for Electron Beam," *J. Vac. Sci. Technol.,* **15,** 883 (1978).
36. H. C. Pfeiffer, "Variable Spot Shaping for Electron-Beam Lithography," *J. Vac. Sci. Technol.,* **15,** 887 (1978).
37. A. M. Carroll and J. L. Freyer, "Measuring Performance of the AEBLE™150 Direct-Write e-Beam Lithography Equipment," *Proc. SPIE,* **537,** 25 (1985).
38. M. Fujinami, N. Shimazu, T. Hosokawa, and A. Shibayama, "EB60: An Advanced Direct Wafer Exposure Electron-Beam Lithography System for High-Throughput, High-Precision, Submicron Pattern Writing," *J. Vac. Sci. Technol.,* **B5,** 61 (1987).
39. T. Takigawa, H. Wada, Y. Ogawa, R. Yoshikawa, I. Mori, and T. Abe, "Advanced e-Beam Lithography," *J. Vac. Sci. Technol.,* **B9,** 2981 (1991).
40. K. Meissner, W. Haug, S. Silverman, and S. Sonchik, "Electron-Beam Proximity Printing of Half-Micron Devices," *J. Vac. Sci. Technol.,* **B7,** 1443 (1989).
41. R. Ward, A. R. Franklin, I. H. Lewin, P. A. Gould, and M. J. Plummer, "A 1 : 1 Electronstepper," *J. Vac. Sci. Technol.,* **B4,** 89 (1986).
42. Y. Sakitani, H. Yoda, H. Todokoro, Y. Shibata, T. Yamazaki, K. Ohbitsu, N. Saitou, S. Moriyama, S. Okazaki, G. Matsuoka, F. Murai, and M. Okumura, "Electron-Beam Cell-Projection Lithography System," *J. Vac. Sci. Technol.,* **B10,** 2759 (1992).
43. H. Y. Liu, M. P. deGrandpre, and W. E. Feely, "Characterization of a High-Resolution Novolak Based Negative Electron-Beam Resist with 4 $\mu C/cm^2$ Sensitivity," *J. Vac. Sci. Technol.,* **B6,** 379 (1988).
44. D. F. Kyser and N. S. Viswanathan, "Monte Carlo Simulation of Spatially Distributed Beams in Electron-Beam Lithography," *J. Vac. Sci. Technol.,* **12,** 1305 (1975).
45. M. Parikh, "Self-Consistent Proximity Effect Correction Technique for Resist Exposure (SPECTRE)," *J. Vac. Sci. Technol.,* **15,** 931 (1978).
46. F. Murai, S. Okazaki, N. Saito, and M. Dan, "The Effect of Acceleration Voltage on Linewidth Control with a Variable-Shaped Electron Beam System," *J. Vac. Sci. Technol.,* **B5,** 105 (1987).
47. E. Kratschmer and T. R. Groves, "Resist Heating Effects in 25 and 50 kV e-Beam Lithography on Glass Masks," *J. Vac. Sci. Technol.,* **B8,** 1898 (1990).
48. M. Fujita, K. Shiozawa, T. Kase, H. Hayakawa, F. Mizuno, R. Haruta, F. Murai, and S. Okazaki, "Application and Evaluation of Direct-Write Electron Beam for ASIC's," *IEEE J. Solid State Circuits,* **23,** 514 (1988).
49. F. Murai, Y. Nakayama, I. Sakama, T. Kaga, Y. Nakagome, Y. Kawamoto, and S. Okazaki, "Electron Beam Direct Writing Technology for 64-Mb DRAM LSIs," *J. Appl. Phys.,* **29,** 2590 (1990).
50. N. Samoto, Y. Makino, K. Onda, E. Mizuki, and T. Itoh, "A Novel Electron-Beam Exposure Technique for 0.1 μm T-Shaped Gate Fabrication," *J. Vac. Sci. Technol.,* **B8,** 1335 (1990).
51. K. L. Lee and M. Hatzakis, "Direct Electron-Beam Patterning for Nanolithography," *J. Vac. Sci. Technol.,* **B7,** 1941 (1989).
52. E. M. Clausen, J. P. Harbison, C. C. Chang, H. G. Craighead, and L. T. Florez, "Electron Beam Induced Modification of GaAs Surfaces for Maskless Thermal Cl_2 Etching," *J. Vac. Sci. Technol.,* **B8,** 1830 (1990).

53. D. L. Spears and H. I. Smith, "High-Resolution Pattern Replication Using Soft X Rays," *Electron. Lett.,* **8,** 102 (1972).
54. D. Fleming, J. R. Maldonado, and M. Neisser, "Prospect for X-Ray Lithography," *J. Vac. Sci. Technol.,* **B10,** 2511 (1992).
55. R. Viswanathan, R. E. Acosta, D. Seeger, H. Voelker, A. Wilson, I. Babich, J. Maldonado, J. Warlaumont, O. Vladimirsky, F. Hohn, D. Crockatt, and R. Fair, "Fully Scaled 0.5 μm Metal-Oxide Semiconductor Circuits by Synchrotron X-Ray Lithography: Mask Fabrication and Characterization," *J. Vac. Sci. Technol.,* **B6,** 2196 (1988).
56. I. Okada, Y. Saitoh, S. Itabashi, and H. Yoshihara, "A Plasma X-Ray Source for X-Ray Lithography," *J. Vac. Sci. Technol.,* **B4,** 243 (1986).
57. A. Heuberger, "X-Ray Lithography," *Sol. State Technol.,* **29,** 93 (Feb. 1986).
58. B. S. Fay and W. T. Novac, "Advanced X-Ray Alignment System," *Proc. SPIE,* **632,** 146 (1986).
59. R. B. McIntosh, G. P. Hughes, J. L. Kreuzer, and G. R. Conti, "X-Ray Step-and-Repeat Lithography System for Submicron VLSI," *Proc. SPIE,* **632,** 156 (1986).
60. C. N. Archie, J. I. Granlund, R. W. Hill, K. W. Kukkonen, J. A. Leavey, L. G. Lesoine, J. M. Oberschmidt, A. E. Palumbo, C. Wasik, M. Q. Barton, J. P. Silverman, J. M. Warlaumont, A. D. Wilson, R. J. Anderson, N. C. Crosland, A. R. Jorden, V. C. Kempson, J. Schouten, A. I. C. Smith, M. C. Townsend, J. Uythoven, M. C. Wilson, M. N. Wilson, D. E. Andrews, R. Palmer, R. Webber, and A. J. Weger, "Installation and Early Operating Experience with the Helios Compact Synchrotron X-Ray Source," *J. Vac. Sci. Technol.,* **B10,** 3224 (1992).
61. A. C. Chen, C. J. Progler, F. F. Couch, T. A. Gunther, H. Fair, and K. A. Cooper, "First X-Ray Stepper in IBM Advanced Lithography Facility," *J. Vac. Sci. Technol.,* **B10,** 2628 (1992).
62. J. Conway, C. Alcorn, D. Patel, J. Ricker, R. Yandow, L. Hsia, and A. Flamholz, "Fabrication of High-Density SRAM Chips Using Mix-and-Match X-Ray Lithography," *Proc. SPIE,* **1924,** 309 (1993).
63. A. M. Hawryluk and L. G. Seppala, "Soft X-Ray Projection Lithography Using an X-Ray Reduction Camera," *J. Vac. Sci. Technol.,* **B6,** 2162 (1988).
64. M. Komuro, N. Atoda, and H. Kawakatsu, "Ion Beam Exposure of Resist Materials," *J. Electrochem. Soc.,* **126,** 483 (1979).
65. R. L. Seliger, J. W. Ward, V. Wang, and R. L. Kubena, "A High-Intensity Scanning Ion Probe with Submicrometer Spot Size," *Appl. Phys. Lett.,* **34,** 310 (1979).
66. G. Stengel, G. Bosch, A. Chalupka, J. Fegel, R. Fischer, G. Lammer, H. Loscher, L. Malek, R. Nowak, C. Traher, P. Wolf, P. Mauger, A. Shimkunas, S. Sen, and J. C. Wolfe, "Ion Projector Wafer Exposure Results at 5× Ion-Optical Reduction Obtained with Nickel and Silicon Stencil Masks," *J. Vac. Sci. Technol.,* **B10,** 2824 (1992).
67. J. L. Bartelt, "Masked Ion Beam Lithography: An Emerging Technology," *Sol. State Technol.,* **29,** 215 (May 1986).
68. S. Okazaki, "Resolution Limits of Optical Lithography," *J. Vac. Sci. Technol.,* **B9,** 2829 (1991).

PROBLEMS

1. A proximity printer operates with a 10-μm mask-to-wafer gap and a wavelength of 436 nm. Another printer uses a 30-μm gap with wavelength 365 nm. Which offers higher resolution?

2. Why must the reticle (mask) used in a wafer stepper be completely free of defects? Why can some defects be tolerated in systems exposing the entire wafer at once?

3. Derive Eq. (6.6),

$$H(\nu) = \frac{2}{\pi}[\cos^{-1}(\nu/\nu_m) - (\nu/\nu_m)\sqrt{1-(\nu/\nu_m)^2}]$$

using a figure and simple geometric considerations.

4. Which projector has a better resolution: a coherent machine or a partially coherent one with a coherency factor 1? The MTF required is better than 0.5.

5. Calculate the wavelength of a 20-keV electron beam.

6. Calculate the current density on the image plane of a LaB_6 cathode at 20 kV. The work function, Richardson constant, and solid angle are 2.4 eV, 40 $A/cm^2 \cdot K^2$, and 10^{-2} rad, respectively.

7. Calculate the total exposure time of a variable-shape electron beam system under the following conditions: resist sensitivity, 2 $\mu C/cm^2$; current density, 10 A/cm^2; field size, 3 mm; chip size, 5 mm $\times$ 10 mm; stage speed, 0.2 s/step; deflection settling, 0.1 μs; shot number per chip, 1×10^8; chip number in a wafer, 100.

8. Compare the chip cost between a stepper and electron direct-writing under the following conditions:

(*a*) Machine costs of the stepper and the e-beam are *S* and *E*, respectively.

(*b*) Throughput of stepper and e-beam are α and β per hour.

(*c*) Reticle unit price is *C*, and *n* pieces are necessary.

(*d*) Chip yields, uptimes, process speeds, running costs, depreciation years, and other conditions are the same.

9. Calculate the actual cost of the chip in the previous question if S = \$2M; E = \$3M; α, β = 50, 10/hr; C = \$3K; n = 5000/y; yields are 80%; running cost = 0; uptimes = 5000 hr/y; 100 chips/wafer; and 5-year depreciation.

10. Suppose that an x-ray resist must see a mask modulation greater than or equal to 0.6 to form useful resist images. What is the minimum gold thickness required on an x-ray mask to satisfy this requirement if the exposure wavelength is 4 Å?

CHAPTER 7

Etching

Y. J. T. Lii

7.1 INTRODUCTION

Devices are built from a number of different layer materials deposited sequentially. The lithography techniques described in Chapter 6 are used to replicate circuit and device features, and the desired pattern is transferred by means of etching. As we enter the era of ULSI manufacturing, the etching process grows more important for fabricating features with sub-half-micrometer dimensions. Dry etching is synonymous with plasma-assisted etching, which encompasses several techniques that use plasmas in the form of low-pressure gas discharges. Integrated circuit fabrication processes that use reactive plasmas are commonplace in today's semiconductor production lines because of their potential for very-high-accuracy transfer of resist patterns, i.e., anisotropic etching, as shown in Fig. 1. By contrast, wet chemical etching results in isotropic etching, where both vertical and lateral etch rates are comparable.

In the early 1970s, photo lithography with Novolak-based resists was used to pattern devices with dimensions less than three micrometers. Oxygen plasmas were explored for stripping of resist. The need to maintain dimensional tolerance during etching became a critical requirement that led to numerous advances in plasma processing and improved reactors during the 1980s. Today, there is a broad base of empirical knowledge and some qualitative understanding of etch mechanisms.[1] These insights and systematic experiments led to the process developments needed to meet the requirements of today's submicrometer device manufacturing.

Etch techniques consist of dry and wet etch methods. Dry etch methods include plasma etching, reactive ion etching, sputtering, ion beam etching, and reactive ion beam etching. For dry etch methods, this chapter concentrates only on plasma and reactive ion etching because they are the most popular methods used in semiconductor industry. This chapter starts with a discussion of the fundamentals of

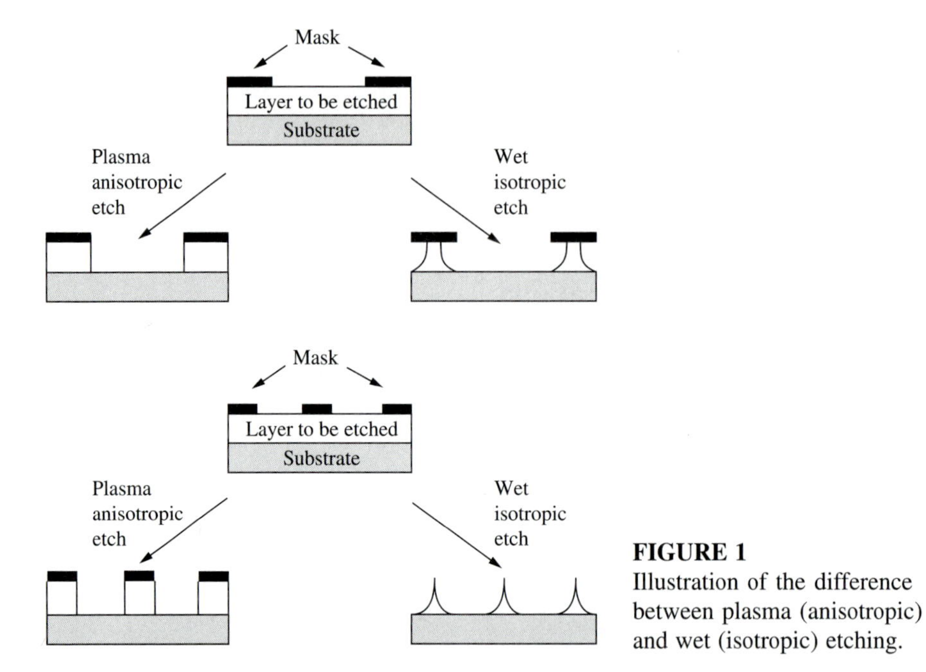

FIGURE 1
Illustration of the difference between plasma (anisotropic) and wet (isotropic) etching.

low-pressure discharge. Next, fluorocarbon plasma chemistry is introduced to review the means for achieving etching directionality and selectivity in reactive etching using glow discharges. Specific plasma processes used to etch different materials important to semiconductor processing are reviewed. Plasma diagnostics, process control, and plasma-induced damage are introduced briefly. In addition, relevant trends in new, magnetically enhanced reactive ion etching systems, in high-density plasma systems, and in integrated etching clustering are discussed. Finally, alternative wet etching methods used in the semiconductor industry are briefly presented.

7.2 LOW-PRESSURE GAS DISCHARGE

Plasma is an ionized gas with approximately equal amounts of positively charged particles (positive ions) and negatively charged particles (usually electrons and a small amount of negative ions). The plasma useful to ULSI processing is a weakly ionized plasma, called a "glow discharge," containing a significant density of neutral particles—more than 90% in most etchers. Although plasma is neutral in a macroscopic sense, it behaves quite differently from a molecular gas, because it consists of charged particles that can be influenced by applied electric and magnetic fields.

Plasma etching systems are extremely complex, and many details of the physical and chemical interactions, both within the plasma and on the surface exposed to the plasma, are not yet understood. The interaction of plasma with surfaces will be discussed in order to describe some of the mechanisms thought to be important in etching. A comprehensive review of this subject can be found in Ref. 2.

7.2.1 AC Plasma Discharge

In plasma etching, a plasma is created by applying a voltage across two electrodes, between which a gas is confined at low pressure. Simple dc power can be used to generate plasma; however, insulating materials cannot be used to cover electrode surfaces, because the insulating materials stop the current flow between the two electrodes and cause surface charging. To deal with the surface charging problem, an ac discharge is used so that the positive charge accumulated during one half-cycle can be neutralized by electron bombardment during the next half-cycle. In plasma etching, an rf field is normally used to generate the gas discharge. One reason for doing so is that the electrodes do not have to be made of a conducting material. The other reason is that electrons can pick up sufficient energy during field oscillation to cause more ionization by electron–neutral atom collisions. As a result, the plasma can be generated at pressures as low as 10^{-3} torr.

The free electrons in the gas are accelerated by the electric field and gain energy. Energy is released when free electrons collide with gas atoms; the gas atoms are then ionized, and more electrons are freed. Electron collisions with neutrals in plasma cause dissociation, ionization, and light emission. These collisions generate atoms, molecules, molecular fragments, and ions. The ions and electrons generated from these collisions raise the density of the plasma. Additional electrons are generated by secondary emission from energetic, positive ions colliding with the electrode surfaces. However, some collisions on the electrode surfaces lower plasma density, such as an ion-ion recombination and an electron-ion recombination. The process continues until equilibrium is reached, that is, until the number of ion-electron pairs created is the same as the number of pairs lost at the electrode surface and the chamber wall. At this point, the plasma is self-sustaining. The density for the plasmas of interest ranges from 10^9 to 10^{12} cm^{-3}. The density of gas molecules at 1 torr is about 10^{16} cm^{-3}, so these discharges are weakly ionized. Therefore, at low pressure, plasma is a partially ionized gas containing electrons, ions, and neutral atoms and/or molecules.

7.2.2 Properties of RF Discharge

When a plasma comes in contact with an electrode or surface, a dark space or *sheath* is formed next to it. The potentials of the rf discharge near the electrodes are important in determining the energies of ions incident on surfaces in the plasma.[3] The potentials near the electrodes, as shown in Fig. 2, result from currents of different particles, specifically electrons and ions. A light particle such as an electron will accelerate more than a heavier particle such as a positive ion in a given electric field. A large number of electrons rush toward electrodes after an electric field is applied. The depletion of electrons from the plasma and a negative charge buildup on the electrodes are inhibited by the development of a negative charge barrier that repels electrons and attracts positive ions. In Fig. 2, V_c is the potential at the rf-powered electrode measured with respect to ground, and V_p is the plasma potential with respect to ground. The potentials across the ion sheaths are $V_p - V_c$ at the rf-powered electrode

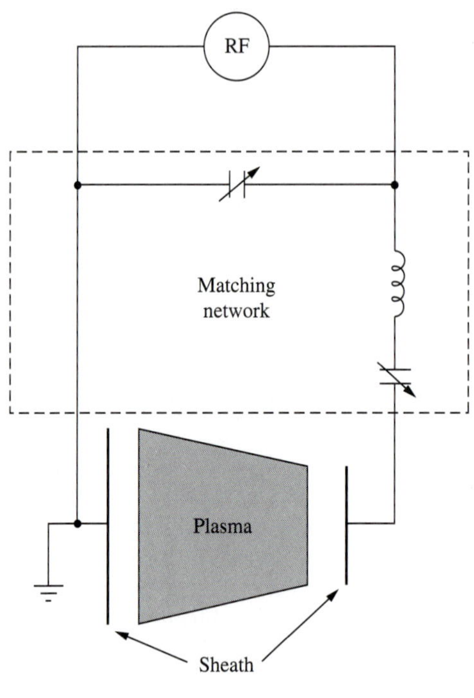

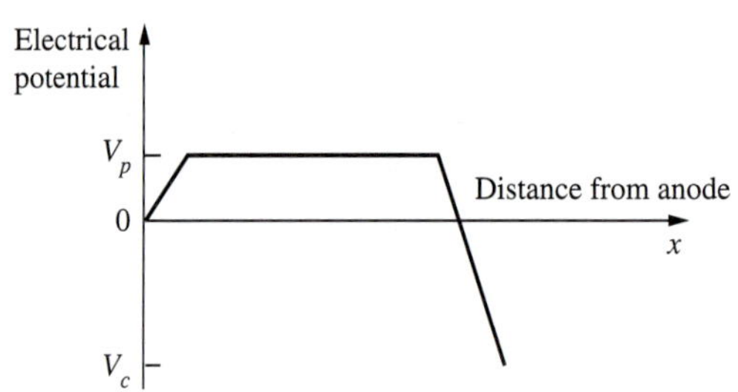

FIGURE 2
Representation of a generic, reactive-ion plasma etcher. RF power is supplied to the electrode supporting the wafer by a generator and matching network. The potential is shown as a function of position in the discharge for the case in which the area of the powered electrode is much less than the area of all grounded surfaces.

and V_p at the grounded surface. The potential of the electrode or surface with respect to that of the plasma determines the maximum energy of the ions bombarding that electrode or surface.

To a first approximation, the sheaths are treated as capacitors. The current density of the positive ions across sheaths is uniform and is equal at both electrodes. The ratio, R, of the area of the rf-powered electrode to the area of all grounded surfaces in contact with the plasma is a key parameter in determining how the applied voltage is distributed among the plasma sheaths. The potential $V_p - V_c$ increases as R decreases. A classic theory of plasma characteristics can be expressed as[2]

$$\frac{V_p - V_c}{V_p} = \left(\frac{A_a}{A_c}\right)^4 \tag{7.1}$$

where A_a and A_c are the grounded surface and rf-powered electrode areas, respectively. The sheath voltage near the small rf-powered electrode is larger than the sheath voltage near the grounded surface. In a typical diode reactive-ion etcher, $V_p - V_c$ can be in the several-hundred-volt range when V_p is less than 100 volts.

7.2.3 Matching Network

A matching network is usually located between the rf power supply and the plasma, as shown in Fig. 2. The purpose of this network is to increase the power dissipation in the discharge and to protect the rf power supply. An rf power supply normally has a purely resistive output with a usual value of 50 ohms. However, an rf plasma normally has a partly capacitive impedance. Therefore, a load equal to the impedance of the rf power-supply output can be simulated by combining the plasma impedance with a variable matching network load. Most plasma etching systems lose power in the matching network; but with careful system design, the power loss can be minimized.

7.2.4 Glow Discharge Chemical Phenomena

Electron energies achieved at low pressure can be so high that they dissociate molecules into free radicals, ions, and new electrons, thus cascading the electrical breakdown of the gas and igniting or sustaining the plasma. At rf frequencies higher than 1 MHz, the ions, because of their low mobility, do not respond to the rf field instantaneously. Electrons undergo primarily inelastic electron-molecule collisions, which supply fresh species to the plasma. The steady-state constitution of a glow discharge is governed by the equilibrium of the rates of production and loss of the various species.

Two main reactions take place in a plasma system:

1. Homogeneous reactions in the bulk plasma, caused mainly by electron impact with neutral species
2. Heterogeneous reactions at the surface contact with the plasma, initiated by ions, electrons, and neutrals

Some examples of the most important types of electron–neutral particle collisions are summarized here:

Simple ionization

$$Ar + e^- \rightarrow Ar^+ + 2e^- \tag{7.2}$$

$$O_2 + e^- \rightarrow O_2^+ + 2e^- \tag{7.3}$$

In molecular gases, ionization may be concurrent with fragmentation; in this case, dissociative ionization is said to occur.

Dissociation

$$CF_4 + e^- \rightarrow CF_3^+ + F + 2e^- \tag{7.4}$$

$$CF_4 + e^- \rightarrow CF_3^+ + F^- + e^- \tag{7.5}$$

Light emission

$$O_2 + e^- \rightarrow O_2^* + e^- \tag{7.6}$$

$$O_2^* \rightarrow O_2 + h\nu \tag{7.7}$$

In the plasma, each formation step balances various loss processes. The equilibrium between formation and loss determines the steady concentrations of species in a discharge. For the charged species in a plasma, these formation and loss processes may be grouped into a few categories as follows:

Electron-ion recombination $O_2^+ + e^- \rightarrow 2O$ (7.8)

Ion-ion recombination $CF_3^+ + F^- \rightarrow CF_4$ (7.9)

Dissociative attachment $e^- + CF_4 \rightarrow CF_3 + F^-$ (7.10)

Atoms and radicals can be lost either by homogeneous reactions or by heterogeneous reactions. The type of loss reaction that will dominate for any species depends on many factors such as pressure, the type of surface, and the ratio of surface area to the volume of discharge.

7.3 ETCH MECHANISMS, SELECTIVITY, AND PROFILE CONTROL

Much of the early work in plasma etching focused on the discovery and optimization of fluorocarbon chemistry.[4] Although this initial work provided many useful industrial etching processes, an understanding of the associated etch mechanisms was typically lacking. Plasma etching is a process in which a solid film is removed by a chemical reaction with ground-state or excited-state neutral species, and plasma etching is often enhanced or induced by energetic ions generated in a gaseous discharge. Some important processing issues discussed in this section include basic etch mechanisms, etch selectivity, and profile control.

7.3.1 Etch Mechanisms

Unlike ion sputtering, which can be compared to sandblasting on an atomic scale, plasma etching methods are based on chemical reactions between the solid to be etched and an active species that arrives on the surface of the solid from the plasma. The microscopic processes of plasma etching of a silicon wafer are shown schematically in Fig. 3.[5] Electrons are accelerated by an applied rf field, and the energetic electrons collide with gas atoms and molecules. Some atoms and molecules become electronically excited, and light emission (glow) is produced from their deexcitation. Other atoms and molecules are dissociated or ionized to form radicals, atoms, and ions by colliding with high-energy electrons. The active species are then transported to the wafer surface where they are adsorbed. The adsorbed species can react with the wafer to form etch products or desorb from the wafer surface without reaction. Etch products then desorb into a gas phase if they are volatile. A brief summary of processes that occur during plasma etching is that a molecular gas, usually relatively inert, is used to generate reactive species, such as F or Cl radicals. By choosing proper chemistry, the radicals react with the wafer surface to form "volatile" products. The chemical compound formed must be volatile with a high vapor pressure

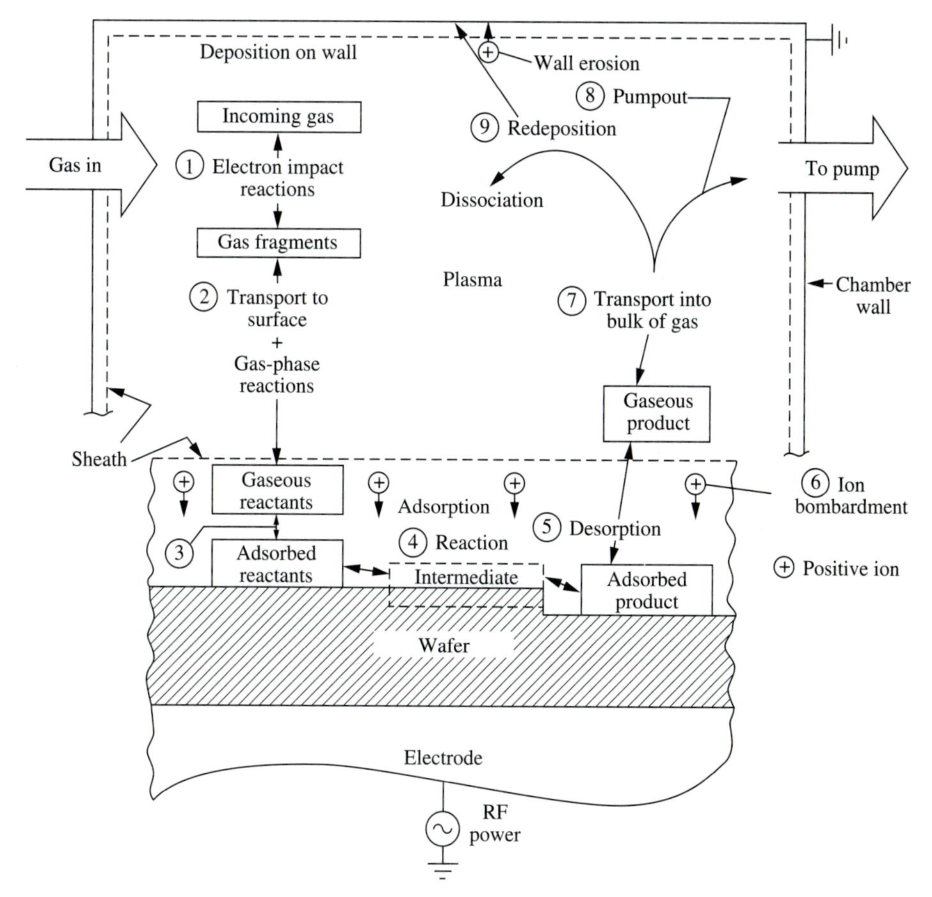

FIGURE 3
Schematic view of microscopic processes that occur during etching of a silicon wafer. (*After Oehrlein and Rembetski, Ref. 5.*)

and stable with a high bonding energy so that it is not dissociated in the plasma and, consequently, can be pumped out of the reactor.

Plasma etching comprises chemical, ion-sputter, and ion-enhanced plasma etching. All these methods are based on the generation of plasma by an rf discharge in a gas at low pressure. But two basic methods can be distinguished—physical methods and chemical methods. The former include sputter etching and the latter include pure chemical etching. There is also a hybrid technique, which uses a physical method to enhance chemical etching. In chemical etching, neutral reactive species generated by the plasma interact with the material's surface to form volatile products. In physical etching, positive ions bombard the surface at high speed; small amounts of negative ions formed in the plasma cannot reach the wafer surface and therefore play no direct role in plasma etching. Chemical and physical etch mechanisms have different basic characteristics. Chemical etching exhibits a high etch rate and good selectivity, produces low ion-bombardment-induced damage, and yields isotropic profiles. Physical etching can yield anisotropic profiles, but it is associated with low etch selectivity and high bombardment-induced damage. Combinations of chemical and physical etching give anisotropic etch profiles, reasonably good selectivity, and moderate bombardment-induced damage.

7.3.2 Fluorocarbon Chemistry

In the early 1980s, molecular gases containing one or more fluorine atoms were used for plasma etching of silicon and silicon compounds. These gases were selected because they are often nontoxic, and fragments produced in a plasma react with silicon or silicon compounds to form volatile compounds at room temperature. For example, CF_4 is extremely stable but dissociates into F atoms and fluorinated fragments (CF_x) in a plasma. Atomic F is the active etchant for Si and SiO_2 by the formation of the volatile SiF_4 and O_2.

$$Si + 4F \rightarrow SiF_4 \tag{7.11}$$

$$SiO_2 + 4F \rightarrow SiF_4 + O_2 \tag{7.12}$$

Energetic electrons colliding with CF_4 are responsible for the generation of radicals, atoms, and ions. However, these radicals and atoms can react with other species, such as gas molecules, to alter the overall chemistry. For example, the etch rates of Si increase dramatically, as shown in Fig. 4,[6] if O_2 is added to CF_4. A maximum

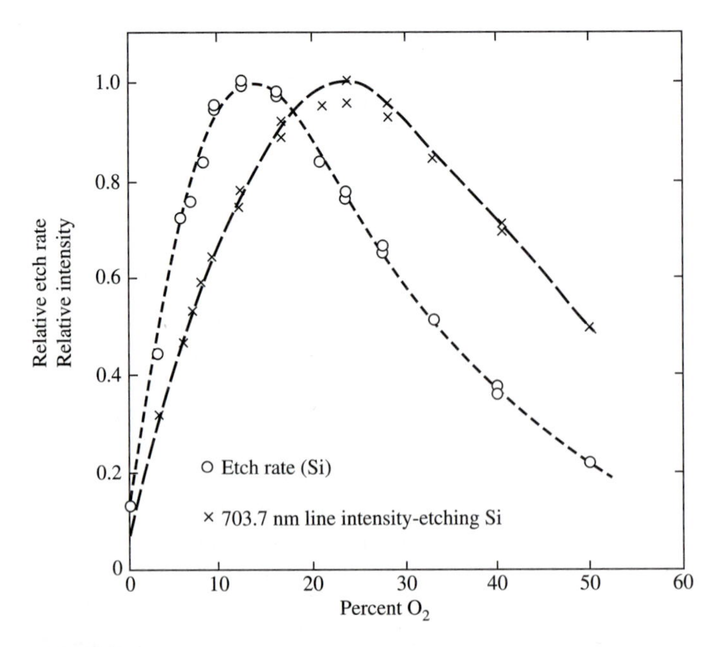

FIGURE 4
Dependence of both the normalized silicon etch rate and the normalized intensity of the emission from electronically excited F atoms on oxygen concentration in the CF_4/O_2 mixture. (*After Mogab, Adams, and Flamm, Ref. 6.*)

etch rate occurs with a CF_4/O_2 mixture that has 12% O_2. A similar effect is observed for SiO_2, where the maximum etch rate is reached at 20% O_2. These effects can be explained by the decreasing recombination rate of F and CF_x ($x \leq 3$). Adding O_2 to the plasma leads to the formation of COF_2, CO, and CO_2, which decreases the concentration of CF_x, thereby decreasing the recombination rate of F and CF_x. This results in a higher F concentration and, therefore, higher Si and SiO_2 etch rates. F concentration increases until the effect of O_2 dilution by CF_4 becomes dominant. This point is reached more quickly with Si etching because oxygen chemsorbs on the Si surface more readily than on the SiO_2 surface thereby blocking the reaction of F with Si.

When H_2 is added to the CF_4 plasma, quite different etch rate changes are observed. Figure 5 shows the etch rates of Si and SiO_2 in a CF_4/H_2 plasma as a function of H_2 concentration.[7] The SiO_2 etch rate is relatively insensitive to the H_2 concentration, but the Si etch rate decreases significantly as the H_2 concentration increases. When the H_2 concentration reaches 40%, the silicon etch stops. A generally accepted explanation of these effects is that two main mechanisms result from H_2 addition. First, H_2 reacts with F to form HF. This reaction consumes the F concentration and decreases the Si etch rate.

$$H + F \rightarrow HF \tag{7.13}$$

Second, the reaction of CF_x with SiO_2 forms etch products SiF_4, CO, CO_2, COF_2, etc. However, the reaction of CF_4 with Si forms a carbon or hydrocarbon polymer on the silicon surface, which blocks silicon etching by F.

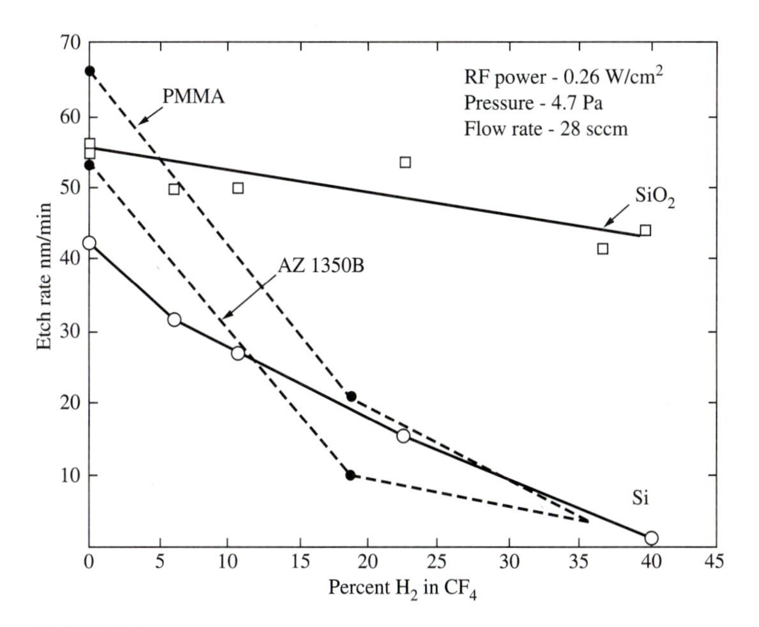

FIGURE 5
Etch rates of silicon (solid line, round points), silicon dioxide (solid line, square points), and PMMA and AZ1350B resists (dashed lines) vs. percentage of H_2 in CF_4. (*After Ephrath, Ref. 7.*)

$$CF_4 + SiO_2 \rightarrow SiF_4 + (CO, CO_2, COF_2) \qquad (7.14)$$

$$CF_4 + Si \rightarrow SiF_x + (C, CF_y, CH_xF_y) \qquad (7.15)$$

Similarly, CHF_3 can be used to replace the CF_4/H_2 mixture for selective silicon dioxide–to-silicon etch. A mixture of CHF_3/CF_4 is often used in industry for selective silicon dioxide–to-silicon etch, because etch selectivity can be controlled by changing the ratio of CHF_3 to CF_4 concentrations.

7.3.3 Anisotropy

Etch profiles are extremely important for submicrometer semiconductor fabrication, and careful control of the surface etch processes is required to ensure directional etching. Etch mechanisms can be understood by analyzing surface etching processes. Plasma consists of neutrals, ions, and electrons, and the effects of each species on etch results are not easily identified. Much of the fundamental understanding of the surface processes that occur during plasma etching comes from well-controlled plasma experiments in which beams of ions, electrons, and neutrals are directed, either together or alternately, at well-characterized surfaces, using apparatus such as that shown in Fig. 6.[8] Etch products are usually analyzed by mass spectrometry. The surface compositions are diagnosed by in-situ or ex-situ x-ray photoelectron spectroscopy, Auger spectroscopy, and secondary ion mass spectrometry. Without radiation or any external energy, the probability of a reaction of Si with F is approximately 10^{-2}, which could be enhanced by ion-beam bombardment. Figure 7 shows the etch rate of silicon with inert Ar ions and with reactive flux to the silicon surface. This etch rate is larger than the sum of the individual etch rates for ion sputtering and neutral etching because chemical etching is enhanced by ion bombardment.[9] Without experiments like this, it is difficult to identify the dominant etch mechanism.

The precise anisotropic etch mechanisms for any given material system are not fully understood, but it is clear that the ions striking the surface must themselves be anisotropic, that is, traveling mainly in a direction perpendicular to the wafer surface. Ions are oriented and accelerated in a sheath. Two proposed mechanisms for anisotropic plasma etching are as follows:

1. Perpendicular ion bombardment creates a damaged surface that is then more reactive toward neutral etchants.
2. Ions help to desorb etch-inhibiting species, such as etch products, from the surface.

In both mechanisms, ion transport must be perpendicular to the surface so that only the etch rate of the bottom surface is enhanced. An illustration of these anisotropic mechanisms is shown in Fig. 8.[4]

Ion–neutral particle collisions from the plasma sheath result in a fraction of the ions hitting the sidewall, and some lateral etching may occur. The number of ion–neutral collisions in the sheath is directly proportional to the sheath thickness and inversely proportional to the ion mean free path. Because the ion mean free path is

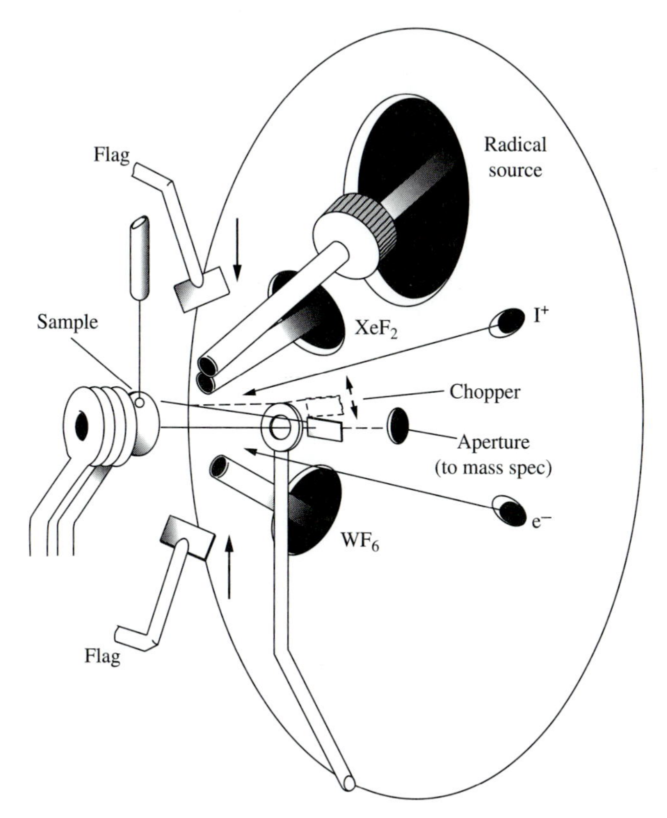

FIGURE 6
Beam-surface experimental apparatus used for the simulation of plasma etching. Such experiments are important for understanding mechanisms of surface-plasma interactions. (*After Winters and Plumb, Ref. 8.*)

usually proportional to pressure, lowering the pressure reduces ion–neutral collisions and enhances anisotropic etching.

In mechanism (2), the formation of sidewall films during etching plays a critical role in the development of the anisotropic etch profile; the films protect the sidewalls from etching by nonperpendicular incoming ions. In other cases, such as during silicon trench etching, thick deposits, often in the upper area of the trench, lead to progressive narrowing of the trench opening and a round bottom.[10] Various mechanisms contribute to sidewall film growth. Three cases are important in the formation of sidewall films:[11,12]

1. Redeposition of etch products or sputtered material from the bottom of the etched structure by adsorption and recombination
2. Deposition from glow discharge
3. Redeposition of sputtered mask material

A schematic description of sidewall deposition phenomena during silicon trench etching by HBr is shown in Fig. 9. Sidewall passivation is governed by the

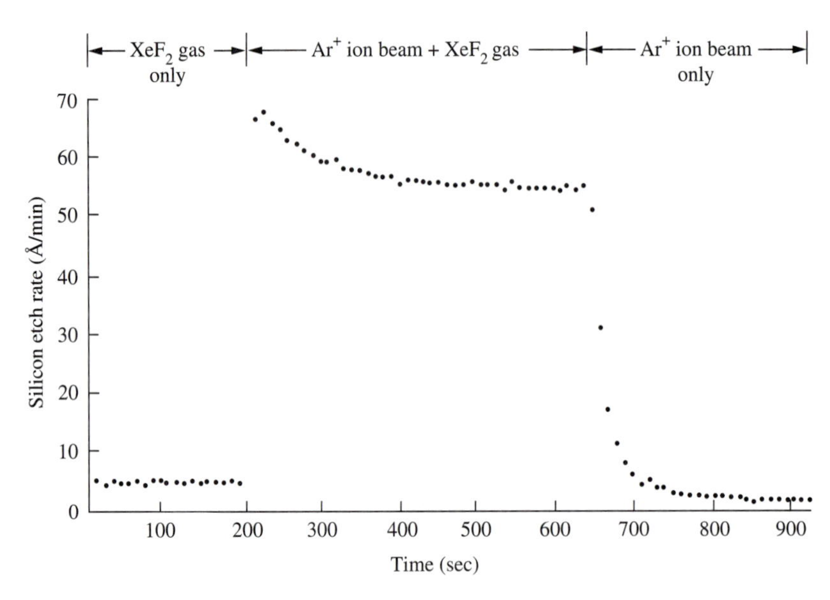

FIGURE 7
Etch rate of silicon as a function of time. This graph illustrates ion-enhanced chemical etching of silicon by an Ar^+ ion beam and XeF_2 flux. (*After Coburn and Winters, Ref. 9.*)

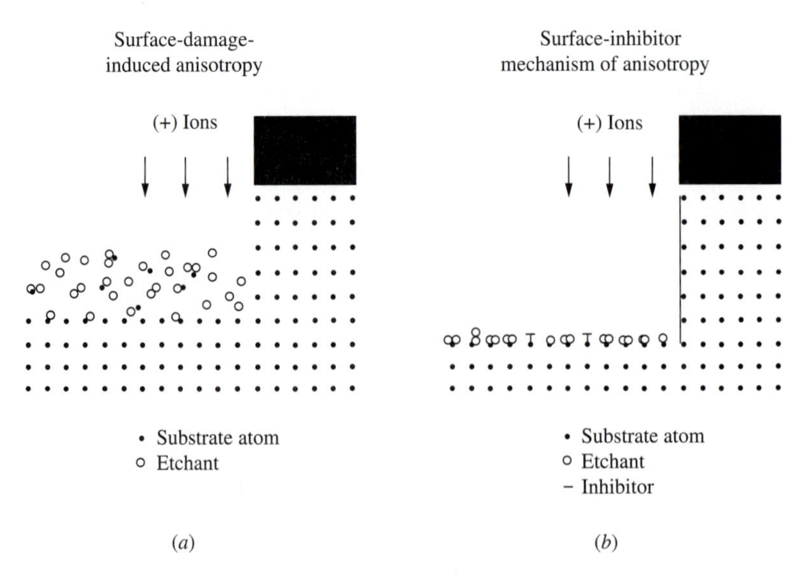

FIGURE 8
Schematic of two proposed mechanisms for ion-enhanced surface etching. Ion-enhanced surface etching results in anisotropic etch profiles. (*After Flamm and Donnelly, Ref. 4.*)

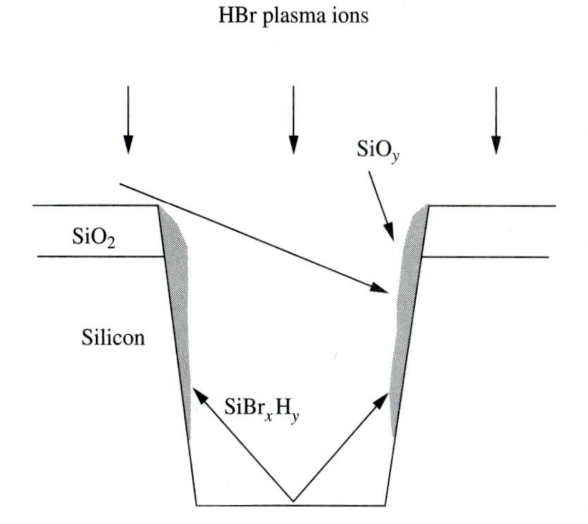

FIGURE 9
Schematic of sidewall formation mechanisms during silicon trench etching by HBr plasma.

equilibrium between deposition and ionic sputtering on vertical surfaces. Sidewall film thickness control becomes critical if tapered sidewalls and dimensional loss is not desired. For example, if the passivation layer becomes too thick during polysilicon gate etching, the gate dimensions increase and problems result.

7.3.4 Selectivity

Selectivity is defined as one film etching faster than another film under the same etching conditions. A higher etch rate ratio (ERR) between different layers is the crucial advantage of reactive ion etching over physical sputtering. The etch rate difference between two different materials is due to different surface etch mechanisms, such as adsorption, reaction, and desorption. During etching, the selectivities with respect to the masking material and the underlying layer require careful control. The required selectivity is defined according to a specified percentage of overetch and film thickness. The gate oxide thickness is in the range of 10 nm for today's 0.8-micrometer, complementary metal-oxide-silicon (CMOS) technologies, and an etch selectivity of polysilicon to thin gate oxide of more than 10 : 1 is required. In the case of a contact hole etch, a large amount of overetch time is required because of the different thicknesses of the oxide layers over the polysilicon gate and over the source and drain. Therefore, a high etch selectivity between silicon and oxide is required for oxide contact hole etch. Thinner masks, such as photoresist, are preferred for optical lithography. A thinner mask film can be used in the etching if a high ERR between the etched film and the mask is obtainable.

Selectivity usually results from two main mechanisms. First, the gas chemistry chosen should be able to generate thermodynamically favorable reactions with the primary film. For example, chlorine-based plasma is usually employed for a polysilicon–oxide etch, and the selectivity arises from the fact that Cl atoms etch SiO_2 very slowly. Etching SiO_2 with chlorine is not a thermodynamically favored reaction. Second, gas chemistry is selected to produce a volatile etch product with the primary film. For example, to etch SiO_2 over aluminum, a fluorine-based chemistry is used because of the high vapor pressure of SiF_4 from the SiO_2 etching. A similar principle has also been applied to SiO_2-over-silicon etching, and the selectivity is usually achieved by forming a passivation layer on silicon. Fluorocarbon polymer forms are used to etch oxide selectively with respect to silicon. Carbon reacts with oxygen during oxide etch, and a fluorocarbon polymer film forms on the silicon surface and provides a good selectivity.

7.3.5 Microscopic Uniformity

Some problems in achieving microscopic uniformity occur because etching rates and profiles depend on feature size and pattern density. Microscopic uniformity problems can be grouped into two categories—aspect ratio–dependent etching (ARDE) and pattern-dependent etching, known as microloading.[13] The cause of the problems is ion and neutral transport.

A typical trench is shown in Fig. 10 with an aspect ratio of 8. Significant ARDE is seen, and trenches with a large aspect ratio etch more slowly than trenches with a small aspect ratio,[14] as shown in Fig. 11. This phenomenon became serious when the

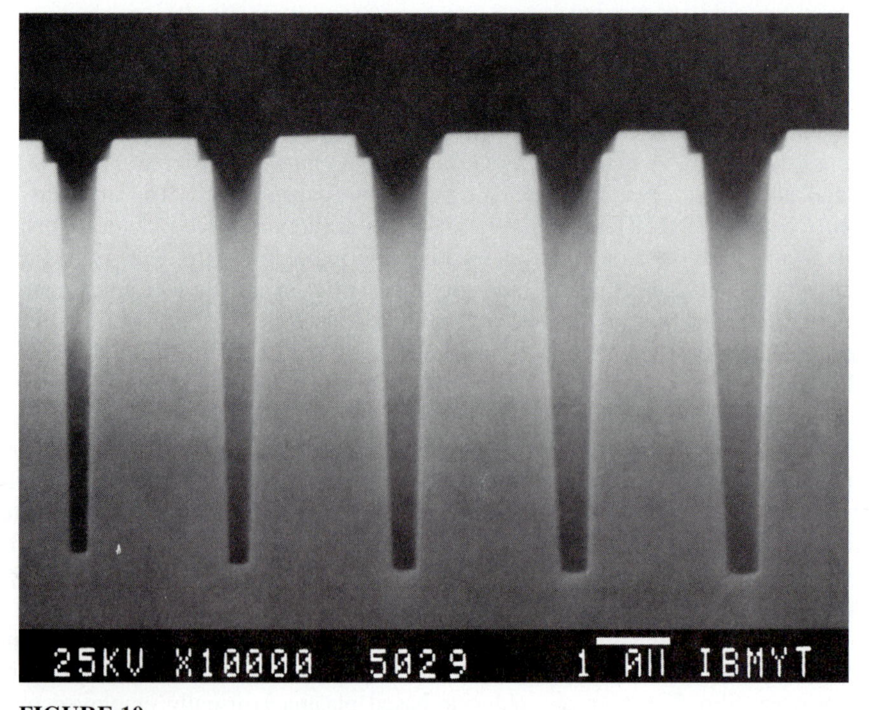

FIGURE 10
Effect of aspect ratio–dependent etching on a silicon trench etch. The etching rate decreases as the trench dimensions are reduced.

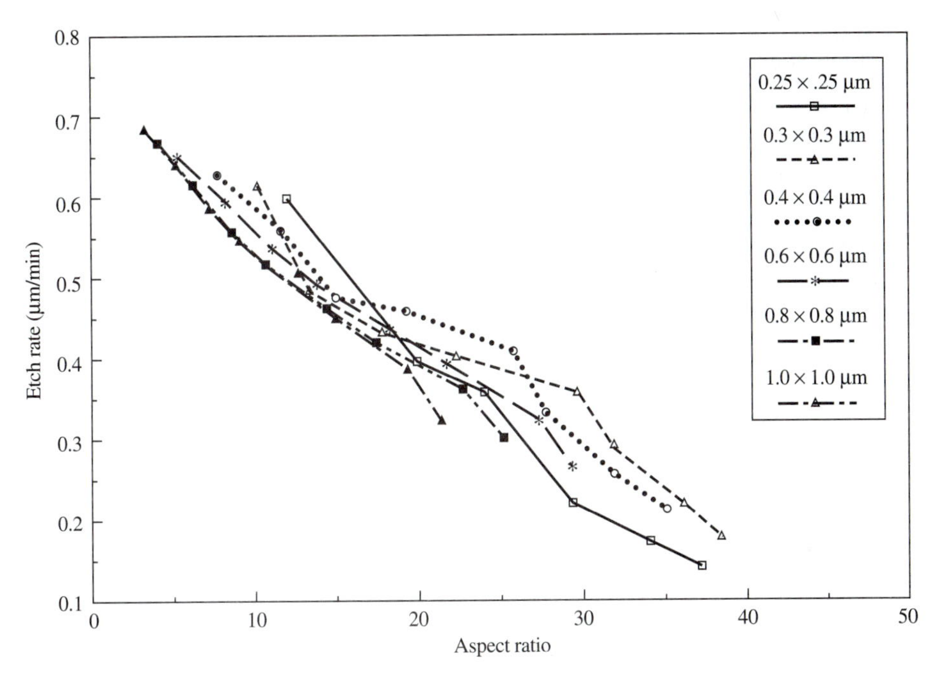

FIGURE 11
Dependence of average silicon trench etch rate on aspect ratio.

era of submicrometer etching began in recent years. Ion bombardment, electron bombardment, reactive neutral species, product desorption, and redeposition all appear to be important in determining the relative etch rates of trenches. Four different mechanisms are the main contributors to ARDE: limited Knudsen transport of neutrals, ion shadowing from ion–neutral collisions in the plasma sheath, neutral shadowing caused by geometrical effects, and ion scattering from insulator charging.[15,16,17]

The term *microloading* usually refers to the dependence of the etching rate on pattern density for identical features.[18] Microloading results from depletion of reactants because the wafer has a local, higher-density unmasked area.[14]

7.4 REACTIVE PLASMA ETCHING TECHNIQUES AND EQUIPMENT

Plasma reactor technology in the electronics industry has changed dramatically since the first application of plasma processing to photoresist stripping. A reactor for plasma etching contains a vacuum chamber, pump system, power supply generators, pressure sensors, gas flow control units, and an end-point detector. Plasma etch tools may be categorized according to the etch mechanism each of them employs. Table 1 shows the similarities and differences in the types of etch equipment that are commercially available. A comparison of pressure operating ranges and ion energies

TABLE 1
Etch mechanisms, pressure ranges, and electrode arrangements of plasma reactors

Etch tool configuration	Etch mechanism	Pressure range (torr)	Electrode arrangement
Barrel etching	Chemical	0.1 ~ 10	Electrodes outside barrel
Downstream plasma etching	Chemical	0.1 ~ 10	Electrodes located in another chamber
Reactive ion etching	Chemical and physical	0.01 ~ 1	Planar diode, triode, cylindrical diode
Magnetic enhanced RIE	Chemical and physical	0.01 ~ 1	Planar diode with magnetic field parallel to the cathode
Magnetic confinement triode RIE	Chemical and physical	0.001 ~ 0.1	Planar diode or triode with magnetic field confinement
Electron cyclotron resonance etching	Chemical and physical	0.001 ~ 0.1	External microwave excitation with biased electrodes
ICP or helicon plasma etching	Chemical and physical	0.001 ~ 0.1	External RF excitation with biased electrodes

for different types of reactors is shown in Fig. 12. Each etch tool is designed empirically and uses a particular combination of pressure, electrode configuration and type, and source frequency to control the two primary etch mechanisms—chemical and physical. High etch rates and tool automation are emphasized for most etchers used in manufacturing.

7.4.1 Barrel Reactor

A barrel reactor usually consists of a cylindrical vacuum chamber with a pair of rf electrodes concentrically located inside. The cross-sectional and side views are shown in Fig. 13. The wafers are placed upright in a boat along the axis of the chamber so that they lie concentrically within the reactor. The process gas pressure is usually between 0.1 to 5 torr, depending on the type of material to be etched. A large batch of wafers can be processed in each run. In the barrel reactor the active radicals are formed between the two electrodes and diffuse through the holes in the electrode to the wafers. The resultant etch profile is isotropic, because there is not much ion bombardment on the wafer surface. The barrel reactor is suitable for less critical process steps, such as the removal of photoresist. Reactants transport to the wafer surface through gas-phase diffusion from the edge of the wafer to the center of the wafer. Therefore, etch uniformities within a wafer and among wafers are very poor, and a large amount of overetch is usually needed to solve this etch uniformity problem.

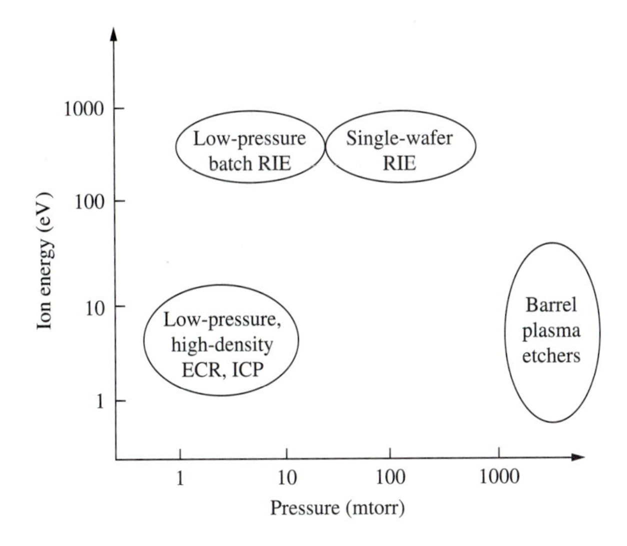

FIGURE 12
Comparison of ion energy and operating pressure ranges for different types of plasma reactors.

7.4.2 Downstream Plasma Etchers

To completely eliminate ion bombardment and photon radiation from the barrel reactor, wafers can be placed in a downstream process chamber. For example, in photoresist stripping, the elimination of ion bombardment is desirable to avoid ion-induced damage or wafer charging. Microwave or rf energy can be used to excite the plasma in the upstream chamber. In addition to its application for photoresist stripping, some single-wafer downstream etchers are used for isotropic plasma etching of silicon, silicon nitride, and silicon dioxide.

7.4.3 Reactive Ion Etchers (RIE)

In the late 1970s, scientists from AT&T Bell Laboratories invented the hexode RIE reactor, which found extensive use in the microelectronic industry. A batch of wafers is mounted on a central, hexagonal-shaped cathode powered by rf, and the

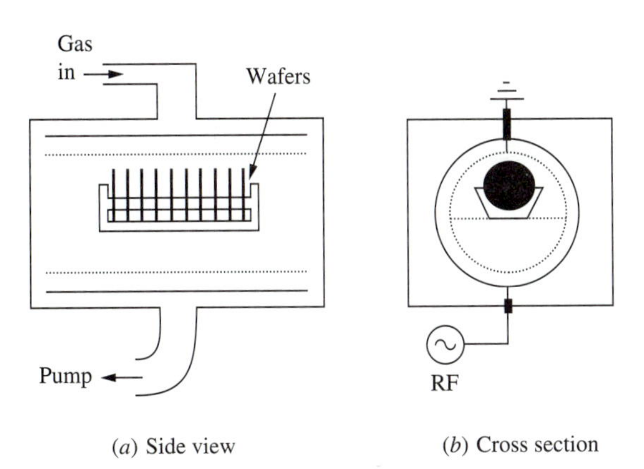

FIGURE 13
Schematic of a barrel plasma reactor showing (*a*) side view and (*b*) cross section.

surrounding bell jar is grounded. As device dimensions shrink, etch uniformity across the wafers becomes a problem for batch RIE systems, and these reactors are not easily automated. Plasma processing is changing from a batch operation to a single-wafer operation, and the parallel-plate diode system has become popular. In this etcher, the wafer is placed on an rf capacitive-coupled bottom electrode, which is considerably smaller than the grounded part of the system. This and the relatively low working pressure (< 500 mtorr) mean that the wafers are subjected to a heavy bombardment of energetic ions from the plasma as a result of the large, negative self-bias at the wafer surface. Anisotropic etching is obtained because ion-enhanced chemical etching has a higher etching rate in the direction perpendicular to the wafer surface than in the direction parallel to the wafer surface.

The etch selectivity of this system is relatively low compared to barrel etch systems because of strong physical sputtering. However, selectivity can be improved by choosing the proper etch chemistry, for example, by polymerizing the silicon surface with fluorocarbon polymers to obtain selectivity of SiO_2 over silicon. Alternatively, a triode-configuration RIE etch, as shown in Fig. 14, can separate plasma generation from ion transport. Ion energy is controlled through a separate bias on the wafer electrode, thereby minimizing the loss of selectivity and the ion-bombardment-induced damage observed in most traditional RIE systems.

7.4.4 Magnetic-Enhanced and Confinement Plasma Etchers

Magnetic-enhanced reactive ion plasma etchers (MERIEs) use a group of permanent magnets located behind the etching wafer or pairs of direct-current electric coils to generate a magnetic field parallel to the wafer surface, as shown in Fig. 15. The magnetic field is perpendicular to the electric field (because of the cathode dc bias) and confines electrons to a circular trajectory near the cathode. Electron confinement reduces the loss of electrons to the wall of the system and increases the frequency of

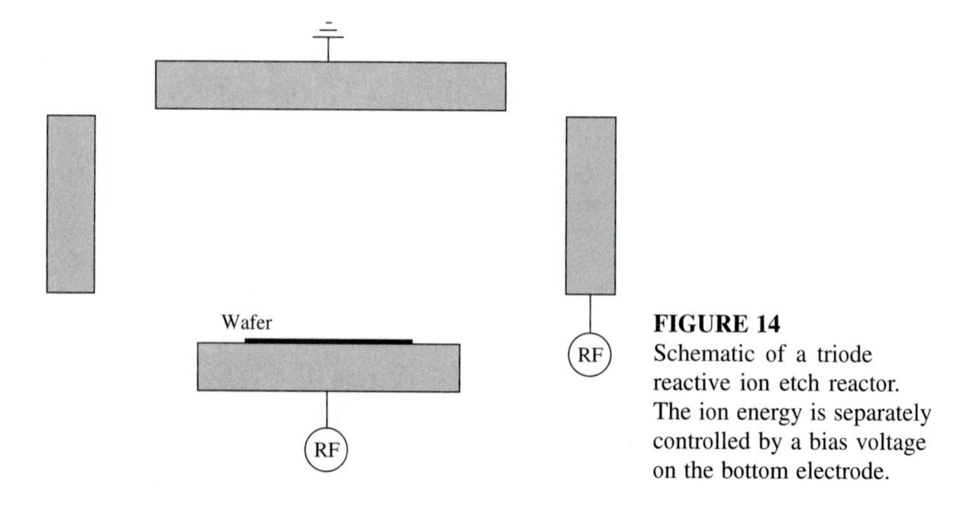

FIGURE 14
Schematic of a triode reactive ion etch reactor. The ion energy is separately controlled by a bias voltage on the bottom electrode.

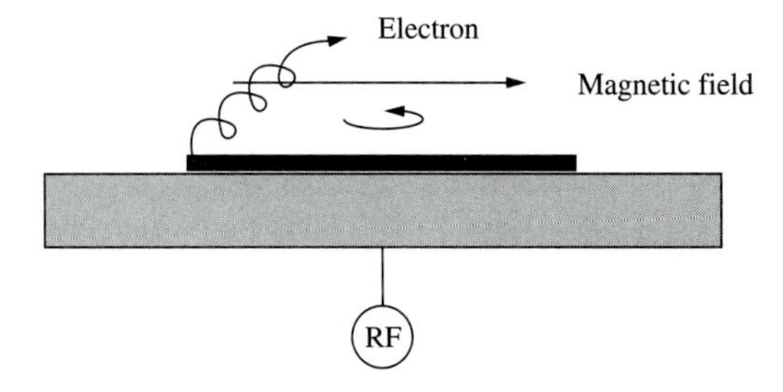

FIGURE 15
Magnetic-enhanced reactive ion etch (MERIE) reactor with a time-average magnetic field applied parallel to the cathode.

electron–neutral collisions. The associated higher frequency of collisions increases ion density, and electron confinement reduces the mobility of electrons toward the cathode, thereby reducing the self-bias voltage,[19] as shown in Fig. 16.[10] To achieve the required time-average plasma uniformity across the wafer surface, the field direction is rotated electrically with a period that is short with respect to the processing time. Magnetic fields also modify the ion-bombardment energies and can be used as another parameter for process optimization.

Uniformity of the plasma and radical fluxes to the wafer are critical issues. Magnetic confinement plasma etchers use permanent-magnet pole pieces and arrange them alternately around the process chamber to create a magnetic-field-free region around the wafer, as shown in Fig. 17. A high density of ions is achieved through the reflection of electrons back into the plasma from the surface magnetic-field bucket. This greatly increases the effective path length of electrons, and higher ion density is generated from the higher frequency of electron–neutral collisions.

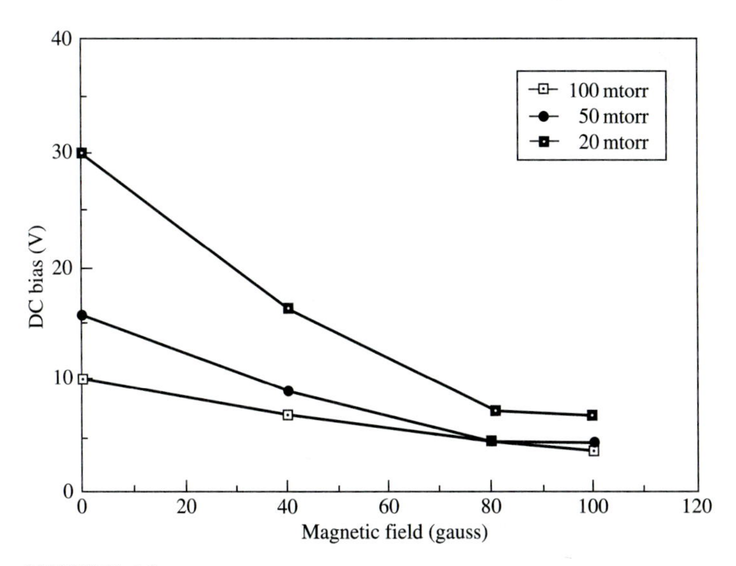

FIGURE 16
Effect of magnetic field on self-bias voltage. (*After Lii, Ng, and Danner, Ref. 10.*)

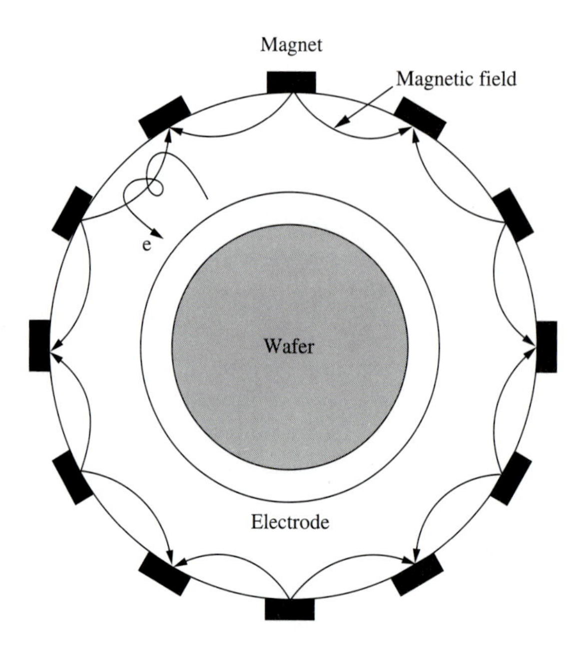

FIGURE 17
Schematic of magnetic-confinement reactive ion etch reactor with a multipolar magnetic bucket surrounding the reactor chamber.

7.4.5 Electron Cyclotron Resonance Plasma Etchers (ECR)

Most parallel-plate plasma etchers, except triode RIE, do not provide the ability to independently control plasma parameters, such as electron energy, ion energy, plasma density, and reactant density. As a result, ion-bombardment-induced damage becomes serious as device dimensions shrink.

ECR etching uses microwave excitation in the presence of a magnetic field to generate a high-density discharge. The Lorentz force causes the electrons to circulate around the magnetic field lines in circular orbits, with a characteristic cyclotron frequency of

$$\omega_{ce} = \frac{eB}{m_e} \tag{7.16}$$

where e is the electron charge, B is the magnetic field, and m_e is the electron mass. When this frequency equals the applied microwave frequency, a resonance coupling occurs between the electron energy and the applied electric field, which results in a high degree of dissociation and ionization (10^{-2} for ECR compared to 10^{-6} for RIE).[20] With a microwave frequency of 2.45 GHz, the required magnetic field is 875 gauss. Figure 18 shows one of the possible ECR plasma etching configurations.[21] Microwave power is coupled via a waveguide through a dielectric window into the ECR source region. The magnetic field, supplied from magnetic coils, decreases as a function of distance from the coils.

Because the gradient in the magnetic field decreases, the electrons are accelerated away from the plasma source, creating a negative potential in the direction of the wafer. Ions diffuse by ambipolar diffusion out of the source region into the the wafer process chamber. The wafer is rf- or dc-biased to control the energy of the ions to achieve the desired etch anisotropy. Plasma uniformity degrades

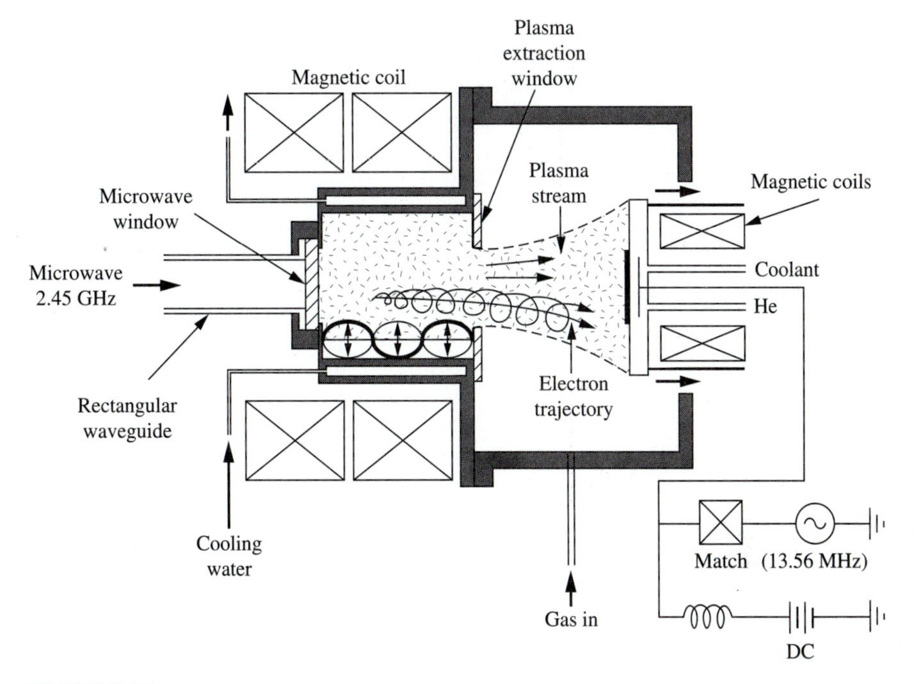

FIGURE 18
Schematic of electron cyclotron resonance etch reactor with a mirror magnetic field around the wafer to confine electron trajectory. *(After Chen et al., Ref. 21.)*

because of the ambipolar diffusion and mirror magnetic field,[21] but a multipole magnetic bucket[22] can be used to improve plasma uniformity. Etch uniformity can also be improved by putting the wafer in an ECR source region surrounded by an optimized magnetic field.[23]

7.4.6 Inductively Coupled and Helicon Wave RF Plasma Etchers

As feature sizes for ULSI continue to decrease, the limits of the conventional rf capacitive-coupled parallel system are being approached. Most ECR plasma reactors are suitable for etching ULSI. However, they are not popular in manufacturing because of their complexity. Other types of high-density plasma sources, such as inductively coupled plasma (ICP) sources or helicon plasma sources, may become the main plasma sources for future ULSI processing. A comparison of recent plasma sources is shown in Table 2.[24] The PMT MORI®source is a helicon wave plasma source. The Lam TCP source is a transformer-coupled plasma source.

An inductively coupled plasma source, shown in Fig. 19, generates high-density, low-pressure plasma that is decoupled from the wafer, and it allows independent control of ion flux and ion energy.[25] Plasma is generated by a flat spiral coil that is separated from the plasma by a dielectric plate on the top of the reactor. The wafer is located several skin depths away from the coil, so it is not affected by the electromagnetic field generated by the coil. There is little plasma density loss because plasma

TABLE 2
Comparison of different commercially available plasma sources[24]

	Plasma source comparison						
	PMT MORI	**Lam TCP**	**Hitachi ECR**	**Drytech Hol. Anode**	**Prototech Helical Res.**	**Lucas Labs Helicon**	**Tegal HRe**
Source freq. (MHz)	13.56	13.56	2450	13.56	13.56	13.56	13.56
Sub. bias freq. (MHz)	13.56	13.56	2/13.56	13.56	13.56	13.56	< 500 kHz
Source B field (gauss)	50–120	—	800	—	—	50–400	†
Ion density (cm^3 at wafer)	3×10^{12}	$0.5–2 \times 10^{12}$	1×10^{11}	1×10^{12}	1.00×10^{12}	3×10^{11}	3×10^{11}
Ion potential (eV)	20+	20+	20+	20+	20+	20+	20+
Electron temp. (eV)	3.5	3.5–6	4	3–5	4	4	4–6
Pressure range (mtorr)	0.5–10	1–25	0.4–10	10–180	0.05–3000	1–10	1–20

† No magnetic field at wafer.

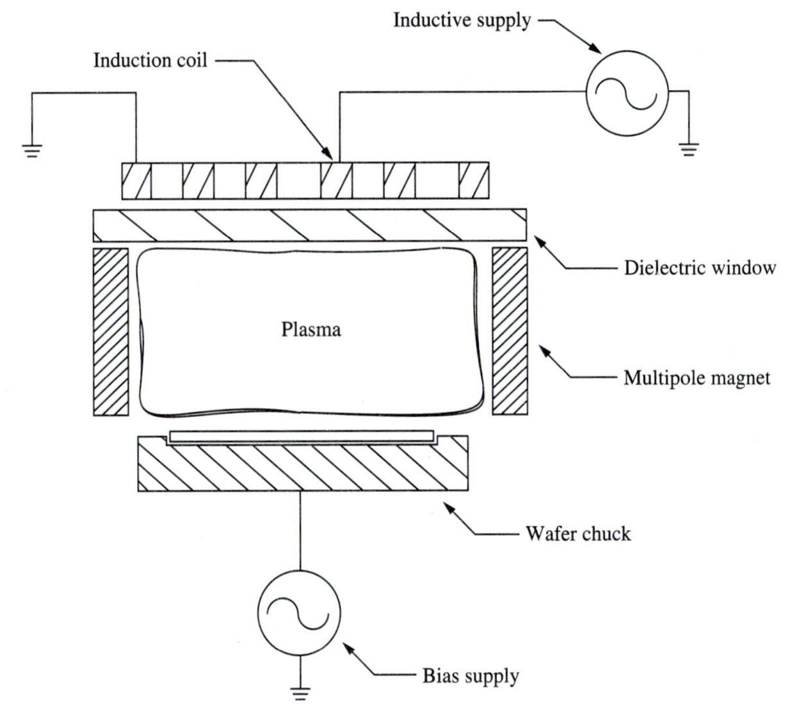

FIGURE 19
Illustration of an inductively coupled plasma reactor. (*After Keller, Forster, and Barnes, Ref. 25.*)

is generated only a few mean free paths away from the wafer surface. Therefore, a high-density plasma and high etch rates are achieved.

A helicon plasma source can also be used to generate a high-density ($> 10^{11}/cm^3$) discharge. A transverse electromagnetic radio-frequency wave (13.56 MHz), excited by a double-loop or single-loop antenna located outside a quartz source tube, is coupled with a steady longitudinal magnetic field B_0 (approximately 100 gauss) generated by a solenoid coil, as shown in Fig. 20.[26] The resonance condition (propagation of a helicon wave) depends on the magnitude of the longitudinal field and the dimension of the reactor. If the wavelength of the helicon wave is the same as the antenna length, the coupling will be resonant. High-density plasma then diffuses into the wafer chamber. In addition, the wafer can be biased separately with a second rf generator.

7.4.7 Clustered Plasma Processing

Microelectronic device wafers are processed in cleanrooms to minimize exposure to ambient particulate contamination. As device dimensions shrink, particulate contamination becomes a more serious problem. To minimize particulate contamination and preserve the integrity of thin film interfaces, clustered plasma tools use a wafer

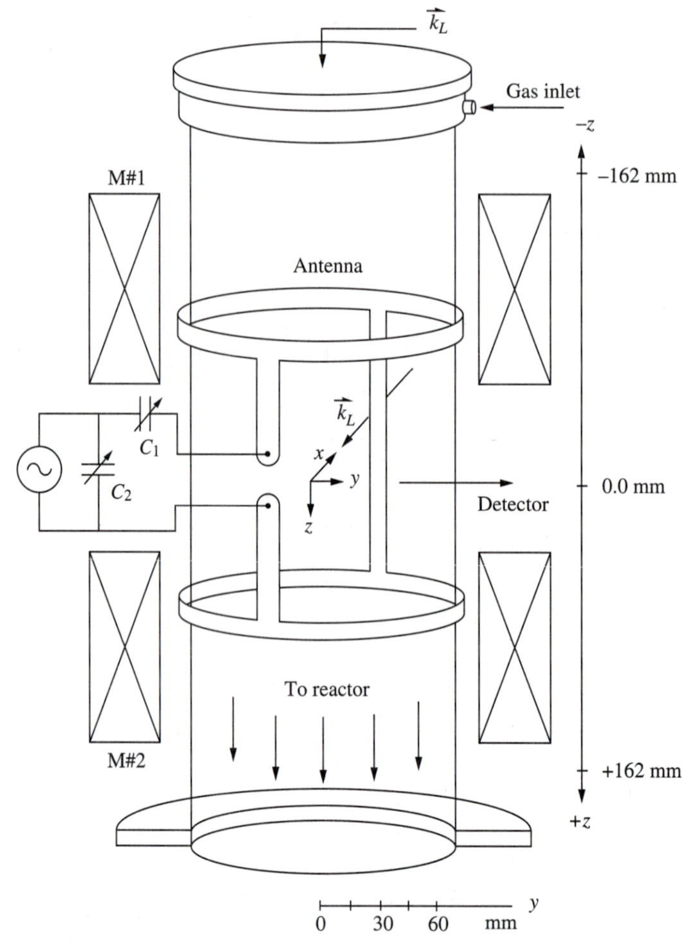

FIGURE 20
Schematic of a helicon wave plasma reactor with a double-loop antenna. (*After Giapis et al., Ref. 26.*)

handler to pass wafers from one process chamber to another in a vacuum environment, as shown in Fig. 21. The wafer can be transferred from one process chamber to another without being exposed to and contaminated by the ambient.[27]

Clustered plasma processing tools can also increase throughput. For example, integration of sequential process steps into multichamber in-situ processing modules, multilayer CMOS gate level etching, dielectric layer deposition with etchback planarization, and metal etching with resist strip are usually accomplished with clustered processing. However, tool reliability becomes very important as tools become even more sophisticated. If they are reliable, clustered tools can provide an economic advantage of high chip yield. The yield is key to reducing manufacturing costs, and clustered tools should provide better yield mainly because the wafer is exposed to less ambient contamination and is handled less.

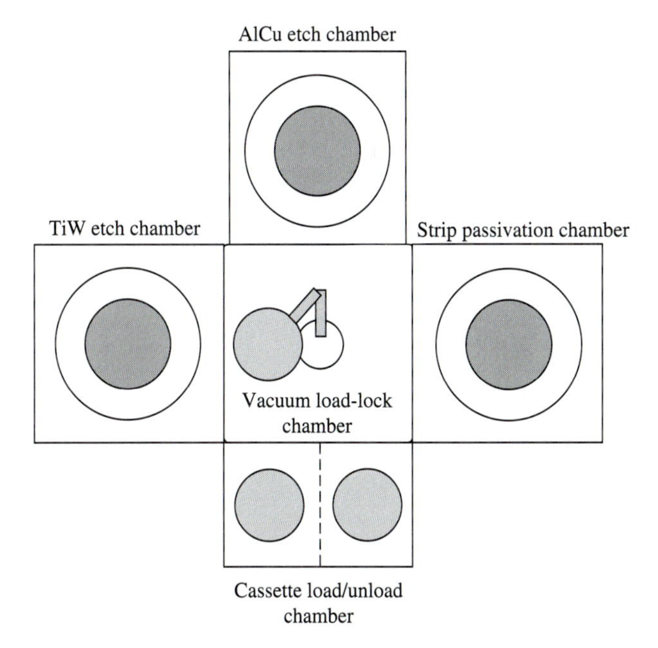

FIGURE 21
Cluster reactive ion etch tool for multilayer metal (TiW/AlCu/TiW) interconnect etching.

7.5 PLASMA PROCESSING PROCESSES

The chemical and physical complexity of plasma-surface interactions makes computer-based modeling and plasma simulation inadequate for developing plasma reactors. As a result, the detailed descriptions required to guide the transfer of processes from one reactor to another or to scale processes from a small to a large reactor are not available. Therefore, the plasma processes for fabricating microelectronic devices have been developed largely by time-consuming, costly, empirical methods.

Excellent critical dimension control and minimized profile microloading are the two most important issues for ongoing research. Extensive research in plasma etching explores ways to optimize the performance of etching processes, characterize and avoid contamination and damage effects, identify the important reactive species and measure their densities, and, in general, improve our limited understanding of plasma etching. See Refs. 28 and 29 on various aspects of plasma etching. Other studies concentrate on the limitations of conventional rf plasma. This research led to the development of enhanced plasma source configurations and processes that are still being examined.

Etch chemistries also play a critical role in the performance of etch processes. Table 3 lists some conventional and new etch chemistries for different etch processes.[30] A truly useful plasma process must have a robust process window.

TABLE 3

Etch chemistries of different etch processes[30]

Material being etched	Conventional chemistry	New chemistry	Benefits
PolySi	Cl_2 or BCl_3/CCl_4 (sidewall passivating gases)	$SiCl_4/Cl_2$	No carbon contamination
	$/CF_4$ (sidewall passivating gases)	BCl_3/Cl_2	
	$/CHCl_3$ (sidewall passivating gases)	$HBr/Cl_2/O_2$	Increased selectivity to SiO_2 and resist
	$/CHF_3$ (sidewall passivating gases)	HBr/O_2	No carbon contamination
		Br_2/SF_6	
		SF_6	Higher etch rate
		CF_4	
Al	Cl_2	$SiCl_4/Cl_2$	Better profile control
	BCl_3 + sidewall-passivating gases	BCl_3/Cl_2	No carbon contamination
	$SiCl_4$	HBr/Cl_2	
Al-Si (1%); Cu (0.5%)	Same as Al	$BCl_3/Cl_2 + N_2$	N_2 accelerates Cu etch rate
Al-Cu (2%)	$BCl_3/Cl_2/CHF_3$	$BCl_3/Cl_2 + N_2 + Al$	Additional aluminum helps etch copper
W	$SF_6/Cl_2/CCl_4$	SF_6 only	No carbon contamination
		NF_3/Cl_2	Etch stop over TiW and TiN
			No carbon contamination
TiW	$SF_6/Cl_2/O_2$	SF_6 only	
WSi_2, $TiSi_2$, $CoSi_2$	CCl_2F_2	CCl_2F_2/NF_3	Controlled etch profile
		CF_4/Cl_2	No carbon contamination
Single-crystal Si	Cl_2 or BCl_3 + sidewall-passivating gases	CF_3Br	Higher-selectivity trench etch
		HBr/NF_3	
SiO_2 (BPSG)	CCl_2F_2	CCl_2F_2	
	CF_4	CHF_3/CF_4	CFC alternatives
	C_2F_6	CHF_3/O_2	
	C_3F_8	CH_3CHF_2	
Si_3N_4	CCl_2F_2	CF_4/O_2	CFC alternatives
	CHF_3	CF_4/H_2	
		CHF_3	
		CH_3CHF_2	

Developing an etch process usually means optimizing many characteristics, such as etch rate, selectivity, profile control, endpoint control, damage, etc., by adjusting a large number of process parameters. Because we do not fully understand etch processes, systematic experiments are required to study the large number of process parameters. The most common method is to use the design experiment and analyze the results with a surface response methodology. In this section, several of the most important etch processes for ULSI fabrication are introduced.

7.5.1 Silicon Trench Etching

As feature sizes in ULSI decrease, a corresponding decrease is needed in the area occupied by both the isolation between circuit elements and the storage capacitor of the DRAM memory cell. This area can be reduced by etching trenches into the silicon substrate and filling them with suitable dielectric or conductive materials. Silicon trenches can also be used as alignment marks for optical or electron beam lithography. Deep silicon trenches, usually with a trench depth larger than 5 μm, are used mainly for forming trench capacitors. Shallow silicon trenches, usually with a trench depth less than 1 μm, are often used for isolation between device elements.

To guarantee a void-free trench refill, anisotropic etching with a high etch selectivity to mask material is desired for submicrometer-deep silicon trench etching. This precludes the use of fluorocarbon chemistry, which tends to undercut the mask material.[18] Fluorine chemistry has a high silicon etch rate, but it has low etch selectivity to the oxide mask and to the isotropic trench profile. However, it is often used for etching shallow silicon trenches with trench depths less than 1 μm, because it satisfies the requirement of etch selectivity of silicon to the photoresist mask. In this section, the deep-trench etching process is discussed because of its stringent requirements of etch rate, anisotropy, and selectivity.

Chlorine-based and bromine-based chemistries have a high silicon etch rate and high etch selectivity to the silicon dioxide mask. However, chlorine-based chemistries tend to give slightly isotropic trench profiles,[31,32] especially in the ion-implanted phosphorus-doped region. Carbon-containing gas has been employed to provide sidewall passivation to avoid undercutting from fast chlorine-silicon reactions.[33] The sidewall passivation layer, however, causes contamination and reduces the etch selectivity of silicon to a silicon dioxide mask. A simple carbon-free etch chemistry leading to minimum thickness of the sidewall passivation is desired to guarantee high accuracy in the pattern transfer. As a general rule, the reactivity of the halogens follows the order F > Cl > Br. Recently, bromine-based chemistries have become popular for etching silicon with anisotropy and extremely high selectivity to the oxide mask.[34]

Sidewall passivation protects the sidewall from lateral etch attack and local ion-enhanced trenching. The desired trench shapes with smooth vertical sidewalls are obtained by controlling the balance between the etch and the inhibition by sidewall passivation. The sidewall shape depends significantly on the wafer temperature. Increasing temperature results in less deposition of sidewall passivation and more lateral etching. Fig. 22[10] refers to the effects of backside cooling under different He pressure. Greater pressure of He results in a cooler wafer temperature.

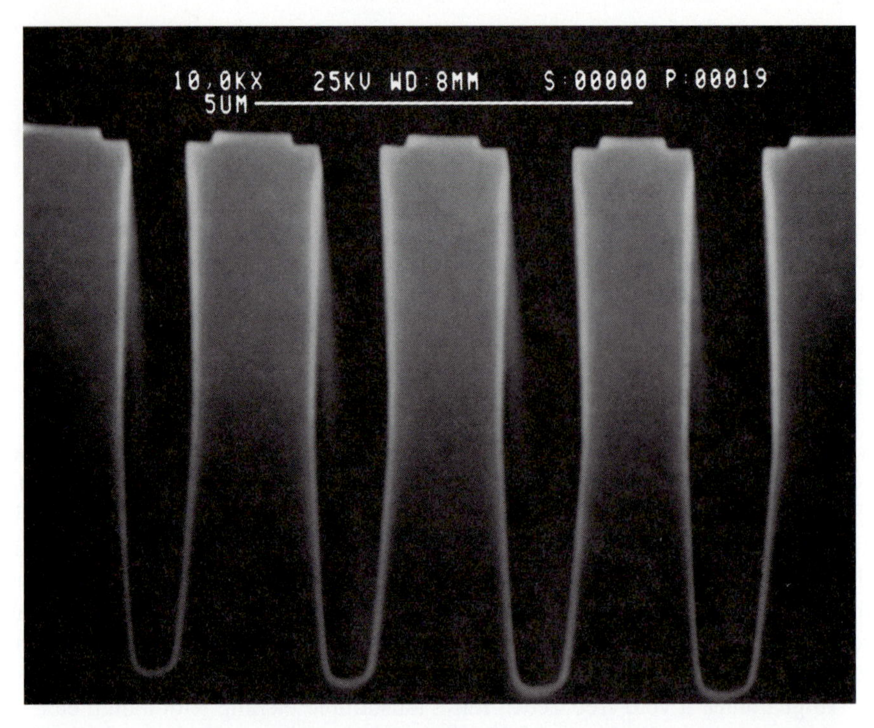

(a)

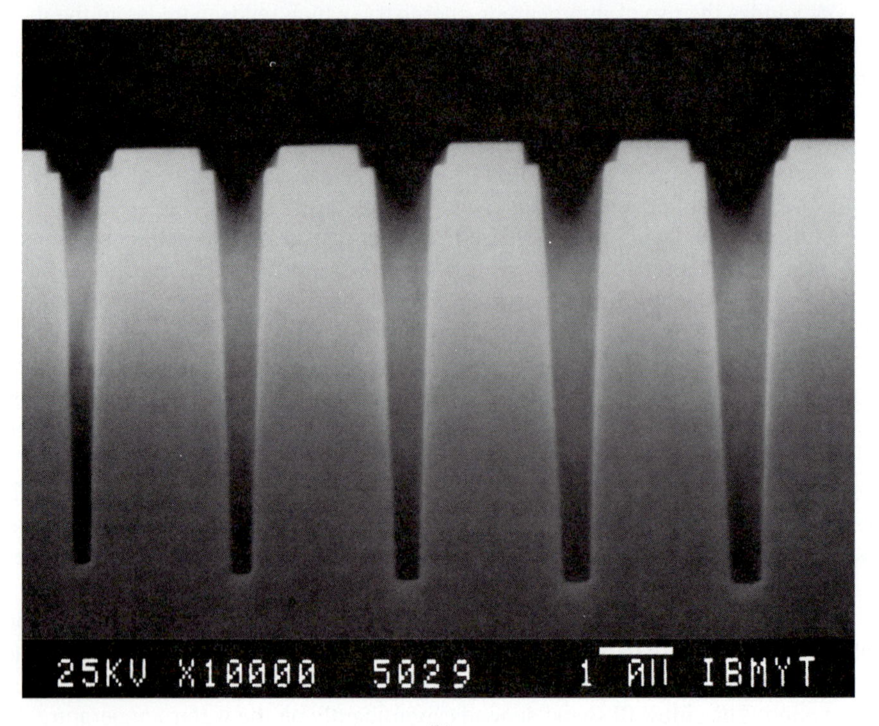

(b)

FIGURE 22
Effect of wafer backside cooling on a silicon trench profile: (*a*) backside He pressure 0.5 torr (*b*) backside He pressure 6.0 torr. (*After Lii, Ng, and Danner, Ref. 10.*)

Aspect ratio–dependent etching (ARDE) is often observed in submicrometer-deep silicon trench etching caused by limited ion and neutral transport within the trench. Trenches with large aspect ratios etch more slowly than trenches with small aspect ratios. Ion scattering results from ion–neutral collisions in the plasma sheath, and the electrical charging on the masks causes ARDE. Some neutrals are transported to the bottom of the trench by Knudsen diffusion, also contributing to ARDE. Low gas pressure reduces the ARDE effect, and chlorine-based chemistry shows less ARDE than fluorine-based chemistry during deep trench etching, because ion-assisted etching is dominant in chlorine-based chemistry.[35]

7.5.2 Polysilicon and Polycide Gate Etching

Low-pressure chemical vapor deposited (LPCVD) polysilicon is usually used as a gate material for MOS devices because of its superior interface property with thin gate oxide at high temperature. Metal silicide over polysilicon has also been used for the MOS gate because of its high electrical conductivity. Anisotropic etching and high etch selectivity between polysilicon and the gate oxide are the most important requirements for gate etching. Achieving high selectivity and etch anisotropy at the same time is difficult for most ion-enhanced etching processes. Therefore, multistep processing is used in which different etch steps in the process are optimized for etch anisotropy and selectivity.

Most chlorine-based and bromine-based chemistries can be used for gate etching to achieve the required etch anisotropy and selectivity. The main etchant today is bromine-based chemistry, because bromine gives higher etch selectivity of polysilicon to gate oxide than chlorine does. Figure 23 shows a scanning electron microscope (SEM) cross section of a polysilicon gate etched by HBr plasma[36] that has a 0.1 μm gate width and a 45 Å gate oxide thickness. The thin native oxide on top of the polysilicon should be removed by fluorine-based plasma before polysilicon etching, i.e., initiation, to avoid any micromasking formation during polysilicon etching. To further improve the selectivity of polysilicon to gate oxide requires removing carbon-containing gases and carbon-containing materials from the etch process and the etch reactor, because carbon tends to reduce the selectivity of polysilicon to gate oxide,[37] as shown in Fig. 24. It is suspected that carbon reacts with the oxygen atoms of silicon dioxide. An enhanced etch attack at the edge of the gate, a phenomenon called "trenching," is shown in Fig. 25. Trenching results in broken gate oxide and is usually observed at low gas pressure. The trenching problem can be solved by increasing etch selectivity, such as adding bromine- and oxygen-containing gases to the etch process.

7.5.3 Oxide and Nitride Etching

An oxide etching process is often used to open a contact window down to a silicon device or metal conductor; therefore, the selectivity of SiO_2 to Si must be high. It is well known that selectivity is achieved by increasing the ratio of CF_x to F atoms

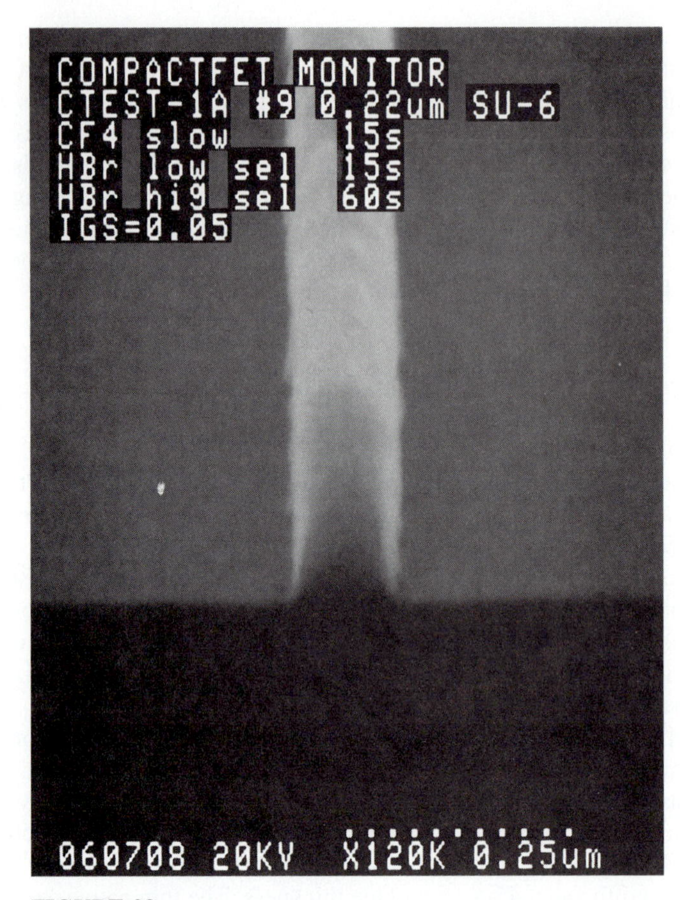

FIGURE 23
SEM cross section of HBr-plasma-etched polysilicon gate with a 0.1 μm gate width and a 45 Å gate oxide thickness. (*After Lii et al., Ref. 36.*)

in the plasma.[7] CHF_3/CF_4 chemistry is a common oxide etchant because it provides high selectivity to Si from carbon polymer deposition on the silicon surface. Therefore, a further silicon surface cleaning must be added to remove the polymer passivation layer after oxide etching to avoid high contact resistance between metal and silicon.

LPCVD or plasma-enhanced CVD silicon nitride are often used as a sidewall around the polysilicon gate in MOS devices for source and drain extension or used as a final passivation layer. Sidewall formation for a polysilicon gate requires anisotropic etching and high selectivity of silicon nitride to silicon dioxide. Si_3N_4 can be etched by both fluorine atoms and CF_x-containing plasma. CF_4 and CF_4/O_2 are widely used to etch silicon nitride but are not selective to silicon or silicon dioxide.[38] Si_3N_4 can be etched anisotropically in an RIE mode with high selectivity to Si and SiO_2 by CHF_3/O_2, CH_2F_2, or CH_3F.[39,40] Details of why the etch mechanism results in high selectivity are not clear. Fluorine-deficient fluorocarbon polymer is thought to contribute to this high selectivity.

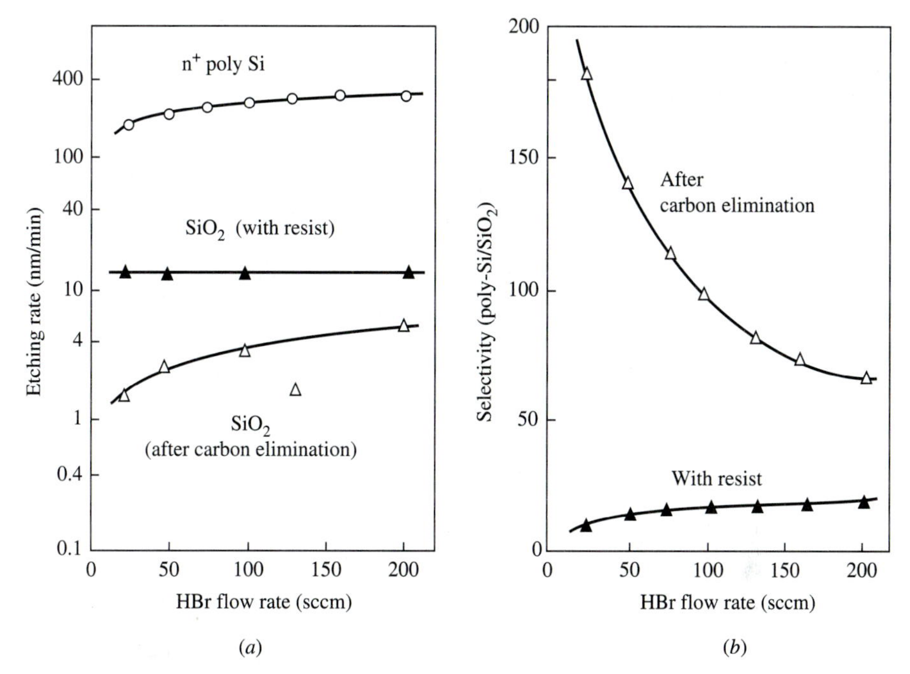

FIGURE 24
Dependence of (*a*) the polysilicon etch rate and (*b*) Si/SiO_2 selectivity on HBr flow rate, carbon reducing Si/SiO_2 selectivity. (*After Nakamura, Iizuka, and Yano, Ref. 37.*)

7.5.4 Aluminum and Tungsten Metal Etching

Etching of a metallization layer is a very important step in semiconductor device fabrication. Aluminum and tungsten are the most popular materials used for interconnection, and anisotropic etching of these materials is usually required. An aluminum etch with fluorine-based chemistry does not work, because of the very low vapor pressure of the etch product AlF_3. Chlorine-based chemistry has been used for Al etching most frequently, and bromine-based chemistry has been investigated recently.[41]

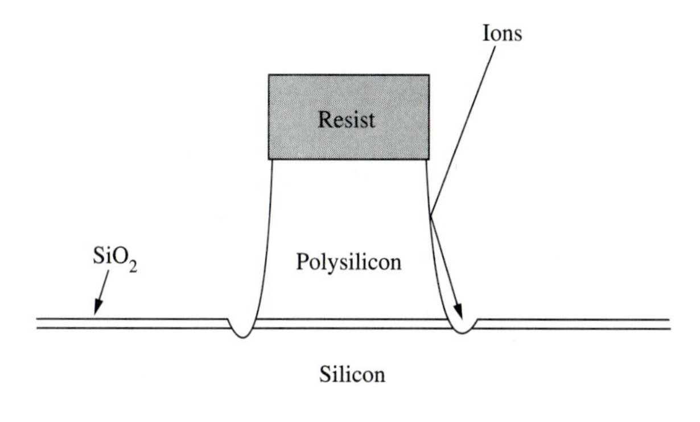

FIGURE 25
Schematic of the trenching mechanism during polysilicon gate etching.

Cl_2 spontaneously etches clean aluminum. Chlorine has a very high chemical etch rate with aluminum and tends to produce an undercut during aluminum etching. Carbon-containing gas is added to form sidewall passivation during aluminum etching to obtain etch directionality. The Al_2O_3 layer, which is always present on air-exposed aluminum, is etched very slowly in chlorine-based gases. Energetic ion bombardment, for example, sputtering, is required to remove Al_2O_3 before aluminum etching. In the semiconductor industry, BCl_3 plasma is often used to remove Al_2O_3 on the aluminum surface.[42] BCl_x also appears to recombine with chlorine at the wafer surface to ensure an anisotropic etch profile during Al etching with a Cl_2/BCl_3 mixture. A small percentage of silicon and copper are usually added to enhance the electromigration resistance of aluminum conductors, but Cu causes micromasking during aluminum etching with chlorine-based plasma because it generates low-volatility etch products. Micromasking can be suppressed by using stronger ion bombardment during etching and addition of aluminum chloride gas.[43] More recently, a barrier metal layer, such as TiW and TiN, has been used to prevent Si migration into Al. TiW and TiN etch readily in chlorine-based plasma.

Corrosion of the aluminum pattern exposure to the ambient is another problem in aluminum etching. Residual chlorine on both the Al sidewall and the photoresist tends to react with water to form HCl, which etches aluminum. In-situ exposure of the wafer to a CF_4 discharge to exchange Cl with F and then to an oxygen discharge to remove the resist, followed by immediate immersion in deionized (DI) water, is successful in eliminating Al corrosion.[44] High-temperature heating of the wafer to evaporate chlorine or hydrolyzing chlorine with hydrogen-containing gases can be used to remove chlorine trapped in the resist and the sidewall.

LPCVD tungsten has been widely used for contact hole filling and first-level metalization because of its excellent deposition conformability. Both fluorine- and chlorine-based chemistry etch W and form volatile etch products.[45,46] Fluorine-based chemistry also etches the silicon dioxide layer; thus it is very difficult to obtain selectivity between W and SiO_2. Chlorine-based chemistries etch tungsten with selectivity to SiO_2. Two tungsten etch processes, blanket W etchback to form a W plug and W pattern etch, are discussed briefly in this section.

To form a W plug, blanket LPCVD W is deposited on top of a TiN barrier layer as shown in Fig. 26. Reducing loading effects during the etchback of W is very important to avoid a W recess in the plug. A two-step process is used for W etchback. In the first step, 90% of the W is etched at a high etch rate and excellent uniformity to achieve high throughput. In the second step, the etch rate is reduced to remove the

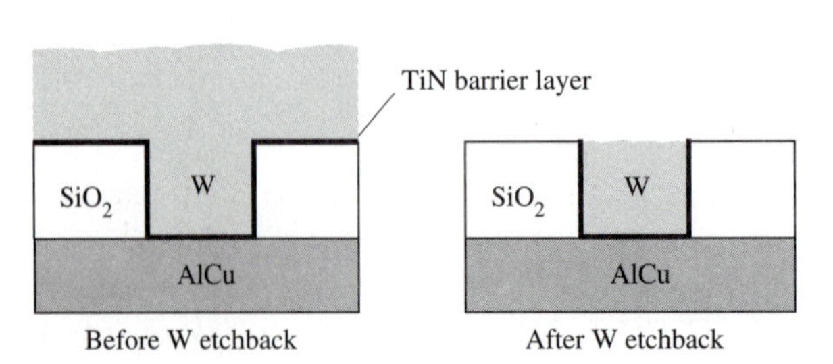

FIGURE 26
Formation of tungsten plug in a contact hole by depositing blanket LPCVD W and then using RIE etchback.

remaining W with an etchant that has a high W-to-TiN-barrier-layer etch selectivity and a minimum loading effect.[47] The loading effect is reduced by decreasing the gas pressure and wafer temperature because of the low chemical etch rate of F with W at these process conditions.

For W interconnect etching, the selectivity between W and the photoresist mask is very low for both fluorine- and chlorine-based chemistries. N_2 is added to the SF_6/Cl_2 mixture to get a slight selectivity increase.[48] Inorganic material, such as SiO_2, can be used as a hard mask during W etching with chlorine-based chemistry because of high etch selectivity of W to the hard mask.

7.5.5 Organic Materials Etching

Oxygen plasmas have been used with a barrel reactor for resist etching since the 1970s. Oxygen atoms are the primary species responsible for etching polymers at a pressure of 1–3 torr. Oxygen atoms initiate polymer etching by extracting hydrogen atoms from a polymer chain to form radicals. These radicals then react with oxygen molecules to form volatile products under high wafer-temperature conditions or under ion bombardment.

Recently, multilayer lithography has become increasingly important for ULSI processing because it provides high resolution that is independent of the underlying materials and substrate topography. The trilayer scheme is a very popular multilayer lithography method. It uses a conventional resist to pattern an intermediate layer, such as silicon dioxide or nitride, that acts as a mask during the subsequent planarization of the organic layer by oxygen RIE. A high degree of anisotropic and selective etching is required in these applications. This implies that both chemical and physical processes are needed, because purely physical etching is not selective, whereas purely chemical etching is not anisotropic. Reactive ion etching contributes both chemical and physical processes. To achieve anisotropic etching, sidewall passivation is required to suppress chemical and physical etching on the sidewalls of the feature.[49] A low wafer temperature and extra carbon- or chlorine-containing gases increase the deposition rate of sidewall films and improve etch anisotropy.[50]

7.5.6 Planarization Etch

With the advent of three or more levels of interconnection, more than 50% of the cost of chip manufacturing will be in the fabrication of electrical interconnects. Between these interconnection levels are inorganic insulators or polymer films, such as doped silicon dioxide glass or polyimide. Because the lithographic exposure achieves maximum resolution over only a limited depth of focus, these films must be flat, or planarized, over the optical exposure area to construct higher levels of interconnection. If the temperature budget does not allow planarization of the oxide layer by means of heat flow, planarization is sometimes accomplished by a repetitive sequence of plasma etching and deposition. An organic material with good flow properties is deposited on top of the oxide layer, and the wafer is etched in a plasma that removes the organic material and the oxide at an approximately equal rate, leaving a uniform oxide layer.

Mixtures of fluorine-based chemistry, such as NF_3, CF_4, and CHF_3, with oxygen are suitable for a nonselective silicon dioxide/organic material etch. Good etch uniformity and a high etch rate are the most important requirements for a planarization etch. The planarization etching step also reduces surface topography because different etch rates are used at different surface angles.

7.6 DIAGNOSTICS, END POINT CONTROL, AND DAMAGE

Dry etching differs from traditional wet chemical etching, and it usually does not have enough etch selectivity to the layer underneath. The plasma reactor must be equipped with a monitor that indicates when the etching process is to be stopped; an end point detection system is needed to avoid etching the underlying layer. Several different types of end-point systems are available. Most of them detect the change in the amount of active reactants or etch products in the discharge. Two of the most popular methods are introduced here. Some etching systems use a combination of these methods for end point detection.

7.6.1 Optical Emission Spectroscopy

Optical emission spectroscopy is a method of analyzing the light emitted by excited atoms and molecules in the gas discharge. The concentration of excited residual products or excited reactants during the etching process is studied by equipping a detector with a filter that lets light of a specific wavelength pass through to be detected, as shown in Fig. 27. Photons passing through a tunable monochromator are detected by a photomultiplier. The presence of neutrals and ions can be determined by correlating these spectra with previously determined spectral series. However, absolute concentrations cannot be obtained unless a complex actinometry technique is used. The emission signal derived from the primary etchant or by-product begins to rise or fall at the end of the etch cycle.

7.6.2 Laser Interferometry

Laser interferometry uses laser light directed toward the wafer surface. If the films on the wafer are transparent, the laser beams reflected from the top and bottom of the etched layer will interfere with each other. As the thickness of the etched layer is reduced, the degree of interference will change, either constructively or destructively. The time between light maxima and minima can be used to determine the etching rate. At the end of the etching process, the interference stops and the signal flattens out. The period of the oscillation is related to the change in film thickness, Δd where

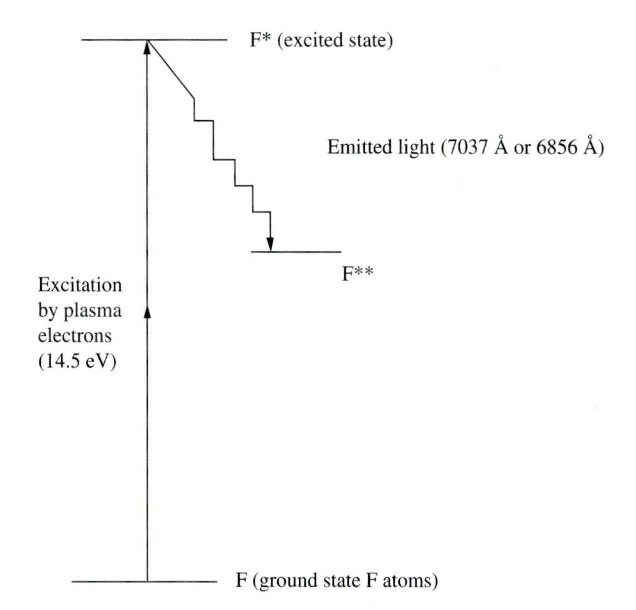

FIGURE 27
Mechanism of optical emission spectroscopy used for end-point detection during plasma etching.

$$\Delta d = \lambda/2n \tag{7.17}$$

λ is the wavelength of the laser light, and n is the refractive index of the etching layer. For example, Δd for polysilicon is 800 Å measured by using a helium-neon laser source for which $\lambda = 6328$ Å. Laser interferometry not only detects the end point but also determines the in-situ etch rate. The disadvantages of this method are that information is obtained from only a limited part of the wafer surface and the output signal is very weak if the etched surface is rough.

7.6.3 Plasma-Induced Damage

As stated above, plasma consists of energized ions, electrons, and excited molecules. When the excited particles recombine, they give off photons having an energy of a few eV. In addition, ion and electron bombardment may all contribute to damage mechanisms. Plasma-induced damage can take many forms, such as a trapped interface charge, material defects migration into bulk materials, and contamination caused by the deposition of etch products on material surfaces.

Etch damage can change the characteristics of sensitive components, such as Schottky diodes. For example, their rectifying capability can be reduced considerably. During oxide contact hole etching, heavy polymer deposition often causes high contact resistance. Sometimes the low-volatility etch products formed during plasma etching, such as silicon bromide from polysilicon etching with HBr, or $CuCl_x$ from AlCu etching with chlorine, are not completely removed after etching and cause device reliability problems. For example, corrosion can occur on dry-etched aluminum lines. Therefore, it is important that the impurities are removed and the damaged area restored before processing is continued. To reduce possible damage from contamination, wet etching is usually used to rinse wafers after plasma etching.

Charge buildup can occur in reactive ion etching because of severe ion bombardment of the wafer surface, and nonuniform plasma stimulates the accumulation of charges on the gate electrode. These stored charges can cause an electric breakdown of the thin gate oxide.[51] In addition, nonuniform electron and ion currents that do not balance locally cause charge buildup and oxide damage.[52] Therefore, it is increasingly important to find new processes and equipment that do not contaminate or cause damage in the underlying layers.

7.6.4 Plasma-Process-Induced Particle Contamination

Plasma processes can generate a large number of particles on the wafer surface during etching. These particles are created by chemical or mechanical means during plasma exposure. Laser light scattering shows that particles are trapped near the plasma/sheath interface during plasma etching.[53] Negatively charged particles are suspended at the plasma/sheath interface because of the electrical potential difference between the sheath boundary and the electrode/wafer surface. These particles range in size from less than a quarter of a micrometer to tens of micrometers at a density of over 10^7 cm^{-3}, and they can fall on the wafer surface when the plasma is turned off. Fluorine-containing gas mixtures produce fewer particles during silicon etching than chlorine- or bromine-containing gases do because fluorine generates etch products that have a high vapor pressure. Also, particle traps are formed by a change of electrode material or topography, which can increase the localized plasma potential. The tool design, especially the electrode, significantly influences the cleanliness of a plasma tool.

Particle contamination control includes the proper operation of plasma shutdown, appropriate process chemistry, and good tool design. Particles suspended near the plasma/sheath boundary can be removed before shutting the plasma process off by using the proper changes in rf power, gas flow, and magnetic field near the end of processing. Particles constitute the etch species in the plasma, so choosing the correct plasma chemistry reduces particle generation, and proper tool design can purge particles to the pump during processing.

7.7 WET CHEMICAL ETCHING

Wet etching is still the practical alternative for a high-throughput, flexible production process. With properly selected chemicals, etch reactions with the primary film are thermodynamically favored over reactions with the other films.[54] The etch-rate ratio usually approaches infinity. Wet etches are especially suitable for blanket etches of polysilicon, oxide, nitride, and metal. Recently, there has been a resurgence of wet etching for certain process steps because plasma etching fails to provide the required etch selectivity, damage-free interface, and particle-contamination-free wafers. Using robotic handling and ultrapure chemicals has improved particle control and process consistency. These improvements have revived wet etching processes for ULSI processing.

7.7.1 Wet Etching Mechanisms and Techniques

Etching of a solid in a solution is a heterogeneous process. Consecutive steps are thought to take place in the wet etch process. These steps are the diffusion of the reacting ions or molecules from solution, the adsorption of the reacting ions or molecules on the surface, the formation of a surface complex, the dissociation of this complex into reaction products, the desorption of the final reaction products, and the diffusion of the etch products into solution. The key criterion is that the etch products must be soluble in the etchant solution.

Most wet etch reactions involve oxidation-reduction. A chemical reaction is considered to be the transition of a system with an initial energy E_r to a final energy E_p. Heat of reaction ΔH is equal to $E_r - E_p$. A heat of reaction less than zero indicates that the reaction is favorable thermodynamically.

Two basic wet etching techniques are immersion etching and spray etching. The choice of one or the other depends on the material to be etched, the etchants, and the economic considerations. Immersion etching is the simplest technique. The masked or unmasked wafer is submerged in the etch solution, and mechanical agitation is usually required to ensure etch uniformity and a consistent etch rate. Spray etching requires less volume of chemicals and is faster than immersion etching. Good process control and etch uniformity are easily obtained from spray etching because fresh etchant is constantly supplied to the wafer surface while the etch products are continuously removed. Fully enclosed systems that eliminate the particle problems incurred when wafers are drawn out of a liquid/air interface are important for future ULSI processing.

7.7.2 Polysilicon and Silicon Nitride Etching

Most etchants may be classified as redox etchants. For silicon etching, H_2O_2 and HNO_3 are used to oxidize the silicon, and HF and HCl are used to dissolve the oxides. Selective isotropic chemical etching of polysilicon is the most widely used etching process for patterning deposited polysilicon films on the oxide. Concentrated KOH at 80 to 100°C is often used for selective polysilicon to oxide etching because of its low SiO_2 etch rate. Dopant concentrations and temperature may affect the etch rate of polysilicon.

Silicon nitride films are etchable at room temperature in concentrated HF or buffered HF and in H_3PO_4 at 150 to 200°C. The etch rate is strongly affected by the oxygen concentration in the nitride films. Selective etching of nitride to oxide is done with 85% H_3PO_4 at 180°C. The etch rate is typically 100 Å/min for CVD nitride, but only a few angstroms per minute for thermal oxide film. Plasma-enhanced CVD nitride has a much higher etch rate than CVD nitride. Sputter-deposited nitride films also have a different etch rate than CVD nitride films.

7.7.3 Silicon Dioxide Etching

SiO_2 can be etched by aqueous fluoride solutions, usually HF with or without the addition of NH_4F. The exact chemical mechanisms are extremely complex. The addition of NH_4F to HF controls the pH value and prevents depletion of the fluoride ions, thus maintaining stable etching performance. Silicon dioxide can also be etched in vapor phase HF. Vapor phase HF oxide etch technology looks promising for submicrometer feature etching because process can be tightly controlled and a desired hydrogen surface passivation can be precisely selected. The main reactions between oxide and HF vapor are

$$SiO_2 + 6HF \rightarrow H_2SiF_6 + H_2O \quad (18)$$

$$H_2SiF_6 + H_2O \rightarrow SiF_4 + 2HF \quad (19)$$

The etch rate of wet oxide etching depends on etchant concentration, agitation, and temperature. In addition, density, porosity, microstructure, and the purity of the silicon dioxide also affect the etch rate. CVD nitride etches are much slower than SiO_2 in HF solution, so an oxide-to-nitride selective wet etch is possible with HF solution.

7.8 SUMMARY AND FUTURE TRENDS

The etch selectivity and etch uniformity within the wafer become very important as film thickness is reduced, device dimensions shrink, and wafer size increases. Half-micrometer processes stretch conventional dry etching approaches to the limit. The first problem observed at quarter-micrometer generation is that conventional RIE will not achieve the clean, vertical sidewall etches demanded for that generation. The second problem is achieving uniformity across a wafer, because etch rates will vary with wafer surface topography. A related problem is known as microloading, in which varying degrees of etch behavior are triggered by alternating dense and sparse etching patterns. The final problem is plasma-induced damage. The concern is that electrostatic charging during the plasma etching process will damage the gate insulators. Process engineers are also concerned about driving unwanted ions and other contaminants into the silicon during etching.

High-density plasma sources with a lower electric potential difference between the plasma and the wafer will, therefore, become necessary to minimize the effects of ion-bombardment-induced damage. The solution lies largely in the high-density plasma sources in which plasma generation is wholly or partly separated from the processing region. The independent control of ion energy and plasma density promises a wider window for process optimization. Such approaches generate considerably denser plasma and avoid creating damaging, high-energy particles. The newer systems offer the necessary selectivity to the resist materials that protect the unetched portions of the wafer as well as to the underlying films that should not be etched. They also help prevent unwanted wafer depositions by using simple gas chemistry. In addition, the combination of a high-density plasma source and a short-gas-residence-time reactor design, i.e., a reactor equipped with high pumping capacity, is important to obtain a high etch rate.[55] Though such solutions appear viable, they remain to be proved in production. Etching will likely pose a major challenge at quarter-micrometer generation.

It is well known that many plasma processes drift from desired conditions and, therefore, require continual correction. Typically, the process drifts to a marginal performance before corrective measures are taken, leading to lost throughput and yield. Future etching systems will need a much higher degree of control and a self-diagnosis capability, so that high process stability can be maintained. Active feedback controls are needed to improve process reliability and reduce process maintenance. Currently, controls are used in plasma etching only for end point detection, with the process completed at a specified time after the end point. With automatic feedback control based on real-time detection of the plasma and wafer state, the optimal etch results could be maintained. Computer modeling of various aspects of plasma etching processes can increase our understanding of etch mechanisms and may enhance system design and process development. Clustered processing will become more popular and will be used to maintain the integrity of sensitive interfaces between thin films.

REFERENCES

1. R. A. Morgan, *Plasma Etching in Semiconductor Fabrication,* Elsevier, New York, 1985.
2. B. Chapman, *Glow Discharge Processes*, Wiley, New York, 1981.
3. J. L. Vossen, "Glow Discharge Phenomena in Plasma Etching and Deposition," *J. Electrochem. Soc.,* **126,** 319 (1979).
4. D. L. Flamm and V. M. Donnelly, "The Design of Plasma Etchants," *Plasma Chem. Plasma Process.,* **1,** 317 (1981).
5. G. S. Oehrlein and J. F. Rembetski, "Plasma-Based Dry Etching Techniques in the Silicon Integrated Circuit Technology," *IBM J. Res. Devel.,* **36,** 140 (1992).
6. C. J. Mogab, A. C. Adams, and D. L. Flamm, "Plasma Etching of Si and SiO_2: the Effect of Oxygen Additions to CF_4 Plasmas," *J. Appl. Phys.,* **49,** 3796 (1978).
7. L. M. Ephrath, "Selective Etching of Silicon Dioxide Using Reactive Ion Etching with CF_4–H_2," *J. Electrochem. Soc.,* **126,** 1419 (1979).
8. H. F. Winters and I. C. Plumb, "Etching Reactions for Silicon with F Atoms: Product Distributions and Ion Enhancement Mechanisms," *J. Vac. Sci. Technol.,* **B9,** 197 (1991).
9. J. W. Coburn and H. F. Winters, "Ion and Electron Assisted Gas-Surface Chemistry—An Important Effect in Plasma Etching," *J. Appl. Phys.,* **50,** 3189 (1979).
10. Y. T. Lii, H. Ng, and D. A. Danner, "Magnetic Enhanced RIE of Silicon Deep Trench," *Proceedings of the 8th Symposium on Plasma Processing,* The Electrochemical Society, Pennington, NJ, 1990, p. 462.
11. Y. T. Lii and J. Jorne, "Redeposition during Deep Trench Etching," *J. Electrochem. Soc.,* **137,** 2837 (1990).
12. H. Hubner, "Calculations of Deposition and Redeposition in Plasma Etch Processes," *J. Electrochem. Soc.,* **139,** 3302 (1992).
13. R. A. Gottscho and C. W. Jurgensen, "Microscopic Uniformity in Plasma Etching," *J. Vac. Sci. Technol.* **B10,** 2133 (1992).

14. Y. T. Lii and T. V. Rajeevakumar, "Quarter Micron Deep Trench Etch for ULSI," *Proceedings of the 9th Symposium on Plasma Processing,* The Electrochemical Society, Pennington, NJ, 1992, p. 328.
15. H. C. Jones, R. Bennett, and J. Singh, "Size Dependent Etching of Small Shapes," *Proceeding of the 8th Symposium on Plasma Processing,* The Electrochemical Society, Pennington, NJ, 1990, p. 45.
16. D. J. Econnomou and R. C. Alkire, "Effect of Potential Field on Ion Reflection and Shape Evolution of Trenches during Plasma Assisted Etching," *J. Electrochem. Soc.,* **135,** 941 (1988).
17. J. W. Coburn and H. F. Winters, "Conductance Consideration in the Reactive Ion Etching of High Aspect Ratio Features," *Appl. Phys. Lett.,* **55,** 2730 (1989).
18. C. J. Mogab and H. J. Levinstein, "Anisotropic Plasma Etching of Polysilicon," *J. Vac. Sci. Technol.,* **17,** 721 (1980).
19. G. Y. Yeom and M. J. Kushner, "Magnetic Field Effects on Cylindrical Magnetron Reactive Ion Etching of Si/SiO_2 in CF_4/H_2 Plasma," *J. Vac. Sci. Technol.,* **A7,** 987 (1989).
20. K. Suzuki, S. Okudaira, N. Sakudo, and I. Kanomata, "Microwave Plasma Etching," *Jpn. J. Appl. Phys.,* **16,** 1979 (1977).
21. C. Chen, M. Chang, C. Yang, and T. Ebata, "Electron Cyclotron Resonance Microwave Plasma Etching of Submicron Polysilicon Gate Structures," *Proceeding of the 8th Symposium on Plasma Processing,* The Electrochemical Society, Pennington, NJ, 1990, p. 368.
22. D. Dane, P. Gadgil, T. D. Mantei, M. A. Carlson, and M. E. Weber, "Etching of Polysilicon in a High-density Electron Cyclotron Resonance Plasma with Collimated Magnetic Field," *J. Vac. Sci. Technol.,* **B10,** 1312 (1992).
23. A. Hall, "Electron Cyclotron Resonance Etching of AlSiCu," *Sol. State Technol.,* **34(5),** 107 (1991).
24. P. Singer, "Meeting Oxide, Poly and Metal Etch Requirements," *Semic. Int.,* 50 (April, 1993).
25. J. H. Keller, J. C. Forster, and M. S. Barnes, "Novel Radio-Frequency Induction Plasma Processing Techniques," *J. Vac. Sci. Technol.,* **A11,** 2487 (1993).
26. K. P. Giapis, N. Sadeghi, J. Margat, R. Gottscho, and T. C. Lee, "Limits to Ion Energy Control in High Density Discharge: Measurement of Absolute Metastable Ion Concentration," *J. Appl. Phys.,* **73,** 7188 (1993).
27. A. S. Bergendahl, D. V. Horak, P. E. Bakeman, and D. J. Miller, "Cluster Tools 2. 16Mb DRAM Processing," *Semic. Intl.,* **13(10),** 94 (1990).
28. J. W. Coburn, "Plasma-assisted Etching," *Plasma Chem. and Plasma Process.* **2,** 1 (1982).
29. D. B. Graves, "Plasma Processing in Microelectronics Manufacturing," *AIChE J.,* **35,** 1 (1989).
30. L. Peters, "Plasma Etch Chemistry: The Untold Story," *Semic. Intl.,* **15(6),** 66 (1992).
31. H. B. Pogge, J. A. Bondur, and P. J. Burkhardt, "Reactive Ion Etching of Silicon with Cl_2/Ar," *J. Electrochem. Soc.,* **130,** 1592 (1983).
32. H. Crazzolara and N. Gellrich, "Profile Control Possibilities for a Trench Etch Process Based on Chlorine Chemistry," *J. Electrochem. Soc.,* **137,** 708 (1990).
33. G. Wohl, A. Weisheit, I. Flohr, and M. Bottcher, "Trench Etching Using a $CBrF_3$ Plasma and Its Study by Optical Emission Spectroscopy," *Vacuum,* **42,** 905 (1991).
34. L. Y. Tsou, "Highly Selective Reactive Ion Etching of Polysilicon with Hydrogen Bromide," *J. Electrochem. Soc.,* **136,** 3003 (1989).
35. M. Sato, S. Kato, and Y. Arita, "Effect of Gas Species on the Depth Reduction in Silicon Deep-Submicron Trench Reactive Ion Etching," *Jpn. J. Appl. Phys.,* **30,** 1549 (1991).

36. Y. T. Lii, C. M. Reeves, D. A. Danner, P. J. Coane, and L. K. Wang, "MRIE 0.1 μm Polysilicon Lines by Using HBr," *Proceedings of the 9th Symposium on Plasma Processing,* The Electrochemical Society, Pennington, NJ, 1992, p. 334.
37. M. Nakamura, K. Iizuka, and H. Yano, "Very High Selective n^+ Poly-Si RIE with Carbon Elimination," *Jpn. J. Appl. Phys.,* **28,** 2142 (1989).
38. J. W. Coburn, "Plasma-Assisted Etching," *Plasma Chem. and Plasma Process.,* **2,** 1 (1982).
39. H. Stocker, "Selective Reactive Ion Etching of Silicon Nitride on Oxide in a Multifacet ("HEX") Plasma Etching Machine," *J. Vac. Sci. Technol.,* **A7,** 1145 (1989).
40. Y. Kawamoto, T. Kure, N. Hashimoto, and T. Takaichi, "Highly Selective Dry Etching Si_3N_4," Electrochem. Soc. Meeting, 84-2, abs. 395 (1984).
41. K. Fujino and T. Oku, "Dry Etching of Al Alloy Films Using HBr Mixed Gases," *J. Electrochem. Soc.,* **139,** 2585 (1992).
42. D. W. Hess, "Plasma Etching of Aluminum," *Sol. State Technol.,* **24,** 189 (1981).
43. M. Sato and Y. Arita, "Al-Cu Alloy Etching Using In-Reactor Aluminum Chloride Formation in Static Magnetron Triode Reactor Ion Etching," *Jpn. J. Appl. Phys.,* **32,** 3013 (1993).
44. P. Riley, S. Peng, and L. Fang, "Plasma Etching of Aluminum for ULSI Circuits," *Sol. State Technol.,* **36(2),** 47 (1993).
45. D. S. Fischl and D. W. Hess, "Plasma Enhanced Etching of Tungsten and Tungsten Silicide in Chlorine Containing Discharges," *J. Electrochem. Soc.,* **134,** 2265 (1987).
46. N. Mutsukura and G. Turban, "Reactive Ion Etching of Tungsten in SF_6–N_2 Plasma," *J. Electrochem. Soc.,* **137,** 225 (1990).
47. I. Miller, R. Frazier, and M. Su, "Controlling Tungsten Etchback on Submicron Devices," *Microelectron. Manuf. Technol.,* 28 (Jan., 1992).
48. R. Rossen, "Magnetically Enhanced Patterned Tungsten Etching," *Microelectron. Manuf. Technol.,* 17 (March, 1991).
49. W. Pilz, J. Janes, K. P. Muller, and J. Pelka, "Oxygen Reactive Ion Etching of Polymers—Profile Evolution and Process Mechanisms," *SPIE, Vol. 1392, Advanced Techniques for Integrated Circuit Processing,* Society of Photo Instrumentation Engineers, 1990.
50. T. Kure, H. Kawakami, S. Tachi, and H. Enami, "Low Temperature Etching for Deep Submicron Trilayer Resist," *Jpn. J. Appl. Phys.,* **30,** 1562 (1991).
51. Y. Kawamoto, "MOS Gate Insulator Breakdown Caused by Exposure to Plasma," *Proceedings of the 7th Symposium on Dry Process,* Institute of Electrical Engineers, Tokyo, 1985, p. 132.
52. C. T. Gabriel and J. P. McVittie, "How Plasma Etching Damages Thin Gate Oxides," *Sol. State Technol.,* 81 (June, 1992).
53. G. S. Selwyn, J. Singh, and R. S. Bennett, "In Situ Laser Diagnostic Studies of Plasma-Generated Particulate," *J. Vac. Sci. Technol.,* **A7,** 2758 (1989).
54. J. L. Vossen and W. Kern, *Thin Film Processes,* Chapter V-1, Academic Press, New York, 1978, p. 401.
55. K. Tsujimoto, T. Kumihashi, and S. Tachi, "Novel Short-Gas-Residence-Time Electron Cyclotron Resonance Plasma Etching," *Appl. Phys. Lett.,* **63,** 1915 (1993).

PROBLEMS

1. Assume that the sheaths are pure capacitors in the plasma model and that positive ions of mass m_i come from the plasma and traverse the sheaths without colliding. The flux i_i is

$$i_i = \frac{KV^{3/2}}{m_i^{1/2} D^2}$$

Derive the equation.

2. What are the major distinctions between traditional reactive-ion etching and high-density plasma etching (ECR, ICP, etc.)? Compare the advantages and limitations of these techniques.

3. Explain why end point detection techniques, such as optical emission and laser interferometry, can detect etch uniformity across a wafer surface?

4. The main etching reaction of chlorine plasma with Si is

$$Si + 4Cl \rightarrow SiCl_4$$

If 90% of input chlorine is consumed by silicon etching, the gas pressure of the etcher is 10 mtorr. In order to achieve a 1 μm/min silicon etch rate for an eight-inch wafer, what is the minimum pumping capacity required?

5. A multiple-step etch process is required for etching a polysilicon gate with thin gate oxide. How do you design an etch process that has no micromasking, has an anisotropic etch profile, and is selective to thin gate oxide?

6. Explain how the mechanisms of using extra Al in the RIE chamber can reduce residue formation during AlCu etching with Cl_2 plasma?

7. What are the major distinctions between bromine-based chemistry and chlorine-based chemistry for Al etching? Compare the advantages and limitations of these chemistries.

8. Based on your understanding of this chapter, briefly summarize the requirements of etching ULSI.

CHAPTER 8

Metallization

R. Liu

8.1 INTRODUCTION

Metallization is the process that connects individual devices together by means of microscopic wires to form circuits. These wires are necessary for a circuit to function, but they introduce parasitic resistance and capacitance, which degrade the performance of devices. Often these wires carry electrical currents in the milliampere range, resulting in very high current density (10^5 A/cm^2) that threatens to break the submicron-size interconnects. Thus, as devices scale down with ever-improved output, the loss within the metal interconnects dissipates a greater portion of the improved performance. Therefore, the essence of metallization practice is to minimize the undesirable effects.

In this chapter metallization issues and techniques will be discussed. The introductory section discusses the major issues. Sections 8.2, 8.3, and 8.4 examine the key metallization processes. Section 8.5 concentrates on the two key techniques for multilevel metallization—interlayer dielectric and global planarization. Section 8.6 covers some aspects of metallization reliability not included in Chapter 12, and Section 8.7 presents a summary and discusses the future materials and processing needs.

8.1.1 Parasitics in Device Performance

The switching speed of a MOSFET is determined by how fast the gate capacitor, plus all the parasitic capacitors, can be charged. Since $Q = CV$ and $i = dQ/dt = C(dV/dt)$, the rate of charging is dominated by (1) the transconductance, which determines the current output of the MOSFET, and (2) the total capacitance. To improve switching speed one has to reduce the total capacitance and maximize the transconductance of the MOSFET.

Figure 1*a* shows the principal parasitics of a MOSFET. C_g is the gate capacitance, C_{os} and C_{od} the overlap capacitance of the source and the drain with the gate, and C_{js} and C_{jd} the junction capacitance of the source and the drain, respectively. Note that C_{os} and C_{od} are part of the physical structure of C_g, but because they do not contribute to the channel current, they are purely parasitic. The effective gate capacitance can be written as $C'_g = C_g + C_{os} + C_{od}$. For a device with a 0.35 μm gate length $C'_g \approx 1.15C_g$, or C_{os} and C_{od} together contribute about 15% of the gate delay.

Junction capacitance is determined by the substrate doping concentration and the junction area. This capacitance is easily estimated using $C_j = \epsilon A/W$, where W is the depletion width given by $W^2 = 2\epsilon V/qN_d$; $\epsilon = 1.05\times10^{-12}$ F/cm, the dielectric permittivity of Si; and N_d, the substrate doping (Chapter 9). For $N_d = 10^{17}$ cm^{-3}, one can estimate a $C_j \approx 3 \times 10^{-16}$ F/(μm)^2, or 0.3 fF/(μm)^2. Although this looks like a very small capacitance, it is not insignificant compared to the gate capacitance C_g. For example, for a MOSFET with a gate length of 0.35 μm and a width of 1 μm, $C_g \approx 1.5$ fF. In this case, the junction capacitance of the source and the drain each contribute a parasitic of about 20% of C_g. The larger the total gate capacitance C'_g and the larger the junction capacitance C_j, the slower the MOSFET can switch on and off. The gate capacitance C_g, however, cannot be reduced because the output current of the MOSFET is proportional to C_g. To improve the switching speed, the parasitic capacitance must be minimized.

The transconductance of the MOSFET can be improved by reducing the total resistance. Figure 1*b* illustrates the resistive components in a MOSFET. The largest resistive component is the channel resistance, R_{ch}, which is similar to the case of capacitive components. The channel resistance is determined by the mobility of the

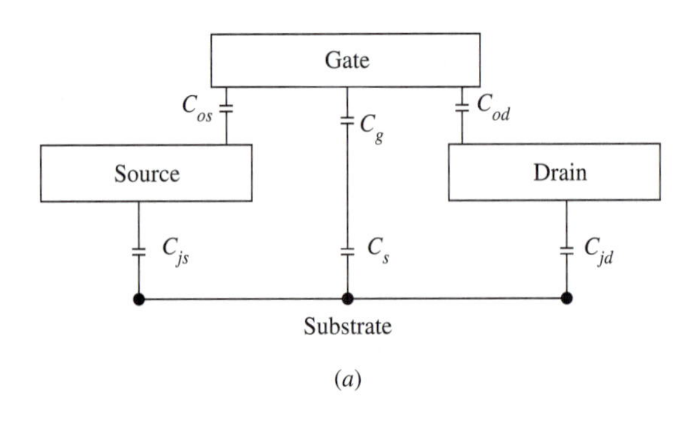

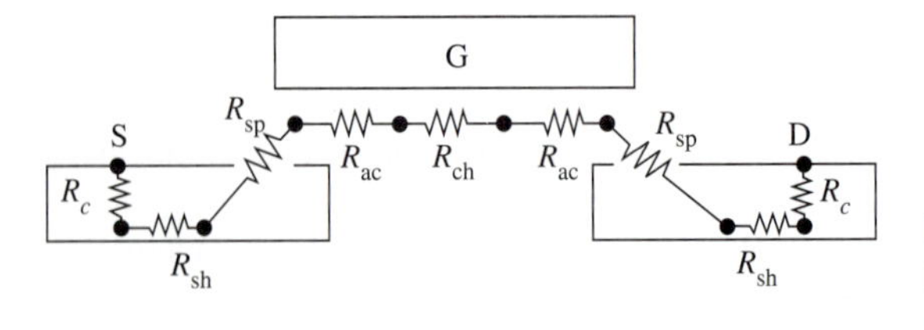

FIGURE 1
Parasitics in a MOSFET: (*a*) Capacitive parasitics; (*b*) resistive parasitics.

carrier in the channel and is part of the device design optimization. All other resistive components are parasitics. The most important are[1] accumulation and spreading resistance $R_{ac} + R_{sp}$, the diffusion sheet resistance R_{sh}, and the contact resistance R_c. The accumulation resistance R_{ac} and the spreading resistance R_{sp} together represent the resistance to the current flowing from the source into the channel inversion layer and then spreading out into the drain. Combined, they make a large contribution to the parasitic resistance, but they are not affected by metallization. R_{sh} and R_c, on the other hand, are affected by junction design and contact metallization. For MOSFETs with gate length L larger than 0.35 μm, R_c seldom reaches 10% of channel resistance. As devices continue to shrink, however, R_c increases because the contact area shrinks. Consequently, at $L < 0.25$ μm, R_c can contribute significant resistance (Section 8.1.3).

Without parasitics, the intrinsic switching speed of a MOSFET can be estimated by $\tau \approx C_g V / I_{Dsat}$, where C_g is the gate capacitance; V, the power supply voltage; and I_{Dsat}, the saturation current. For an nMOSFET with $L = 0.35$ μm operating at 3.3 V, $\tau \approx 12$ ps, corresponding to a cutoff frequency $f_T \approx 50$ GHz. Adding the (mostly capacitive) parasitics, the speed degrades by about 60% to 20 ps.

Further scaling of devices will decrease C_g, C_{os}, and C_{od}, as well as C_{js} and C_{jd} by about $1/S$, where S is the scaling factor. Parasitic resistance, however, will increase as contact size shrinks and decreases the transistor current drive. In addition, the power supply voltage scales down and further reduces the output current. Therefore, without reducing the parasitic capacitance and resistance, higher switching speed will be difficult to achieve.

8.1.2 Parasitics in Circuit Performance

The switching speed of the transistor sets the upper limit of the circuit speed. The speed of a circuit, however, is determined by the capacitive and resistive loads on the transistors and is thus only proportional to the switching speed of the transistors. In the simplest case, a CMOS inverter is connected to another identical inverter, as shown in Fig. 2*a*. Figure 2*a*(*i*), (*ii*), (*iii*) shows a simple inverter circuit, two inverters connected by a wire of length L, and the resistance and capacitance of each component.

The switching delay of a single inverter can be estimated from the total capacitance of the circuit. The width of the pMOS is usually about twice that of the nMOS so as to match the output current. Therefore, the total capacitance is $\sim 3C_n$, where $C_n = C_{os} + C_{od} + C_{js} + C_{jd} + C_g$ is the capacitance of an nMOS. The switching delay of the inverter is then $t_{in} \approx 3\tau$, or about three times the switching speed of the nMOSFET (with parasitics). The capacitance of the wire adds another delay to the signal before it reaches the next inverter. This delay is given by $t_w = C_w V / i$, where C_w is the wire capacitance. The wire capacitance can be calculated using $C_w = \epsilon_{ox} L w k / d$, where L, w are the length and the width of the wire, respectively; d is the thickness of the oxide separating the wire from the ground plane; $k \approx 3$, a geometry factor accounting for fringe fields; and $\epsilon_{ox} = 0.345$ pF/cm, the dielectric permittivity for SiO_2. For $w = d(= 0.5$ μm$)$, we can calculate $C_w \approx 1000$ fF/cm.

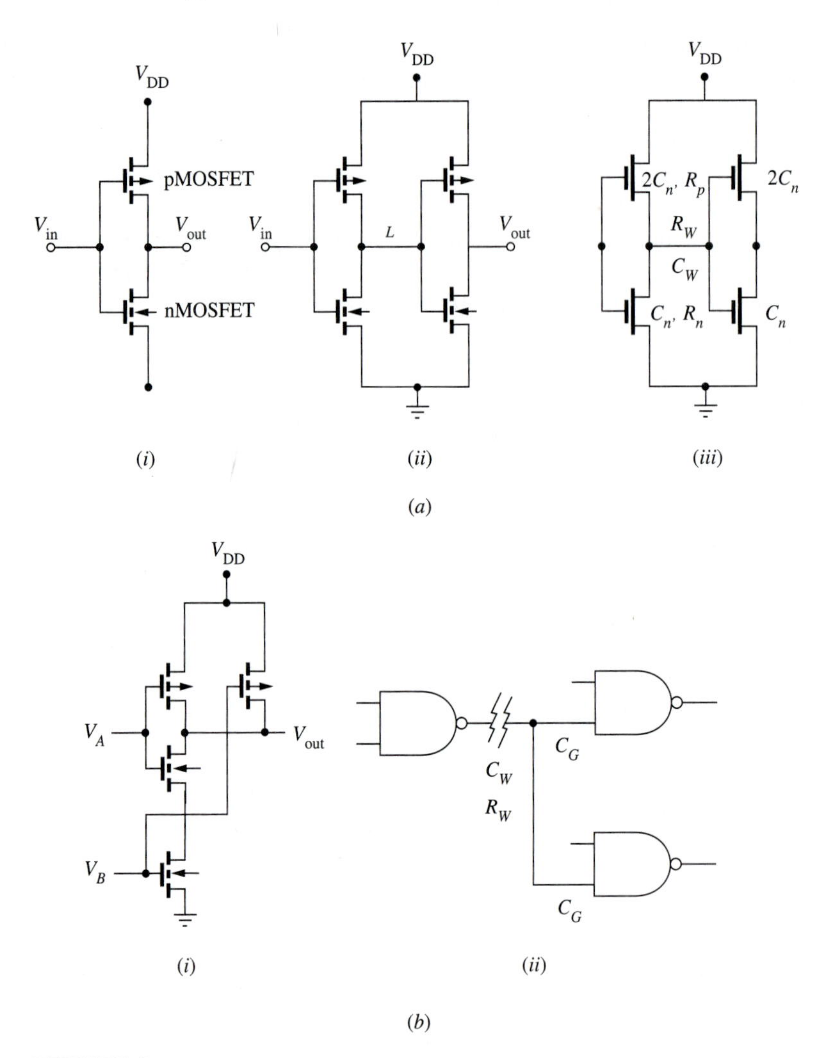

FIGURE 2
Parasitics in simple circuits: (*a*) CMOS inverters: (*i*) a CMOS inverter; (*ii*) two inverters connected by a wire of length *L*; (*iii*) capacitive and resistive components. (*b*) NAND gates: (*i*) a two-input NAND gate; (*ii*) a NAND gate connected through a long wire to two more NAND gates (fan-out 2).

The total capacitance of the inverter and the wire is $C = C_w + 3C_n$. For a minimum dimension nMOS the transistor width is about 0.5 μm, which gives a $C_n \approx 1.2$ fF. If the wire is 10 μm in length, then $C_w \approx 1$ fF, and the total capacitance is 4.6 fF. The current output of the nMOS is about 0.2 mA at 3.3 V, which gives a switching delay of $t = CV/i = 77$ ps. About 17 ps of the 77-ps delay is due to the capacitance load of the wire. The output of the second invert is further delayed by its capacitance. The total delay of the two inverters is then 77 ps + 60 ps = 137 ps. Without the wire capacitance, the delay would be 120 ps. Therefore, the capacitance of a 10-μm wire contributes about a 10% delay.

If the wire length L connecting the two inverters is 1 mm instead of 10 μm, then C_w = 100 fF, and the total capacitance becomes 107.2 fF. The capacitive load is now dominated by the wire capacitance, and the switching delay becomes $\Delta t = CV/i = 1769$ ps = 1.769 ns. It is clear that to drive a large capacitive load, more current from the driving circuit is needed to obtain high-speed performance.

Figure 2*b* shows the parasitic capacitance and resistance seen by a logic gate. The two-input NAND gate requires two nMOSFETs and two pMOSFETs and thus has a total capacitance of 6 C_n. As this logic gate drives other logic gates at the end of a long wire (1 to 2 mm in length), the switching speed of the gate is much reduced. The current from the driver gate not only has to charge up the equivalent capacitance of 12 C_n but also has to charge up the extra capacitance of the long wire. Figure 3 shows the effects of *RC* delay and *IR* drop on the switching speed. The switching speed as a function of wire length without considering the effect of *IR* drop in the line is plotted in Fig. 3*a*. Note that even when the *RC* load on the wire is ignored, the switching speed of the logic gate is still about a factor of 10 slower than that for a MOSFET. This is unavoidable, since the extra load from the fan-out gates requires the charging of a larger total capacitance.

With the presence of a long wire, both a capacitive load C and a resistive load R_L are added. The resistive load R_L, however, is usually much smaller than the channel and the parasitic resistance of the nMOS and pMOS. For example, for a 1-mm Al wire 0.5 μm × 0.5 μm in cross section, $R_w \approx 120\ \Omega$, whereas the resistance of a 1-μm wide pMOSFET is > 5 kΩ. Therefore, R_L has little effect on circuit delay.

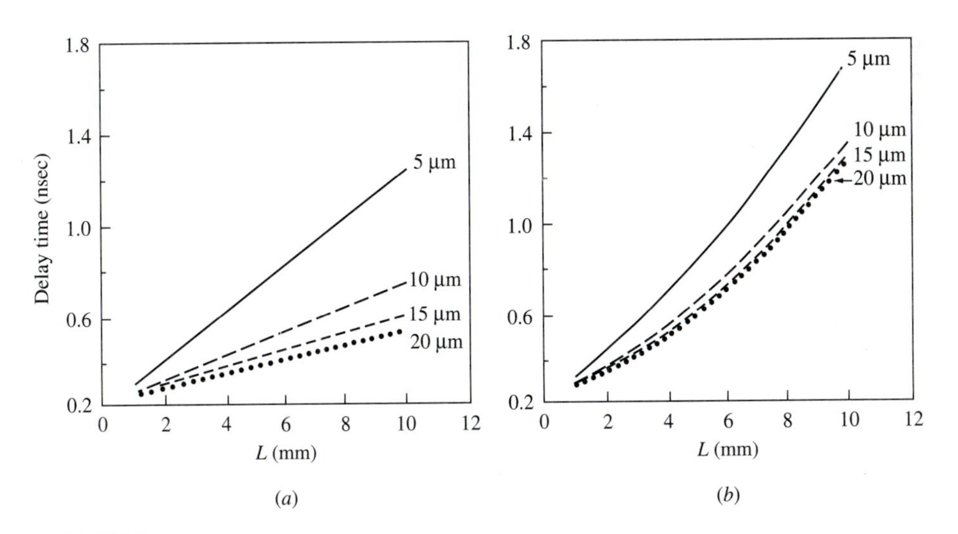

FIGURE 3
RC delay time for a CMOS inverter driving a metal line with length L and three fan-outs. The metal line width = 0.5 μm, and the dielectric thickness = 0.5 μm. Labels are for the transistor width of nMOSFETs. The width of pMOSFETs is 2× that of nMOS. (*a*) Delay time when no line resistance is considered. (*b*) Delay time when the 60 mΩ/□ Al line resistance is considered. Note that the benefit of wide transistors is diminished owing to the *IR* drop. (*Courtesy of H. I. Cong.*)

When only the capacitive load of the long wire is considered, the switching delay increases linearly with the length of the wire, since the capacitance increases linearly with the wire length. To maintain a fast switching speed, one can send more current down the wire by increasing the width of the driving transistor. Thus, when the *IR* drop along the wire is ignored, it appears that the *RC* delay can be compensated by a higher current from wide transistors.

Consider the situation when the *IR* drop along the wire cannot be ignored. The *IR* drop along the wire adds a nonlinear component to the curve shown in Fig. 3*a*, now replotted in Fig. 3*b*. The reason for this nonlinearity is that the *IR* drop depends linearly on the wire length and, as a result, the delay time now increases quadratically with the wire length. Even more significant is the diminishing benefit of wide transistors. The higher current provided by the wide transistor now brings a larger *IR* drop along the wire, which results in a drop in the drain voltage seen by the driver transistor. With a large enough *I*, the *IR* drop reduces the drain voltage of the driver transistor sufficiently so that it operates in the linear region and behaves like a resistor. Once this condition is reached, a further increase in transistor width only keeps the transistor operating further in the linear region with no increase in its current output. Thus, when the wire resistance becomes significantly high, the *IR* drop imposes a self-limit to circuit speed.

8.1.3 Contact Issues

Ideally, the metal wires make contact with the semiconductor device without adding parasitic resistance. In practice, there is always some contact resistance and the best one can achieve is to minimize it. A common practice is to dope the Si or polysilicon heavily to produce a low-resistance ohmic contact. The heavy doping in Si reduces the thickness of the Schottky barrier and allows carriers to tunnel through the barrier.[2]

The quality of a metal-semiconductor contact is measured by the specific contact resistance, ρ_{sc}, which has dimensions of Ω-cm^2. Figure 4 shows the specific contact resistance on Si as a function of Si doping.[2] The resistance of a metal wire contacting the source, drain, or gate of a MOS device is simply ρ_{sc}/A, where A is the contact area. The specific contact resistance of a metal to n$^+$ Si may reach $\sim 1\times10^{-7}$ Ω-cm^2, whereas that to p$^+$ Si seldom goes below 2×10^{-7} Ω-cm^2. Thus, for a contact window of 1 μm to n$^+$ Si the contact resistance is $\sim$ 13 Ω. This is insignificant compared with the channel resistance of several kΩ.

Consider, however, a MOSFET with a channel length of 0.18 μm and with contact windows of 0.2 μm in diameter. The channel resistance is about 1 kΩ/μm, but now the contact resistance is $\sim$ 300 Ω. With these design rules the contact resistance is a significant parasitic that will reduce the performance of the MOSFET.

An innovative way to reduce the contact resistance is to increase the metal-semiconductor contact area without increasing the contact window size. This is done by metallizing the entire source/drain and gate area with a self-aligned silicide, or *salicide* (Section 8.3). The metal wiring now contacts the silicide instead of Si. Because a metal-to-metal contact usually has low specific contact resistance, $< 1 \times 10^{-8}$ Ω-cm^2, the contact resistance can be reduced by more than an order of

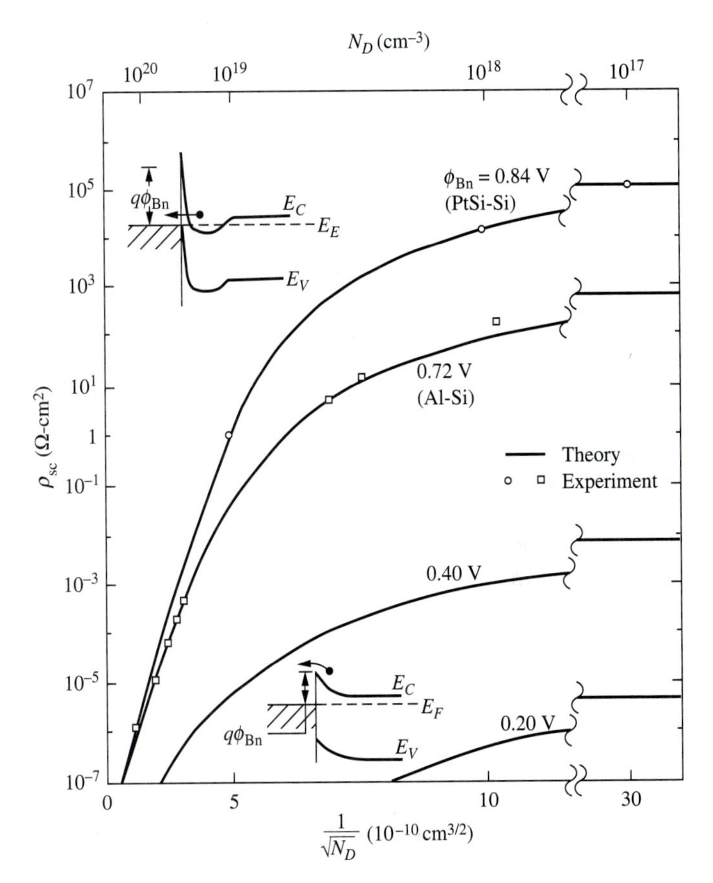

FIGURE 4
Specific contact resistance to Si. (*From Ref. 2 with permission.*)

magnitude without increasing the contact size. The use of silicides, however, causes other complications, which will be discussed in Section 8.3.

The specific contact resistance of metal-semiconductor interfaces represents the lower limit of contact resistance. In practice, both etching residues in the contact windows and damage of the Si by contact window etching may cause contact resistance 10 to 100 times higher than this limit. To minimize contact resistance, details of contact reactive ion etching (RIE) and contact cleaning are painstakingly designed to alleviate the etching damage and eliminate residues.

Vias, which connect different levels of metal wirings, also are affected by etching damage and residues but to a lesser degree. However, the tolerance for contact resistance in the via level is much smaller (usually $< 1\ \Omega$). Consequently, the via contact presents a challenge as difficult as that of the Si contact.

8.1.4 Multilevel Metallization

Nearly all MSI (medium-scale integration), LSI (large-scale integration), and earlier generations of VLSI circuits used only one level of Al wiring. Today, as well as in the foreseeable future, virtually all VLSI and ULSI circuits are made with two to five levels of Al or W wirings. Multilevel metallization provides greater flexibility in circuit design and, even more importantly, a substantial reduction in die size and thus chip cost. Curiously, multilevel metallization has been introduced reluctantly; many innovative ideas have been practiced to circumvent the use of more metal wiring layers.

The reluctance to adopt multilevel wiring is readily traced to the delayed development of processing and design tools. The development of a reliable, low-temperature, plasma-enhanced, interlayer dielectric (ILD) process, and the development of ILD planarization, prove to be the bottlenecks even today. In addition, the capability and user friendliness of CAD tools for logic and custom circuits continue to lag behind the need for faster multilevel wiring design.

Figure 5 compares a two-level metal and a three-level metal design of a segment of a standard cell ASIC (application-specific integrated circuit). The details of the standard cells are irrelevent to this discussion and are not shown. A significant advantage of the third level of metal is to reduce the routing channels between cells because routing in the horizontal direction can now be done over the cell. Other types of nonmemory circuits can, in general, realize a similar advantage with the additional level of wiring. The reduction in chip area is 20 to 35%, depending on the particular design.

The performance of the circuit is increased because of the smaller die size and the added design flexibility. The economics of an additional level of metal, however,

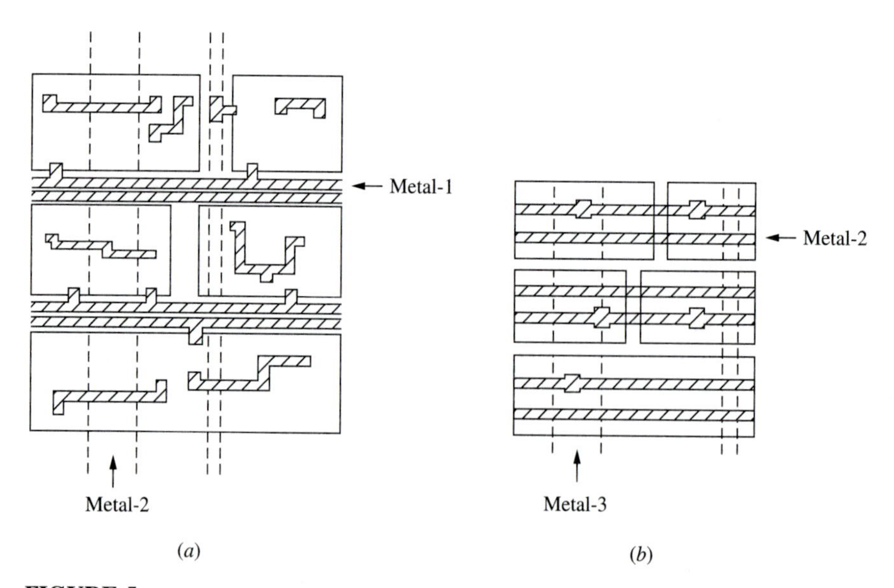

FIGURE 5
Two-level metal and three-level metal routing of wires for a segment of standard cell design: (*a*) 2LM requires routing channels between functional cells; (*b*) 3LM can route over cells and reduce area by ~ 30%. Metal-1 (not shown) is used only for local interconnection within the cell.

are not always simple. For a simple CMOS process with, for example, 12 masks using two levels of metal, the addition of the third level of metal adds approximately 15% to the processing cost. If the new metal level does not result in lower yield, then the chip yield will increase because of die size reduction. This will more than compensate for the increased processing cost.

Each added level of metal, however, creates more processing difficulty. If there is a significant loss in yield, or expensive new processes are needed to achieve high yield, then the economic benefit is not ensured. Therefore, to utilize multilevel metallization both design and processing engineering are vital to achieve minimum die size, low cost, and high-yield processing.

8.1.5 Metallization Reliability

With each new level of integration the current output for the devices increases, yet the cross section of the Al wire decreases. In addition, these thinner wires are buried under more layers of other wiring and dielectric materials and are subjected to increased mechanical stress. The reliability of the Al wiring therefore becomes an important concern. The fundamental mechanism and parameters that affect electromigration and stress migration are discussed in Chapter 12. In Section 8.6 we will examine the impact of scaling and multilevel interconnect on the reliability.

8.2 METAL DEPOSITION TECHNIQUES

Metals are deposited by a variety of techniques but principally by physical vapor deposition (PVD) or by chemical vapor deposition (CVD). These techniques and their applications will be discussed in the following sections.

8.2.1 Physical Vapor Deposition

The most common forms of PVD are evaporation, e-beam evaporation, plasma spray deposition, and sputtering. Evaporation and e-beam evaporation were used extensively in earlier generations of MSI and LSI but have since been replaced by sputtering. Today, all VLSI and ULSI devices are fabricated by sputtered metal, and we will focus only on this method. Details of evaporation principles and techniques can be found in Refs. 3 and 4.

Sputtering prevailed because of (1) the high deposition rate afforded by modern cathode and target design, (2) the capability to deposit and maintain complex alloy compositions, (3) the capability to deposit high-temperature and refractory metals, (4) the capability to maintain well-controlled, uniform deposition on large (200 mm diameter and larger) wafers, and (5) the capability, in multichamber systems, to clean the contact before depositing metal. No other deposition technique can offer all of these advantages, and for Al alloys and contact metals, sputtering has been and probably will continue to be the technique of choice.

Fundamentals of the sputtering process

The basic sputtering process employs a plasma of Ar gas. The Ar plasma can easily be generated by a glow discharge, as shown in Fig. 6*a*. A glow discharge typically consists of several different dark and glowing regions. Near the cathode is a very thin layer of ion space charge, followed by a dark space or gap (the Crookes dark space) where most of the potential between the cathode and the anode drops, followed by a negative glow (the glow discharge), a negative dark space (the Faraday dark space), and positive columns at the anode. The Ar plasma (the negative glow region) is sustained by the secondary electrons generated from ion bombardment of the cathode. These electrons accelerate away from the cathode, acquiring nearly the full energy of the dc potential across the electrodes when crossing the cathode dark space, and collide with and excite the Ar atoms in the glow discharge region, creating the plasma. After several collisions, the electrons lose their energy and continue to

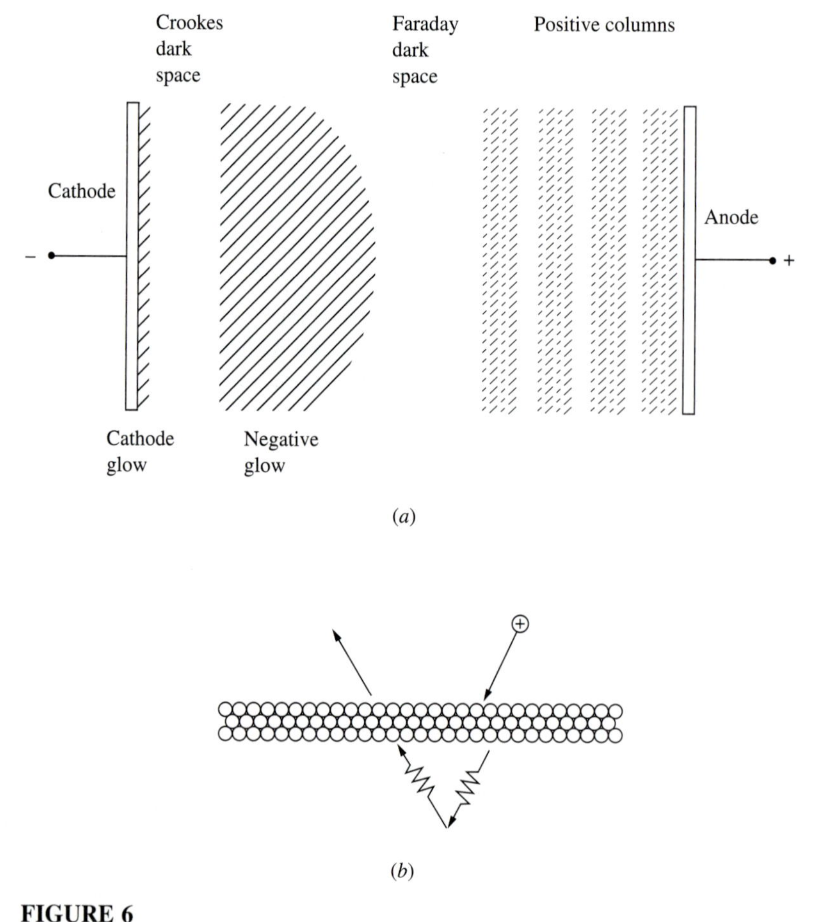

FIGURE 6
Sputtering process: (*a*) Characteristic regions of a glow discharge; (*b*) momentum transfer through phonon interaction.

drift toward the anode. Since they cannot excite any Ar atoms, nor can they create new electron-ion pairs, this region is dark (the Faraday dark region). After that, the electrons are gradually accelerated again by the anode, creating columns of faint glows and dark regions called the positive column. The alternating faintly glowing and dark regions are created by electrons acquiring enough energy to excite Ar atoms and then falling below the minimum energy level again.

Since most of the potential drop in a glow discharge is across the Crookes dark space, if one moves the anode forward beyond the position of the positive column the negative glow is virtually unaffected. As a result, virtually all sputtering systems are designed with the anode very close to the negative charge. The closer the target is to the wafer, the higher the deposition rate. The details of basic sputtering techniques can be found in Refs. 5 and 6.

The mechanism of sputtering is quite simple. The charged ions in the Ar plasma diffuse into the Crookes dark zone, acquire virtually all the energy from the voltage drop, and hit the cathode (target) surface. The momentum of the Ar ion transfers to the target material through phonon interactions. The result is the ejection of one or more atoms from the surface of the target, as shown in Fig. 6*b*. The ejected atom, being neutral, flies through the plasma (with a small chance of being ionized on the way) and lands on the wafer. The angular distribution of sputtered materials has been studied[6] and found to follow the cosine law. Therefore, like evaporation processes, sputter deposition occurs essentially along a line-of-sight path with a cosine distribution, resulting in poor step coverage if the surface topography of the wafer is abrupt.

The sputtering yield from the target varies widely with the material and with the energy and the incident angle of the ions. The sputtering yield for a number of materials peaks at incident angles of 45 to 60°. Although this does not have a direct bearing on the discussion of PVD, it provides the basis of the use of Ar sputter-etch, for example, for ILD processing (Section 8.5). Sputtering yield as a function of target material and ion energy can be found in Ref. 6.

In sputter deposition, the most important concern is to increase the ion bombardment rate on the cathode so that a reasonable deposition rate can be achieved. The glow discharge, on the other hand, relies on the secondary electrons from the target and is intrinsically inefficient. Consequently, means to increase the secondary electron production, as well as to increase the efficiency of ionization, are designed to improve the sputtering rate. One common design is the use of an $\mathbf{E} \times \mathbf{B}$ field to enhance the trapping of electrons. The cathode design using a magnetic field is called *magnetron.* The invention of magnetron sputtering greatly improved the sputter deposition rate and made sputtering the leading technique of PVD deposition.

Sputtering can also be achieved by applying an rf field instead of a dc field to create the plasma. In an rf field, both electrons and ions receive an ac field, but because of the high frequency only electrons can respond to the rf field. Consequently, the ions act as if the field were dc. The benefit of rf sputtering is that the electrons are oscillating with the field, resulting in no charge accumulation at the target even if it is an insulator. Therefore, rf sputtering is used for depositing insulating or semiconducting materials.

Figure 7 shows a magnetron sputtering system. The particular material to be sputtered is made into a disc that is thermally bonded to the cathode. Quite a large

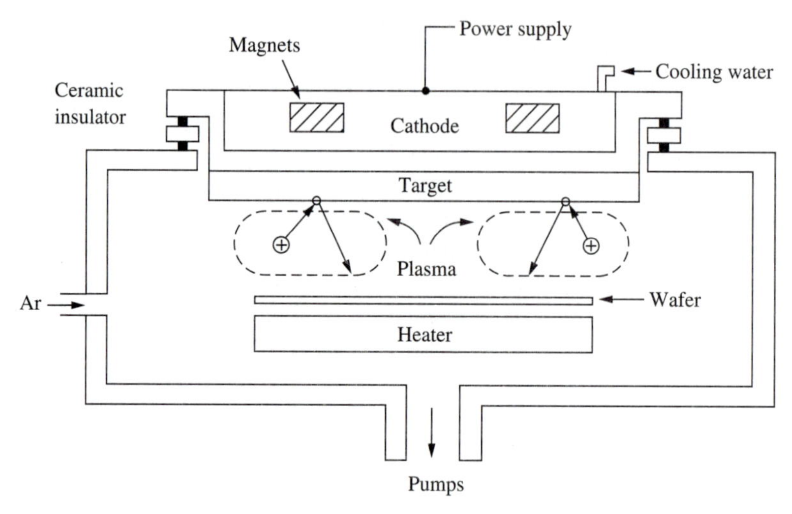

FIGURE 7
A dc-magnetron sputtering system.

amount of power is supplied to the Ar plasma to maximize the sputtering rate (in the range of 3 kW to 20 kW). Most of this power is absorbed by the sputtering target, which needs to be cooled through thermal contact with the cathode. Usually, the cathode is water-cooled. Apart from the high rate, another critical criterion for sputter deposition is the uniformity of the deposited film, especially on large wafers. To achieve high rate and uniformity, new sputtering cathodes are designed with rotating, rare earth, high-strength permanent magnets. This improved magnetron cathode is the basis for modern high-rate, single-wafer deposition systems.

Film properties

Sputtered film properties have been studied thoroughly.[7] The substrate temperature and the Ar pressure are the most important parameters. At low substrate temperature, the surface mobility of the sputtered atoms is low, resulting in porous films with poor physical properties. At higher temperatures with $T/T_m \approx 0.5$ to 0.7 (T_m is the melting temperature of the metal), surface diffusion occurs during sputtering and results in films with good physical properties comparable to the bulk. The films are columnar in this region because the grains grow unidirectionally. The grain size increases with increasing temperature because of lower density of nucleation at higher temperature.

At higher temperatures, $T/T_m > 0.7$, bulk diffusion occurs during deposition, and the film consists of equiaxially larger grains. This pattern is caused by recrystallization during film deposition. Although large-grained films are desirable (e.g., for electromigration resistance), in practice high-temperature sputtering has seldom been used until recently (see Section 8.4.3) because of equipment limitations. Most Al films are deposited between 200 and 300°C, which corresponds to $T/T_m \approx 0.5$ to 0.6. Good-quality Al films with grain size between 0.5 and 1.0 μm are achieved at these temperatures.

The Ar pressure plays an important role in the film stress and electrical and physical properties. During high-rate sputtering, Ar gas can be incorporated in the sputtered film[8] to nearly 1 at% (atomic percent). Ar incorporation changes the film properties substantially. Even worse, since Ar is not tightly bonded with metal, it can condense during subsequent thermal treatment into bubbles, resulting in blistering or delamination of the film.

Generally, better film properties are achieved at a higher substrate temperature and a lower Ar pressure. In practice, the rate of deposition is a function of plasma density, which in turn is a function of the Ar pressure. Therefore, until the higher-plasma-density magnetron, which allows high-rate sputtering at low pressure, was invented, high-quality films were hard to achieve. Today, with even more efficient magnetron designs, good film properties for Al are achieved routinely.

Alloy sputtering

For electromigration (EM) resistance, Cu is alloyed to Al to enhance the EM lifetime. For contacts, the high solubility of Si in Al causes the Al to spike into the Si substrate, leading to source/drain junction leakage. This is caused by Si diffusion from the source/drain to the Al metal to satisfy the high solubility of Si in Al at low temperature.[9] To prevent Al spiking, Si is often alloyed into Al to reduce Si diffusion. Therefore, in practice, Al is always used in alloy forms and not as a pure metal.

Sputtering is especially suitable for alloy deposition. It is known that the sputtering yield is a strong function of the target material. This is also true for alloy sputtering. It is generally difficult to equalize the sputtering rate of various elements. However, when an alloy target is used, the differential sputtering rate for different elements in the alloy results in a composition that is enriched in the element with the lower yield. Consequently, the exposed target surface has a higher concentration of the element of low yield. This self-compensating effect is important because it ensures that the film deposited on the wafer has essentially the same alloy composition as the target material.

Reactive sputtering

The ions in the Ar plasma generally have energy near or higher than the binding energy of the target material. Consequently, refractory metals can be readily sputter-deposited. In addition, compounds such as SiC, TiN, etc., which are difficult to deposit by other PVD methods, can be sputtered readily.

Conductive compounds such as TiN are especially important for ULSI application as barrier and glue metal layers[10] (see Section 8.4 on CVD W plug) and as antireflective coatings over Al alloys. TiN can be sputtered by two techniques—by using a compound TiN target or by reactive sputtering. Both are used in ULSI fabrication.

Reactive sputtering of TiN is done by introducing nitrogen into the sputtering chamber in addition to the Ar plasma. The plasma provides enough energy to dissociate the nitrogen molecules into atomic nitrogen, which subsequently can interact with Ti to form TiN.[11] The film properties of TiN vary widely with deposition conditions[12] as shown in Fig. 8. With no N_2 flow, pure Ti is sputtered, as expected. With increasing N_2 flow, the film resistivity increases until a maximum is reached

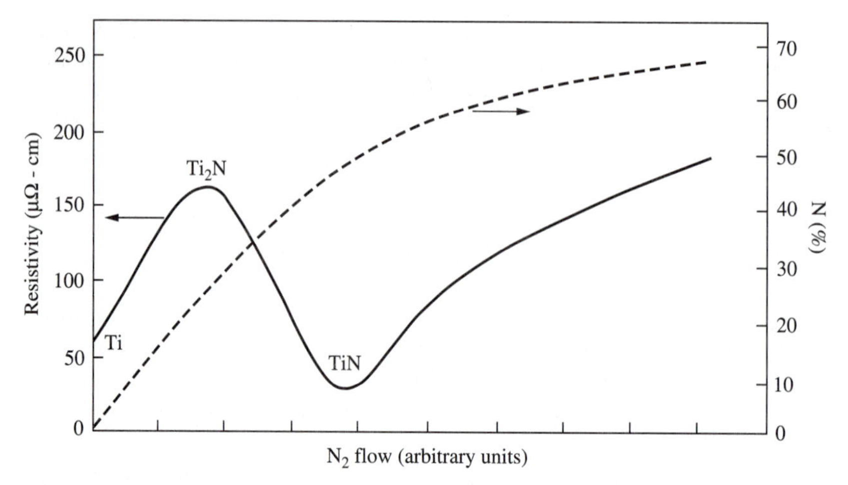

FIGURE 8
Properties of TiN as a function of N_2 flow rate.

at a Ti/N ratio of ~ 2 : 1. Then the resistivity decreases again and reaches a minimum when the Ti/N ratio is about 1. This is the condition when stoichiometric TiN is deposited. A further increase in the N_2 flow results in nitrogen-rich TiN with high resistivity.

TiN is a high-melting-point, hard metal. In addition, the thermal expansion coefficient of TiN is different from Si, and as a result TiN deposited at 300 to 400°C can have a high compressive stress in the range of 1000 MPa. This brittle, high-stress material naturally is a potential source for particle generation, e.g., from flaking from the walls of the sputtering chamber. Careful design of the sputtering equipment and a carefully controlled deposition process are the prerequisites for particle-free reactive TiN deposition.

Step coverage

During sputtering, the atoms ejected from the sputtering target follow the cosine distribution.[5] To avoid contamination from residue gas, sputtering requires high vacuum and high-purity Ar with a pressure of 1 to 10 mtorr. At this pressure the mean free path is on the order of several centimeters, or approximately the distance between the target and the wafer. Therefore, atoms ejected from the target experience few collisions with the ambient and fly to the wafer essentially following a line-of-sight path.

The line-of-sight trajectory of the atomic flux causes poor coverage on vertical or steeply slanting surfaces. For metal wiring this is undesirable, since the resistance of the wire is increased by the thin sections. An even more serious difficulty is deposition of metal into contact windows and vias. Because of the severe geometric restrictions, the sidewalls and the bottoms of contacts and vias receive only 10% or less of the metal deposited on the top surface. Consequently, for contacts with a high aspect ratio, the very thin metal in the contacts causes severe reliability problems.

The planar-type sputtering cathode improves the metal step coverage only marginally. Adding a collimator, thus artificially modifying the cosine distribution of the atomic flux, improves the bottom coverage in high-aspect-ratio contacts, but at the expense of worsened sidewall coverage. Increasing the Ar pressure, thus increasing the number of collisions, causes a deterioration of the film property. Higher substrate temperature improves the surface mobility of Al atoms and provides better step coverage, but the nucleation and growth steps need to be carefully controlled (see Section 8.4.3).

For critical contact application, the poor step coverage of sputtering eventually forces the adoption of CVD metal-plug processes.

8.2.2 Directional Enhancement—Collimation

The metal coverage at the bottom of a contact can be increased by employing directional enhancement. This is done simply by inserting a collimator between the sputtering cathode and the wafer.[13] The collimator replaces the Si wafer as the electrical ground for the plasma. Neutral species sputtered off the target at a high angle are then intercepted by the collimator and deposit on the side wall of the collimator instead of reaching the wafer. Consequently, only atoms that are ejected within a small angle to the normal of the target will reach the wafer. These atoms have a much higher chance of reaching the bottom of the contact hole than those with larger angles.

Figure 9*a* shows the bottom coverage obtained by collimated sputtering compared with that from conventional sputtering. Simulation results using a cosine source distribution and a line-of-sight path agree well with available experimental

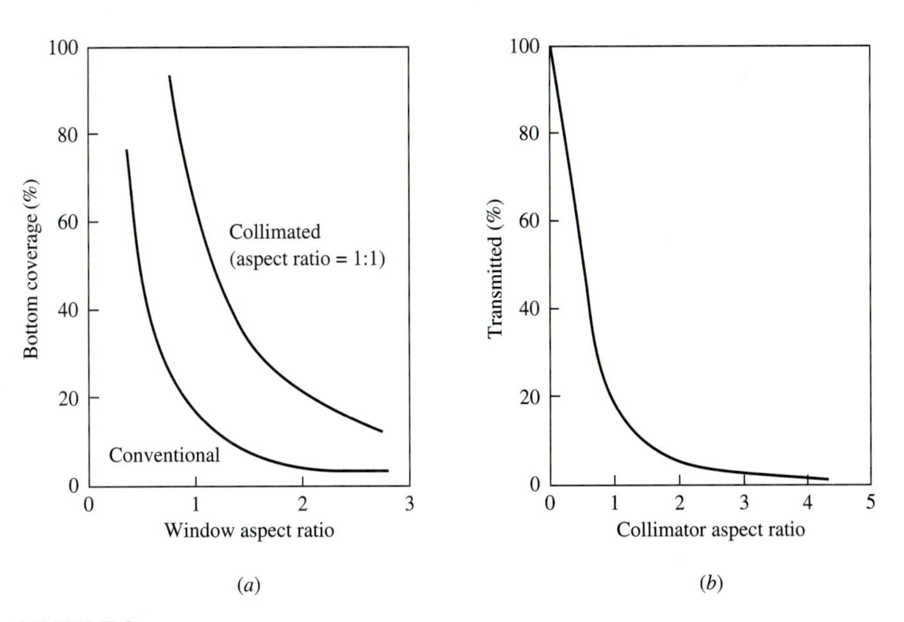

FIGURE 9
Properties of collimated sputtering: (*a*) Bottom coverage in windows; (*b*) transparency of the collimator as a function of its aspect ratio.

data, indicating that the collimator works simply as an interceptor that reduces the flux from large-angle ejectors. The improvement in the bottom coverage is significant,[13] from $< 10\%$ to $> 25\%$ for a contact hole aspect ratio of 2 : 1.[13,14]

The sidewall coverage, however, is degraded, as expected. The lack of sidewall coverage makes collimated sputtering undesirable for interconnect metal such as Al. But for contact and barrier material such as Ti and TiN, where bottom coverage is crucial, collimation provides a simple extension of an existing process and can be easily integrated into IC processing. As a result, collimated sputtering for Ti and TiN is quickly adopted in CMOS processing for 0.35-μm and 0.25-μm designs.

Collimation, however, also means discarding a large portion of sputtered material from the target, since only material ejected within a small angle passes through the collimator. Figure 9*b* shows the transparency of a collimator as a function of its aspect ratio. For a collimator aspect ratio (defined by length/diameter) of 1 : 1, approximately 70% of sputtered material is intercepted by the collimator, and for an aspect ratio of 1.5 : 1, nearly 90% is intercepted.[15] The low transparency of the collimator increases the cost of deposition because (1) the deposition rate is drastically reduced and (2) the target utilization is reduced and, therefore, both the material cost and the maintenance cost increase. Thus, even though collimation is easily retrofittable to IC processing, the higher cost makes this process less desirable if a low-cost alternative (such as CVD) can be developed.

8.2.3 Chemical Vapor Deposition

A number of metals and metal compounds can be deposited by chemical reaction or thermal decomposition of precursors. Techniques for depositing Al,[16–20] Cu,[21–24] WSi_2[25–27] TiN,[28–31] and W,[32–37] have all been developed and applied to devices and/or prototype circuits. However, except for WSi_2 and W, CVD metals are not widely used yet because of immature process and equipment development.

Metal CVD techniques

The nature of metal CVD is not different from Si or SiO_2 CVD. A precursor chemical containing the desired metal and a chemical reaction (including decomposition), which has the elemental metal or the metal compound as the final product, are the key ingredients. Usually the wafer needs to be heated to 100 to 800°C to provide the initial thermal energy to overcome the reaction barrier. The kinetics of the CVD process have been discussed in detail in Chapter 5, and the metal CVD follows the same general principle.

CVD metal can also be deposited in a plasma. The advantage of plasma deposition is that the energy required to overcome the initial reaction barrier can be supplied by the plasma, and the deposition temperature can be reduced. This reduction is especially important for metal or metal compounds used for vias and multilevel interconnects since their tolerance for high temperature is limited.

CVD W

The chemical vapor deposition of W and WSi_2 is widely used for circuit fabrication.[38] WSi_2 is used on top of gate polysilicon as a *polycide* structure and for local interconnect.[26,27] CVD W is used both as a contact plug and as a first-level metal.

The basic chemistry for CVD W is straightforward:

$$WF_6 + 3H_2 \rightarrow W + 6HF \tag{8.1}$$

$$2WF_6 + 3Si \rightarrow 2W + 3SiF_4 \tag{8.2}$$

$$WF_6 + SiH_4 \rightarrow W + SiF_4 + 2HF + H_2 \tag{8.3}$$

$$2WF_6 + 3SiH_4 \rightarrow 2W + 3SiF_4 + 6H_2 \tag{8.4}$$

In addition, WF_6 can be reduced by Al and Ti through

$$WF_6 + 2Al \rightarrow W + 2AlF_3 \tag{8.5}$$

$$2WF_6 + 3Ti \rightarrow 2W + 3TiF_4 \tag{8.6}$$

In practice, the process is more complex, with several intermediate reaction products such as WF_4 and WF_5 involved.[39]

During CVD W deposition, the wafer is held on a heated chuck that is between 400 and 500°C, opposite a shower head where the WF_6, H_2, or SiH_4 gases are injected. Usually a two- or three-step process is used. SiH_4 is introduced without any flow of WF_6 to initiate the deposition of a very thin layer (a few nanometers) of amorphous Si as a prenucleation layer. This step is followed by a $SiH_4 + WF_6$ silane reduction nucleation process, and then the high-rate $H_2 + WF_6$ hydrogen reduction deposition. At the nucleation stage less than 100 nm of W is deposited, and the bulk of the W deposition is by hydrogen reduction.

This complex procedure was developed to ensure that during the initial nucleation stage the Si from the source/drain area did not participate in the chemical reaction. As seen from Eqs. (8.2), (8.5), and (8.6), WF_6 reacts readily with Si, Al, and Ti. The free energies of the WF_6 reduction processes [Eqs. (8.1)–(8.6)] can be calculated from standard thermochemical tables.[40] The ΔG values of these reactions are shown in Table 1. It is clear that hydrogen reduction is the least favorable process, whereas reduction of WF_6 by Si, Al, and Ti are highly favorable. When WF_6 reacts with the Si from the source/drain region, it causes junction leakage.[41,42] Therefore, precautions must be taken (by introducing SiH_4 first) to avoid the consumption of Si from the substrate. WF_6 also reacts with lower-layer Al wiring and forms AlF_3 [Eq. (8.5)], which increases the via contact resistance.[43] The most common problem, however, is the reaction of WF_6 with the Ti glue layer [Eq.(8.6)]. This reaction causes the delamination of the TiN and "volcanoes." (See Section 8.4.)

TABLE 1
Free energy changes for CVD W deposition

Reaction	**ΔG at 500°K (KJ/g-atom W)**
$WF_6 + 3H_2 \rightarrow W + 6HF$	−113
$WF_6 + 1.5Si \rightarrow W + 1.5SiF_4$	−478
$WF_6 + 1.5SiH_4 \rightarrow W + 1.5SiF_4 + 3H_2$	−870
$WF_6 + 1.5Ti \rightarrow W + 1.5TiF_4$	−1037
$WF_6 + 2Al \rightarrow W + 2AlF_3$	−1162

Even with elaborate deposition procedures, junction leakage caused by the local attack of WF_6 can still be a problem. "Wormholes" with diameters of only a few nanometers have been observed to penetrate several hundred nanometers into the Si.[44] Consequently, a barrier metal layer such as TiN or TiW is almost always used before CVD W deposition. The barrier metal layer serves more than one function. In addition to reducing the junction leakage, this metal layer also provides adhesion of the CVD W on SiO_2 and prevents the generation of particles from poorly grown CVD W on SiO_2.

W deposited by a CVD process has an electrical resistivity 1.5 to 2 times that of the bulk W. This increase is due mainly to the incorporation of F, which can reach > 1 atomic percent, in the W film. As a result, CVD W has a resistivity 3 to 4 times that of Al and is unsuitable for interconnect purposes except for short runs. CVD W also grows into ragged morphology with poor reflectivity, even though the grain size is not large (usually < 200 nm), because the exposed W surface consists of the as-grown crystal facets with no preferred orientation, resulting in random reflection of light. Therefore, for a CVD W thicker than 200 nm, patterning of submicron W lines is difficult.

CVD TiN

TiN is widely used as a barrier-metal layer for CVD W deposition and can be deposited by sputtering from a compound target or by reactive sputtering. CVD TiN can provide a better step coverage than PVD methods and is potentially more economical than collimated sputtering.

CVD TiN can be deposited[28,45,46] using $TiCl_4$ and NH_3; H_2/N_2; or NH_3/H_2:

$$6TiCl_4 + 8NH_3 \rightarrow 6TiN + 24HCl + N_2 \tag{8.7}$$

$$2TiCl_4 + 2NH_3 + H_2 \rightarrow 2TiN + 8HCl \tag{8.8}$$

$$2TiCl_4 + N_2 + 4H_2 \rightarrow 2TiN + 8HCl \tag{8.9}$$

The deposition temperature is 400 to 700°C for the NH_3 reduction and is > 700°C for the N_2/H_2 reaction. Generally, the higher the deposition temperature, the better the TiN film and the less Cl incorporated in the TiN. However, even the best TiN produced by this process contains $\sim 0.5\%$ of Cl. The lower-temperature processes result in even higher ($\sim 5\%$) Cl in the TiN,[45] causing concerns about the corrosion reliability of the Al wiring. CVD TiN by $TiCl_4$ processes at near 700°C has good electrical resistivity, $\sim 100\ \mu\Omega$-cm or less, but has high stress, ~ 2000 MPa. Because of the high deposition temperature (≥ 650°C) these processes are unsuitable for via applications.

TiN can also be deposited by using metal-organic precursors.[47–50] These precursors are usually in the form of a Ti-alkylamine complex. One example is the use of tetrakis-(dimethylamido)-Ti [also written as TDMAT, or $Ti(NMe_2)_4$] and NH_3; the reaction[51] follows the $TiCl_4 + NH_3$ reaction closely:

$$6Ti[N(CH_3)_2]_4 + 8NH_3 \rightarrow 6TiN + 24HN(CH_3)_2 + N_2 \tag{8.10}$$

Similar reactions using tetrakis-(diethylamido)-Ti (TDEAT) have also been reported.[50]

TiN deposition by metal-organic precursors can be done at low temperature ($\leq$ 450°C) and imposes no danger of Cl incorporation. The low deposition temperature makes the metal-organic CVD (MOCVD) process suitable for use on all metal levels. However, the deposited films have lower densities and are unstable.[52] Because of carbon and oxygen inclusion the films have high resistivity, $\geq 500\ \mu\Omega$-cm, compared with the 40 to 100 $\mu\Omega$-cm for reactive sputtered TiN. In addition, the step coverage[52] for high-aspect-ratio vias is only marginal. As a result, low-temperature MOCVD TiN is not an ideal barrier material yet. TiN deposited by TDEAT at 425°C[50] has better step coverage and lower resistivity (180 $\mu\Omega$-cm). Recently, mixed amorphous/crystalline CVD TiN has been deposited[53] using TDMAT with good step coverage (~70%).

Low-temperature deposition of TiN using $TiCl_4$ and NH_3 with plasma assistance has also been reported.[29,54] Similar to plasma-enhanced dielectric processes, with an rf plasma TiN can be deposited at about 500 to 600°C. Plasma-enhanced CVD TiN contains lower Cl and has better physical and electrical properties than plain CVD. However, the Cl content is still in the 1% range, and the film properties degrade quickly below 500°C. Recently, high-density ECR (electron cyclotron resonance) plasma-enhanced deposition of TiN has been reported. By using a $TiCl_4$ precursor, good quality TiN with resistivity about 40 $\mu\Omega$-cm has been achieved at 400°C.[55] By using a TDMAT precursor, TiN has been deposited by downstream ECR plasma at < 400°C.[56,57] TiN with a resistivity of 200 to 300 $\mu\Omega$-cm and low C and oxygen content has been reported.

CVD Ti

Although an excellent barrier material, TiN directly in contact with Si causes high contact resistance. A Ti layer below the TiN is usually used to ensure good contact resistance. CVD of Ti, however, is difficult because Ti is highly electronegative and precursors that can reduce to elemental Ti are thermodynamically unfavorable. Plasma-assisted CVD is necessary to impart the extra energy needed. Recently, ECR plasma-assisted CVD of Ti has been reported.[55,58] At 400°C the ECR-enhanced CVD Ti reacted directly with Si to form $TiSi_2$. Both good contact resistance and junction leakage were observed.

CVD Cu

Copper can be deposited by CVD readily. Because of its low resistivity and good electromigration resistance, Cu metallization has received much attention recently. There are two types of metal-organic precursors for Cu: the divalent Cu^{II} and the monovalent Cu^{I} compounds. The Cu^{II} precursors have the form of Cu^{II} (β-diketonate)$_2$. The most commonly used Cu^{II} precursor is bis-hexafluoroacetyl-acetonate-Cu^{II}, or Cu(hfac)$_2$. The structure $(CF_3COCHCF_3CO)_2Cu$ is shown in Fig. 10. Cu(hfac)$_2$ is a solid at room temperature and sublimes at low heat (35 to 130°C). In a CVD system, the Cu(hfac)$_2$ compound is heated, and a carrier gas (Ar or H_2) is passed through the bed of Cu(hfac)$_2$ powder to bring the vapor into the reaction chamber. Although Cu(hfac)$_2$ can decompose directly without a reaction with another chemical, the deposition rate is low and the Cu film is inferior. Higher deposition rates and better quality films are achieved by hydrogen reduction:[27,58,59]

$$\mathrm{Cu(hfac)_2 + H_2 \rightarrow Cu + 2H(hfac)} \tag{8.11}$$

Although the Cu(hfac)$_2$ + H_2 reaction starts at ~175°C, it does not reach the surface

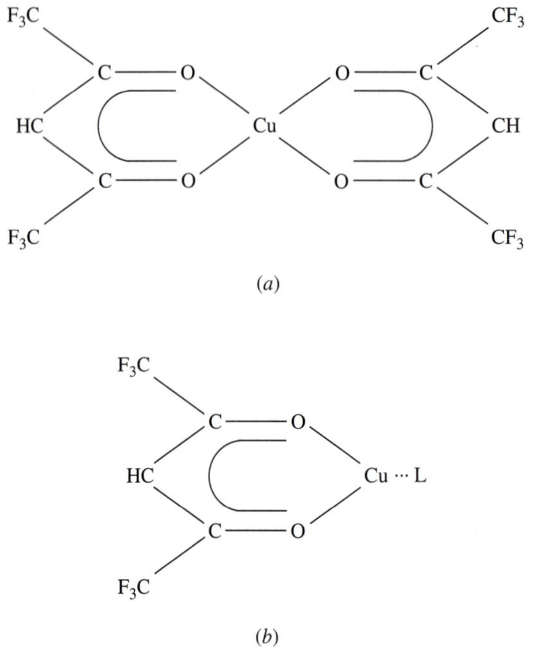

(a)

(b)

FIGURE 10
Molecular structures of (*a*) $Cu^{II}(hfac)_2$; (*b*) $Cu^{I}(hfac)L$.

reaction-limited stage until 320°C.[59] To get lower resistivity Cu films, one must adjust the deposition temperature to 350 to 450°C[59,60,61] Because Cu is likely to be used with low-permittivity polymeric dielectric, the high deposition temperature is undesirable.

Copper from $Cu(hfac)_2$ CVD can be deposited as blanket films or selectively. Study of the kinetics of deposition[60] using $Cu(hfac)_2$ and H_2O discovered an incubation period for growth that differs on Pt and on SiO_2. The difference in nucleation time can be explained[60,61] by the catalytic action of Pt to help the chemisorption and decomposition of $Cu(hfac)_2$ through

$$Cu^{II}(hfac)_2 \rightarrow Cu^{I}(hfac) + hfac \tag{8.12}$$

$$Cu^{I}(hfac) \rightarrow Cu + hfac \tag{8.13}$$

$$hfac + H \rightarrow Hhfac \tag{8.14}$$

$$Hhfac + Hhfac \rightarrow (Hhfac)_2\uparrow \tag{8.15}$$

Equation 8.13 indicates that $Cu^{I}(hfac)$ is a more readily usable compound as a Cu precursor. However, $Cu^{I}(hfac)$ is not stable, and better precursors need to be found.

Various Cu^{I} precursors have been tested[22] in search of a better CVD process. The most promising is the Cu^{I} β-diketonates. This group of precursors has the form of $Cu^{I}(\beta\text{-diketonate})L$ where L is a neutral ligand weakly bonded to the Cu. Examples of L are 2-butyne, vinyltrimethylsilane (vtms), trimethylphosphine $(PMe)_3$, etc. Hexafluoroacetylacetonate (hfac) is the most commonly used β-diketonate. The structure of $Cu^{I}(hfac)L$ is similar to $Cu^{II}(hfac)_2$, with the L ligand replacing one of the hfac rings (Fig. 10*b*).

Depending on the L ligand, CVD using Cu^{I} compounds may be selective, blanket, or both. $Cu^{I}(hfac)(PMe)_3$ was found to deposit Cu on metal selectively only,[62] but Cu^{I}(hfac)(2-butyne) deposits Cu only blanketly.[63] Both precursors can deposit clean Cu at high rate (200 to 1000 nm/min), but both are solids at room temperature. Recently, a liquid precursor Cu^{I}(hfac)(vtms) was synthesized,[64] and low-resistivity Cu[23,65,66] at $< 200°C$ was achieved.

Cu^{I}(hfac)L deposits pure Cu through a disproportionation mechanism:[62,63]

$$2Cu^{I}(hfac)L \rightarrow Cu^{0} + Cu^{II}(hfac)_2 + L \tag{8.16}$$

Both $Cu^{II}(hfac)_2$ and L are stable gaseous compounds and desorb from the surface readily at the deposition temperature. Detailed reaction kinetics for Cu^{I}(hfac)(vtms) has been studied.[66,67] Because this material is liquid at room temperature and good Cu films can be achieved below 200°C, it is a suitable precursor for CVD Cu.

Copper lacks a dense oxide and is thus vulnerable to corrosion. Also, like Au, Cu can diffuse through SiO_2 and cause deep impurity levels in Si, reducing bipolar gain and causing junction leakage. In addition, Cu has no volatile compounds at room temperature and thus cannot be etched by the familiar RIE process at moderate temperatures ($\leq 200°C$). Therefore, the use of Cu as interconnect depends on issues well beyond the deposition techniques.

CVD Cu has resistivity $\sim$ 10 to 20% above the bulk. To prevent corrosion and Cu diffusion into Si, a cladding layer (e.g., TiN or Ta) is needed. The lower resistivity of Cu is therefore not totally realizable. In addition to better conductivity, Cu has better electromigration resistance than Al because of its higher atomic weight. Therefore, Cu metallization and CVD Cu are still being intensely studied for future ULSI application.

CVD Al

Al can be deposited by several CVD precursors; the most extensively studied is tri-isobutyl-Al, $(C_4H_9)_3Al$ or TIBA.[17] The chemistry is a three-step decomposition:

$$TIBA + H_2 \rightarrow DIBAH + C_4H_8 \tag{8.17}$$

$$DIBAH + H_2 \rightarrow AlH_3 + 2C_4H_8 \tag{8.18}$$

$$AlH_3 \rightarrow Al + \tfrac{3}{2}H_2 \tag{8.19}$$

where DIBAH is the abbreviation for di-isobutyl Al hydride, or $(C_4H_9)_2$ AlH. The first reaction is at 40 to 50°C, before the gas reaches the wafer; and the second reaction occurs on the wafer with the wafer held at 150 to 300°C. In practice, DIBAH is the chemical precursor that decomposes on the wafer surface. However, the direct use of DIBAH is impractical owing to its low vapor pressure and low deposition rate.

The Al deposited by this process contains virtually no hydrocarbon residue and has a resistivity of $\sim 2.8\ \mu\Omega$-cm, close to that of bulk Al. When deposited on a good nucleation layer such as TiN, (111)-textured Al with high reflectivity can be achieved.[68]

A problem with the application of CVD Al to ULSI circuits is that there is no Cu in the CVD Al to provide electromigration resistance. A simple solution to this problem is to deposit $\sim$ 50% of the total thickness of Al by CVD and the rest by sputtering an Al-Cu alloy.[69] A subsequent heating to 250 to 400°C allows the Cu in the alloy to redistribute to the entire Al wiring. The maximum Cu concentration is determined by the solubility of Cu in Al at the annealing temperature. At 300°C, a Cu content of ~ 0.35 to 0.40 at% has been reported.[69] Recently, deposition of CVD Al with simultaneous Cu doping using dimethyl aluminum hydride (DMAH) and cyclopentadienyl copper triethylphosphine (CpCuTEP) has been reported.[70] Cu content as high as 2.5% was achieved.

One disadvantage of CVD Al is that the utilization rate of the precursor is low since Al is a light element. In a TIBA molecule, less than 15% of the molecular weight is from the Al. In practice, only a fraction of the precursor reaches the wafer surface and, consequently, a sizable amount of used precursor must be reprocessed or disposed of, which adds to the cost of the process.

The fire and explosion hazard of the precursor material for CVD Al is another issue that must be addressed. TIBA is pyrophoric (self-igniting) in air and reacts violently with water. Other precursors are generally more hazardous than TIBA. One should note that Al alkyls have been in use by the polymer industry for a number of years, and standard industrial safety and hygiene procedures are readily available. However, semiconductor manufacturers do not have this experience yet.

Even though the process and the equipment development of CVD Al are immature compared with CVD W, the advantage of lower cost and better reliability still makes CVD Al desirable. CVD Al is not vulnerable to corrosion and does not require elaborate cladding. Al CVD can be deposited at low temperature ($\leq 250°C$) and is compatible with low-permittivity dielectric materials. Thus, interest in CVD Al continues.[70,71,72]

CVD $TaSi_2$

Sputtered $TaSi_2$ is used for polycide gate structure.[73] CVD $TaSi_2$ is a lower-cost replacement but has not been used widely because of its complex CVD behavior. CVD $TaSi_2$ can be deposited at 600°C using the silane reduction of $TaCl_5$:[74]

$$TaCl_5 + 2SiH_4 \rightarrow TaSi_2 + 5HCl + 1.5H_2 \tag{8.20}$$

A competitive reaction to Eq. (8.20) is

$$5TaCl_5 + 3SiH_4 + 6.5H_2 \rightarrow Ta_5Si_3 + 25HCl \tag{8.21}$$

During LPCVD, Ta_5Si_3 is the more favored reaction. $TaSi_2$ is formed only at the $TaSi_x$/polysilicon interface where an extra supply of Si is available. In the presence of Si, the Ta-rich silicide converts to $TaSi_2$ through

$$Ta_5Si_3 + 7Si \rightarrow 5TaSi_2 \tag{8.22}$$

$TaCl_5$, however, reacts rapidly with Si through the displacement reaction

$$4TaCl_5 + 13Si \rightarrow 4TaSi_2 + 5SiCl_4 \tag{8.23}$$

This reaction is self-limiting at 150 to 250 nm when Si diffusion through the $TaSi_2$ is insufficient to support the reaction. The displacement reaction, however, deposits $TaSi_2$ nonuniformly, resulting in a jagged[74] silicide/polysilicon interface, which causes gate oxide breakdown. Therefore, in practice, a polysilicon layer of

$\sim$ 150 nm is deposited first, followed by the silane reduction process. Finally, an anneal at $> 600°C$ is needed to convert the Ta-rich silicide to $TaSi_x$ with $x \approx 2.2$.

LPCVD of $TaSi_2$ can also be achieved by dichlorosilane reduction[75] at 650°C, that is,

$$TaCl_5 + 2SiH_2Cl_2 + 2.5H_2 \rightarrow TaSi_2 + 9HCl \tag{8.24}$$

This reaction suffers from the same Si displacement reaction as the silane reduction. Consequently, the same three-step process, Si deposition + reduction reaction + sintering, is needed.

CVD WSi_2

WSi_2 and $MoSi_2$ are used widely on top of gate polysilicon to form a low-resistance polycide gate. Both sputtered and CVD WSi_2 are used for ULSI device fabrication. CVD WSi_2 is readily deposited using the silane reduction of WF_6 at 300 to 400°C,[25,26,76,77]

$$WF_6 + 2SiH_4 \rightarrow WSi_2 + 6HF + H_2 \tag{8.25}$$

A comparison of Eq. (8.25) with Eqs. (8.3) and (8.4), however, immediately reveals the ambiguity of the chemical reactions since the reactants are the same. In a CVD reactor, the flow rates of WF_6 and SiH_4 control the outcome of the reaction. As can be expected, a higher SiH_4/WF_6 ratio results in WSi_2 deposition. WSi_2 does not form at a SiH_4/WF_6 ratio below 3. Instead, W with Si incorporated in the grain boundaries is deposited at low flow rates.[76] In practice, a SiH_4/WF_6 ratio greater than 10 is used to ensure the deposition of $WSi_x (x \approx 2.2–2.6)$. Although a displacement reaction similar to $TaSi_2$ [Eq. (8.23)] also occurs, it is self-limiting at < 20 nm because of the lower deposition temperature for WSi_2. Therefore, the differences in chemistry and the deposition temperature make CVD WSi_2 more desirable than $TaSi_2$.

Unlike CVD W, which is used for plug/interconnect, WSi_2 is used on gate polysilicon, which has to endure high temperatures (800 to 1000°C) during source/drain drive and reoxidation. CVD WSi_2 films have high stress ($\sim$ 1000 MPa),[77] and silane reduction produces CVD films with relatively poor step coverage. During high-temperature processes, peeling and cracking of the WSi_2 at step edges have been a problem. In addition, F incorporation at high concentrations of 10^{20} atoms/cm^3 is common. The F diffuses to the gate oxide and causes flat band shifts. Therefore, even though silane-reduced WSi_2 has been in manufacturing for some time, a better process is needed.

WSi_2 can be deposited by dichlorosilane (SiH_2Cl_2, or DCS) reduction at higher temperatures (500 to 600°C)

$$WF_6 + 3.5SiH_2Cl_2 \rightarrow WSi_2 + 1.5SiF_4 + 7HCl \tag{8.26}$$

$$WF_6 + 3.5SiH_2Cl_2 \rightarrow WSi_2 + 1.5SiCl_4 + 6HF + HCl \tag{8.27}$$

Depending on the surface reaction kinetics, SiF_xCl_y, Cl_2, and F_2 are all possible reaction products.

WSi_2 deposited by dichlorosilane reduction[78,79] contains much less F (10^{18} atoms/cm^3) than from silane reduction, and the Cl content is also low (10^{18} atoms/

cm^3). The resistivity and the film stress are comparable to that from silane reduction. However, it has a deposition rate that is three to five times as high and a better step coverage. As a result the DCS process reduces the peeling and cracking problems at the step edge as well as the device instability problems. Therefore, DCS will gradually replace silane in CVD WSi_2.

CVD $TiSi_2$

$TiSi_2$ has a resistivity three to four times lower than WSi_2 and $TaSi_2$ and is the most desirable for polycide. Because of the advantage of self-aligned silicide (salicide), CVD $TiSi_2$ has not been widely regarded as important. However, self-aligned structures are reaching a limitation in thickness and thermal stability (see Section 8.3.2). The CVD process, especially the selective deposition on the source/drain and the gate, may become important for future ULSI devices. $TiSi_2$ can be deposited by LPCVD using $TiCl_4$ and silane reduction[80] at 650 to 700°C:

$$TiCl_4 + 2SiH_4 \rightarrow TiSi_2 + 4HCl + 2H_2 \tag{8.28}$$

Excessive Si substrate consumption (300 to 600 nm), however, occurs through the displacement reaction:

$$TiCl_4 + 2Si + 2H_2 \rightarrow TiSi_2 + 4HCl \tag{8.29}$$

The displacement reaction is controlled by lowering the deposition temperature to $< 600°C$ (to slow the diffusion) and by increasing the silane flow. Lowering the deposition temperature, however, lowers the deposition rate too. In addition, film properties degrade at lower temperatures. By employing plasma enhancement (PECVD), low-resistivity (14 to 25 $\mu\Omega$-cm) $TiSi_2$ has been deposited[81] at 590°C. In PECVD, the reaction is probably more complex than suggested by Eqs. (8.28) and (8.29). Spectroscopic studies[82] detected HCl, SiH_3Cl, SiH_2Cl_2, and SiCl as by-products.

Selective deposition of $TiSi_2$ follows essentially the chemistry in Eq. (8.29), and observations[83] confirmed the Si consumption from the displacement reaction. Thermodynamic calculations[84] show that Si consumption can be minimized at the expense of lower selectivity. Recently, selective $TiSi_2$ deposition using a high silane flow[85] to reduce the Si consumption has been reported. In this approach, the displacement reaction [Eq. (8.29)] provides the selectivity but the silane reduction reaction [Eq. (8.28)] checks the rate. Consequently, $TiSi_2$ could be selectively deposited without excessive Si etching.

8.2.4 Electro- and Electroless Plating

Electro- and electroless plating have been used extensively in printed circuit board and multichip module applications but have not been applied in manufacturing ULSI chips. Plating provides a low-cost method for depositing thick metal layers over large areas and is ideal for printing circuit boards and for multilayer board fabrication. Because of their tolerance to higher temperature, on-chip interconnects have been made exclusively by PVD and CVD processes. For future ULSI devices, however, low-temperature, low-cost processes compatible with low-permittivity dielectric materials are important. Plating of metal (especially Cu and barrier materials) is an ideal low-cost process for this application.

Selective electroless Ni plating using nickel sulfate for submicrometer contact plugs by hypophosphite reduction has been reported:[86]

$$Ni^{2+} + (H_2PO_2)^- + H_2O \rightarrow Ni^o + 2H^+ + H(HPO_3)^- \quad (8.30)$$

Electroless plating of Cu for on-chip interconnects has been reported.[87] A Ti underlayer was used to provide a conducting surface, and the Ti was activated by Pd deposition in $PdCl_2$ solution before the Cu deposition. Cu can be deposited using $CuSO_4$ and formaldehyde reduction:[88]

$$Cu^{2+} + 2HCHO + 4OH^- \rightarrow Cu^o + 2HCOO^- + 2H_2O + H_2 \quad (8.31)$$

The electroless-deposited Cu has resistivity between 2.0 and 2.5 $\mu\Omega$-cm, which is lower than Al (2.7 $\mu\Omega$-cm). After a 400°C anneal, the resistivity drops[87] further by about 10%. Electroless Cu with resistivity close to bulk ($<$ 1.9 $\mu\Omega$-cm compared to 1.7 $\mu\Omega$-cm) has been reported.[89]

8.3 SILICIDE PROCESS

Si forms many stable metallic and semiconducting compounds with metals. Several metal silicides show low resistivity and high thermal stability, making them suitable for ULSI application. The formation kinetics, film properties, and potential application have been thoroughly examined by Murarka[90] and by Nicolet and Lau.[91] In this section we address some specific applications and discuss the issues regarding the use and choice of silicides.

Silicides can be formed by several means: (1) sputtering and sinter, (2) CVD, and (3) self-alignment. The first process is not selective nor self-aligned, and is used only for gate silicide. The last process is self-aligned and can be used for source/drain/gate or for source/drain only.

8.3.1 Polycide Gate Structure

Polycide is the condensed name for polysilicon-silicide, which is widely used for gate structures that contain a non-self-aligned silicide. In this structure, a silicide is blanket-deposited over the polysilicon and the composite is patterned together. A well-designed polycide gate can achieve a gate sheet resistance $\leq$ 10 $\Omega/\square$ compared to that for polysilicon of $\geq$ 40 $\Omega/\square$. Because of the considerably lower resistivity, this poly structure can be used for local routing over short distances. This is especially beneficial when only one or two levels of metals are used in a circuit, since the poly routing can help to reduce the die size by as much as 20%. In fact, the invention of polycide structure effectively delayed the need for multilevel metallization by several years.

A typical polycide formation sequence is shown in Fig. 11*a*. For sputter deposition, a high-temperature, high-pressure sintered, high-purity compound target is

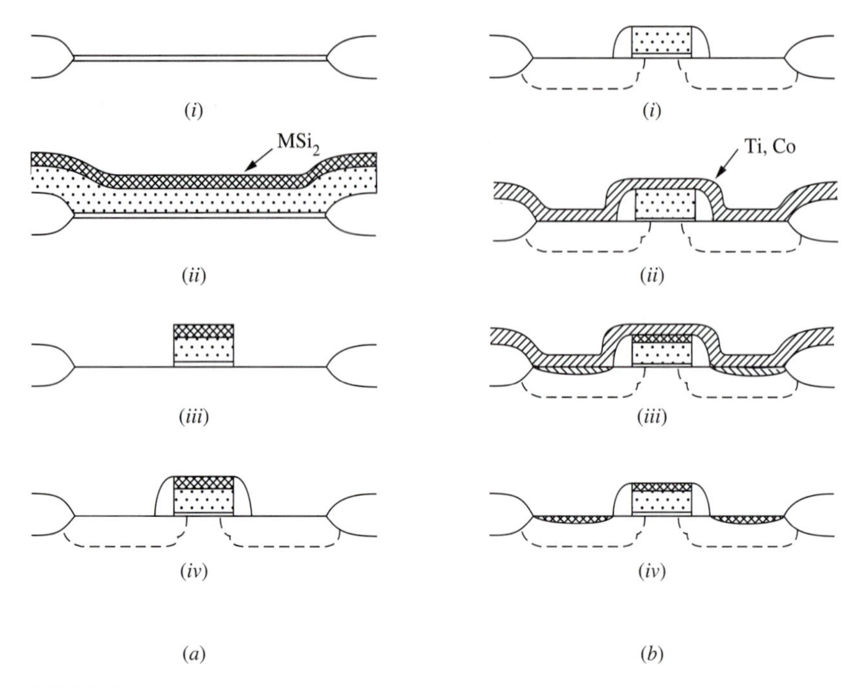

FIGURE 11
Polycide and salicide processes: (*a*) Polycide structure: (*i*) gate oxide; (*ii*) polysilicon and silicide deposition; (*iii*) pattern polycide; (*iv*) lightly doped drain (LDD) implant, sidewall formation, and S/D implant. (*b*) Salicide structure: (*i*) gate patterning (polysilicon only), LDD, sidewall, and S/D implant; (*ii*) metal (Ti, Co) deposition; (*iii*) anneal to form salicide; (*iv*) selective (wet) etch to remove unreacted metal.

used to ensure the quality of the silicide. The as-deposited film usually has a higher Si content in the form of MSi_x with $x \approx 2.4$ to 2.6 instead of 2.0. During the high-temperature sintering process the excess Si precipates out to the underlying poly-Si, reducing the silicide to a near stoichiometric composition with $x \approx 2.0$ to 2.2. Silicides with a lower Si content tend to oxidize badly during sintering, resulting in peeling and blistering problems. Silicides can also be deposited with a CVD process.

The most commonly used polycides are WSi_2,[26,92] $TaSi_2$,[73] and $MoSi_2$.[93] All are refractory, thermally stable, and resistant to processing chemicals. These properties make them better choices for polycide even though their electrical resistivity (50 to 100 $\mu\Omega$-cm) is higher than for $TiSi_2$ and $CoSi_2$.

$TiSi_2$ and $CoSi_2$ have the lowest resistivity among all silicides (10 to 15 $\mu\Omega$-cm); however, they are not good candidates for polycides. Both are vulnerable to processing chemicals, especially to HF solutions. The most difficult obstacle, however, is in polycide patterning. $TiSi_2$ is attacked by Cl-based chemistry during polysilicon etch, and this makes linewidth control difficult at the critical gate level. $CoSi_2$, on the other hand, lacks a volatile reaction by-product and is difficult to pattern by dry etching. $TiSi_2$ and $CoSi_2$, however, are ideal for self-aligned silicide or salicide application.

8.3.2 Self-Aligned Silicide (Salicide)

A self-aligned silicide process is illustrated in Fig. 11*b*. In this process, the polysilicon gate is patterned without any silicide, and a sidewall spacer is formed to prevent shorting the gate to the source and drain during the silicidation process. A metal layer, either Ti or Co, is blanket-sputtered on the entire structure, followed by silicide sintering. Silicide is formed, in principle, only where the metal is in contact with Si. A wet chemical wash then rinses off the unreacted metal, leaving only the silicide. This technique eliminates the need to pattern the composite polycide gate structure and adds silicide to the source/drain area to reduce the contact resistance.

In practice, both $CoSi_2$ and $TiSi_2$ are formed by multistep processes:

$$\mathrm{Co + Si \rightarrow CoSi} \qquad (450°\mathrm{C}) \tag{8.32}$$

$$\mathrm{CoSi + Si \rightarrow CoSi_2} \qquad (700°\mathrm{C}) \tag{8.33}$$

The selective etch by wet chemical is done after the formation of CoSi. If rapid thermal processing (RTP) is used for silicide formation, step 1 [Eq. (8.32)] can be skipped and $CoSi_2$ is formed directly at 700 to 750°C. The wet etch then is applied directly to $CoSi_2$.

To form $TiSi_2$, two RTPs in nitrogen are required:

$$\mathrm{Ti + 2Si \rightarrow TiSi_2} \quad (\mathrm{C\text{-}49}) \tag{8.34}$$

$$\mathrm{TiSi_2(C\text{-}49) \rightarrow TiSi_2} \quad (\mathrm{C\text{-}54}) \tag{8.35}$$

The first RTP is done at 620 to 680°C, and the second RTP is done at a temperature higher than 750°C. The second step [Eq. (8.35)] transforms the high-resistivity, metastable C-49 phase $TiSi_2$ to the low-energy low-resistivity orthogonal $TiSi_2$ (C-54 phase). The resistivity of C-49 $TiSi_2$ is about three to four times higher than that for C-54 $TiSi_2$. This phase transition occurs at 700 to 750°C. Direct heating of Ti in contact with Si to $\geq$ 750°C forms the stable C-54 $TiSi_2$ directly. However, severe shorting between the source/drain and gate occurs (see the discussion of processing issues later in this section).

Since $TiSi_2$ and $CoSi_2$ have the lowest resistivity, $\sim$ 10 to 15 $\mu\Omega$-cm, even a substantially thinner layer of $TiSi_2$ or $CoSi_2$ can provide a lower sheet resistance than polycide structures. For example, 50 nm of $TiSi_2$ has a sheet resistance of 2 to 3 $\Omega/\square$ whereas even 150 nm of WSi_2 can only give a sheet resistance of $\geq$ 6 $\Omega/\square$. Therefore, by using a self-aligned silicide, not only is the gate patterning simplified but also one can more fully use the benefit of a low-resistance poly-Si in circuit design.

Equally important is lowering the source/drain contact resistance. In most designs using 0.5 μm design rules, the source/drain contact resistance imposes no threat as a critical parasitic resistance. For wide transistors, multiple contacts are required to ensure that the entire width of the transistor is used. A self-aligned silicide can act as a shunt for the entire source/drain area, allowing a single contact and additional design freedom.

Effects on transistor performance

A self-aligned silicide on the source/drain area lowers the contact resistance because the metal-Si contact area is increased from the area of the contact window(s) to the entire source/drain area. Consider the contact resistance of a nMOS transistor with a source/drain area of 1 μm $\times$ 1 μm and a contact window size of 0.4 μm. For the specific contact resistance, let us use 1×10^{-7} Ω-cm^2 for the metal/n$^+$ contact, and 0.2×10^{-8} Ω-cm^2 for the metal/metal contact. The contact resistance for the silicide/Si contact is

$$R_{c1} = 10^{-7}\ \Omega\text{-cm}^2/(10^{-8}\ \text{cm}^2) = 10\ \Omega$$

The contact resistance for the metal/silicide contact is

$$R_{c2} = 0.2 \times 10^{-8}\ \Omega\text{-cm}^2/(0.12 \times 10^{-8}\text{cm}^2) = 1.6\ \Omega$$

The total contact resistance is then $R_{c1} + R_{c2} = 11.6\ \Omega$.

If no silicide is formed in the source/drain, then only R_{c1} exists:

$$R_{c1} = 10^{-7}\ \Omega\text{-cm}^2/(0.12 \times 10^{-8}\ \text{cm}^2) = 80\ \Omega$$

Therefore, even for a nominal-size transistor, the self-aligned silicide significantly decreases the contact resistance.

Even 80 Ω of contact resistance is only a small fraction of the channel resistance. To reduce this quantity further does not improve the device performance. Let us consider, however, a wider transistor used to supply more current to subsequent stages. Assume that the transistor is 10 μm wide and that this nMOSFET is capable of delivering 4 mA of current at $V_g = 3.3$ V. Let us consider the effect of the R_c on the drain side and on the source side, individually:

1. The effect of R_c on the drain side. The effect of R_c on the drain side is to reduce the drain voltage V_d by the amount $R_c I_d$. If the transistor is operating in the saturated region (as in most digital circuits), then the reduction in V_d has little effect. If the transistor is operating in the linear region (as in analog circuits), then the output current I_d is reduced from

$$I_d = V_d / R_{\text{channel}} \tag{8.36}$$

to

$$I_d = (V_d - R_c I_d) / R_{\text{channel}} \tag{8.37}$$

or

$$I_d = V_d / (R_c + R_{\text{channel}}) \tag{8.38}$$

Therefore, the effective resistance in the linear region has been increased by the amount R_c. For this transistor, the channel resistance in the linear region, R_{channel}, can be calculated as V_{dsat}/I_d, or approximately 250 Ω. If only a single contact is made, the contribution of R_c, 80 Ω, is about 30%.

2. The effect of R_c on the source side. The effect of R_c on the source side is more severe. A resistance on the source side of the MOSFET not only decreases the drain voltage by the same amount as on the drain side but also raises the potential at the source by R_cI_d. The effective gate voltage V_g is then reduced by the same amount:

$$V'_g = V_g - R_cI_d \tag{8.39}$$

A reduction in effective gate voltage is equivalent to a reduction in transconductance:

$$g_m = \Delta I_d/\Delta V_g \tag{8.40}$$

$$g'_m = g_m/(1 + R_c g_m) \tag{8.41}$$

An nMOSFET with a $g_m = 300\ \mu S/\mu m$ and $R_c = 80\ \Omega$ has a $g'_m \approx 0.98 g_m$, or only a 2% loss in current drive for a 1-μm wide transistor; but for a 10-μm transistor, a 20% loss for the current drive results. Therefore, for wide transistors, either multiple contacts or a self-aligned silicide is needed to reduce the parasitic resistance.

The contact resistance will degrade the device performance more severely on further scaling of design rules. At a 0.25-μm design rule, the contact windows will be about 0.3 μm, giving a $R_c \approx 140\ \Omega$ compared with the 80 Ω in the example for 0.35-μm design rules. At a 0.18-μm design rule (corresponding to 1 Gbit DRAM), R_c will be $\sim 250\ \Omega$ without self-aligned silicide. Even for a 1-μm wide transistor, the g'_m will be reduced to only about $0.9 g_m$. If self-aligned silicide is used, R_c drops to about 18 Ω in the 0.25-μm design and to about 25 Ω in the 0.18-μm design.

Effects on layout area

Figure 12 illustrates the benefit of self-aligned silicide in reducing layout area. Consider the current distribution along a wide transistor with no silicide in the source/drain area. Because of the need for shallow junctions, the diffusion resistance increases with each generation of scaling. As a consequence, at design rules of 0.35 μm or below, a diffusion R_{sh} of $< 100\ \Omega/\square$ is not achievable for n^+ diffusion. Consider the configuration in Fig. 12*a*, where two contact windows at one end of a wide transistor are used. The current distribution within the transistor is not uniform, and most of the current is crowded into the lower half of the transistor, as shown in Fig. 12*b*. This observation results in a current reduction of approximately 20 to 30%. The higher the diffusion R_{sh}, the more severe the current crowding becomes. If a self-aligned silicide is used, then the current crowding is avoided since the diffusion R_{sh} is largely replaced by the low silicide R_s.

The current crowding problem can be avoided by using multiple contacts. As shown in Fig. 12*d*, strategically placed contacts can reduce the current crowding. The use of more contacts, however, requires metallization to connect these contacts. The result is that valuable routing area is used for source/drain contacts, requiring a larger die size. This penalty can be reduced when more levels of metallization are used, since more degrees of design freedom are available to minimize the area penalty.

Processing issues

There are three important processing issues in self-aligned silicides: (1) the forming of a bridge between gate and source/drain; (2) the difficulty in forming

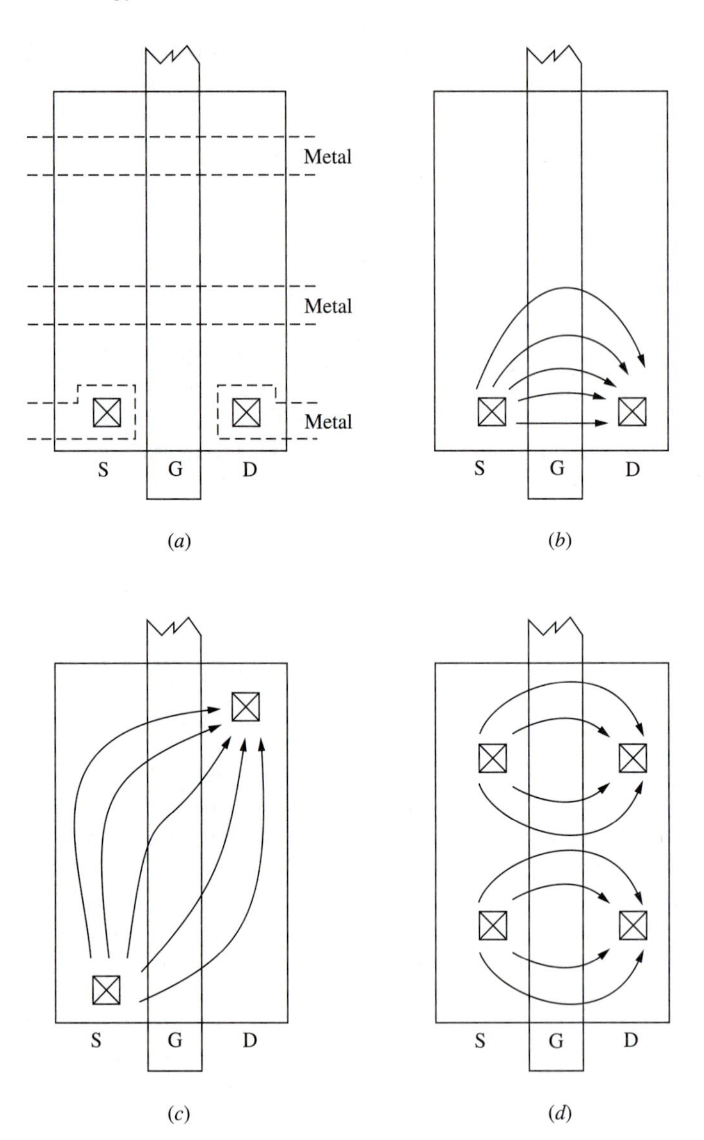

FIGURE 12
Effects of salicide on layout area: (*a*) If only one contact is made in the S/D, then the rest of the area can be used for routing; (*b*) if the S/D is not silicided, then current crowding occurs on the lower half of the transistor; (*c*) a different contact arrangement reduces the current crowding but takes away some routing area; (*d*) multiple contacts allow the use of full transistor width but leave no room for other metal runners over the transistor. If the S/D is silicided, then no current crowding will occur and only one contact, as in (*a*), is needed.

silicide on thin strips of polysilicon and the diffusion region; and (3) the thermal stability of thin silicides.

The forming of a bridging short between the gate and source/drain is most apparent for $TiSi_2$. Figure 13 illustrates the formation of $TiSi_2$ along the sidewall of the gate, resulting in a short between the gate and the source/drain. The mechanism for this is "Si pumping" during the formation of $TiSi_2$. During silicide formation of

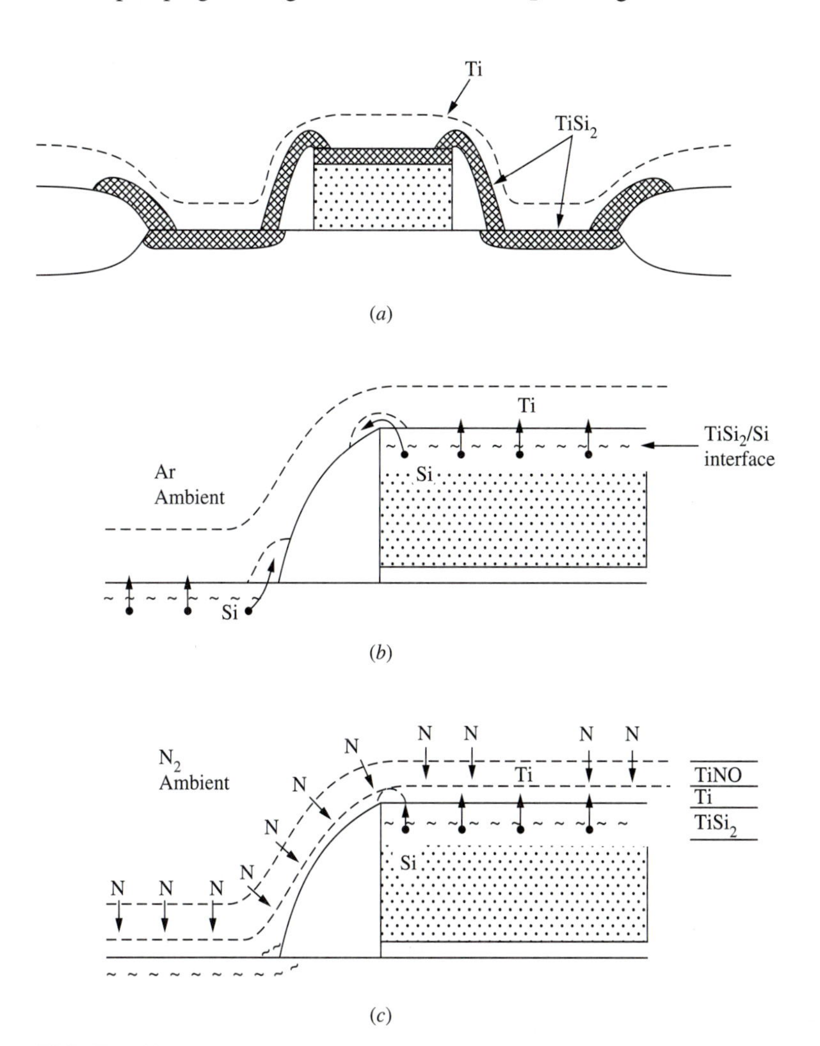

FIGURE 13
"Bridging" during the formation of Ti salicide: (*a*) $TiSi_2$ bridges over sidewall spacer and a short circuit is formed between S/G and D/G; (*b*) during $TiSi_2$ formation, Si diffuses into Ti to form $TiSi_2$; under favorable conditions Si can diffuse a long distance and form $TiSi_2$ on the sidewall spacer; (*c*) in a nitrogen atmosphere, nitrogen diffuses into Ti and blocks the diffusion of Si, reducing or eliminating bridging; TiN and/or Ti–O–N are formed on top of $TiSi_2$.

$TiSi_2$, Si is the moving species, which diffuses into the Ti metal, as shown in Fig. 13*b*. As a result, $TiSi_2$ formation is no longer confined to where the Ti metal contacts Si. Naturally, the extent of bridge formation is determined by the temperature and time of the sintering process. Unfortunately, at least 600°C is needed to cause Ti to react with Si for $TiSi_2$ sintering and at this temperature extensive diffusion of Si in Ti already occurs. This problem can be reduced by processing in a nitrogen ambient and using RTP.[94–96] Within a finite temperature range between 620 and 680°C, nitrogen can diffuse into the Ti metal before extensive diffusion of Si occurs and stop the "Si pumping," as depicted in Fig. 13*c*. At temperatures higher than 680°C, Si diffuses rapidly in Ti and bridging occurs.

For $CoSi_2$, the diffusing species during silicide formation is alternately Co and Si. As a result, the problem of bridge formation is much less severe. $CoSi_2$ can be formed by using either a furnace in a reducing atmosphere[97] or by RTP.[98]

At a polysilicon line width below about 0.5 μm, the formation of $TiSi_2$ becomes difficult,[99,100] resulting in high poly resistance. The transformation of the high-resistance C-49 phase $TiSi_2$ to the low-resistance C-54 phase $TiSi_2$ was found to be the limiting factor.[99] After the first RTP, small-grained (0.1 to 0.2 μm) C-49 $TiSi_2$ was formed on both wide ($>$ 10 μm) and narrow (0.4 μm) polysilicon lines. After a 725°C anneal, the $TiSi_2$ on the wide polysilicon lines transformed to large-grained (1 to 10 μm) C-54 phase. However, on narrow lines only smaller-grained ($<$ 1 μm) C-54 $TiSi_2$ was observed, and only on a few lines. At a higher-temperature (775°C) anneal, C-54 phase was formed on both wide and narrow lines. At 800°C, $TiSi_2$ started to agglomerate, and the line resistance rapidly degraded.

An explanation[99] for the difficulty of the C-49 to C-54 phase transformation can be constructed based on the limitation of grain size on narrow lines. Generally, thin films do not grow into large grains several times the width of the line. Large-grained structures (such as Al) can be obtained by growing a large-grained thin film and then patterning it into lines. In the case of $TiSi_2$ the polysilicon lines are already defined. The nucleation of C-54 phase at 700 to 750°C is slow, and the density of nuclei is low. On wide lines, the nuclei grow into large grains, and only a few nuclei are needed to cover the entire line with large-grain C-54 $TiSi_2$. On narrow lines, however, the grain size is limited by the linewidth, and only a fraction of the line is covered by the C-54 $TiSi_2$ grown from the few nuclei available. This problem can be solved by heating the $TiSi_2$ to $>$ 750°C to accelerate the nucleation of the C-54 phase. Unfortunately, thin Ti silicide agglomerates at $>$ 750°C, rendering this solution unusable.

Several efforts have been made to reduce the narrow-line effects.[100–102] Pre-amorphization of the polysilicon[100,101] was used to improve the process margin. Selective W was used as a strapping[102,103] over $TiSi_2$ to improve the resistivity. $CoSi_2$ was proposed to replace $TiSi_2$ for very fine lines,[104,105] since $CoSi_2$ transformation occurs at a lower temperature (600 to 700°C) and thus full formation of low-resistance $CoSi_2$ is achieved before the silicide agglomerates.

Because of these processing issues, the use of salicide for 0.25-μm and smaller devices requires considerable process development and in some cases new processes have to be used. The salicide process, however, delivers unique advantages for logic and MPU (microprocessor unit) circuits at low cost. Continued scaling of devices will demand even more development work for this difficult process.

8.3.3 Local Interconnect

Local interconnects are short polysilicon, polycide, or metal wires that connect adjacent devices to make simple functional groups such as logic gates, inverters, flip-flops, etc. Often, these short wires only connect part of the circuit to help handle cross-over of wiring. Figure 14 illustrates the use of a polysilicon line to connect the drains of the driving transistors to the gates of the opposite transistors in a SRAM cell. In this particular case, low resistance is not required and high resistance polysilicon can be used. For memory circuits such as the SRAM, more than one layer of polysilicon is often used as a resistive load and for local interconnect. This type of application is called *multi-poly technology.* Sometimes both multi-poly and multilevel metal are used with the poly for local connections and with metal (Al) for longer (global) interconnects.

For logic and microprocessors using a $TiSi_2$ salicide structure, one can take advantage of the TiN formed during the silicidation process and pattern it for use as a local interconnect.[106] As shown in Fig. 15, instead of the wet selective etch, the local interconnect area is masked and dry or dry/wet etched to remove only the TiN and unreacted Ti from unmasked areas. This process forms the salicide and the local interconnect in a single deposition/etching step. Good selectivity of TiN relative to $TiSi_2$ is the necessary condition for this approach.

TiN local interconnect can also be used with $CoSi_2$ salicide, although through a more complex process. After the salicide process is completed, a TiN layer is deposited and masked. A dry etching of the TiN completes the patterning of the local interconnect. This is a higher-cost process, since the TiN has to be deposited separately. The advantage is that good selectivity between TiN and $CoSi_2$ is guaranteed. Other silicide, polycide, or metal (such as W) can also be used in the same way as long as selective etching can be done.[107,108]

Local interconnect has the advantage of adding a (limited) degree of design freedom at relatively low cost. The process described in Fig. 15 requires only one mask to acquire this design freedom; no window mask is needed!

8.3.4 Silicide and Shallow Source/Drain Junctions

A severe disadvantage of a salicide in the source/drain region is the junction leakage caused by silicide. To form a silicide, a substantial portion of the Si from the junction is consumed. For example, to form 50 nm of $TiSi_2$ requires the consumption of approximately 50 nm of Si from the junction. For a shallow junction about 0.15-μm deep this represents one-third of the junction depth if the silicide is uniformly formed. Silicide tends to be not totally uniform, and even a minor nonuniformity (say, 20 nm) means that in some places 70 nm out of the 150-nm junction are consumed, leaving a very shallow effective junction.

In addition to the consumption of Si, dopants in the junction redistribute during silicide formation. This phenomenon[109] is especially damaging in $TiSi_2$. During the formation of silicide, because of the diffusion of Si into the Ti metal, a large number of vacancies are generated in the Si matrix, which enhances the diffusion of dopants

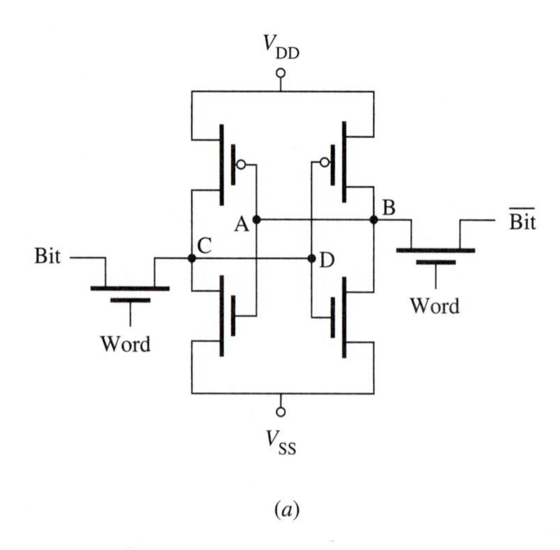

(*a*)

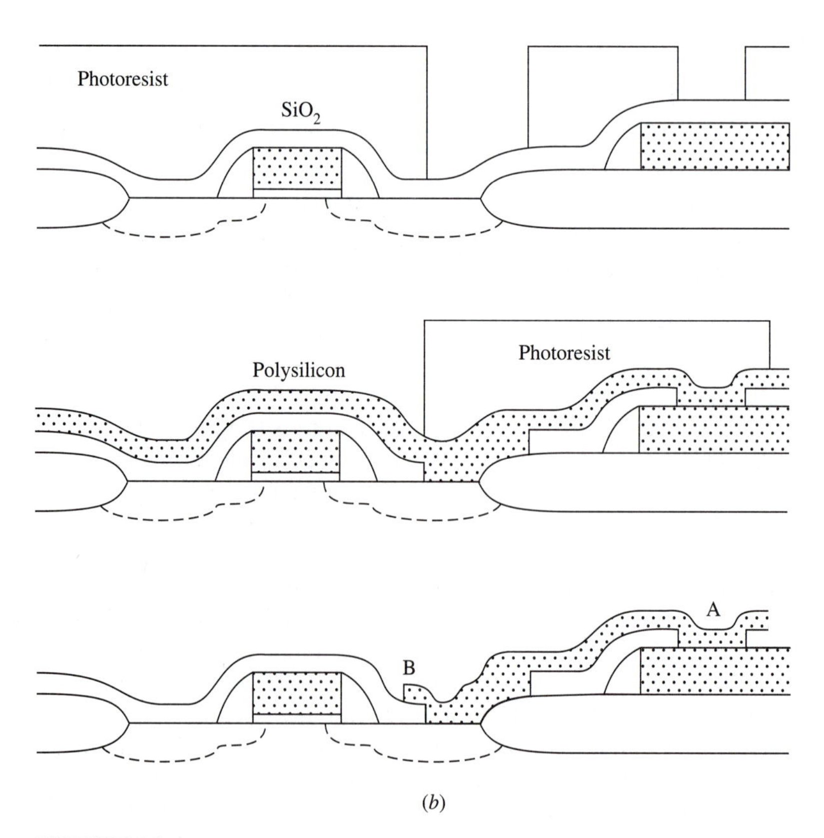

(*b*)

FIGURE 14
Polysilicon local interconnect for SRAM: (*a*) A circuit diagram for a six-transistor CMOS SRAM cell; A, B and C, D are connected with polysilicon; (*b*) a two-mask process for implementing the polysilicon local interconnect: The first mask is used to open contact windows through a thin dielectric layer (interpoly dielectric or IPD) and the second mask is used to define the local interconnect.

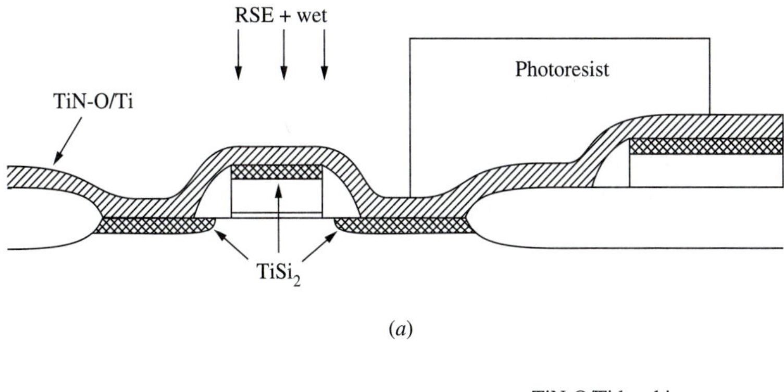

(*a*)

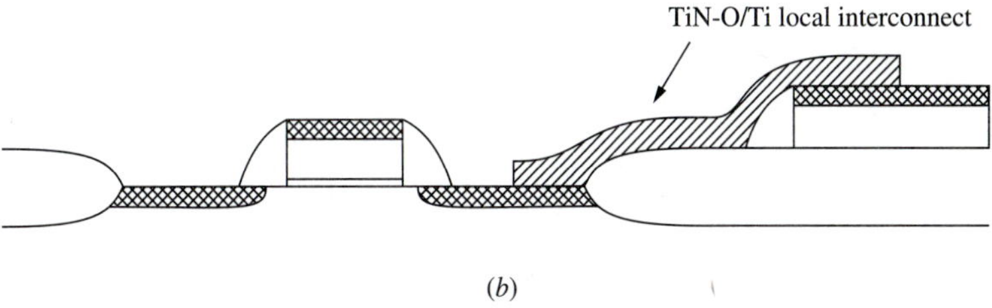

(*b*)

FIGURE 15
One-mask process for TiN local interconnect: (*a*) Save the Ti–O–N/Ti after $TiSi_2$ formation in N_2; (*b*) after photolithography and etching (dry + wet), a local interconnect is formed between the drain and the polysilicon of another device.

in the junction. In the first place, the Si containing the highest dopant concentration is consumed by the silicide. In addition, the enhanced diffusion absorbs more dopant from the remaining Si junction into the silicide, further depleting the junction of dopants, as shown in Fig. 16. The depleted junctions are more vulnerable to leakage,[110] and the low dopant concentration at the silicide/Si interface causes high contact resistance (see Fig. 4).

A technique to solve the junction leakage and dopant depletion problem is to implant the junction after the silicide is formed. To do this, the dopant has to be implanted either into the silicide or through the silicide. Good junctions formed by implanting into $CoSi_2$ and driving the dopants out by annealing have been demonstrated.[111] By using this technique, junctions as shallow as 0.12 μm can be made. For $TiSi_2$ the same process is difficult because of the very high affinity of As and B for $TiSi_2$; only very limited success has been reported.[112] It is difficult to form very shallow junctions ($\leq 0.15\ \mu$m) by implanting dopant through the silicide, since the peak of the implant has to be in the silicide and only the tail portion can be used. The junction depth then becomes a sharp function of the silicide thickness, which in the case of $TiSi_2$ is difficult to control precisely.[113]

Once the junction depth becomes less than $\sim 0.1\ \mu$m, further scaling (with salicide) becomes extremely difficult. Even when the junctions can be formed, the

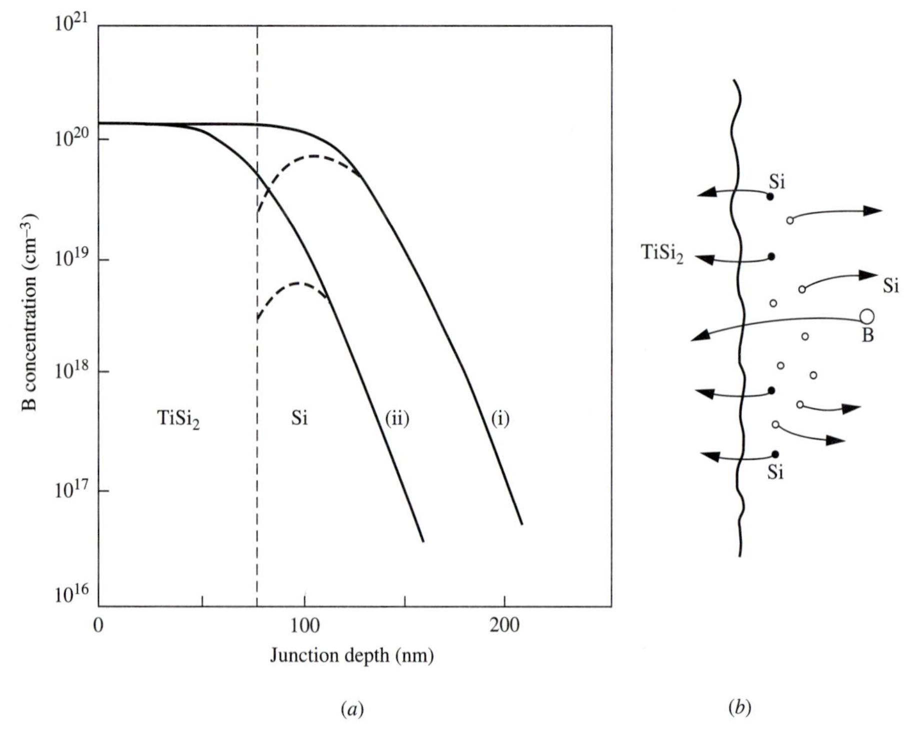

FIGURE 16
Redistribution of B during $TiSi_2$ formation: (*a*) B redistribution for 200-nm junction (*i*), and for 150-nm junction (*ii*). Solid curves are B concentration profiles before $TiSi_2$ and dashed curves are after $TiSi_2$. (*b*) Mechanism of enhanced diffusion during $TiSi_2$ formation: Si moves into Ti to form $TiSi_2$, leaving a vacancy behind; the vacancy diffuses rapidly into Si, enhancing the diffusivity of B. There is no back-diffusion flux from $TiSi_2$ even as the B concentration in $TiSi_2$ is higher because B binds tightly with Ti.

very shallow effective junction (< 0.05 μm) is vulnerable to other processing and environmental factors such as mechanical stress.[114] Eventually innovations, such as raised source and drain junctions, will be necessary to replace the conventional structures (see Section 8.7).

8.4 CVD W PLUG AND OTHER PLUG PROCESSES

One of the most difficult problems in metallization is to ensure enough metal continuity at contact windows and vias. The step coverage of sputtered Al decreases rapidly with increasing contact window aspect ratio, and at small design rules the step coverage at contacts and vias drops below 20%. No metal reliability can be guaranteed at this level of step coverage, and consequently various forms of metal plugs have been developed.

8.4.1 Selective CVD W Plug

On a Si contact, the selective process starts from a Si reduction process as shown in Eq. (8.2). This process provides a nucleation layer of W grown on Si (but also consumes about 20 nm of Si) but not on SiO_2. The real W plug is grown by a hydrogen reduction process as shown in Eq. (8.1). The hydrogen reduction process deposits W rapidly on the nucleation layer, forming the plug. This process, however, does not have perfect selectivity, and as a result, spurious nucleation and W growth can occur on the SiO_2 (Fig. 17*a*). The mechanism of the loss of selectivity has been studied and reported numerous times,[115,116] but so far no foolproof technique has been identified to guarantee total selectivity.

Another factor that is unfavorable to a selective W plug is the difficulty in filling contact windows of different heights, as shown in Fig. 17*a*. Since the contact to the gate is always shallower (by an amount that equals the gate height + field oxide) than to the source/drain, selective W cannot fill both contact windows simultaneously. Therefore, selective W is suitable only for via contacts.

Various innovative techniques have been proposed to solve the loss of selectivity. The most promising deposits a nucleation layer such as TiN and then selectively removes this material from the SiO_2 area, leaving it only in the contacts,[117] as shown in Fig. 17*b*. This has several advantages: (1) It solves the different window height problem, because W grows from the sidewall as well as from the bottom of the contacts; (2) the selectivity loss is less severe, since now the plug grows from the sidewall and much thinner W is needed; and (3) the adhesion of the W plug to the contact is better because of growth from the sidewall. This process, however, shifts the difficulty of selective W deposition to the selective etching of TiN. Until a simple, low-cost selective W process is found, the majority of manufacturers will use the more expensive but more mature blanket W-plug process.

8.4.2 Blanket CVD W Plug

By depositing a metal nucleation layer on the entire wafer, CVD W can be blanket-deposited on the wafer and in the contact windows. The W on the SiO_2 is then etched off by RIE, leaving only the thicker W in the contact, as shown in Figs. 17*c* and 17*d*. Since this process relies on the removal of all CVD W except in the contacts, the uniformity of the W deposition and RIE etchback is critical for the control of this process. If this control is not exercised, then the W plug will be substantially recessed after the etchback, resulting in poor Al step coverage again.

The most commonly used nucleation layer is TiN. Although TiW also serves as a good nucleation layer, the RIE selectivity between TiW and W is poor, resulting in both localized etching along the sidewall and difficult end point detection. As a result, the process control of the plug process for TiW is inferior to that for TiN.

Usually, a silane reduction process [Eq. (8.2)] is used as the first step in blanket W deposition, to serve as a nucleation layer and also to reduce junction damage. After the silane reduction, hydrogen reduction as shown in Eq. (8.1) is used to grow the blanket W layer rapidly.

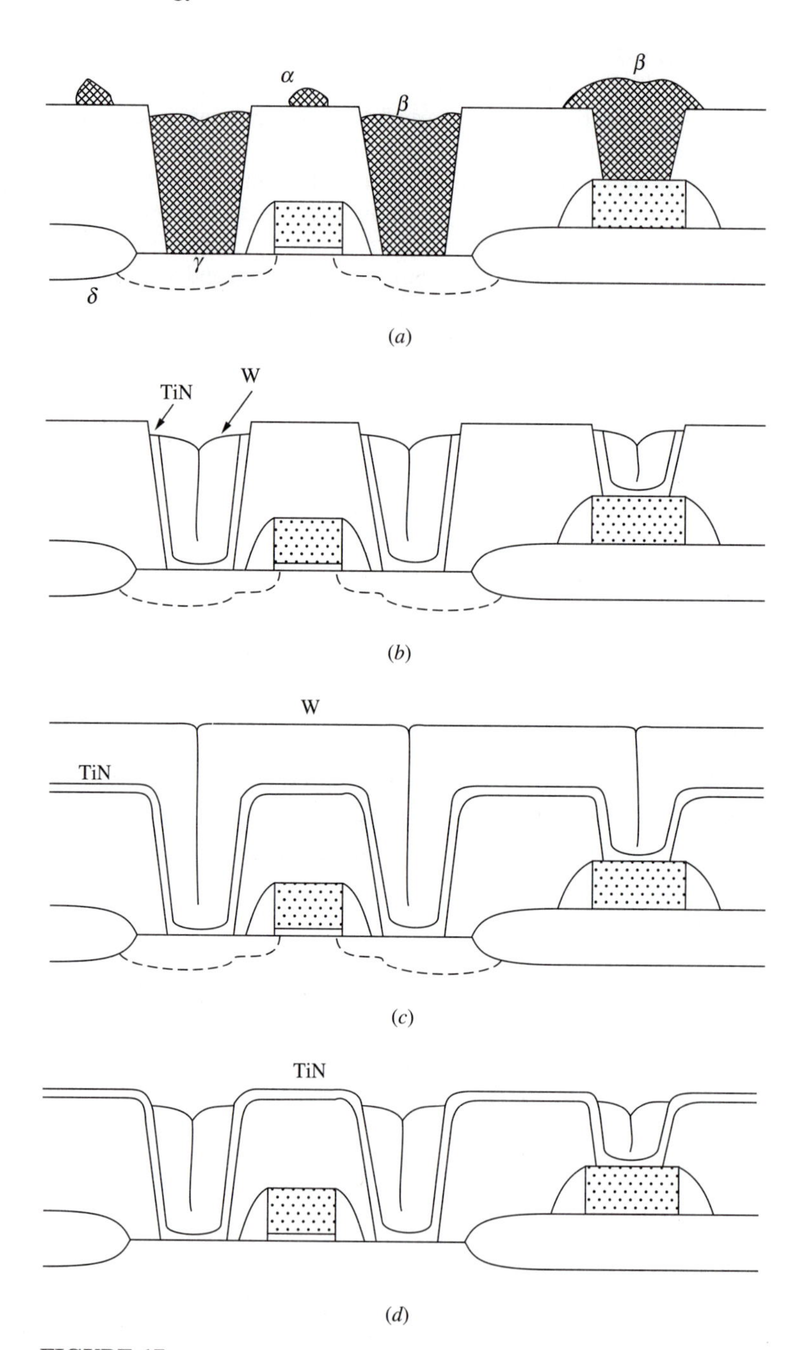

FIGURE 17

CVD W plug processes: (*a*) Selective W with no barrier layer: α: loss of selectivity; β: uneven fill; γ: junction leakage; δ: wormholes. (*b*) Selective W with barrier liner; does not have the problems shown in (*a*), but liners are hard to form. (*c*) Blanket deposition of W on TiN barrier. (*d*) After W etchback to form W plugs.

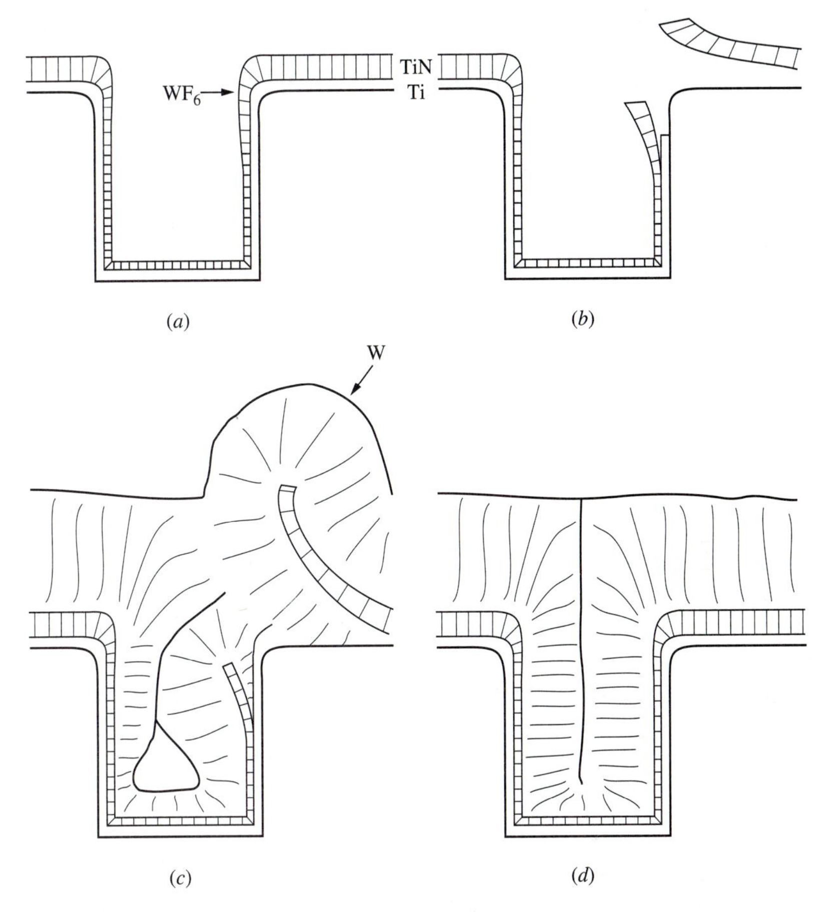

FIGURE 18
Tungsten volcano from barrier failure: (*a*) High stress and weak TiN at sharp corner attacked by WF_6; (*b*) Ti reacts with WF_6 and TiN peels back; (*c*) W deposits on both sides of peeled TiN, forming the volcano; (*d*) window with rounded corner stays intact.

TiN does not produce good contact resistance to n^+ and p^+ Si, and a thin layer of Ti (~ 30 to 50 nm) under the TiN is used. Ti, however, reacts readily with WF_6 (Table 1) and is more effective than H_2, Si, or SiH_4 in reducing WF_6. If there is any defect in the TiN film, or if the TiN is not properly "stuffed" with oxygen, then WF_6 can penetrate through the weak spots of the TiN film and react with the underlying Ti. Figure 18 shows the formation of "volcanoes" at the edge of contact windows. Sputter-deposited TiN has high stress, and the highest stress point is at the edge of a step (Fig. 18*a*). Because of the columnar growth of TiN, the material at the corner also is more porous and thus more vulnerable to WF_6 penetration. Once a reaction starts, the Ti is exposed to WF_6 and reaction proceeds rapidly. Losing support, the TiN peels off (Fig. 18*b*). Meanwhile, W nucleates on both sides of the peeling TiN (Fig. 18*c*) and grows into thick W films, forming a hump usually $> 1\ \mu m$ in size.

Because these humps or "volcanoes" are large, the etchback RIE process cannot remove them completely, resulting in intra- and interlevel metal shorts.

The W volcano problem is more severe for smaller contact windows because higher stress develops at surfaces with larger curvatures. The barrier properties of the TiN are critical for the prevention of W volcanoes. The deposition process also plays a significant role, especially at the nucleation stage. Rounding the edge of the contact window and relieving the local stress (Fig. 18*d*) reduce the occurrence of W volcanoes.

The equipment for blanket CVD W deposition and etchback has been developed in the last decade, and this process is widely applied to both memory and logic circuits.

8.4.3 Aluminum Plug

Recently, reflow Al plugs formed at high temperatures have become popular. The concept of depositing Al and reflow by a high-temperature anneal by RTP[118] or by excimer laser[119] has been investigated before. Reflow by RTP causes junction damage because Al stays in the molten state for about one second and reacts with the barrier material. Heating the Al by an excimer laser, on the other hand, melts the Al in a microsecond and no visible damage to the barrier layer is done. However, Al is highly reflective even in the UV range; as a result, high fluence of laser light is needed to melt Al. Consequently, the laser dose required to fill contact holes is close to that which causes runaway laser ablation.[120] Antireflective coatings reduce the laser fluence needed for melting and increase the process margin.[121] However, the alloying of metals such as Ti into Al causes high resistivity, whereas Cu causes etching problems. Therefore, despite some reported success,[119] laser reflow of Al has not been widely used.

In situ reflow of Al by heating the wafer to a high ($> 500°C$) temperature for a few minutes after cold Al deposition has been reported.[122] Because of the high temperature, the barrier metal has to be robust; usually TiN is used. Figure 19*a* illustrates the formation of an Al plug using this reflow method. The driving force for the Al flow process is the reduction of surface area to minimize the surface energy. To completely fill contact windows with aspect ratio ≥ 1, the wafer needs to be heated to about 550°C for about 3 minutes. At this high temperature significant bulk diffusion of Al occurs. Since surface energy provides the driving force for Al flow, it is imperative that the Al surface remain oxide-free throughout the reflow process. High vacuum, on the order of 1×10^{-8} torr, with low residual oxygen and water vapor is necessary. Also, proper wetting of Al on the barrier layer is important to prevent the agglomeration of Al. Although TiN is a good barrier layer material, the wetting of Al on TiN is poor. Therefore, composite wetting/barrier layers such as Ti/TiN, $TiSi_2$/TiN, or Si/TiN are used to enhance the Al reflow.

Al plugs can also be formed by high-temperature deposition or by multistep deposition.[123–125] Figure 19*b* shows a multistep Al plug process. A thin Al layer is sputtered at low temperature to serve as a continuous nucleation layer. The majority

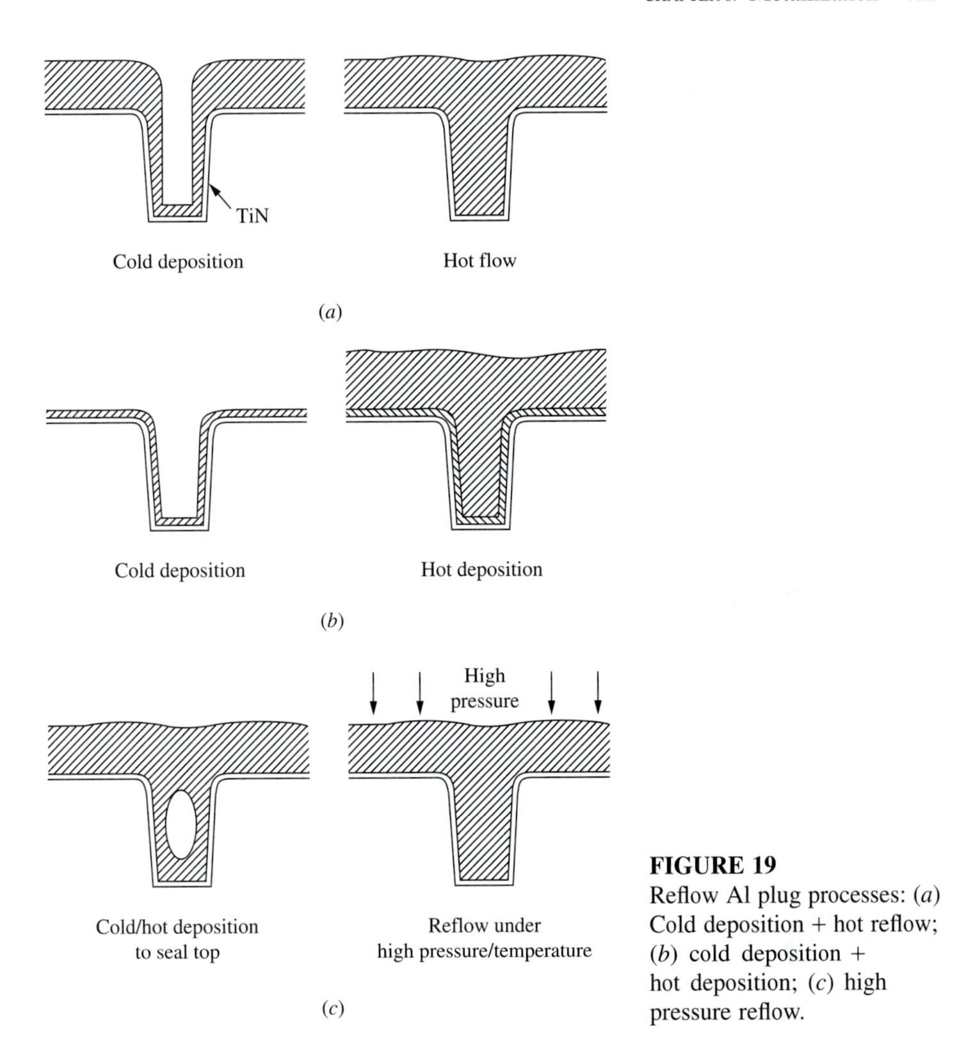

FIGURE 19
Reflow Al plug processes: (*a*) Cold deposition + hot reflow; (*b*) cold deposition + hot deposition; (*c*) high pressure reflow.

of the Al is then deposited at high temperature, 400 to 500°C. In this process, most of the mass transport occurs through surface diffusion. Therefore, the process can be done at a lower temperature (by about 50 to 100°C). The lower processing temperature puts a less stringent requirement on the barrier metal. However, because of the dependence on surface diffusion, the vacuum requirement is more demanding than for the Al reflow process.

For via applications the flow process needs to occur at < 450°C. This can be achieved by using special alloys. For example, for an Al-Ge alloy, the flow temperature can be reduced to below 400°C.[126] Ge, like Si, tends to precipitate out of Al during cooling and affects the contact resistance in small vias. Low-temperature Al flow is important for the opportunity it provides for low-cost fabrication of multilevel wiring. Recently, a low-temperature, high-pressure Al flow process[127] has been reported. In this process, a thick layer of Al is deposited first to seal the top of the contact vias. Ar at high pressure (600 MPa) is then introduced into the specially designed chamber where the wafer is held at 350 to 450°C. The combined effects of temperature (which makes Al soft) and pressure force a massive migration of Al, which flows and fills the vias, as shown in Fig. 19*c*.

Because of the dependence on surface and bulk diffusion, the flowed Al plug processes are sensitive to the shape and the aspect ratio of the contacts. The dependence on the wetting layer also causes processing problems, for example, the formation of $TiAl_3$ when Ti is used as the wetting layer and the formation of Si nodules when Si is used for wetting. The processing temperature range is small because Al does not flow if the wafer temperature is low, yet the barrier fails if the temperature is too high. Well-controlled and very uniform heating of the wafer is necessary. Therefore, although flowed Al plugs are used for the fabrication of some products at low cost, the technique has not reached maturity yet. This is still an area of intense R&D interest.

8.4.4 Other Metal Plug Processes

Contact plugs can be formed by other CVD processes such as CVD Al[20] and CVD Cu.[23,24] Both of these are still in their infancy. Polysilicon-based plug processes have been investigated. In this approach, polysilicon is blanket-deposited and then removed by RIE etchback to form a Si plug in the contact windows. A metal is deposited next and then annealed at low temperature to form a metal silicide with the Si plug. The excess metal is then stripped off by a selective chemical etch, and a silicide plug is left in the contact windows. By using this approach, silicide plugs of $CoSi_2$[128] and Ni_3Si[129] have been reported. Electroless plating of Co, CoW, and Ni to form selective plugs have been reported.[86] These processes, even though promising, have not been tested in large circuits to verify their compatibility with ULSI processing.

8.5 MULTILEVEL METALLIZATION

8.5.1 Introduction

The most difficult problems in multilevel metallization are (1) to deposit a void-free interlevel dielectric (ILD) that can fill the small gaps between metal lines without particle or reliability problems and (2) to smooth out the surface morphology and to reduce the accumulated topography to allow sufficient margin for the depth of focus for lithography. In this section we will focus in these two areas—ILD deposition and ILD planarization. We will also discuss the merit of more advanced architecture—stacked contacts, borderless contacts, and packaging limitations.

8.5.2 Interlevel Dielectric (ILD) Deposition

Plasma-enhanced SiO_2

Even though high-quality CVD oxides can be deposited without plasma assistance at high temperatures (650 to 850°C), they are not suitable for ILD because high temperature is not compatible with Al wiring. Oxides deposited at lower temperatures (300 to 400°C) are porous and water-absorbing and, in general, unsuitable for ILD use. With plasma enhancement (PECVD), oxides can be deposited at $\leq$ 400°C with acceptable quality. The details of CVD oxide and PECVD deposition have been described in Chapter 5. here we will examine the merit of PECVD oxides for ILD.

A particularly important precursor for plasma oxide deposition is TEOS, or tetraethylorthosilicate (also called tetraethoxysilane) $Si(OC_2H_5)_4$, which without plasma dissociates pyrolytically at about 650 to 750°C. TEOS dissociates cleanly into SiO_2 and complex by-products[130] with little C residue in the oxide. The chemistry of TEOS dissociation follows roughly the reaction

$$Si(OC_2H_5)_4 \rightarrow SiO_2 + \text{by-products} \tag{8.42}$$

In the presence of plasma, TEOS can react with oxygen according to

$$Si(OC_2H_5)_4 + O_2 \rightarrow SiO_2 + \text{by-products} \tag{8.43}$$

In Eq. (8.43), oxygen is added to help the oxidation of TEOS. With plasma enhancement, good-quality oxide can be achieved at $<$ 400°C. What makes TEOS an important precursor for CVD oxide is its capacity to cover the steps conformally. The molecular structure of TEOS consists of a silicate (SiO_4) core with an ethyl group attached to the oxygen atoms at each corner of the tetrahedral silicate structure. This structure is particularly difficult to chemisorb on the surface because Si cannot reach the surface directly. Therefore, it forms weak bonds with the surface and is mobile there before chemical reaction occurs. (Alternative theories propose that fragments of TEOS are mobile on the surface.) Consequently, oxides deposited with a TEOS precursor have good step coverage.

Silane-based chemistry can also be used for plasma oxide deposition. Commonly used chemistries are reactions with N_2O or with O_2:

$$SiH_4 + 2N_2O \rightarrow SiO_2 + 2H_2 + 2N_2 \tag{8.44}$$

$$SiH_4 + O_2 \rightarrow SiO_2 + 2H_2 \tag{8.45}$$

Silane-based oxides have poor step coverage, contain a large amount of hydrogen, and are seldom used. However, with advanced high-density plasma sources excellent step coverage and film properties can be achieved with silane-based chemistry. (See Section 8.5.2 on high-density plasma deposition).

Although the step coverage of PETEOS is conformal, it has a tendency to build up at the upper corners of metal lines, as shown in Fig. 20*a*. This overhang is caused by the shadow effect of LPCVD. At low pressure, the mean free path of the precursor molecule is larger than the feature size. Consequently, the bombardment of the surface by the precursor molecules follows a line-of-sight path, which produces

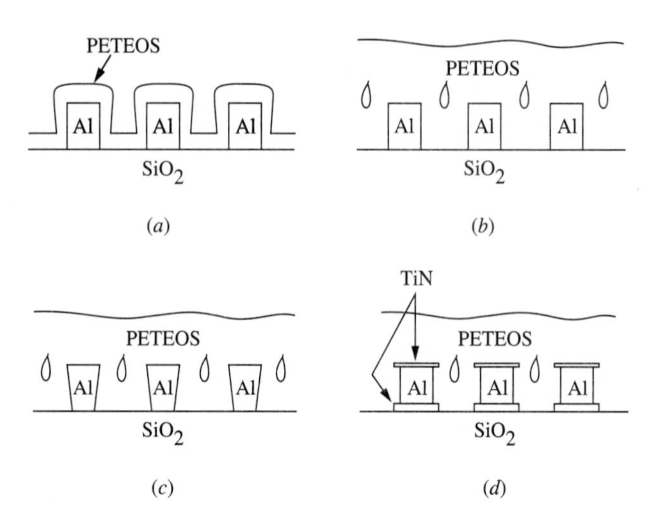

FIGURE 20
Keyholes in plasma-enhanced CVD oxide: (*a*) PECVD TEOS tends to build up at corner of metal lines; (*b*) keyhole voids; (*c*) re-entrant metal etching profiles cause larger keyholes even when the oxide is perfectly conformal; (*d*) metal etching profile control is more difficult for multilayered metal structures, resulting in keyholes.

the shadow effect. As a result, when PECVD TEOS is used to fill a small gap between two metal lines a void can form in the center, as shown in Fig. 20*b*. Note that even without the shadow effect and the overhang, metal profiles may be imperfect, as shown in Fig. 20*c*, which also results in voids. These are especially serious when multilayered metal structures for which uniform etching profiles are difficult to maintain are used (Fig. 20*d*). Voids in ILD not only pose a reliability concern, for fear of trapped chemicals in the void, but also may cause breaks in metal lines if a void is etched open during the planarization process, as shown in Fig. 21.

The task of depositing ILD without voids therefore cannot be solved by a process that is 100% conformal. Consequently, multistep processes using ion bombardment to round off corners to enhance the ILD filling capability have been developed.[131] As shown in Fig. 22, the multistep process (also called dep/etch/dep process) alternately deposits and etches the ILD to create a desired profile. The Ar etch step is used to cut the corners of the ILD and form a less severe geometry for the next PECVD process. Since a sidewall on the metal is created, this process is not sensitive to the etching profile of the metal lines. The multistep process, however, is more costly to fabricate because it requires multiple deposition and etching steps.

The repeated dep/etch/dep process can fill metal gaps with a height/space ratio (the aspect ratio, or AR) of about 1.0 without any voids. In principle the multistep process of deposition and etching can be tailored to fill gaps with higher aspect ratios, as shown in Fig. 23 By increasing the number of PECVD deposition and Ar etch steps, higher aspect ratio gaps can be filled. However, the cost for the ILD process also increases, with diminishing returns for each additional layer. Less expensive

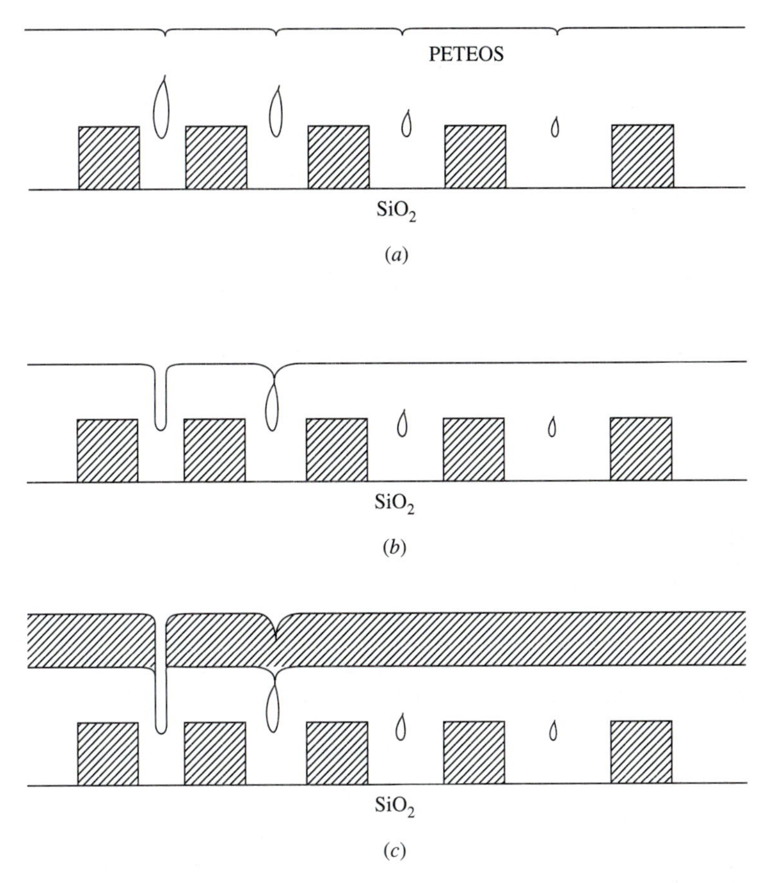

FIGURE 21
Effects of keyholes in ILD on metal step coverage: (*a*) ILD before planarization; (*b*) after ILD planarization, some keyholes open into trenches; (*c*) after metal deposition.

processes that can fill a wider range of gaps are more desirable. These will be discussed in the next four sections.

Atmospheric CVD TEOS/O_3 oxide

Silicon dioxide can be deposited by TEOS at low temperatures (150 to 400°C) with ozone[132] without plasma. Because the decomposition of TEOS is induced by O_3 and a plasma is not required, this process can be done at atmospheric or nearly atmospheric pressure to get better step coverage. At atmospheric pressure, the mean free path is $\leq 0.1\ \mu m$ and the deposition is conformal. The oxide can be deposited in a batch-type atmospheric pressure (APCVD) or in a single wafer or minibatch system at subatmospheric pressure (SACVD) for better process control. The chemistry of the two processes is the same.

The mechanism of TEOS/O_3 deposition is still poorly understood. Figure 24 shows a typical deposition rate vs. deposition temperature curve, and Fig. 25 shows

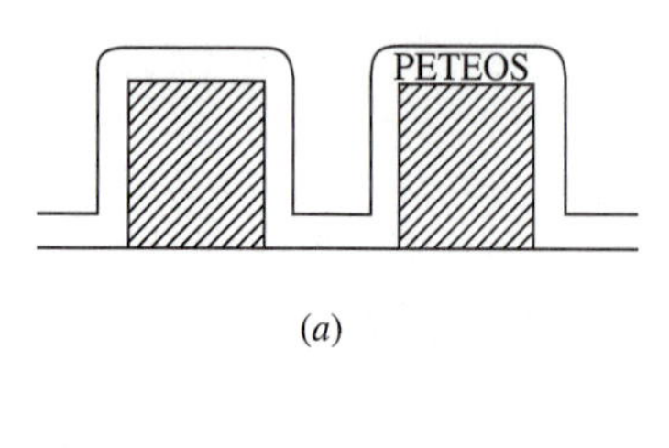

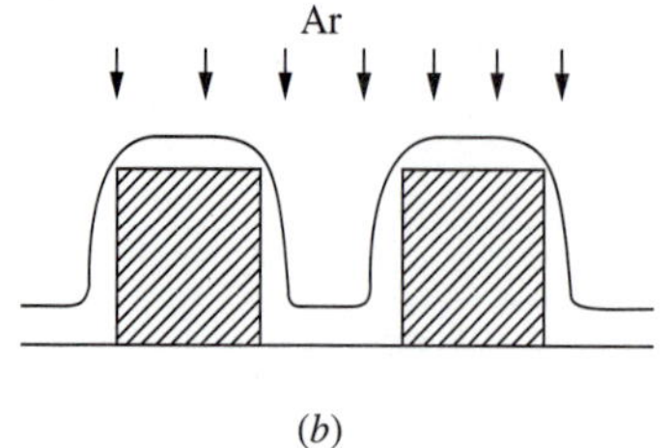

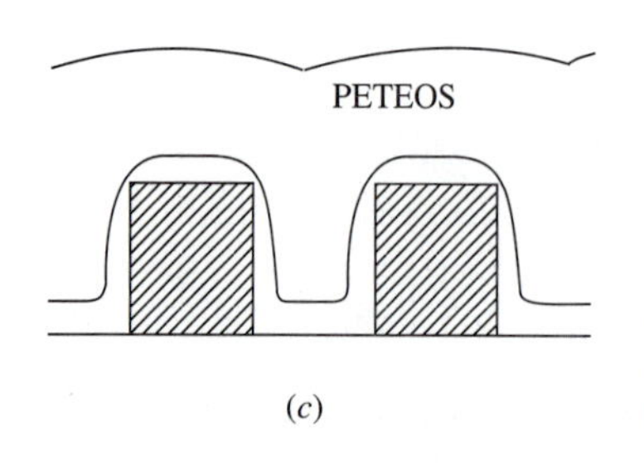

FIGURE 22
A multistep ILD process: (*a*) PECVD TEOS deposition; (*b*) Ar sputter etch to remove overhang; (*c*) new layer of PECVD TEOS to complete void-free gap fill.

a typical deposition rate vs. O_3 concentration curve. The deposition rate follows roughly an Arrhenius relation with $1/T$ at low temperature up to about 250°C, where it peaks and then declines at higher temperatures. The concentration of O_3 has a similar effect on the deposition rate. The deposition rate and the oxide properties are strong functions of the underlying Si surface.[133] For example, on thermal oxide

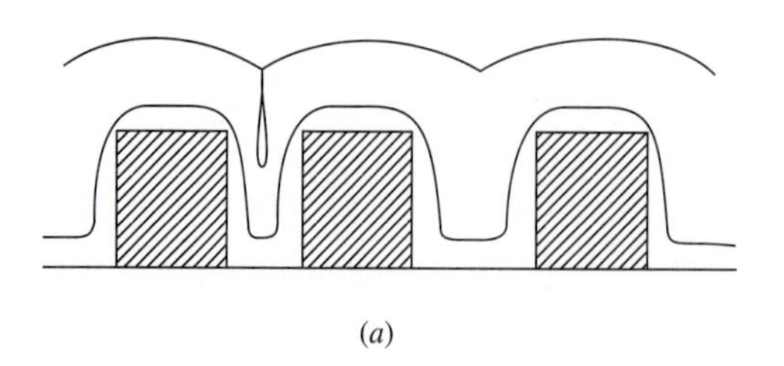

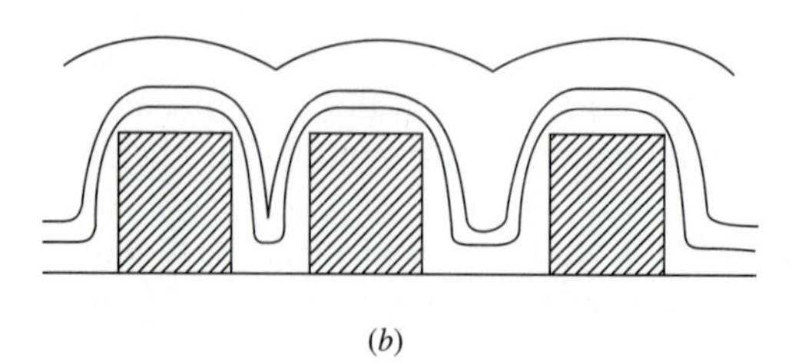

FIGURE 23
ILD gap-filling processes by repeated deposition and Ar etch: (*a*) dep/etch/dep; (*b*) dep/etch/dep/etch/dep.

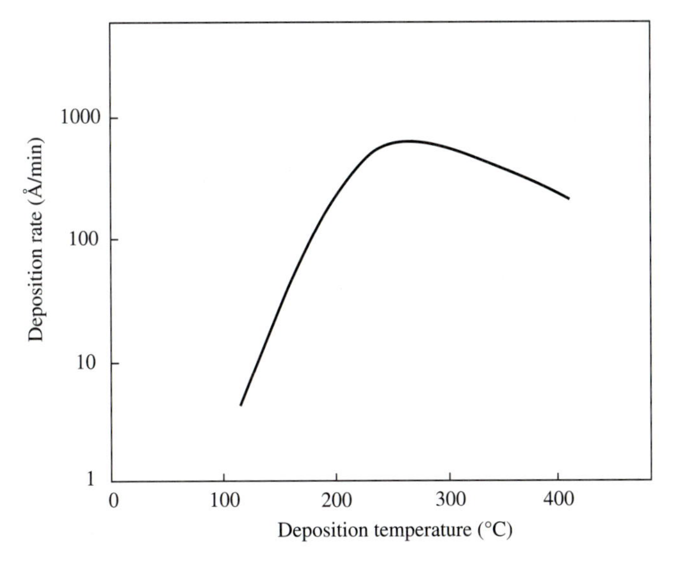

FIGURE 24
Deposition rate of O_3-TEOS vs. wafer temperature.

the growth rate is only about 70% that on Si. This sensitivity to surface condition suggests that a heterogeneous reaction on the surface is responsible for the decomposition of TEOS. The behavior in Fig. 24 also suggests that there is more than one reaction at higher deposition temperatures.

A peculiar characteristic of TEOS/O_3 deposition is the exhibition of flow-like step coverage,[134] as shown in Fig. 26. This led to a model[134] for the formation of liquid-like polymers of TEOS on the surface:

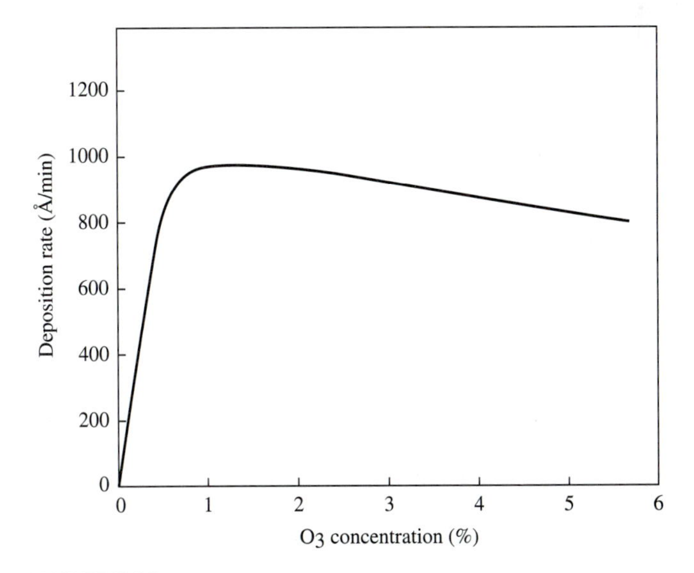

FIGURE 25
Deposition rate of O_3-TEOS vs. O_3 concentration.

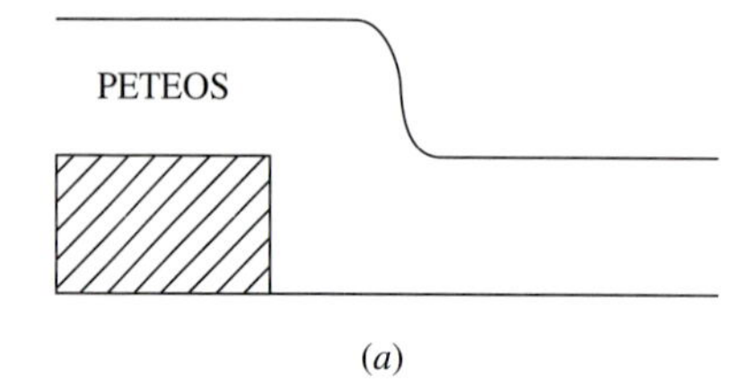

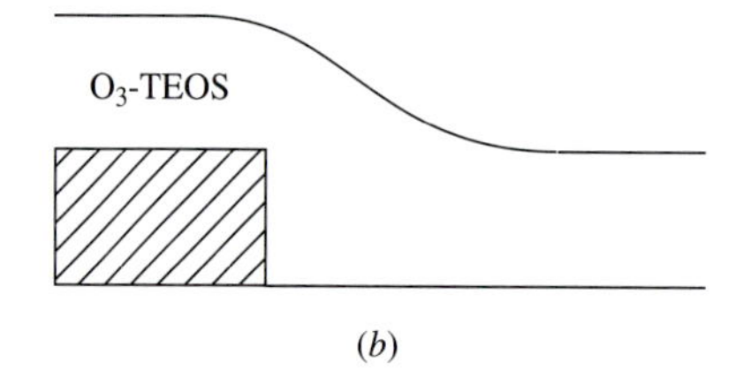

FIGURE 26
Flow-like characteristics of O_3-TEOS deposition: (*a*) PECVD TEOS; The step coverage is conformal; (*b*) O_3-TEOS APCVD; The step is flow-like.

$$n\mathrm{Si(OC_2H_5)_4} \rightarrow \mathrm{Si}_n\mathrm{O}_{(n-1)}\mathrm{(OC_2H_5)}_{2(n+1)} + 2n\mathrm{C_2H_4} + n\mathrm{H_2O} \qquad (8.46)$$

In this model, the TEOS oligomer eventually decomposes into SiO_2:

$$\mathrm{Si}_n\mathrm{O}_{(n-1)}\mathrm{(OC_2H_5)}_{2(n+1)} \rightarrow n\mathrm{SiO_2} + 2n\mathrm{C_2H_4} + n\mathrm{H_2O} \qquad (8.47)$$

Note that, although the mechanism for TEOS/O_3 deposition is not known in detail, the foregoing liquid model is not widely accepted. The flow-like behavior has also been explained by the more recent surface reaction model. These models[135,136] propose that the dissociation of TEOS is through the oxidation by atomic O from O_3. Although the details are beyond the scope of this book, Fig. 24 can be understood with a simplified model. If we assume that both the dissociation of O_3 (into $O + O_2$) and the oxidation of TEOS by atomic O occur on the surface, the Arrhenius behavior of Fig. 24 can be explained. The activation energy matches that for O_3 dissociation, and one may conclude that this is the rate-limiting factor.[135,137] At higher temperatures, desorption of TEOS from the surface becomes significant, which lowers the deposition rate. The flow-like structure can be explained[137] using the well-known fact that the vapor pressure of a droplet of radius r follows the Kelvin equation:[138]

$$\log_e p/p_0 = 2\gamma V/rRT \qquad (8.48)$$

where γ = surface tension
V = molar volume
p_0 = vapor pressure of a planar surface

The shoulder portion of the deposition (Fig. 26) has a higher vapor pressure for TEOS, which results in a lower deposition rate because of a higher desorption rate. For the knee portion the situation is the opposite: A convex geometry causes a low vapor pressure (for TEOS) and a higher deposition rate. Further experimental verification of these models is necessary to establish a detailed mechanism for TEOS/O_3 deposition.

The characteristics of O_3-TEOS can be summarized: (1) It can be deposited by APCVD or SACVD at 250 to 400°C at rates depending on both temperature and O_3 concentration; (2) the oxide has good step coverage and can fill higher aspect ratio gaps than PECVD TEOS; (3) the film properties are sensitive to surface conditions; and (4) the film properties are functions of deposition temperature and O_3 concentration. Higher temperature and higher O_3 concentration lead to higher-quality oxides but at the cost of lower deposition rates.

O_3-TEOS oxide is somewhat porous, with a shrinkage of 1 to 3% after a 450°C anneal. When left in air, the film absorbs moisture gradually. The water absorption can cause gate oxide damage and via poisoning and has to be avoided. Figure 27 shows the process sequence of a structure where the O_3-TEOS oxide is etched back to provide a gap filler. In this approach, a thin layer of O_3-TEOS oxide is first deposited to fill the gap (Fig. 27*a*). If it is left on top of the metal, then moisture absorption and re-release during metal deposition will lead to oxidation and poisoning of the metal contact (Fig. 27*b*). Instead, an RIE process is used to remove the O_3-TEOS oxide (Fig. 27*c*). Finally, a thick PECVD TEOS is deposited to complete the ILD process (Fig. 27*d*). This process can fill narrow gaps with aspect ratios of about 1.3 to 1.5. For higher aspect ratio gaps new deposition processes are needed.

Spin-on-glass (SOG)

Spin-on-glasses are widely used for low-cost fabrication of IC circuits. There are two basic types of SOG: organic and inorganic. The inorganic SOGs are silicate-based, and the molecular structure is shown in Fig. 28*a*. The structure of an organic

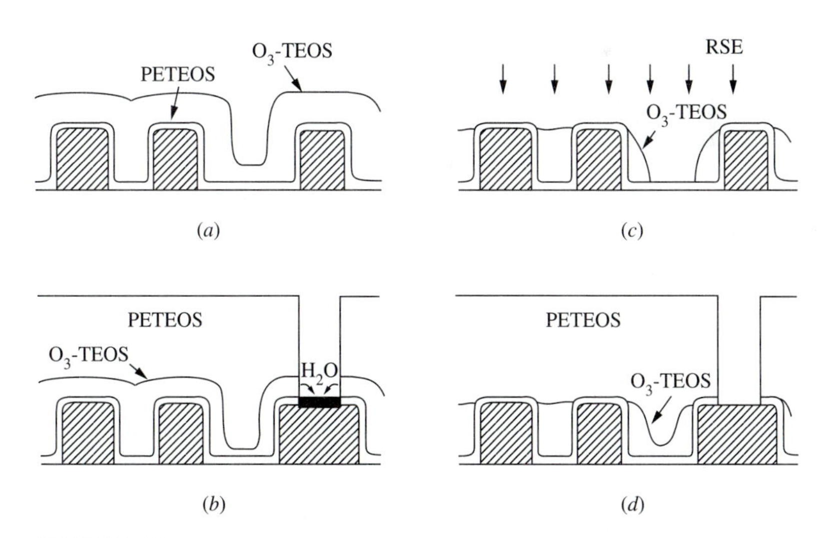

FIGURE 27
O_3-TEOS/PETEOS ILD processes: (*a*) PETEOS liner before O_3-TEOS deposition; (*b*) if the O_3-TEOS is left over metal, the exposed portion may absorb moisture and outgas, oxidizing the metal contact; (*c*) to avoid contact problems, remove O_3-TEOS from contact area by RSE; (*d*) a process that leaves no exposed O_3-TEOS.

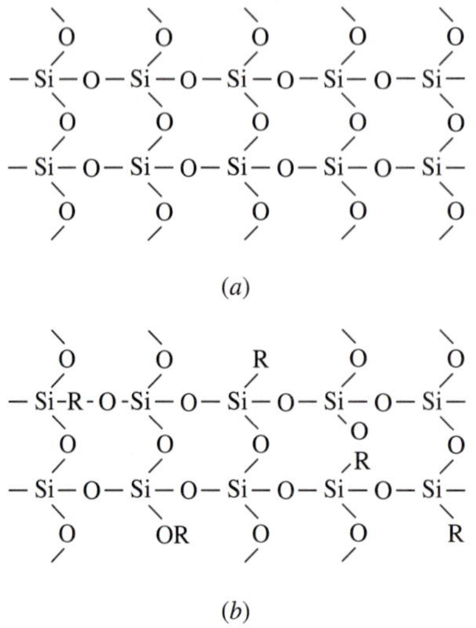

```
   \      \      \      \      \
    O      O      O      O      O
   /      /      /      /      /
 - Si - O - Si - O - Si - O - Si - O - Si -
   \      \      \      \      \
    O      O      O      O      O
   /      /      /      /      /
 - Si - O - Si - O - Si - O - Si - O - Si -
   \      \      \      \      \
    O      O      O      O      O
   /      /      /      /      /
```

(*a*)

```
   \      \             \      \
    O      O      R      O      O
   /      /      /      /      /
 - Si-R-O-Si - O - Si - O - Si - O - Si -
   \      \      \      \      \
    O      O      O      O      O
   /      /      /     /R     /
 - Si - O - Si - O - Si - O - Si - O - Si -
   \      \      \      \      \
    O      OR     O      O      R
   /             /      /
```

(*b*)

FIGURE 28
Schematic of the molecular structures of SOGs: (*a*) silicate SOG; (*b*) siloxane SOG.

SOG is shown in Fig. 28*b*. The organic, or siloxane-based, SOGs are featured with radical groups replacing or attaching to oxygen atoms. From these two basic structures, numerous commercial modifications have been made to adjust the molecular weight, the viscosity, and the final film properties for specific applications.

After full curing, the silicate SOG behaves like SiO_2. It does not absorb water in significant quantity and is thermally stable. However, the volume shrinkage during curing is large, and as a result the silicate SOG develops high stress ($\sim$ 400 MPa) and cracks easily during curing. Applying only a thin layer (100 to 200 nm) avoids the cracking problem. To build up thicker layers to fill gaps, multiple application and curing are needed. The ability to fill narrow gaps is also limited since low-viscosity SOGs (which have better filling properties) usually have higher shrinkage.

Silicate SOGs can be modified by adding P (up to 4 at%) to produce a softer SOG. The P is incorporated into the silicate by replacing Si atoms and breaking Si–O bonds. Therefore, adding P to a silicate SOG creates a structure similar to the siloxane SOG (Fig. 28*b*). The modified SOG (called P-silicate) is softer and has less shrinkage. The stress after curing is reduced to $\sim$ 200 MPa, and the SOG is more crack-resistant. It also has a better gap-filling capability and can be applied in thicker layers. Adding P, however, increases the water absorption and the modified SOG is both mechanically and electrically less stable than the silicate SOG.

Siloxane SOG, on the other hand, retains carbon even after curing. It has lower stress ($\sim$ 150 MPa) and can be applied in thicker layers. The organic SOG, however, is sensitive to water absorption, cannot tolerate even moderate temperature ($>$ 400°C) cycles, and has poor mechanical strength. Organic SOGs may also become unstable under plasma exposure. The amount of organic component in the siloxane is an adjustable parameter, and the SOG can be tailored to specific applications.

Because silicate SOGs are more SiO_2-like, they are used mainly for gap filling even when multiple applications and curing are necessary. The organic SOGs are more suitable for sacrificial SOG, e.g., the planarizing layer for etchback.[139] (See

Section 8.5.3.) Because of water absorption and subsequent via poisoning by SOGs, they are used only for gap fill and the excess is removed. The processing sequence shown in Fig. 27*a*, *c*, *d* applies to SOG with the simple replacement of SOG for O_3-TEOS oxide.

Because both the material and the method of application are inexpensive, SOG has become a popular low-cost ILD for gap fill. The etchback step (Fig. 27*c*) is expensive and a potential source of particles. Therefore, a nonetchback SOG process is important. Recent work has reported spin-on materials that can be left on metal without jeopardizing via contacts.[140,141] Fluorine treatment of a siloxane SOG to use as a nonetchback ILD is also reported.[142] No large-scale testing has been completed for these new materials and therefore the acceptance is still experimental. Although some fully cured SOGs have lower dielectric constants (3.0 to 3.6 compared with 3.9 to 4.5 for SiO_2) than oxides, in general SOGs absorb water and both the dielectric and mechanical properties are inferior to those of PECVD oxides. Lower cost, good gap-filling capability, and better ILD planarity still make SOG an attractive alternative to PECVD oxide. Currently, SOGs are used for manufacturing devices from $> 1\ \mu m$ to $0.35\ \mu m$ in size.

Bias-sputtered quartz

SiO_2 can be deposited by rf sputtering from a quartz target at room temperature. The sputter-deposited oxide is porous, has poor step coverage, and is not suitable for ILD use. Adding a dc bias, however, changes both the film quality and the step coverage drastically. The ion bombardment during sputter deposition provides sufficient energy to densify the oxide to a level similar to thermal oxide. The energetic ions sputter off the overhang from feature edges and create shadow-free deposition. Consequently, a complete filling of gaps between metal lines is achieved. Figure 29*a* shows the resputtering rate of SiO_2 as a function of ion incident angle.[143] The sputtering rate is low at normal incident angle and stays low up to about 30°. It has a broad peak at about 70° and then falls off sharply to zero at 90°. In bias quartz sputtering the deposition rate and bias are adjusted so that the deposition rate is between the minimum and the peak of the resputtering rates. If the deposition rate falls below the minimum resputtering rate, then the substrate is eroded and there is no deposition. If the deposition rate is higher than the peak of the resputtering rate, then bias sputtering is not effective, resulting in poor step coverage. Figure 29*b* shows the layer profiles over different features for biased sputter deposition. Note the building of sharp ridges over metal features. As the SiO_2 grows in thickness the ridges on smaller features disappear first. Bias-sputtered deposition (Fig. 29*b*) can fill narrow, high-aspect-ratio gaps solidly when parameters are properly adjusted.

Bias-sputtered quartz has not been used widely for ILD applications for the following reasons: (1) the low deposition rate; (2) particulates from chamber walls and shielding; (3) metal contamination from chamber parts; (4) radiation damage from energetic electrons (which produce x-rays); and (5) high compressive stress (~ 1000 MPa) of the film, which cracks easily. However, the benefits of bias sputtering during deposition have been recognized. The combination of bias sputtering with plasma-enhanced CVD of SiO_2 led to the most promising ILD process (see next section on HDP oxide).

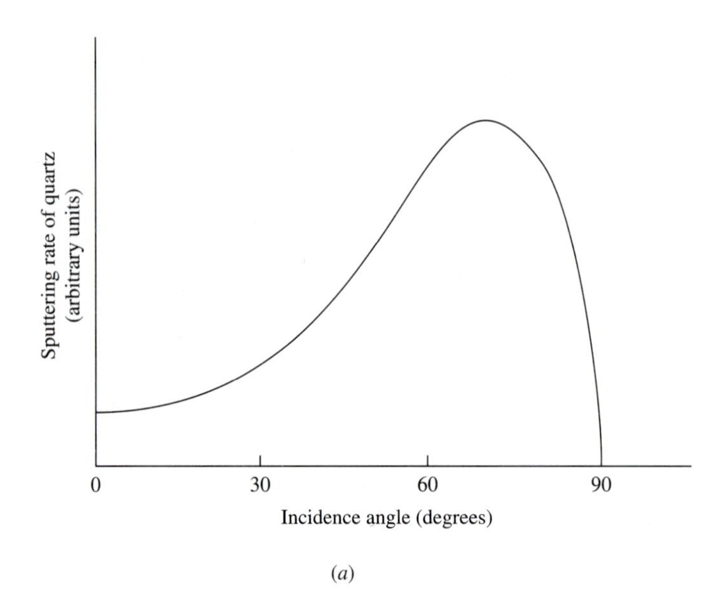

(*a*)

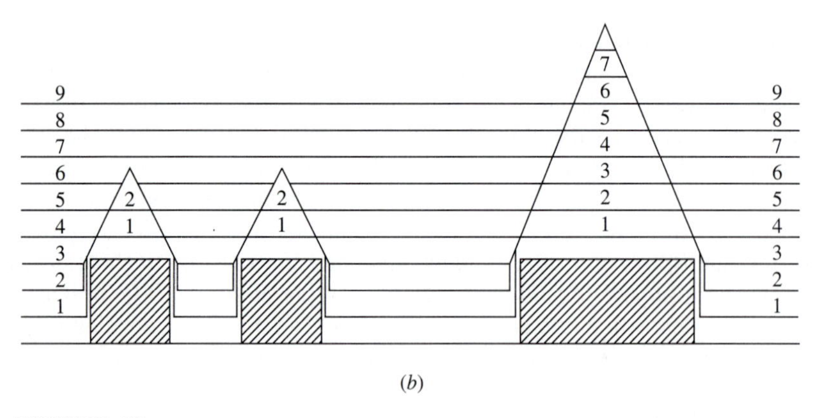

(*b*)

FIGURE 29
Bias-sputter deposition: (*a*) The sputtering rate of quartz vs. the incidence angle of Ar^+; (*b*) profiles of bias-sputter-deposited layers.

High-density plasma (HDP) enhanced SiO_2

Many of the shortcomings of bias-sputtered quartz can be overcome by combining bias sputtering with plasma-enhanced deposition. CVD oxide has a much higher deposition rate and is free from particulates from chamber walls. In addition, CVD oxide is free from metallic contamination and can produce films with lower stress. However, to achieve high deposition rate and effective bias sputtering without radiation damage one needs a high density of low-energy ions. Recent advances in high-density plasma sources (see Chapter 5) provide the necessary ingredient for combined CVD SiO_2 and bias sputtering for high-quality ILD at low temperature.[144,145]

Modern high-density plasma sources are capable of supplying low-energy ions with density $\geq 10^{12}$ cm^{-2}. The high density of plasma provides oxides of exceptional quality even at low deposition temperatures. In addition, by applying a bias, the re-sputtering effect, similar to bias-sputtered quartz, allows the filling of narrow gaps. As a result, even hard-to-fill gaps caused by the undercutting of Al wiring during the RIE of multilayered structures can be filled with a good quality oxide.

TEOS/O_2 chemistry similar to PECVD TEOS oxide has been used to deposit SiO_2 using electron cyclotron resonance (ECR) plasma at low temperature.[146] More commonly, silane and oxygen are used following the chemistry in Eq. (8.45). Silane is preferred over TEOS because (1) good step coverage is ensured by bias sputtering and therefore the conformal coverage for TEOS is not a significant advantage, (2) silane chemistry leaves no C residue, (3) silane is a gaseous precursor and is easier to handle than the liquid TEOS precursor, and (4) the poor film properties of silane-based oxide in PECVD are remedied by the higher-density plasma.

One drawback of high-density plasma is that, because it is so efficient, the plasma is sustained at a pressure one order of magnitude lower than for PECVD. Lower pressure (mtorr) means a lower deposition rate, about 200 nm/min. Note that this is not one order of magnitude lower than the rate of PECVD oxide, because higher plasma density helps to accelerate the oxidation of silane. To increase the deposition rate one must increase the pressure, the flow rate, or the microwave power. It is desirable to maintain a low pressure to avoid gas phase reactions and subsequent particulate generation. Usually, the maximum microwave power has already been applied, and it is hard to increase the plasma density further. By increasing the silane flow rate, a deposition rate as high as 650 nm/min has been achieved.[147] At this rate low-cost single wafer processing is possible.

In HDP oxide deposition, bias sputtering is achieved by an rf bias separately applied to the substrate to avoid charge accumulation. During deposition, the substrate is self-biased to a dc voltage that serves to accelerate the Ar ions for bias sputtering. This independent rf supply also provides an additional adjustable parameter for controlling the film properties, deposition rates, etc. Because of the high density of plasma, heating of the wafer during deposition is significant. Consequently, the thermal chuck that is used to maintain the wafer temperature needs to be carefully designed to provide effective cooling.

The high plasma density also raises the question of device damage. Although the energy of the ionized species in the plasma is low (< 40 eV), the substrate rf self-bias can be several hundred volts. Consequently, the oxide is deposited under constant ion bombardment with significant energy. Studies have been conducted to investigate the plasma damage.[148,149] These results indicate that there is no direct gate oxide damage (10 to 15 nm) during ECR oxide deposition. Larger-scale and long-term effects on gate oxide have not been reported yet.

High-density plasma oxide deposition equipment has just become commercially available. Because of the ideal properties of this process for ILD and for gap filling, the HDP oxide will become the dominating ILD process for devices 0.35 μm and below.

Device and contact degradation by ILD

Device and contact degradation can be caused by the ILD process through three different paths: (1) hydrogen or hydroxyl ions in low-quality oxides, (2) plasma damage, and (3) moisture absorption and the subsequent oxidation of metal contacts. The first two can cause V_t shift and hot carrier degradation, whereas the oxidation of metal contacts can cause high via contact resistance and early via electromigration failure.

PECVD oxides have a large number of unsaturated bonds. In addition, the oxides are more defective and have a density only $\sim$ 90% of that of thermal oxide. If they are not to be used as an ILD, a high-temperature anneal (800 to 950°C) in O_2 or steam can densify these oxides to nearly 100% of thermal oxide. However, if they are to be used as an ILD, the maximum temperature for annealing is restricted to $\leq$ 450°C and the oxide property remains inferior. An exception to this is the oxide deposited by high-density plasma, for which good oxide quality is achieved at low temperature.

Hydrogen and hydroxyl groups ($-$OH) are known to cause dangling bonds in the gate oxide and to accelerate hot carrier aging.[150] These species diffuse rapidly through PECVD oxides and are not effectively gettered by phosphosilicate glass (PSG). Consequently, some ions can reach the gate oxide where traps are created. The detailed mechanisms of hot carrier aging are discussed in Chapter 12.

Both PECVD oxide and HDP oxide employ a plasma and can potentially cause plasma damage to the gate oxide. Since the ILD is deposited after the first level of metal, some transistors are subjected to a high "antenna ratio." An example of the antenna effect is when a long Al wire is attached to the gate of a MOSFET. During a plasma process (RIE or deposition) the entire length of this wire can collect charges and charge up the gate oxide of the MOSFET. Since the ratio of the wire length to the width of the MOSFET can be large (100 to 1,000) significant charge accumulation can occur. Plasma damage effects are discussed in detail in Chapter 12.

O_3-TEOS and SOG are porous and water-absorbing. When exposed to aqueous solutions, or even cleanroom air, these oxides absorb water quickly from the atmosphere. The absorbed water can outgas in the vacuum chamber before contact metal deposition. Because the vias are opened directly to the metal, and the pumping in the via is always poor, the water vapor is chemisorbed on the metal and oxidizes the metal just before the contact metal deposition. This oxidation causes high contact resistance, known as via poisoning. To avoid this problem, these materials are used only as a gap filler, as shown in Fig. 27, to prevent air exposure.

8.5.3 Planarization

The demand for more levels of metal wiring for the ever-increasing complexity of circuits can only be fulfilled if ILD planarity can be maintained. The planarity of ILD serves two purposes—to provide a smooth surface to ensure good metal step coverage and to provide a flat-enough field, within the lithography depth of focus, that contact vias and metal wires can be patterned.

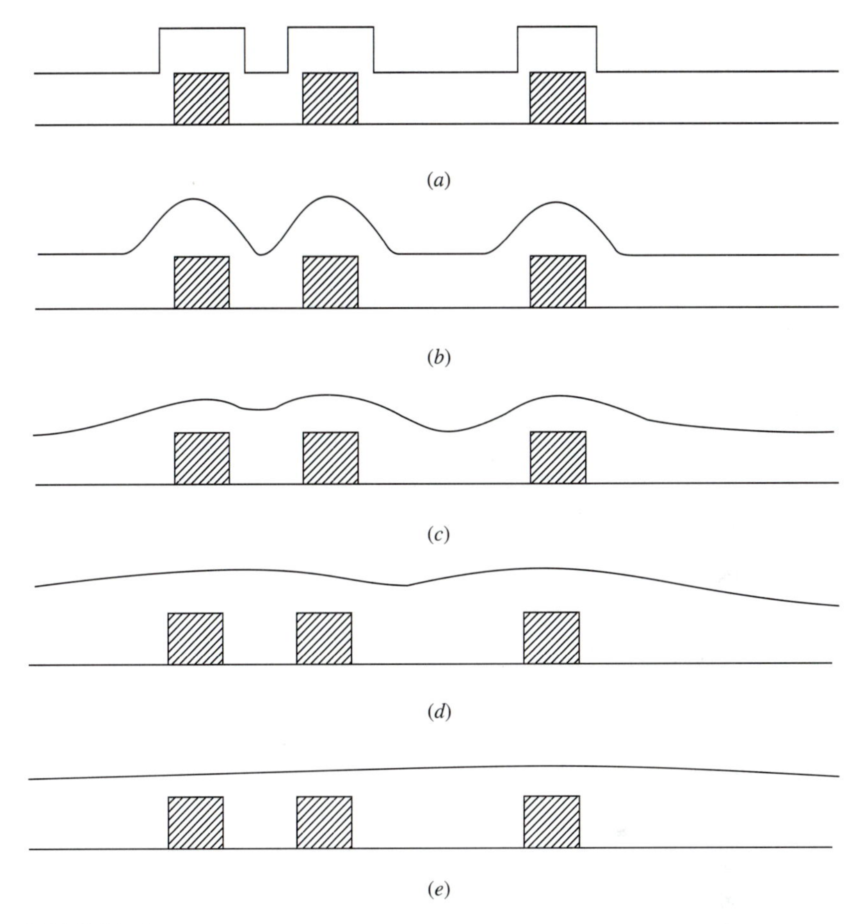

FIGURE 30
Qualitative definition of planarization: (*a*) No planarization; (*b*) smoothing; (*c*) partial planarization; (*d*) local planarization; (*e*) global planarization.

It is useful to define planarization terms unambiguously to avoid confusion. There is currently no universally recognized standard for planarization. Here we will define planarization both qualitatively and quantitatively. We follow the terminology defined by Wolf,[151] as shown in Fig. 30. Planarization is divided qualitatively into four categories:

1. *Smoothing:* Corners rounded and sidewalls sloped, but the step height not significantly reduced (Fig. 30*b*).
2. *Partial Planarization:* Smoothing plus a reduction in step height locally, as in Fig. 30*c*.
3. *Local Planarization:* Complete filling up of smaller gaps (1 to 10 μm); total step height to flat areas not significantly reduced (Fig. 30*d*).
4. *Global Planarization:* Local planarization plus a significant reduction in the total step height, as shown in Fig. 30*e*.

There are two ways to define planarization quantitatively. The first is to measure the reduction in step height. The metric used, the planarization factor β, is defined by $\beta = 1 - h_f/h_i$, where h_i is the initial step height before planarization and h_f the final step height. This definition does not consider the lateral extent of the planarization and is used for characterizing local planarization. For global planarization, the planarity is defined by the relaxation distance,[152] or its equivalent, the planarization angle θ. As shown in Fig. 31, at a single step edge the dielectric becomes thicker but eventually falls back to the original thickness. This behavior can be approximated by three straight line segments. The horizontal span of the tapered region is defined as the planarization relaxation distance R. The angle between the taper and the x-axis is defined as the planarization angle θ. R and θ are related simply by $R(\tan\theta) = h$, where h is the step height.

Both R and θ are used in the literature as metrics for planarity. Figure 32 shows the ranges of R and θ for various ILD deposition and planarization processes. PECVD, O_3-TEOS, and HDP oxides are mainly for ILD and gap fill and have relaxation distances in the range of < 5 μm and $\theta \approx 30$ to 45°. Local planarization is achieved with SOG or resist etchback with R between 1 and 100 μm and θ between 1 and 45°. Chemical/mechanical polishing (CMP), on the other hand, provides true global planarization with R in millimeters and $\theta < 0.01°$.

Lithography depth of focus

The limiting factors of resolution vs. depth of focus for optical lithography have been discussed in Chapter 6. In simple forms the resolution and the depth of focus can be written as

$$\text{Resolution} = K_1\lambda/\text{NA}^2 \tag{8.49}$$

$$\text{DOF} = \pm K_2\lambda/(\text{NA})^2 \tag{8.50}$$

where K_1 and K_2 are factors controlled by both the optical system and the photoresist and NA is the numerical aperture of the lens. To achieve ever higher resolution, the wavelength has been decreasing and the NA increasing for each generation of new technology. Consequently, the depth of focus for lithography systems is shrinking faster than improvements in the resolution are appearing. This decrease, plus the

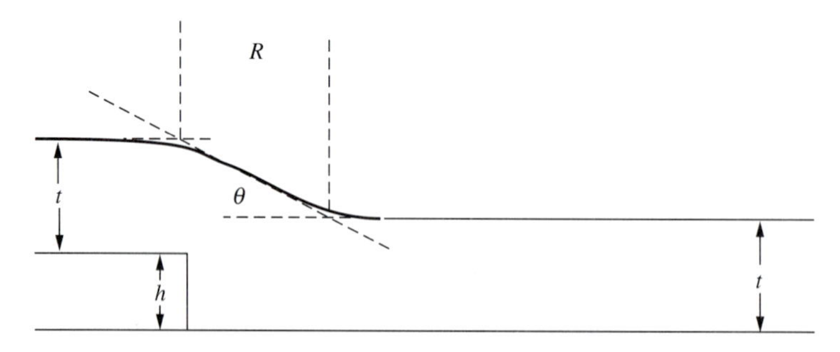

FIGURE 31
Quantitative definition of planarization: R = the relaxation distance, θ = planarization angle.

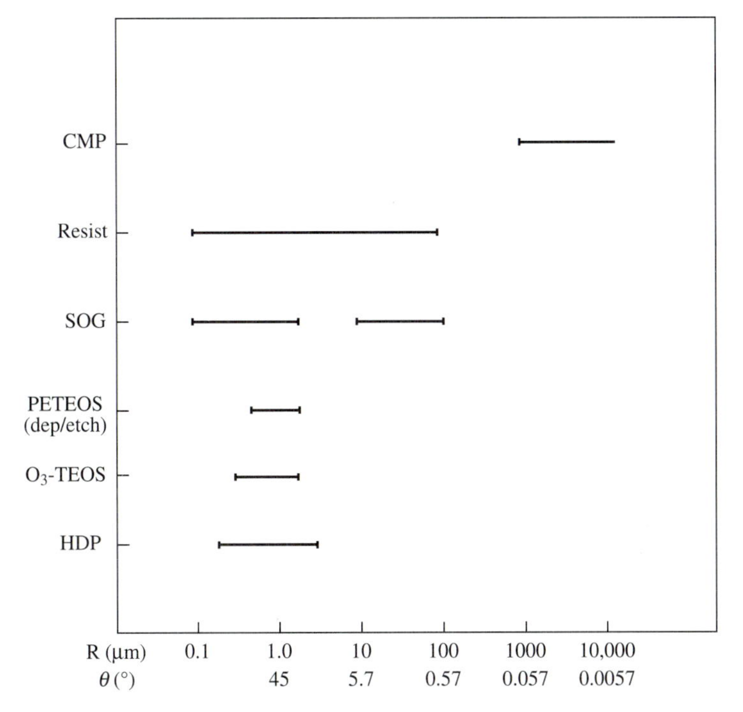

FIGURE 32
The range of planarization for various processes.

demand for more levels of metal wiring, is rapidly pushing the surface topography beyond the depth of focus capability. Figure 33 shows how the depth of focus is exceeded quickly by the buildup of topography when multiple levels of metals are used. The film thickness in Fig. 33 is typical for a 0.5-μm design. At window-2 and metal-2, the total topography has already exceeded the depth of focus limit if 0.5-μm resolution is desired from I-line lithography. Without using global planarization to reduce the topography, window-2 and metal-2 design rules have to be relaxed to allow the use of I-line lithography. At window-3 and metal-3, the design rules have to be grossly relaxed to stay within the depth of focus limits. The increase in die size causes a substantial penalty in yield and cost.

Etchback planarization

The surface topography caused by previous polysilicon or Al wiring can be smoothed by applying a thick layer of dielectric and a sacrificial layer of planarizing material such as photoresist or SOG and etching back, as shown in Fig. 34. The photoresist or SOG can smooth the surface features from a few microns to a few tens of microns. The entire structure is then etched back to the final ILD thickness by RIE, as shown in Fig. 34*b*. The smoothing of the ILD surface helps to achieve better Al step coverage as well as better metal linewidth control. The latter occurs through the removal of sharp edges at steps, which helps to reduce reflection from metal surfaces during lithography. Since this planarization is only local, it does not reduce the

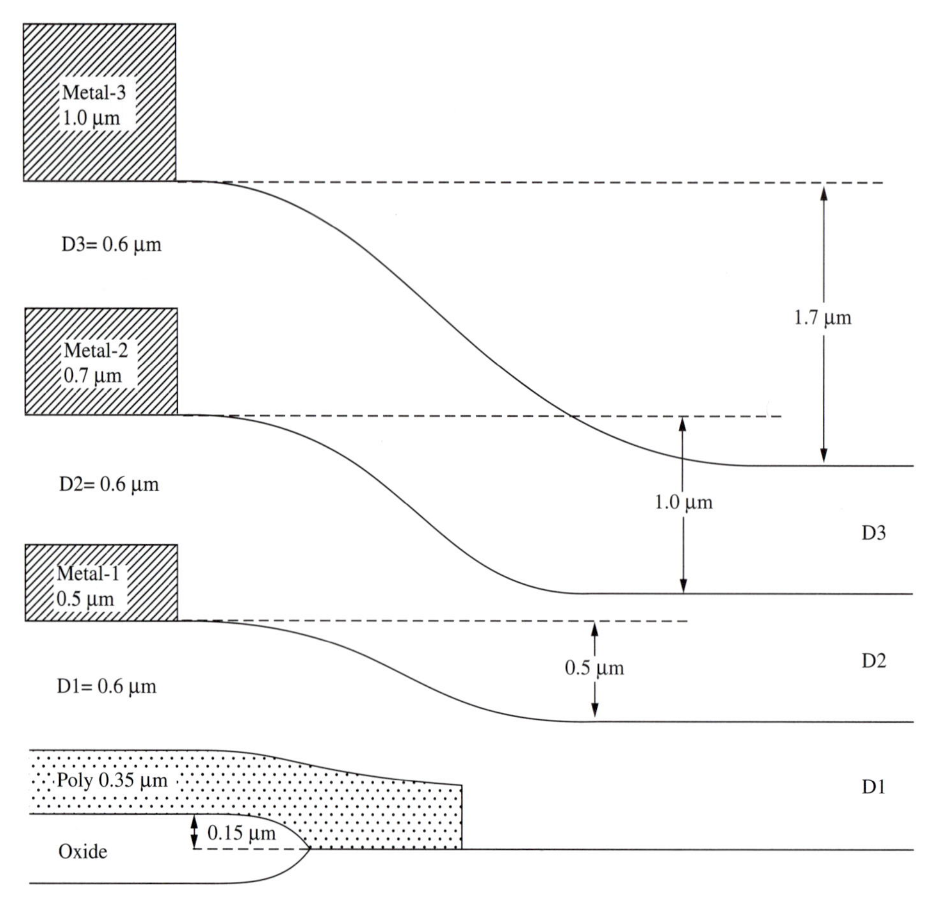

FIGURE 33
The buildup of topography by multilevel metallization.

total topography within a lithography field. The buildup of total surface topography increases just as rapidly as without planarization (Fig. 34*d*). Consequently, this process can be applied only to the first and at most the second dielectric and is difficult to extend to ILD planarization for three or more levels of metallization.

An etchback process can be used for global planarization in combination with a dummy resist, metal, or oxide mask.[153] (The dummy mask is sometimes called a planarization block mask, or PBM.) For example, if the metal pattern is designed so that there is no open area greater than a few micrometers in size, then the etchback process is adequate to provide uniform topography for the entire lithography stepper field. In practice, such a restriction cannot be part of the design rules, since it would compromise design freedom excessively. A method that does not pose any restriction to the design rules is to add electrically inactive "dummy" features whenever an open area greater than a set number (e.g., 4 μm) is detected. This approach of adding dummy metal features is shown in Fig. 35. The dummy metal features are incorporated into the metal mask and the processing sequence is not changed.

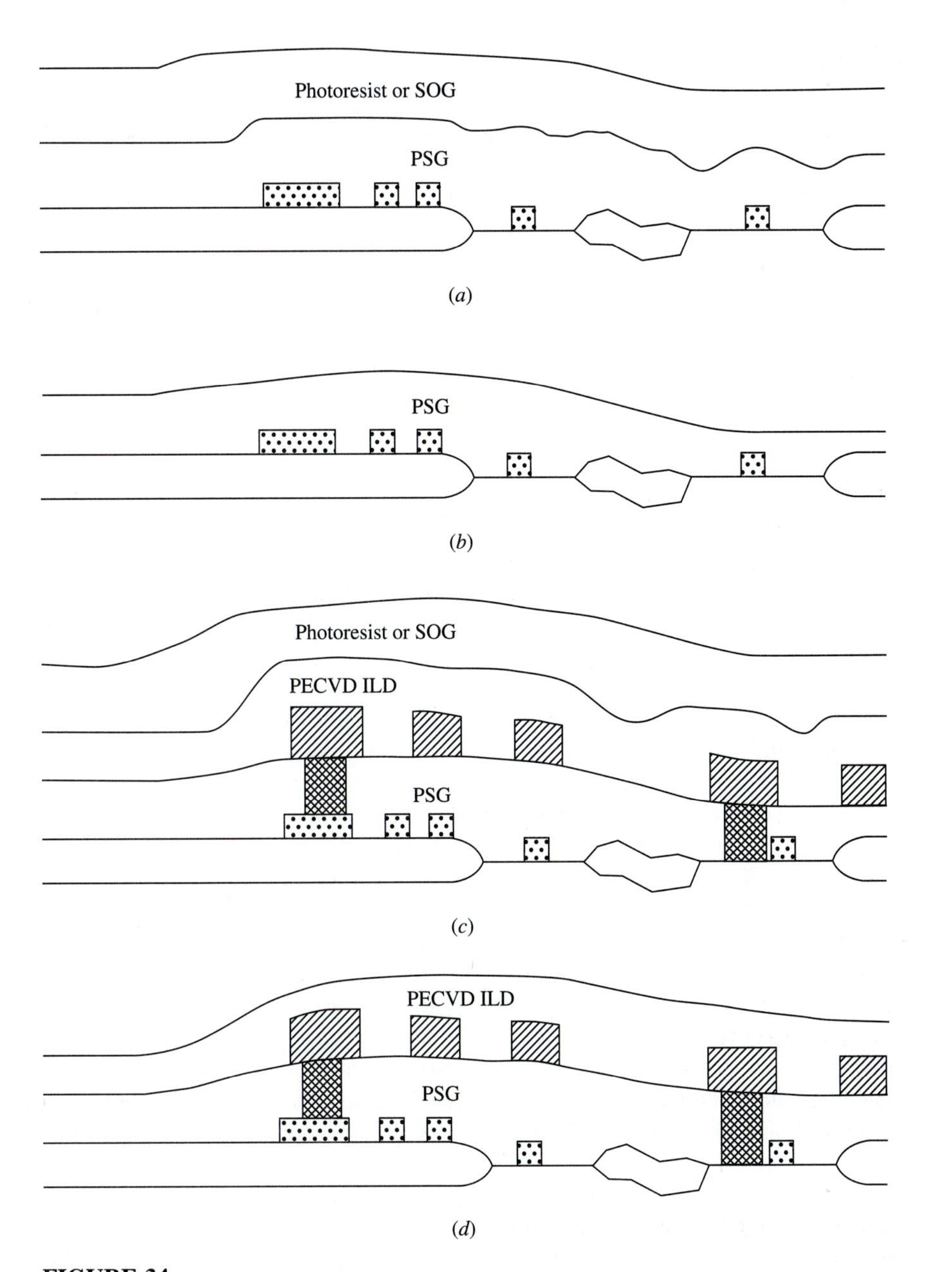

FIGURE 34
Etchback planarization: (*a*) Sacrificial layer over PSG; (*b*) topography smoothed by etchback; (*c*) after W plug, metal-1, ILD, and the deposition of a new sacrificial layer; (*d*) etchback provides a smooth morphology again, but does not remove built-up topography.

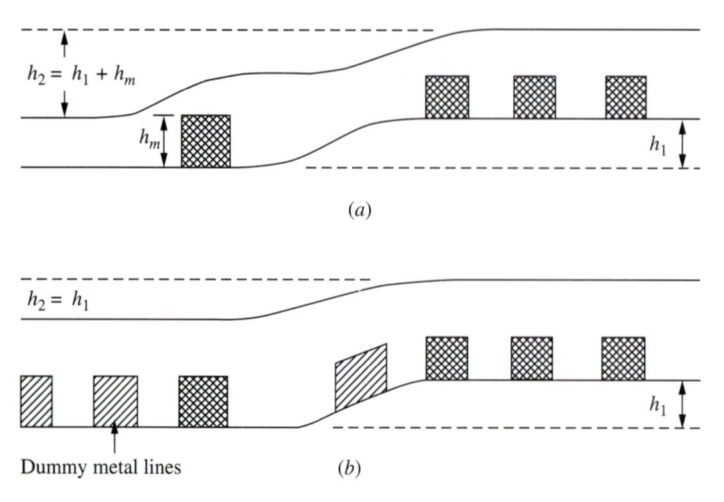

FIGURE 35
Global planarization using etchback and dummy metal features: (*a*) etchback with no dummy metal features results in local planarization only; (*b*) dummy metal + etchback provides global planarization.

Generating dummy features of any kind, however, requires substantial CAD development and CAD processing time. As a result, the design cost and the design cycle time are increased even when the processing cost is not always increased. Metal dummy features are also undesirable from the performance point of view, since wiring capacitance is increased. Even when the dummy features are electrically inactive, they must be held either at ground or at certain voltages to avoid interference with the circuit. ULSI circuits are designed through CAD simulation that does not account for nearby dummy features that can float to any voltage. To ground and connect the dummy metal features not only requires more CAD work but also complicates the design work. Therefore, even though dummy metal lines are easy to process, from a design point of view they are unacceptable.

An alternative is to use dummy oxide features instead of metal features. This is a more complex process because the oxide features need to be deposited, patterned, and etched independent of the metal features. The benefit is that it does not impose capacitance and grounding problems and is a more acceptable solution than metal dummy features.

One dummy oxide feature approach is shown in Fig. 36. After the lower-level metal patterning and etching are completed, a regular ILD is applied. A dummy oxide mask, which has to be generated by a CAD tool and contains information about densely patterned and open areas, is then used to pattern the ILD. As shown in Fig. 36*b*, the area containing densely packed metal lines is then etched by roughly the amount corresponding to the height of the underlying metal. This approach leaves a final topography between dense and open areas of roughly the same height, as shown in Fig. 36*c*. After this, a new layer of ILD and a sacrificial layer of photoresist or SOG are applied, and the entire structure etched back to the final ILD thickness. This approach is relatively simple, the oxide mask is noncritical and does not cost a great deal of CAD time, and the processing is straightforward. A drawback is its

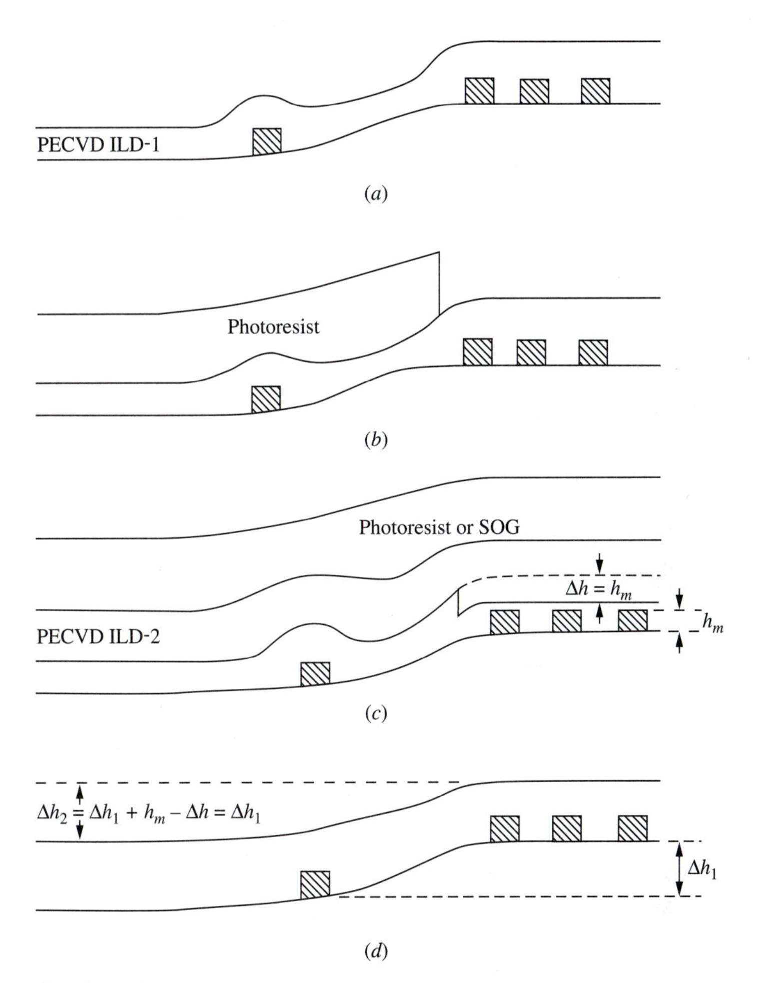

FIGURE 36
Global planarization by oxide mask and etchback: (*a*) Deposit ILD-1 for dummy features, the thickness ≈ metal height; (*b*) lithography for dummy features, (*c*) RIE ILD-1, apply a thicker ILD-2 and a sacrificial etchback material, (*d*) etchback planarization. Note that this process provides both local smoothing and global planarization, but does not remove topography from previous layers.

inability to handle gray areas where metal lines are separated by 5 to 20 μm and are somewhat irregular. To handle all possible designs, a refinement of this process is necessary, which will make the process more complex and the CAD work more elaborate.

Global planarization can also be achieved with an oxide etching process. As shown in Fig. 37, the topography introduced by metal lines can be removed by etching trenches into the next layer ILD and then smoothing out the structure by SOG or resist etchback. First, a thick layer of ILD is applied over the lower-level metal. Then a reverse-tone metal mask is used to pattern the ILD. RIE is used to etch trenches into the ILD, followed by wet etching to smooth the features somewhat, as in Fig. 37*c*.

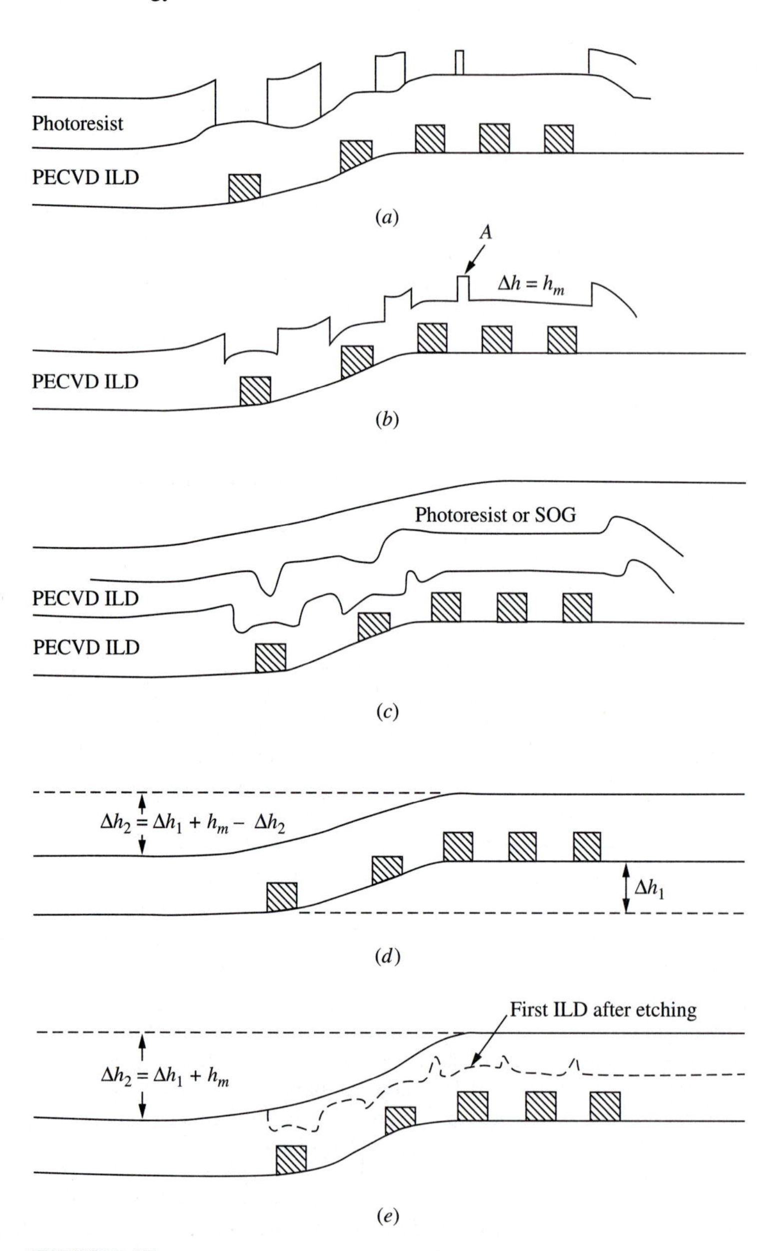

FIGURE 37
Global planarization by reverse-tone oxide mask and etchback: (*a*) Reverse-toned mask has larger features to relax alignment difficulties; (*b*) after oxide etch, the features are still abrupt; (*c*) wet etch removes sharp corners and thin "walls" such as A; then deposit ILD-2 and a sacrificial layer; (*d*) etchback planarization; (*e*) topography in areas with certain metal spacings is not reduced.

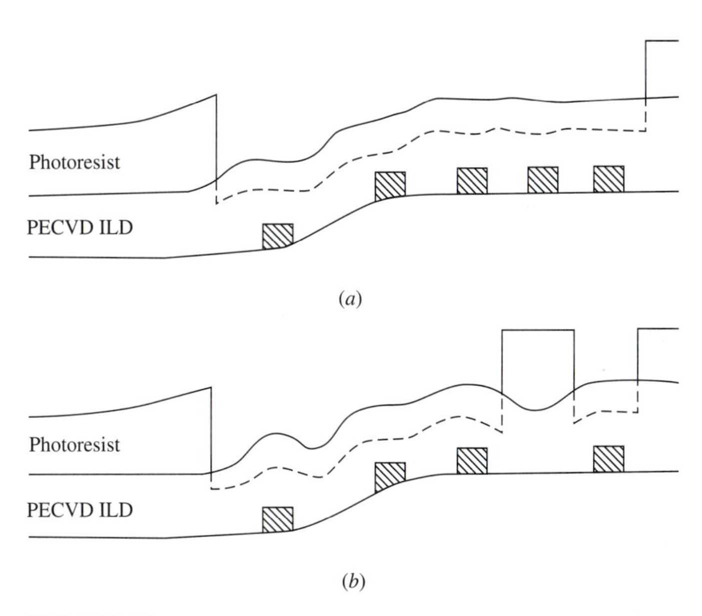

FIGURE 38
Using CAD to refine reverse tone mask: (*a*) When metal spacings are less than a certain value, no pattern is put between neighboring lines; (*b*) when two lines are separated by a certain distance, an extra line is put in. The dashed lines show the trenches after RSE of ILD-1. With the CAD refinement, the situation as in Fig. 37*e* will not occur.

Finally, a sacrificial photoresist or SOG is applied and the entire structure blanket-etched back to the final ILD thickness. An immediate problem is the misalignment between the trenches and the metal lines. In addition, as can be seen from Fig. 37*e*, for clustered metal lines separated by a certain distance the topography reduction is poor. At this separation an array of "walls" is left after RIE and wet etching, rendering planarization ineffective.

The reverse tone oxide patterning technique can provide a global planarization but is imperfect. This problem can be reduced by applying some CAD surgery to the reverse tone oxide mask. A thorough design with a CAD tool can decide where to put trenches and the proper size of the trenches. By applying these techniques, a reverse tone oxide mask can produce good global planarization, as shown in Fig. 38.

Even though the etchback approach can provide global planarization when combined with dummy feature masks or reverse tone masks, it does not provide totally flat topography and requires considerable process development. The additional mask required, the CAD work, and the additional oxide etch also increase the processing cost. These processes are rapidly being replaced by a low-cost, truly global planarization process, discussed next.

Chemical/mechanical polishing (CMP)

A true global planarization can be achieved by chemical/mechanical polishing (CMP) of the ILD.[154] This process has been used to prepare Si wafers for more than

30 years, as well as for glass polishing and for bonded silicon-on-insulator (SOI) wafers. Although its application to ILD planarization is more recent, both the equipment and the technique are well known to the semiconductor industry.

The requirements and the process for polishing a dielectric are quite different from those for polishing Si. For dielectric polishing, the goal is to remove topography and yet maintain good uniformity across the entire wafer. The amount of material removed is about 0.5 to 1 μm, compared to Si polishing where several tens of micrometers of material are removed. The uniformity requirement for ILD polishing is much more stringent than for Si polishing, since nonuniform ILD films lead to window etching and plug formation difficulties. Consequently, even with more than 30 years of experience in Si wafer polishing, the industry found that new processes and new equipment had to be developed for ILD polishing. Recently, CMP has been applied to polishing metal for W plug formation and embedded metal structures. Metal polishing involves chemistries significantly different from those for oxide polishing and requires further research and development.

The schematic of the CMP equipment is shown in Fig. 39. The essence of a CMP process is an automated rotating polishing platen and a wafer holder, which can both exert a force on the wafer and rotate the wafer independent of the rotation of the platen. The polishing is accomplished by a polishing slurry consisting of colloidal silica suspended in a KOH solution. An automatic slurry feeding system is used to ensure the uniform wetting of the polishing pad and the proper delivery and recovery of polishing slurry. For a unit designed for industrial use, automatic wafer loading and a cassette-to-cassette handler are also incorporated.

The basic polishing mechanism for an SiO_2 dielectric is the same as for glass polishing and has been summarized recently.[155,156] The mechanical removal rate of the glass is given by the Preston equation,[157]

$$R = K_p p v \tag{8.51}$$

where R = rate of removal
p = applied pressure
v = relative velocity between the wafer and the polishing pad
K_p = proportionality constant

K_p has units of $(\text{pressure})^{-1}$ and is known as the Preston coefficient. K_p is a function

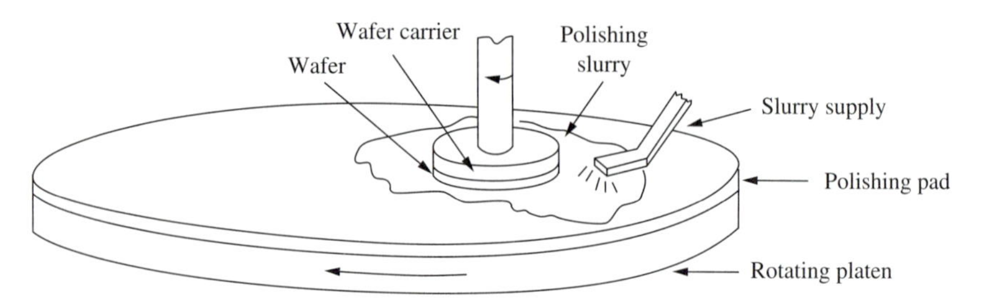

FIGURE 39
A schematic of a CMP polisher.

of the mechanical properties of the glass (hardness, Young's modulus), the polishing slurry, and the composition and the structure of the polishing pads.

Although Eq. (8.51) is essentially mechanical in nature, the microscopic action of polishing is both chemical and mechanical. The exact mechanism for polishing is still not fully understood. Figure 40 shows the current understanding of the chemical/mechanical events during polishing. The chemical reactions can be divided into four stages: (1) the formation of hydrogen bonds with the oxide surfaces of both the wafer and the slurry particles (hydroxylation), as in Fig. 40*a*; (2) the formation of hydrogen bonds between the wafer and the slurry (Fig. 40*b*); (3) the formation of molecular bonds between the wafer and the slurry (Fig. 40*c*); and (4) the breaking of the oxide bonds with the wafer (or the slurry) surface when the slurry particle moves away (Fig. 40*d*).

From the chemical/mechanical events shown in Fig. 40, there are three important implications: (1) Polishing is not a mechanical abrasion of slurry against the wafer surface; (2) both the presence of water and the pH of the solution affect the formation of hydrogen bonds; and (3) the size and the composition of the slurry particles are important. Relying on chemical/mechanical action rather than mechanical abrasion avoids a mechanically damaged surface layer on the ILD. The microscopic nature of CMP distinguishes it from mechanical abrasion, which is unacceptable to ULSI applications. Because of the chemical nature of bond breaking, the polishing slurry has a large effect on the polishing rate. The most commonly used slurry for SiO_2 polishing is silica with a particle size about 10 to 90 nm. In general, slurries made from oxides with higher oxygen bond strengths give higher polishing rates. The highest polishing rate achieved, with cerium oxide, is several times higher than that of silica.

Figure 41 illustrates the arrangement of a patterned wafer on a polishing pad. Instead of the wafer being held directly on the chuck, a backing film is sandwiched between the wafer and the chuck. The film provides elasticity between the chuck and the wafer. A composite set of two pads is used to achieve the desired rigidity/elasticity for polishing. Figure 42*a* illustrates the need for elasticity between the chuck and the wafer. Without a backing film any defect or particle on the chuck or on the back of the wafer will cause a thin spot. In addition, wafer breakage is more likely. If the polishing pad is too soft, as in Fig. 42*b*, then polishing is conformal to the wafer surface and no planarization is achieved. Planarization is caused by the capability of the polishing pad to bridge over low spots on the wafer and thus preferentially removing material from the high spots, as shown in Fig. 42*c*.

Therefore, microscopically, the removal of material in contact with the polishing pad is mainly chemical in nature. Planarization occurs because only high spots on the wafer touch the polishing pad. Because the macroscopic removal of material depends on the mechanical contact between the wafer and the polishing pad, the Preston equation [Eq. (8.51)] still applies well to CMP.

The detailed mechanism for planarization has not been fully elucidated and is an active field of research. A phenomenological model[158] using three parameters, corresponding to high features, low features, and horizontal erosion, is successful in simulating the erosion profiles of isolated and clustered features. Recently, a fluid model, which treats the slurry as a hydrodynamic layer, has been developed.[159] This

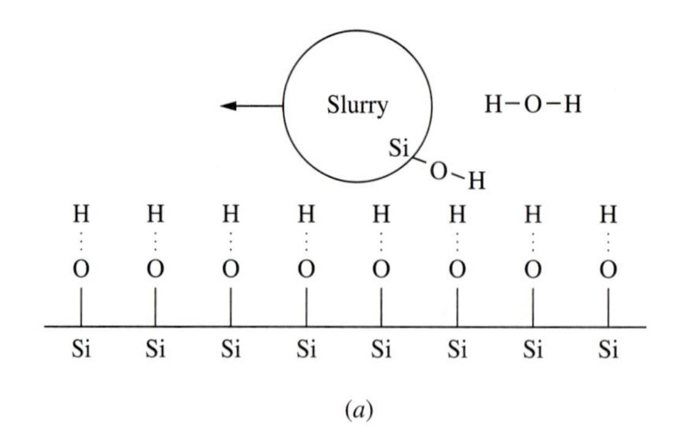

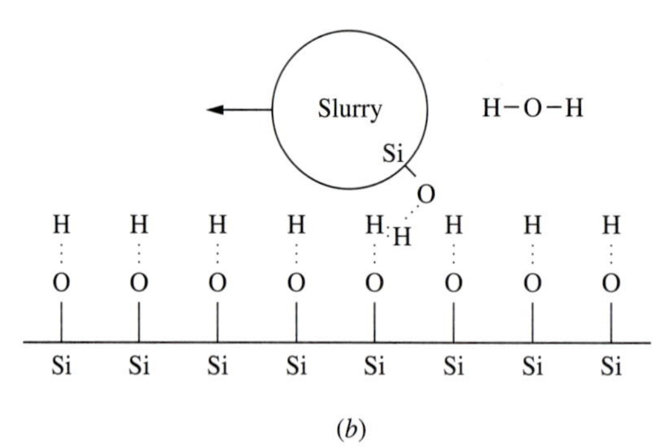

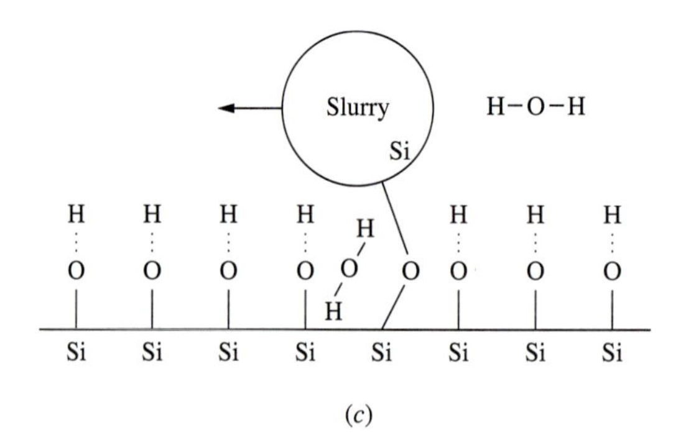

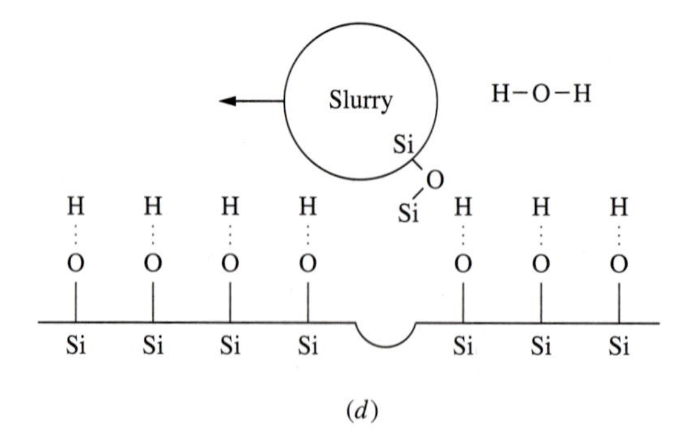

FIGURE 40
The mechanism of chemical/mechanical polishing: (*a*) In aqueous solution oxide forms hydroxyls; (*b*) hydrogen bond is formed between the slurry particle and the wafer; (*c*) Si–O bonds are formed by releasing a water molecule; (*d*) the Si–Si bond breaks when the slurry particle moves away.

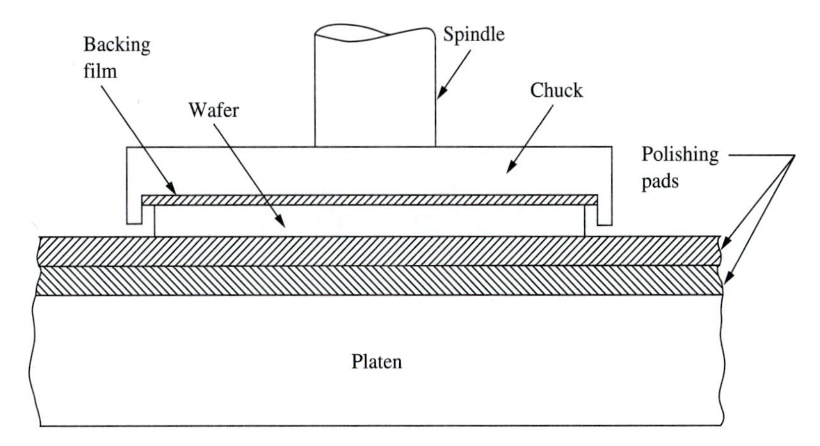

FIGURE 41
Details of the CMP wafer carrier and polishing pads.

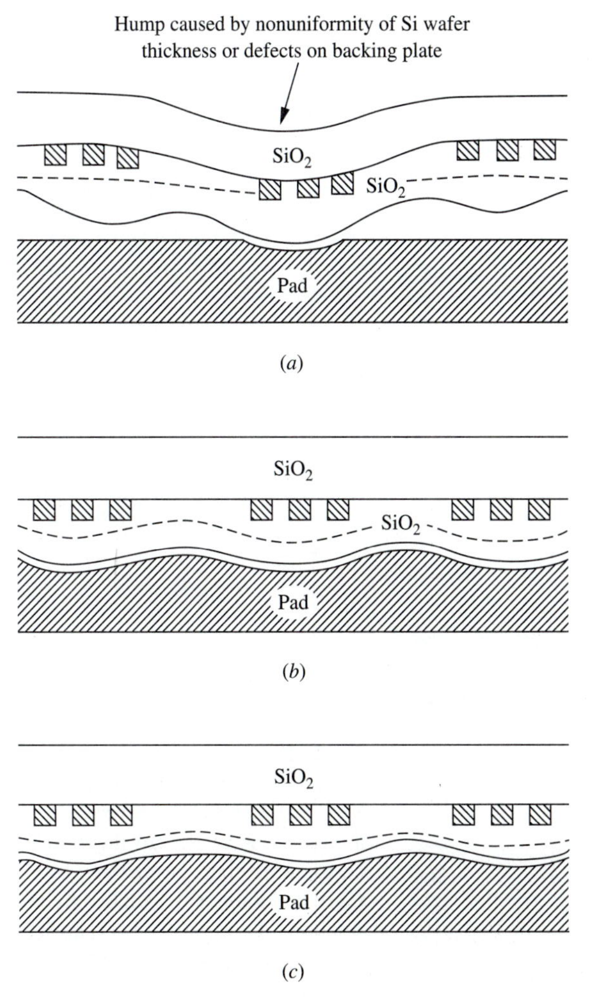

FIGURE 42
The effect of rigidity of the backing chuck (film) and the polishing pad: (*a*) A rigid chuck may bend the wafer and cause a thin spot; (*b*) a soft pad conforms to the surface and does not planarize; (*c*) a moderately rigid pad polishes the high spot on the wafer. The dashed lines show the final polished profile.

model showed limited success in simulating the erosion profiles. Therefore, qualitatively the planarization by CMP is understood, but quantitative understanding is still lacking.

The mechanism for CMP of metal is less studied and less understood and is more complex than oxide polishing. Figure 43 illustrates one proposed mechanism for metal polishing.[160] This model employs both a chemical etching and a passivation mechanism. Metals form an oxide layer on the surface, as shown in Fig. 43*a*. During polishing, this oxide layer is removed by a mechanism similar to that for oxide polishing (i.e., hydroxylation + bond formation with slurry + bond breaking from wafer). Once the oxide is removed, the metal is etched by the chemicals in the slurry solution, as shown in Fig. 43*b*. Simultaneously, the exposed metal forms a new passivation layer through oxidation by the solution, as in Fig. 43*c*. In reality, the three processes (the removal of metal oxide, metal etching, and repassivation) occur simultaneously

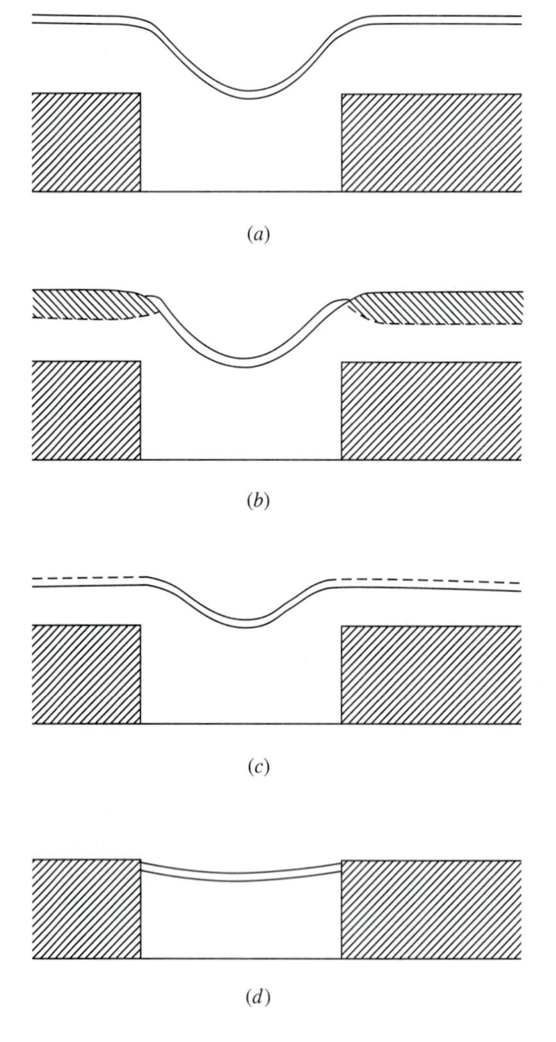

FIGURE 43
Mechanism for metal polishing: (*a*) Free metal surface is covered by a passivation oxide; (*b*) the oxide on the high spots is removed by CMP and the metal is etched by the solution (shaded area); (*c*) regrowth of oxide; (*d*) processes in (*b*) and (*c*) repeat until no high spot is left.

and the polishing rate is determined by the balance of these processes. Therefore, for metal CMP the polishing slurry must contain all three components: the fine slurry particles, a corrosion (etching) agent, and an oxidant. Figure 43*d* shows the eventual end point of metal CMP. Planarization is achieved by the mechanical rigidity of the polishing pad similar to oxide polishing.

For metal CMP, the most extensive experience is with W CMP. Fine alumina powder is the most commonly used slurry. The slurry solution consists of potassium ferricyanide, $K_3Fe(CN)_6$, and potassium dihydrogen phosphate, KH_2PO_4, with the pH adjusted to about 5.0 to 6.5.[160] In this solution, the ferricyanide serves both as the etchant and the oxidant. The KH_2PO_4 provides the desired pH for etching. The etching reaction is as follows:

$$W + 6[Fe(CN)_6]^{3-} + 4H_2O \rightarrow (WO_4)^{2-} + 6[Fe(CN)_6]^{4-} + 8H^+ \quad (8.52)$$

Competing with the etching reaction is the passivation reaction:

$$W + 6[Fe(CN)_6]^{3-} + 3H_2O \rightarrow WO_3 + 6[Fe(CN)_6]^{4-} + 6H^+ \quad (8.53)$$

Because of the delicate balance between etching, passivation, and oxide removal, the control of the slurry is important. If etching [Eq. (8.52)] becomes dominant, then W is etched isotropically and no planarization is achieved. To improve planarity, a weak base, ethylenediamine, is added to the slurry to adjust the pH closer to neutral. Potassium ferricyanide, however, is highly toxic, and the disposal of used slurry causes severe environmental impacts. This compound is now replaced by ferric nitrate $Fe(NO_3)_3$.

A major difficulty in the application of CMP is to guarantee uniform polishing over the entire wafer (uniformity). To achieve uniform polishing, three conditions have to be met: (1) Every point on the wafer has to travel at the same relative velocity to the polishing pad, (2) the polishing slurry is uniformly distributed under the entire wafer, and (3) the wafer itself should be symmetrical. Condition 3 is difficult to achieve for 150-mm wafers with a primary or a secondary flat and is easier to achieve for 200-mm wafers with a notch instead of flats. Condition 2 is hard to guarantee because new slurry always gets to the edge of the wafer first. Condition 1 is a controllable parameter and is examined in the following.

Consider the configuration in Fig. 44. Let the rotation speed of the wafer, which is at a distance R from the center of the platen, be ω_2, and let the rotation speed of the platen itself be ω_1. Consider three points A, B, and C on the wafer. A and B are at the opposite edge of the wafer whereas C is at the center of the wafer. The relative velocities of points C, A, and B to the platen are given by

$$v_C = \omega_1 R \quad (8.54)$$

$$v_A = \omega_1(R - r) + \omega_2 r \quad (8.55)$$

$$v_B = \omega_1(R + r) - \omega_2 r \quad (8.56)$$

It is easy to see that these three points will move at the same relative velocity to the platen if $\omega_2 = \omega_1$. Figure 45*a* shows the trajectories of the points A, B, and C on the platen after a complete revolution. Under the condition of $\omega_2 = \omega_1$, the loci of points A and B are circles and both A and B travel the same distance as C.

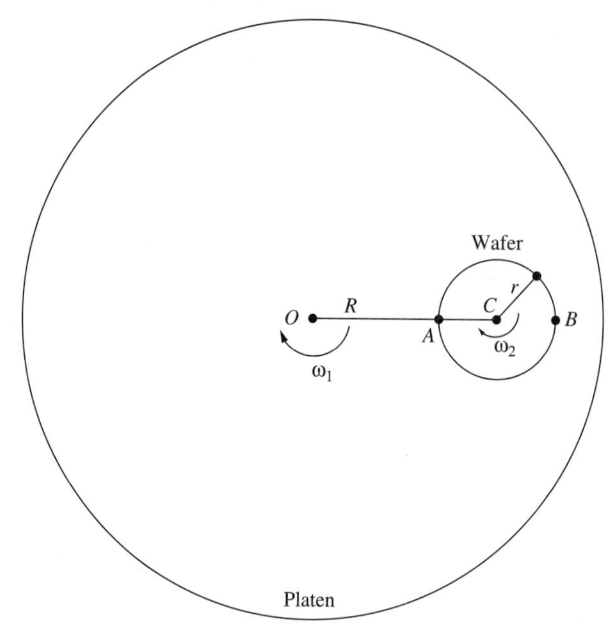

FIGURE 44
The relative motion of a wafer to the platen. Point C, the center of the wafer, is fixed in space. Both the platen and the wafer rotate.

This is equivalent to $v_A = v_B = v_C$. (It has also been proven mathematically that when $\omega_2 = \omega_1$ every point on the wafer moves at the same relative velocity.[161]) If $\omega_2 = -\omega_1$, however, the loci of points A and B travel a longer distance than C, and the edge of the wafer will polish at a higher rate than the center, as in Fig. 45*b*. If the wafer rotates faster than the platen, i.e., when $\omega_1 < \omega_2$, then a point near the edge of the wafer travels a longer distance than a point near the center, as shown in Fig. 45*c*. Thus, we can conclude that when the wafer rotates at a higher speed than the platen, the edge of the wafer will polish at a higher rate, according to the Preston equation. Figure 45*d* shows the case when the wafer rotates at a speed of only $\frac{1}{2}$ of the platen. Under the condition of $\omega_1 = 2\omega_2$, a point at the edge of the wafer follows a cardiod curve whose averaged travel path over a complete turn is comparable to that at the center of the wafer.

From the foregoing illustration the best uniformity is achieved with $\omega_1 \geq \omega_2$. The edge of the wafer polishes at a higher rate than the center when $\omega_1 < \omega_2$. In practice, the polishing surfaces are not rigid and other factors contribute to the polishing uniformity. Depending on the values of R and r, a condition between $\omega_1 = 2\omega_2$ and $\omega_2 = 2\omega_1$ is used.

The other condition that must be met for uniform polishing is the uniform distribution of the polishing slurry under the wafer. Currently, the details of how slurry is transported from the edge to the center of the wafer are not clear. Since slurry transport is mostly through the surface of the pad, the edge of the wafer gets a more abundant supply of slurry than the center of the wafer. Therefore, no matter whether

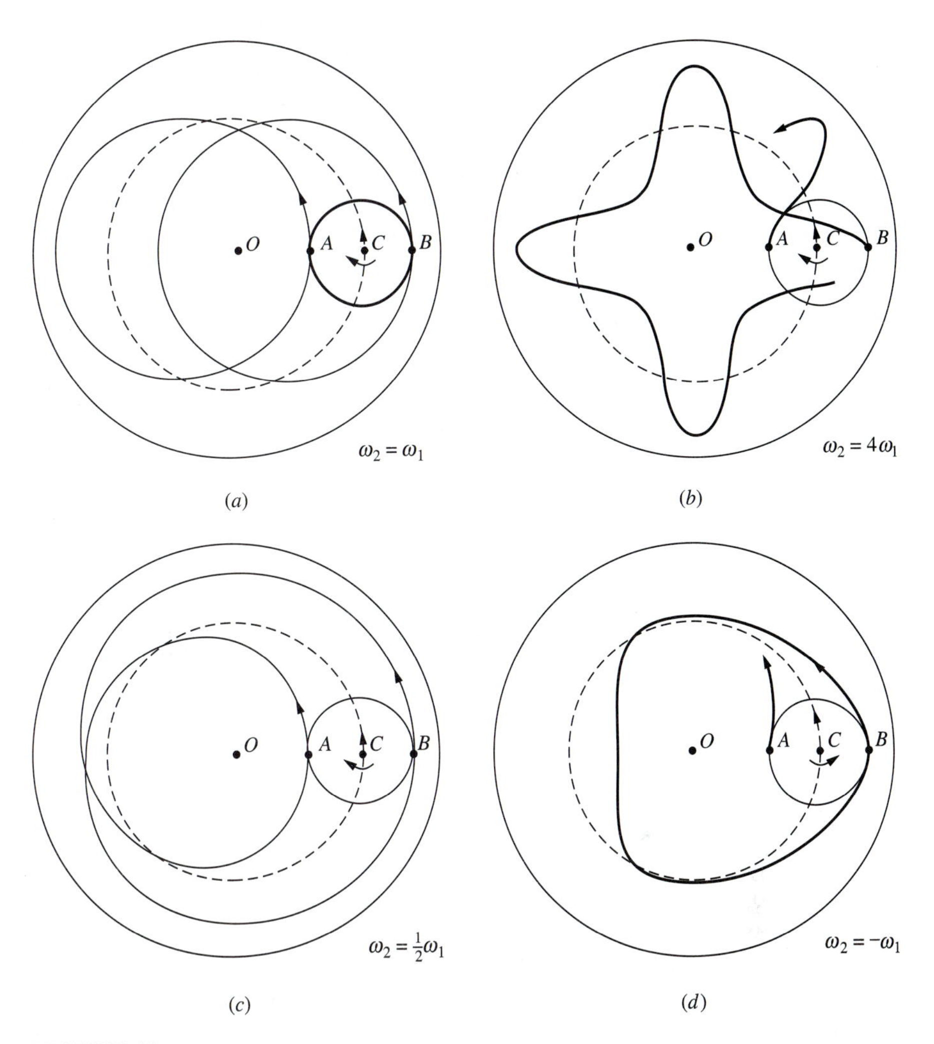

FIGURE 45
The loci of different points on a wafer: (*a*) When the wafer rotates at the same angular speed as the platen; (*b*) when the wafer rotates at a higher speed than the platen; (*c*) when the wafer rotates at $\frac{1}{2}$ the speed of the platen; (*d*) when the wafer rotates at the same speed as the platen, but in the opposite direction.

from the relative velocity aspect or from the slurry supply aspect, the edge of the wafer has a polishing rate equal to or higher than the center. To compensate for this, some equipment uses a slightly convex curvature on the carrier to exert a higher pressure toward the center of the wafer.

Consistent polishing rate and uniformity (reproducibility) are not routinely achieved for CMP yet. The lack of reproducibility is attributable to multiple sources and is still poorly understood. The Preston equation [Eq. (8.51)] connects the polishing rate to K_p (the Preston coefficient), the pressure, and the relative velocity. From our earlier analysis uniform relative velocity is achievable. The pressure p and the Preston coefficient are both functions of the *consumables* and the pattern on the wafer. The polishing pads (and to a lesser degree the backing film) and the polishing slurry are commonly referred to as the consumables. The real pressure for CMP depends both on the wafer pattern and on the contact with the polishing pad. A soft pad gives a larger contact area than a harder pad and generates a lower pressure under the same load force on the wafer carrier. Therefore, a soft pad gives a lower polishing rate. Similarly, wafers with different patterns generate different contact areas under the same load force and are polished at different rates.

The polishing pad also changes with use. Polishing pads are made of polyurethane with 1-mm perforations punched into the pads to help transport the slurry and facilitate polishing. The surface of the polishing pad is intentionally kept rough with many asperities about 1 to 10 μm in size to help hold/transport the slurry. After polishing a number of wafers, the surface asperities are flattened,[162] resulting in a state called *glazing*. When a pad is glazed, it loses some of its capacity to hold polishing slurry. In addition, glazing increases the contact area and thus causes a drop in pressure. Both changes reduce the polishing rate. A glazed pad can be rejuvenated by a process called *conditioning*. A diamond-embedded wheel is applied to the pad while the platen continues to rotate. The detailed mechanism for conditioning is not known. There is no consensus on whether the pad material is actually removed by the diamond wheel lapping. The net effect of the conditioning, however, is the regeneration of rough surfaces, which can hold/transport the slurry and provide a smaller contact area. Pad conditioning provides a key for consistent polishing rate. It is applied after the polishing of each wafer, and in some equipment concurrent pad conditioning and polishing are used.

The polishing slurry is the other important consumable, and consistency in the slurry is a key to reproducibility. For dielectric polishing, the most commonly used slurry is colloidal silica or fumed silica suspended in a KOH-based solution. The polishing rate depends on the pH of the solution, the concentration (in wt %) of the solid particles in the slurry, and the size and the distribution of the slurry particles. The polishing rate increases with increasing pH, with increasing slurry particle concentration, and with increasing particle size.[163]

To summarize: although the material removal rate in CMP can be described by the Preston equation, out of the three parameters in this equation only the relative velocity can be directly controlled. The pressure and the Preston coefficient are complex functions of processing, consumables, and wafer pattern variables. A manufacturable CMP process can only evolve from the understanding and control of these parameters.

A serious drawback for CMP is the difficulty in end point detection. Without a clear signal about when the process is completed, CMP is done using empirical polishing rates and timed polish. Since the process control is immature, the empirical method fails from time to time, causing yield drops. Attempts to install an end point mechanism include capacitive measurement and optical measurement.[164] A commercial end point apparatus with limited capability has recently been introduced.[165]

The need to remove the polishing slurry from the wafer and to ensure a low level of particles on the wafer led to the specialty of "post-CMP clean." The slurry particles used in CMP range from 10 to 100 nm. These small particles attach to the wafer mainly by electrochemical forces. To establish a cleaning chemistry, one must understand the details of particle attachment. So far, the post-CMP clean uses brute force—second platen buffing, soft scrubbing, ultrasonic cleaning, and chemical etching. Experience indicates that this approach is adequate for 0.5-μm device fabrication. For future, more demanding, applications, fundamental understanding of the fine particle issues will be necessary.[166]

One risk of the CMP process is the abrupt introduction of a globally planarized ILD, which creates uneven via heights over a previously unplanarized or locally planarized dielectric. Figure 46 shows the difference in via depth after CMP of an ILD over a lower level of metal on a locally planarized dielectric. Depending on the thickness of the metal wire and the ILD, the difference in via depth can be as much as 100%. Vastly different via depths increase the difficulty in the already difficult via

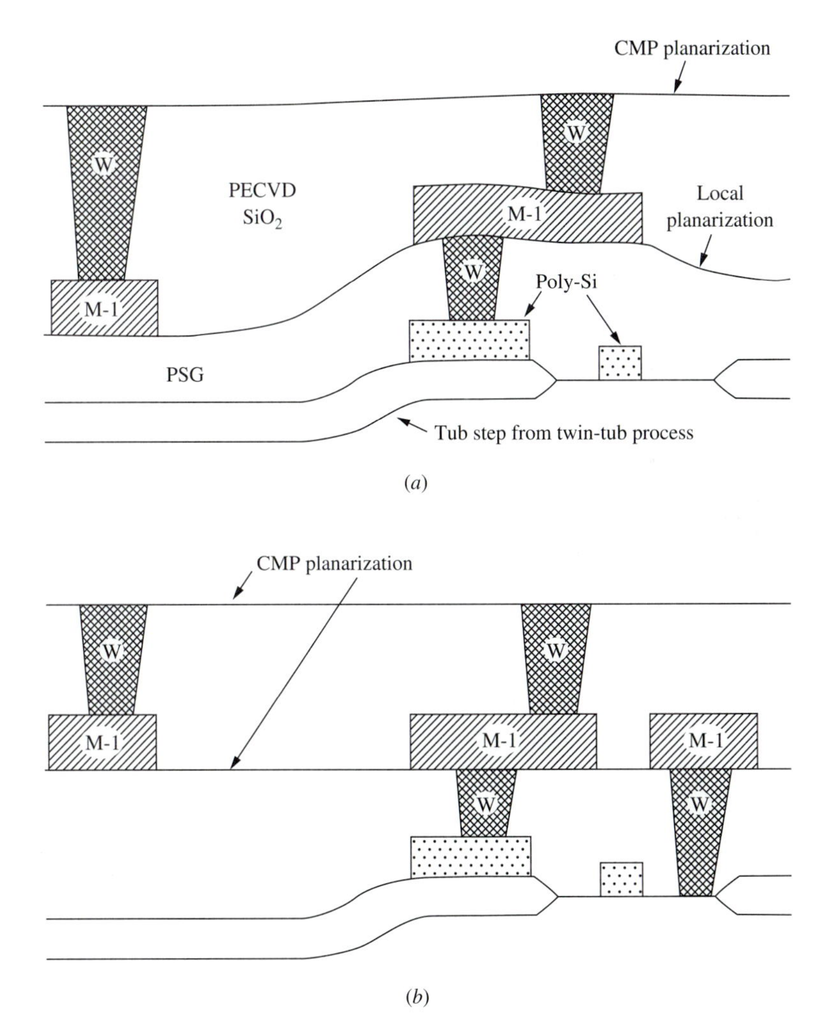

FIGURE 46
Via depth differences by different planarization approaches: (*a*) Large difference in via depths when CMP is used after local planarization; (*b*) vias of similar depth when CMP is used twice.

lithography and etching processes. This problem can be solved by applying CMP or global planarization on earlier levels of the dielectric. The ideal approach is to apply a global planarization at the source/drain/gate contact level. Unless a totally flat structure is used, such as recessed isolation and recessed gate, the contact windows are of different depths. A global planarization at this level does not add more difference in window depth compared to a local planarization process, as shown in Fig. 46*b*. When CMP or global planarization is applied to this dielectric (the poly-metal dielectric or PMD), uniform via depth can be achieved without any penalty. The choice of how many global planarization processes should be used and where they should be applied is a complex process integration issue of cost/yield analysis.

Damascene and dual damascene

The damascene process derives its name from the ancient art of the Middle East involving inlaying metal in ceramic or wood for decoration. In ULSI application, a damascene process refers to a similar structure, as shown in Fig. 47. After the via plug process the ILD is deposited without planarization, since the surface is already flat. Trenches for metal lines are then defined and etched in the ILD and

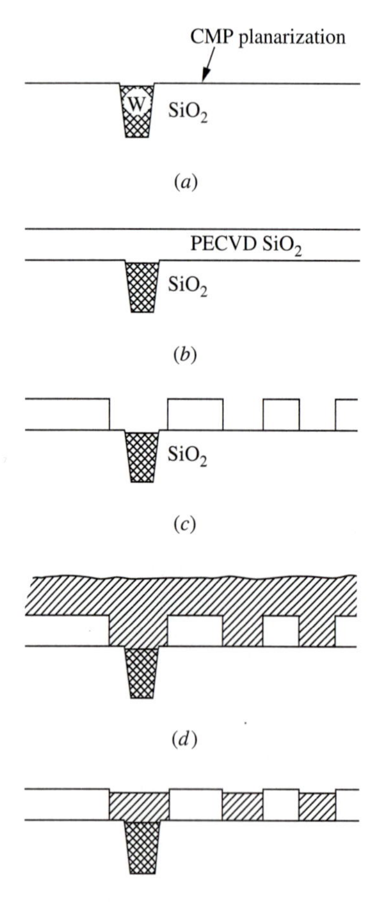

FIGURE 47
Damascene process: (*a*) Metal plugs formed after the SiO_2 is planarized by CMP; (*b*) deposited PECVD SiO_2 ILD; (*c*) trench patterning and RIE for metal lines; (*d*) metal deposition to fill trenches; (*e*) CMP metal to complete metal definition.

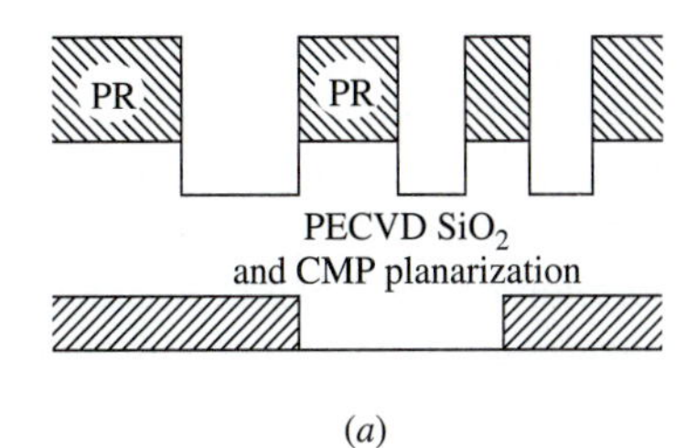

(*a*)

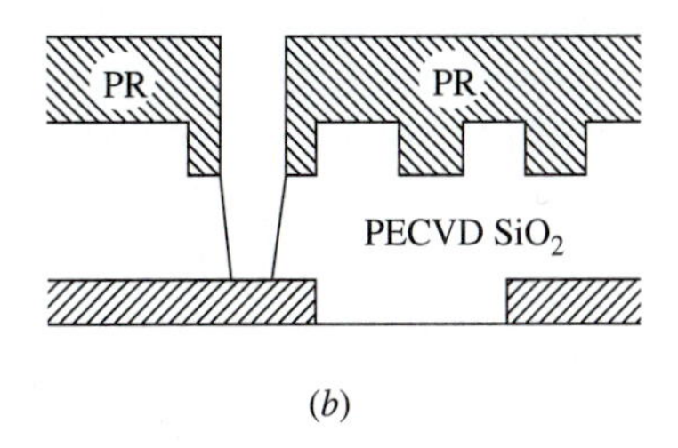

(*b*)

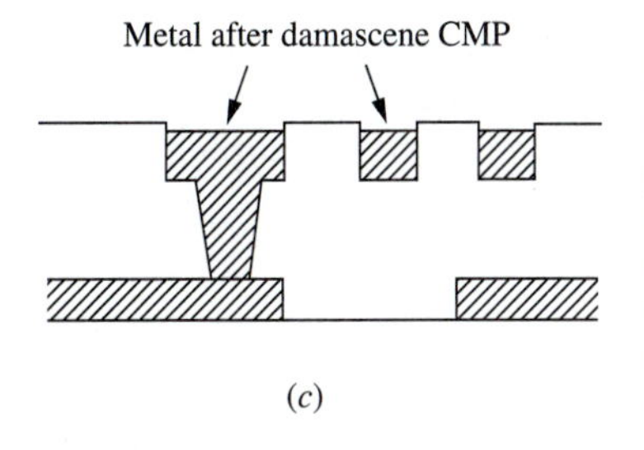

(*c*)

FIGURE 48
Dual damascene process: (*a*) After ILD planarization, define trench pattern; (*b*) strip photo resist, apply new photoresist, define via pattern, and RIE vias; (*c*) deposit metal in both vias and trenches, and CMP to remove excess metal. Note that the via plug is of the same material as the upper-level metal in this process.

filled with metal, either by a CVD process or by a metal flow process. The excess metal on the surface is removed and a planar structure with metal inlays in the dielectric is achieved. This process has several advantages over the traditional metal/ILD/planarization approach: (1) The surface at any time is totally flat; (2) the process eliminates the difficulty in filling small gaps between metal wires; and (3) it eliminates the difficulty in metal etching, especially for Cu and other hard-to-etch metals. On the other hand, the damascene process is more complex since (1) it still requires the via plug process and the CMP of both metal and dielectric, and (2) it requires a flat topography to start with and thus forces the use of recessed isolation, etc.

A dual damascene process is illustrated in Fig. 48. In this process vias and trenches are defined using two lithography and RIE steps, but the via plug is filled in the same step as the metal line, as shown in Fig. 48*c*. Dual damascene reduces the number of processing steps by reducing the barrier layer depositions from two to one and by eliminating the CVD W plug processes. One special benefit of dual damascene is that the via plug is now of the same material as the metal line and the risk of via electromigration failure is reduced (see Section 8.6.1).

Damascene has been demonstrated on a number of applications. The most commonly applied process is the first metal or local interconnect.[167–170] Some early damascene structures were achieved with RIE,[167] but CMP is exclusively used today. Metal interconnects using damascene of Cu[171–174] and damascene of Al[175–177] have been reported.

Stacked contact and borderless contacts

As the technology scales to smaller dimensions, one of the most difficult obstacles to continued scaling is the level-to-level alignment in lithography, especially in complex structures of metals and contacts. Consequently, the metal packing density becomes limited by design rules governing the separation of one level of contacts from another, and by design rules for nesting tolerance or for borders used around contacts.

Figure 49 illustrates the area penalty imposed by contact-to-contact separations and by borders. From a fully bordered, no-stacking structure, Fig. 49*a*, to a borderless, fully stacked structure, Fig. 49*e*, a total area reduction of 62% is achieved. Intermediate area savings of significant portions can be realized by various combinations of borders and stacking, Fig. 49*b*–*d*. Unless the circuit is bonding-pad limited, the use of stacked contacts and borderless contacts can save more area than going to a new generation of design rules.

However, the practice of stacking contacts and borderless contacts is still in its infancy, mainly owing to processing difficulties. Figure 50 illustrates the problem with stacked contacts. If the lower contact is made with a nonplug process, then the poor Al step coverage prohibits a good contact, as shown in Fig. 50*a*. If a metal plug is used in the lower-level contact, as in Fig. 50*b*, then a good contact can be made when the second contact is stacked on top of the first one. If the metal-plug control

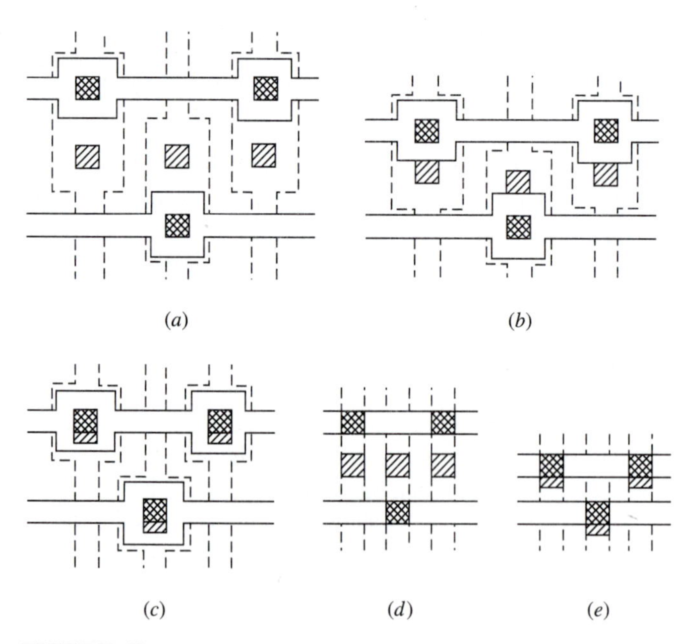

FIGURE 49

Stacked contacts and borderless contacts: (*a*) No stacking, design rules on all lines; (*b*) no stacking, no design rule separating metals on different levels; (*c*) stacked, with border; (*d*) no stacking, borderless; (*e*) stacked and borderless.

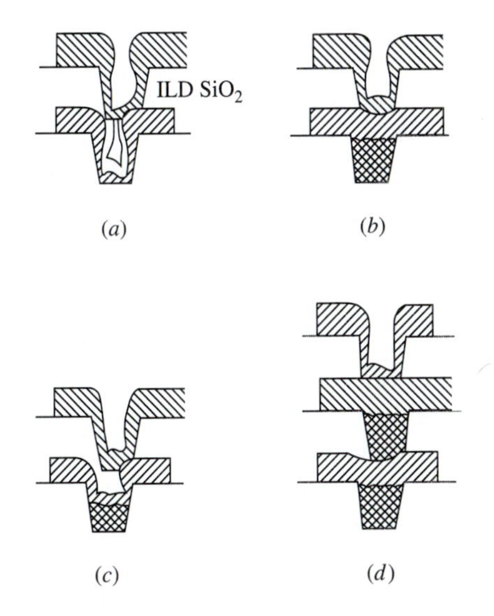

FIGURE 50
Stacked contacts: (*a*) No metal plugs in the lower contact via results in poor contact from the upper via; (*b*) a good metal plug ensures a good via contact; (*c*) a poor metal plug can still cause a poor via contact; (*d*) good via plugs allow multiple stacking of vias.

is poor, as shown in Fig. 50*c*, then poor Al step coverage can occur even with a metal plug, and a good stacked contact is difficult to make. If the metal-plug control can be maintained, as in a damascene process, then multiple stacking of contacts is possible, as shown in Fig. 50*d*.

Figure 51 illustrates the problems with borderless contacts. For a nonplug process, Fig. 51*a*, borderless contacts provide very little contact area between the upper-level via and the lower-level metal and the contact is unreliable. With any misalignment, even making contact is not ensured. If the lithography misalignment is significant, the contact area is still small, and current crowding can occur at the via contact, as in Fig. 51*b*. To make reliable borderless contacts, one can tighten the lithography alignment tolerance or relax the via design rule, as in Fig. 51*c*. True borderless contacts remain difficult to make reliably because of these problems. The use of dual damascene will help attain borderless contacts, because the filling of via contact holes and the metal wiring are accomplished in a single operation.[178]

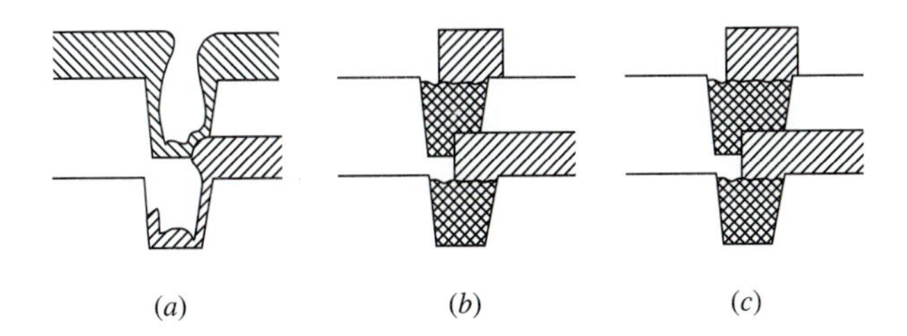

FIGURE 51
Borderless stacked vias: (*a*) Nonplug process gives poor contact; (*b*) misalignment can also cause poor contact; (*c*) a larger upper via provides better contacts to both levels of metal.

8.5.4 Bonding Pad Limitations

For many ASIC applications, the power, ground, and I/O signals require hundreds of bonding pads. With the increasing capability to pack more components into a chip, the circuit sophistication increases with each generation of design rules, resulting in further demand for additional bonding pads. For many designs, the number of bonding pads becomes the limitation on chip size and, until bonding technology advances further, improvement in packing density cannot be realized by simply shrinking design rules or adding more levels of metal wiring.

Currently, bonding pitches of 80 to 120 μm are used for wire bonding in Si chip packaging. Smaller pitches are still in the exploratory stage. Consider a 400-pin device with all bonding pads distributed along the perimeter of the chip. If we assume a 100-μm bonding pad pitch, then to accommodate 100 pins per side the bonding pads will occupy 100×100 μm or 1 cm of length on each side. In this case, no further increase in circuit packing density can be achieved until the chip area becomes greater than 1 cm^2. For current technologies in volume production (0.5 to 1.0 μm minimum design rule), 1 cm^2 is the limit of optimum cost at the current achievable defect level. As a result, the strongest limitation in cost reduction is the number of bonding pads. Until packaging technology advances to reduce the bonding pitch, sophisticated circuits cannot enjoy the benefit provided by further shrinking the design rule or adding more levels of metallization.

A technique to circumvent the bonding pad limit is to use an array of pads instead of using only perimeter pads. This has the benefit of providing not only more bonding pads per chip but also shorter metal wires and thus faster circuits. This point will be discussed further in Chapter 10. It is clear, however, that packaging technology must scale and advance together with the other ULSI processing technologies. Otherwise, it will become the limitation, quickly rendering all other efforts of scaling useless.

8.6 METALLIZATION RELIABILITY

The reliability of metal wires under high current density operation and under mechanical stress has been a focus of intensive studies. These two very important phenomena are discussed in detail in Chapter 12. In this section we will discuss the contact electromigration associated with multilevel metallization, and metal corrosion.

8.6.1 Contact Electromigration

With multiple levels of wiring, it is necessary to pass current from one level of metal to another through vias. When the metal design rules scale down, the size of the contact vias also shrinks accordingly, and the current density in the vias can be as high as, and sometimes even higher than, that in the metal conductors. In practice via electromigration has been observed for two reasons: (1) The metal step coverage in the vias is poor, resulting in very high current density; and (2) a different material is used in the via (such as a W plug), causing localized current crowding[179,180] and high current density. Either of these two can cause severe contact electromigration.

The first case is illustrated in Fig. 52*a*. Consider the case of a contact via that carries a current of 0.4 mA from a transistor. If the step coverage in the via is about 10%, which is normal for a sputtered film in a hole with an aspect ratio of 1 : 1, then the amount of Al on the via wall is only about 50 nm for a 500-nm layer of Al deposition. The current density in the via can be calculated as 4×10^{-4}A/0.075 $\times 10^{-8}$cm^2 = 5×10^5A/cm^2, or about six times that of a fully filled via. This high current density is three to five times higher than the EM design limits and will cause early EM failure. Since the current is from a nominal-size transistor, the via electromigration problem cannot be solved without an increase in the step coverage. This is accomplished by using W or Al plugs.

Figure 52*b* illustrates the EM-induced voiding when a CVD W plug is used in a via and Al wiring is in direct contact with the W plug. Even with a plug in the via, current crowding still occurs when current goes through the plug and into the next

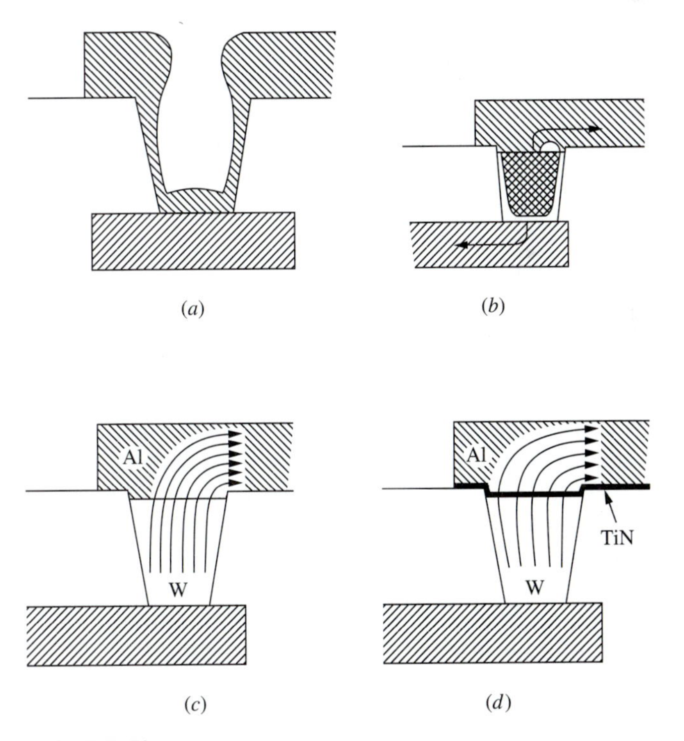

FIGURE 52
Contact electromigration: (*a*) Poor Al step coverage can cause high current density and Joule heating; (*b*) Al/W-plug contacts cause voiding at the corner of the Al/plug interface; (*c*) current crowding occurs at the inner corner of an Al/W-plug contact; (*d*) adding a TiN layer between the W plug and the Al wire reduces the current crowding.

level of Al wire. This happens because the via intersects the Al wire at a right angle, and the electrical current has to turn 90° when entering the Al wire. As a result, various current paths offer different resistances, and current tends to go through the least resistive path—the inner corner, as shown in Fig. 52*b*. The current crowding drives more current through the inner corner of the plug and the adjacent Al wire. If the plug is made of Al, then no void will form since the Al flux leaving the high-current spot will be compensated by Al flux from inside the plug and from Al farther up the line. For a W plug, however, there is no Al flux that can come from the plug to replenish the Al flux leaving the hot spot. Consequently, voids will develop at the inner corner where current leaves the W plug into the Al wire, as shown in Fig. 52*c*.

There are two approaches to reduce the EM voiding at the via contact: (1) Replace the W plug with an Al plug to ensure the continuity of the Al flux, or (2) reduce the current crowding at the corner of the via contact. The second approach is illustrated in Fig. 52*d*. By applying a layer of TiN between the W plug and the Al wire, the current crowding at the inner corner can be reduced. The high resistance of the TiN helps to distribute the current from the via to a wider area in the Al wire, thus reducing the current crowding. To calculate the current distribution accurately three-dimensional analysis is needed, although the calculation is straightforward. A simple two-dimensional calculation is given in Problem 18 at the end of this chapter. Interestingly, the worst current crowding occurs when an Al plug is used. Voiding does not occur, however, because the Al flux through the via is continuous. Voids form from electromigration only when dissimilar materials are used and when current crowding occurs.

8.6.2 Metal Corrosion

Aluminum grows a passivating oxide in air and is naturally protected against corrosion. Aluminum wiring used in ULSI circuits, however, contains Cu, which has no passivating oxide, and the Al-Cu alloy used is more vulnerable to corrosion. The corrosion of Al wires comes from four sources: (1) Cl transported through the plastic packaging and passivation materials, (2) Cl from etching compound and etching by-products, (3) phosphorous acid formed from excess P in the phosphosilicate glass (PSG), and (4) electrochemical (galvanic) corrosion from dissimilar materials. Chlorine plays an important role in the corrosion of Al through these reactions:

$$\mathrm{Cl^- + H_2O \rightarrow HCl + OH^-} \tag{8.57}$$

$$\mathrm{3HCl + Al \rightarrow AlCl_3 + 1.5H_2} \tag{8.58}$$

$$\mathrm{AlCl_3 + 3H_2O \rightarrow Al(OH)_3 + 3HCl} \tag{8.59}$$

Note that the reactions in Eqs. (8.58) and (8.59) are cyclic. After the initial formation of HCl no additional Cl is needed. The presence of the Cl ion is only to facilitate the net reaction [from Eq. (8.58) and Eq. (8.59)]:

$$\mathrm{Al + 3H_2O \rightarrow Al(OH)_3 + 1.5H_2} \tag{8.60}$$

Therefore, only a small amount of Cl is needed to cause severe local corrosion of the Al lines, because the Cl^- ions recycle themselves during the corrosion process.

Since most chemicals used for Al dry etching (RIE) contain Cl, $AlCl_3$ or a similar compound is formed on the Al surface after the plasma etching. Upon exposure to the moisture in air, the reaction in Eq. (8.59) and subsequently the cycling of Eq. (8.58) and Eq. (8.59) can proceed rapidly. Therefore, Al corrosion can occur very quickly after metal etching. To prevent etching-induced corrosion, chlorine compounds and elemental Cl must be removed from the metal surface immediately after plasma etching. Usually, a water rinse or a water vapor treatment is given right after etching. Even with this precaution corrosion can still occur from time to time. Modern metal-etching equipment now uses multiple chambers with an in-situ water rinse or water vapor treatment to ensure the removal of all Cl compounds.[181]

Phosphorus-doped glasses usually contain 4 to 5 wt % of P. P in the PSG can leach out if the concentration is higher than about 6 wt %. The P then interacts with the H_2O in the glass to form metaphosphoric acid (HPO_3), which subsequently etches the Al and causes corrosion. Note that the P content in PSG may vary locally in a wafer. Consequently, the P content can exceed 6 wt % locally even when the average P content is well below 4 to 5 wt %.

Modern metal structures use multilevels of dissimilar materials such as Ti/TiN/Al-Cu/TiN or Ti/Al-Cu/TiN, and the chance of electrochemical corrosion is increased. For example, the emf (electromotive force) between Al and W is about 1.5V, and in W-plug/Al-wiring structures the Al is in direct contact with W. In this kind of structure it is important that the dielectric material does not form an electrolyte to allow a galvanic reaction to occur. Therefore, water-absorbing materials should be avoided. Note that electrochemical corrosion can also be accelerated by the presence of Cl^- (from, e.g., plasma etching). In the presence of Cl^- and moisture, the galvanic action,

$$Al \rightarrow Al^{3+} + 3e^- \qquad \text{(Anode: Al)} \tag{8.61}$$

$$3H^+ + 3e^- \rightarrow 1.5H_2 \qquad \text{(Cathode: W)} \tag{8.62}$$

can proceed rapidly. Since Al is more electronegative than W it becomes the anode and is corroded away.

Copper is electropositive (relative to hydrogen) and is not vulnerable to electrochemical corrosion. However, in air copper oxide grows linearly with time, indicating the lack of a protective oxide. This lack of a passivating oxide makes Cu more vulnerable to chemical corrosion. Since there is practically no ULSI circuit made with Cu wiring today, experience in corrosion is nearly nonexistent. Most proposals for using Cu metallization involve some protective layers.[174]

8.7 SUMMARY AND FUTURE TRENDS

8.7.1 Summary

In summary, the demand for multilevel metallization has forced advances in all fronts, from film deposition to new architecture. Precise control of metal deposition and uniformity on large wafers is achieved by new designs of magnetron sputtering, and reactive sputtered and collimated deposition of Ti/TiN become standard processes. Polycides and self-aligned salicides provide more design freedom at low cost. CVD W has been widely adopted and will be followed by CVD TiN and CVD Cu for future applications. Plasma-enhanced CVD oxides are standard processes for 0.8-μm and 0.5-μm devices and will eventually be replaced by O_3-TEOS and HDP oxide for ILD gap fill. The advent of CMP and damascene builds the foundation for global planarization and totally planar structures and allows the building of four, five, and more levels of metals.

The rapid expansion of new applications in telecommunication, networked computation, multimedia processors, personal digital assistants, etc., in the next decade will no doubt further emphasize the need for high levels of integration. Higher levels of integration and high levels of performance will further complicate metallization while the conventional Al-alloy and SiO_2-dielectric approach their limits of *RC* delay and *IR* drop. Therefore, new materials and structures that can alleviate the parasitic limits and add higher sophistication to circuits are needed to ensure continued growth in the early 2000s.

8.7.2 New Conductor Materials

New metallization materials to reduce *RC* delay and *IR* drop or to improve electromigration resistance and thus allow the use of higher drive current are urgently needed. Since Al is already a good conductor the improvement in *R* seems limited. Ag, Cu, and Au are the only three metals that have higher electrical conductivity than Al. Fortunately, all three are heavier than Al and have better electromigration resistance, but even for Ag the improvement in *R* cannot exceed a factor of 2.

Cu is by far the most studied substitute for Al. CVD deposition of Cu has been discussed in Section 8.2.3. The use of Cu wiring for multilevels of interconnects has been demonstrated by dual damascene,[24,173,182,183] but the commercial application of this material to devices has not yet begun. Cu can diffuse through SiO_2, and a barrier layer, such as Si_3N_4, is needed. The most difficult obstacle, however, is its vulnerability to corrosion because Cu does not possess a tight, self-passivating oxide as does Al. To protect Cu from corrosion, passivation layers are needed that, unfortunately, take precious space and lower the conductance per unit volume. Consequently, Cu metallization can only reduce the effective *R* by 20 to 30%. Properly done, however, Cu can improve the electromigration resistance of the metal lines[184,185] and allow the use of higher current, which in turn improves the circuit speed. Investigations of Au[186] and Ag[187] for interconnects have been reported only recently. Much work is needed to make these new materials viable.

8.7.3 Low-Dielectric-Constant Materials

Low-permittivity insulators provide a larger margin for improving the *RC* delay than new conductors. In addition, a lower dielectric constant also reduces the line capacitance and thus cuts down the cross talk between conductors. An ideal low-permittivity dielectric, however, is still beyond reach today, because few materials can substitute for the properties of SiO_2. A dielectric material must meet at least the following criteria: (1) low permittivity, (2) high breakdown field, (3) low leakage, (4) no moisture absorption or excessive permeability to moisture, (5) thermal stability to $> 400°C$, (6) good adhesion to metal and to dielectric, (7) no metallic contaminants, (8) low defect density, (9) no significant outgassing, and (10) resistance to plasma. Unfortunately, few inorganic insulators have low permittivity, and few organic insulators show good thermal stability, good adhesion to metal, no outgassing or moisture absorption, and resistance to plasma.

Among the inorganic dielectric materials, BN has a low dielectric constant of 2.9. However, BN is hygroscopic[188] and is unsuitable as an ILD. SiBN with a composition of $Si_{0.10}B_{0.39}N_{0.51}$ has a dielectric constant of 3.6 and is stable in room air.[188] Other reliability factors for SiBN have not been fully studied yet. Recently, fluorinated silicon dioxide, or SiOF, has shown promise as a viable candidate.[189,190] SiOF has a dielectric constant of about 3.0 to 3.6 and can be deposited by PECVD[189] using TEOS and C_2F_6, or by a room temperature process.[190] The PECVD process is particularly attractive because it represents a modification of a process already widely used, with the benefit of a 15 to 30% lower dielectric constant.

Among the organic materials, most of the experience is with polyimides. These have been used extensively in packaging, multichip modules, and for some ULSI circuits as ILDs and as passivation. Reference 191 gives an overview of the experience and the issues on the applications of polyimides. With a dielectric constant of 3.0 to 3.6 and relatively good thermal and mechanical properties the application of polyimides to ULSI circuits is within reach. Some polyimides (e.g., PMDA-ODA from DuPont) can withstand temperatures to about 400°C without losing mechanical stability. Integration of polyimides with CVD W and Al-Cu wiring has been reported.[192] Multilevel metal structures using Cu/polyimide and dual damascene have been demonstrated.[24]

Organic materials with dielectric constants lower than 3.0 include parylene, BCB, and Teflon™. Parylene [poly-(*p*-xylylene)] has a dielectric constant of about 2.6 and can be vapor-deposited. Figure 53 shows the chemical structure of parylene, a polymer of *p*-xylylene. It consists of a benzene ring with 2 CH_2 radicals attached in the *para* positions. Parylene does not absorb moisture, but it is stable only to about 300°C. Fluorinated parylene (parylene-F) demonstrates a higher thermal stability to about 350 to 400°C. Investigation of this material for ULSI application has been reported recently.[193,194] Both the thermal stability and the resistance to plasma etching are promising. Much more characterization and probably synthesis of new materials will be needed to demonstrate full compatibility with ULSI processing. BCB, or

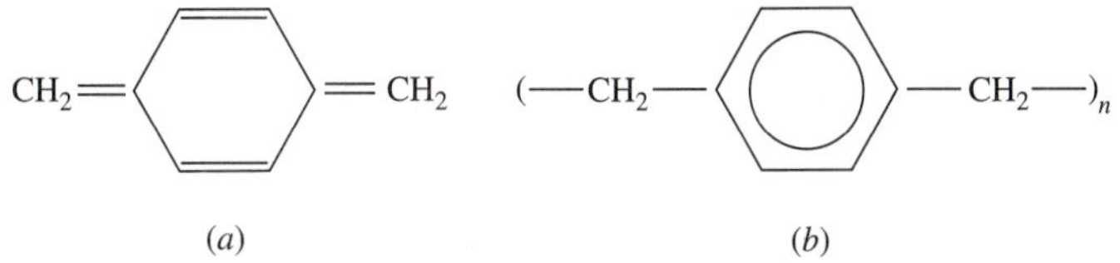

FIGURE 53
Molecular structures for (*a*) para-xylylene, which is a precursor for parylene; (*b*) parylene (poly-*p*-xylylene) polymer.

benzocyclobutane, has properties similar to parylene (no moisture absorption, dielectric constant of about 2.6, and stable to ~300°C). Chemicals related to the Teflon™ family have dielectric constants ≤ 2.0 but are thermally and mechanically less stable. In addition, Teflon has poor adhesion to metals. Therefore, this family of chemicals is hard to use for ILD. Recently, fluorinated polyimides with dielectric constants about 2.6 have been synthesized. Newly synthesized polymers, including some Si-containing polymers with dielectric constants below 3.0, have been listed in Ref. 195.

To summarize: the barrier to using dielectric materials with dielectric constants of 3.0 to 3.6 seems surpassable in a few years. Both organic polyimides and the inorganic SiOF may serve this purpose. To go below 3.0, the barrier is much higher. The existing polymers (commercially available) are probably all insufficient to fulfill the stringent requirements for ILD. A substantial number of new materials are being synthesized. The characterization and the trial application of these materials to ULSI circuits will be important.

8.7.4 New Contact Structures

The benefit of area reduction by stacked contacts and borderless contacts is substantial, rivaling that of design rule reduction. With the progress and acceptance of fully plugged contacts, structures with stacked contacts will become widely accepted. Completely borderless contacts are difficult to make, especially for small design rules. But partially relaxed border design rules that allow some contacts to be nearly borderless can still reduce the design area and increase circuit density. In the next decade, the stacked contact and variations of borderless contacts will be fully utilized to achieve higher packing density.

Packaging with arrays of pins instead of perimeters-only pins has already been designed. Because the pin count can now increase as the square of the die size, instead of linearly with the die size, this approach will alleviate the bonding pad limitation faced by many designs. In addition, the use of array bonding allows many signals to travel a shorter distance than in perimeter bonding and will improve circuit performance. Advances in packaging technology will further encourage the use of multilevel metallization.

8.7.5 Raised Source/Drain Structures

At 0.20 μm and below, silicides in the source/drain area are needed to reduce the contact resistance to an acceptable level (Section 8.3.4). However, it is difficult to make shallow junctions (0.10 μm or below) with silicide. Even if such junctions can be made, they will be so vulnerable[114] to stress, window RIE, etc. that it is impractical to make traditional contacts to them. Therefore, shallow junctions for 0.2-μm or below devices need to be made differently.

Raised source/drain junctions have been demonstrated.[196] By using selective growth of Si, 100 nm or more of epitaxially grown Si can be added on the

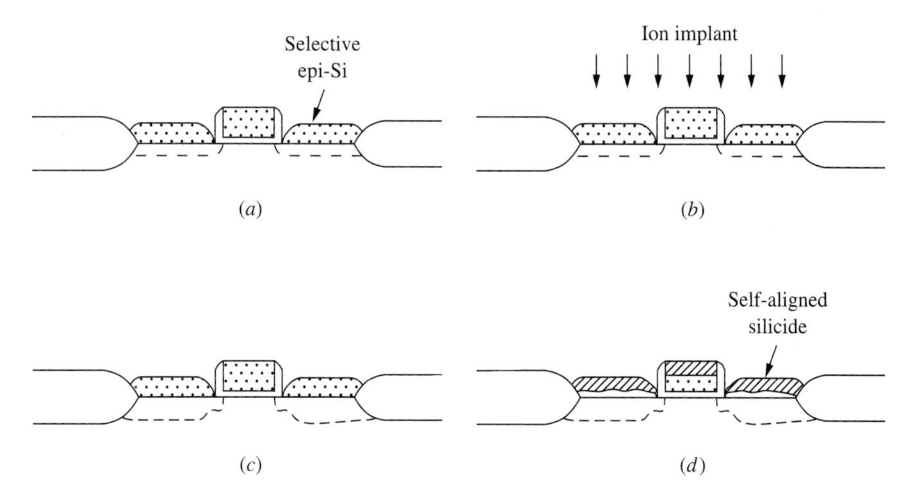

FIGURE 54
Raised source/drain junctions: (*a*) Selective growth of epitaxial Si on the source/drain; (*b*) implant of dopants into the epi-Si; (*c*) driving the dopants in to form the junction; (*d*) formation of salicide on the epi-Si and on poly-Si.

source/drain, as shown in Fig. 54*a*. Shallow junctions can then be formed by implanting into the epitaxial Si, as in Fig. 54*b*, and driving into the substrate, as in Fig. 54*c*. Finally, a salicide is formed on the epitaxial Si, Fig. 54*d*. Since the Si consumed for forming the salicide is from the epitaxial Si, the silicide does not erode the shallow junction. Because of the thicker Si, the sheet resistance of the source/drain diffusion is also lower. In addition, a thicker silicide can be used since the Si consumed does not come from the shallow junction. This allows a lower sheet resistance for the poly lines, for more design freedom. A drawback of raised source/drain junctions is the addition to the sidewall capacitance (Miller capacitance). To reduce this parasitic, a thicker sidewall spacer should be selected so that the Miller capacitance is only a small fraction of the gate capacitance.

8.7.6 Low-Cost Processing

The trend of using three, four, or more levels of metal wiring for ASICs and microprocessors started in the 1980s and will continue throughout the 1990s and into the next century. Multilayered Al structures are already widely used, and CMP of ILD and metal are beginning to be accepted as mainstream processes. The cost of processing the metallization structure is now comparable to the cost of processing CMOS devices. To keep the processing cost (per unit area) constant or to lower it, process complexity must be reduced.

The popular W plug and multilayer Al structure require repeated metal deposition and etching. For example, five metal depositions, two etchings, and one lithography are needed to complete one level of metal. Compare this to a simple Al wiring where only one metal deposition, one lithography, and one etching are used. Clearly,

TABLE 2
Processing complexity of various multilevel metal structures

Structure	Process	Dielectric deposition	Metal deposition	RIE or planarization	Lithography	Total
Simple Al, non-plug	ILD deposition	2	0	1	0	3
	ILD planarization	0	0	1	0	1
	Via pattern	0	0	1	1	2
	Metal deposition	0	1	0	0	1
	Metal pattern	0	0	1	1	2
	Total =	2	1	4	2	9
Full-plug layered-metal	ILD deposition	2	0	1	0	3
	ILD planarization	0	0	1	0	1
	Via pattern	0	0	1	1	2
	Plug (barrier + CVD W + etch)	0	3	1	0	4
	Metal deposition	0	3	0	0	3
	Metal pattern	0	0	1	1	2
	Total =	2	6	5	2	15
Al-plug layered-metal	ILD deposition	2	0	1	0	3
	ILD planarization	0	0	1	0	1
	Via pattern	0	0	1	1	2
	Metal deposition	0	3	1	0	4
	ARC deposition	0	1	0	0	1
	Metal pattern	0	0	1	1	2
	Total =	2	4	5	2	13
Dual damascene-layered-metal	ILD deposition	1	0	0	0	1
	Trench pattern	0	0	1	1	2
	Via pattern	0	0	1	1	2
	Metal deposition	0	3	0	0	3
	Metal polish	0	0	1	0	1
	Total =	1	3	3	2	9

the current process is much too complex. If sputtered Al can also be used for the plug, then a much simpler process can be developed. However, since a barrier metal and an antireflective coating (ARC) are needed, the all-Al process only saves one metal deposition and one etching. Dual damascene, on the other hand, can reduce the cost by eliminating the metal lithography and the ARC layer. Table 2 compares the processing complexity of several different approaches. If all processes are mature (i.e., ILD and metal CMP can be done at low cost, and CVD metals are available), then dual damascene will provide the simplest process and the lowest cost.

Other cost reduction approaches are (1) high throughput equipment, (2) standardization of equipment, and (3) standardization of process modules. For example, using RTP and cluster tools, the cycle time can be reduced[197] substantially and fabrication cost is reduced. These measurements are more general than for just metallization. However, being a repetitive process, multilevel metallization will benefit the most from standardization.

REFERENCES

1. K. K. Ng and W. T. Lynch, "The Impact of Intrinsic Series Resistance on MOSFET Scaling," *IEEE Trans. Electron Dev.,* **ED-34,** 503 (1987).
2. S. M. Sze, *Physics of Semiconductor Devices,* 2nd ed., Wiley, New York, 1981, p. 304.
3. R. Glang, "Vacuum Evaporation," in L. I. Maissel and R. Glang, eds., *Handbook of Thin Film Technology,* McGraw-Hill, New York, 1970, p. 1-3 to p. 1-130.
4. D. B. Fraser, "Metallization," in S. M. Sze, ed., *VLSI Technology,* McGraw-Hill, New York, 1983, pp. 350–357.
5. J. L. Vossen and J. J. Cuomo, "Glow Discharge Sputter Deposition," in J. L. Vossen and W. Kern, eds., *Thin Film Processes,* Academic, New York, 1978, pp. 12–75.
6. R. K. Watts, "Planar Magnetron Sputtering," in J. L. Vossen and W. Kern, eds., *Thin Film Processes,* Academic, New York, 1978, pp. 131–174.
7. J. A. Thornton, "Influence of Apparatus Geometry and Deposition Conditions on the Structure and Topography of Thick Sputtered Coatings," *J. Vac. Sci. Technol.,* **11,** 666 (1974).
8. L. I. Maissel, "Application of Sputtering to the Deposition of Films," in L. I. Maissel and R. Glang, eds., *Handbook of Thin Film Technology,* McGraw-Hill, New York, 1970, p. 4-1 to p. 4-44.
9. M. Hansen and K. Anderko, *Consititution of Binary Alloys,* McGraw-Hill, New York, 1958, pp. 132–134.
10. C. Y. Ting and M. Wittmer, "The Use of Titanium-Based Barrier Layers in Silicon Technology," *Thin Solid Films,* **96,** 327 (1982).
11. B. N. Chapman, *Glow Discharge Processes,* Wiley, New York, 1980, p. 369.
12. M. Wittmer, "Properties and Microelectronic Applications of Thin Films of Refractory Metal Nitrides," *J. Vac. Sci. Technol.,* **A-3,** 1797 (1985).
13. S. M. Rossnegal and D. Mikalsen, "Collimated Magnetron Sputter Deposition," *J. Vac. Sci. Technol.,* **A-9,** 261 (1991).
14. R. V. Joshi and S. Brodsky, "Collimated Sputtering of TiN/Ti Liners into Sub-Half-Micron High Aspect Ratio Contacts/Lines," *Proceedings of the 9th International VLSI Multilevel Interconnection Conference,* 1992, p. 253.
15. F. H. Baumann, R. Liu, C. B. Case, and W. Y.-C. Lai, "Application and Constraints of Collimated Ti/TiN Sputtering for 0.25 μm and sub-0.25 μm Contact Structures," *Proceedings of the 10th International VLSI Multilevel Interconnection Conference,* 1993, p. 412.
16. W. M. Bolton, R. E. Breining, and F. O. Deutscher, "Deposition of Aluminum," U.S. Patent 2,990,295, 1961.
17. M. L. Green, R. A. Levy, R. G. Nuzzo, and E. Coleman, "Al Films Prepared by Metal-Organic Low Pressure Chemical Vapor Deposition," *Thin Sol. Films,* **114,** 367 (1984).
18. H. W. Piekaar, L. F. Tz. Kwakman, and E. H. A. Granneman, "LPCVD of Aluminum in a Batch-Type Load-Locked Multi-Chamber Processing System," *Proceedings of the 6th International IEEE VLSI Multilevel Interconnection Conference,* 1989, p. 122.
19. W. Y.-C. Lai, K. P. Cheung, D. P. Favreau, C. J. Case, and R. Liu, "CVD Aluminum for Submicron VLSI Metallization," *Proceedings of the 8th International IEEE VLSI Multilevel Interconnection Conference,* 1991, p. 89.
20. T. Ohta, N. Takeyasu, E. Kondoh, Y. Kawano, and H. Yamamoto, "New Structure of Selective-CVD-Al for Submicron Plug Applications and Process Using UHV Cluster Tool," *Proceedings of the 11th International VLSI Multilevel Interconnection Conference,* 1994, p. 329.

21. R. W. Moshier, R. E. Sievers, and L. B. Spendlove, "Vapor-Plating Metals from Fluorocarbon Keto Metal Compounds," U.S. Patent 3,356,527, 1967.
22. M. E. Gross, "A Thermoanalytical Survey of Precursors for Copper Metal-Organic Chemical Vapor Deposition," *J. Electrochem. Soc.,* **138,** 2422 (1991).
23. A. V. Gelatos, S. Poon, R. Marsh, C. J. Mogab, and M. Thompson, "CVD of Copper From a Cu(+1) Precursor and Water Vapor and Formation of TiN-Encapsulated Submicron Copper Interconnects by Chemical-Mechanical Polishing," *1993 Symposium on VLSI Technology—Digest of Technical Papers,* 1993, p. 123.
24. B. Luther, J. F. White, C. Uzoh, T. Cacouris, J. Hummel, W. Guthrie, et al., "Planar Copper-Polyimide Backend of the Line Interconnections for ULSI Devices," *Proceedings of the 10th International VLSI Multilevel Interconnection Conference,* 1993, p. 15.
25. B. L. Crowder and S. Zirinsky, "1 μm MOSFET VLSI Technology: Part VII—Metal Silide Interconnection Technology—A Future Perspective," *IEEE Trans. Electron Dev.,* **ED-26,** 369 (1979).
26. K. C. Saraswat, D. L. Brors, J. A. Fair, K. A. Monnig, and R. Beyers, "Properties of Low Pressure CVD Tungsten Silicide for MOS VLSI Interconnections," *IEEE Trans. Electron Dev.,* **ED-30,** 1497 (1983).
27. T. Ohba, S. Inoue, and M. Maeda, "Selective CVD Tungsten Silicide for VLSI Applications," *Tech. Dig. IEDM,* 213 (1987).
28. A. Kohlhase, M. Mändl, and W. Palmer, "Performance and Failure Mechanisms of TiN Diffusion Barrier Layers in Submicron Devices," *J. Appl. Phys.,* **65,** 2464 (1989).
29. J. Laimer, H. Stori, and P. Rodhammer, "Plasma Assisted Chemical Vapor Deposition of Titanium Nitride in a Capacitively Coupled Radio-Frequency Discharge," *J. Vac. Sci. Technol.,* **A-7,** 2952 (1989).
30. A. Sherman, "Growth and Properties of LPCVD Titanium Nitride as a Diffusion Barrier for Silicon Device Technology," *J. Electrochem. Soc.,* **137,** 1892 (1990).
31. E. O. Travis, W. M. Paulson, F. Pintchovski, B. Boeck, L. C. Parrillo, M. L. Kottke, K.-Y. Fu, M. J. Rice, J. B. Price, and E. C. Eichman, "A Scalable Submicron Contact Technology Using Conformal LPCVD TiN," *Tech. Dig. IEDM,* 47 (1990).
32. J. J. Cuomo, "The Kinetics of Chemical Vapor Deposition of Tungsten," *Proceedings of the 3rd International Conference Chemical Vapor Deposition,* 1972, pp. 270–291.
33. P. A. Gargini and I. Beinglass, "WOS: Low Resistance Self-Aligned Source, Drain and Gate Transistors," *1981 Tech. Dig. IEDM,* 54 (1981).
34. E. K. Broadbent and C. L. Ramiller, "Selective Low Pressure Chemical Vapor Deposition of Tungsten," *J. Electrochem. Soc.,* **131,** 1427 (1984).
35. M. L. Green and R. A. Levy, "Structure of Selective Low Pressure Chemically Vapor-Deposited Films of Tungsten," *J. Electrochem. Soc.,* **132,** 1243 (1985).
36. S. Mehta, S. Mittal, A. Haranahalli, and D. Ranadive, "Blanket CVD Tungsten Interconnect for VLSI Devices," *Proceedings of the 3rd International IEEE VLSI Multilevel Interconnection Conference,* 1986, p. 418.
37. A. S. Sachdev, J. A. Fair, C. Fuhs, and W. Coney, "Blanket Tungsten Applications in VLSI Processing," in R. S. Blewer, ed., *Tungsten and Other Refractory Metals for VLSI Applications,* Materials Research Society, Pittsburgh, PA, 1987, pp. 69–72.
38. P. E. Riley, T. E. Clark, E. F. Gleason, and M. M. Garver, "Implementation of Tungsten Metallization in Multilevel Interconnection Technologies," *IEEE Trans. Semi. Manuf.,* **3,** 150 (1990).
39. C. M. McConica and K. Krishnamani, "The Kinetics of LPCVD Tungsten Deposition in a Single Wafer Reactor," *J. Electrochem. Soc.,* **133,** 2542 (1986).
40. JANAF Thermochemical Tables, 3rd ed., *J. Phys. Chem. Ref. Data,* **14,** Suppl. 1, (1985). D. R. Bradbury, I. E. Turner, K. Nauka, and K. Y. Chiu, "Selective CVD

Tungsten as an Alternative to Blanket Tungsten for Submicron Plug Applications on VLSI Circuits," *Tech. Dig. IEDM,* 273 (1991).
41. G. E. Georgiou, J. M. Brown, M. L. Green, R. Liu, D. S. Williams, and R. S. Blewer, "The Influence of Selective Tungsten Deposition on Shallow Junction Leakage," in V. A. Wells, ed., *Tungsten and Other Refractory Metals for VLSI Applications II,* Materials Research Society, Pittsburgh, PA, 1987, p. 227.
42. R. V. Joshi, K. Y. Ahn, and P. Fryer, "The Role of Process Parameters Selective W Deposited by SiH_4, H_2, and WF_6 Chemistry in Terms of Shallow Junction Leakages," in R. S. Blewer and C. M. McConica, eds., *Tungsten and Other Refractory Metals for VLSI Applications IV,* Materials Research Society, Pittsburgh, PA., 1989, pp. 85–92.
43. S. Kang, R. Chow, R. H. Wilson, B. Gorowitz, and A. G. Williams, "Application of Selective CVD W for Low Contact Resistance Via Filling to Aluminum Multilayer Interconnection," *J. Electron. Matls.,* **17,** 213 (1988).
44. Y. Kusumoto, K. Takakuwa, H. Hashinokuchi, T. Ikuta, and I. Nakayama, "A New Approach to the Suppression of Tunneling," in V. A. Wells, ed., *Tungsten and Other Refractory Metals for VLSI Applications II,* Materials Research Society, Pittsburgh, PA, 1987, pp. 103–109.
45. N. Yokoyama, K. Hinode, and Y. Homma, "LPCVD Titanium Nitride for ULSIs," *J. Electrochem. Soc.,* **138,** 190 (1991).
46. M. J. Buiting, A. F. Otterloo, and A. H. Montree, "Kinetical Aspects of the LPCVD of Titanium Nitride from Titanium Tetrachloride and Ammonia," *J. Electrochem. Soc.,* **138,** 500 (1991).
47. T. Suzuki, T. Ohba, Y. Furumura, and H. Tsutikawa, "LPCVD-TiN Using Hydrazine and $TiCl_4$," *Proceedings of the 10th International VLSI Multilevel Interconnection Conference,* 1993, p. 418.
48. K. Ishihara, K. Yamazaki, H. Hamada, K. Kamisako, and Y. Tarui, "Characterization of CVD TiN Films Prepared with Metal-Organic Source," *Jpn. J. Appl. Phys. 1,* **29,** 2103 (1990).
49. R. M. Fix, R. G. Gordon, and D. M. Hoffman, "Synthesis of Thin Films by Atmospheric Pressure Chemical Vapor Deposition Using Amido and Imido Titanium (IV) Compounds as Precursors," *Chem. Mater.,* **2(3),** 235 (1990).
50. I. J. Raaijmakers, R. N. Vrtis, G. S. Sandhu, J. Yang, E. K. Broadbent, D. A. Roberts, and A. Lagendijk, "Conformal Deposition of TiN at Low Temperature by Metal Organic CVD," *Proceedings of the 9th International VLSI Multilevel Interconnection Conference,* 1992, p. 260.
51. L. H. Dubois, B. R. Zegarski, and G. S. Girolami, "Infrared Studies of the Surface and Gas Phase Reactions Leading to the Growth of Titanium Nitride Thin Films from Tetrakis(dimethylamido)titanium and Ammonia," *J. Electrochem. Soc.,* **139,** 3603 (1992).
52. J. T. Hillman, M. J. Rice, Jr., D. W. Studiner, R. F. Foster, and R. W. Fiordalice, "Comparison of Titanium Nitride Barrier Layer Produced by Inorganic and Organic CVD," *Proceedings of the 9th International VLSI Multilevel Interconnection Conference,* 1992, p. 246.
53. K. A. Littau, M. Eizenberg, S. Ghanayem, H. Tran, Y. Maeda, A. Sinha, M. Chang, G. Dixit, M. K. Jain, M. F. Chrisholm and R. H. Haveman, "CVD TiN: A Barrier Metallization for Sub-Micron Via and Contact Applications," *Proceedings of the 11th International VLSI Multilevel Interconnection Conference,* 1994, p. 440.
54. N. J. Ianno, A. U. Ahmed, and D. E. Englebert, "Plasma Assisted Chemical Vapor Deposition of TiN from $TiCl_4/N_2/H_2$ Mixture," *J. Electrochem. Soc.,* **136,** 276 (1989).
55. T. Akahori and A. Tanihara, "Preparation of TiN Films by ECR Plasma CVD," *Proceedings of the International Conference on Solid State Devices and Materials,* Yokohama,

1991, pp. 180–182. T. Akahori, A. Tanihara, and M. Tano, "Preparation of TiN Films by Electron Cyclotron Resonance Plasma Chemical Vapor Deposition," *Jpn. J. Appl. Phys.*, **30,** 3558 (1991). ——, "TiN/Ti Films Formation Using ECR Plasma CVD," *Proceedings of the 10th International VLSI Multilevel Interconnection Conference,* 1993, p. 405.

56. A. Intemann, H. Koerner, and F. Koch, "Film Properties of CVD Titanium Nitride Deposited with Organometallic Precursors at Low Pressure Using Inert Gases, Ammonia, or Remote Activation," *J. Electrochem. Soc.,* **140,** 3215 (1993).
57. A. Weber, R. Nikulski, C.-P. Klages, M. E. Gross, W. L. Brown, E. Dons, and R. M. Charatan, "Low Temperature Deposition of TiN Using Tetrakis(dimethylamido)-Titanium in an Electron Cyclotron Resonance Plasma Process," *J. Electrochem. Soc.,* **141,** 849 (1994).
58. J. T. Hillman, R. F. Foster, J. Faguet, R. Arora, M. S. Ameen, C. Arena, and F. Martin, "Titanium Chemical Vapor Deposition," *Proceedings of the 11th International VLSI Multilevel Interconnection Conference,* 1994, p. 365.
59. N. Awaya and Y. Arita, "Selective Chemical Vapor Deposition of Copper," *Dig. Tech. Papers, Symp. on VLSI Technol.,* 103 (1989). Y. Arita, N. Awaya, K. Ohno, and M. Sato, "CVD Copper Metallurgy for ULSI Interconnections," *Tech. Dig. IEDM,* 39 (1990).
60. B. Lecohier, B. Calpini, J.-M. Philippoz, H. van den Bergh, D. Laub, and P. A. Buffat, "Copper Film Growth by Chemical Vapor Deposition," *J. Electrochem. Soc.,* **140,** 789 (1993).
61. D.-H. Kim, R. H. Wentorf, and W. N. Gill, "Film Growth Kinetics of Chemical Vapor Deposition of Copper from $Cu(HFA)_2$," *J. Electrochem. Soc.,* **140,** 3267 (1993).
62. H.-K. Shin, K.-M. Chi, M. J. Hampden-Smith, T. T. Kodas, J. D. Farr, and M. Paffett, "Hot-Wall Chemical Vapor Deposition of Copper from Copper I Compounds 2—Selective, Low-Temperature Deposition of Copper from Copper (I) β-Diketonate Compounds, (β-Diketonate)CuL_n, via Thermally Induced Disproportionate Reactions," *Chem. Mater.,* **4,** 788 (1992).
63. A. Jain, K.-M. Chi, T. T. Kodas, M. J. Hampden-Smith, J. D. Farr, and M. F. Paffet, "Chemical Vapor Deposition of Copper from (Hexafluoroacetylacetonato)(Alkyne)-Copper (I) Complex via Disproportionation," *Chem. Mater.,* **3,** 995 (1991).
64. J. A. T. Norman, B. A. Muratore, P. N. Dyer, D. A. Roberts, and A. K. Hochberg, "New OMCVD Precursors for Selective Copper Metallization," *J. de Phys. IV,* **1,** C2-271 (1991).
65. N. Awaya and Y. Arita, "The Characteristics of Blanket Copper CVD for Deep Submicron Via Filling," *Dig. Tech. Papers, Symp. on VLSI Technol.,* 125 (1993).
66. A. Jain, K.-M. Chi, T. T. Kodas, and M. J. Hampden-Smith, "Chemical Vapor Deposition of Copper from Hexafluoroacetylacetonato Cu(I) Vinyl-Trimethylsilane," *J. Electrochem. Soc.,* **140,** 1434 (1993).
67. L. H. Dubois and B. R. Zegarski, "Selectivity and Copper Chemical Vapor Deposition," *J. Electrochem. Soc.,* **139,** 3295 (1992).
68. K. P. Cheung, C. J. Case, R. Liu, R. J. Schutz, R. S. Wagner, L. F. Tz. Kwakman, D. Huibregtse, H. W. Piekaar, and E. H. A. Granneman, "Improved CVD Al Deposition Using in-situ Sputtered Nucleation Layers," *Proceedings of the 7th International IEEE VLSI Multilevel Interconnection Conference,* 1990, p. 303.
69. L. F. Tz. Kwakman, D. Huibregtse, H. W. Piekaar, E. H. A. Granneman, K. P. Cheung, C. J. Case, W. Y.-C. Lai, R. Liu, R. J. Schutz, and R. S. Wagner, "The Incorporation of Cu in CVD Al by Diffusion from In-Situ Sputtered Sources," *Proceedings of the 7th International IEEE VLSI Multilevel Interconnection Conference,* 1990, p. 282.

70. E. Kondoh, Y. Kawano, N. Takeyasu, T. Katagiri, H. Yamamoto, and T. Ohta, "Interconnection Formation by Simultaneous Copper Doping in Chemical Vapor Deposited Aluminum (Al-Cu CVD)," *Tech. Dig. IEDM,* 277 (1993).
71. T. Amazawa and Y. Arita, "A 0.25μm Via Plug Process Using Selective CVD Aluminum for Multilevel Interconnections," *Tech. Dig. IEDM,* 265 (1991).
72. T. Ohta, N. Takeyasu, E. Kondoh, Y. Kawano, and H. Yamamoto, "New Structure of Selective CVD Al Filling Direct on Al for Submicron Plug Applications and Process Using UHV Cluster Tool," *Proceedings of the 11th International VLSI Multilevel Interconnection Conference,* 1994, p. 329.
73. A. K. Sinha, W. S. Lindenberger, D. B. Fraser, S. P. Murarka, and E. N. Fuls, "MOS Compatibility of High-Conductivity $TaSi_2/n^+$ Poly-Si Gates," *IEEE Trans. Electron Dev.,* **ED-27,** 1425 (1980).
74. D. S. Williams, E. Coleman, and J. M. Brown, "Low Pressure Chemical Vapor Deposition of Tantalum Silicide," *J. Electrochem. Soc.,* **133,** 2637 (1986).
75. K. Hieber and F. Neppl, "Possible Applications of Tantalum Silicide for Very Large Scale Integration Technology," *Thin Sol. Films,* **140,** 131 (1986).
76. M. Suzuki, N. Kobayashi, K. Mukai, and S. Kondo, "Characterization of Silane-Reduced Tungsten Films Grown by CVD as a Function of Si Content," *J. Electrochem. Soc.,* **137,** 3213 (1990).
77. Y. Shioya, K. Ikegami, M. Maeda, and K. Yanagida, "High Temperature Stress Measurement on Chemical-Vapor-Deposited Tungsten Silicide and Tungsten Films," *J. Appl. Phys.,* **61,** 561 (1987).
78. T. Hara, T. Miyamoto, H. Hagiwara, E. I. Bromley, and W. R. Harshbarger, "Composition of Tungsten Silicide Films Deposited by Dichlorosilane Reduction of Tungsten Hexafluoride," *J. Electrochem. Soc.,* **137,** 2955 (1990).
79. S. G. Telford, M. Eizenberg, M. Chang, A. K. Sinha, and T. R. Gow, "Chemically Vapor Deposited Tungsten Silicide Films Using Dichlorosilane in a Single-Wafer Reactor—Growth, Properties, and Thermal Stability," *J. Electrochem. Soc.,* **140,** 3689 (1993).
80. P. K. Tedrow, V. Ilderem, and R. Reif, "Low Pressure Chemical Vapor Deposition of Titanium Silicide," *Appl. Phys. Lett.,* **46,** 189 (1985).
81. J. Lee and R. Reif, "Plasma Enhanced Chemical Vapor Deposition of Blanket $TiSi_2$ on Oxide Patterned Wafers II—Silicide Properties," *J. Electrochem. Soc.,* **139,** 1166 (1992).
82. J. Lee and R. Reif, "Plasma Enhanced Chemical Vapor Deposition of Blanket $TiSi_2$ on Oxide Patterned Wafers I—Growth of Silicide," *J. Electrochem. Soc.,* **139,** 1159 (1992).
83. A. Bouteville, C. Attuyt, and J. C. Remy, "Selective RTLPCVD of $TiSi_2$ without Substrate Consumption," *Appl. Surf. Sci.,* **53,** 11 (1991).
84. J. Engqvist, C. Myers, and J.-O. Carlsson, "Selective Deposition of $TiSi_2$ from H_2-$TiCl_4$ Gas Mixtures and Si: Aspects of Thermodynamics Including Critical Evaluation of Thermodynamical Data in the Ti-Si System," *J. Electrochem. Soc.,* **139,** 3197 (1992).
85. K. Saito, T. Amazawa, and Y. Arita, "Selective Titanium Silicide Chemical Vapor Deposition with Surface Cleaning by Silane and Ohmic Contact Formation to Very Shallow Junctions," *J. Electrochem. Soc.,* **140,** 513 (1993).
86. G. E. Georgiou, P. E. Bechtold, H. S. Luftman, and T. T. Sheng, "Selective Electroless Plated Ni Contacts to CMOS Junctions with $CoSi_2$," *J. Electrochem. Soc.,* **138,** 3618 (1991).
87. P.-L. Pai and C. H. Ting, "Copper as the Future Interconnection Material," *Proceedings of the 6th International IEEE VLSI Multilevel Interconnection Conference,* 1989, p. 258.

88. J. Van den Meerakker, "On the Mechanism of Electroless Plating I—Oxidation of Formaldehyde at Different Electrode Surfaces," *J. Appl. Electrochem.*, **11,** 387 (1981).
89. R. Contolini, S. Mayer, A. Bernhardt, and G. E. Georgiou, "A Copper Via Plug Process by Electrochemical Planarization," *Proceedings of the 10th International VLSI Multilevel Interconnection Conference,* 1993, p. 470.
90. S. P. Murarka, *Silicides for VLSI Applications,* Academic Press, Orlando, FL, 1983.
91. M.-A. Nicolet and S. Lau, "Silicides," in N. G. Einpruch and G. B. Larabee, eds., *VLSI Microstructure Science,* Vol. 6, Academic Press, NY, 1983, p. 329.
92. K. C. Saraswat, F. Mohammadi, and J. D. Meindl, "WSi_2 Gate MOS Devices," *Tech. Dig. IEDM,* 462 (1979).
93. T. Mochizuki, K. Shibata, T. Inoue, and K. Ohuchi, "A New MOS Process Using $MoSi_2$ as a Gate Material," *Jpn. J. Appl. Phys.,* **17-1,** 37 (1977).
94. M. E. Alperin, T. C. Holloway, R. A. Haken, C. D. Gosmeyer, R. V. Karnough, and W. D. Parmanto, "Development of the Self-Aligned Titanium Silicide Process for VLSI Applications," *IEEE Trans. Electron. Dev.,* **ED-32,** 141 (1985).
95. M. Delfino, E. K. Broadbent, A. E. Morgan, B. J. Burrow, and M. N. Norcott, "Formation of $TiN/TiSi_2/p^+/n^-$ Si by Rapid Thermal Annealing (RTA) Silicon Implanted with Boron through Titanium," *IEEE Electron. Dev. Lett.,* **EDL-6,** 591 (1985).
96. T. Brat, C. M. Osburn, T. Finstad, J. Liu, and B. Ellington, "Self-Aligned Ti Silicide Formed by Rapid Thermal Annealing," *J. Electrochem. Soc.,* **133,** 1451 (1986).
97. S. P. Murarka, D. B. Fraser, A. K. Sinha, J. J. Levinstein, E. J. Lloyd, R. Liu, D. S. Williams, and S. J. Hillenius, "Self-Aligned Cobalt Disilicide for Gate and Interconnection and Contacts to Shallow Junctions," *IEEE Trans. Electron Dev.,* **ED-34,** 2108 (1987).
98. L. Van den Hove, R. Wolters, K. Maex, R. F. De Keersmaecker, and G. J. Declerck, "A Self-Aligned $CoSi_2$ Interconnection and Contact Technology for VLSI Applications," *IEEE Trans. Electron. Dev.* **ED- 34,** 554 (1987).
99. J. B. Lasky, J. S. Nakos, O. J. Cain, and P. J. Geiss, "Comparison of Transformation to Low-Resistivity Phase and Agglomeration of $TiSi_2$ and $CoSi_2$," *IEEE Trans. Electron Dev.,* **ED-38,** 262 (1991).
100. I. Sakai, H. Abiko, H. Kawaguchi, T. Hirayama, L. E. G. Johansson, and K. Okabe, "A New Salicide Process (PASET) for Sub-Half Micron CMOS," *Dig. Tech. Papers, Symp. on VLSI Technol.,* 66 (1992).
101. T. Horiuchi, H. Wakabayashi, T. Ishigami, H. Nakamura, T. Mogami, T. Kunio, and K. Okumura, "A New Titanium Salicide Process (DIET) for Sub-Quarter Micron CMOS," *Dig. Tech. Papers, Symp. on VLSI Technol.,* 121 (1994).
102. Y. Matsubara, M. Sekine, N. Nishio, T. Shimura, K. Noguchi, T. Horiuchi, Y. Yamada, and T. Kitano, "A Novel Low Resistance Salicide Technology (SWAN) for Quarter-Micron CMOS," *Dig. Tech. Papers, Symp. on VLSI Technol.,* 103 (1993).
103. M. Sekine, K. Inoue, H. Ito, I. Honma, H. Miyamoto, K. Yoshida, H. Watanabe, K. Mikagi, Y. Yamada, and T. Kikkawa, "Self-Aligned Tungsten Strapped Source/Drain and Gate Technology Realizing the Lowest Sheet Resistance for Sub-Quarter Micron CMOS," *Tech. Dig. IEDM,* 493 (1994).
104. K. Goto, T. Yamazaki, A. Fushida, S. Inagaki, and H. Yagi, "Optimization of Salicide Processes for Sub-0.1μm CMOS Devices," *Dig. Tech. Papers, Symp. on VLSI Technol.* 119, (1994).
105. W. M. Chen, J. Lin, and J. C. Lee, "A Novel $CoSi_2$ Thin Film Process with Improved Thickness Scalability and Thermal Stability," *Tech. Dig. IEDM,* 691 (1994).
106. T. E. Tang, C.-C. Wei, R. A. Haken, T. C. Holloway, L. R. Hite, and T. G. W. Blake, "Titanium Nitride Local Interconnect Technology for VLSI," *IEEE Trans. Electron Dev.,* **ED-34,** 682 (1987).

107. D. C. Chen, S. S. Wong, P. V. Voorde, P. Merchant, T. R. Cass, J. Amano, and K. Y. Chiu, "A New Device Interconnect Scheme for Submicron VLSI," *Tech. Dig. IEDM,* 118 (1984).
108. K. K. Young, H. K. Hu, C. H. Lin, D. Peters, and K. Y. Chiu, "TiW Local Interconnect Technology for Advanced ULSI," *International Electronic Devices and Materials Symposium, Taipei, R.O.C.,* 1990, p. 231.
109. R. Beyers, D. Coulman, and P. Merchant, "Titanium Disilicide Formation on Heavily Doped Silicon Substrates," *J. Appl. Phys.,* **61**, 5110 (1987).
110. R. Liu, D. S. Williams, and W. T. Lynch, "A Study of the Leakage Mechanisms of Silicided n^+/p Junctions," *J. Appl. Phys.,* **63,** 1990 (1988).
111. R. Liu, F. A. Baiocchi, L. A. Heimbrook, J. Kovalchick, D. L. Malm, D. S. Williams, and W. T. Lynch, "Formation of Shallow p^+/n and n^+/p Junctions with $CoSi_2$," in *Proceedings of the 1st International Symposium on ULSI Science and Technology,* S. Broydo and C. M. Osburn, eds., Vol. 87-11, The Electrochemical Society, Pennington, NJ, 1987, pp. 446–462.
112. V. Probst, H. Schaber, P. Lippens, L. Van den Hove, and R. De Keersmaecker, "Limitation of $TiSi_2$ as a Source for Dopant Diffusion," *Appl. Phys. Lett.,* **52,** 1803 (1988).
113. C. Y. Lu, J. M. Sung, R. Liu, N. S. Tsai, R. Singh, S. J. Hillenius, and H. C. Kirsch, "Process Limitation and Device Design Trade-Offs of Self-Aligned $TiSi_2$ Junction Formation in Submicrometer CMOS Devices," *IEEE Trans. Electron Dev.* **ED-38,** 246 (1991). H. Jiang, C. M. Osburn, Z.-G. Xiao, G. McGuire, and G. A. Rozgonyi, "Ultra Shallow Junction Formation Using Diffusion from Silicides III—Diffusion into Silicon, Thermal Stability of Silicides, and Junction Integrity," *J. Electrochem. Soc.,* **139,** 211 (1992).
114. R. Liu, D. S. Williams, and W. T. Lynch, "Mechanisms for Process-Induced Leakage in Shallow Silicided Junctions," *Tech. Dig. IEDM,* 58 (1986).
115. K. C. Saraswat, S. Swirhun, and J. P. McVittie, "Selective CVD of W for VLSI Technology," *Proceedings of the 2nd International Symposium on VLSI Science and Technology,* J. E. Bean and G. A Rozgonyi, eds., Vol. 84-7, The Electrochemical Society, Pennington, NJ, 1984, pp. 409–419.
116. D. R. Bradbury and T. I. Kamins, "Effect of Insulator Surface on Selective Deposition of CVD W Films," *J. Electrochem. Soc.,* **133,** 1214 (1986).
117. K. K. Choi, S. B. Hwang, H. L. Park, and C. G. Ko, "A New Selective W-CVD Process Using Poly-Si Glue Layer," *Proceedings of the 9th International VLSI Multilevel Interconnection Conference,* 1992, p. 286.
118. A. Kamgar, R. V. Knoell, F. A. Baiocchi, K. J. Orlowsky, K. P. Cheung, and R. Liu, "Impact of Al Melting on Diode Integrity," *Proceedings of the 6th International IEEE VLSI Multilevel Interconnection Conference,* 1989, p. 190.
119. R. Mukai, N. Sasaki, and M. Nakano, "High-Aspect-Ratio Via-Hole Filling with Aluminum Melting by Excimer Laser Irradiation for Multilevel Interconnection," *IEEE Electron Dev. Lett.,* **EDL-8,** 76 (1987).
120. D. Pramanik and S. Chen, "Characterization of Laser Planarized Aluminum for Submicron Double Level Metal CMOS Circuits," *Tech. Dig. IEDM,* 673 (1989).
121. W. Y.-C. Lai, R. Liu, K. P. Cheung, and R. Heim, "The Use of Ti as an Antireflective Coating for the Laser Planarization of Al for VLSI Metallization," *Proceedings of the 6th International IEEE VLSI Multilevel Interconnection Conference,* 1989, p. 501, ——, "A Study of Pulsed Laser Planarization of Aluminum for VLSI Metallization," Ibid., 1989, p. 329.
122. C. S. Park, S. I. Lee, J. H. Park, J. H. Sohn, D. Chin, and J. G. Lee, "Al-PLAPH (Aluminum Planarization by Post-Heating) Process for Planarized Double Metal CMOS

Applications," *Proceedings of the 8th International IEEE VLSI Multilevel Interconnection Conference,* 1991, p. 326.
123. F. S. Chen, Y. S. Liu, G. A. Dixit, R. Sundaresan, C. C. Wei, and F. T. Liou, "Planarized Aluminum Metallization for Sub-0.5 μm CMOS Technology," *Tech. Dig. IEDM,* 51 (1990).
124. M. Ionoue, K. Hashizume, and H. Tsuchikawa, "The Properties of Aluminum Thin Films Sputter Deposited at Elevated Temperature," *J. Vac. Sci. Technol.,* **A-6,** 1636 (1988).
125. C. J. Case and R. Liu, unpublished.
126. K. Kikuta, T. Kikkawa, and H. Aoki, "0.25 μm Contact Hole Filling By Al-Ge Reflow Sputtering," *Dig. Tech. Papers, Symp. on VLSI Technol.,* 35 (1991).
127. G. A. Dixit, M. F. Chrisholm, M. K. Jain, T. Weaver, L. M. Ting, S. Pourch, K. Mizobuchi, R. H. Havemann, C. D. Dobson, A. I. Jeffryes, P. J. Holverson, P. Rich, D. C. Butler, and J. Hems, "A Novel High Pressure Low Temperature Aluminum Plug Technology for Sub-0.5μm Contact/Via Geometries," *Tech. Dig. IEDM,* 105 (1994).
128. C. S. Wei, V. Mureli, M. L. A. Dass, and D. B. Fraser, "The Use of Selective Silicide Plugs for Submicron Contact Fill," *Proceedings of the 6th International IEEE VLSI Multilevel Interconnection Conference,* 1989, p. 136.
129. T. Iijima, A. Nishiyama, Y. Ushiku, T. Ohguro, I. Kunishima, K. Suguro, and H. Iwai, "A Novel Selective Ni_3Si Contact Plug Technique for Deep-Submicron ULSIs," *Dig. Tech. Papers, Symp. on VLSI Technol.,* 70 (1992).
130. A. C. Adams and C. D. Capio, "The Deposition of Silicon Dioxide Films at Low Pressure," *J. Electrochem. Soc.,* **126,** 1042 (1979). A. C. Adams, "Dielectric and Polysilicon Film Deposition," in S. M. Sze, ed., *VLSI Technology,* McGraw-Hill, NY, 1983, p. 107.
131. G. C. Schwartz and P. Johns, "Gap-Fill with PECVD Silicon Dioxide Using Deposition/Sputter-Etch Cycles," *Proceedings of the 3rd International Symposium on ULSI Science and Technology,* The Electrochemical Society, Pennington, NJ, 1991, pp. 720–729.
132. K. Maeda, Y. Nishimoto, N. Tokumasu, and K. Fujino, "Dielectric Film Deposition by Atmospheric Pressure and Low Temperature CVD Using TEOS, Ozone, and New Organometallic Doping Sources," *Proceedings of the 6th International IEEE VLSI Multilevel Interconnection Conference,* 1989, p. 382.
133. K. Fujino, Y. Nishimoto, N. Tokumasu, and K. Maeda, "Dependence of Deposition Characteristics on Base Materials in TEOS and Ozone CVD at Atmospheric Pressure," *J. Electrochem. Soc.,* **138,** 550 (1991).
134. K. Fujino, Y. Nishimoto, N. Tokumasu, and K. Maeda, "Silicon Dioxide Deposition by Atmospheric Pressure and Low-Temperature CVD Using TEOS and Ozone," *J. Electrochem. Soc.,* **137,** 2883 (1990).
135. J. Huang, K. Kwok, D. Witty, and K. Donohoe, "Dependence of Film Properties of Subatmospheric Pressure Chemical Vapor Deposited Oxide on Ozone-to-Tetraethylorthosilicate Ratio," *J. Electrochem. Soc.,* **140,** 1682 (1993).
136. J. A. Mucha and J. Washington, "Infrared Studies of Ozone-Organosilicon Chemistries for SiO_2 Deposition," *Mat. Res. Conf. Proc.,* **334,** 31 (1994).
137. J. A. Mucha, "The Effects of Ethoxy-Groups on the Kinetics of O_3-Induced SiO_2 Deposition," submitted for publication (1995).
138. See for example C. J. Adkins, *Equilibrium Thermodynamics,* McGraw-Hill, London, pp. 211–216, 1968.
139. L. Forester, A. L. Butler, and G. Schets, "SOG Planarization for Polysilicon and First Metal Interconnect in a One Micron CMOS Process," *Proceedings of the 6th International IEEE VLSI Multilevel Interconnection Conference,* 1989, p. 72.
140. L. Forester, H. Meynen, and B. Coenegrachts, "Characterization of a Direct-on-Metal Spin-on Glass Planarization Process," *Proceedings of the 10th International VLSI Multilevel Interconnection Conference,* 1993, p. 175.

141. B. T. Ahlburn and K. A. Scheibert, "A Non-Etchback Spin On Glass for 0.5 μm Devices Using Hydrogen Silsesquioxane as a Replacement for Methylsiloxane," *Proceedings of the 11th International VLSI Multilevel Interconnection Conference,* 1994, p. 120.
142. L.-J. Chen, S.-T. Hsia, and J.-L. Leu, "Fluorine-Implanted-Treatment (FIT) SOG for the Non-Etchback Intermetal Dielectric," *Proceedings of the 11th International VLSI Multilevel Interconnection Conference,* 1994, p. 81.
143. C. Y. Ting. V. J. Vivalda, and H. G. Schaefer, "Study of Planarized Sputter-Deposited SiO_2," *J. Vac. Sci. Technol.,* **15(3),** 1105 (1978).
144. S. Matsuo and M. Kiuchi, "Low Temperature Chemical Vapor Deposition Method Utilizing an Electron Cyclotron Resonance Plasma," *Jpn. J. Appl. Phys. Lett.,* **22,** L210 (1983).
145. K. M. Kerney, "ECR Finds Applications in CVD," *Semiconductor International,* 67 (Mar. 1989).
146. C. S. Pai, J. M. Miner, and P. D. Foo, "Electron Cyclotron Resonance Microwave Discharge for Oxide Deposition Using Tetraethoxysilane," *J. Electrochem. Soc.,* **139,** 850 (1992).
147. R. Chebi and S. Mittal, "A Manufacturable ILD Gap Fill Process with Biased ECR CVD," *Proceedings of the 8th International IEEE VLSI Multilevel Interconnection Conference,* 1991, p. 61.
148. B. M. Somero, R. P. Chebi, E. O. Travis, H. B. Haver, and W. K. Morrow, "An Integrated, Low Temperature Interlayer Dielectric Using ECR Technology for Advanced CMOS ASIC and Logic Applications," *Proceedings of the 9th International VLSI Multilevel Interconnection Conference,* 1992, p. 72.
149. A. Bose, M. M. Garver, and R. A Spencer, "Advanced Inter-Metal Dielectric Deposition—A Comparative Analysis Between ECR-CVD and O_3/TEOS," *Proceedings of the 10th International VLSI Multilevel Interconnection Conference,* 1993, p. 89.
150. N. Lifshitz and G. Smolinsky, "Water-Related Charge Motion in Dielectrics," *J. Electrochem. Soc.,* **136,** 2335 (1989). ———, "Hot Carrier Aging of the MOS Transistors in the Presence of Spin-On Glass as the Interlevel Dielectric," *IEEE Electron Dev. Lett.,* **EDL-12,** 140 (1991).
151. S. Wolf, *Silicon Processing for the VLSI Era, Volume 2—Process Integration,* Lattice Press, Sunset Beach, CA, 1990, pp. 199–201.
152. T. H. Daubenspeck, J. K. DeBrosse, C. W. Koburger, M. Armcost, and J. R. Abernathey, "Planarization of ULSI Topography over Variable Pattern Densities," *J. Electrochem. Soc.,* **138,** 506 (1991).
153. A. Schiltz and M. Pons, "Two-Layer Planarization Process," *J. Electrochem. Soc.,* **133,** 178 (1986).
154. B. Davari, C. W. Koburger, R. Schulz, J. D. Warnock, T. Furukawa, M. Jost, Y. Taur, W. G. Schwittek, J. K. DeBrosse, M. L. Kerbaugh, and J. L. Mauer, "A New Planarization Technique Using a Combination of RIE and Chemical Mechanical Polish (CMP)," *Tech. Dig. IEDM,* 61 (1989).
155. L. M. Cook, "Chemical Processes in Glass Polishing," *J. Non Cryst. Solids,* **120,** 152 (1990).
156. S. Sivaram, R. Leggett, A. Maury, K. Monnig, and R. Tolles, "Overview of Planarization by Mechanical Polishing of Interlevel Dielectrics," *Proceedings of the 3rd International Symposium on ULSI Science and Technology,* J. M. Andrews and G. K. Celler, eds., Vol. 91-11, The Electrochemical Society, Pennington, NJ, 1991, pp. 607–616.
157. F. W. Preston, "The Theory and Design of Plate Glass Polishing Machines," *J. Soc. Glass Tech.,* **11,** 214 (1927).
158. J. Warnock, "A Two-Dimensional Process Model for Chemimechanical Polish Planarization," *J. Electrochem. Soc.,* **138,** 2398 (1991).

159. S. R. Runnels, "Feature-Scale Fluid Based Erosion Modeling for Chemical-Mechanical Polishing," *J. Electrochem. Soc.,* **141,** 1900 (1994).
160. F. B. Kaufman, D. B. Thompson, R. E. Broadie, M. A. Jaso, W. L. Guthrie, D. J. Pearson, and M. B. Small, "Chemical-Mechanical Polishing for Fabricating Patterned W Metal Features as Chip Interconnects," *J. Electrochem. Soc.,* **138,** 3460 (1991).
161. W. J. Patrick, W. L. Guthrie, C. L. Standley, and P. M. Schiable, "Application of Chemical Mechanical Polishing to the Fabrication of VLSI Circuit Interconnections," *J. Electrochem. Soc.,* **138,** 1778 (1991).
162. I. Ali, S. R. Roy, and G. Shinn, "Chemical-Mechanical Polishing of Interlayer Dielectric: A Review," *Sol. State Technol.,* 63 (Oct. 1994).
163. R. Jairath, J. Farkas, C. K. Huang, M. Stell, and S-M. Tzeng, "Chemical-Mechanical Polishing: Process Manufacturability," *Sol. State Technol.,* 71 (July 1994).
164. W. Y.-C. Lai, G. L. Miller, R. J. Schutz, G. Smolinsky, and E. R. Wagner, "Capacitive End-Point Detection for Oxide Planarization by Chemical-Mechanical Polishing," *Proceedings of the 10th International VLSI Multilevel Interconnection Conference,* 1993, p. 147.
165. P. Singer, "Chemical-Mechanical Polishing: A New Focus on Consumables," *Semic. Intl.,* 48 (Feb. 1994).
166. H. Aoki, T. Nakajima, K. Kikuta, and Y. Hayashi, "Novel Electrolysis-Ionized-Water Cleaning Technique for Chemical-Mechanical Polishing (CMP) Process," *Dig. Tech. Papers, Symp. on VLSI Technol.,* 79 (1994).
167. E. K. Broadbent, J. M. Flanner, W. G. Wilbert, G. M. Van den Hoek, and I.-W. Huang Connick, "High Density High Reliability Tungsten Interconnection by Filled Interconnect Groove Metallization," *IEEE Trans. Electron Dev.,* **ED-35,** 952 (1988).
168. C. W. Kaanta, W. J. Cote, J. E. Cronin, K. Holland, P. Lee, and T. Wright, "Submicron Wiring Technology with Tungsten and Planarization," *Tech. Dig. IEDM,* 209 (1987).
169. C. W. Kaanta, S. G. Bombardier, W. J. Cote, W. R. Hill, G. Kerszykowski, H. S. Landis, D. J. Poindexter, C. W. Pollard, G. H. Ross, J. G. Ryan, S. Wolff, and J. E. Cronin, "Dual Damascene: A ULSI Wiring Technology," *Proceedings of the 8th International IEEE VLSI Multilevel Interconnection Conference,* 1991, p. 144.
170. J. Givens, S. Geissler, O. Cain, W. Clark, C. Koburger, and J. Lee, "A Low-Temperature Local Interconnect Process in a 0.25μm-Channel CMOS Logic Technology with Shallow Trench Isolation," *Proceedings of the 11th International VLSI Multilevel Interconnection Conference,* 1994, p. 43.
171. K. D. Beyer, W. L. Guthrie, S. R. Markarewicz, E. Mendel, W. J. Patrick, K. A. Perry, W. A. Pliskin, J. Riseman, P. M. Schiable, and C. L. Standly, "Chemical-Mechanical Polishing Method for Producing Coplanar Metal/Insulator Films on a Substrate," U.S. Patent 4,944,836, 1990.
172. A. Krishnan, C. Xie, N. Kumar, J. Curry, D. Duane, and S. P. Murarka, "Copper Metallization for VLSI Applications," *Proceedings of the 9th International VLSI Multilevel Interconnection Conference,* 1992, p. 226.
173. N. Misawa, S. Kishii, T. Ohba, Y. Arimoto, Y. Furumura, and H. Tsutikawa, "High Performance Planarized CVD-Cu Multi-Interconnection," *Proceedings of the 10th International VLSI Multilevel Interconnection Conference,* 1993, p. 353.
174. H. M. Dalal, R. V. Joshi, H. S. Rathore, and R. Fillipi, "A Dual Damascene Hard Metal Capped Cu and Al-Alloy for Interconnect Wiring of ULSI Circuits," *Tech. Dig. IEDM,* 273 (1993).
175. Y. Hayashi, K. Kikuta, and T. Kikkawa, "A New Abrasive-Free, Chemical-Mechanical-Polishing Technique for Aluminum Metallization of ULSI Devices," *Tech. Dig. IEDM,* 976 (1992).

176. T. Shinzawa, K. Sugai, A. Kobayashi, Y. Hayashi, T. Nakajima, S. Kishida, H. Okabayashi, T. Yako, K. Tsunenari, and Y. Murao, "Adhesion-Layerless Submicron Al Damascene Interconnection Using Novel Al-CVD," *Dig. Tech. Papers, Symp. on VLSI Technol.* 77 (1994).
177. K. Kikuta, Y. Hayashi, T. Nakajima, K. Harashima, and T. Kikkawa, "Aluminum-Germanium-Copper Multilevel Damascene Process Using Low Temperature Reflow Sputtering and Chemical Mechanical Polishing," *Tech. Dig. IEDM,* 101 (1994).
178. S. Roehl, L. Camiletti, W. Cote, E. Eckstein, K. H. Froehner, P. I. Lee, D. Restaino, G. Roeska, V. Vynorius, S. Wolff, and B. Vollmer, "High Density Damascene Wiring and Borderless Contacts for 64M DRAM,"*Proceedings of the 9th International VLSI Multilevel Interconnection Conference,* 1992, p. 22.
179. R. N. Hall, D. M. Brown, R. H. Wilson, and D. W. Skelly, "Electromigration Reliability Studies of Intermetal Contact Having CVD Tungsten Via Plugs," *Proc. Tungsten and Other Refractory Metals for VLSI Applications III,* V. A. Wells, ed., Materials Research Society, Pittsburgh, PA, 1987, pp. 231–237.
180. F. Matsuoka, H. Iaai, K. Hama, H. Itoh, R. Nakata, T. Nakakubo, K. Maoguchi, and K. Kanzaki, "Electromigration Reliability for Tungsten Plug Filled Via Hole Structure," *IEEE Trans. Electron Dev.,* **ED-37,** 562 (1990).
181. S. Fujimura, K. Shinagawa, M. T. Suzuki, and M. Nakamura, "Resist Stripping in an O_2 + H_2O Plasma Downstream Reactor," *J. Vac. Sci. Technol.,* **A-9,** 357 (1991).
182. J. S. H. Cho, H. K. Kang, I. Asano, and S. S. Wang, "CVD Cu Interconnection for ULSI," *Tech. Dig. IEDM,* 297 (1992).
183. S. Lakshminarayanan, J. Steigerwald, D. Price, M. Bourgeois, T. P. Chow, R. J. Gutmann, and S. P. Murarka, "Dual Damascene Copper Metallization Process Using Chemical-Mechanical Polishing," *Proceedings of the 11th International VLSI Multilevel Interconnection Conference,* 1994, p. 49.
184. J. S. H. Cho, H.-K. Kang, C. Ryu, and S. S. Wong, "Reliability of CVD Cu Buried Interconnect," *Tech. Dig. IEDM,* 265 (1993).
185. T. Nitta, T. Ohmi, T. Hoshi, S. Sakai, K. Sakaibara, S. Imai, and T. Shibata, "Evaluating the Large Electromigration Resistance of Copper Interconnects Employing a Newly Developed Accelerated Life-Test Method," *J. Electrochem. Soc.,* **140,** 1131 (1993).
186. K. Urabe, T. Katoh, and Y. Murao, "Fully Planarized Multilevel Gold Metallization with Gap Filled Wires," *Proceedings of the 10th International VLSI Multilevel Interconnection Conference,* 1993, p. 366.
187. Y. Ushiku, H. Ono, T. Iijima, N. Ninomiya, A. Nishiyama, H. Iwai, and H. Hara, "Planarized Silver Interconnect Technology with a Ti Self-Passivation Technique for Deep Submicron ULSIs," *Dig. Tech. Papers, Symp. on VLSI Technol.,* 121 (1993).
188. M. Maeda, T. Makino, E. Yamamoto, and S. Konaka, "A Low Permittivity Interconnection Using an SiBN Interlayer," *IEEE Trans. Electron Dev.,* **ED-36,** 1610 (1989).
189. J. Ida, M. Yoshimaru, T. Usami, A. Ohtomo, K. Shimokawa, A. Kita, and M. Ino, "Reduction of Wiring Capacitance with New Low Dielectric SiOF Interlayer Film for High Speed/Low Power Sub-Half Micron CMOS," *Dig. Tech. Papers, Symp. on VLSI Technol.,* 59 (1994).
190. T. Homma, R. Yamaguchi, Y. Murao, "Room Temperature Chemical Vapor Deposition SiOF Film Formation Technology for the Interlayer in Submicron Multilevel Interconnections," *J. Electrochem. Soc.,* **140,** 687 (1993).
191. R. M. Geffken, "An Overview of Polyimide Use in Integrated Circuits and Packaging," in *Proceedings of the 3rd International Symposium on ULSI Science and Technology,* J. M. Andrews and G. K. Celler, eds., Vol. 91-11, The Electrochemical Society, Pennington, NJ, 1991, pp. 667–677.

192. R. V. Joshi, L. Hsu, H. Dalal, P. Klymco, M. Jaso, and H. Ng, "A Novel Application of Polyimide-W-Al/Cu for VLSI Interconnect," *Proceedings of the 8th International IEEE VLSI Multilevel Interconnection Conference,* 1991, p. 75.
193. S. Babrad, X. Zhang, B. J. Howard, C. Chiang, G. Cuan, K. Huang, R. Olson, H. Bakhru, C. Steinbruchel, T.-M. Lu, and J. F. McDonald, "Aliphatic Tetrafluorinated Parylene (PA-f) as a Conformal Insulator for Submicron Multilevel Interconnections," *Proceedings of the 9th International VLSI Multilevel Interconnection Conference,* 1992, p. 86.
194. X. Zhang, R. Tacito, D. S. Yaney, C. Chiang, C. Steinbruchel, and J. F. McDonald, "A Submicron Parylene Etch Process for Multilevel Interconnection," *Proceedings of the 11th International VLSI Multilevel Interconnection Conference,* 1994, p. 87.
195. J. Paraszczak, D. Edelstein, S. Cohen, E. Babich, and J. Hummel, "High Performance Dielectrics and Processes for ULSI Interconnection Technologies," *Tech. Dig. IEDM,* 261 (1993).
196. C. Mazure, J. Fitch, and C. Gunderson, "Facet Engineered Elevated Source/Drain by Selective Epitaxy for 0.35 Micron MOSFETs," *Tech. Dig. IEDM,* 853 (1992).
197. M. Moslehi, L. Velo, A. Paranjpe, R. Chapman, Y. J. Lee, S. Huang, J. Kuehne, C. Schaper, T. Breedjik, D. Yin, D. Anderson, and D. Davis, "Microelectronic Manufacturing Science and Technology (MMST): Single-Wafer RTP-Based 0.35 μm CMOS IC Fabrication," *Tech. Dig. IEDM,* 649 (1993).

PROBLEMS

1. (*a*) Estimate the capacitance per unit length of two parallel Al wires 0.5 μm × 0.5 μm in cross section and separated by a SiO_2 dielectric layer $d = 0.5$ μm. Use $f = 1.5$ for fringe field effects ($\varepsilon_{ox} = 0.345$ pF/cm). (*b*) Estimate the capacitance per unit length for one Al wire at the distance $d = 0.5$ μm over a ground plane, where $f = 3$.

2. If we assume no external resistance, what is the "intrinsic" *RC* delay for the configuration in Problem 1(*b*), (*a*) for an Al wire length of 1 mm, and (*b*) for a wire of 2-mm length? The resistivity of Al is $\rho = 2.7$ μΩ-cm.

3. Show that the intrinsic *RC* delay for two parallel Al wires is independent of the metal height. How does the intrinsic *RC* scale with the scaling factor $S \approx \sqrt{2}$ (*a*) if the die size also scales down as $1/S$, (*b*) if the die size stays constant, and (*c*) if the die size scales up as S?

4. For an nMOSFET with $L = 0.35$ μm, $W = 1$ μm, and a gate oxide thickness = 9 nm, operating at $V_d = V_g = 3.3$ V, the on-current $I_{on} \approx 0.5$ mA is achieved. (*a*) If a $V_{dd} = 3.3$ V is applied to the drain of the MOSFET through a long 1 mm Al wire with a cross section of 0.5 μm × 0.5 μm, what is the V_d? Will the MOSFET operate in the saturation mode? (*b*) Do the same calculations for a wide MOSFET with $W = 10$ μm. (*c*) Do the same calculations for a wider MOSFET with $W = 40$ μm and a 2 mm Al wire. (Let $V_{dsat} = 0.8$ V; the resistivity for Al is $\rho = 2.7$ μΩ-cm.)

5. When a MOSFET is operating in the saturation region it can be considered as a current source. Let us define the node delay time as the time required to charge the capacitance of a node, $\Delta t_n = C_n \Delta V / i_n$, where C_n is the node capacitance, i_n the charging current, and $\Delta V = V_{dd}$ for full-swing operation. (*a*) Consider the case $C_n = 20$ fF, $V_{dd} = 3.3$

V, and a MOSFET as in Problem 3 with $W = 10\ \mu m$; what is the node delay time? (*b*) Consider the case with $W = 40\ \mu m$, and V_{dd} is connected to the drain of the MOSFET through a 2 mm Al wire, as in Problem 3(*c*). Ignore the *RC* delay in the Al wire, and assume a $V_{sat} = 0.8$ V; what is the node delay time now? (*c*) What is the node delay time for $W = 80\ \mu m$?

6. The rule of thumb for the electromigration limit in Al wiring is to restrict the current density below 1×10^5 A/cm^2. Consider the device and metallization parameters in the table below. At what generation of technology will the current density in the Al wire exceed the design limit from the current of a minimum dimension transistor?

	nMOSFET				Metallization	
Technology	**L(μm)**	**W(μm)**	**G_{ox}(nm)**	**I_{on}/μm**	**w(μm)**	**h(μm)**
0.5 μm (3.3 V)	0.50	0.70	12	0.3 mA	0.70	0.50
0.35 μm (3.3 V)	0.35	0.50	8	0.5 mA	0.50	0.50
0.25 μm (2.5 V)	0.25	0.35	6	0.5 mA	0.35	0.40
0.18 μm (1.8 V)	0.18	0.25	4	0.5 mA	0.25	0.35

7. The mean free path of a molecule in a gas is given by $\lambda = kT/(\sqrt{2}\pi d^2 p)$, where p is the gas pressure and d the diameter of the molecule. Estimate the mean free path for (*a*) a sputtering system with an Ar pressure of 4 mtorr ($d = 0.36$ nm); (*b*) a CVD W deposition system with hydrogen pressure of 30 mtorr ($d_{WF_6} = 0.56$ nm, $d_H = 0.28$ nm); (*c*) an O_3-TEOS system with a N_2 pressure of 600 torr ($d_{N_2} = 0.37$ nm, $d_{TEOS} = 2$ nm); and (*d*) an evaporation system with an H_2O pressure of 10^{-6} torr, ($d = 0.47$ nm).

8. A collimator with an aspect ratio of 1 : 1 allows only about 20% of the sputtered material to pass. Let us assume that the conventional sputtering is 100% transparent and that the bottom coverage in the contact holes is 10% for conventional sputtering and 33.33% for collimated sputtering. (*a*) Calculate the thickness of collimated TiN barrier metal needed if 100 nm is used by conventional sputtering. (*b*) Estimate the cost of collimated TiN relative to conventional TiN sputtering, assuming that the only relevant factor is the equipment throughput. (*c*) How can the economy of collimated sputtering be improved?

9. Consider three CVD metal deposition processes: (*a*) CVD Al using TIBA, or tri-isobutyl-Al [$(C_4H_9)_3Al$]; (*b*) CVD W using WF_6; and (*c*) CVD Cu using Cu(hfa)$_2$. If we assume a utilization rate of 5% in all cases, how many grams of each chemical precursor are needed to deposit a 0.5-μm coating of metal on 200-mm wafers? For a 5000 wafer-start/week factory and three applications of the metal per wafer, how much precursor material (in weight) is needed per year (50 weeks)? How much waste material needs to be disposed of or burned? (The densities of Al, W, and Cu are 2.7, 19.3, and 8.9 g/cm^3, and the atomic weights for W, Cu, Al, and F are 184, 63.5, 27, and 19, respectively. Assume deposition occurs only on the wafer.)

10. Derive Eq. (8.41).

11. An nMOSFET with $L = 0.18\ \mu m$ and $W = 1\ \mu m$ can produce a $I_{on} = 0.5$ mA at $V_d = 1.8$ V, with characteristics as shown in the *I-V* curve. (*a*) If we assume there is no current crowding, what is the output current if a single 0.2-μm (round) contact is

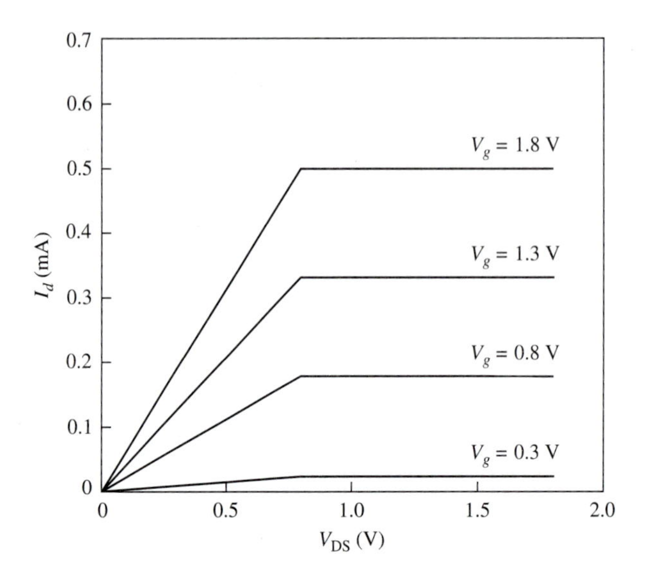

made on both the source and the drain of the transistor? Use $\rho_{sc} = 1 \times 10^{-7}\ \Omega\text{-cm}^2$, $2 \times 10^{-7}\ \Omega\text{-cm}^2$, and $5 \times 10^{-7}\ \Omega\text{-cm}^2$ and compare the results. (*b*) Repeat (*a*) for $W = 10\ \mu\text{m}$. (*Hint:* Solve graphically.)

12. Consider a wide nMOSFET with $L = 0.25\ \mu\text{m}$ and $W = 40\ \mu\text{m}$. Calculate the IR drop along the width of the transistor assuming $I = 10^{-12}\text{A}/\mu\text{m}$. Also calculate the RC delay along the width of the transistor assuming a $C = 1\ \text{fF}/\mu\text{m}$ (*a*) for a polysilicon gate with $R_s = 100\ \Omega/\square$; (*b*) for a polycide gate with $R_s = 15\ \Omega/\square$; and (*c*) for a salicide gate with $R_s = 3\ \Omega/\square$.

13. Consider a MOSFET with $L = 0.25\ \mu\text{m}$ and $W = 10\ \mu\text{m}$ with a single pair of contacts at one end of the source and drain, as shown in Fig. 12*b*. The current crowding effect can be roughly simulated by a set of parallel resistors with values $R_{ch} + 2kR_{sh}$, where R_{ch} is the channel resistance per unit width, R_{sh} the diffusion sheet resistance, and k the number of squares for the kth parallel current path. (*a*) If $R_{ch} = 3.5\ k\Omega/\mu\text{m}$ and $R_{sh} = 100\ \Omega/\square$, what is the ratio I_d/I_{do}, where the I_{do} is the drain current without current crowding (i.e., current flows through the entire width of the MOSFET)? (*b*) What is I_d/I_{do} if $R_{sh} = 200\ \Omega/\square$? (*Hint:* Divide the wide transistor into 10 parallel 1-μm wide transistors.)

14. To refine the model in Problem 13, consider a similar MOSFET with $L = 0.25\ \mu\text{m}$ and $W = 20\ \mu\text{m}$. Let the current distribution along the width of the transistor be $i_d(x) = i_{do}x_o/(x_o + \alpha x)$, where x is the lateral distance in μm from the center of a contact window, $X_o = 1\ \mu\text{m}$, and $\alpha = 2R_{sh}/R_{ch} = 200/3500 = 0.057$ for $R_{sh} = 100\ \Omega/\square$ and $R_{ch} = 3.5\ k\Omega/\mu\text{m}$. Calculate the drain current ratio I_d/I_{do} for (*a*) a single contact to the source/drain at the center of the MOSFET; (*b*) a single contact to the source/drain at one end of the transistor; (*c*) 19 contact windows, separated by 1 μm, to the source/drain; and (*d*) repeat (*a*), (*b*), (*c*) for silicided source/drain with $R_{sh} = 3\ \Omega/\square$.

15. Let us assume that the cost for a lithography step is \$10; for etching, metal deposition, and dielectric deposition is \$5; the cost for sacrificial material (resist, SOG) is \$2; and for CMP is \$8, per wafer, per application. Assume that the lithography depth of focus is 1.0

μm and poly and metal thicknesses are $M_0 = M_1 = M_2 = M_3 = M_4 = 0.6$ μm. What is the most economical four-level-metal metallization process? What is the most economical process if the cost per application for CMP is $20? You may use any combination of global and local planarization processes.

16. Show that during a CMP process with $\omega_2 = 2\omega_1$ the locus of a point at the edge of the wafer on the platen is an ellipse with the major axis $= R + r$ and the minor axis $= R - r$. (*Hint:* Solve graphically.)

17. Show that during CMP, the locus of a point on the edge of the wafer for $\omega_2 = \frac{1}{2}\omega_1$ traces a shorter path than that for $\omega_2 = -\frac{1}{2}\omega_1$. (*Hint:* Solve graphically.)

18. The current crowding at a metal-via contact can be calculated by a finite element method using commercial tools. Consider a very simple 2-D case with a resistance network as shown below. Assign $R_{21} = 2R_{11}$ and $R_{31} = 3R_{11}$ to account for the difference in the length of current paths for I_1, I_2, and I_3. (*a*) Let $R_{12} = R_{22} = R_{32} = 4R_{11}$ for a CVD W plug. Find the ratios I_1/I_2 and I_1/I_3. (*b*) Repeat for a TiN overcoat between the W plug and Al; let $R_{12} = R_{22} = R_{32} = 10R_{11}$. (*c*) Repeat for an Al plug where $R_{12} = R_{22} = R_{32} = R_{11}$.

19. To use Cu for wiring one must overcome several obstacles: (1) the diffusion of Cu through SiO_2, (2) adhesion of Cu to SiO_2, and (3) corrosion of Cu. One way to overcome these obstacles is to use a cladding/adhesion layer (e.g., Ta or TiN) to protect the Cu wires. Consider a cladded Cu wire with a square cross section of 0.5 μm × 0.5 μm and compare it with a layered TiN/Al/TiN wire of the same size, with the top and bottom TiN layers 40 nm and 60 nm thick, respectively. (*a*) What is the maximum thickness of the cladding layer if the resistance of the cladded Cu wire and the TiN/Al/TiN wire is the same? (*b*) What is the maximum cladding layer thickness if 20% lower resistance is desired? (*c*) What is the maximum cladding layer thickness if 40% lower resistance is desired? (*d*) Keeping the cross section as a square and decreasing the wire size to 0.3 μm × 0.3 μm, repeat (*b*). What is the sheet resistance of the structures (for both Al and Cu) for (*b*) and for (*d*)? (The resistivity of Al is 2.7 μΩ-cm, and the resistivity of Cu is 1.7 μΩ-cm.)

20. Consider a long wire with resistance R and capacitance (to ground) C driven by a transistor with maximum output current I_o. The voltage across the capacitor at time t can be expressed by $V(t) = V_o + (V_{dd} - V_o)\{1 - \exp[-(t - t_o)/RC]\}$ where $V_o = V_{dd} - i_o R$ is the voltage across the capacitor at time t_o. The total time required to charge this line to a voltage $V_f \approx 0.9V_{dd}$ can be derived as $t = t_o + RC[\log_e(V_{dd} - V_o)/(V_{dd} - V_f)]$, and $t_o = CV_o/I_o$. (*a*) For a 2-mm long Al wire with a cross section of 0.5 μm × 0.5 μm at 0.5 μm from the ground plane, as in Problem 2(*b*), connected to a transistor with a maximum output of 2.5 mA, what is the time required to charge this line to a $V_f = 3.0$ V ($V_{dd} = 3.3$ V)? (*b*) What is the fraction of this time caused by the wire RC delay? (*c*) What is the time required to charge this line to 3.0 V if it is Cu? (*d*) What is the minimum time required to charge this wire to 3.0 V if the driving transistor can deliver unlimited current? (*e*) In case (*d*), what fraction of the delay is caused by the wire RC delay? (*f*) If Cu wire is used in (*d*) instead of Al, what is the minimum delay time? What fraction of it is due to the wire RC delay?

CHAPTER 9

Process Integration

C. Y. Lu and W. Y. Lee

9.1 INTRODUCTION

Integrated circuit (IC) technology has produced dramatic advances over the past 20 years. Since the early 1970s technical progress has propelled the integrated circuit industry from small-scale integration (SSI), at less than 30 devices per chip, to medium-scale integration (MSI), 30 to 10^3 devices per chip, to LSI, 10^3 to 10^5 devices per chip, to VLSI, 10^5 to 10^7 devices per chip, and now to ULSI, 10^7 to 10^9 devices per chip. The present period started in the early 1990s, roughly corresponding to the introduction of four-Mbit dynamic random-access memory (DRAM) devices. The minimum line width span for ULSI, from 0.80 to 0.15 micrometers, puts us squarely in the submicrometer regime. In this chapter, we discuss various integration issues for different technologies and circuits. These technologies have much in common; they differ only in special blocks of processes particular to their applications. The mainstream of ULSI technology is in complementary metal-oxide-semiconductor (CMOS)-based silicon processes.[1] These processes are shown schematically in Fig. 1. Bipolar technology, the major technology in the past, is becoming a niche application technology in ULSI integration. In Sections 9.3 through 9.5 we treat the CMOS, bipolar, and bipolar CMOS (BiCMOS) technologies in turn. For each technology, we describe a typical process flow in some detail, along with any special process modules needed. These descriptions are not intended to standardize the processes because there are many options; rather, we present baseline process sequences that are useful primarily as vehicles for illustrations.

Throughout the years, a considerable number of process modules have been developed for process integration. In Section 9.2 we describe some of the important process modules, or building blocks, that can be used, in general, for all circuits, and we discuss the special processing modules relevant to their particular technologies in

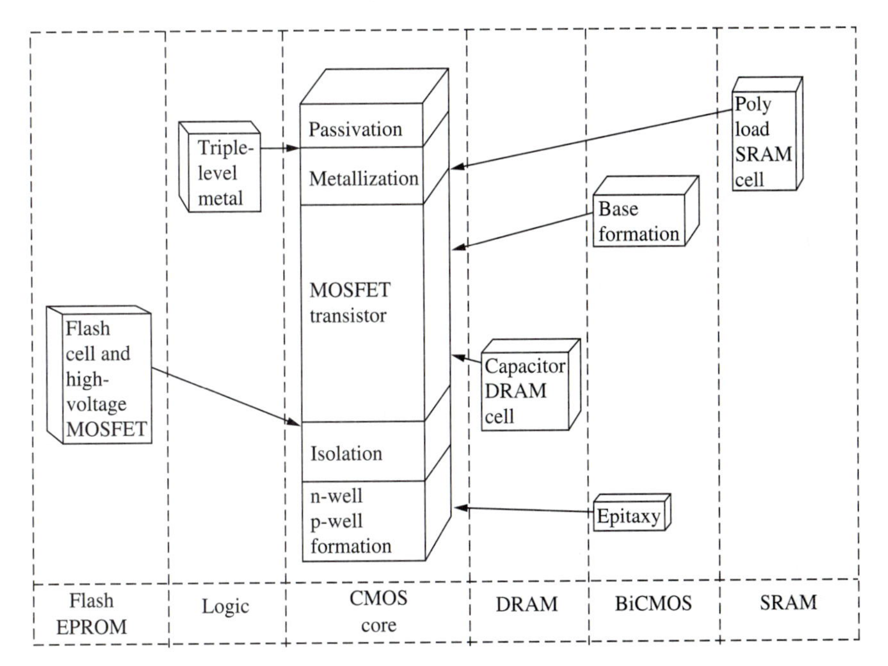

FIGURE 1
Process harmonization for CMOS core-based technologies. (*After Chatterjee, Ref. 1.* ©1991 IEEE.)

Sections 9.3 through 9.6. At the end of these sections we give special consideration to some of the fabrication and device issues of that particular technology. Standard memory technologies of DRAM and static random-access memory (SRAM) are covered in detail in Section 9.6. Section 9.7 describes some major considerations for process integration in ULSI fabrication. Finally, Section 9.8 summarizes the chapter and explores future trends for this industry.

The purpose of this chapter is to present fundamentals of integrating silicon processing steps to create silicon devices in the ULSI context. Since device performance is an important criterion for process optimization, we discuss how changes in processing steps affect the important device parameters. A detailed treatment of device physics is beyond the scope of this book, and we assume that readers are familiar with these fundamentals. However, readers may refer to several excellent textbooks on this subject, such as those by Sze[2] and Shur,[3] in the reference list at the end of the chapter.

9.2 BASIC PROCESS MODULES AND DEVICE CONSIDERATIONS FOR ULSI

9.2.1 Introduction

As the complexity and performance of integrated circuits increase, more processing steps are needed to fabricate them. Four or five mask levels were quite adequate for primitive ICs in the 1970s, whereas 16-Mb DRAM (ULSI) memory chips require more than twenty mask levels. Often it was found that, as dimensions shrank, it became more difficult to make smaller-geometry devices with pattern transfer techniques alone; single-unit process steps that served well before were no longer sufficient because the devices thus obtained would have degraded performance. It takes a group of operations, sometimes involving a dozen individual steps, to replace the original unit process step in order to maintain or improve the device performance and manufacturability. An example is the salicidation process described in subsection 9.2.6. This process replaces the simple source and drain diffusion to reduce device series resistances and increase circuit density. Another example is the intermetal layer dielectric spin-on-glass (SOG) planarization steps, which smooth out the topography caused by smaller line and space pitches. As Fig. 1 shows, a modular approach to managing the design of multiple technologies, with a maximum of common modules, or building blocks, and a common set of manufacturing equipment, is essential for ULSI technology development.[1] Because of their importance, we describe some common process modules below. For special-purpose ICs, additional modules are covered in subsequent sections.

9.2.2 Well Structures

The CMOS device is the most popular choice in ULSI applications. The first consideration in the CMOS process is to determine how the substrates will be formed for the two types of transistors. The n-channel transistors need a p-substrate, and the p-channel transistors require an n-substrate. The three approaches to forming the two different substrates are referred to as p-well, n-well, and twin-well processes (see Fig. 2).

The p-well (also called p-tub) process involves implanting or diffusing p-type dopant into an n-substrate at a concentration high enough to overcompensate the n-substrate and to give good control over the desired p-type doping (Fig. 2*a*). The concentration of n-substrate dopant must be high enough to ensure that the p-channel device characteristics are adequate. The concentration of p-well dopant must typically be five to ten times higher than that in the n-substrate to ensure this control. However, excessive p-well doping produces deleterious effects in the n-channel transistor, such as increased back-gate bias sensitivity, reduction in mobility, and increased source/drain to p-well capacitance.

The n-well (also called n-tub) process is an alternative approach in which an n-well is formed in a p-type substrate. Because the n-channel device is formed in the p-type substrate, this approach is compatible with standard n-channel MOS (NMOS) processing (Fig. 2*b*). In this case the n-well overcompensates the p-substrate, and the p-channel transistor suffers from excessive doping effects.

The twin-well (twin-tub) process allows two separate wells to be implanted into very lightly doped silicon. This allows the doping profiles in each well region to be tailored independently so that neither type of device will suffer from excessive doping effects. The lightly doped silicon is usually an epitaxial layer grown on a heavily doped silicon substrate (Fig. 2*c*). The substrate can be either n-type or p-type.

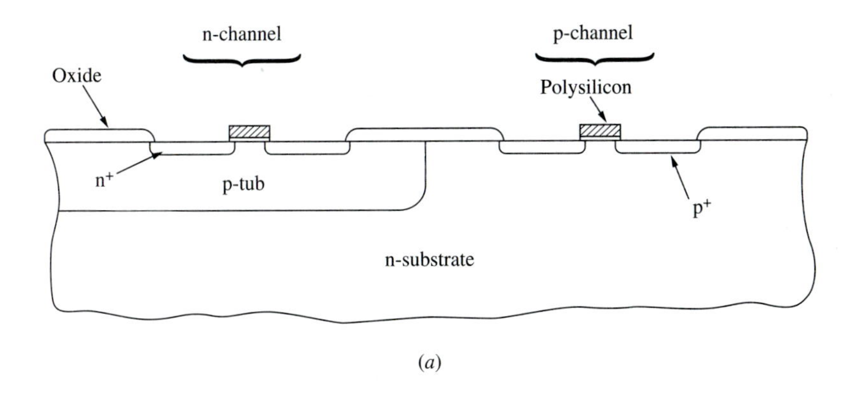

(*a*)

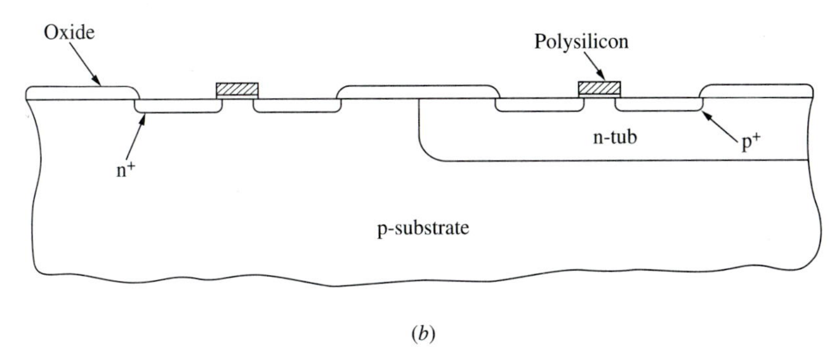

(*b*)

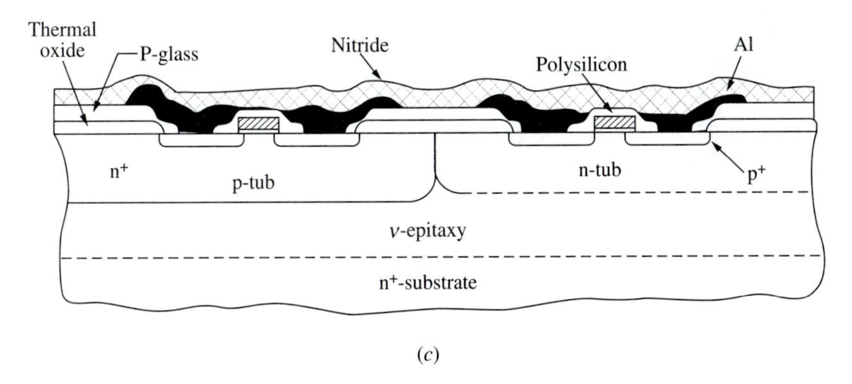

(*c*)

FIGURE 2
Various CMOS structures: (*a*) p-well, (*b*) n-well, (*c*) twin-well. (*After Parrillo et al., Ref. 4.* ©1980 IEEE.)

Twin well

Figure 3 shows the twin-well process module. Although the process is a little complicated, both its n- and p-channel devices can be optimized without compromising their performance.[4,5] First, the wafer is oxidized and capped with a nitride layer, which is selectively removed from the n-well regions by a patterning step. Phosphorus for n-well dopant is then implanted. After the photoresist mask is stripped, a thick (350 nm) oxide is grown over the n-well regions. The nitride

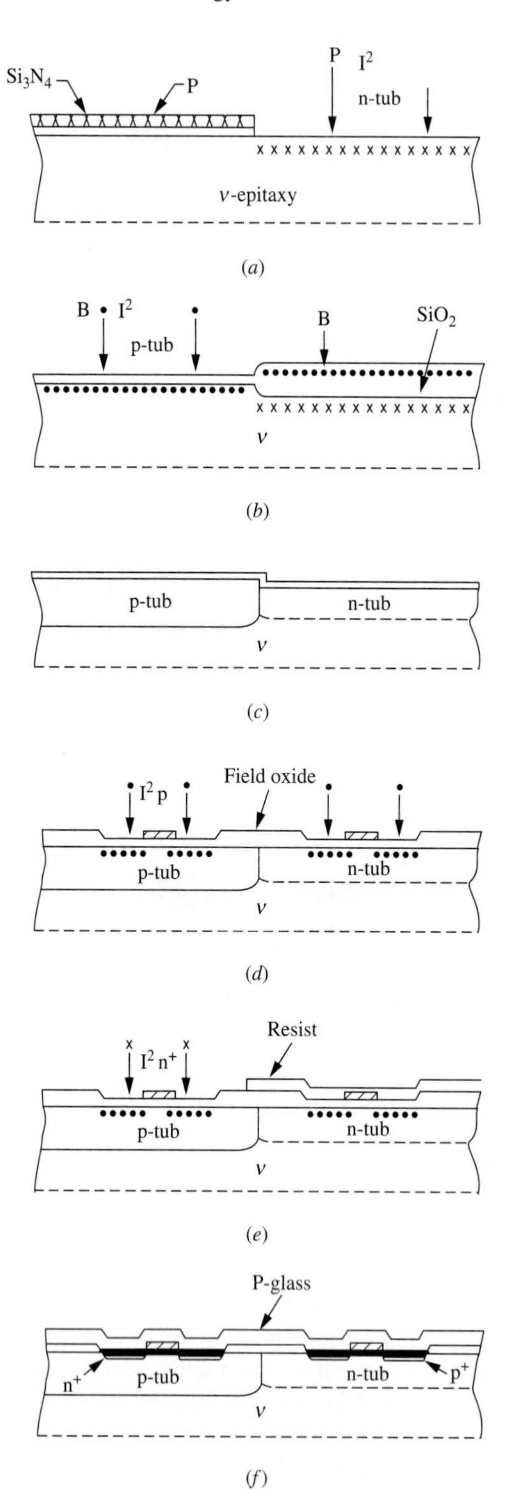

FIGURE 3
Twin-well CMOS structure at several stages of the process: (*a*) n-well ion implant (I^2); (*b*) p-well implant; (*c*) twin-well drive-in; (*d*) nonselective p^+ source/drain implant; (*e*) selective n^+ source/drain implant using photoresist mask; (*f*) P-glass deposition. (*After Parrillo et al., Ref. 4.* ©1980 IEEE.)

layer is then etched off, and p-well regions are exposed, while the n-well regions are covered with the implant-blocking oxide. Thus, during the p-well implant, boron is implanted self-aligned to the n-well edge; the n-well oxide protects the n-well from counter-doping. Next, both wells are driven in simultaneously (at 1100°C for 500 min). At the end of the drive-in step, the concentration on the surface may be 1×10^{16} cm^{-3} for the p-well, and 3×10^{16} cm^{-3} for the n-well. The higher concentration of the n-well is designed to improve the punch-through performance of the p-channel metal-oxide-semiconductor field-effect transistor (pMOSFET) and to eliminate the need for a separate masking step to increase the surface impurity concentration and, therefore, the threshold voltage for the n-well field isolation (channel stop).

Retrograde well

Conventional wells are formed by implanting dopants and diffusing them to the desired depth. However, they diffuse laterally as well as vertically, which reduces packing density. If a high-energy implant is used to place the dopants at the desired depth without further diffusion, much less lateral spread will occur. Such high-energy implants also bury the peak of the impurity profile to a certain depth within the substrate, and the concentration of the impurity decreases as it approaches the wafer surface. Wells with such a profile are called retrograde wells.[6]

Besides the potential benefit of packing density, such wells also have the following advantages:

A retarded field is created in the parasitic bipolar transistor in the CMOS structure, thereby reducing the possibilities of latch-up.[7]

Susceptibility to vertical punch-through is reduced.

The conductivity in the bottom of the well is increased, which also provides some further latchup protection.[7]

Sometimes boron at high energies (such as 400 keV) is implanted after the field oxide is grown; this may provide higher field threshold voltage and less channel encroachment by the field dopants.

9.2.3 Isolation

For MOS circuits, threshold voltage for the field-oxide areas must be higher to isolate individual devices. Many isolation methods have been proposed. We consider three of the most well-known structures suitable for ULSI applications: LOCOS, poly-buffered LOCOS, and shallow trench isolation.

LOCOS

The localized oxidation isolation method (LOCOS) is certainly the most dominant isolation process used in IC technologies.[8] It utilizes the property that oxygen diffuses through Si_3N_4 very slowly. When silicon is covered with silicon nitride, Si_3N_4, no oxide can grow. In addition, nitride itself oxidizes at a very slow rate; thus it will remain as an integral oxidation barrier layer throughout the entire oxidation step. Figure 4 shows the major process steps. After a wafer with a bare silicon surface is cleaned, a 20 to 60 nm SiO_2 layer is thermally grown on its surface. Next, a 100

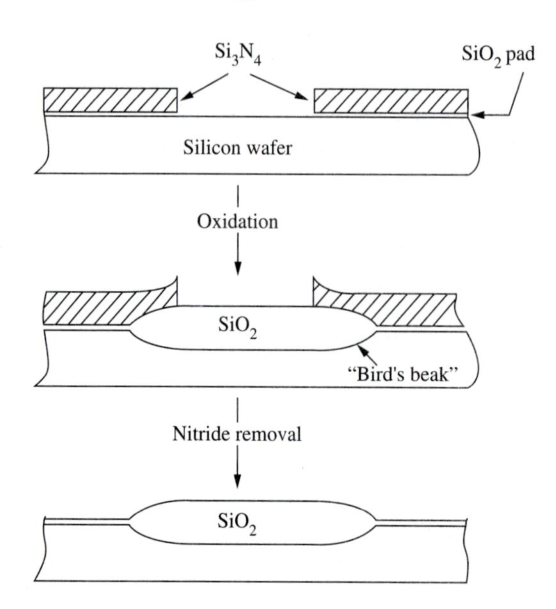

FIGURE 4
Cross-section depiction of process sequence for local oxidation of silicon LOCOS process.

to 200-nm-thick layer of chemical vapor deposition (CVD) silicon nitride, which functions as an oxidation mask, is deposited. The active regions are then defined with a photolithographic step so that they are protected by the photoresist patterns. The composite oxide/nitride layers are then plasma-etched as a stack. With the photoresist patterns in place, the wafer is subsequently implanted with a 10^{12} to 10^{13} cm^{-2} dose of boron and with energies in the range of 60 to 100 keV. This channel stop is now self-aligned to the n-channel devices. The dosage is chosen to provide adequate field-inversion voltage but not so high as to cause high n^+ junction-to-substrate capacitance and excessively low n^+ breakdown voltage. The nitride films have a drawback; they exhibit very high tensile stress when deposited on silicon. The underlying layer of oxide, called a pad or buffer oxide, is used to cushion the transition of stresses between the silicon substrate and the nitride film. In general, the thicker the pad oxide is, the less likely it is that the edge force transmitted to the substrate could cause crystal dislocations. On the other hand, a thicker pad oxide will render the nitride layer less effective as an oxidation mask by allowing lateral oxidation to take place. The lateral regions at the nitride edges, which span the transition from the thick field oxide to the thinner pad oxide, are called a *bird's beak* because they are shaped like the beak of a bird.

After the channel-stop implant, the field oxide with a thickness of 500 to 900 nm is thermally grown by wet oxidation at temperatures of 900 to 1000°C for 4 to 8 hours. This and subsequent high-temperature steps have the following device consequences: Because of preferential impurity segregation, boron redistributes at the interface and becomes more depleted. The lateral diffusion of boron also causes it to encroach into the NMOS active areas, and the resultant transistor will behave as if it is a narrower device. Another phenomenon, named the Kooi effect after its discoverer, occurs during the growth of the field oxide.[8] It was found that a thin layer of silicon oxy-nitride can form on the silicon surface near the bird's beak as a result of the reaction of NH_3 with the silicon substrate.[9, 10] The NH_3 is generated from the reaction between water vapor and the masking nitride during the field oxidation step and then diffuses through the pad oxide and reacts with the substrate to form silicon oxy-nitride spots or ribbons (thus it is sometimes called the white-ribbon or black-belt effect). When the gate oxide is grown, its growth rate may be more impeded at these locations than elsewhere, and is thus thinner at these spots, causing low-voltage breakdown and reliability problems with the gate oxide.

In the next stage of the process sequence, the masking nitride layer is stripped with phosphoric acid at 180°C using a reflux boiler. Then the pad oxide is etched. An extra "sacrificial" gate oxide layer is then grown to eliminate the black belts and white ribbons of the Kooi effect by lifting the micromasking oxy-nitride skin off the interface and washing it away when this sacrificial oxide is etched off.[10]

Poly-buffered LOCOS

Although conventional LOCOS has enjoyed long staying power over a number of years, it runs into serious limitations in the submicrometer regime because of its relatively long bird's beak (approximately 0.4 to 0.6 μm on each side). Many schemes have been modified to correct this shortcoming. The simplest way is to etch back a portion of the field oxide after it is grown. The most promising approach is the poly-buffered LOCOS, or PBLOCOS.[11,12] We know that the bird's beak can be substantially reduced if we increase the nitride and reduce the pad thicknesses, but severe dislocations may be generated in the substrate because of the stress exerted by the nitride film during oxidation.[9] However, it was found that if we substitute the oxide pad with a poly-buffered pad layer (50 nm poly/10 nm oxide) and use a much thicker nitride (approximately 400 nm), good results can be achieved. The polysilicon acts as a cushion against damaging stresses, and a bird's beak of 0.1 to 0.15 μm is possible. With some modification, PBLOCOS can even achieve a bird's beak of almost 0 μm without leakage and gate oxide degradation problems.[12]

Shallow trench

No matter how we modify the LOCOS process, it will probably be difficult to reduce the bird's beak length to much less than 0.1 μm per side with totally flat topology. Therefore, for sub-quarter-micrometer technology, new approaches to isolation with totally flat topology are needed. In shallow-trench isolation,[13] trenches about 0.3 to 0.8 μm deep are anisotropically etched into the silicon substrate through dry etching. Active regions are those that are protected from the etch when the trenches are created. Next, a CVD oxide is deposited on the wafer surface and then etched back so that it remains only in the recesses, with its top surface at the same level as the original silicon surface. Etchback is performed using a sacrificial photoresist method. This technique has the advantages of having no bird's beak and no encroachment. Also, when two devices are separated by a trench, the electrical field lines have to travel a longer distance and change direction twice, so they are considerably weakened. Therefore, trenches of submicrometer dimensions are adequate for isolation to prevent punch-through and latch-up phenomena.[14] However, because it is a complicated process, and because up to now LOCOS and PBLOCOS have been satisfactory methods, shallow-trench isolation has not been used extensively in the industry.

9.2.4 Drain Engineering

Lightly doped drain (LDD)

In scaling down devices, thinner gate oxide and higher-doped channels are needed to boost the punch-through voltage of the short channeled devices. Both measures drastically increase the electric field (both the horizontal and vertical components) near the drain regions where charge carriers, accelerated by this field and becoming hot, can overcome the oxide barrier and inject into the gate. Hot carriers induce damage to the gate oxide and may be trapped there, thereby degrading the device performance. By reducing this drain field, the lightly doped drain (LDD) alleviates this problem.[15]

In the LDD structure, the drain is formed by two implants (Fig. 5). One of these is self-aligned to the gate electrode and the other is self-aligned to an oxide spacer at the edge of the gate. The spacer is formed by depositing an oxide layer about 200 nm thick and then etching it with a blanket anisotropic reactive ion etch. All the oxide in the field will be cleared except at regions next to sharp steps where the deposited oxide is the thickest. This oxide spacer has a lateral dimension roughly equal to 75 to 85% of the oxide's original thickness and can be controlled to a high degree of precision.

The first, lighter dose forms a lightly doped section of the drain and source at the edges near the channel. In the NMOS devices, this dose is normally 1 to 5×10^{13} cm^{-2} of phosphorus. The second, heavier dose forms low-resistivity regions of the source and drain that merge into the lighter doped regions. In NMOS devices, this implant is typically arsenic with a dose of 1 to 5×10^{15} cm^{-2}. These two implants are separated and displaced in a self-aligned manner by the spacers. As a result, the electric field can be reduced to an acceptable level near the gate edge because the impurity concentrations there are much lighter than those in a conventional drain. Since the heavier, second dose is further away from the gate edge than would be the case in a conventional drain structure, its depth can be made somewhat greater without adversely affecting the device operation. The increased junction depth also has the advantages of lowering both the sheet and contact resistances and providing better protection against junction spiking by the aluminum contact.

Large-angle-tilt implanted drain (LATID)

In the sub-half-micrometer regime, the usefulness of the LDD structure itself may be reaching a limit, because the spacer interface degradation from hot carrier injection is not subjected to gate control. Better results can be obtained by implementing the LATID.[16,17] It is made by implanting phosphorus for a total flux of 1 to 8×10^{13} cm^{-2} with a tilt angle θ (20 to 60 degrees) from the source and drain sides and by rotating the wafer stepwise using a reduced spacer length L_s (0.05 to 0.15 μm) (Fig. 6). This drain is thus fully compatible with the conventional process and the usual LDD masking steps. In the example given, $L_s = 0.08$ μm and the n^- dose $= 10^{13}$ cm^{-2} for an optimized 0.25-μm device structure. Devices of better reliability can be obtained using this scheme of off-axis implant, because there is better gate control action over the LDD regions.

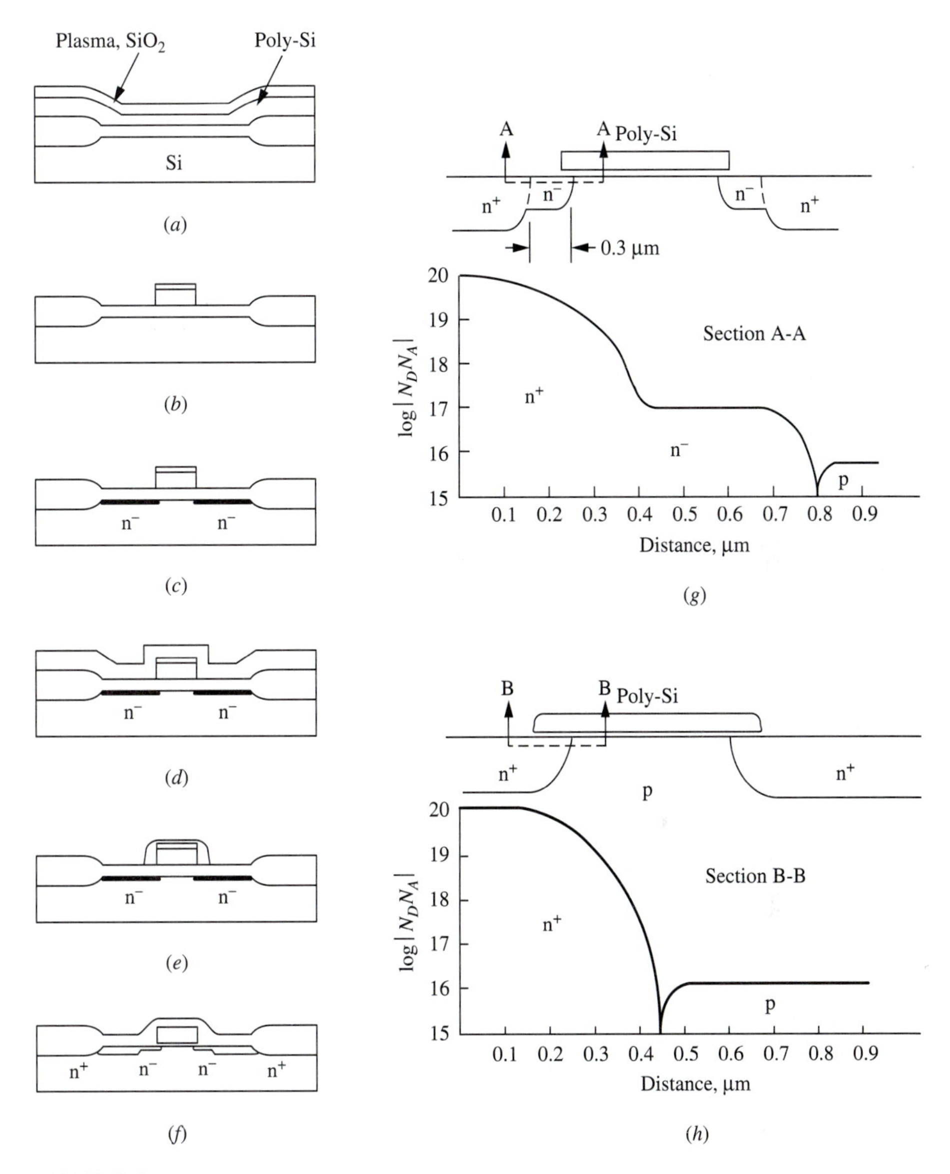

FIGURE 5
(*a–f*) Process sequence used to form lightly doped drain (LDD) structures. (*g*) Doping profile in an LDD structure taken through section A-A. (*h*) Doping profile in a conventional drain structure taken through section B-B. (*After Ogura et al., Ref. 15.* ©1980 IEEE.)

Double-implant lightly doped drain (DI-LDD), or halo

To improve the punch-through voltage of a MOSFET, a "halo" implant can be applied to the drain of the device.[18] For an n-channel device, boron may be implanted with a dosage of about 1×10^{13} cm^{-2} at the drain side. Because of the faster diffusion of boron, its junction is pushed past that of the n$^+$ even if its concentration is

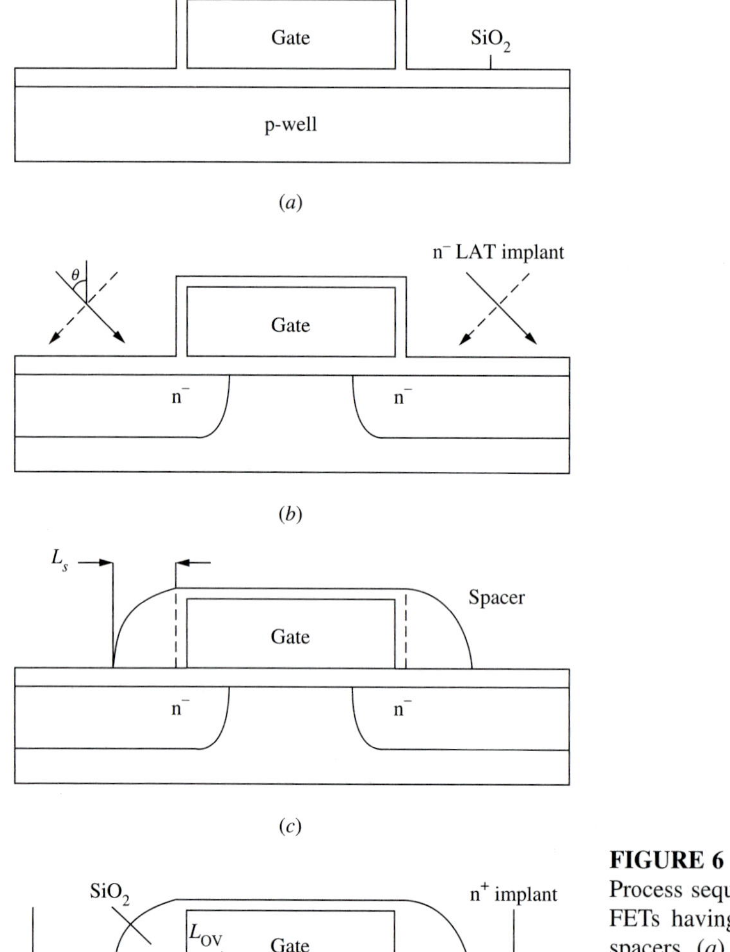

FIGURE 6
Process sequence for LATID FETs having sidewall spacers. (*a*) Gate etching and sidewall oxidation. (*b*) Phosphorus n^- LAT implant. (*c*) Formation of sidewall spacers. (*d*) Arsenic n^+ implant. (*After Hori, Odake, and Yasui, Ref. 16.* ©1992 IEEE.)

much less. The p-type impurities form a ring-shaped halo around the n^+ junction. It has the advantage that its enhanced concentration would increase the punch-through voltage, and this feature takes much less space because it is self-aligned. Also, since the source side is not affected, the body factor[2,3] of the device will not be increased.[18]

9.2.5 Shallow Junctions

As devices shrink, the vertical source and drain junction depths of MOSFET devices must also shrink to reduce short channel effects and improve isolation properties.[19]

It is much harder to make a shallow p^+/n junction than a shallow n^+/p junction. Because of its lighter mass, the boron-implant impurity profile does not follow a strict Gaussian distribution; rather, it has a long tail formed by excessive channeling of boron atoms in the silicon lattice, which makes it hard to produce p^+/n junctions shallower than 0.25 μm by conventional method. To reduce p^+/n junctions to 0.25 μm, BF_2 is more suited than boron because of its higher molecular weight and lower tendency of channeling in silicon. But even at very low implant energies such as 15 keV, the channeling tail is still considerable. This method can only be regarded as a stop-gap measure and not as a satisfactory way to obtain shallower junctions.

Extensive work has been done to attain very shallow junctions with high activation and low defect. The process steps of a typical example follow:

1. Do preamorphization of the silicon substrate with a Si^+ or Ge^+ implant; Ge is preferable to Si because it is more efficient in amorphization. Preamorphization randomizes the crystal surface and reduces channeling.[20,21]
2. Implant boron or BF_2 with very low energy (< 10 keV); boron is preferable to BF_2 because it is more difficult to anneal out damages in the presence of fluorine.[22]
3. Anneal with rapid thermal annealing (RTA).

It is important to set the right energies and depths of the implants to minimize high diode leakage, which is governed by the following considerations:

The depth of the amorphous/crystal (a/c) interface (X_a) must be shallow since the end-of-range defects of the a/c interface cannot be annealed out completely.

The initial boron implant range R_p must be considerably smaller than the a/c depth so that no channeling will occur and so that the boron tail will not receive enhanced diffusion as it would if it were in the interstitial-rich region under the a/c interface.[20]

The junction can then be pushed out during annealing to about 70 nm beyond the a/c interface, so that all the crystal defect regions are completely enclosed by the p^+/n region. This technique minimizes excessive diode leakage.[20]

Since the enhanced diffusion is only a transient phenomenon aided by the presence of interstitials, by the time the junction reaches the a/c interface, most of the interstitials will be gone and no enhanced diffusion will occur.

The following lists a few examples from the literature[20] of conditions under which shallow p^+/n junctions were achieved. X_j is the final p^+/n junction depth.

Case 1:

Ge preamorphization for $X_a = 40$ nm
BF_2 implant at 10 keV, $2 \times 10^{15}\text{cm}^{-2}$
RTA anneal at 950°C for 10 sec, $\rightarrow X_j = 80$ nm

Case 2:

Ge preamorphization, implant at 27 keV, 3×10^{14} cm^{-2}
Boron implant at 1.35 keV, 5×10^{14} cm^{-2}
RTA anneal at 1050°C for 10 sec, $\rightarrow X_j = 110$ nm

Case 3:

Si preamorphization, implant at 15 keV, 2×10^{15} cm^{-2}
BF_2 implant at 15 keV, 2×10^{15} cm^{-2}
Low-temperature anneal at 600°C, 1 hour
RTA anneal at 1000°C for 10 sec $\rightarrow X_j = 120$ nm

9.2.6 Salicidation

In the salicidation, or self-aligned silicidation, process, the entire source and drain regions and the top of the poly-silicon gate of a MOSFET are covered with a low-resistivity metallic silicide film. This approach is attractive because such a film is formed using a self-aligned process that does not require any additional masking steps.[23,24] The conventional salicidation scheme of post-junction-formation silicide (PJS) process sequence of $TiSi_2$ is described as follows (see Fig. 7):

1. After poly-Si sidewall oxide spacers are formed, the source/drain (S/D) regions are implanted to form junctions (Fig. 7*a*).

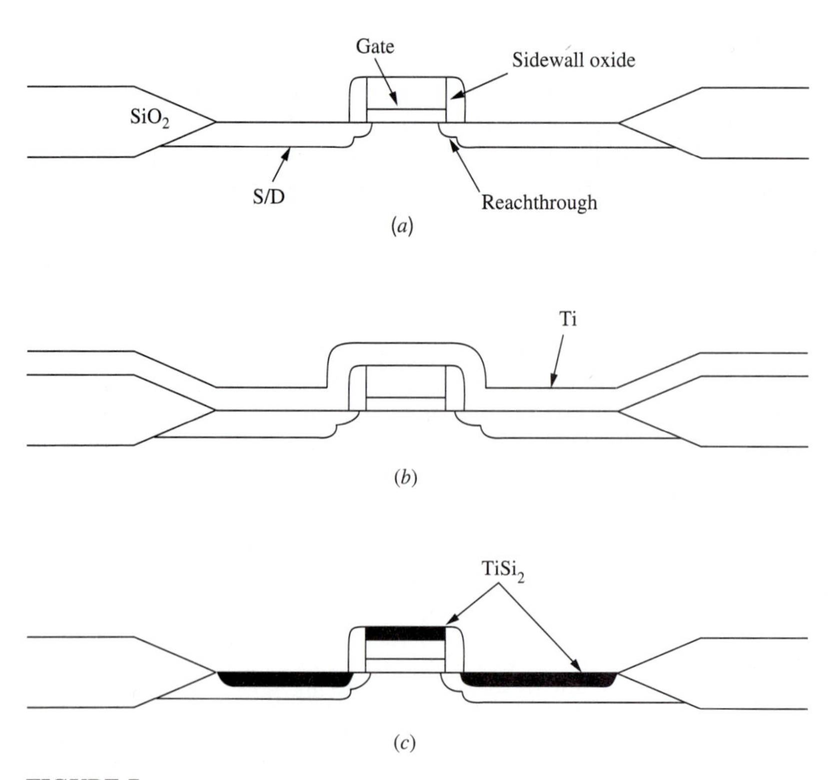

FIGURE 7
Fabrication of MOS transistors with low-resistivity gates and junctions using the self-aligned $TiSi_2$ process. (*After Tang et al., Ref. 28.* ©1987 IEEE.)

2. A HF dip is applied to ensure a clean interface on the wafer and promote uniform reaction. Immediately afterward, 50 to 100 nm of metal titanium film is deposited (Fig. 7*b*).
3. The wafer is then heated to 600 to 800°C in nitrogen, which causes the silicidation reaction to occur wherever metal is in contact with silicon or poly; everywhere else, the metal remains unreacted. Special precaution must be taken to avoid oxygen contamination in the furnace tube, since Ti reacts readily with oxygen to form unwanted titanium oxides. Nitrogen diffuses into and reacts with Ti over the oxide regions to form a stable layer of TiN, which acts as a diffusion barrier. Silicide formation is faster than TiN formation at higher temperature; therefore, at higher anneal temperatures silicon diffuses laterally from the substrate and gate regions and converts Ti over the oxide spacer regions into silicides. This can cause a silicide film to form on top of the oxide sidewall, which can short the gate to the substrate. However, since a lower temperature is chosen, no advance effect occurs at this step.
4. The unreacted metal is then selectively removed by an etchant that does not attack silicide, the silicon substrate, or the oxide. As a result, the poly-gate and the source and drain exposed regions are now completely covered by a silicide film, but there is no silicide elsewhere.
5. A subsequent annealing of 800 to 900°C is then used to further reduce the sheet resistance to its final values, typically 10 to 1.5 $\Omega/\square$ (Fig. 7*c*).

In order to integrate the salicide process with a shallow junction in ULSI CMOS application, several novel schemes have been proposed, such as dopant drive-out (DDO),[25] an implant through metal film (ITM),[26] and an implant through prior-formed silicide film (ITS).[27] For a detailed discussion of these various approaches, see Chapter 8 of this book or Ref. 24 at the end of this chapter.

9.2.7 Local Interconnect (LI)

An interconnection scheme derived from the salicidation process described above uses TiN to replace buried contacts by making direct connections from the diffusion to the poly gates and, as a result, increases the packing density. This TiN local interconnect approach requires no extra deposition steps since it utilizes a layer that is normally discarded.[28] Figure 8 illustrates how the salicidation process has been modified. After the silicide reaction step, the conductive TiN is patterned and etched. Following resist strip, an 800°C anneal in argon reduces the TiN and $TiSi_2$ resistivities to their final values. This process is capable of realizing LI linewidth of less than 0.8 μm and specific contact resistance as low as 2×10^{-8} Ω-cm^2. LI also allows junctions to metal contact to be extended over isolation regions so that minimum-geometry junctions can be realized to reduce circuit areas and capacitances. An example of how LI technology increases circuit packing density is that the cell area of a six-transistor SRAM cell, using the same design rules, is reduced by 25%.[28]

Local interconnect is not limited to using TiN. For the same purpose, a polysilicon layer, with or without a silicide on top, can be used as landing pads on top of source or drain diffusion junctions. They can be extended to make metal connections on top of field oxide or directly to other diffusions or gates to form a sub-level

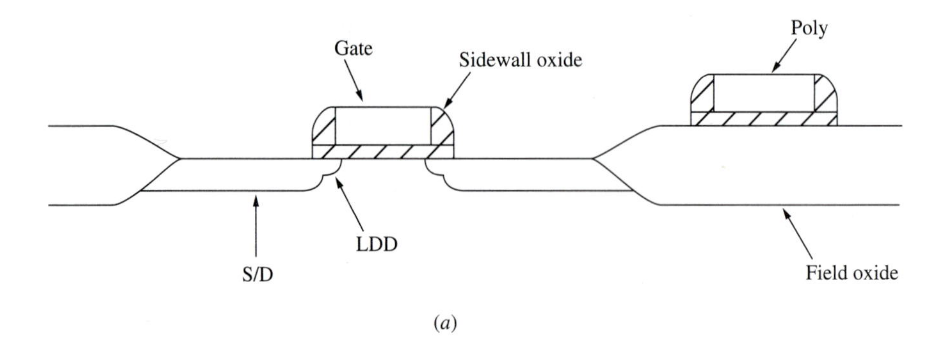

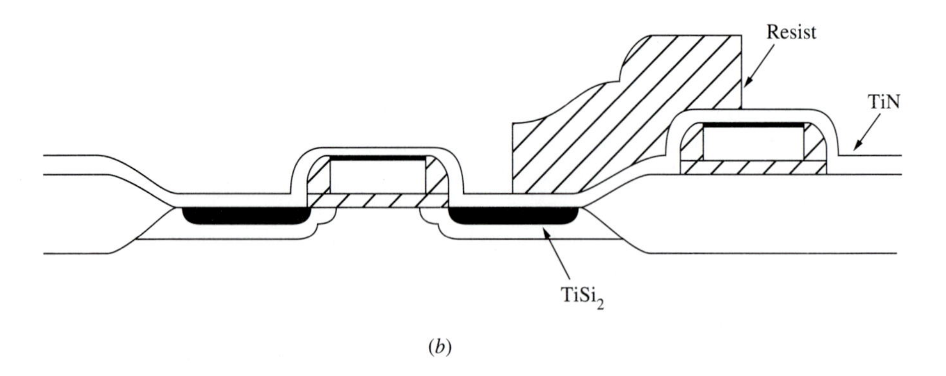

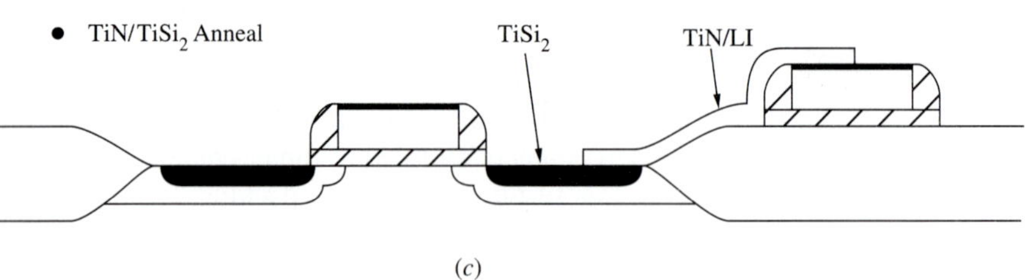

FIGURE 8
Modification of the self-aligned titanium silicide process to realize a level of local interconnect using titanium nitride. (*After Tang et al., Ref. 28.* ©1987 IEEE.)

local interconnect. A representative example is described by the folded extended window structure MOSFET (FEWMOS) with raised source and drain.[29] For its implementation, as depicted in Fig. 9, the polysilicon (or polycide) and its top deposited dielectric layers are patterned as a stack, and then oxide sidewall spacers are formed at this poly/dielectric stack by anisotropic etchback. Then a doped polysilicon layer is deposited and contacted with the exposed source and drain regions. This poly layer is then patterned to form a metal-contact landing pad and local interconnect. This doped polysilicon landing pad on top of the source and drain areas can serve as a

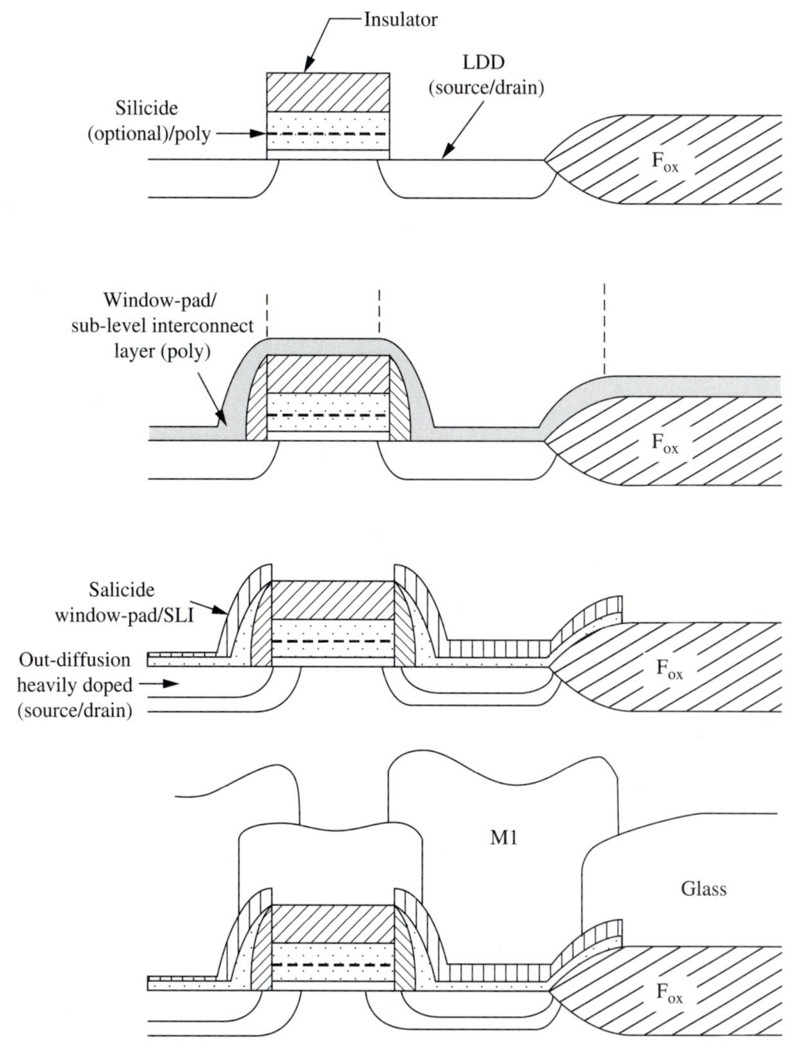

FIGURE 9
The process sequence of a FEWMOS structure. (*After Lu et al., Ref. 29.* ©1988 IEEE.)

diffusion source to form ultra-shallow junctions in bulk silicon with raised source and drain areas. This raised source and drain landing pad can also minimize metallization spiking with shallow junction as a barrier. This device and LI structures could provide a fundamental solution to the shallow-junction issues of scaled MOSFET devices in ULSI.[29]

9.2.8 Self-Aligned Structures and Contact

Self-alignment is a technique in which multiple levels of regions on the wafer are formed using a single mask, thereby eliminating the alignment tolerances required

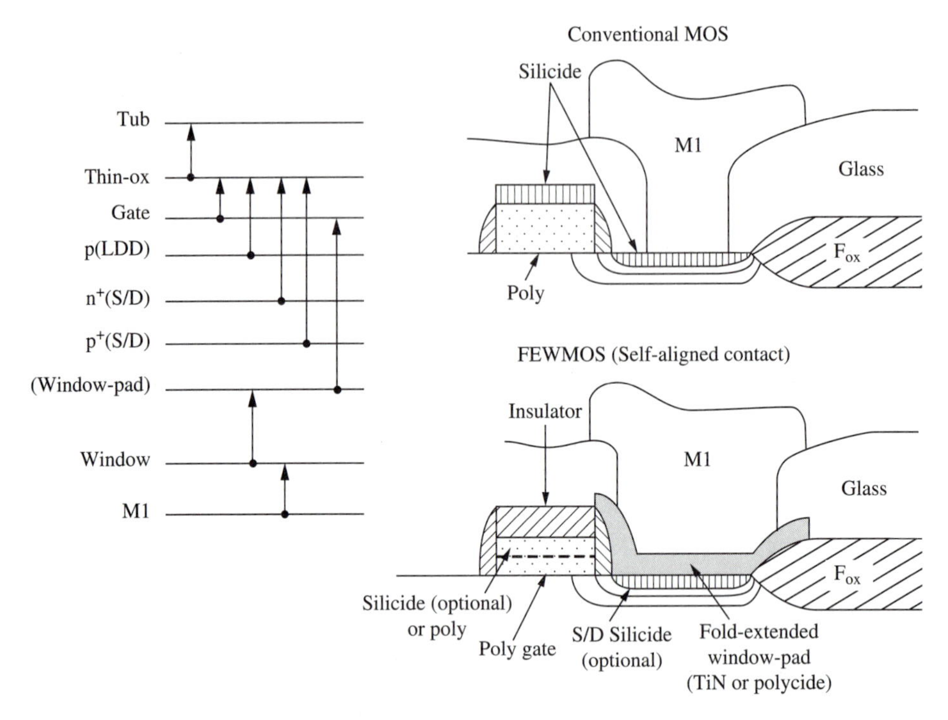

FIGURE 10
Cross-section comparison of the conventional LDD MOS structure and the FEWMOS LDD structure, and the alignment sequence of various levels. (*After Lu et al., Ref. 29.* ©1988 IEEE.)

by additional masks. This powerful approach is being used more often as circuit sizes decrease. There are many examples of this technique; one of the earliest and most widely applied is the self-aligned source and drain implant to the poly gate. Self-aligned contacts are often used in memory cells where contacts are limited only by the spacers and field oxide bird's beaks or a contact window landing pad. Therefore, the mask contact window can be oversized to underneath contact area, and no contact borders are needed, resulting in significant space saving[29, 30] (see Fig. 10).

9.2.9 Multilevel Interconnect

A key requirement of submicrometer technology is a high-density, high-reliability metallization system to interconnect the many active devices on a single substrate. Usually more than one level of interconnect is needed in ULSI circuits. There are three main challenges in the implementation:

Planarization of intermetal dielectric layers.
Filling of high-aspect-ratio gaps between metal space, contact holes, and vias.
Integration of new conductor materials that exhibit low resistance, high reliability, and process compatibility.

The important subject of multilevel metallization is treated thoroughly in Chapter 8.

9.3 CMOS TECHNOLOGY

9.3.1 Introduction

The pairing of complementary n- and p-channel (CMOS) transistors to form low-power ICs was originally proposed in the early 1960s.[31] For much of its history, the CMOS transistor was considered to be only a runner-up for the design of MOS ICs. It was not until the severe limitations in power density and dissipation occurred in NMOS circuits that CMOS became the dominate technology for IC manufacturing. CMOS technology employs both NMOS and PMOS transistors to form logic elements. The advantage of CMOS technology is that its logic elements draw significant current only during the transition from one state to another and draw very little current between transitions, allowing power to be conserved.

The circuit diagram of a CMOS inverter is illustrated in Fig. 11*a*. The cross section of the inverter structure (Fig. 11*b*) shows the n-channel transistor formed in a p-region called a tub or well. The p-channel transistor is formed in the n-substrate. The gates of the transistors are connected to form the input. If we define the threshold voltage of n-channel and p-channel transistors as V_{Tn} (e.g., 0.6 V) and V_{Tp} (e.g., -0.6 V), respectively, we can track the output voltage V_o, shown in Fig. 11*c*, as a function of the input voltage V_i. The transistor current for the individual transistors is shown in Fig. 11*d* as a function of the input voltage. At $V_i = 0$, the n-channel transistor is off, since $V_i < +0.6$ V; but the p-channel is on, since $V_i = 0 > V_{Tp}$ (-0.6 V). As the gate voltage of the n-channel transistor is raised above V_{Tn}, it begins to conduct current, which flows into the p-channel transistor. The p-channel transistor is on at this point since the magnitude of its gate voltage is well above the threshold voltage. Continuing to increase V_i will bring the gate voltage of the p-channel device closer to the p-channel threshold voltage and eventually turn the device off. The important point to remember is that in either logic state, where V_o is either V_{DD} or V_{SS}, one of the two transistors is off; since the transistors are in series, negligible current is conducted through the inverter. Significant current is conducted through the inverter only when both transistors are on during the transition, as shown in Fig. 11*d*. Therefore, in either logic state, very little current flows from V_{SS} to V_{DD} and very little power is consumed. The low power consumption of CMOS is one of its most important attributes for high-density applications.

9.3.2 Fabrication Process Sequence

There are many variations in the design of a CMOS process.[32] The version presented here is the mainstream approach and illustrates the general sequence of process steps and mask layers used in ULSI applications.[5]

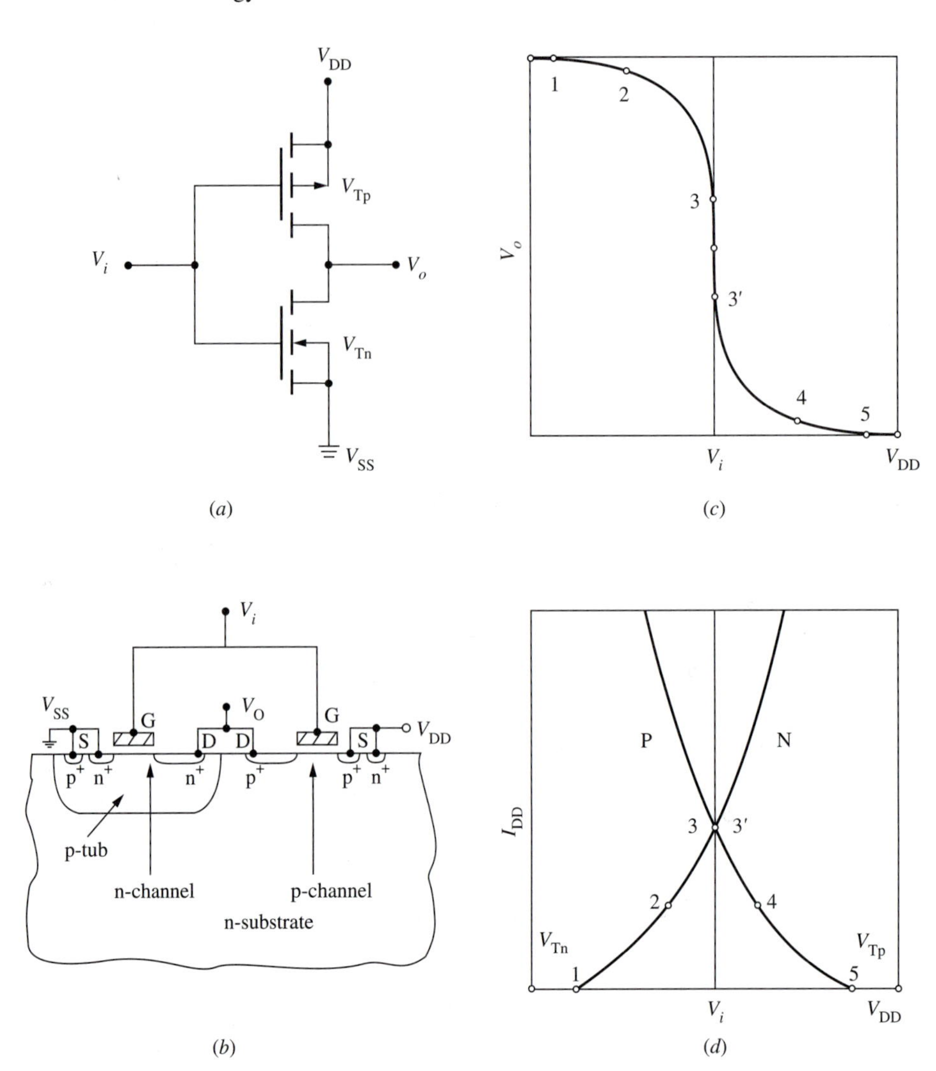

FIGURE 11
CMOS inverter. (*a*) Circuit schematic; V_{DD} and V_{SS} are the highest and lowest circuit potentials, respectively. (*b*) Device cross section. (*c*) Output (V_o) versus input (V_i) voltage of inverter. (*d*) Current through inverter as a function of input voltage (solid curve); *I-V* characteristics of n- and p-channel transistors (dashed curves). The numbers correspond to different points on the inverter transfer characteristic. (*After* B. Hoefflinger and G. Zimmer, "New CMOS Technologies," in J. Carrol, ed., *Solid-State Devices 1980,* from *10th European Solid-State Device Research Conf., Sept. 1980,* Institute of Physics Conference Series **57,** 1980.)

The AT&T Twin-Tub VI 0.6-μm CMOS process[5] sequence is shown in Fig. 12. The CMOS devices are fabricated with p^--epi on p^+ substrate wafers. Conventional twin-well (twin-tub) processing using high-pressure oxidation (HIPOX) to grow a LOCOS field oxide is used.[33] A selective arsenic implant and a deep phosphorus implant are performed to form the n-well followed by a selective well steam oxida-

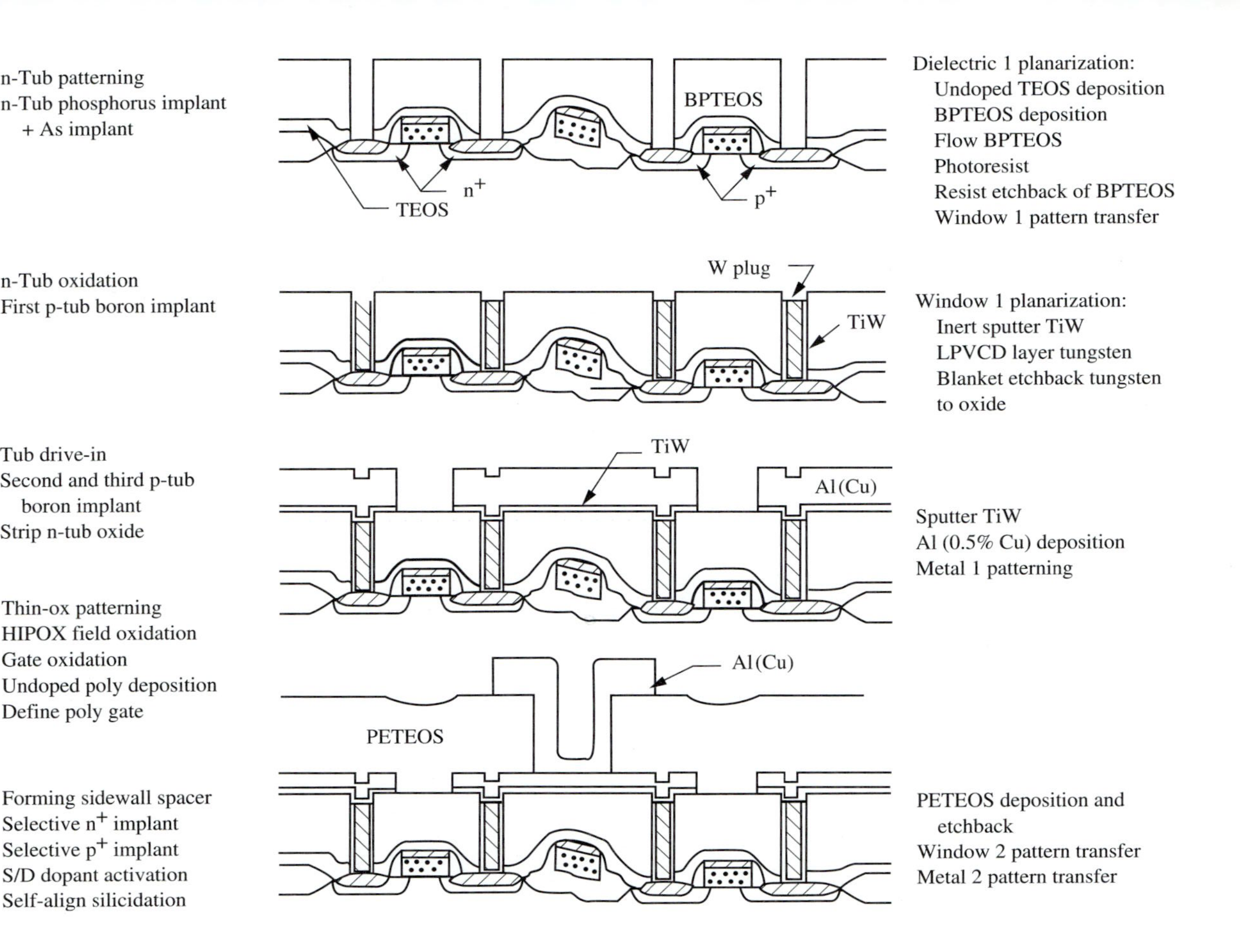

FIGURE 12
Process sequence of the AT&T Twin-Tub VI CMOS process. (*After Lu et al., Ref. 5.* ©1989 IEEE.)

tion at 1050°C. By using the well oxide as a mask, a shallow p-well boron implantation is done followed by the 1050°C-well drive-in in an N_2 ambient. After the well drive-in, a deep boron implant is performed to prevent device punch-through. The final depth of both wells is about 1.8 μm. No channel stop implant is needed.[33]

After the 12.5-nm gate oxidation, a layer of undoped low-pressure chemical vapor deposition (LPCVD) polysilicon is deposited as the gate material. For the gate doping, either a blanket arsenic implant is performed followed by a selective low-energy boron implant to counter-dope the poly on p-channel device regions, or the polysilicon is left undoped. After gate patterning, a thin layer of dry oxide is grown on the poly and the substrate. The lightly-doped drain implantation may be performed to reduce hot-carrier effect. A sidewall spacer is then formed by depositing LPCVD tetraethylorthosilicate (TEOS) oxide and then anisotropically etching it back by reactive ion etch (RIE).

The n^+ region is then patterned and formed by a high-dose arsenic implant followed by a lower-dose phosphorus implant. The high-dose arsenic implant amorphizes the Si substrate and, hence, eliminates the phosphorus implantation tail. After n^+ photoresist mask stripping, the p^+ region is patterned and doped by a 30-keV BF_2 implant or by a 20-keV boron implant with a Si substrate preamorphization by Si implant. If a BF_2 implant is chosen to form the p^+ junction, the high-temperature junction drive of an 875°C anneal could be shared by n^+ and p^+ junctions to anneal the implantation damage and push the junctions away from the residual damage. However, if a preamorphized boron implant is used, the 875°C anneal is too high for the p^+ shallow junction; therefore, this high-temperature (875°C) anneal should be performed immediately after the n^+ implant. After silicon preamorphizing and boron implanting, an additional, brief 800°C anneal is performed to drive both junctions slightly. In order to fully activate the dopants without pushing the junction too deep, a high-temperature rapid thermal anneal (RTA) is used. For the BF_2-implanted p^+n junction, the RTA is essential for low parasitic series resistance ($\leq$ 2.2 kΩ/μm) of p-channel devices. However, for a preamorphized boron p^+ junction, the RTA is not as critical to obtain low parasitic resistance for pMOSFETs. In general, the process with the BF_2 implant is simpler, gives a shallower junction, and has better junction leakage properties. This activation RTA step also alleviates the gate depletion problems for implant doped gates.[34]

Following the junction formation, a brief HF cleaning is performed to remove the residue oxide on the source/drain/gate regions, and a 60-nm titanium film is sputtered followed by Si implant for ion mixing. Due to shallow junctions, this Si-implant energy needs to be adjusted to be optimum for the least junction leakage and to maximize the ion-mixing effects for salicide formation. RTA is used to form the $TiSi_2$ of phase C49 at 625°C. The unreacted Ti is then etched away by a wet Ti etch solution followed by a 900°C RTA to change C49 $TiSi_2$ into a low-resistivity and stable phase (C54) of $TiSi_2$. A nominal $TiSi_2$ thickness of 60 nm and a sheet resistance of 2.5 Ω/□ has been obtained. The actual junction depths obtained from SIMS (secondary-ion mass spectroscopy) analysis are 140 and 180 nm underneath the $TiSi_2$ interface for the n^+p and p^+n junctions, respectively. About 30 nm of Si was consumed in the $TiSi_2$ process. Then passivation glass is deposited, window holes are opened, and metallization is performed.

9.3.3 Special Considerations

Latch-up

A major problem in CMOS circuits is the inherent self-destructive phenomenon known as latch-up, in which a very-low-resistance path is established between the V_{DD} and V_{SS} power lines, allowing large currents to flow through the circuit.[35] This can cause the circuit to cease functioning or even destroy itself from heat damage. The susceptibility to latch-up arises from the presence of complementary parasitic bipolar transistors contained in the CMOS structures (see Fig. 13); when in close proximity to one another, they can interact electrically to form pnpn device structures, which behave like thyristors. Such devices normally do not conduct; however,

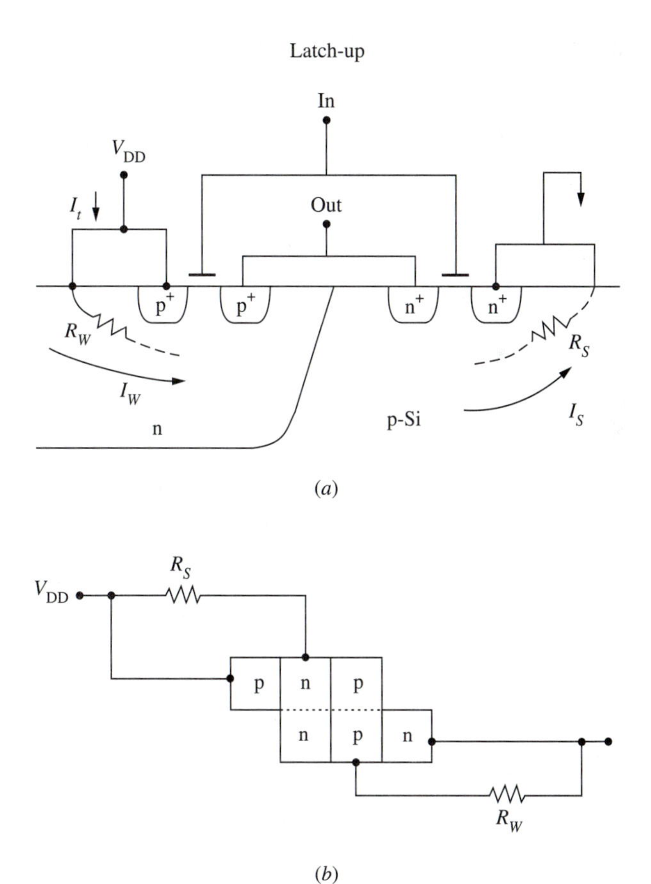

FIGURE 13
(*a*) Cross section of a CMOS inverter. R_w and R_s represent the resistances associated with the well and the substrate, respectively. (*b*) Circuit and schematic representation of the cross-coupled parasitic npn and pnp transistors.

a triggering current may forward bias a pn junction, and then the structure will fire into its highly conductive mode and remain so even after the trigger ceases.[2,3]

There are various measures that can minimize the risks of latch-up:[35]

To reduce the current gain of the CMOS vertical parasitic bipolar transistor, create a deeper well or a retrograde well.

To reduce the resistances of the parasitic current paths, use a heavily doped substrate with devices fabricated on a lightly doped epi-layer.

To put current sinks in the parasitic current paths, implement changes through design rules, such as putting p^+ guard rings around n-channel devices and increasing the separations between the n- and p-channel devices.

Dual polysilicon gates

For ULSI CMOS circuits in the submicrometer range, a major issue is the scaling of the PMOS transistors. For ease of fabrication, n^+ doped polysilicon gates are used for both n-channel and p-channel devices. Asymmetry results because of work-function differences for the two types of devices; this creates a large native p-channel threshold voltage for pMOSFET,[36] which requires adjustment to the appropriate magnitude with a boron compensation implant. This implant could be large enough to counter-dope the surface region of the n-well, creating the "buried channel" structure. Such a structure can produce drain-to-source punch-through current at short channel lengths because there is less voltage control of the gate. This situation has been tolerated down to 0.35-μm pMOSFET or in the case where PMOS is used as loads. A great improvement can be made if a p^+-doped gate is used for the p-channel devices. This makes both the n-channel and p-channel transistors completely symmetrical in their threshold voltages, device dimensions, and dopings; and the p-channel transistor becomes a surface channel device. Such a device has been made successfully by a number of workers[5,34,37] and has even been put in production, although at the cost of more process complexity. To produce such a device, undoped polysilicon gates are patterned first, and they are doped at the same time the source and drain of each type of device are implanted.[5] Some special processing problems need to be considered: one problem is that, at high processing temperatures, boron tends to diffuse through the gate oxide very rapidly into the channel region, especially in the presence of hydrogen-related species such as water vapor.[38] If this happens, the threshold of the device will change, or it may even become a depletion mode device. Therefore, special care must be taken to ensure that the furnace operations are free of hydrogen.[39]

Another concern is that, since the individually doped poly gates are formed by implantation, the doping species must be sufficiently activated, otherwise a depletion layer may form at the poly/oxide interface during device operation. This can result in a significant reduction of the transconductance of the device.[34]

In the dual-polygate CMOS fabrication process, BF_2 implantation was usually used simultaneously for pMOSFETs' source, drain, and gate implant doping; however, fluorine incorporated into p^+-gate devices can jeopardize the PMOS stability even under moderate annealing conditions.[40,41] It was demonstrated that fluorine can severely influence the p^+-gate PMOS device characteristics in addition to the hydrogen effects. These threshold voltage instability and anomalous edge effects were proven to be caused by the enhanced boron penetration through the thin gate oxide[40,41] and by the generation of negative-charge interface-traps in the gate oxide.[40,42] These effects have an impact on any ULSI process steps for dual-polygate MOSFET device fabrication that involve fluorine in addition to the H_2/H_2O-containing ambient anneal.[39]

Another concern in the dual-polygate CMOS process is the interdiffusion of dopants between the two different doping-type regions through the neighboring horizontal junctions. The interdiffusion is especially significant through the shorting overlayer of metal silicide, if a salicidation structure is used. Dopant lateral diffusions in n^+/p^+ dual-polygate CMOS were studied in $CoSi_2$,[37] $TaSi_2$,[43] and $TiSi_2$,[44] and it was found that lateral diffusion occurs even at 800°C annealing in $TiSi_2$ and at higher temperatures in other silicide layers. Such lateral diffusion of dopants can compensate the normal dopant concentration or even convert the dopant type of the underlying polysilicon of the neighboring gate that shares the common polycide runner, resulting in large variations in threshold voltages of n- and p-channel MOSFETs.

9.4 BIPOLAR TECHNOLOGY

9.4.1 Introduction

Bipolar technology was the only available technology in early IC production. With better understanding and improvements in the oxide-silicon interface, bipolar technology was surpassed by PMOS and NMOS technologies. Subsequently, CMOS has become the mainstream technology since the early 1970s. Although they have long held a distant second place to CMOS ICs with their high circuit density, bipolar ICs have nevertheless been progressing rapidly, approaching densities of 100 k gates per chip.[45] Bipolar IC density still falls far short of ULSI density, but bipolar ICs offer unique advantages; they have faster speed and higher current-driving capability than CMOS ICs, and they are supplanting CMOS ICs as the random logic of choice in top-of-the-line digital electronic equipment.

Interestingly, the advances in bipolar ICs result from adaptations of the strategies of rival technologies; most of the new developments in MOS process equipment and technology are also applicable to bipolar technology. New developments in ion implantation, epitaxial growth, and polysilicon deposition have made extremely thin-base transistors possible, with their base diffusion extending less than 150 nm into the silicon.

9.4.2 Standard Bipolar Process

The three types of planar structures for the early bipolar IC all used a reverse-biased pn junction to provide isolation between collectors on the same chip. The three structures are (1) standard buried-collector transistor (SBC),[46] (2) collector-diffused isolation transistor (CDI),[47] and (3) triple-diffused transistors (3D).[48]

A transistor fabricated with the SBC process has a collector region consisting of a heavily doped n^+ buried layer and a lightly n-doped epitaxial layer. The collectors are isolated from one another by p-diffusions that extend all the way through the epi-layer to contact the p-substrate. The base and emitter regions are formed

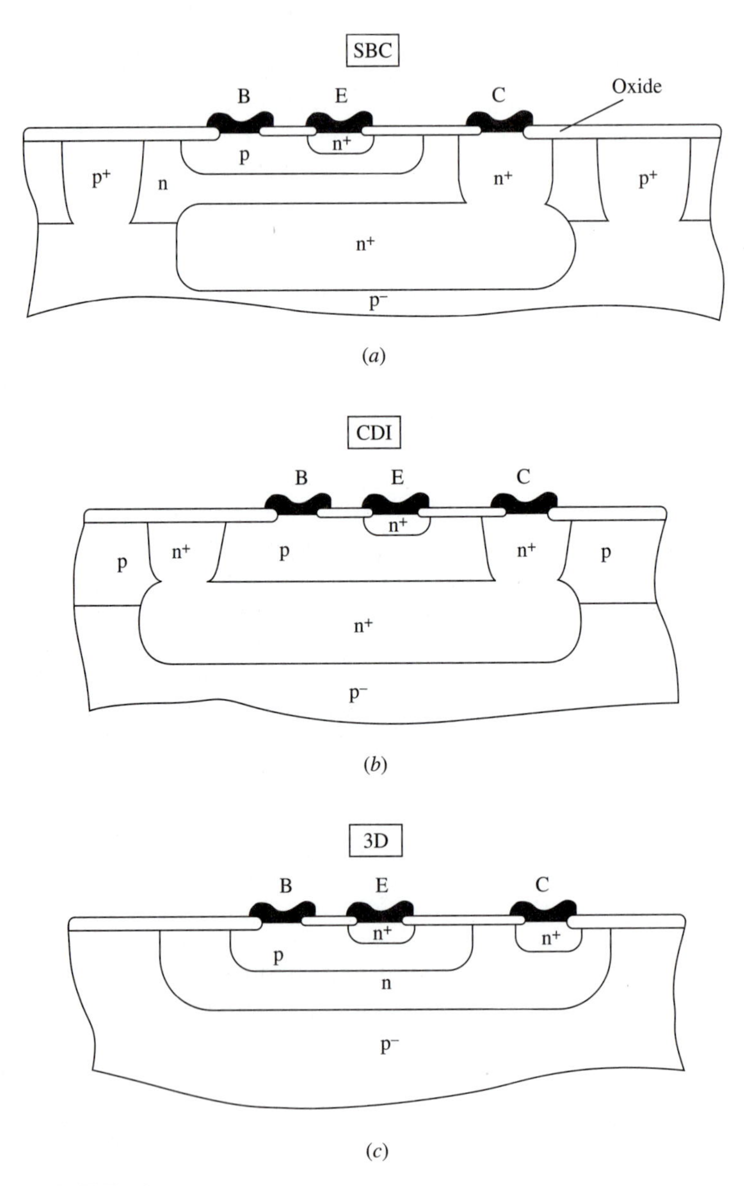

FIGURE 14
Cross sections of three junction-isolated processes: (*a*) Standard buried-collector (SBC) process, (*b*) collector-diffused isolation (CDI) process, and (*c*) triple-diffused (3D) process. (*After Wolf, Ref. 52.*)

by diffusion after the isolation structures have been created (Fig. 14*a*). In the CDI process, an n^+ ring completely encircles the device and merges with the bottom buried n^+ layer so that the transistor is completely isolated by the collector/substrate junction (Fig. 14*b*). The 3D process has no buried or epitaxial layers (Fig.14*c*). Instead, the collector as well as the emitter and base regions are formed by means of implanting ions directly into the substrate. Figure 14 shows the cross-sections of these junction-isolated processes.

The large sizes of these junction-isolated transistors make it difficult for them to meet the functional density requirement of ULSI applications and also cause them to have high parasitic capacitances. By replacing junction isolation with oxide isolation, it is possible to shrink the transistor size and hence the propagation-delay times. In this section, only the oxide-isolated structures that have self-aligned poly emitters will be discussed, since these structures are more suitable for ULSI applications.[49–52]

9.4.3 Self-Aligned Bipolar Structures and Fabrication

The most widely used state-of-the-art bipolar approach employs a double-polysilicon process to form an emitter that is self-aligned to the base, as shown in Fig. 15. The first layer of polysilicon (poly 1) is doped p^+ and serves as the base electrode, whereas the second layer (poly 2) is doped n^+ and makes contact with the transistor's emitter region. The emitter and the extrinsic base regions are formed by means of dopant outdiffusion from poly 2 and poly 1, respectively.

Figure 16 shows the sequence of steps that form this self-aligned structure.[51,52] After the field oxide is grown and the oxide is removed from the surface of the active regions, poly 1 is deposited and heavily p-doped with boron. This layer is then covered with a CVD oxide (Fig. 16*a*). The emitter mask is used to pattern the emitter-area regions, and a dry-etch process is used to produce openings in the CVD oxide and poly 1 (Fig. 16*b*). A thermal oxide is then grown over the etched structure, and a relatively thick oxide (approximately 0.1 to 0.4 μm) is grown on the vertical

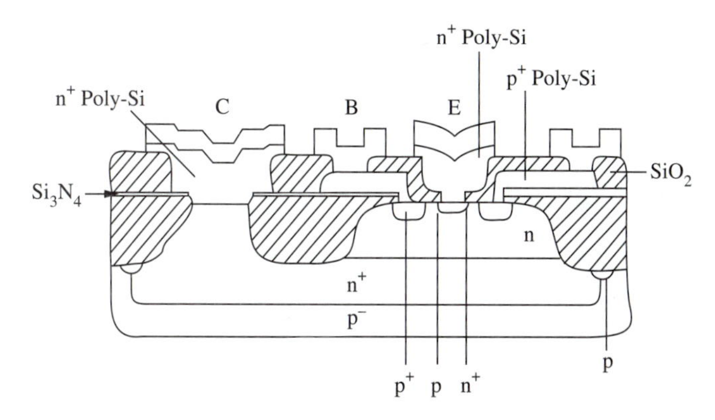

FIGURE 15
Cross section of a double-polysilicon self-aligned bipolar transistor. (*After Wieder, Ref. 49.* ©1986 IEEE.)

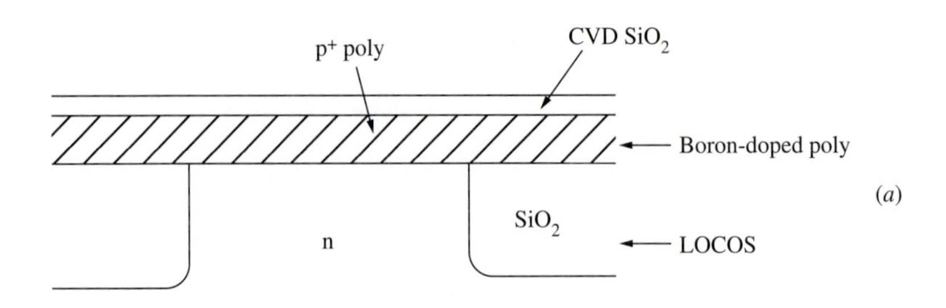

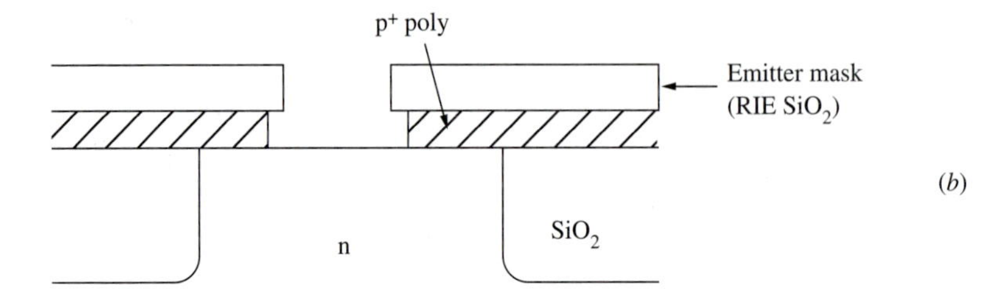

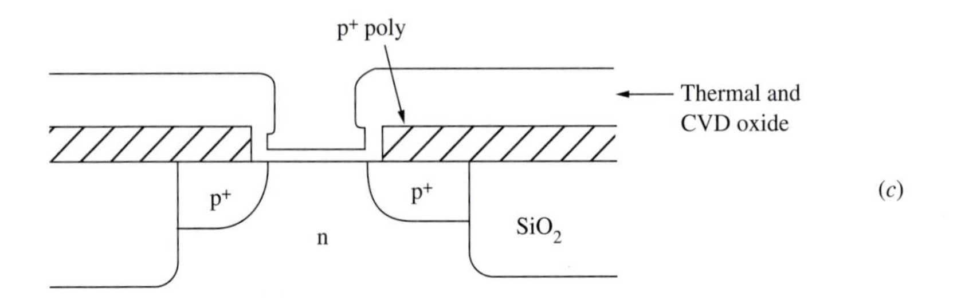

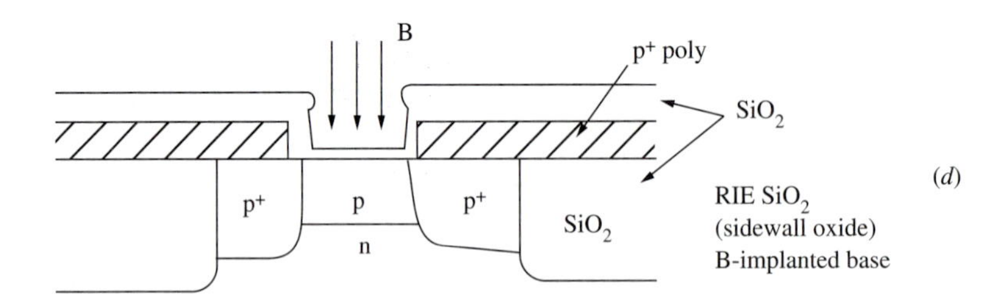

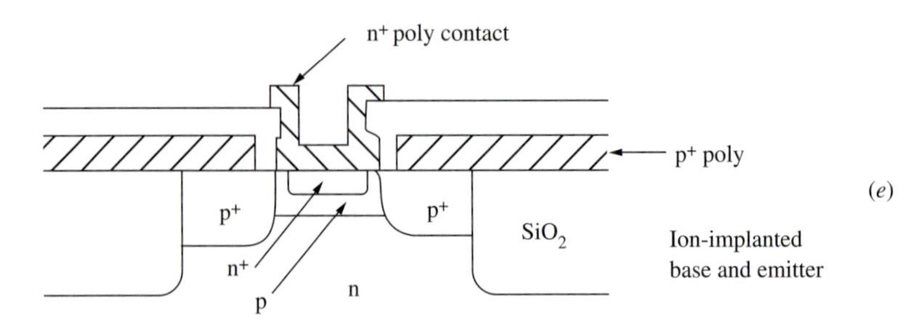

FIGURE 16
Process sequence for the fabrication of double-polysilicon self-aligned npn bipolar transistor. (*After Tang et al., Ref. 50.* ©1980 IEEE.)

sidewalls of the heavily doped poly. The thickness of this oxide determines the spacing between the edges of the base and emitter contacts. The extrinsic base regions are also formed during the thermal-oxide growth step as a result of the outdiffusion of boron from the poly 1 into the substrate (Fig. 16*c*). Because boron diffuses laterally as well as vertically, the extrinsic base region will be able to make contact with the intrinsic base region that is formed next, under the emitter contact.

The parasitic series resistance of the extrinsic base region, R_{BI}, will now consist of the sum of the resistances of the poly 1 electrode, R_{BIpoly}, and of the path between the edges of the base and emitter contact, R_{BIsub}. Since the latter distance is so small, R_{BIsub} is negligibly small. The doping of poly 1 can be much heavier than that of the extrinsic base region in the substrate, and the value of R_{BIpoly} can therefore be made much smaller than R_{BI} in a nonself-aligned bipolar transistor.

Following the oxide-growth step, the intrinsic base region is formed using ion implantation of boron (Fig. 16*d*). This serves to self-align the intrinsic and extrinsic base regions. After the contact is cleaned to remove any oxide layer, poly 2 is then deposited and implanted with As or P. A shallow emitter region is then formed through dopant outdiffusion from poly 2 (Fig. 16*e*). The use of rapid thermal anneal for the base and emitter outdiffusion steps has also been reported to facilitate the formation of shallow emitter-base and collector-base junctions.

This self-aligned structure allows the fabrication of emitter regions smaller than the minimum lithographic dimension. When the sidewall-spacer oxide is grown, it fills the contact hole to some degree because the thermal oxide occupies a larger volume than the original volume of polysilicon. Thus, a 0.8-μm-wide opening will shrink to about 0.4-μm if a 0.2-μm-thick sidewall oxide is grown on each side. The relatively conventional process steps of borophosphosilicate glass (BPSG) deposition, contact opening, reflow, and metallization follow.

9.4.4 Special Considerations

High-performance bipolar transistor

An optimal bipolar device requires a minimum number of masking steps, very few critical alignments, and has very high yields. It should also have the following characteristics:

Low junction capacitances
Low series emitter and collector resistances
Low base transit time
Good dc gain
High breakdown voltages

Scaling

Bipolar performance is improved by shrinking the device in both the horizontal and vertical directions. The planar surface geometry shrinking has been spurred by advances in photolithography, etching, and various self-alignment techniques. Vertical scaling is very important, because junction depth must be as shallow as possible to minimize capacitances and transit time, while maintaining adequate breakdown voltages and dc gain. To get a feel for this problem, note that the transit time is proportional to the electrical base-width squared, W^2, and a small increase in W can have a large and detrimental effect on transit time. However, reducing the base junction X_j has the undesired effect of reducing the punch-through voltage, because it would be easier for the collector depletion region to touch the emitter. Increasing the base impurity concentration will reduce the depletion spread but will increase capacitances and reduce gain. A very shallow emitter junction would give lower emitter efficiency, lowering the dc current gain. This problem is alleviated by using the poly emitters,[53] as explained later.

Vertical impurity profile

Figure 17 gives a general doping profile of a bipolar transistor along a cross section perpendicular to the surface, passing through the emitter, base, and collector. The collector region consists of a heavily doped n^+ buried layer and a lightly doped epitaxial layer. The base and emitter regions are formed by implants and subsequent diffusions after isolation structures are created. The lightly doped side of the collector determines the breakdown voltage of the collector-base junction, BV_{CBO}, caused by avalanche breakdown, except in the case of punch-through. The breakdown voltage of the emitter-base junction, BV_{EBO}, is limited by the lightly doped base

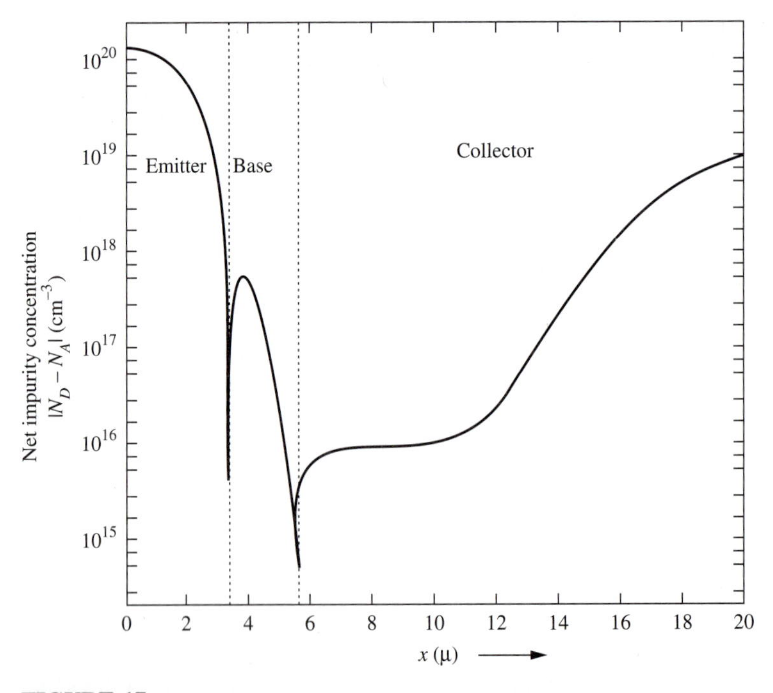

FIGURE 17
The impurity distribution in an SBC bipolar transistor. $|N_d - N_A|$ is the net impurity concentration, and x is the depth into silicon surface. (*After Wolf, Ref. 52.*)

region. Since the base is more heavily doped than the collector, BV_{EBO} is much smaller than BV_{CBO}. (For a general discussion on bipolar device physics, see Ref. 54.) The combined structure of the n^+ buried collector and the lightly doped epi allows high BV_{CBO} and low R_c.

Poly emitter structure

The poly emitter has the following advantages:

The lateral and vertical dimensions of the emitter can be scaled more easily.
Since the emitter is formed by outdiffusion of dopant from the polysilicon in direct contact with the substrate, it does not have the severe yield problem that occurs in implanted emitters because of the residual damage of the implant.
Aluminum contact spiking in the emitter-base junction is minimized.

The current gain of transistors fabricated with a poly emitter is three to seven times that of transistors made with conventional emitters (e.g., emitter depth is approximately 200 nm, and base depth is approximately 100 nm). Transistors with poly emitters also have a higher cut-off frequency, f_T, of 17 to 30 GHz and less than 50 ps gate delay.[55] An interfacial oxide tunneling model has been postulated to explain this increase in current gain: An unavoidable, thin native oxide always forms on the substrate before polysilicon deposition; this acts as a tunneling barrier to hole injection into the polysilicon, and thus increases the emitter efficiency.[56] Another reason for the higher current gain may be enhanced impurity segregation at the substrate/poly interface. SIMS analysis confirms that such a dopant pileup does occur.[57] This impurity-segregation effect gives rise to a potential-energy barrier that hinders the injection of holes into the emitter during forward bias and decreases the slope of the hole concentration in the emitter, leading to a smaller hole current and an increased current gain.[58]

In processing, the nonuniform, variable nature of the unintentionally grown interfacial oxide has caused poor reproducibility, especially as devices are being scaled down. A manufacturable process flow is described as follows: The emitter driving-in step is done by RTA at 1050°C, preceded by a careful cleaning and a HF dip just before poly deposition. This results in a reproducible native oxide of approximately 0.5 to 1 nm, and allows a shallow (approximately 50-nm) As-doped emitter region to be regrown without disrupting the continuous grain boundary at the poly/mono silicon interface. By following this procedure the emitter resistance can be reduced to below 30 Ω-μm^2. Better results can also be obtained by depositing poly at 275°C using a plasma-enhanced (PE) CVD process.

Base regions

A major obstacle to small base width, W_B, in advanced bipolar devices is the channeling tail of the intrinsic-base boron implant. One way to remedy this is to eliminate the direct implantation of boron into the substrate. Instead, after the sidewall spacer oxide is grown, poly 2 is deposited and implanted with boron. After the intrinsic base is formed by outdiffusion, As^+ is implanted. This approach has produced npn transistors with a base width of 100 nm, an emitter junction depth of 50 nm, and cut-off frequencies of 16 GHz.[59]

The connection between the extrinsic and intrinsic base regions in the double-poly self-aligned bipolar structure is critical. The space between the edge of the base and the emitter contacts is determined by the thickness of the sidewall oxide spacer. The amount of overlap between the two base regions depends on the spacer thickness and the lateral diffusion from the p^+ poly. Control of these two parameters is vital in order to fabricate the structures with high yield. If the overlap between the intrinsic and extrinsic regions is too large, low values of BV_{EBO}, β, and f_T result. On the other hand, if the space between the edges of the two regions is too great, the result will be anomalously high R_B values and low values of perimeter punch-through voltage between the emitter and collector through the lightly doped link region of the base.

Injection of minority carriers into the base occurs at the sidewalls as well as at the bottom of the emitter. It can be seen that the current gain resulting from transistor action from the sidewall injection is much smaller than that from the bottom injection. This is because both the base width and the doping concentration are smaller in the base region beneath the emitter than they are along its sidewalls. For devices with emitter dimensions smaller than 1 μm, the perimeter-to-area ratio may increase to such a degree that common emitter current gain, β, can be significantly degraded by the perimeter injection effect.

Super-high-performance SiGe HBT

One of the promising new developments for bipolar technology is the heterojunction bipolar transistor (HBT) using an alloy of silicon and germanium as base material. The Ge_xSi_{1-x} alloy has a narrower band-gap for the base region than silicon and allows an appreciable increase in emitter efficiency. However, these devices are not even in the LSI stage yet.

9.5 BiCMOS TECHNOLOGY

9.5.1 Introduction

BiCMOS is a technology that integrates both CMOS and bipolar device structures on the same chip. CMOS can offer low power and high density to a digital IC, but it is usually slower than the emitter-coupled logic (ECL)-based IC. Bipolar transistors, on the other hand, can deliver large drive currents, operate with small logic swings, and have high noise immunity; however, they exhibit high power consumption, poor density, and limited circuit options. BiCMOS offers the benefits of both bipolar and CMOS circuits.[60] By appropriately trading off the characteristics of each technology, speed and power can be balanced. This desirable result, however, is attained with the penalty of adding more process complexity. So far BiCMOS has penetrated the application-specific IC (ASIC) and very-high-speed SRAM market. Even though it may not become the dominant IC technology, it will carve out a sizable niche in the high-performance digital and analog arena.

9.5.2 Different Process Design Approaches

There have been two major approaches to BiCMOS process design. The low-end BiCMOS process incorporates a moderate-speed bipolar transistor into an aggressive CMOS process. The high-end BiCMOS process extends an aggressive bipolar process to include MOS devices. In low-end BiCMOS, the overall objective is to improve performance with a minimum increase in cost. CMOS attributes, large component density and low power dissipation, are retained while performance is boosted by the application of bipolar devices used selectively on the chip. The high-end BiCMOS usually incorporates large memory blocks using MOS elements of high-speed ECL peripheral circuits[54] to implement mass memory with fast access time.

9.5.3 Fabrication Process Sequence

Since BiCMOS is a mixture of two technologies in various proportions, one can be dazed by the sheer variety. First we consider the simplest kind of BiCMOS, and then we add on more features and finally discuss the full-fledged technology.

Triple-diffused (3D) BiCMOS

The simplest way to add an npn bipolar transistor is to use the PMOS n-well as the collector of the bipolar device and introduce an additional mask level for the p-base region.[61] Figure 18 shows a cross section of a typical structure. The NMOS device is built in a 15 to 10-μm-thick p-epitaxial layer on top of a p^+ substrate, whereas the PMOS transistor is built in an implanted n-well 4 to 5 μm deep. The p^+ substrate is used to reduce latch-up susceptibility by providing a low-impedance current path through a vertical pnp parasitic device. Poly gates are used for both n- and p-channel devices. The p-type base region of the bipolar device is 1 μm deep, with a doping level of 1×10^{17} cm^{-3}. Many of the processing steps can be

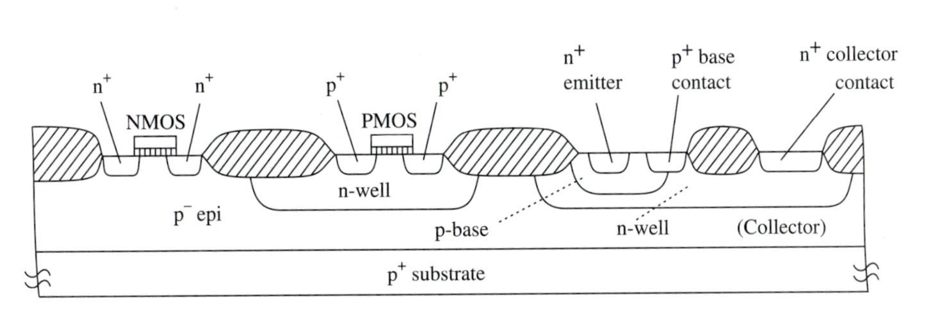

FIGURE 18
BiCMOS structure formed by the simple addition of an npn bipolar transistor to the basic n-well CMOS process. (*After Haken et al., Ref. 62.*)

shared between the MOS and bipolar devices. For example, the n^+ source/drain (S/D) doping step is used for the emitter and collector contact of the bipolar structure. The p-channel S/D implant is the same as that used for the p^+ base contact. This simple process is called the 3D BiCMOS process.

Standard buried-collector (SBC) BiCMOS

From a bipolar performance standpoint, the simple approach described above has a number of limitations. The most significant of these is that the lightly doped n-well leads to a high collector resistance, which limits the usefulness of the bipolar transistor. Figure 19 illustrates how this problem is overcome by introducing a buried n^+ layer under the n-well.[60] The buried n^+ layer is first formed in the substrate by arsenic implantation and then capped with a 2-μm-thick n-type epi layer. This n^+ layer not only reduces the collector resistance, but also reduces the susceptivity to latch-up. The p-wells for the NMOS device and the bipolar transistor are then fabricated in the thin n-epitaxial layer. The resistivity of the n-epi is typically 1 Ω-cm and is chosen so that it can support both the PMOS and the bipolar transistors. The collector resistance is further reduced by adding a deep n^+ connection to the buried n^+ layer with an additional mask; this may be either a deep n^+ implant or a poly plug. The p-well step is also used to form the junction isolation structures of the bipolar transistor collector. Thus, with two added mask levels, this approach merges the process steps needed to achieve low bipolar collector resistance with those required to reduce CMOS latch-up susceptibility.[60]

Twin-well BiCMOS

The previous approach produces a bipolar device that, despite much-improved characteristics, still has a number of drawbacks. Its packing density is limited by the p^- substrate doping level and by the spacings that must be used to prevent punch-through from one collector to another. Raising this doping level would increase collector-to-substrate capacitance. Also, the p-type epi-layer has to be counter-doped to form the n-well regions for the PMOS devices. Counter-doping the 2-μm-thick 1 Ω-cm material can cause process control difficulties and can reduce PMOS perfor-

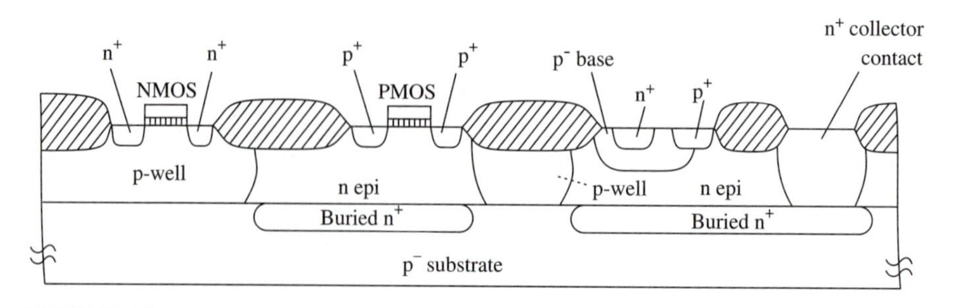

FIGURE 19
BiCMOS structure showing the introduction of a buried n^+ layer and deep n^+ top side contact to reduce collector resistance. (*After Haken et al., Ref. 62.*)

mance by degrading mobility. Figure 20 (*a*)–(*e*) shows how bipolar packing density can be improved by self-aligning buried p-layers to the buried n^+ regions at the cost of higher collector-to-substrate capacitance.[62]

A lightly doped (10 Ω-cm) p^--type substrate is used for the starting wafer. An oxide pad and nitride composite layers are first deposited and patterned for the buried n^+ layer. Antimony is implanted into the windows and annealed at 1250°C to diffuse the antimony and remove defects (see Fig. 20*a*). At the same time, a thick oxide is grown over the buried n^+ regions. This thick oxide serves as a blocking mask for the self-aligned boron-buried p-layer implant, as illustrated in Fig. 20*b*. All oxide is removed from the surface prior to epitaxy, and a short HCl etch is performed in situ to remove native oxides before growing a thin (1 to 1.5 μm) n-epitaxial layer with a resistivity of more than 15 Ω-cm.

The wafer is then oxidized and capped with a nitride layer, which is selectively removed from the n-well regions by a patterning step. After the n-well dopant of phosphorus is implanted, a thick oxide (350 nm) is grown over the n-well regions, and nitride is then stripped from the p-wells. The subsequent p-well implant is self-aligned to the n-well edges (see Fig. 20*c*). A well drive is then performed at 1000°C with the oxide cap in place. All oxide is again removed from the surface; another thin oxide pad and nitride composite layer are deposited, and the active areas are defined (see Fig. 20*d*). Normal CMOS processing steps are then followed using a LOCOS-type isolation process to isolate adjacent active regions. Next, standard NMOS and

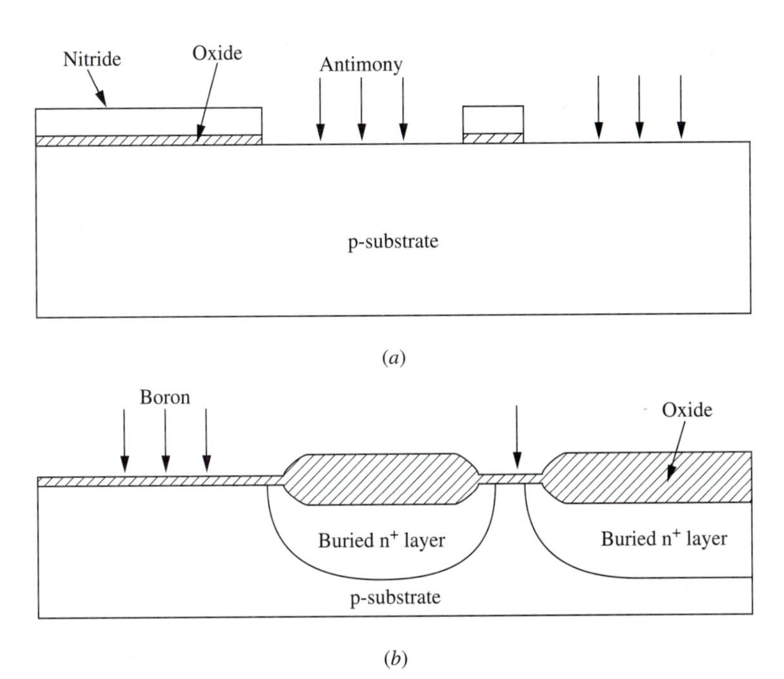

FIGURE 20
(*a*) Device cross section for BiCMOS process flow showing buried n^+ implant. (*b*) Device cross section for BiCMOS process flow showing buried p implant self-aligned to buried n^+. (*Continued*)

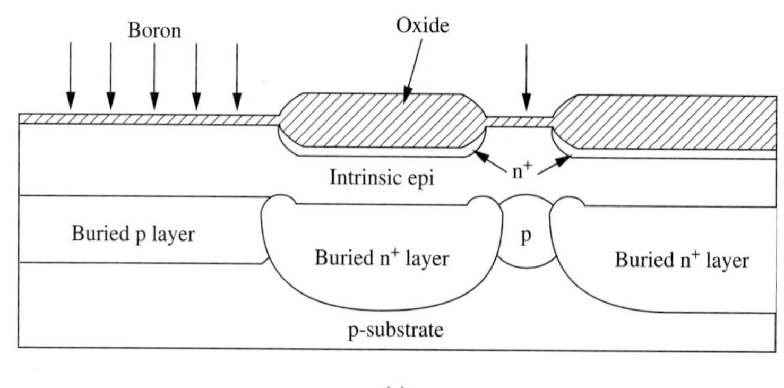

(*c*)

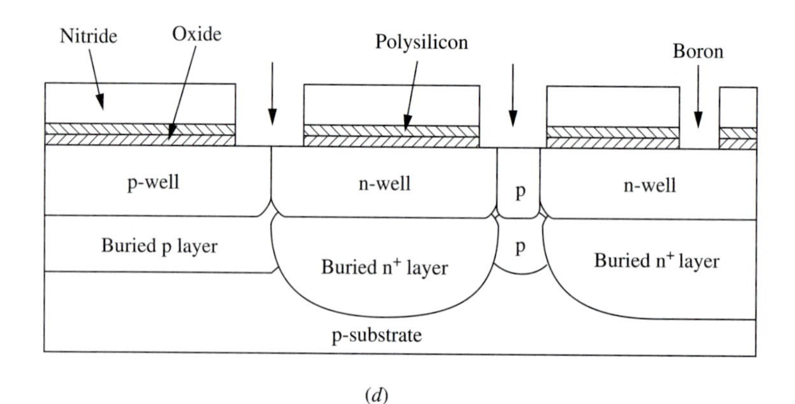

(*d*)

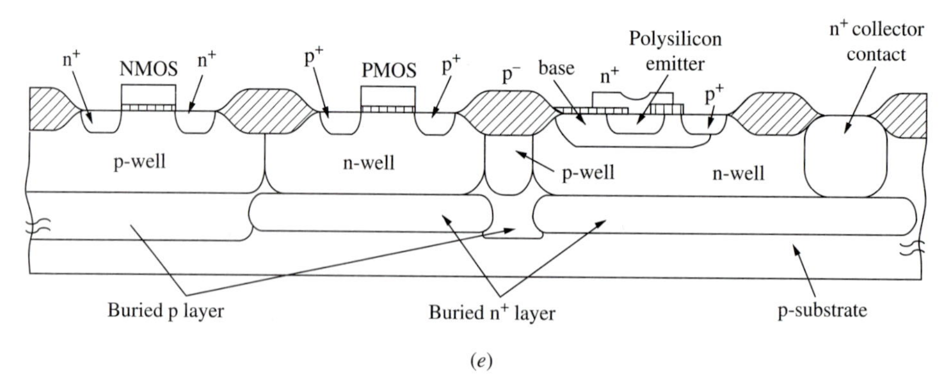

(*e*)

FIGURE 20 (***continued***)
(*c*) Device cross section for BiCMOS process flow showing p-well implant self-aligned to oxide-masked n-well. (*d*) Device cross section for BiCMOS process flow showing channel-stop implant after etch of field isolation regions. (*e*) Optimized BiCMOS device structure. Key features include self-aligned p and n^+ buried layers for improved packing density, separately optimized n and p-wells (twin-well CMOS) formed in an epitaxial layer with intrinsic background doping, and a polysilicon emitter for improved bipolar performance. (*After Haken et al., Ref. 62.*)

PMOS channel implants are performed to set the MOS device threshold voltages. A contact plug mask is patterned and a high dose of phosphorus is implanted into the npn collector region to reduce its resistance and prevent the device from entering saturation. After gate oxidation, a poly layer is deposited and n^+ doped to form the MOS electrodes (see Fig. 20*e*).

The process described above shows how a twin-well CMOS structure can be implemented without heavily counter-doping the epi-layer. Instead of using an epi-layer in which the doping level is determined by the bipolar and PMOS requirements, a near-intrinsic epi-layer is deposited. Up to this point, four additional mask levels, buried n^+, deep n^+ contact, p-base, and emitter, are required to merge this high-performance bipolar process with a baseline CMOS flow, resulting in a total mask count of fourteen compared to a typical ten-mask CMOS process. Figure 21 compares typical CMOS and bipolar technology process flows to a high-performance BiCMOS technology process flow that has these additional masking steps. All flows are assumed to have LDD MOSFETs, twin-wells, and double-level metal.[62]

Further enhancement

For the cost of an extra mask level, the bipolar performance can be further improved by replacing the diffused emitter (which is formed by the n^+ source/drain implant) with a polysilicon emitter (see Section 9.4). Higher bipolar performance is obtained because shallower emitters and narrower base-widths are achieved; this reduces both transit time and parasitic capacitance by reducing emitter junction area.

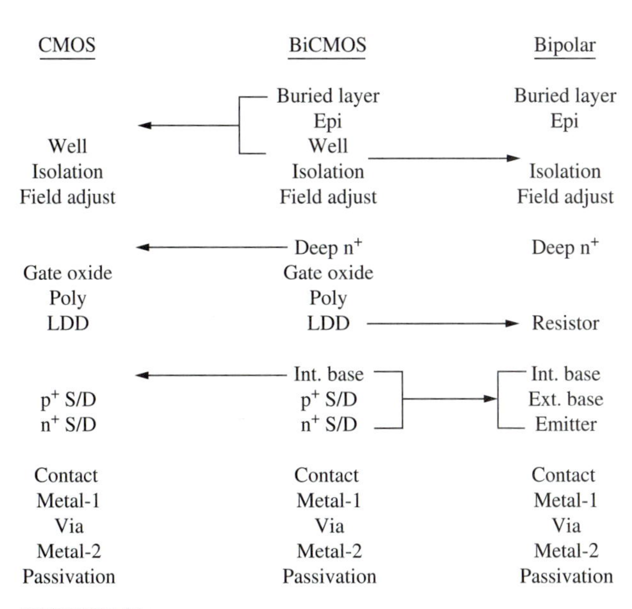

FIGURE 21
Schematic representation of CMOS, bipolar, and BiCMOS process steps. (*After Alvarez, Ref. 60.*)

After poly gate patterning, the NMOS and PMOS source/drain are implanted. In this process, LDD structures are used for both NMOS and PMOS transistors. The n^- and n^+, and the p^- and p^+ S/D implants are also used for the collector and base contacts, respectively, of the npn device.

The npn intrinsic-base implant of boron is performed next, followed by the deposition of an interpoly undoped oxide layer. Next, the emitter opening is patterned in the deposited oxide layer. A second polysilicon layer is then deposited and patterned to form the poly emitter. Care should be exercised to minimize damage and contamination of the crystal surface during RIE oxide etching of the emitter windows. It is also important to minimize the thickness of the interfacial oxide, which may be grown on the silicon immediately before the poly 2 deposition; excessive thickness of the interfacial oxide results in low current gain and high emitter resistance. After the poly 2 is doped, a final high-temperature anneal is used to drive arsenic from the poly layer into the silicon to form the emitter. Since metal contacts are placed on top of the emitter to reduce resistance, an undercoating barrier material is used to prevent aluminum diffusion through the poly into the shallow emitter-base junction.

To further enhance performances, a salicide process may be implemented in which the gates, emitters, and diffusions are silicided in a self-aligned manner to reduce their sheet resistances from 20 to 50 $\Omega/\square$ to 1.5 $\Omega/\square$. The p^+ extrinsic base can also be self-aligned to the sidewall of oxide spacers next to the emitters from the outdiffusion of an extra poly layer. Local-interconnect technology may be used to connect the gates and diffusions without the need for contact holes and metal borders. As examples of these approaches, the cross sections of a high-performance twin-well submicrometer BiCMOS[49] and a trench-isolated BiCMOS[63] are shown in Figs. 22 and 23, respectively.

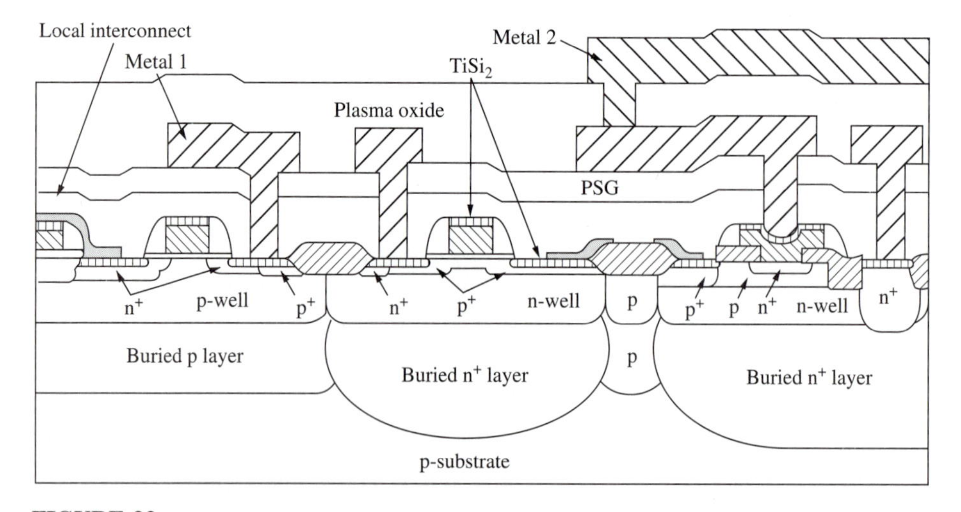

FIGURE 22
Device cross section for BiCMOS process flow following completion of double-level metal interconnect. (*After Haken et al., Ref. 62.*)

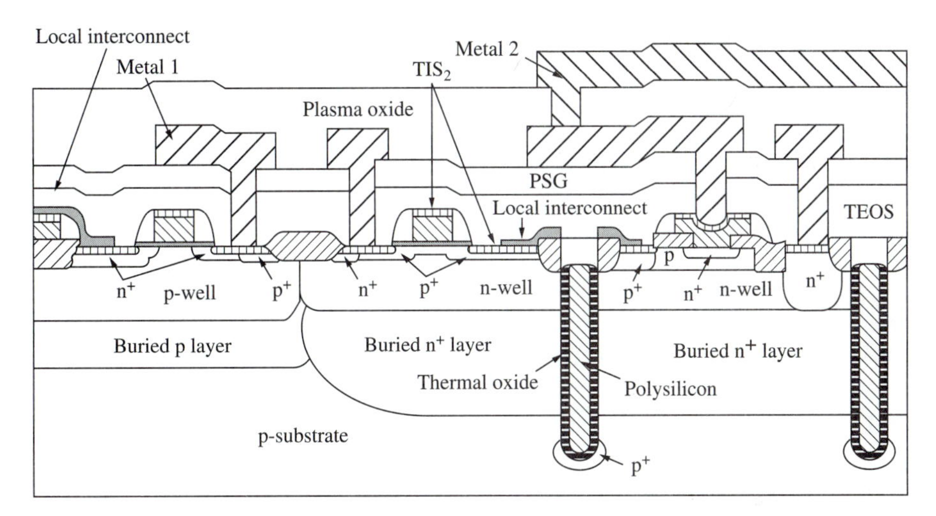

FIGURE 23
Cross section of high-performance submicrometer BiCMOS process showing both trench-isolated bipolar devices and silicided gates, emitters, and diffusions for low sheet resistance. (*After Havemann et al., Ref. 63.* ©1987 IEEE.)

9.5.4 Special Considerations

Buried layers

For BiCMOS, process requirements for the CMOS well and the bipolar collector need to be considered. One major process design trade-off involves the characteristics of the epi and well profiles. For bipolar devices, minimum epi thickness is determined by the emitter breakdown voltage (BV_{CEO}), the collector capacitance, and the manufacturing controllability. On the other hand, if the epi is too thick, f_T will be reduced and R_C will increase. From the CMOS perspective, the use of a p^+ buried layer beneath the NMOS devices dictates that the epi must be thick enough to keep the NMOS back-bias body effect[2,3] and junction capacitance from being excessive.

The n-well in the twin-well structure not only affects the PMOS devices but also serves as the collector of the bipolar devices. Thus, besides giving adequate CMOS characteristics, the n-well should be doped heavily enough to prevent the Kirk effect,[2] yet lightly enough to provide an adequately high BV_{CBO} for the bipolar devices.

Epitaxy and autodoping

The growth of a lightly doped, thin epi layer over two types of buried layers presents a challenge to the epitaxial deposition process.[64] Autodoping for both types of dopants must be minimized in both the vertical and lateral directions to avoid the need for excessive counter-doping in the wells. It has been shown that arsenic autodoping can be suppressed through the use of reduced-pressure epitaxy, but unfortunately, boron autodoping increases at lower pressures.[65] A novel two-step epi process has been reported in which a high-temperature (1150°C), low-pressure (10 torr) SiH_2Cl_2-epi cap is deposited first, followed by a low-temperature (900°C), low-pressure SiH_4 epi-layer deposition. This process produces an abrupt epi substrate transition and minimal arsenic and boron autodoping. Lateral autodoping can also be reduced if the time and temperature of the prebake period (the period between in-situ etching of the surface and the actual deposition) is increased, since the buried-layer dopants diffuse to the surface and evaporate into the gas stream during this cycle.[64]

9.6 MOS MEMORY TECHNOLOGY

9.6.1 DRAM Processing Technology

Dynamic random access memory is so named because its cells can retain information only for a limited time before they must be read and refreshed at periodic intervals. A DRAM cell consists of one transistor and one storage capacitor. For bit densities of up to one megabit, planar-type storage capacitors are used. However, as component density has increased, the amount of charges needed for a sufficient noise margin remains fixed. Therefore, in order to increase the specific capacitance, two different routes have been taken (see Fig. 24): the first solution is to store charges vertically in a trench; the second solution, which allows the cell to shrink in size without losing storage capacity, is to stack the capacitor on top of the access transistor. The majority of manufacturers have taken the stacking approach. In Fig. 24, we can see that as the memory density increases, the capacitor structure becomes more intricate and grows in the vertical direction.

Representative process flow

We now describe a typical process flow for a 16-Mb DRAM. A stacked capacitor structure is chosen because most products use it.[52] To fabricate this device, we use a modified CMOS process with a minimum geometry of 0.5 μm. A cross-sectional view of this cell is depicted in Fig. 25.

Up to the gate-oxide growth, the process uses conventional CMOS fabrication steps. Two hundred nm of poly 1 is deposited and doped with $POCl_3$, followed by a deposition of 250 nm of TEOS oxide. The poly 1 photomask is applied and the composite film is plasma-etched as a stack. Selective LDD implants are performed. Next, a layer of 150 nm of TEOS is deposited and plasma-etched, so that spacers are formed at each poly step. At this point, the gates of the peripheral devices, as well as those of the cell transfer devices, are formed and covered by a thick layer of oxide, both on top and at the sides of the poly 1 gates. The peripheral NMOS source and drain mask is then applied, and an arsenic dose of 5×10^{15} cm^{-2} is implanted.

The photomasking level of the cell contact comes next. This layer consists of oversized contact openings at the drain regions of the transfer devices. Plasma-etch is used to rid the silicon substrate of any thin oxide layer. Three hundred nanometers of poly 2 layer is then deposited, doped, and patterned. This will be the lower

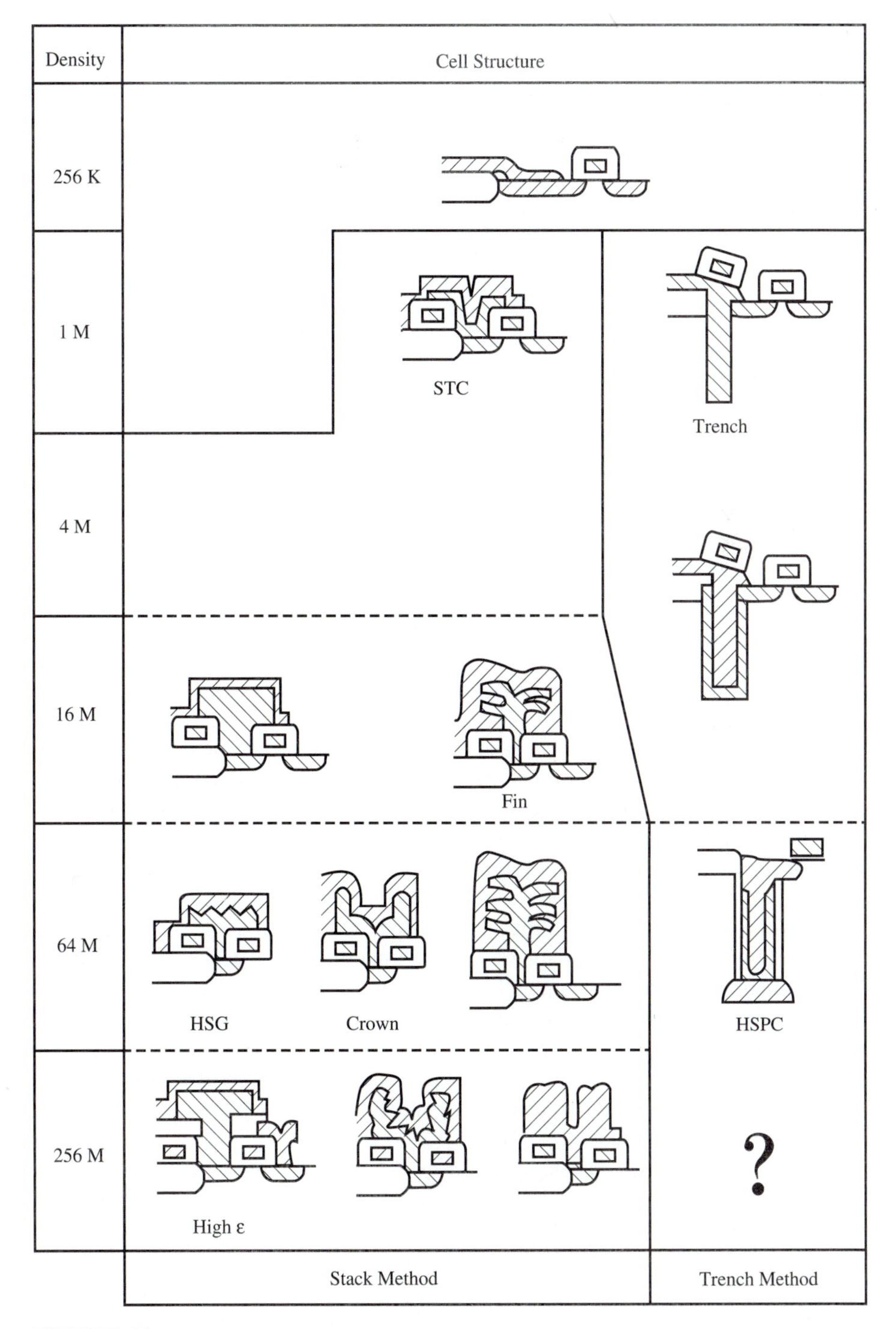

FIGURE 24
Two methods for fabricating high-density DRAM storage capacitors. STC = stack capacitor; Fin = finlike stack capacitor; HSG = hemisphere-grain stack capacitor; Crown = crownlike stack capacitor; HSPC = high-substrate plate capacitor. (*After* M. Ogirima, *1993 Symposium on VLSI Technology, Dig. of Tech. Papers.* Kyoto, Japan, 1993, p. 1. ©1993 IEEE.)

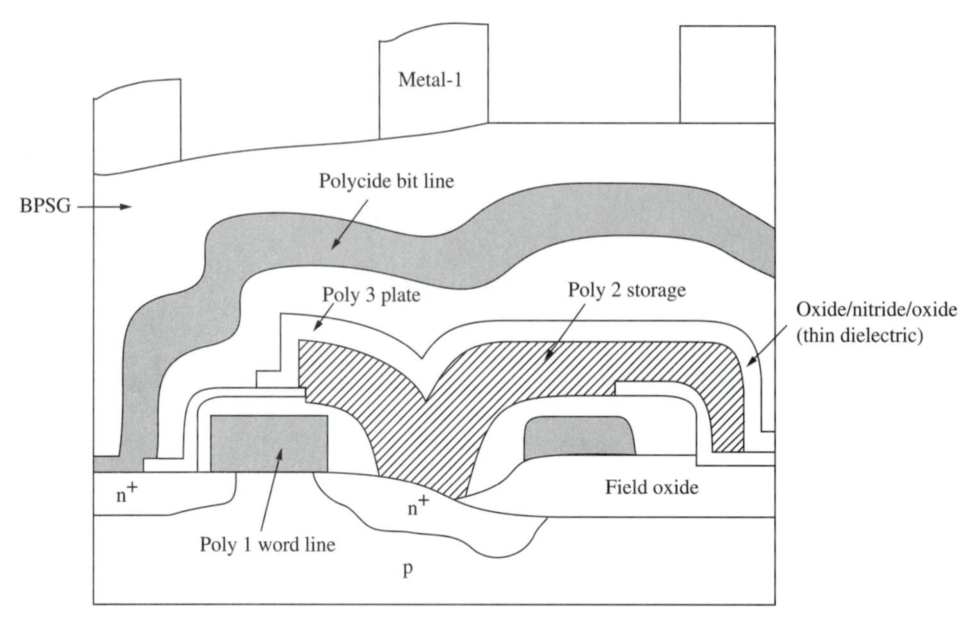

FIGURE 25
16-Mb DRAM cell cross section—stacking capacitor structure.

capacitor electrode. Note that, because the poly 1 gates are completely protected by a thick oxide, no alignment borders are needed for the cell contacts, and the contact mask could be oversized. The boundaries of the contacts are limited by the poly 1 spacers and the bird's beak of the field oxide, not by the mask dimensions. This self-aligned feature significantly reduces the cell area and increases its density.

For the capacitor dielectric, a composite film of oxide/nitride/oxide with an equivalent oxide thickness of 6 nm is deposited. The native oxide on the poly 2 is used as the first oxide layer, on top of which 5 nm of nitride is deposited. The nitride layer is then reoxidized in steam at 920°C to grow a top oxide layer 2 nm thick. The oxidation step seals any potential pinholes in the oxide/nitride/oxide composite film, making the film very reliable and high yielding. A counter electrode of 150 nm of polysilicon film is deposited next, and then it is doped and patterned. This electrode extends throughout the memory array area and has openings cut out over the source regions of each transfer device.

At this point, the p^+ masking and BF_2 implants are implemented for the peripheral devices. This step is deliberately placed after the nitride reoxidation step in order to reduce p^+ junction depth. Three hundred and fifty nanometers of BPSG is then deposited over the capacitors and reflowed at 850°C in order to smooth out the topography and reduce step height (see Fig. 25). Contact openings are then made in the source areas by using a photomasking step and etching through the BPSG layer. Seventy nanometers of poly is then deposited and doped by phosphorus implantation, and a layer of tungsten silicide is chemical-vapor deposited. Using another masking step, the composite poly/silicide (polycide) layers are patterned to define the bit lines for the cells.

Next come the conventional steps of the back-end processing of metallization. The word lines in the memory arrangement are stitched with metal lines at equal intervals in order to reduce their resistances.

Special considerations for deep-trench capacitors

An alternative way to obtain larger specific capacitance is to bury the capacitors in deep trenches. The process technology making these structures possible is anisotropic etching of silicon by RIE. The requirements for etching deep, narrow trenches—3 to 4 microns for the 0.8-μm DRAM, and 5 to 7 microns for the 0.5-μm DRAM technologies—are very exacting for devices with acceptable yield:[52]

1. The etched trench should have sidewalls that are smooth and tapered with an angle of roughly 87°. Too much taper would waste cell space, but too little could cause void formation during the refilling step by CVD oxide or polysilicon.
2. The sidewalls should have no bowing or kinks, and the trench should have a smooth bottom with rounded corners, so as to prevent stress-induced defects from being generated during oxide growth and to avoid large fields at sharp corners.
3. Etch damage in the substrate must be minimized, so as not to have excessive diode leakage.
4. The trench depth must be uniform both across the wafer and from wafer to wafer.

As a common capacitor plate for all the cells, in-situ doped polysilicon is deposited to fill the trenches. Excessive material on the surface is then etched back or planarized by the chemical mechanical polishing (CMP) process.

Sometimes in deep trenches an inversion problem, due to positive charges either at the trench surface or in the trench-refill material, forms an undesirable leakage path.[66] The polysilicon that fills the trench must also be highly doped to prevent depletion effects.

9.6.2 SRAM Processing Technology

Static random access memory is called *static* because it does not require periodic refresh signals to retain stored data. The bit states in a SRAM are stored in a pair of cross-coupled inverters, which form a circuit known as a flip-flop. Figure 26*a* is a circuit schematic of a six-transistor SRAM cell. Figure 26*b* is a circuit schematic of a four-transistor cell where the active loads are replaced by passive poly resistors. As SRAM has evolved, it has increased in density. At present, with a 0.5-μm minimum line width, it has reached a size of four Mb.

BiCMOS SRAM offers densities and power dissipation levels close to those of CMOS SRAM, but with higher operating speeds. Typically, the memory array in a BiCMOS SRAM consists of an NMOS cell in a p-well. The decoders, word line, and write drivers are implemented with BiCMOS logic gates. The sense amplifiers, which require high input sensitivity, use pure bipolar circuits.

CMOS SRAM process flow

Now we examine a 1-Mb CMOS SRAM with 0.8-μm technology as an example of CMOS SRAM process flow.[67] The processing steps that precede the gate oxidation and follow the metal contact hole formation are fairly conventional and are omitted.

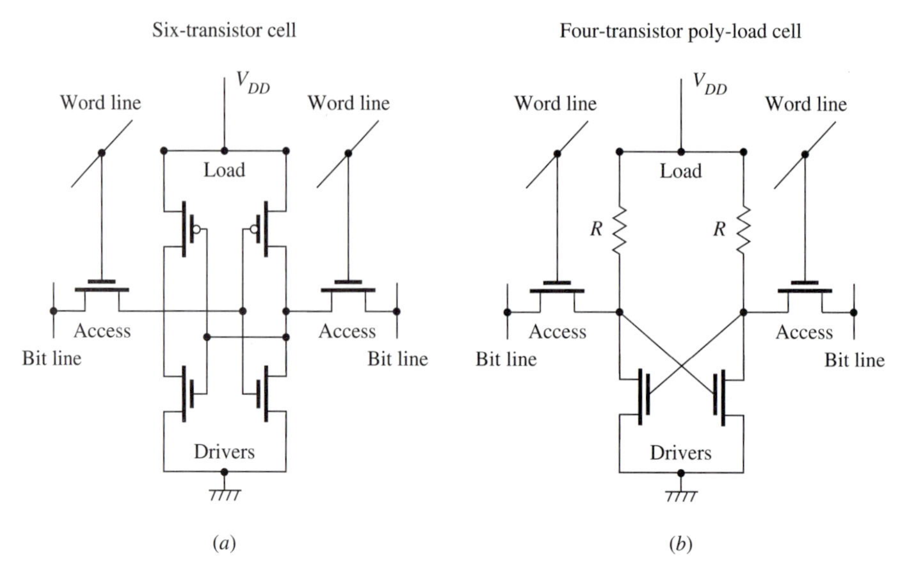

FIGURE 26
(*a*) The full-CMOS six-transistor cell has a lower power consumption and a greater immunity to transient noise and temperature fluctuations than poly load four-transistor cells. (*b*) The four-transistor cell requires less space on the chip than the six-transistor cell because it is fabricated in a separate poly layer above the transistors.

Figure 27 shows the layout of a SRAM memory cell. For maximum density in a memory device, the cell must be laid out as small a size as possible. The size is determined by the cell's topography and by the design rules of the IC fabrication technology. For this process, we use an advanced six-transistor cell. Using a PMOS thin-film transistor (TFT) on isolator load, the memory cell size is 5.2 μm $\times$ 7.9 μm. The SRAM is fabricated with a single aluminum and a triple-level polysilicon, twin-well CMOS technology. The first polysilicon level forms the gate electrodes of the cell driver NMOS transistors and the word lines. The second polysilicon level, crystallized from amorphous silicon, is used for the PMOS TFT loads and the V_{CC} lines. The third interconnect layer, tungsten silicide, is used as ground lines and upper word lines. The bit lines are formed by aluminum. A cross-sectional view of the memory cell is shown in Fig. 28. The p^+ drain of the PMOS TFT is directly connected to the n^+ polysilicon (poly 1) through a contact hole formed above the buried contact region.

Both n-channel and p-channel MOSFETs have an LDD drain structure. The gate oxide thickness of the bulk silicon CMOSFET is 20 nm. The gate oxide thickness for the TFT is 40 nm, on which 60 nm of amorphous polysilicon is deposited. The body of the TFT is annealed in an oxidizing atmosphere and crystallized to polysilicon with a grain size in the range of 100 to 500 nm. An offset at the drain of the TFT reduces its gate-drain electric field and thus reduces its off current.

Special considerations for buried contact, poly load, and thin-film transistor load

In order to establish contacts from the drain regions of the load devices to those of the n-channel devices, poly lines are used to connect the diffusion directly and

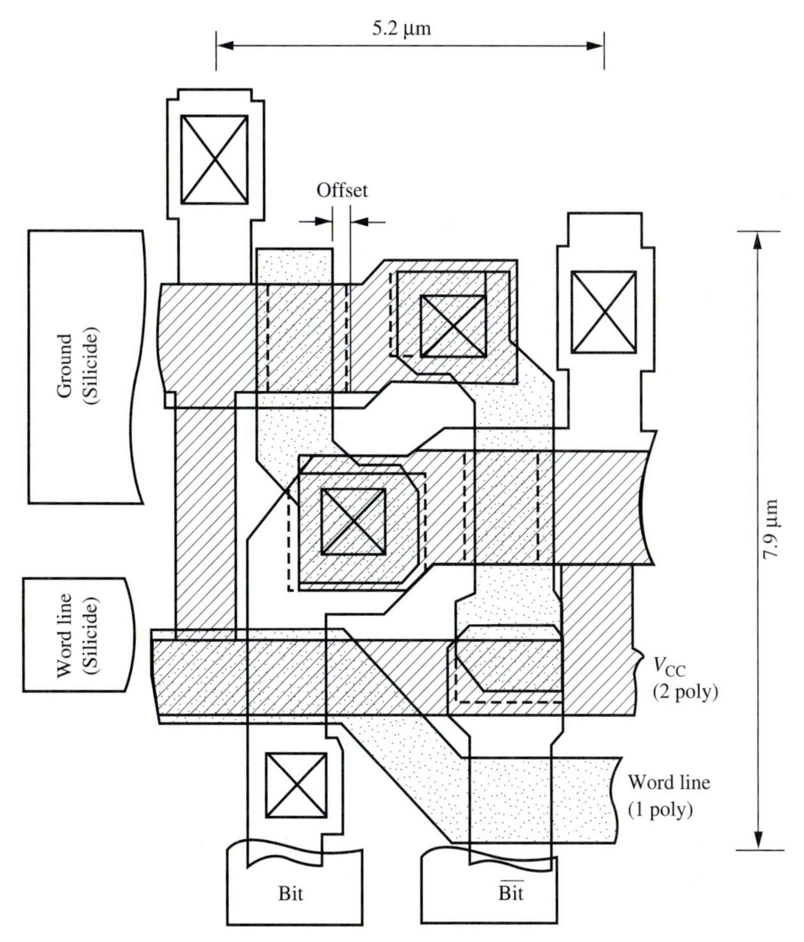

FIGURE 27
Typical SRAM memory cell layout with a PMOS TFT load. (*After Ando et al., Ref. 67.* ©1989 IEEE.)

serve as local interconnects. The poly lines are from the same layer as that of the poly gate. These contacts are buried deep under the interlayer dielectric and are thus called *buried contact.* Because it is borderless, a buried contact takes up less space and also can be used as a local interconnect between the drain of the access transistors and the gate of the driver transistors.

A representative structure of the buried-contact module is depicted in Fig. 29. Immediately after gate oxide growth, a thin layer of doped polysilicon of 50 nm is deposited. A mask layer is applied that opens the buried contact areas but protects the poly gate areas. Plasma etching is used to etch this composite layer until the substrate is exposed. At this stage, the gate oxide is protected by the thin poly so that it is sealed from contamination. Three hundred and fifty nanometers of doped polysilicon is then deposited before a second poly gate mask is applied. This mask again defines the poly gate areas and the buried contacts, but allows a small offset of the contacts to account for misalignment from the previous mask. During the second RIE etching, this offset

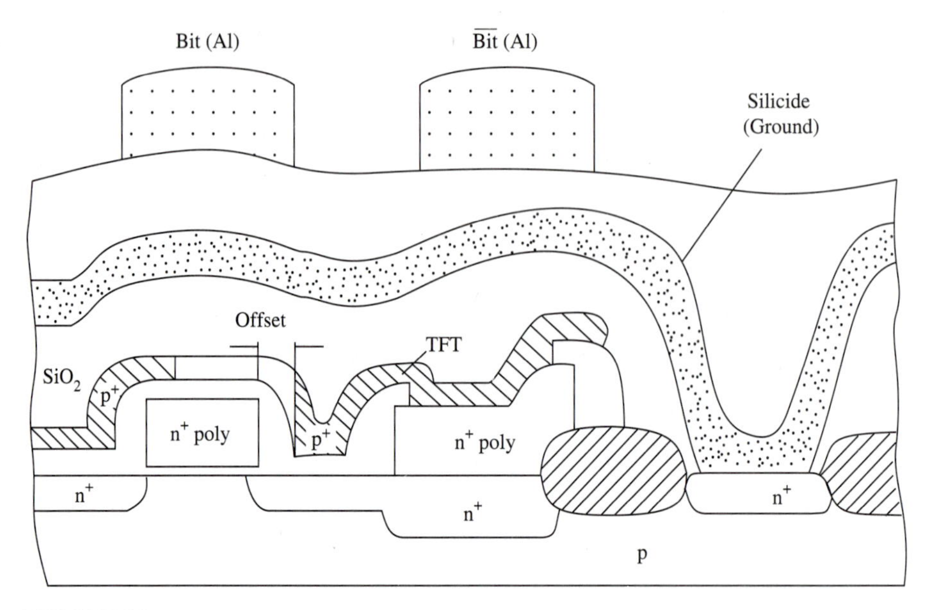

FIGURE 28
Cross section of the SRAM memory cell with a PMOS TFT load. (*After Ando et al., Ref. 67.* ©1989 IEEE.)

causes the digging of a shallow dimple in the substrate because there is no gate oxide in this area to protect it from the poly overetching. Arsenic is then implanted for the source and drain of the device. At the dimple, electrical continuity is achieved when lateral diffusion from this implant joins with the outdiffusion of impurities from the poly layer on top of the buried contacts.

The SRAM cell that uses poly film resistor as a load is easier to implement than the six-transistor cell with polysilicon thin-film transistor as a load (see Fig. 26*b*), but

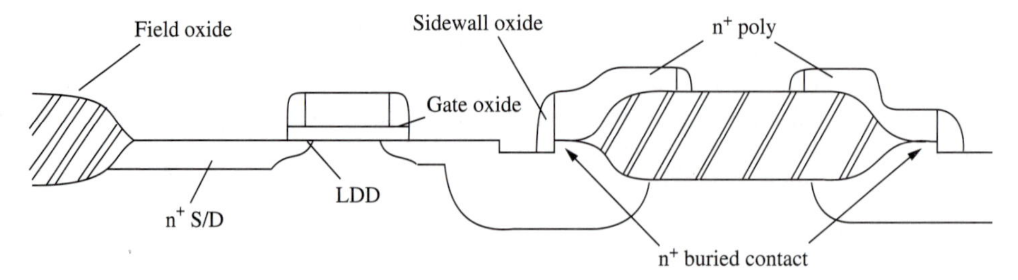

FIGURE 29
Limitations of buried contacts: (1) Phosphorus outdiffusion limits device isolation and buried-contact-to-gate design-rule scaling. (2) The use of n-type polysilicon limits buried contacts in CMOS to n-channel transistors only. (*After Tang et al., Ref. 28.* ©1987 IEEE.)

it has a leakage current three orders of magnitude larger than that of a six-transistor cell. It is usually considered to be practical up to one-megabit memories.[68] Poly load resistors are fabricated from undoped polysilicon films implanted with arsenic in a dose from 1×10^{13} to 1×10^{15} cm^{-2}. The sheet resistivity can be controlled from 10^4 $\Omega/\Box$ up to 10^{12} $\Omega/\Box$. For a 1-Mb SRAM, a poly film with a sheet resistance of 26 $G\Omega/\Box$ will be needed. This can be obtained with an arsenic implant dose of 3×10^{13} cm^{-2}.

At densities of SRAM above four megabits, polysilicon thin-film transistors must be used as loads. They are stacked above the n-channel devices (see Fig. 28). Such an arrangement has smaller cell area, smaller leakage current, and better alpha-particle soft-error immunity than a poly-resistor load.

9.6.3 Soft-Error Rate (SER) Consideration

Soft errors are single, nonrecurring read errors on a single bit of a memory array. It is not a permanent error in that the cause is not a process defect. Although soft errors can be caused by such circuit-related problems as supply-voltage noise, inadequate noise margins, and sense-amplifier imbalance, there is one specific physical failure mode that will cause soft errors. The cause of this failure mode was first identified as alpha particles originating from the decay of radioactive impurities in IC packages.[69]

The alpha particles, when incident on the memory chip, generate electron-hole pairs in the substrate and are collected by the depletion regions of the pn junction. Since a bit of information is stored on a DRAM or a SRAM cell by the presence or absence of charges in the potential well of the storage capacitor, additional carriers, because of the alpha particles collected by these junctions, can upset the original state of the junction and cause it to flip to the opposite state; thus an error results (see Fig. 30).

Soft errors can occur in both DRAM and SRAM, and the situation is aggravated as the memories are scaled down. One general way of reducing SER is to create a deep doping gradient that produces a potential-energy barrier to the diffusing electrons, preventing some of them from reaching the depleted regions of the storage nodes. The well-substrate junction acts as a reflective barrier for diffusing minority carriers created outside the well. Putting the memory on a heavily doped substrate with a lightly doped epitaxial layer on the surface is also beneficial. Since the minority carriers generated in the heavily doped substrate have a much shorter lifetime than those generated in the lightly doped regions, they are recombined before they can do harm.

9.6.4 Other Memory Technologies

Another family of semiconductor memory that uses submicrometer design rules is the nonvolatile memory, which includes electrically erasable programmable read-only memory (EEPROM), erasable programmable ROM (EPROM), and the flash EEPROM. However, because their combined production volume is eclipsed by that of DRAM and SRAM, these memories will not be discussed here. Among these ROM devices, flash EEPROM could emerge as a mainstream memory technology in the next decade.

Effects of an Alpha Particle

0 | 1

- Potential well filled with electrons
- p-type silicon in inversion
- One million electrons

- Potential well empty
- p-type silicon in deep depletion

0 | 1

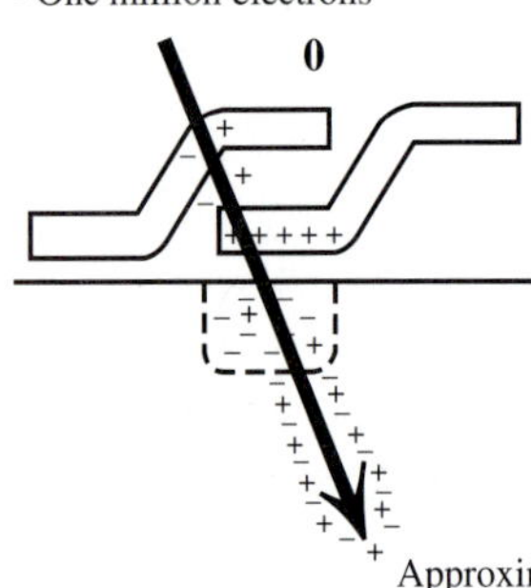

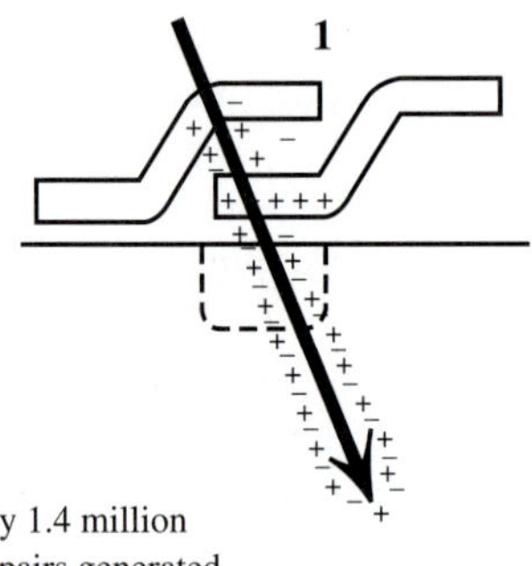

Approximately 1.4 million electron-hole pairs generated to a depth of approximately 25μ

- Natural alphas up to approximately 8 MeV in energy
- A typical 5 MeV: –25 μ range in Si
  –1.4 × 10^6 e-h pairs (3.5 eV/e-h pair)

0 | 1

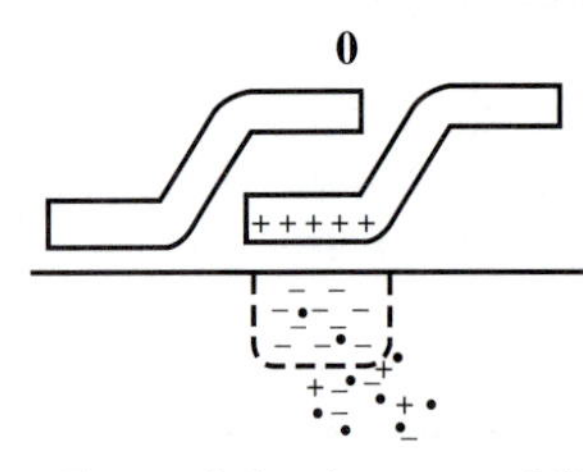

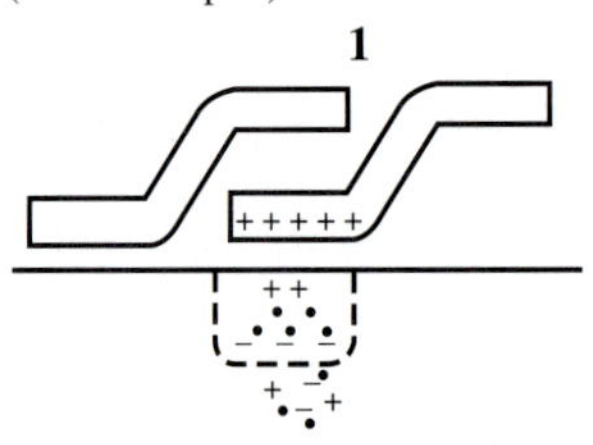

- Electron-hole pairs generated diffuse
- Electrons reaching depletion region are swept by electric field into well; holes are repelled
- Collection efficiency = fraction of electrons collected

0 | 1 → 0

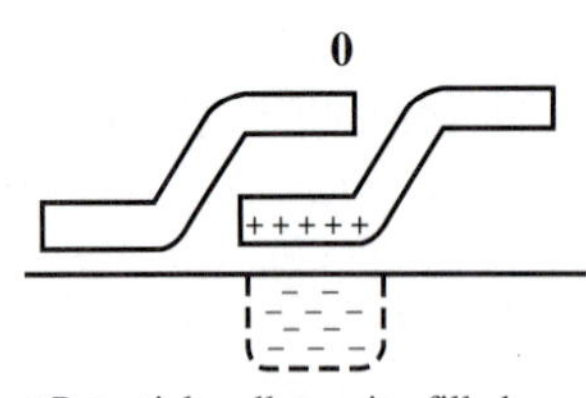

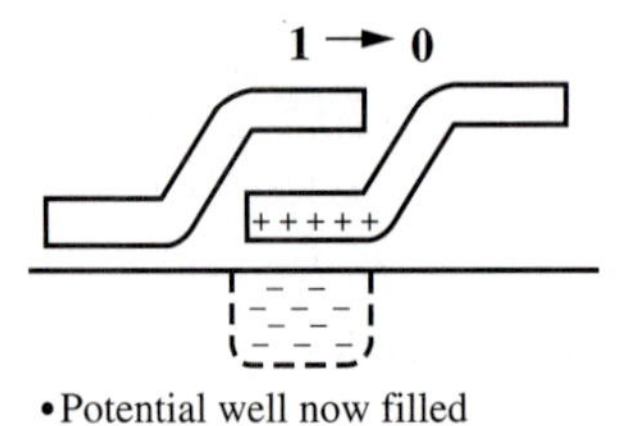

- Potential well remains filled
- No appreciable collection
- If (collection effiency) × (number of electrons generated) > critical charge, a soft error results
- A single alpha can cause an error
- No permanent damage results

- Potential well now filled

FIGURE 30
Stages of soft-error creation by alpha particles in dynamic memories. (*After May and Woods, Ref. 69.* ©1985, 1989 IEEE.)

9.7 PROCESS INTEGRATION CONSIDERATIONS IN ULSI FABRICATION TECHNOLOGY

9.7.1 Introduction

With further shrinking of dimensions, the circuit processes become more stringent; old requirements are tightened, and new requirements have to be considered. In this section, we take a close look at some of the most important issues involved in the process integration of ULSI circuits.

9.7.2 Design Rules

A complete IC process usually involves ten to twenty masking steps according to the complexity of the process. Each mask introduces some changes in the surface features, whether they are in the silicon substrate or in the films deposited on it. Minimum-dimension rules, governed by processing or physical requirements, apply to these mask layers. When these minimum spacings are violated, havoc ensues in the circuit's operation. Circuit designers use the minimum design rules, or "design rules" for short, to construct the various layers, and they treat these rules as sacred. Many factors are involved in determining these rules. Some of the most important factors are described below:

Line-width tolerances

The line-width dimension on a finished wafer usually differs from the corresponding line-width dimension on the mask. This is because the photo and etching steps used to pattern this line generally shrink or expand the original line by a biased amount, Δ. Because this is a statistical process, this shrinkage or expansion introduces a variation to the uncertainty of this line width. This bias and its variation depend very much on the processing equipment and procedure.

Junction depth and lateral diffusion

After all the heating cycles are completed, each kind of junction has certain dimensions, both vertical and lateral, and each dimension has its own variation value. For instance, the n-well junction depth is about 2 μm, and the source/drain junctions are 0.1 to 0.3 μm deep. The lateral diffusion spread is sometimes estimated as 80% of the vertical junction depth.

Depletion width

When a voltage is applied to a pn junction, the depletion width change depends on the doping profile of the junction and the amount of voltage applied.[2,3]

Thickness tolerances

The thicknesses of the deposited films usually do not affect design rules; however, they will affect the rules if they give a lateral extent of a surface feature. For instance, an oxide spacer has a lateral dimension and its thickness will displace an implanted junction by that amount. The bird's beak at the edge of the field oxide is another example; the thickness of pad oxide usually affects the bird's beak length.

Masking tolerances

Masking tolerances result from two separate causes: Within the same mask, the dimensions may vary because of the mask-making process. Alignment tolerances also vary from one mask level to another. For a modern optical stepper tool, the former value is approximately 0.025 μm and the latter value is approximately 0.1 μm.

In determining the design rule of a particular feature, all the biased amounts of the factors mentioned above are added up algebraically, and their standard deviations are added in a root-mean-square manner. Consequently, the different tolerances are "nested" within each other. For example, if we want to find the minimum distance between two isolated n^+ junctions, we have the following calculations, where an uppercase letter represents the bias and a lowercase letter represents the deviation of each factor:

$$\text{Total bias} = 2D + 2B + 2A$$

$$\text{Total deviation} = 1.4 \times (d^2 + b^2 + e^2)^{1/2}$$

where D is the bird's beak, B is the lateral n^+ junction extent, and A is the process bias of the active area; and d and b are the deviations of D and B, respectively, and e is the mask run-out tolerance.

9.7.3 Compactness

IC density is of prime importance. Density can be achieved to some extent by the brute force approach, such as using better steppers or etchers. However, tighter design rules can be obtained by using more innovative process integration. One powerful method is self-alignment; the alignment tolerance between two mask levels can consume a great deal of chip area. A common example is the self-aligned source and drain implant with respect to the polysilicon gate. Another useful idea is the self-aligned contact. The object is to make a contact to the substrate between the poly gate and the field oxide or between two poly gates.[29] The main process flow, depicted in Fig. 9, is described as follows:

Usually, we need to create a clearance space between a gate and the contacts to its S/D in order to avoid a shorting problem; but the following method enables us to put a contact hole next to the edge of the gate making a connection to the n^+ junction. A polysilicon film is deposited with a thick CVD oxide on top, and then the composite film is etched at the gate stack. Another 200 nm of CVD oxide is then deposited, and the wafer is blanket-etched in plasma so that only oxide spacers are left at the edge of every poly step. At this stage, the poly gates are completely enclosed by oxide, both on the top and on the side, and the substrate areas where we want to make contacts are bare; so another poly film is deposited and patterned to form a contact landing pad. The advantage of this method is that no alignment borders are needed for these contacts. The sizes of these contacts on the mask and their alignment to the prior structure become noncritical; the actual contact dimensions on the substrate are limited by the extent of the spacers and by the bird's beak of the field oxide.

9.7.4 Step Heights

At the various processing stages, there are features on the chip that differ in height. The range of variation of this vertical profile at any mask level is termed *step height.* A major limitation in processing submicrometer feature-sized ICs is the very small depth of focus of the optical steppers used to pattern the circuit features. The focus margin (FM) or the usable depth of focus (DOF) is calculated by the following equation:

$$\mathrm{FM} = \mathrm{DOF} - \mathrm{CDOF} - \mathrm{FB} - H_s$$

where the focus budget, FB, equals 0.95 μm and is contributed by the imperfection of the optical system and mask; H_s is the step height for the specified layer; CDOF is the correction of depth of focus resulting from photoresist thickness variation; and DOF is a function of exposure geometry and is approximately 1.7 μm for 0.5-μm lines.

As an example, the focus margins for a hypothetical 16-Mb DRAM process of 0.5-μm geometry have been calculated and are shown in Table 1. Two of the layers, poly 4 and metal 1, have potential problems because their usable focus margins are negative; this is because their step height is excessive. These layers need to be planarized better so that their focus margins are within acceptable limits.

9.7.5 Process Window and Sensitivity

When a piece of equipment performs an operation, its performance is controlled by many parameters. A change in the magnitude of any of the process parameters

TABLE 1
Focus margins for DRAM process layers

Layer	H_S μm	Resolution μm	DOF μm	CDOF μm	FM μm
Active area	0	0.5	1.7	0.0	0.75
Poly 1	0.14	0.5	1.7	0.0	0.61
Cell contact	0	0.5	1.7	0.0	0.75
Poly 2	0.45	0.50	1.7	0.0	0.3
Poly 3	0.45	0.7	2.7	0.0	1.3
BL contact	0	0.5	1.7	0.0	0.75
Poly 4	1.00	0.5	1.7	0.0	−0.25
M 1 contact	0.506	0.6	2.25	0.0	0.79
Metal 1	1.3	0.6	2.25	0.91	−0.91
Via	0.5	0.6	2.25	0.0	0.8
Metal 2	0.5	0.7	2.7	0.91	0.34

induces a corresponding response in its performance. The window of operation for this equipment is defined as the specification limits of the response induced by this change in parameter. For example, when exposing the photoresist of a masking layer using a stepper, the exposure dosage affects the line width of the resist. Higher dosage gives smaller line width. Therefore, in using this machine, there is a specific range of dosage used to obtain the acceptable line width. The incremental change of line width that occurs with a change of dosage is termed the *sensitivity* of the process. In designing a manufacturing process, it is desirable to have as large a process window and as small a sensitivity as possible, so that large variations in equipment performance can be tolerated. It may be helpful to plot the process variables versus the process response in a response contour surface-plot.

9.7.6 Process Parameter Monitoring

For ULSI, manufacturing technology is as important as processing technology. In fact, without a reliable way to fabricate ICs in quantity, it is doubtful that good-quality circuits can be made. An important cornerstone in IC manufacturing is statistical process control (SPC). Throughout the fabrication process, various processing parameters are collected, and their deviations from the targeted values, as well as their distribution variation over a period of time, are analyzed. Since these measurements are taken on-line, any abnormal deviation from prespecified control limits would immediately alert workers to find the underlying causes and correct the problem before a large amount of defective units are produced. Some examples of processing parameters are poly critical dimension (CD), gate oxide thickness, and poly sheet resistance. For the product, the parameters are the various wafer acceptance tests (WAT) of electrical measured data. For the equipment, the parameters are such items as the temperature control of a furnace or an implanter's dosage uniformity. For any given parameter, it is convenient to define the following quantities:[70]

$$C_p = |\text{USL} - \text{LSL}|/6s$$

$$K = (x - \text{TGT})/(|\text{USL} - \text{LSL}|)/2$$

$$C_{pk} = C_p(1 - |K|)$$

Where,

LSL (USL) = lower (upper) specification limits
TGT = target value, usually centered between LSL and USL
x = sample mean value
s = sample standard deviation

The spec of a given parameter usually has a TGT of $\pm 3s$. ULSI manufacturing demands very tight control; usually a parameter requires six-sigma values to be contained within its specification limits. Thus the quantities defined above must be limited to the following values:

$$C_p \geq 2$$

$$K \leq 0.25$$

$$C_{pk} \geq 1.5$$

9.8
SUMMARY AND FUTURE TRENDS

In this chapter, we have presented the highlights and an overview of the most important developments in process integration for different technologies. A subject this broad requires volumes for a complete discussion of every detail. Although a chapter can only give a cursory treatment of the subject, we consider our task fulfilled if this whets the reader's appetite to seek other reading materials such as the references listed at the end of this chapter.

In the previous sections, we describe various current processing technologies. Silicon technology is treated exclusively because it has the widest application and is the only one adequate for ULSI. Bipolar devices have long been eclipsed by CMOS technology, although bipolar and CMOS technology have merged to form BiCMOS technology. Analog and high-voltage circuits are not covered in this chapter because they are involved at lower levels of integration. Thin-film transistor (TFT) display and silicon-on-insulator (SOI) systems are also omitted because they have not yet reached the ULSI level.

What future is in store for us? Prediction is always risky, often not because it may be too wild, but because it may be too conservative. Did Dr. William Shockley imagine his invention of the bipolar junction transistor would lead to submicrometer devices? Or did Dr. J. S. Kirby, while applying black wax for masking by hand, and Dr. Noyce, who used Al metallization for the first IC, foresee that their inventions would lead to a chip with one gigabit of memory? Nevertheless, we are safer to limit our predictions to around the year 2005, to see what new technology is waiting for us.

Semiconductor processing equipment is playing an increasingly dominant role in the advancement of IC technologies; it is the engine that drives progress. We can say with some certainty that optical lithography can reach 0.20 μm by utilizing wave-front engineering and using DUV (deep ultraviolet, wavelength less than about 200 nm) lithography. Pattern overlay accuracy with a total budget of 0.10 μm is feasible now for 0.5- to 0.35-μm technology, and may reach 0.02 μm. For patterning, highly anisotropic etching with zero etch bias is possible with the electron cyclotron resonance (ECR) type of plasma etcher, which offers low ion energies, low damage, high etch rates, and high selectivities. X-ray or electron-beam direct-write lithography may take over the job down to dimensions of 0.08 μm. The myriad problems that have plagued X-ray or electron beam direct-write lithography may be solved in time for the production of gigabit devices. Gate oxide annealed in N_2O ambient is found to yield devices with higher mobilities, better reliability and lifetime, less susceptibility to degradation, and a better barrier against impurity diffusion; thus a high-quality gate dielectric of 4 to 5 nm thickness is in sight. CVD-deposited metal film can give better step coverage and lower resistivity and is capable of filling via holes with large aspect ratios. The use of polyimide as inter metal dielectric (IMD) material coupled with pillar or reverse pillar technique can offer nearly perfect planarity and low dielectric constant to as many levels of interconnect as needed.[71]

For advanced device capabilities, it has been demonstrated by several groups that high-speed devices with 0.1 μm channel length were achieved with transconductance of 750 mS/mm for nMOSFETs, and 400 mS/mm for pMOSFETs.[72] Shallow-trench isolation coupled with retrograded wells should be capable of providing good isolation with 0.25-μm device spacings without the danger of latch-up.

If we consider the present generation to be 0.5-μm minimum geometry technology, we can project the future generation of technology to be 0.35 μm, 0.25 μm, 0.17 μm, and 0.12 μm, spaced evenly with the same ratio.[73] Assuming we maintain our present speed of progress of four years for every generation, before 2003 we could have one gigabit of DRAM memory cells per chip. Then another milestone will have been crossed; the era of giga-scaled integration (GSI) will have arrived. For product trends, fast synchronous DRAM memories with short data access time will be used as cache memory, and flash EPROM will be used as all programmable read-only memories. Highly integrated chips will be expected to incorporate as much as 10^8 transistors in the year 2000. On the basis of this capacity, a large-scale system-on-a-chip that integrates a microprocessor unit (MPU) core, an application-specific-standard-product (ASSP) core, Mb memory, and analog functions will become feasible.[74]

Because of various practical considerations, the drive for smaller minimum dimensions will probably be slowed after 0.10 μm; however, this by no means is the end of progress.[75] Better process module development can further increase the circuit density. A great deal of advancement can and will be made in semiconductor manufacturing technology.

Density and yield are the essence of semiconductor microelectronics and the key ingredients in the competitive market place. Price per circuit function is the driving force, the major components of which are performance, reliability, and manufacturing costs. Costs may only be reduced by factory automation and a reduction of packaging subassemblies through larger-scaled integration.

We predict that IC manufacturing is moving toward full-scale automation. This means complete computer control of processing equipment and data handling (such as real-time data collection and analysis, recipe download, and data upload); robotic material-handling systems to perform material transport and stocker control; and real-time lot scheduling and tracking, optimized batching, scheduling, and dispatching. Equipment tools will be mostly single-wafer processed, and cluster-oriented.

The wafer process environment will have to be much cleaner than it is now. As a rule, dust particles of one tenth the minimum geometry will be fatal defects, so we have to worry about 10-nm-sized particles. Our present practice of building cleanrooms of class one and beyond has long reached the point of diminished returns. It is common sense that contamination control is only as good as the weakest link; and most of our present sources of contamination are from the processing equipment, which so far we know little about. Many believe that, rather than making the whole building clean, it is more important to consider the immediate wafer environment. And if the wafer environment is sufficiently tight, workers should be able to work in shirtsleeves without any bad effects on the products. This would be a cleaner, much less expensive factory and more fitting for human workers.

As the IC industry reaches its maturity, we will see more consolidation of manufacturing facilities and standardization of processes, moving from a more exploratory age to a more routine one. But the electronic industry will not remain stagnant for long. As sure as the automobile followed the horse-and-buggy, somewhere a breathtaking breakthrough will be made. Some new material and device principles will be discovered, and progress will continue to new heights, which we cannot imagine now, in the twenty-first century and beyond.

REFERENCES

1. P. Chatterjee, "ULSI-Market Opportunities and Manufacturing Challenges," *IEDM Tech. Dig.,* 11 (1991).
2. S. M. Sze, *Semiconductor Devices: Physics and Technology,* Wiley, New York, 1985, *Physics of Semiconductor Devices,* 2nd ed., Wiley, New York, 1981.
3. M. Shur, *Physics of Semiconductor Devices,* Prentice Hall, Englewood Cliffs, NJ, 1990.
4. L. C. Parrillo, R. S. Payne, R. E. Davis, G. W. Reutlinger, and R. L. Field, "Twin-Tub CMOS—A Technology for VLSI Circuits," *IEDM Tech. Dig.,* 752 (1980).
5. C. Y. Lu, J. J. Sung, H. C. Kirsch, N. S. Tsai, R. Liu, A. S. Manocha, and S. J. Hillenius, "High-Performance Salicide Shallow-Junction CMOS Devices for Submicrometer VLSI Applications in Twin-Tub VI," *IEEE Trans. Electron Dev.,* **36(11)**, 2530 (1989).
6. R. D. Rung, C. J. Delloca, and L. C. Walker, "A Retrograde p-Well for Higher Density CMOS," *IEEE Trans. Electron Dev.,* **28(10)**, 1115 (1981); S. Odanaka, T. Yabu, N. Shimizu, H. Umimoto, and T. Ohzone, "A Self-Aligned Retrograde Twin-Well Structure with Buried p^+-Layer," *IEEE Electron Dev. Lett.,* **10(6)**, 280 (1989).
7. A. G. Lewis, R. A. Martin, T. Y. Huang, J. Y. Chen, and M. Koyanagi, "Latchup Performance of Retrograde and Conventional n-Well CMOS Technologies," *IEEE Trans. Electron Dev.,* **34(10)**, 2156 (1987).
8. E. Kooi, *The Invention of LOCOS,* IEEE, Inc., New York, 1991.
9. E. Kooi, J. G. van Lierop, and J. A. Appels, "Formation of Silicon Nitride at a Si/SiO_2 Interface Using Local Oxidation of Silicon and during Heat Treatment of Oxidized Silicon in NH_3 gas," *J. Electrochem. Soc.,* **123(7)**, 1117 (1976).
10. T. T. Sheng, C. Y. Lu, R. D. Chang, and S. T. Chiang, "From White Ribbon to Black Belt: A Direct Observation of the Kooi Effect Masking Film by TEM," *J. Electrochem. Soc.,* **140(11)**, L163 (1993).
11. R. H. Havemann and G. P. Pollack, U.S. Patent 4,541,167 (1985).
12. J. M. Sung, C. Y. Lu, L. B. Fritzinger, T. T. Sheng, and K. H. Lee, "Reverse L-Shape Sealed Poly-Buffer LOCOS Technology," *IEEE Electron Dev. Lett.,* **11(11)**, 549 (1990).
13. H. Mikoshiba, T. Homma, and K. Hamano, "A New Trench Isolation Technology as a Replacement of LOCOS," *IEDM Tech. Dig.,* 578 (1984).
14. S. H. Goodwin and J. D. Plummer, "Electrical Performance and Physics of Isolation Region Structures for VLSI," *IEEE Trans. Electron Dev.,* **31(7)**, 861 (1984).
15. S. Ogura, P. J. Tsang, W. W. Walker, D. L. Critchlow, and J. F. Shepard, "Design and Characteristics of the Lightly Doped Drain-Source (LDD) Insulated Gate Field-Effect Transistor," *IEEE Trans. Electron Dev.,* **27(8)**, 1359 (1980).

16. T. Hori, Y. Odake, and T. Yasui, "Deep-Submicrometer Large-Angle-Tilt Implanted Drain (LATID) Technology," *IEEE Trans. Electron Dev.,* **39**(**10**), 2312 (1992).
17. Y. Okumura, M. Shirahata, A. Hachisuka, T. Okudaira, H. Arima, and T. Matsukawa, "Source-to-Drain Nonuniformly Doped Channel (NUDC) MOSFET Structures for High Current Drivability and Threshold Voltage Controllability," *IEEE Trans. Electron Dev.,* **39**(**11**), 2541 (1992).
18. C. Codella and S. Ogura, "Halo Doping Effect in Sub-Micron DI-LDD Device Design," *IEDM Tech. Dig.,* 230 (1985).
19. J. R. Brews, W. Fichtner, E. H. Nicollian, and S. M. Sze, "Generalized Guide to MOSFET Miniaturization," *IEEE Electron Dev. Lett.,* **1**(**1**), 2 (1980).
20. A. Tanaka, T. Yamaji, A. Uchiyama, T. Hayashi, T. Iwabuchi, and S. Nishikawa, "Optimization of Amorphous Layer Thickness and Junction Depth on the Preamorphization Method for Forming Shallow-Junction in Silicon," *IEDM Tech. Dig.,* 785 (1989); S. N. Hong, G. A. Ruggles, J. J. Wortman, and M. C. Ozturk, "Material and Electrical Properties of Ultra-Shallow p^+-n Junctions Formed by Low-Energy Ion Implantation and Rapid Thermal Annealing," *IEEE Trans. Electron Dev.,* **38**(**3**), 479 (1991).
21. M. C. Ozturk, J. J. Wortman, C. M. Osburn, A. Ajmera, G. A. Rozgonyi, E. Frey, W. K. Chu, and C. Lee, "Optimization of the Ge Preamorphization Conditions for Shallow-Junction Formation," *IEEE Trans. Electron Dev.,* **35**(**5**), 659 (1988).
22. C. Carter, W. Maszara, D. K. Sadana, G. A. Rozgonyi, J. Liu, and J. Wortman, "Residual Defects Following Rapid Thermal Annealing of Shallow Boron and BF_2 Implants into Preamorphized Silicon," *Appl. Phys. Lett.,* **44**(**4**), 459 (1984).
23. C. K. Lau, Y. C. See, D. B. Scott, J. M. Bridges, S. M. Parna, and R. D. Daires, "Titanium Disilicide Self-Aligned Source/Drain Plus Gate Technology," *IEDM Tech. Dig.,* 714 (1982).
24. C. Y. Lu, J. J. Sung, R. Liu, N. S. Tsai, R. Singh, S. J. Hillenius, and H. C. Kirsch, "Process Limitation and Device Design Trade-Offs of Self-Aligned $TiSi_2$ Junction Formation in Submicrometer CMOS Devices," *IEEE Trans. Electron Dev.,* **38**(**2**), 246 (1991).
25. R. Liu, F. A. Baiocchi, L. A. Heimbrook, J. Kovalchick, D. L. Malm, D. S. Williams, and W. T. Lynch, "Formation of Shallow p^+/n and n^+/p Junction with $CoSi_2$," *Electrochem. Soc. Proceedings of the 1st International Symposium on ULSI Sciences and Technology,* **87–11** 1987, p. 446.
26. E. Nagasawa, H. Okabayahi, and M. Morimoto, "Mo and Ti-Silicided Low-Resistance Shallow Junction Formed Using the Ion Implantation through Metal Technique," *IEEE Trans. Electron Dev.,* **34**(**3**), 581 (1987).
27. D. L. Kwong, Y. H. Ku, S. K. Lee, E. Louis, N. S. Alvi, and P. Chu, "Silicided Shallow Junction Formation by Ion Implantation of Impurity Ions into Silicide Layers and Subsequent Drive-In," *J. Appl. Phys.,* **61**(**11**), 5084 (1987).
28. T. E. Tang, C. C. Wei, R. A. Haken, T. C. Holloway, L. R. Hite, and T. G. W. Blake, "Titanium Nitride Local Interconnect Technology for VLSI," *IEEE Trans. Electron Dev.,* **34**(**3**), 628 (1987).
29. C. Y. Lu, D. S. Yaney, K. H. Lee, M. S. Twiford, N. S. Tsai, T. Kook, L. B. Fritzinger, M. L. Chen, and T. S. Yang, "A Folded Extended Window MOSFET for ULSI Applications," *IEEE Electron Dev. Lett.,* **9**(**8**), 388 (1988).
30. W. T. Lynch, "Self-Aligned Contact Schemes for Source-Drains in Submicron Devices," *IEDM Tech. Dig.,* 354 (1987).
31. F. M. Wanlass and C. T. Sah, "Nanowatt Logic Using Field-Effect MOS Triodes," *ISSCC Tech. Dig.,* 32 (1963).
32. J. Y. Chen, *CMOS Devices and Technology for VLSI,* Prentice Hall, Englewood Cliffs, NJ, 1989.
33. S. J. Hillenius and L. C. Parrillo, U.S. Patent 4,554,726 (1985).

34. C. Y. Lu, J. M. Sung, H. C. Kirsch, S. J. Hillenius, T. E. Smith, and L. Manchanda, "Anomalous C-V Characteristics of Implanted Poly MOS Structures in n^+/p^+ Dual-Gate CMOS Technology," *IEEE Electron Dev. Lett.,* **10**(**5**), 192 (1989).
35. R. R. Troutman, *Latch-up in CMOS Technology: The Problem and Its Cure,* Kluwer Academic Publishers, Norwell, MA, 1986; D. B. Estreich and R. W. Dutton, "Modeling Latch-Up in CMOS Integrated Circuits," *IEEE Trans. Computer-Aided Design of Integrated Circuits and Syst.,* **1**(**4**), 157 (1982).
36. S. J. Hillenius, and W. T. Lynch, "Gate Material Work Function Considerations for 0.5 μm CMOS," *Proceedings of the IEEE International Conference on Computer Design: VLSI in Computers, Port Chester, NY,* 1985, p. 147.
37. S. J. Hillenius, R. Liu, G. E. Georgiou, R. L. Field, D. S. Williams, A. Kornblit, D. M. Boulin, R. L. Johnston, and W. T. Lynch, "A Symmetric Submicron CMOS Technology," *IEDM Tech. Dig.,* 252 (1986).
38. M. Ghezzo and D. M. Brown, "Diffusivity Summary of B, Ga, P, As, and Sb in SiO_2," *J. Electrochem. Soc.,* **120**(**1**), 146 (1973).
39. L. Manchanda, "Hot-Electron Trapping and Generic Reliability of p^+-Polysilicon/SiO_2/Si Structures for Fine-Line CMOS Technology," *Proceedings of the International Reliability Physics Symposium,* 1986, p. 183.
40. J. J. Sung and C. Y. Lu, "A Comprehensive Study on p^+ Polysilicon-Gate MOSFET's Instability with Fluorine Incorporation," *IEEE Trans. Electron Dev.,* **37**(**11**), 2312 (1990).
41. J. R. Pfiester, F. K. Baker, T. C. Mele, H. H. Tseng, P. J. Tobin, J. D. Hayden, J. W. Miller, C. D. Gunderson, and L. C. Parrillo, "The Effects of Boron Penetration on p^+ Polysilicon Gated PMOS Devices," *IEEE Trans. Electron Dev.,* **37**(**8**), 1842 (1990).
42. C. Y. Lu and J. M. Sung, "Negative Charge Induced Degradation of pMOSFETs with BF_2 Implanted p^+-Poly Gate," *Electron. Lett.,* **25,** 1685 (1989).
43. L. C. Parrillo, S. J. Hillenius, R. L. Field, E. L. Hu, W. Fichtner, and M. L. Chen, "A Fine-Line CMOS Technology That Used p^+ Poly/Silicide Gates for NMOS and PMOS Devices," *IEDM Tech. Dig.,* 418 (1984).
44. W. Lin, M. L. Chen, R. H. Doklan, and C. Y. Lu, "Dopant Diffusion in n^+/p^+ Poly Gate CMOS Process," *Solid-State Electronics,* **32**(**11**), 965 (1989).
45. G. Wilson, "Advances in Bipolar VLSI," *Proc. IEEE,* **78**(**11**), 1707 (1990).
46. W. C. Till and J. T. Luxton, *Integrated Circuits: Materials, Devices, and Fabrications,* Prentice Hall, Englewood Cliffs, NJ, 1982.
47. B. T. Murphy, V. J. Glinski, P. A. Gray, and R. A. Pederson, "Collector Diffusion Isolated Integrated Circuits," *Proc. IEEE,* **57**(**9**), 1523 (1969).
48. J. Buie, "Improved Triple Diffusion Means Densest ICs Yet," *Electronics,* 101 (Aug., 1975).
49. A. Wieder, "Submicron Bipolar Technology—New Chances for High Speed Applications," *IEDM Tech. Dig.,* 8 (1986).
50. D. D. Tang, T. H. Ning, R. D. Isaac, G. C. Feth, S. K. Wiedmann, and H. N. Yu, "Subnanosecond Self-Aligned I^2L/MTL Circuits," *IEEE J. Solid-State Circuits* **15**(**4**), 444 (1980).
51. G. P. Li, T. H. Ning, C. T. Chuang, M. B. Ketchen, D. D. Tang, and J. Mauer, "An Advanced High-Performance Trench-Isolated Self-Aligned Bipolar Technology," *IEEE Trans. Electron Dev.,* **34**(**11**), 2246 (1987).
52. S. Wolf, *Silicon Processing for the VLSI Era, Vol. 2—Process Integration,* Lattice Press, Sunset Beach, CA, 1990, p. 510.
53. T. H. Ning and R. D. Isaac, "Effect of Emitter Contact on Current Gain of Silicon Bipolar Devices," *IEEE Trans. Electron Dev.,* **27**(**11**), 2051 (1980).
54. S. M. Sze, *Semiconductor Devices: Physics and Technology,* chapter 4, "Bipolar Devices," Wiley, New York, 1985, p. 109.

55. T. C. Chen, K. Y. Toh, J. D. Cressler, J. Warnock, P. F. Lu, D. D. Tang, G. P. Li, C. T. Chuang, and T. H. Ning, "A Submicrometer High-Performance Bipolar Technology," *IEEE Electron Dev. Lett.,* **10(8)**, 364 (1989).
56. H. C. De Graaff and J. G. De Groot, "The SIS Tunnel Emitter: A Theory for Emitters with Thin Interface Layers," *IEEE Trans. Electron Dev.,* **26(11)**, 1771 (1979).
57. P. Ashburn and B. Soerowirdjo, "Comparison of Experimental and Theoretical Results on Polysilicon Emitter Bipolar Transistors," *IEEE Trans. Electron Dev.,* **31(7)**, 853 (1984).
58. C. C. Ng and E. S. Yang, "A Thermionic-Diffusion Model of Polysilicon Emitter," *IEDM Tech. Dig.,* 32 (1986).
59. T. Yuzubira, T. Yamaguchi, and J. Lee, "Submicron Bipolar-CMOS Technology Using 16 GHz f_T Double Poly-Si Bipolar Devices," *IEDM Tech. Dig.,* 748 (1988).
60. A. R. Alvarez, "Introduction to BiCMOS," in A. R. Alvarez, Ed., *BiCMOS Technology and Applications,* Kluwer Academic Publishers, Narwell, MA, 1989, p. 1.
61. K. Miyata, "BiCMOS Technology Overview," *IEDM Short Course on BiCMOS Technology,* IEEE Inc., New York, 1987, p. 1.
62. R. A. Haken, R. H. Havemann, R. H. Eklund, and L. N. Hutter, "BiCMOS Process Technology," in A. R. Alvarez, Ed., *BiCMOS Technology and Applications,* Kluwer Academic Publishers, Narwell, MA, 1989.
63. R. Havemann, R. Eklund, R. Haken, D. Scott, H. Tran, P. Fung, T. Ham, D. Favreau, and R. Virkus, "An 0.8 μm 256K BiCMOS SRAM Technology," *IEDM Tech. Dig.,* 841 (1987).
64. J. Borland, M. Gangani, R. Wise, S. Fong, Y. Oka, and Y. Matsumoto, "Silicon Epitaxial Growth for Advanced Device Structures," *Sol. State Technol.,* 111 (Jan., 1988).
65. M. Graef and B. Leunissen, "Antimony, Arsenic, Phosphorus, and Boron Autodoping in Silicon Epitaxy," *J. Electrochem. Soc.,* **132(8)**, 1942 (1985).
66. W. P. Noble, A. Bryant, and S. H. Voldman, "Parasitic Leakage in DRAM Trench Storage Capacitor Vertical Gated Diodes," *IEDM Tech. Dig.,* 340 (1987).
67. M. Ando, T. Okazawa, H. Furuta, M. Ohkawa, J. Monden, N. Kodama, K. Abe, H. Ishihara, and I. Sasaki, "A 0.1-μA Standby Current, Ground-Bounce-Immune 1-Mbit CMOS SRAM," *IEEE J. Solid State Circuits,* **24(6)**, 1708 (1989).
68. N. C. C. Lu, L. Gerzberg, C. Y. Lu, and J. D. Meindl, "Modeling and Optimization of Polycrystalline-Silicon Resistors," *IEEE Trans. Electron Dev.,* **28(7)**, 818 (1981); N. C. C. Lu, L. Gerzberg, and J. D. Meindl, "Scaling Limitations of Monolithic Polycrystalline-Silicon Resistors in VLSI Static RAMs and Logic," *IEEE Trans. Electron Dev.,* **29(4)**, 682 (1982).
69. T. C. May and M. H. Woods, "Alpha-Particle-Induced Soft Errors in Dynamic Memories," *IEEE Trans. Electron Dev.,* **26(1)**, 2 (1989); B. Chappell, S. E. Schuster, and G. A. Sai-Halasz, "Stability and SER Analysis of Static RAM Cells," *IEEE Trans. Electron Dev.,* **32(2)**, 463 (1985).
70. P. E. Fieler, "Understanding Motorola's Six Sigma Program," *International Reliability Physics Symposium Tutorials, 2.1,* IEEE Inc., New York, 1990.
71. J. L. Yeh, G. W. Hills, and W. T. Cochran, "Reverse Pillar and Maskless Contact—Two Novel Recessed Metal Schemes and Their Comparisons to Conventional VLSI Metallization Schemes," *Proceedings of the VLSI-Multilevel Interconnection Conference, Sunnyvale, CA,* 1988, p. 95.
72. Y. Taur and Y. Mii, "0.1 μm CMOS and Beyond," *Proceedings of the 1993 International Symposium on VLSI Technology, Systems, and Applications,* Taipei, Taiwan, 1993, p. 1.
73. C. Hu, "Future CMOS Scaling and Reliability," *Proc. IEEE,* **81**, 682 (1993).
74. H. Komiya, "Future Technological and Economic Prospects for VLSI," *ISSCC Tech. Dig.,* 16 (1993).

75. J. D. Meindl, V. K. De, and B. Agrawal, "Prospects for Gigascale Integration (GSI) Beyond 2003," *ISSCC Tech. Dig.*, 12 (1993).

PROBLEMS

1. The white ribbon, or the Kooi effect, in the LOCOS field oxidation process can cause thin spots in the gate oxide, resulting in low gate oxide yield. State how this effect can be remedied.

2. Compare the relative merits of the n-well, p-well, and the twin-well processes for the CMOS circuits.

3. Calculate the capacitance for the capacitor of a DRAM that has a total electrode surface area of 3.14 μm^2 and uses oxide-nitride-oxide as a dielectric, with consecutive layers 1 nm, 6 nm, and 1.2 nm thick.

4. The critical dimensions (CDs) of the poly gate were measured for five lots of a CMOS process:

Lot number	Poly gate CD (μm)
L233100	0.781
L233101	0.752
L233102	0.748
L233103	0.777
L233104	0.790

The gate dimension on the mask is 0.70 μm. What are the average value and standard deviation for the poly gate dimensions?

5. From experiments, we know the following values:

Items	Bias	Deviation
Mask registration	0.00	0.028
Mask deviation	0.00	0.025
Misalignment between two mask layers	0.00	0.100
Contact layer process variations	0.04	0.100

The processing spec. for the poly gate CD is 0.750 ± 0.100 μm. What is the minimium design rule of the poly gate in problem 4 for the contact spacing in a MOS transistor?

6. What are the C_p and C_{pk} values from the latest factory lots in problems 4 and 5?

7. Formulate a detailed process flow of the fabrication of a submicrometer n-channel MOS capacitor in order to monitor the capacitor's threshold voltage.

8. What are the essential requirements for device isolation?

9. Why is the "hot-electron" problem more severe for submicrometer devices, and what are the measures we use to solve this problem in processing?

10. Explain how alpha particles increase the soft-error rate (SER) for a memory array, and describe the process procedures for reducing SER.

CHAPTER 10

Assembly and Packaging

T. Tachikawa

10.1 INTRODUCTION

The packaging of the ULSI chip (also called a die or bar) is a broad subject that ranges from preassembly wafer preparation to fabrication technologies for the packages. The purposes of the packaging are to provide electrical connection, expand the chip electrode pitch for the next level of packaging, protect the chip from mechanical and environmental stress, and provide a proper thermal path for the heat that the chip dissipates.

Packaging significantly affects and, in many cases, dominates the overall cost, performance, and reliability of the packaged chip and the system in which the package is applied. Enhanced performance and downsizing of electronic appliances result mainly from miniaturization of the geometry and higher integration of the chip itself, but they also have been aided by package improvements. Leading-edge ULSI devices require superior package performance. Memories have been the driving force in the advancement of leading-edge wafer fabrication technology; they lead to reduced package size and the resulting increased density. Logic and microprocessor devices, with high input/output (I/O) terminal numbers, higher speed, and high power dissipation, create additional challenges for packaging technology. Packaging designers need to achieve, simultaneously, larger I/O pin counts, a better thermal path, and excellent electrical performance in a smaller package.

Progress in the wafer fabrication process also leads to challenges in assembly techniques. For example, the introduction of new metallization material to reduce stress-induced migration requires a corresponding breakthrough in interconnection techniques. The ULSI package design must meet various demands from packaged chip testing, handling systems, and next-level packaging. Generally, a high-density ULSI package that contains a comparatively larger chip requires a smaller external terminal spacing. This jeopardizes the post-packaging processes described above.

Mechanical, thermal, and electrical integrities in a single-chip package do not always guarantee integrities in next-level packaging. They sometimes impose burdens, such as printed wiring board (PWB) design restrictions, mounting and soldering difficulties, and solder joint reliability problems.

Packaging cost is important because it accounts for a considerable portion of the total cost of the packaged ULSI chip. The cost, typically 1 cent/pin in small-scale integration (SSI) and in medium-scale integration (MSI) packaging, has been increasing, often reaching as much as 10 cent/pin. Packaging cost is likely to be a more dominant cost than the cost of the chip itself. ULSI packaging is the technology that must bridge the gap between the chip and the next-level packaging at a reasonable cost and achieve the required performances through a variety of engineering trade-offs.

This chapter describes all of these packaging issues. It covers single-chip packages, assembly techniques, and mechanical, electrical, and thermal design considerations, paying special attention to relations with the next-level packaging, particularly at the PWB level.

10.2 PACKAGE TYPES

A wide variety of package types can be used for ULSI devices. However, they can be divided into two basic types: hermetic-ceramic and plastic packages. In a hermetic package, the chip resides in an environment decoupled from the external environment by a vacuum-tight enclosure. The package is usually ceramic-based and designed for high-performance applications that allow some cost penalties. In a plastic package, on the other hand, the chip is not perfectly decoupled from the external environment because it is encapsulated with resin materials, typically epoxy-based resin. The outside ambient affects the chip over a long time and gradually penetrates the plastic. Plastic packages have become more popular as their applications expand and they perform better. Because the production process is typically automated batch-handling, they are also cost competitive. This popularity persists because of improved plastics and other process developments. Design improvements that build performance features into the package also drive the trend.

Figure 1[1] illustrates a typical hermetic package. The chip resides in a cavity of the package. Package base material is a formed ceramic on which metallization wirings and external leads are placed. The chip and the package are interconnected by fine Al wire. Hermetic sealing is complete with the cap, usually ceramic or metal, lidded to the package. The package excludes environmental contaminants and has little mechanical or chemical effect on the chip, since the package components do not touch the chip surface. Al_2O_3 is the usual ceramic material; however, AlN would be a good choice where higher power dissipation is required.

Figure 2[2] shows a typical plastic package. The chip is attached to the paddle of the lead frame. The frame, made of etched or stamped thin metal (usually Fe-Ni or Cu alloys), serves as a skeleton around which the package is assembled, and it provides external leads in the completed package. Interconnections are fine gold

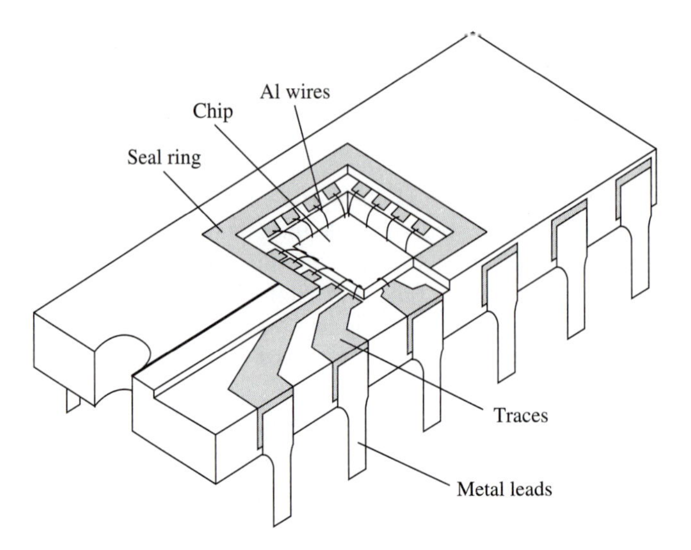

FIGURE 1
A typical hermetic package. A silicon chip is placed in the cavity of a ceramic-based package and wedge-bonded to make electrical connections to the terminals on the package. Metallized traces, usually of W, form electrical paths on the package. Lid sealing makes the cavity hermetic. (*After Prokop and Williams, Ref. 1.*)

wire. Encapsulation is carried out by the transfer molding method using epoxy resin. The resin covers the chip and forms the package's outer shape at the same time. External leads are formed into the final shape after molding. The package cannot thoroughly decouple the chip from the environment, so the mechanical and chemical effects from the plastic touching the chip require attention.

10.2.1 Through-Hole and Surface-Mount Packages

Next-level packaging, particularly at the PWB level, classifies single-chip packages into through-hole (TH) and surface-mount (SM) types. The TH types include the dual-in-line package (DIP) and the pin-grid-array (PGA) package. Both are available in hermetic-ceramic and plastic types. TH packages provide easier placement for soldering and stronger solder joints at the PWB level. However, they sacrifice PWB design flexibility, restrict mounting density because of the fine-pitch drilling process, and add drilling cost.

The prevalence of SM packages is remarkable. Over 50% of the packages used today are SM types, and this type has been accepted for ULSI packaging. Although SM packages include both hermetic-ceramic and plastic types, the lower-cost plastic package is preferred because package cost increases approximately in proportion to the increase in pin count. SM packages enhance PWB design flexibility because through-holes are unnecessary. They allow finer-pitch soldering than TH packages

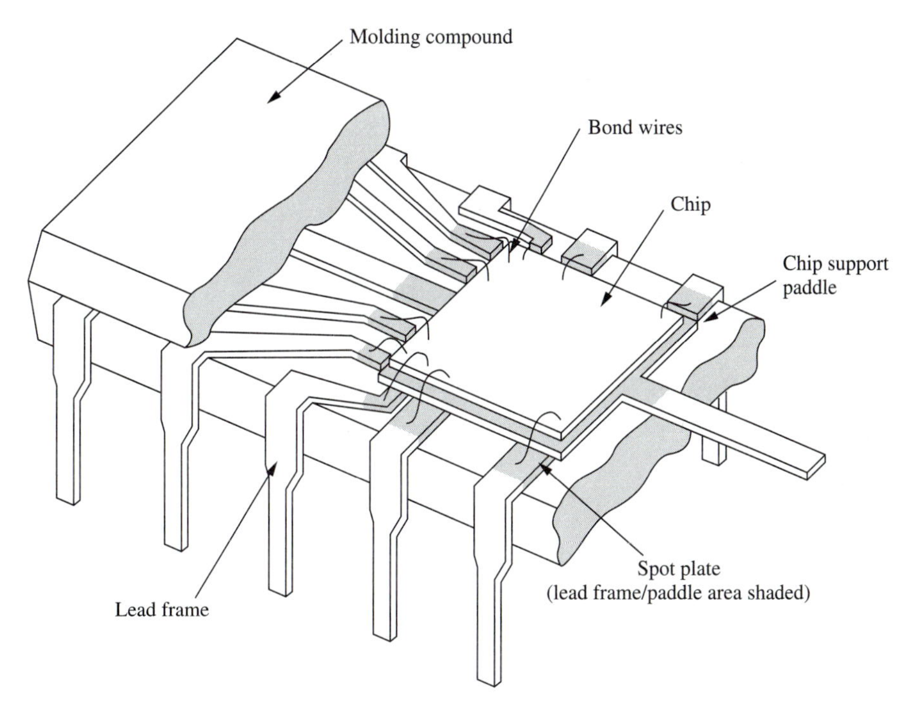

FIGURE 2
A typical plastic package. The package is a composite structure consisting of a silicon chip, a metal lead frame, and a plastic molding compound. The chip support paddle and inner tips of the lead frame are Ag-plated. External leads are solder-plated after molding. (*After Howell, Ref. 2.*)

do, because the pitch is determined by PWB (Cu foil) etching process capability, not by the drilling requirements. Generally, SM packages provide better area efficiency. Finer-pitch SM packages, especially higher-pin-count packages for ULSI devices, challenge both single-chip packaging and surface-mount techniques. The key issues are maintaining lead integrity (such as coplanarity and skew), uniform solder printing, precise mounting, and controlled reflow soldering. SM package leads come in two different geometric forms—*J-leads,* such as those shown in Fig. 6, and *gull-wing leads,* such as those shown in Fig. 25. The small-outline J-lead (SOJ) package and the quad flat J-lead (QFJ) package are examples of the former, and the small-outline package (SOP) and the quad flat package (QFP) are examples of the latter.

In memory ULSI applications, where comparatively smaller pin counts are required, the J-lead package is a frequent choice. The soldering portions of J-leads are located beneath the plastic body, so they do not need extra space for soldering as gull-wing leads do, thus decreasing the overall area the package occupies. However, in logic and microprocessor ULSI applications, where higher pin counts are needed, gull-wing leads predominate. Gull-wing leads are easier to form than J-leads, and the better geometric integrity of the leads that results is particularly suitable for higher-pin-count and finer-pitch areas.

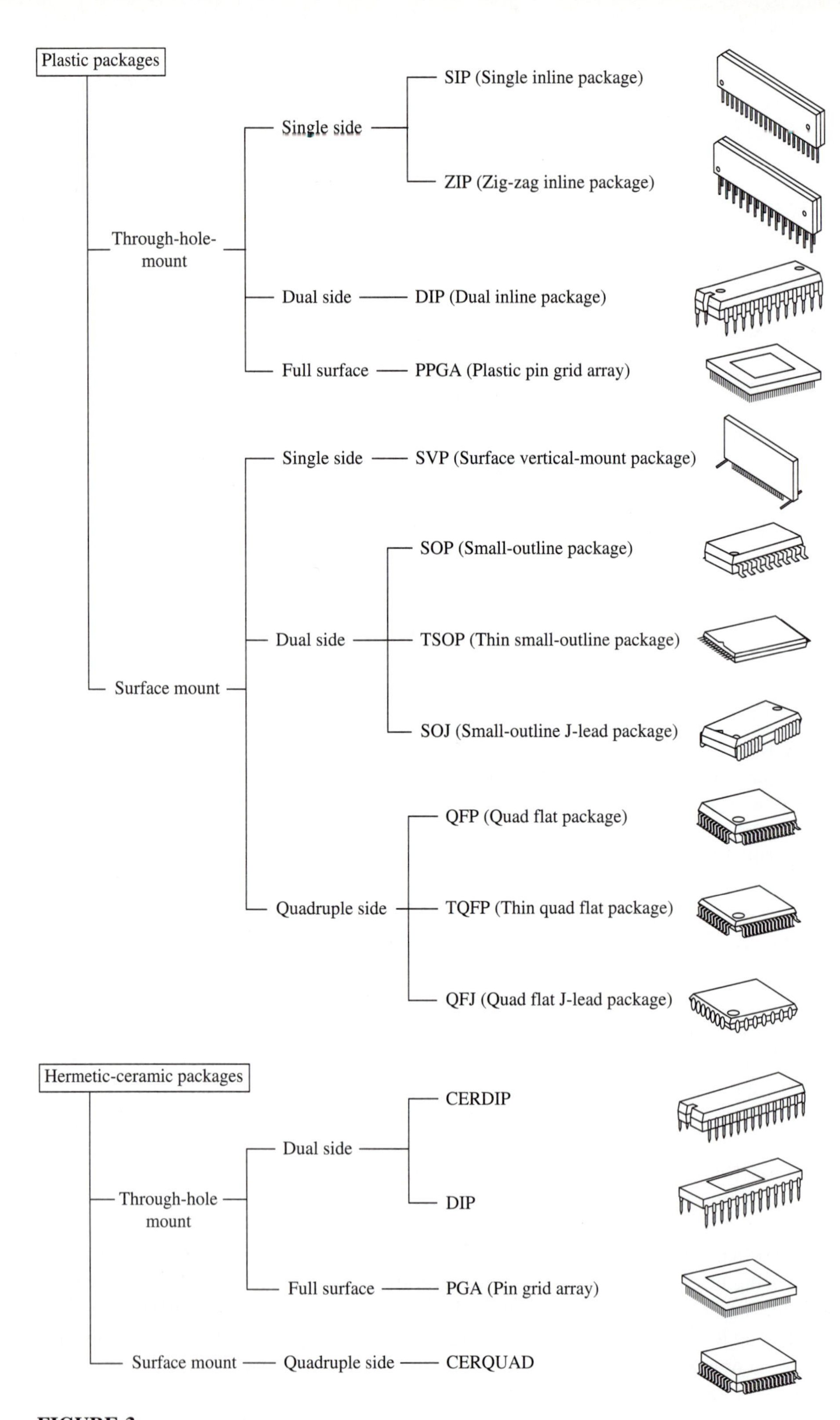

FIGURE 3
A variety of package designs. Not all ceramic package varieties are shown. (*Courtesy of Mitsubishi Electric Corp.*)

Many package variations exist in the industry. Figure 3 illustrates some of these packages. (List of packages is based on standards and registrations of the Joint Electron Device Engineering Council (JEDEC).)

10.2.2 ULSI Package Trends

This section describes the recent trends for ULSI device packages. We shall consider device trends and their effects on packaging, newer packages for specific devices, and packaging level integrations.

Device trends and packaging

Memory chips have quadrupled in complexity, in terms of bits per chip, every three years. Today, 16-Mbit DRAM is in industrial use, and the subsequent generations, 64-Mbit DRAM and 256-Mbit DRAM, are at the research and development stage. The increase in complexity has accompanied a significant increase in chip size. Figure 4[3] shows the relationships among chip size, cell size, and DRAM density for several design generations. Each new generation increases the chip size approximately 1.5 times over that of the previous generation. Nevertheless, the trend for downsizing of electronic components requires that new chips be placed in the same

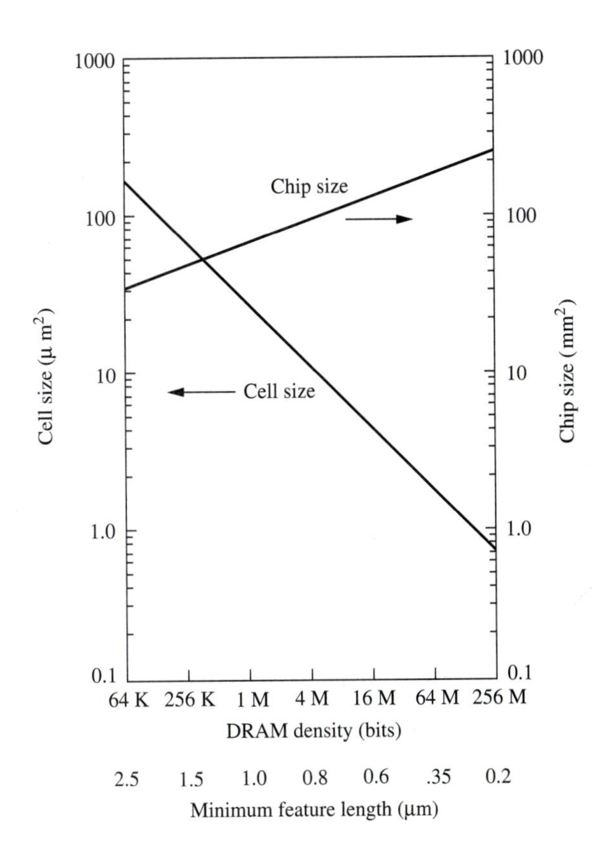

FIGURE 4
Relationship between chip size and DRAM generation. Bit density quadruples as the generation changes, every three years, because of the miniaturization of chip design features. (*After Prince, Ref. 3.*)

size, or smaller, package as the previous generation. Memory packaging has so far succeeded in meeting the requirement by increasing the ratio of the chip-to-package area. The ratio is 70% in 16-Mbit DRAM.

Meanwhile, logic and microprocessor chips continuously increase in operating speeds, usually resulting in increased power consumption and higher I/O numbers in accord with the growth of circuit density. Figure 5[4] shows this trend in microprocessor devices, exploiting the SPECmark value that features operating speed efficiency. In CMOS logic devices, the relationship between operating speed and power consumption is expressed by[5]

$$P = \frac{(C \cdot V^2 \cdot \upsilon \cdot S_{\text{eff}})}{2} \tag{10.1}$$

where P is the power consumption, C is the load capacitance, V is the source voltage, υ is the switching cycle number, and S_{eff}, typically 0.4 through 0.6, is the circuit integration parameter.

Higher-speed operation means an increase in the υ value. Today, some state-of-the-art microprocessors consume as much as 30 watts. Consequently, package design, structure, and process techniques are affected by these increasing power requirements.

Rent's rule describes experimentally the relationship between I/O number and gate count in logic devices:[6]

$$\text{Number I/O} = \alpha(\text{Gate Count})^{\beta} \tag{10.2}$$

where α and β are constants determined by the device design. The rule has applied for logic devices that have partitioned roles in the system in which they are used.

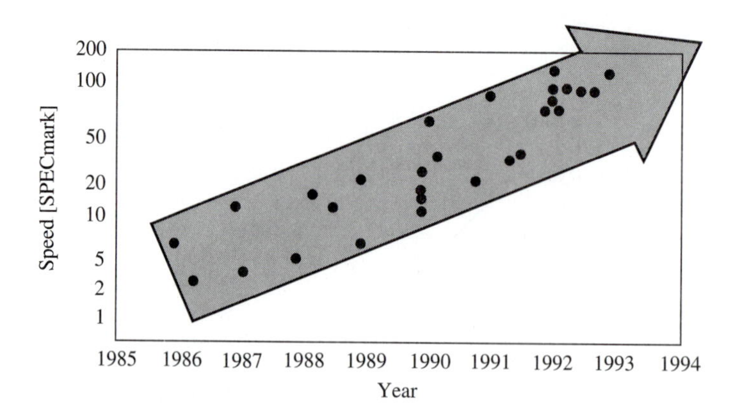

FIGURE 5
Trend of microprocessor devices. Each dot shows when a device of a given speed appeared in the industry. SPECmark is a typical benchmark value that expresses the operation speed of a microprocessor executing a standard program consisting of several calculations. (*After Nikkei Microdevices, Ref. 4.*)

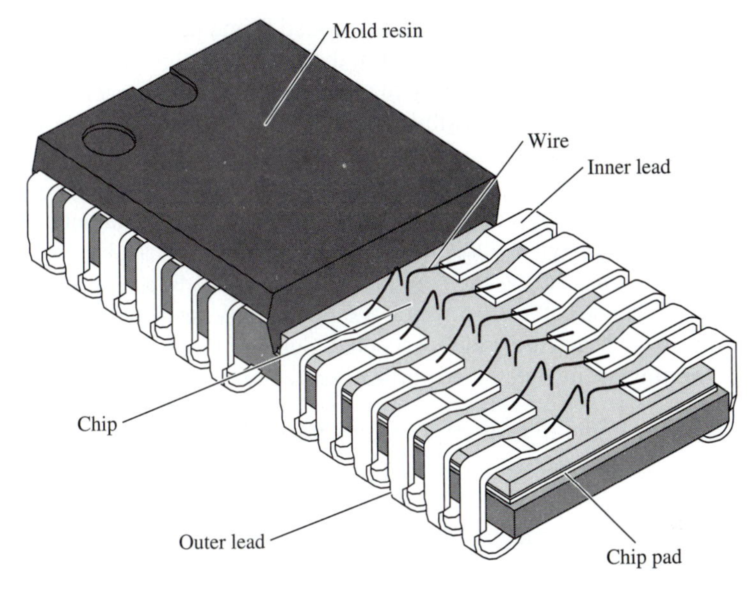

FIGURE 6
A typical LOC package structure. Tips of the lead frame extend above the chip center where the bonds are made. (*After Ueda, Ref. 9.*)

Gate counts, as large as 400,000 gates today and one million gates in the near future, require higher package pin counts in the range of 500 to 700 pins. These increases in package pin counts are caused not only by Rent's rule, but also by the increase in the number of power/ground pins. More power/ground lines are necessary for larger power consumptions and high-speed operations without electromagnetic noise. In leading-edge microprocessors, the bus-line-width has grown to 16 bits, 32 bits, and 64 bits. This growth also dictates the increase in package pin counts. To meet these requirements, leading-edge packages for ULSI devices that have more than 1000 pins are in the research stage.

Packages for memory ULSI devices

Memory devices usually must be low cost, since many of them are used in electronic appliances. The need for low-cost memory devices has been driving the trend for cheaper plastic packages. A variety of SM plastic packages have been developed and are in industrial use, such as SOJ, SOP, and thin SOP (TSOP). Except for TSOP, these have bodies typically 2 mm thick. TSOP packages have 1-mm-thick plastic bodies suitable for compact appliances. As discussed earlier, the chip occupancy continues to grow, and the stringent requirements this imposes have led to considerable changes in package structures. The lead-on-chip (LOC) structure,[7] in which wire interconnections within the package are made above the chip-circuitry surface, is notable.[8] In conventional packages for older-generation devices, such as that shown in Fig. 2, the interconnections were made only in the periphery and outside the chip area, exploiting an additional area for the interconnections. Figure 6[9] shows a typical LOC package. Here the tips of the lead frame extend over the chip surface, and Au wires are stitch-bonded to the lead frame tips to connect them with the chip bonding pads, which are located in the interior of the chip area. The LOC structure increases the chip occupancy to more than 70% of the package area. The structure provides chip design flexibility because it allows the pads to be located on the chip in almost any position. It also allows the placement of substitutes for additional inner lead portions (called bus-bars) that work as alternatives to power and ground Al metallizations on the chip circuitry. The bus-bars are numerous in some high-speed memory designs, since they can enhance the electrical performance of the devices without increasing the chip size. In current ULSI memories, LOC and chip-on-lead (COL) structures, in which the electrically insulated chip resides on the tips of the inner leads, are widely accepted. Some of these packages are vertical SM types, forming a category called surface vertical-mount packages (SVPs).[10]

Packages for logic and microprocessor ULSI devices

Figure 7 shows packages of choice for logic and microprocessor ULSI devices in terms of I/O pin count capability and corresponding external terminal spacing choices. PGA, a typical through-hole, high-pin count package, is widely used up to approximately 500 pins, because the package is easy to solder to the PWB and generally achieves a well-balanced performance among various requirements. Usual pin pitches are 2.54 mm (100 mil) and 1.78 mm (70 mil) for through-hole use and 1.27 mm (50 mil) for an SM package to prevent increasing the package size in higher pin count applications. The quad flat package (QFP) is the prevalent type used in logic and microprocessor packaging that requires up to approximately 300 pins. QFPs that vary in pin count, geometric dimensions, and package internal structure are now in industrial use, and their external lead pitches are shrinking continuously. The 1.27 mm pitch is now rather obsolete, because pitches are decreasing to 0.8 mm, 0.65 mm, and 0.5 mm. Today, 0.4 mm and 0.3 mm are the finest pitches in use. QFPs that can accomodate thinner bodies suitable for smaller electronic components are being standardized as low profile QFP (LQFP) and thin QFP (TQFP), segregated by their overall heights. The quad tape carrier package (QTP) enables more than 500 pins and introduces finer pitches of, typically, 0.25 mm or 0.2 mm. QTP is one of the choices to meet the highest pin count requirements. However, the pitches challenge both single-chip and next-level packagings stringently.

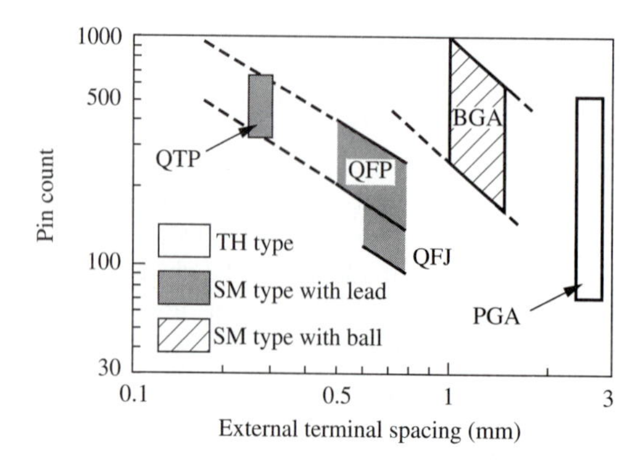

FIGURE 7
Packages of choice for logic and microprocessor devices in terms of I/O pin count and external terminal spacing.

Newer packages, such as the ball-grid array (BGA) package,[11] which has external terminals on one surface of the package and not on the sides, have been proposed as candidates to solve the problem. These packages provide larger external terminal pitches of 1.27 mm or 1.0 mm, for example, since they are located in a matrix. We discuss the details in Section 10.7.

Packaging-level integrations

As circuit integration proceeds on the chip, it also proceeds in the packaging through interactions among several levels of packaging. Generally, packaging exclusive of the final system construction is classified into three levels, as shown in Fig. 8. Final system requirements determine a specific selection of the packaging method, or how to combine the levels.[12, 13, 14] Types 1 through 4 show the major methods that have been used in the industry. Type 1 is the most common choice. In type 1, the chip is first packaged as a single chip and then packaged at the third level, typically at the PWB level. Types 2 and 3, usually called multi-chip-module (MCM) technologies, are used in high-performance systems, typically in mainframes. In type 2, the chip is single-chip-packaged as in type 1, and the packaged chips are then packaged at the second level onto a smaller substrate; this forms a functionally larger and geometrically smaller unit and utilizes the finer multilayer wiring of the substrate. The substrate is attached to a larger mother board in the third-level packaging. Type 3 is similar to type 2 in that it uses a smaller substrate as an intermediary stage. However, bare chips are attached directly to the substrate, usually providing a superior electrical and geometrical performance but with some drawbacks, such as more difficult testability, increased cost, and lower yield. Type 4 is the simplest way of packaging, where bare chips are attached directly to the system board.

ULSI devices with complex functions will affect packaging significantly; that is, type 1 has penetrated into higher-performance applications, such as the engineering work station trend, and type 4, which originated in consumer products, is now seen in some personal computers. The single-chip packaging technologies discussed in the following sections are in transition, but they continue to be fundamental techniques.

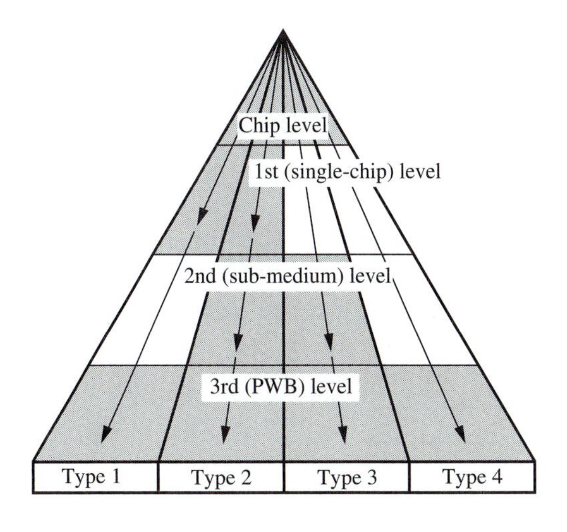

FIGURE 8
Circuit-level integrations for final system construction.

10.3
ULSI ASSEMBLY TECHNOLOGIES

This section covers the basic assembly operations in use today for ULSI devices. We consider the steps from wafer preparation through wire bonding, including some design considerations that are closely related to assembly process technologies. Figure 9 shows a generic assembly flowchart that applies to both plastic and ceramic packages.

10.3.1 Wafer Preparation

Wafer preparation starts with the back-grinding process at the end of the wafer fabrication process. Large-diameter, and hence thick, wafers need to be thinned before assembly begins. For example, 6-inch-diameter wafers are thinned from 650 μm to approximately 400 μm thickness. Thicker wafers place more demands on the dicing process. Moreover, they place other demands on package design and assembly processes,[15, 16] such as chip interconnection, encapsulation, and final packaged chip reliability. In thinner SM ULSI packages that need higher area or volume densities, thinner wafers are especially beneficial to reduce the thermal stresses resulting from mismatching of the thermal coefficient of expansion (TCE) between the larger ULSI chip and the plastic molding materials described later. Back-grinding is done by fully

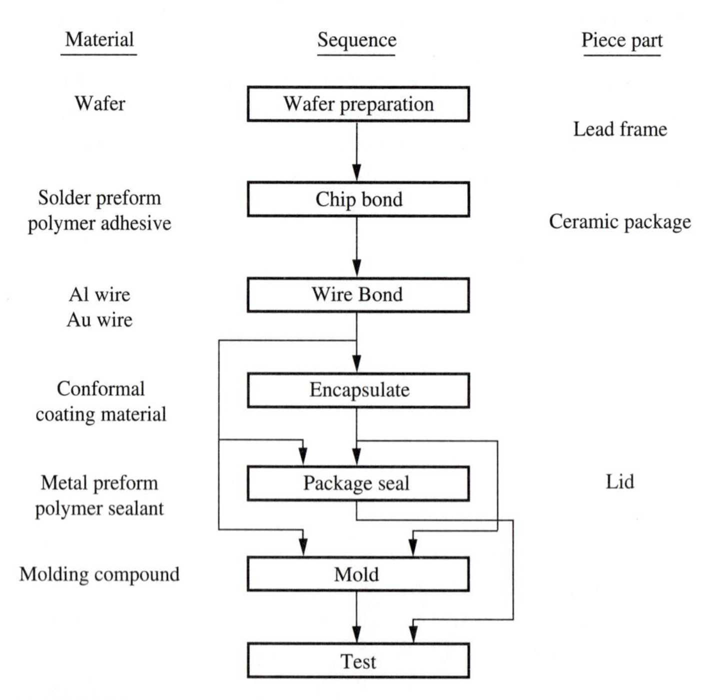

FIGURE 9
Generic assembly sequence for plastic and ceramic packages.

automated machines. This operation must be done carefully. Poor operation degrades the mechanical strength of the chips and may lead to chip-crackings, which begin in the back after the chips are packaged. If required, back-side metallization is done after back-grinding. Au, multilayer metallization of Au-Ni-Ag, and Ti-Ni-Au, in order from the silicon side, are the usual choices for metallization. Proper combinations of the back-side metallization materials and chip-bonding operations (described later) provide superior adhesion strength and electrical connections.

A diamond-blade dicing saw separates wafers into individual chips. Wafers are typically adhesive-mounted to a tape that has been preassembled to a frame before dicing. An advanced diamond blade, typically 25 μm thick rotating at a speed of 20,000 rpm, cuts the wafer from 90 to 100% saw-through, allowing dicing streets as narrow as 60 μm with dissimilar materials such as passivation, metallization, and gold usually serving as test patterns. Established methods such as scribe-and-break or 50% saw-through dicing-and-break are not usually acceptable for ULSI devices and have been replaced by 90 through 100% saw-through dicing, which leads to better-quality cuts without damage to the chip surface. Wafer-dicing machines are completely automatic and contain features such as alignment systems, an integral wafer-cleaning station, a drying oven, and quality monitors. The sawed-apart wafers, still mounted in the tape-frame fixtures, are then loaded into the chip-bonding operation, where an automatic bonder sorts good chips either by recognizing the bad-chip marks or by reading the positional mapping data from the wafer test operations.

10.3.2 Chip Interconnection

Chip interconnection consists of two steps. In the first step, chip bonding, the back of the chip is mechanically attached to an appropriate medium, such as a ceramic substrate, a multilayer ceramic-package piece part, or a paddle of a metal lead frame. The chip bonding operation is necessary as preparation for the next interconnect operation, wire bonding. Chip bonding also provides a thermal path from the chip to the medium and sometimes allows electrical connections to be made to the back of the chip. In the second step, the bond pads on the circuit side of the chip are electrically interconnected to the package by wire bonding. Fine metal wires (Au or Al) are used for wire bonding for SSI up to USLI packaging. Alternatives also used today for ULSI devices are tape-automated bonding (TAB) and flip-chip solder bonding, described in Section 10.7.

10.3.3 Chip Bonding

Chip bonding is the first operation in which the chip meets the package. The bonding technology is determined by selecting among a variety of trade-offs, such as thermal stresses caused by TCE mismatch between the chip and the package materials, power dissipation, electrical capabilities, resulting reliabilities, and cost. This selection has become more difficult and sophisticated in ULSI, because the growing chip size and more stringent demands on the performance of the packaged chip allow little room to make engineering trade-offs. Some features that were negligible in SSI,

MSI, and even in VLSI have led to degradations of performance and reliability in ULSI. Chip-bonding technology has become more difficult; hence, before describing the chip-bonding process itself, we discuss the major structural problems of ULSI packaging.

The first problem that increased ULSI chip size presents is the TCE mismatch between the chip and the package medium to which the chip is attached. The TCE mismatch causes various thermal stresses in several operations that the chip undergoes. The first stress occurs when the chip is cooled to room temperature after chip bonding. Then wire bonding and encapsulation temperatures create major additional stresses during assembly operations. Final test operations, including burn-in, soldering at the next packaging level, and power cycling in actual use mixed with cooling conditions, place different and integrated stresses on the devices. Sometimes these stresses cause chip cracking or chip delamination from the package medium, resulting in degraded thermal performance and reliability. In plastic packages, the stress, combined with plastic moisture absorption, might lead to package cracking rather than chip cracking; this also results in degraded reliability.

Figure 10[17] shows TCEs of materials for semiconductor packaging. Materials that have TCEs close to that of a silicon crystal are preferable in package construction. Choosing only the packaging materials that have TCEs the same as that of silicon would be ideal, reducing the thermal stresses to zero. Realistically, however, no material in industrial use allows such a choice. In ceramic packages, Al_2O_3 substrates that provide good TCE matching have been preferred for many years. The established chip-bonding process for the combination of a silicon chip and Al_2O_3 is the Au-Sn eutectic method. Although this method, called hard soldering, provides

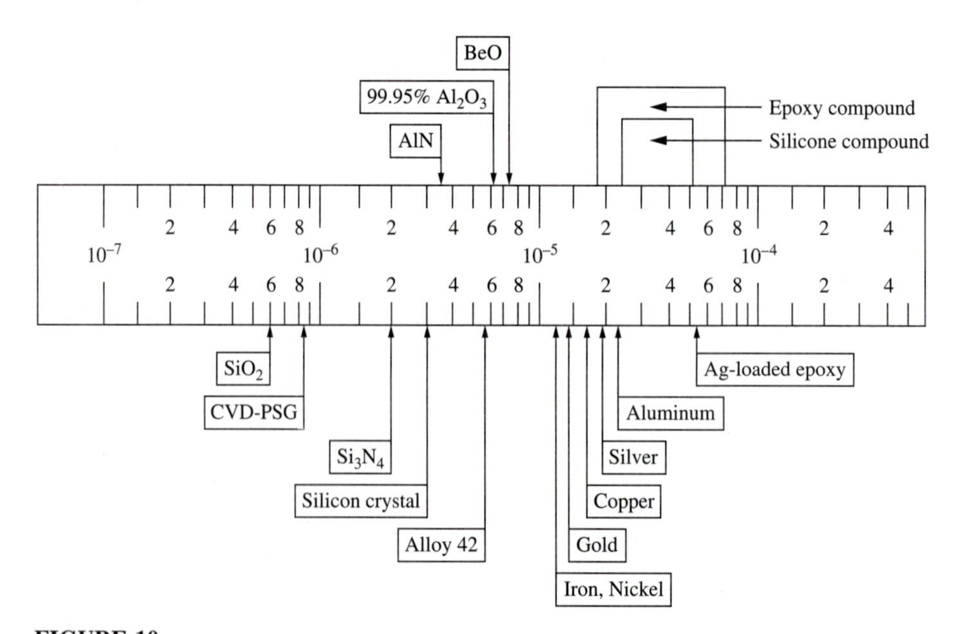

FIGURE 10
Thermal coefficients of expansion (TCE) of materials used in semiconductor devices, in °C^{-1}. (*After Kakei, Ref. 17.*)

little stress buffer effect, it has an effect even in ULSI chip bonding where the TCE mismatch is smaller and the established bonding technique is appreciated. In plastic packages, the Alloy 42 (42% Ni–58% Fe alloy) lead frame has been widely used for the same reason. Chip-bonding techniques began with Au-Si hard soldering for SSI; later, resin chip-attach materials and Pb-Sn solders (soft solders) were used to bond MSI chips up to approximately 5 mm square because they were better stress buffers. However, in ULSI packaging, where enhanced thermal and electrical capabilities are sometimes more important than the higher reliabilities that come with better TCE matching, Alloy 42 no longer can be the single candidate. Most of the heat that a chip dissipates usually flows to the package medium through the chip attachment. The medium also determines the allowable transient heat-peak value. In ULSI plastic packages, especially in logic and microprocessor applications, a significant number of packages now use Cu alloys that provide approximately ten times the thermal conductivity of Alloy 42. Cu alloys that have smaller permeability are valued highly for their electrical superiority as well. The adoption of a Cu alloy challenges the chip-bonding technique to solve the TCE mismatch problem. Today, improved, epoxy resins with Ag flakes, which have smaller elastic moduli (E), are commonly used along with improved process control techniques. Controlled chip-attach material thickness (usually as thick as 10 to 20 μm) and a smaller E value compensate TCE mismatch to some extent.

The second problem is the TCE mismatch between the chip and the molding compound in plastic packages. This mismatch does not exist in ceramic packages, because no packaging materials come into contact with the chip surface. Although the α_1 value (that is, the TCE below the glass transition temperature, T_g) of the improved molding compound has been reduced to 10 to 20 ppm/°C, the mismatch is still significant because the ULSI chips are large. The resulting thermal stresses on the chip start as compressive stresses after the molding operations, where the chip is cooled from approximately 175°C to room temperature. Of course, the stress is complicated, since the package at this stage has a composite structure consisting of a chip, a metal lead frame, and plastic molding compound. Reduction of the TCE of molding compounds, typically to 7 ppm/°C, provides better matching with the chip, but such an excessive reduction induces a serious side effect—an intolerable TCE mismatch between the molding compound and the Cu alloy lead frame. In ULSI plastic package constructions, Cu alloy lead frame materials, which usually have a TCE of 17 ppm/°C, are becoming more popular, but they require a larger TCE of the molding compound, from 10 to 15 ppm/°C. Countermeasures are required, such as increasing adhesion strengths of contacting package components, optimizing geometric dimensions, and modifying structure.

The third and usually the last TCE mismatch problem is between the completed package and the board to which the package is attached. This problem is discussed in Section 10.5. Table 1 summarizes the mechanical characteristics of packaging materials that influence TCE mismatches.

In the chip-bonding operation, prepared chips are attached to the package media, typically paddles of a lead frame or ceramic package cavities, using an automatic chip bonder. The chip bonder usually has features such as a good-chip sorting function, a pick-up and place unit, a chip-attach material supply station, and a lead frame loader and unloader.

TABLE 1
Mechanical characteristics of packaging materials

Material	Thermal coefficients of expansion ($\times 10^{-6}$/K)	Elastic modulus (GPa)	Thermal conductivity (W/m-k)
Silicon	2.6–3.6	10.2–16.3	150
SiO_2	0.6–0.9	≈ 7.1	1.4
SiN	2.8–3.2	≈ 32.6	3.0
Polyimide	20–50	3–4	0.12–0.2
Au	1.4	78	318
Al	23	70.5	229
Cu	16.7	112.7	394
Cu alloys	16.3–17.3	117–132	151–351
Ni-Fe alloys (Alloy 42)	4.14	147	15
Kovar	4.4	143	16.7
Mo	3.7–5.3	343	130
37Pb-63Sn	24.7	32.1	49
95Pb-5Sn	29.8	7.4	63
80Au-20Sn	12.3	59.2	24
Al_2O_3	6.7–7.1	262–378	19–25
AlN	4.4–4.9	326–357	70–170
Glass epoxy	16	0.26	0.36
CuW	6.0–7.0	347	210
Epoxy	12–22	13.2–17.3	0.67
Ag epoxy	33	1.96–6.86	1.26

The major choices for a chip-attach process include hard (or eutectic) solder, polymers, and Ag-filled glasses. A comparison of the chip-attach processes for hermetic-ceramic and plastic packages shows two emerging technologies for ULSI—eutectic for hermetic-ceramic packages and polymers for plastic packages. The eutectic process is essentially contamination-free, has excellent shear strength, and ensures low-moisture packages. The major disadvantages are that it is difficult to automate and it places high thermal stresses on the chip because of the high process temperature. Typical polymer chip-attach materials for ULSI in plastic packages are epoxies, with and without Ag flakes to enhance electrical and thermal conduction. Epoxies with Ag flakes are more popular. Formerly, epoxies were high in hydrolyzable ions, which could directly affect reliability. Epoxy suppliers have improved their materials, controlling the residual sodium and chloride ions to less than 10 ppm each. Epoxies are widely used today for ULSI.

Although Ag-filled glasses and polyimides, with and without Ag, are choices in some ULSI packages, the following sections describe only the eutectic and epoxy chip-attach processes, because they are the most widely used today.

Eutectic chip bonding

Figure 11 illustrates the fundamental aspects of a chip bond. Eutectic chip bonding metallurgically attaches the chip to a substrate material, typically a metal lead frame made of Alloy 42, or to a ceramic substrate, usually 90 to 99.5% Al_2O_3. Metallization is often required on the back of the chip to make it wettable by the

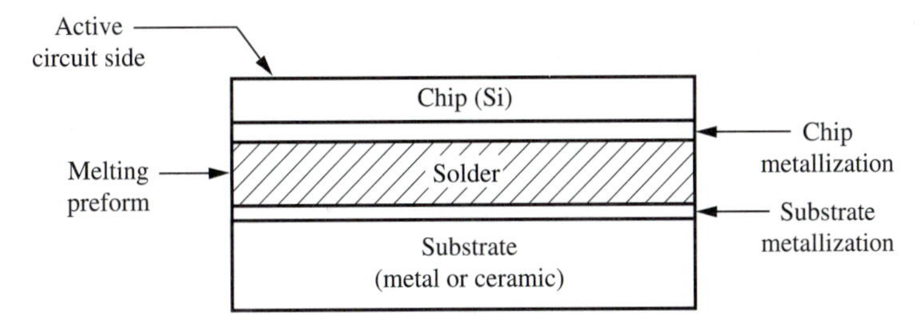

FIGURE 11
The basic structure of a silicon device chip-bonded with a metal preform.

chip-bonding preform. The preform is a thin sheet, less than 0.05 mm thick, of the appropriate solder-bonding alloy. Table 2[18] lists the compositions and melting points for preform materials. The melting point must be higher than the temperature of the next operation, usually wire bonding, otherwise remolten preform makes the operation difficult. The substrate material is usually metallized with plated Ag (lead frame) or Au (lead frames or ceramic). Solder chip bonding to ceramic packages, which are to be hermetically sealed, is usually performed with a Au or Au–2% Si preform. The TCE mismatch between silicon and the ceramic substrate is relatively small, so thermal stress caused by a TCE mismatch is not a critical issue. Still, the bonding must be done carefully to create uniform wetting; partial wetting induces undesirable stress concentrations on the bond and may lead to chip cracking. In the presence of some mechanical scrubbing and a temperature of about 370°C, the preform reacts to dissolve the silicon. The Au-Si eutectic composition is reached and then exceeded, because the bonding temperature is set to more than 400°C. As the composition of the composite structure becomes more silicon-rich, it freezes and the chip bond is completed. In other applications that require lower temperatures or more ductile chip-bonding solders, Pb-Sn-based solders are widely used.

Epoxy chip bonding

Silver-filled epoxy adhesives are major choices as polymer-based, chip-attach materials. The silver filler, usually flakes, makes the epoxy both electrically conduc-

TABLE 2
Compositions and melting points for chip-attach preforms[18]

Composition	Temperature (°C) Liquids	Solids
80% Au, 20% Sn	280	280
92.5% Pb, 2.5% Ag, 5% In	300	—
97.5% Pb, 1.5% Ag, 1% Sn	309	309
95% Pb, 5% Sn	314	310
88% Au, 12% Ge	356	356
98% Au, 2% Si	800	370
100% Au	1063	1063

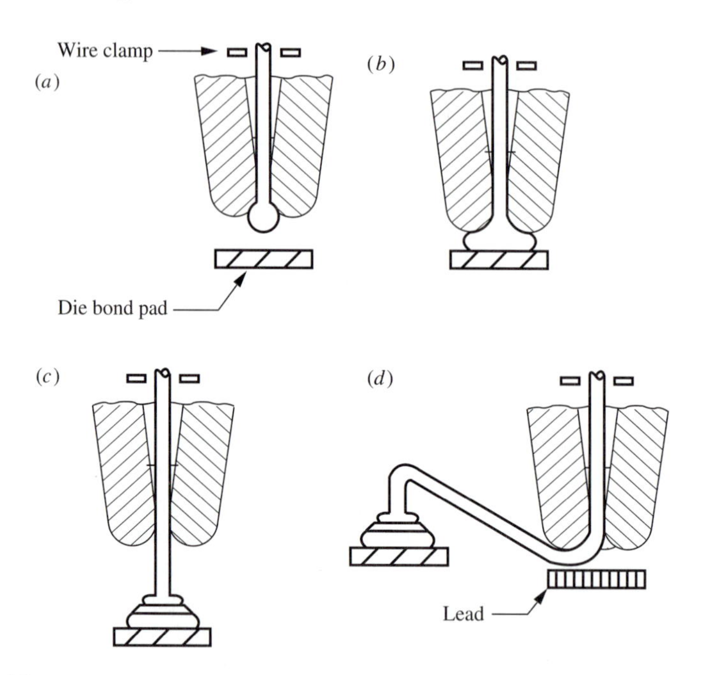

FIGURE 12
Tailless ball-and-wedge bonding cycle. (*a*) The capillary targets the chip's bond pad and is positioned above the chip, with the ball formed on the end of the wire and pressed against the face of the capillary. (*b*) The capillary descends, bringing the ball in contact with the chip. The inside cone, or radius, grips the ball in forming the bond. In a thermosonic system, ultrasonic vibration is then applied. (*c*) After the ball is bonded to the chip, the capillary rises to the loop-height position. The clamps are open and wire is free to feed out of the end of the capillary. (*d*) The lead of the devices is positioned under the capillary, which is then lowered to the lead. Wire is fed out the end of the capillary, forming a loop.

tive, to provide lower resistance between the chip and the substrate, and thermally conductive, to allow a good thermal path between the chip and the rest of the package. Silver-filled epoxies are less expensive than the high-gold-content hard solders, are more flexible in absorbing the thermal stress between the chip and the substrate, and make it easy to automate the attach processes. In ULSI manufacturing, epoxy is fed onto the substrate material through a multinozzle or a well-designed, single-nozzle dispenser to ensure the required bond line thickness is created without voids. The back of the chip often does not require metallization because epoxy provides better adhesion to the natural silicon oxides on the back, whereas the substrate is usually metallized because the chip-attach portion is plated when other necessary substrate portions (such as stitches for wire bonding) are plated. The automatic chip-attach operation proceeds at a speed of 1 to 2 sec per chip. Epoxy chip bonds do not need mechanical scrubbing, unlike eutectic chip bondings; hence, there is a lower probability of damage to the chip edges in an epoxy chip bond. Because epoxies are thermosetting polymers, they must be cured at elevated temperatures to complete the chip bond. Typical cure conditions range from 125 to 175°C and require 1 to 2 hours.

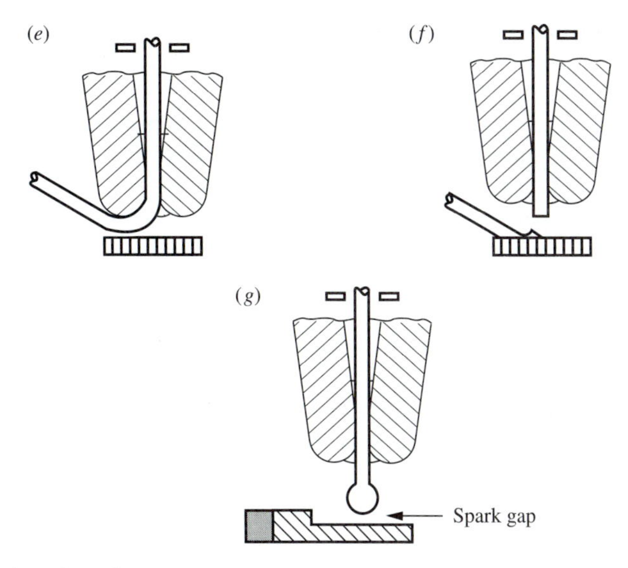

FIGURE 12 *(continued)*
(*e*) The capillary deforms the wire by pressing against the lead, producing a wedge-shaped bond that has a gradual transition into the wire. In a thermosonic machine, ultrasonic vibration is then applied. (*f*) The capillary rises off the lead, leaving the stitch bond. At a preset height, the clamps are closed while the capillary is still rising with the bonding head. This prevents the wire from feeding out of the capillary, and pulls at the bond. The wire breaks at the thinnest cross section of the bond. (*g*) A new ball is formed on the tail of the wire that extends from the end of the capillary. A hydrogen flame or an electronic spark may be used to form the ball. The cycle is completed and ready for the next ball bond. (*After Kulicke and Soffa Industries, Ref. 19.*)

10.3.4 Wire Bonding

Wire bonding is the most common method for connecting the pads on the chip to those on the package. It is performed on chips after they have been chip-bonded to the appropriate piece part. Au or Al is usually chosen as the wiring material because they bond well to pads on the chip and to the metallized portions on the package. In addition, these wires form metallurgically stable connections as long as the bonding conditions are properly controlled, and the manufacturing processes of fine (typically 25 to 30 μm in diameter) wires are mature and well-prepared. Typically, gold wire is ball-and-wedge bonded by thermosonics or thermocompression; that is, it is ball-bonded to the chip bond pad (Al) and wedge-bonded to the package substrate (Au or Ag), as shown in Figure 12.[19] One advantage of ball-and-wedge bonding is that the wedge bond can be performed anywhere on a 360° arc around the ball bond. There are two reasons for this. First, the symmetrical geometry of the capillary tip allows the form action of a symmetrical first bond from which the wire can dress in any direction. Second, the basic metallurgical connection of the ball-bond is performed through thermocompression, not ultrasonic vibration. In the ultrasonic bonding of Al wire, the wire direction needs to be the same as that of the vibrating horn (the machine head that holds the capillary tip or wedge), whereas thermocompression allows the wire to dress in any direction regardless of the horn direction. As the horn is placed in a designated direction, the versatility of ball-and-wedge bonding allows the wire to dress in any direction around the ball. Au wire ball-and-wedge bonding is widely used in ULSI packaging, especially in plastic packaging operations. The state-of-the-art, fully automatic ball-and-wedge bonders run at speeds over 6 wires per sec. Usually, ball bonds are Au-Al and wedge bonds are Au-Au or Au-Ag metallurgical diffusions performed at bonding temperatures ranging from 300 to 350°C in thermocompression bonding and from 150 to 250°C in thermosonic bonding.

The major issue in establishing high-quality, reliable wire bonding is process control of variables such as bonding temperature, loads, and ultrasonic vibration magnitudes. Visual inspections performed according to MIL-STD-883 METHOD 2010 can detect defects such as chip cracks, missing wires, and poor wire shapes that might cause short circuits. In addition, two types of process tests are used today—the wire-pull test and the ball-shear test. Wire-bond pull tests have long been used to evaluate both ball and wedge bonds. The test is done by engaging the wire with a hook near the center of the wire span and pulling to destruction. Failure modes in the test are shown in Fig. 13.

The ball-shear test (Fig. 14) has also been around a long time. It evaluates the quality of ball bonds by pushing the ball to destruction; this characterizes the Au-Al intermetallic formation, which usually cannot be characterized by the wire-pull test.

Metallurgical diffusion, in its earlier stage, follows the equations

$$X^2 = Dt \tag{10.3}$$

$$D = D_o \exp(-Q/RT) \tag{10.4}$$

where X is the diffusion thickness, D is the diffusion constant, t is the storage time, D_o is the frequency factor, Q is the activation energy, R is the gas constant, and T is the storage absolute temperature. D_o and Q are constants determined by the combination of the two materials.

Gold and aluminum form a variety of intermetallics; of these, Au-Al is formed first and gradually changes to Au-Al_4, degrading the bond strength. Newer chip

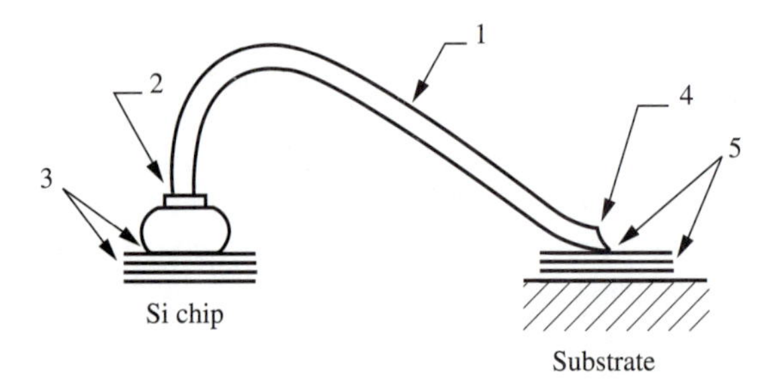

FIGURE 13
Failure modes in wire-bond pull tests. (1) Break in wire. (2) Wire break above the ball. (3) The ball-bond failure; it should be determined whether the location is in the Au-to-Al or in the Si chip buried layers. (4) Break in the wire at the wedge-bond heel. (5) Wedge-bond interface failure; it should be determined whether the location is in the Au-to-Ag or in the metallized layers.

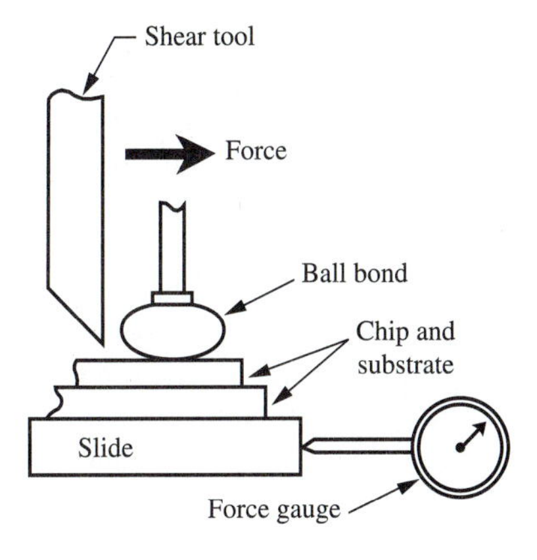

FIGURE 14
Setup of the shear test. Force is applied by the shear tool and the force needed to shear is recorded on the force gauge.

metallization materials, such as Al-Si-Cu, Al-Cu, and Cu, that are either being used in the industry or still being studied, must be accompanied by improvements in wire bonding techniques; otherwise, they invite poor interconnections, leading to degraded reliability.

Al wire wedge bonding (Fig. 15) is another major choice in ULSI wire bonding techniques. It is widely accepted in hermetic-ceramic packages.

ULSI design rules

The establishment of, and adherence to, good chip design rules is absolutely essential to achieve high yields in ULSI package assembly. Rules must be generated for the particular package type used and must be compatible with the assembly equipment. These decisions should be made early, preferably before the chip layout is started and definitely before the layout is completed. Bonding-pad design rules are especially important. In memory ULSI, pads are often placed only on the two facing sides of the chip; or they are located on the chip-center in LOC packages,

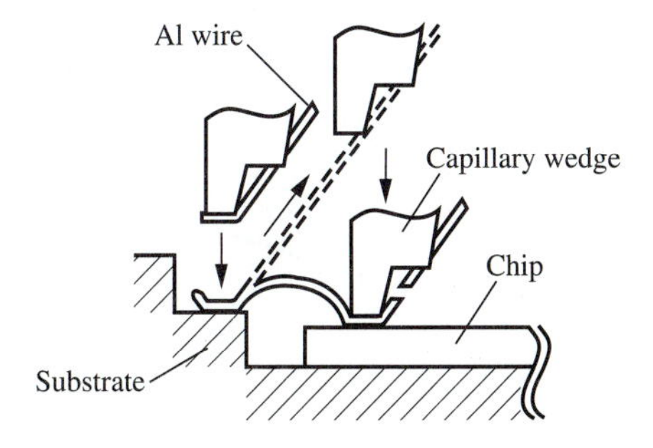

FIGURE 15
Al wire wedge bonding. A capillary wedge holding the wire descends and presses the wire against the pad on the substrate. The first bond is made by applying ultrasonic vibration. Then the capillary wedge rises to the loop-height position and moves to target the pad on chip, forming a loop. The capillary wedge is lowered to and pressed against the pad on chip. After the second bond is produced, wire clamps (not shown in the figure) pull and break the wire, and the bonding cycle is completed.

to increase the chip occupancy or the area efficiency of the PWB where the packages are placed. In logic and microprocessor ULSIs, the growing I/O count and the shrinking active device size present stringent constraints on chip design. The space required for interconnection not only represents a large fraction of the chip area, but sometimes determines the chip-size, and considerable extra space is needed around the active area for pad placement. To avoid this problem, the effective bonding-pad sizes and spacings (also called pitch) should be reduced. Today, possible spacings range from approximately 100 to 150 μm. Figure 16 shows the relationships between chip size and the allowable pin count for several spacings. Typical design rules for wire-bonding pads are illustrated in Fig. 17; these cover recommended pad sizes and pitch, clearances between pads and the edge of the chip, and clearances to internal adjacent metal conductors of other critical design features.

The chip designer must understand another essential restriction that affects package selection and bonding-pad layout. Bond pads on packages, metallized pads in ceramic packages, or lead frame tips in plastic packages need particular spacings that are determined by the capabilities of the manufacturing process. Lead frames are usually chemically etched or mechanically stamped using a progressive die to produce designated patterns from 100- to 150-μm-thick flat metal. For lead frames, the aspect ratio in the etching process or the strength of the die parts determines the spacings.[20] For ceramic packages, screen printing of the metallized-conductor is the constraint. Today, the finest spacing in industrial use is 220 μm, both for lead frames

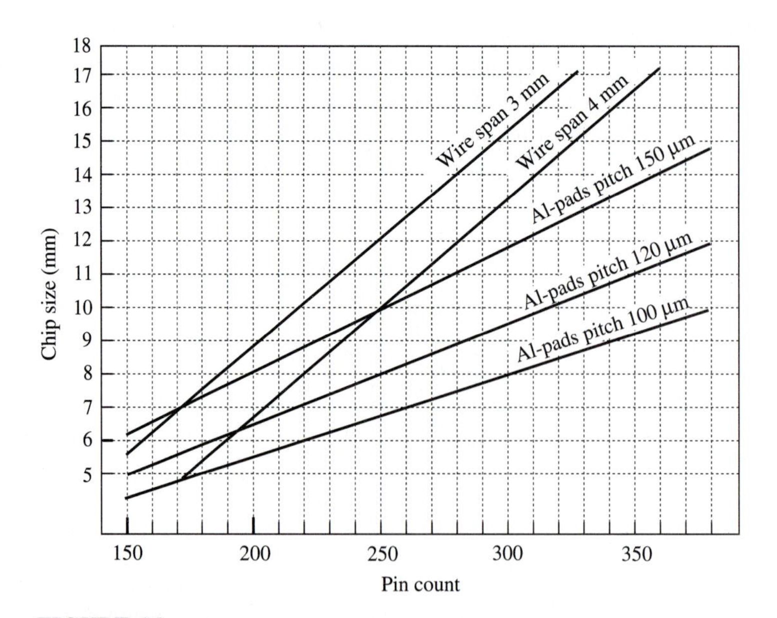

FIGURE 16
Pin count and required minimum chip size for several wire spans and Al-pad pitches. Where the lines of the two constraints, wire span and Al-pad pitch, cross, the upper line represents the bottleneck restriction.

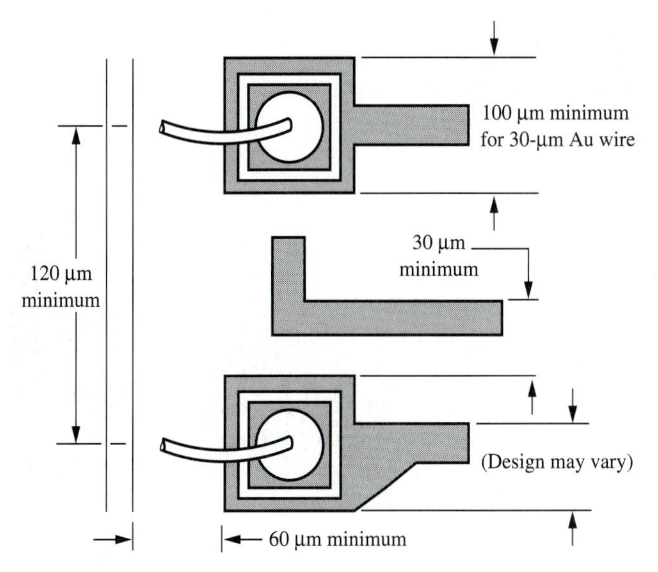

FIGURE 17
Typical bonding pad sizes and spacing. (*Courtesy of Mitsubishi Electric Corp.*)

and metallized conductors. The maximum allowable I/O number or package bond-pad number that can be placed around a chip that has specific geometric dimensions is determined by the package bond-pad spacings and the possible wire spans that connect the chip and the package. Wire span (from the pad on the chip to the substrate or to the lead frame bond pad) is usually about 1 to 4 mm, and the resulting pin count limitations are shown in Fig. 16 for several wire spans. Longer wire could lead to potential edge shorts to the chip and shorts to adjacent wires, or to the conductor on the lead frame or metallized conductors of the package, since longer wire is more likely to droop or deform. In plastic packages, longer wire can also be deformed by the resin flow during the molding operation. This wire span is often stringently restrictive in high-I/O chip design; hence, terraced packages, often combined with staggered pads on the chip, are sometimes used as a solution (Fig. 18).

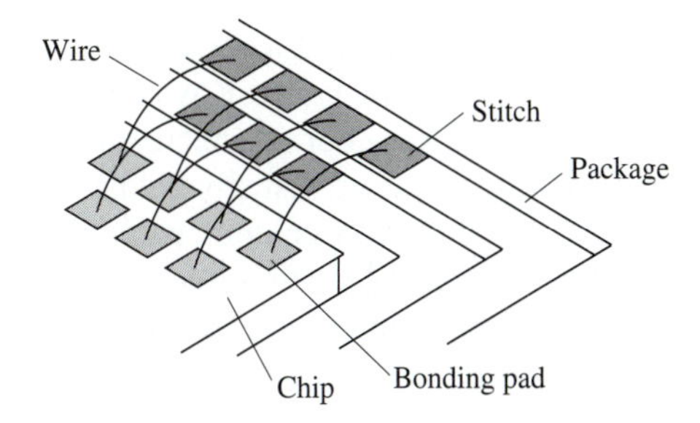

FIGURE 18
A terraced package and staggered pads on the chip—an arrangement that results in a lower-pitch bonding.

These design rules can be integrated into chip-design CAD systems, so that the chip designer can choose a proper lead frame (or a substrate); the location of all lead frame wedge-bond targets and the optimal location for the ball bond on the chip can appear as design templates in the system.

10.4 PACKAGE FABRICATION TECHNOLOGIES

Single-chip packages are usually based either on refractory-ceramic technology or molded plastics. Ceramic packages are usually used for state-of-the art devices where maximum reliability or high-power treatment is required. For more mature products where low cost is critical but a hermetic seal is still required, ceramic dip (CERDIP) technology is employed. This technology uses a combination of lead frames, dry-pressed refractory ceramic parts, and glass sealing. Plastic packaging usually is used for more mature products where cost is important and a hermetic seal is not required. However, rapid and continuous improvements in plastic technology, such as highly reliable plastic resins, proper molding process controls, and progress in relevant techniques, have been broadening its applications to leading-edge ULSI devices. This section discusses processing technologies that generally apply to a variety of package types.

10.4.1 Ceramic-Package Technology

Ceramic-package fabrication consists of the operations that are done before the chip assembly operations. Figure 19[21] illustrates multilayer ceramic technology. A dispersion, or slurry, of ceramic powder and liquid vehicle (solvent and plasticized resin binder) is first prepared and then cast into thin sheets by passing a leveling blade (doctor blade) over the slurry. After they dry, the sheets are cut to size, and via holes (holes through the dielectric layers, through which interconnections are made) and cavities are mechanically punched into the sheet. Then custom wiring paths, usually a slurry of tungsten powder, are screened onto the surface and the via holes are filled with metal. Several of these sheets are press-laminated together in a precisely aligned fixture, and the entire structure is fired at 1600°C to form a monolithic sintered body. The refractory ceramic technology is a complex process requiring careful process control.

After the laminate is sintered, it is ready for the finishing operations of lead attachment and metallization plating. Nickel is plated over the tungsten in preparation for lead brazing. The lead material is either an Fe-Ni-Co alloy called Kovar, or an Fe-Ni alloy, usually Alloy 42; the brazing material is a silver-copper eutectic alloy. All exposed metal surfaces are electroplated or chemically plated (usually gold over nickel) for bondability and environmental protection. Multilayer ceramic packages can be made of sizes up to 100×100 mm in lateral dimensions to a tolerance of $\pm 0.5\%$, and they can consist of up to 30 layers in the most advanced processes.

Ceramic-package technology is very effective for constructing complex packages with many signal, ground, power, bonding, and sealing layers. It does have,

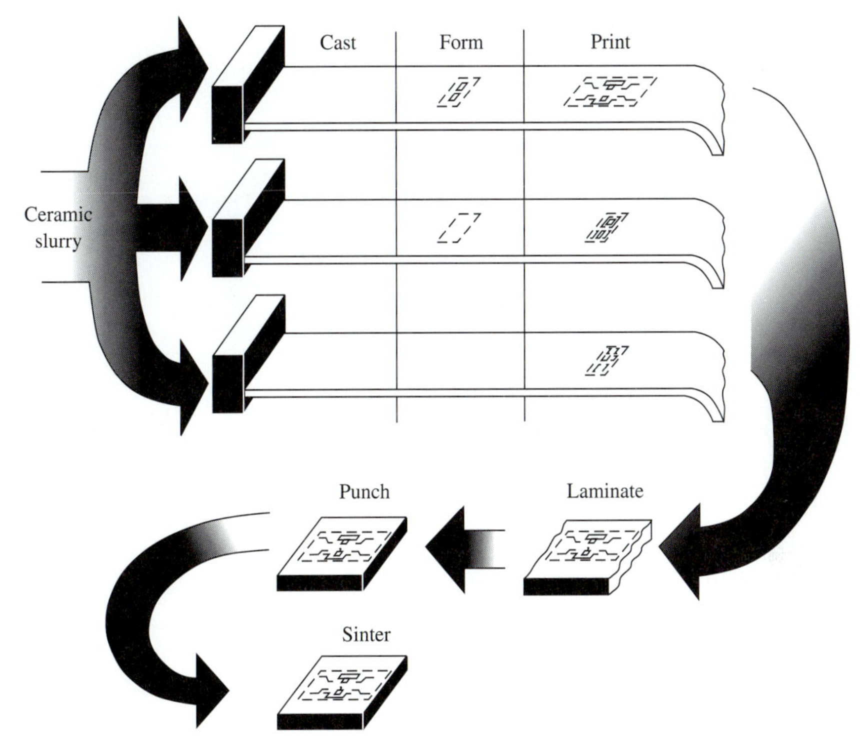

FIGURE 19
Process sequence to create a laminated refractory-ceramic product starting from a ceramic slurry. (*After Gardner and Nufer, Ref. 21.*)

however, three drawbacks: hard-to-control tolerances caused by high shrinkage during processing, high dielectric constant ($\epsilon_i/\epsilon_0 = 9.5$), and the modest conductivity of alumina. The tolerance problem makes it difficult to use edges as accurate references, and the high dielectric constant affects signal-line capacitive loading. Although they are very expensive, beryllia (BeO) or aluminum nitride (AlN) modules could be used instead of Al_2O_3 modules to produce greatly superior thermal performance and a significantly lower dielectric constant.

Ceramic packages can have brazed pins or leads and can have edge or array pinouts. The package designer has the option of locating the chip cavity either on the side with the pins or on the side away from the pins. Devices requiring good thermal-dissipation characteristics are packaged with the cavity on the leaded side. This provides a direct thermal path to the surface (facing away from the PWB), which is exposed to the cooling air flow. In this arrangement, the heat transfer can be enhanced by attaching a heat fin to the package. The disadvantage of this arrangement is increased package size, since the pins cannot be placed at the cavity area. In contrast, when thermal dissipation is comparatively small, the cavity can be located on the opposite side, providing a smaller outer package size. Figure 20 shows an example of refractory multilayer-ceramic packages for ULSI use. After chip interconnections are formed, hermeticity is completed by lidding operations, in which a metal or ceramic cap is attached with glasses or Au-Sn eutectic materials.

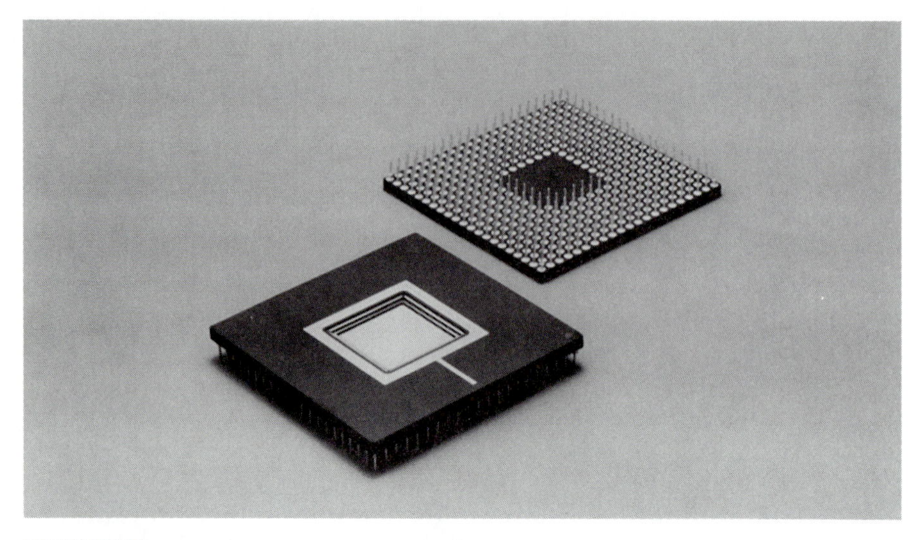

FIGURE 20
Refractory multilayer-ceramic pin-grid-array (PGA) package. (*Courtesy of Mitsubishi Electric Corp.*)

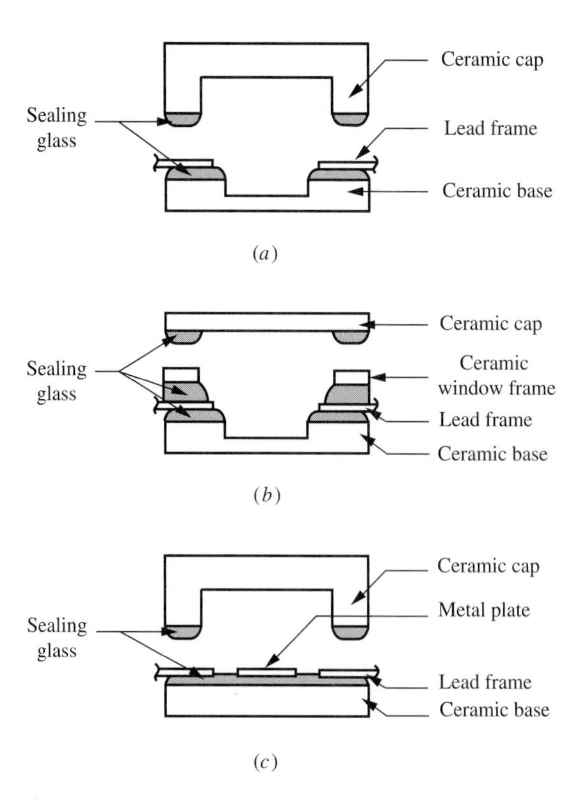

FIGURE 21
Structure of CERDIP and CERQUAD (or "quad CERPAC"). (*a*) Construction details for the standard CERDIP structure consisting of a base, a cap, a lead frame, and a glass-frit seal. (*b*) Construction details for a package consisting of a base, lead frame, window frame subassembly, and a ceramic cap using polymer or soldering sealing. (*c*) Construction details for a CERQUAD consisting of a base, a planar lead frame with an integral chip pad, a domed ceramic cap, and a glass-frit seal. This arrangement provides a good TCE match to the chip and a simpler means of providing an electrical connection from the chip pad to an external I/O lead. (*Courtesy of Kyocera International, Inc.*)

10.4.2 Glass-Sealed Refractory Technology

Figure 21 illustrates a low-cost ceramic technology. This technology relies on glass-sealing a lead frame (after chip interconnections are formed) between two pressed ceramic units using low-temperature glass. The glasses used for glass-sealing are $PbO\text{-}ZnO\text{-}B_2O_3$ types. Both crystallizing and noncrystallizing glasses are used. Sealing is usually performed above 400°C in an oxidizing ambient to avoid deoxidizing the metallic components of the glass and, consequently, degrading the electrical insulations. Since Au-Al intermetallic growth would be severe at these temperatures, Al bonding wire is used. The lead frames, typically Alloy 42, usually have a strip of Al deposited or clad on their lead tips so that an all-aluminum bonding system is possible. The hermetically glass-sealed package has the potential of being automated, and it competes with plastic technology for low-cost packaging. But in ULSI packaging, where delicate submicrometer chips are concerned, this glass-sealing technology must be used carefully. The high sealing temperature, above 400°C, would cause additional thermal diffusions at the transistor junctions, so that the electrical characteristics of the chips might shift slightly.

Both TH types (CERDIPs) and SM types (CERQUADs) are available. The laminated ceramic technology described in Section 10.4.1 can provide DIP and quad packages, but the terms CERDIP and CERQUAD specifically apply to glass-sealed refractory packaging.

10.4.3 Plastic Molding Technology

Plastic encapsulation involves a variety of techniques. In glob-top-coating, for example, the post-bonding chip is coated with liquid plastic resin and the plastics are cured for cross linking.

In ULSI plastic packaging, a premolding technique is sometimes used. This technology is the plastic equivalent of the refractory ceramic-cavity package. First the package is molded together with a lead frame, forming a plastic body and a cavity equivalent; then the chip and interconnects are added. The premold package has considerable future potential, but in ULSI packaging, postmold technology is still dominant. Only the postmold technology is discussed here. The postmolding technology is a transfer molding method using thermosetting (cross-linking) epoxy resins to mold around the lead frame-chip assembly, forming the package body. This process is relatively harsh, because the chip and its wire bonds are exposed to viscous molding material. However, improvements in molding materials, in molding process control, and in overall integrated automation of the assembly process, especially the automation of the chip-bonding, wire-bonding, and molding steps, have established the reliability to meet ULSI device objectives at a lower cost.

Molding material

The polymers most commonly used for IC packaging are epoxies. The original epoxy resin used to mold ICs was made by condensing epichlorohydrin with bisphenol-A to produce a material called Epoxy-A. An excess of epichlorohydrin was used to leave epoxy groups on each end of the low-molecular-weight polymer. Today, NOVOLAC epoxies are generally preferred; their higher functionality makes them heat resistant because each repeating group contains an epoxy group. These resins, called Epoxy-B, are made by reacting epichlorohydrin with NOVOLAC phenolic resin and a base. The synthesis of the resins produces NaCl as a by-product. Both Na and Cl ions reduce device reliability, so these by-products must be carefully washed from the resins before they are mixed into molding compounds. For ULSI plastic packages, lower-molecular-weight epoxy resins like biphenyl are used in addition to NOVOLAC epoxies, because they reduce moisture absorption from the ambient. Molding compounds are mixtures of epoxy resin, fillers, small amounts of pigments, mold-release agents, antioxidants, plasticizers, and flame retarders, usually formed into tablet shape. The selections and the ratios of these materials have significantly changed over the past few years[22, 23] to meet the challenges of ULSI devices. These changes greatly affect the rheological, chemical, and thermophysical properties of the molding compounds, and they improve the moldability, the in-process yield, and the reliability of packaged chips.

Thermal coefficient of expansion (TCE) is the most important feature of the molding compound. Fillers, usually amorphous or crystalline SiO_2, added to the resin decrease the resultant TCE of the compound. In ULSI packaging, higher filler loading, close to 80% by weight, reduces α_1 (TCE below T_g) of the compound to as low as 10 to 17 ppm/°C, providing better TCE matching to the lead frame (α = 17ppm/°C in Cu alloys and 5 ppm/°C in Alloy 42) and to the chip (α = 4 ppm/°C). Highly loaded fillers improve the mechanical strength and the thermal conductivity of the resin; however, they also increase the elastic modulus (E) and cause poor moldability as side effects. Because a larger E results in a larger stress in the package, the TCE is reduced by modifying the epoxy resin in two ways—adding a flexible material and changing the resin chemistry. Sometimes a hardener is added. Silicone rubber mixed with epoxy resin forms a "sea-island" structure; this has been widely used, as well as new, lower-molecular-weight and more flexible chemistry, such as biphenyl. Reducing the moisture-absorption ratio of the molding compound is essential in SM packages; this subject is detailed in Section 10.5.

ULSI memory, logic, and microprocessor devices challenge designers to formulate newer molding compounds that reduce TCE and E, reduce moisture absorption, improve adhesion strength to the chip and the lead frame, improve moldability, and offer various physical and chemical features from the increasing variations in plastic packages.

Molding process

Thermoset molding materials, that is, epoxy compounds, are usually transfer-molded in large multicavity molds. After it enters the pot, the molding compound is usually preheated, melts under pressure and heat, and flows to fill the mold cavities containing lead frame strips with their attached chips. The lead frame often has long, fragile fingers, and the chip is interconnected to these thin leads with 25- to 30-μm diameter Au wire. To avoid damaging this fragile structure, the viscosity and velocity of the molding compound must fall within certain ranges. Commercial molding compounds are designed to meet these requirements when molded at approximately

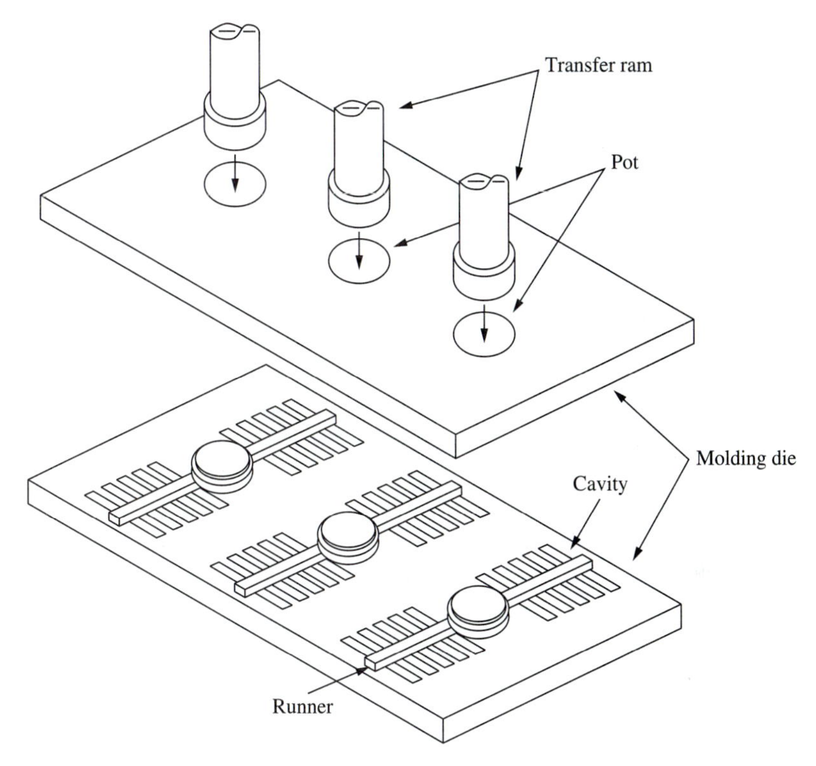

FIGURE 22
Schematic of a multipot transfer-mold system featuring plural ram, pot, and runner sets. Small mold-compound tablets, each large enough to fill a few cavities, are used in this arrangement; these provide better process control compared to a single-pot system.

175°C and at pressures of about 6 MPa, while the upper and lower molding dies are pressed at pressures of about 18 MPa to avoid resin flashes. To control the velocity of the molten molding compound, each device cavity has a gate to slow the material flow. The gate also influences the viscosity of the fluid; the mechanics of this fluid are relatively complex because the materials are non-Newtonian. In addition, partial cross-linking can occur during the molding process, which affects the material viscosity.

For molding ULSI devices, which are usually more difficult because of the larger package sizes and longer Au wire lengths, the multipot approach (see Fig. 22) is preferred because it provides more stable process control and operates at cycles ranging from 1 to 5 minutes. Products usually have insufficient cross-links after the molding operation and require post-cure storage at approximately 175°C for several hours.

10.5 PACKAGE DESIGN CONSIDERATIONS

In this section, we discuss some basic package design features. The package designer must simultaneously realize the electrical, thermal, and mechanical performances required in one specific package, make engineering trade-offs, and consider the manufacturing process capabilities described in Sections 10.3 and 10.4.

10.5.1 Electrical Design Considerations

As the speed of ULSI devices increases and their noise margins decrease, electrical design for packages must be carefully considered. A poorly designed package may spoil all other efforts, such as enhancing drivability of MOS transistors, increasing switching speed by reducing gate and line capacitances, lowering voltage interface at the chip level, and processing pipeline or parallel signals at the system level. Packages with a design geometry much larger than that of silicon can significantly affect the electrical performance of the packaged chips. Several electrical performance criteria are important: minimum signal delay, signal-reflection control, and noise reduction, including simultaneous switching noise and cross talk. These criteria, often discussed in PWB design, also apply to the package. They are mutually dependent and require trade-offs, since they are related through simple package-geometry variables.

Signal delay

Signal delay time, t_d, is given by

$$t_d = \frac{l}{v} = \frac{l}{c/\sqrt{\epsilon_r}} \tag{10.5}$$

where l (m) is the signal line length, v (m/sec) is the velocity of signal, c (m/sec) is the velocity of light, and ϵ_r is the dielectric constant of the surrounding material.

High-speed operation requires smaller t_d. The ratio of t_d to the cycle time usually dominates the system performance. In package construction, a short signal line (bonding wire length plus lead length) in small dielectric material, typically polyimide resin, is preferable. Table 3 lists the dielectric constants of common packaging materials. An excessively small dielectric-constant value of the surrounding

TABLE 3
Dielectric constants of packaging materials

Material	Dielectric constant (at 1 MHz)
Al_2O_3	9.6–10.2
AlN	8.7
Mold compounds	3.9–4.3
Polyimide	3.5
Si	11.7
GaAs	12.9
SiO_2	3.8
Glass ceramic	3.9–7.8
Glass epoxy	4.2

material, however, induces signal reflections that degrade operating speed; hence, an optimum dielectric value exists.

Signal reflection

A mismatched impedance causes signal reflections when a signal is transmitted from a driver to a receiver through a transmission line. In CMOS ULSI devices, multiple reflections occur at the driver and receiver ends when the output impedance of the output buffer is smaller than that of the transmission line. These reflections cause a ringing phenomenon that may slow operation or cause the circuit to malfunction.[24] These reflections cannot be treated lightly when the following relationships exist.

$$l > \frac{c}{\sqrt{\epsilon_r} \cdot \upsilon_0} \tag{10.6}$$

$$\upsilon_0 = \frac{0.35}{t_r} \tag{10.7}$$

where l (m) is the signal line length, c (m/sec) is the velocity of the light, ϵ_r is the dielectric constant of the surrounding material, υ_0 (1/sec) is the critical frequency, and t_r (sec) is the signal rise (or fall) time.

The equations show that a shorter signal line is beneficial to high-speed operation requiring a smaller package size. Larger package constructions that have longer signal lines can no longer be dealt with as lumped-element circuits but must be considered distributed-element circuits. In particular, longer wires, longer via-hole connects or both, which have larger impedance-mismatch potentials, should be avoided if possible, or matched-impedance designs should be used instead. Leading-edge packages, multilead frame plastic packages, and some other multilayered packages that contain strip, microstrip, or coplanar constructions provide better impedance matching. Figure 23 illustrates the strip, microstrip, and coplanar structures. The characteristic impedances, $Z_0(\Omega)$, of strip and microstrip structures are expressed by[25]

$$Z_0 = \frac{60}{\sqrt{\epsilon_r}}\left[\ln\left(\frac{4b}{0.67\pi W(0.8 + t/W)}\right)\right] \quad \text{(strip line)} \tag{10.8}$$

$$Z_0 = \frac{87}{\sqrt{\epsilon_r + 1.41}}\left[\ln\left(\frac{5.98h}{0.8W + t}\right)\right] \quad \text{(microstrip line)} \tag{10.9}$$

where ϵ_r is the dielectric constant of the dielectric material, t (m) and W (m) are the thickness and width of the conductor, and h (m) and b (m) are the thicknesses of the dielectric material beneath and surrounding the conductor, respectively.

For a typical coplanar structure such as a symmetrical double-strip, coplanar waveguide, the characteristic impedance, $Z_0(\Omega)$, is expressed by[26]

$$Z_0 = \frac{\eta_0}{\pi\sqrt{(\epsilon_r + 1)/2}}\left[\ln\left(2 \cdot \frac{1 + \sqrt{S/(2W + S)}}{1 - \sqrt{S/(2W + S)}}\right)\right] \quad \text{(coplanar line)} \tag{10.10}$$

where η_0 (equal to 120π Ω) is the characteristic impedance of free space, S (m) is the gap between the two lines, and W (m) is the width of the line. Assumptions are

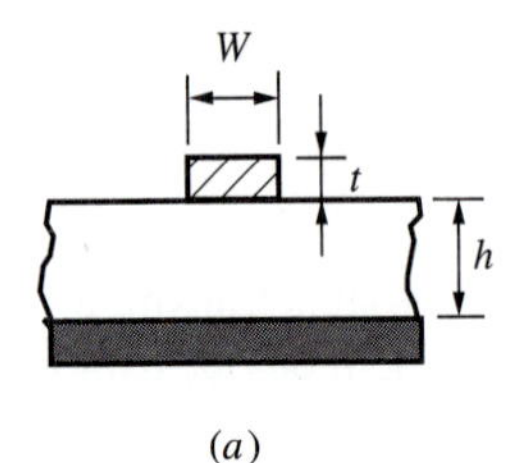

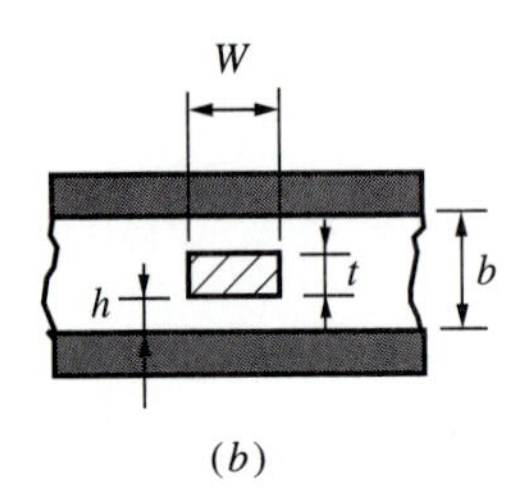

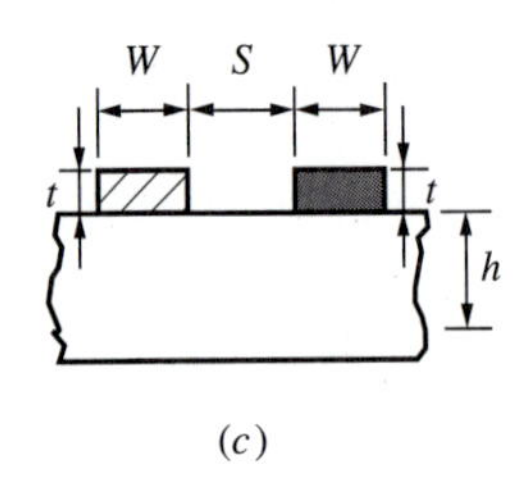

FIGURE 23
Cross-sectional sketches of several structures for impedance control: (*a*) microstrip line, (*b*) strip line (*after Kaupp, Ref. 25*), and (*c*) coplanar structure.

that the conductor lines are infinitely thin ($t = 0$), the dielectric material is infinitely thick ($h = \infty$), and the gap widths range is $0.173 \leq S/(2W + S) < 1$.

Noise

Two typical types of noise, cross-talk noise and simultaneous switching noise (ΔI noise), are discussed here.

Cross-talk noise (see Fig. 24*a*) occurs when a line is undesirably affected by another line that is placed very close to it because of the electromagnetic coupling between the two lines. The noise, coupled by C_m, L_m, or both (*m*: mutual) between the two lines, increases in proportion to the signal-voltage or current gradient and the AC-coupling strength. Cross-talk noise is a more serious problem in ULSI packaging intended to handle higher speeds, larger signal counts, and the resulting narrow signal-line spacings. Major countermeasures in package design are shorter parallel-signal runs, closer ground (or power) planes, and lower dielectric-constant materials.

Simultaneous switching noise, one of the most practical electrical-design problems, particularly in CMOS ASIC devices, occurs when many output buffers switch simultaneously. Figure 24*b* illustrates the mechanism. When an output buffer switches from high to low, transition current (i) flows from the power line (V_{cc}) into the load capacitance (C_l), inducing the noise voltage given by

$$V_n = L_g \frac{di}{dt} \tag{10.11}$$

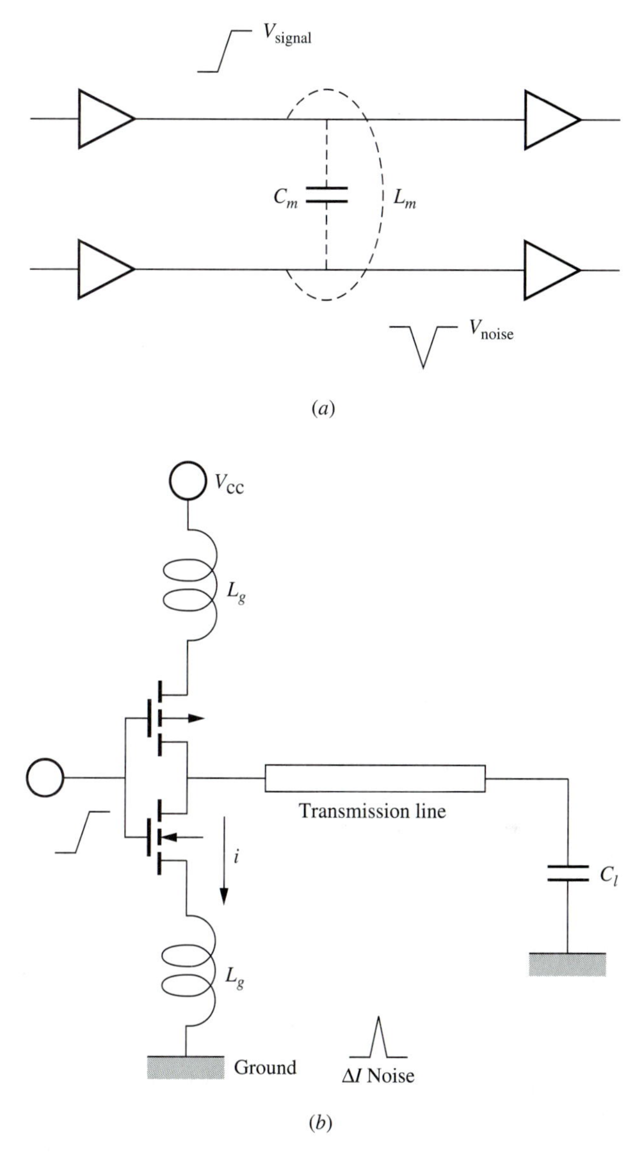

FIGURE 24
(*a*) Cross-talk noise that appears in one of the two adjacent lines, and (*b*) simultaneous switching (ΔI) noise that occurs when the output buffers switch simultaneously.

where V_n (V) is the induced voltage, L_g (H) is the inductance of the power lead, and di/dt is the derivative of current with respect to time.

In addition, when a line switches from low to high, an electric charge stored in the load capacitance flows into the ground line through the transmission line, inducing the same noise voltage as shown in Eq. (10.11). If j lines are switching simultaneously, then V_n is given by

$$V_n = \sum_j L_{g_j} \frac{di_j}{dt} \tag{10.12}$$

This noise voltage, V_n, leads to misoperation of other signal lines because it causes a bounce in the ground level. To reduce V_n, the L_g of the package must be reduced.[27] In actual package design, mutual inductance, that is, interaction between a line and the adjacent line, plays a significant role in reducing the L_g. But here we discuss only self-inductance, L, for simplification.

The Biot–Savart law derived from Maxwell's fundamental equation gives the lead inductance value L for a simplified columnar conductor model, assuming $l \gg a$:

$$L = \frac{l}{2\pi}\left(\frac{\mu}{4} + \mu_0\left[\ln\left(\frac{2l}{a}\right) - 1\right]\right) \tag{10.13}$$

$$= L_i + L_e \tag{10.14}$$

where l (m) is the length, a (m) is the radius, μ (kg-m/C^2) is the magnetic permeability of the conductor, and μ_0 (kg-m/C^2) is the permeability of a vaccum. L is the sum of L_i, the internal inductance, which depends on the internal magnetic flux, and L_e, the external inductance, which depends on the external magnetic flux. Either L_i or L_e, or both, if possible, must be smaller to suppress ΔI noise. Cu alloy reduces L_i considerably, because it has a smaller μ value, whereas shorter and wider geometric dimensions of the lead reduce L_e. Multiple grounds control the noise effectively, since, if m ground leads are used, the total inductance becomes approximately L_g/m. However, multiple grounds have a large impact on packaging density. Inductance can be reduced significantly by using large-area power and ground planes within the package. Figure 25 illustrates an example of a multilayer lead frame[28, 29, 30] used to provide the planes.

The electromagnetic-field simulator has become a powerful analytical tool for predicting the impact of package design on the various kinds of noise described above. Through the simulation, electrical parameters such as inductances, capacitances, resistances, and impedances of the package portions are calculated from the geometric dimensions of the package portions and the electrical properties of the components. These parameters allow a description of the package as an equivalent circuit and are used to perform overall packaged-chip circuit simulations, typically using Simulation Program with Integrated Circuit Emphasis (SPICE).

10.5.2 Thermal Design Considerations

The objective of thermal design is to keep the operating junction temperature of a silicon chip low enough to minimize the failure rate caused by temperature-activated failure mechanisms. The design must prevent the chip from exceeding the acceptable thermal limit for a particular application.[31] Only in the simplest possible applications can this be done by considering the packaged silicon device alone. Usually the packaged-device environment must be established for most of the following variables[32]: PWB temperature, total power dissipation on the board, local-neighbor power dissipation, degree of forced-air cooling, lateral and vertical space

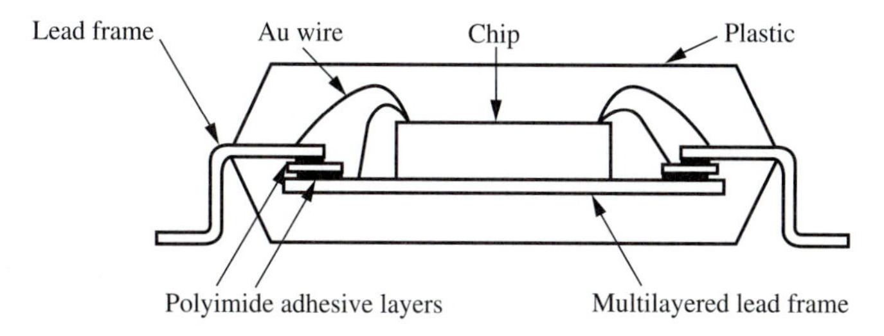

FIGURE 25
Multilayer lead frame package that provides large-area power and ground planes. The ground plane, which also serves as the chip support paddle, and the power plane are attached to the lead frame with electrically insulating adhesives. Many bonds are made from pads on the chip to these planes.

between boards that can be used by the package, conductivity of the PWB, and ideal performance of the isolated package. The thermal modeling of a single chip in a package is discussed below.

We start from the definition of thermal resistance. When heat P_{XY} (watts) flows from point X to point Y, the thermal resistance R_{XY} (°C/watt) between the two points is given by

$$R_{XY} = \frac{(T_X - T_Y)}{P_{XY}} \tag{10.15}$$

where T_X, and T_Y, in °C, are the temperatures at points X and Y and $T_X > T_Y$. Similarly, thermal resistance of a package (see Fig. 26) is expressed as

$$\theta_{ja} = \frac{(T_j - T_a)}{P} \tag{10.16}$$

where θ_{ja} (°C/W) is the junction-to-ambient thermal resistance, T_j (°C) is the average chip or junction temperature, T_a (°C) is the ambient temperature, and P (watts) is the power.

Figure 26 is a simplified heat-transfer model of a packaged chip, where heat is transferred from the chip to the surface of the package by conduction and from the

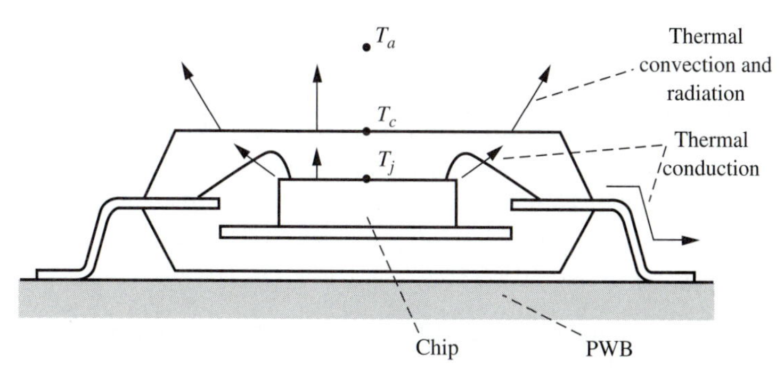

FIGURE 26
A simplified heat-transfer model of a packaged chip.

package surface to the ambient by convection and radiation. In most applications, the temperature difference between the case (or the package surface) and the ambient is small, and radiation can be neglected. Conduction heat transfer through the package terminals (see Fig. 26) can be significant, particularly in high-I/O ULSI packages. However, if we neglect it for simplification, the overall thermal resistance in this model can be considered as the sum of two thermal components, θ_{jc} and θ_{ca}, defined as

$$\begin{aligned}\theta_{ja} &= \theta_{jc} + \theta_{ca} \\ &= \frac{(T_j - T_c)}{P} + \frac{(T_c - T_a)}{P}\end{aligned} \tag{10.17}$$

where θ_{jc} (°C/watt) is the junction-to-case thermal resistance, θ_{ca} (°C/watt) is the case-to-ambient thermal resistance, and T_c(°C) is the average case temperature.

θ_{jc} is relatively insensitive to the ambient and is mainly a function of package materials and geometry. θ_{ca} depends on the package geometry, the package orientation in the application, and the conditions of the ambient in the operating environment (whether heat transfer is free or by forced-convection). Heat transfer is classified into three categories—conduction, convection, and radiation. How these work in ULSI packages is summarized below.

Conduction

Conduction dominates the heat transfer from the chip to the package surface. Fourier's equation at one dimension and at a stationary state is given by

$$Q = (T_1 - T_2) \cdot \kappa \cdot \frac{S}{L} \tag{10.18}$$

where Q (watts) is the amount of heat flow, T_1 and T_2 (°C) are the temperatures at the two ends of the thermal path, κ (watts/m-K) is the thermal conductivity, S (m^2) is the cross-sectional area, and L (m) is the length of the insulated heat flow path.

In actual package, Eq. (10.18) can be written as

$$P = (T_j - T_c) \cdot \kappa \cdot \frac{S}{L} \tag{10.19}$$

So, we have

$$\theta_{jc} = \frac{(T_j - T_c)}{P} = \frac{L}{(\kappa \cdot S)} \tag{10.20}$$

ULSI packages have a high packaging density (i.e., smaller S), so high-thermal-conductivity components such as Cu alloy lead frames, AlN substrates, and thermo-conductive molding compounds are particularly important because they increase the overall package κ value. Thinner package constructions, which reduce the L value, are also important.

Convection

Heat transfer from the package surface to the ambient results mostly from convection, given by Newton's cooling law,

$$Q = h \cdot A \cdot (T_c - T_a) \tag{10.21}$$

where Q (watts) is the amount of heat flow, h (W/m^2-K) is the heat transfer coefficient, A (m^2) is the surface area, and T_c and T_a (°C) are the temperatures of the surface and the ambient.

From Eqs. (10.17) and (10.21), we have

$$\theta_{ca} = \frac{(T_c - T_a)}{Q} = \frac{1}{(h \cdot A)} \tag{10.22}$$

θ_{ca} is reduced through increased conduction[33] and larger package surface area in ULSI packaging. The application system constructions are forced-air convections, liquid coolants in place of air coolings, and additional heat sinks attached to the package surface.

Radiation

Radiation helps transfer some heat from the package surface to the ambient, but usually the contribution is small. Blackbody radiation follows the Stefan–Boltzmann law

$$E_b = \sigma \cdot T^4 \tag{10.23}$$

where E_b (W/m^2) is radiation energy flux density, σ is Stefan–Boltzmann's constant (5.67×10^{-8} W/m^2-K^4), and T (K) is the absolute temperature.

Since actual materials are not perfectly black, actual radiation E_1 becomes

$$E_1 = \epsilon \cdot E_b = \sigma \cdot \epsilon \cdot T^4 \tag{10.24}$$

where ϵ is the emissivity.

Heat transfer caused by radiation between two plates, that is, between a package of temperature T_1 and emissivity ϵ_1, and an ambient of temperature T_2 and emissivity ϵ_2, is expressed as

$$Q = \sigma \cdot f \cdot A(T_1^4 - T_2^4) \tag{10.25}$$

where Q (watts) is the amount of heat flow, A (m^2) is the package surface area, and f is expressed by

$$f = \frac{1}{1/\epsilon_1 + 1/\epsilon_2 - 1} \tag{10.26}$$

When $T_1 - T_2 \ll T_1$ and $T_1 - T_2 \ll T_2$, Q could be expressed as follows, by placing $(T_1 + T_2)/2 = T_m$

$$\begin{aligned} Q &= 4\sigma \cdot f \cdot T_m^3 \cdot A(T_1 - T_2) \\ &= hr \cdot A(T_1 - T_2) \end{aligned} \tag{10.27}$$

where $hr = 4\sigma \cdot f \cdot T_m^3$ and is the radiation transfer coefficient.

Therefore, the thermal resistance of the radiation is

$$\theta_{\text{rad}} = \frac{(T_1 - T_2)}{Q} = \frac{1}{hr \cdot A} \tag{10.28}$$

In actual applications, black-dyed packages and external heat sinks are sometimes preferred, since they increase hr values.

10.5.3 Mechanical Design Considerations

In this section, we discuss mechanical design issues. We focus on plastic packages because TCE mismatches result in more complex problems in packages where the molding compound comes into contact with the chip and the lead frame surfaces.

Thermal stress of plastic mold

Figure 27 shows a simplified plastic package model in which the silicon chip and the lead frame are assumed to be structurally homogeneous, having the same TCE. When the construction is held at a temperature T (°C), the compressive stresses (σ_c) in the plastic resin and in the chip (including the lead frame) are respectively given by

$$\begin{aligned} \sigma_{cr} &= E_r \frac{\lambda - \alpha_r \cdot \Delta T \cdot L}{L} \\ \sigma_{cc} &= E_c \frac{\lambda - \alpha_c \cdot \Delta T \cdot L}{L} \end{aligned} \tag{10.29}$$

where E (GPa) is the elastic modulus, λ (m) is the elongation of the construction, ΔT is the temperature difference between the temperature T and that of the molding environment, L (m) is the package length, and the subscripts r and c correspond to resin and chip.

From Eq. (10.29), we see that

$$\frac{\sigma_{cr}}{E_r} + \alpha_r \cdot \Delta T = \frac{\sigma_{cc}}{E_c} + \sigma_{cc} \cdot \Delta T \tag{10.30}$$

Also, from equilibrium of the forces, we find that

$$\sigma_{cr} \cdot A_r + \sigma_{cc} \cdot A_c = 0 \tag{10.31}$$

where A (m^2) is the lateral cross-sectional area of the package.

By combining Eq. (10.30) and Eq. (10.31), stresses can be given by

$$\begin{aligned} \sigma_{cr} &= A_c \cdot E_r \cdot E_c \cdot (\alpha_r - \alpha_c) \cdot \frac{\Delta T}{(A_r \cdot E_r + A_c \cdot E_c)} \\ \sigma_{cc} &= A_r \cdot E_r \cdot E_c \cdot (\alpha_r - \alpha_c) \cdot \frac{\Delta T}{(A_r \cdot E_r + A_c \cdot E_c)} \end{aligned} \tag{10.32}$$

E_r and α_r are actually temperature-dependent functions; hence, Eq. (10.32) becomes

$$\sigma_{cc} = \int_{T_m}^{T} E_c \cdot \frac{\alpha_r(T) - \alpha_c}{1 + A_c E_c / A_r E_r(T)} \tag{10.33}$$

Since $\alpha_r(T) > \alpha_c$, it is apparent that the chip receives a compressive stress when the package is cooled from T_m, and it receives a tensile stress when heated. Stress reduction is essential because stress induces various kinds of defects such as device-parameter shift, chip cracking, Al-metallization slide, and Au wire fatigue. The latter results from relative dislocation between the wire and the plastic and might lead to wire breakage after temperature cycling. The countermeasures are to control

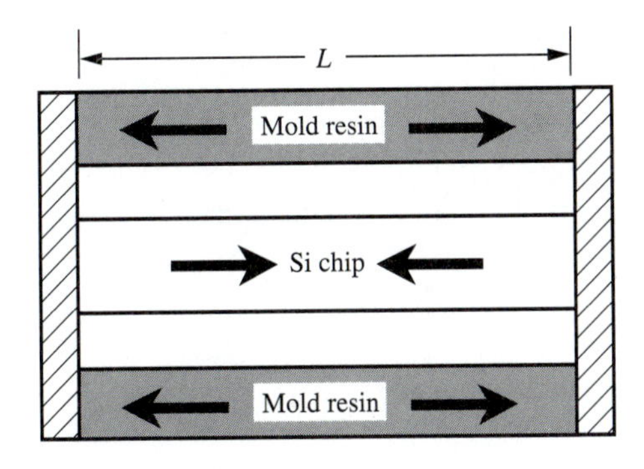

FIGURE 27
A simplified plastic package model, assuming that the silicon chip and the lead frame are structurally homogeneous. Arrows show the directions of stresses.

package-geometry dimensions and to carefully select package materials and process variables; choosing molding compounds that provide a smaller α_r and a reduced E_r is fundamental.

Moisture-induced cracking

Moisture-induced cracking resistivity[34] is another important issue, especially in thinner plastic package design. Moisture-induced cracking occurs when a moisture-saturated package is exposed to a reflow soldering temperature of approximately 215°C to 260°C.[35] Figure 28 illustrates the mechanism.[36] Absorbed water concentrates on the boundary surface between the chip pad and molding resin because the adhesion there is usually poor, and it flashes to steam when the package is reflow soldered. The steam pressure is sufficient to rupture thinner packages, because they have thin sections of plastic over and under the chip/lead frame surfaces. Recent studies[37, 38] on this phenomenon express the maximum stress of resin ($\sigma_{\max}$) that occurs at the center in terms of the vapor pressure P by[39]

$$\sigma_{\max} = k \cdot P \cdot a^{x_1}/h^{x_2} \tag{10.34}$$

where k is the proportional constant, a is the length of the shorter side of the chip pad, h is the resin thickness under the chip pad, and x_1 and x_2 are indices that depend on the package type.

Equation (10.34) shows that a smaller chip pad and thicker resin under the chip pad improve antimoisture capability. These measures, however, often jeopardize the ability to put larger ULSI chips in thinner packages. Therefore, additional measures are being implemented: improving molding resin (lower E and TCE), improving adhesion between the molding resin and the chip pad (dimples or holes in the chip pad), baking the package and packing it in a moisture-free container for shipment by the supplier and/or prebaking it before mounting by the user (reduced P), and lowering soldering temperatures at the next level of assembly operations (reduced P).

TCE mismatch between package and PCB[40]

When a ceramic package (typically TCE = 6 ppm/°C) is surface-mounted onto a plastic board (typically TCE = 15 ppm/°C), the TCE mismatch sometimes causes a large thermal stress on the lead-solder joints that could lead to joint breakage caused by solder fatigue from temperature cycling operations. A similar problem occurs in ULSI plastic packages, typically in a thinner small-outline package (SOP), with a chip occupancy as large as approximately 80% and an overall

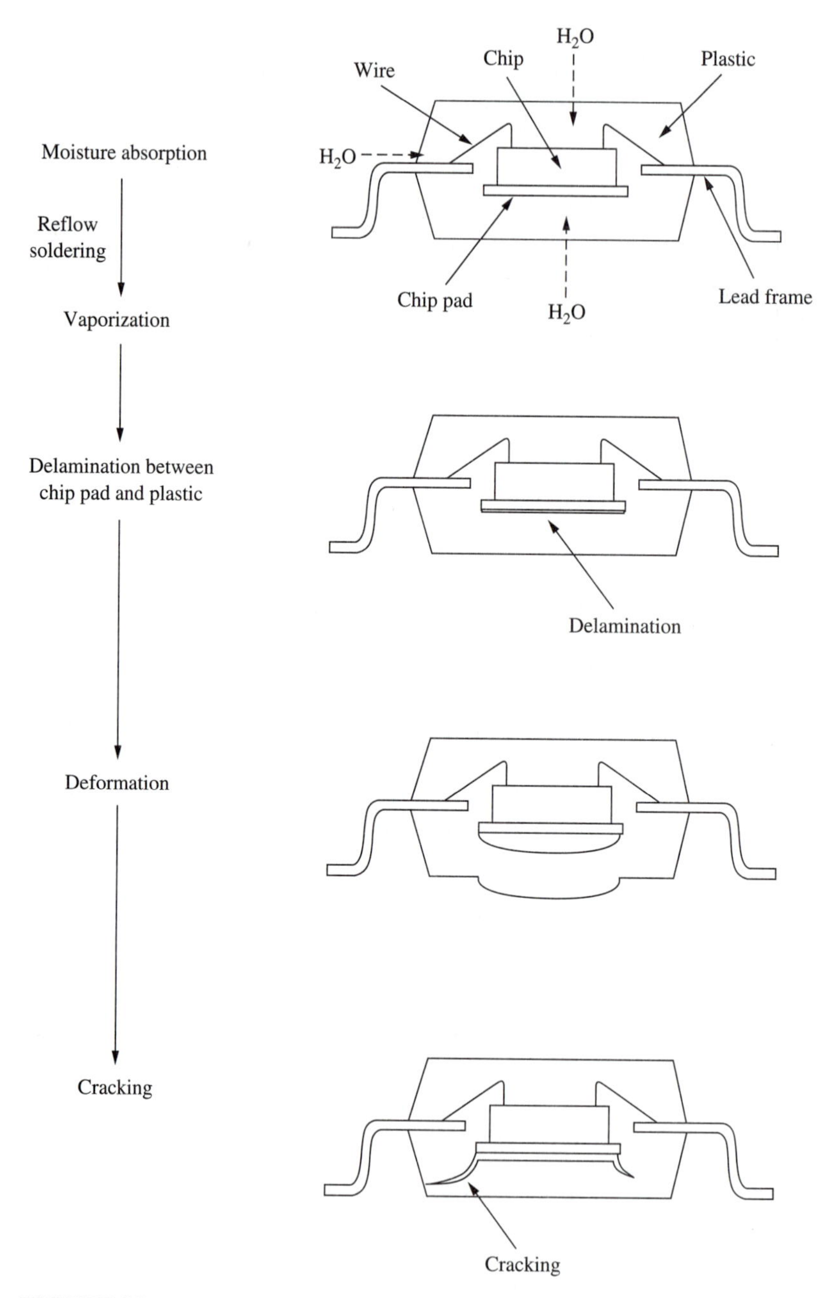

FIGURE 28
Mechanism of moisture-induced cracking. Plastic package absorbs moisture from the room ambient. The moisture flashes into steam during reflow soldering of the package to the PWB. Cracking resulting from the vapor pressure is apt to occur at the edge of chip support paddle.

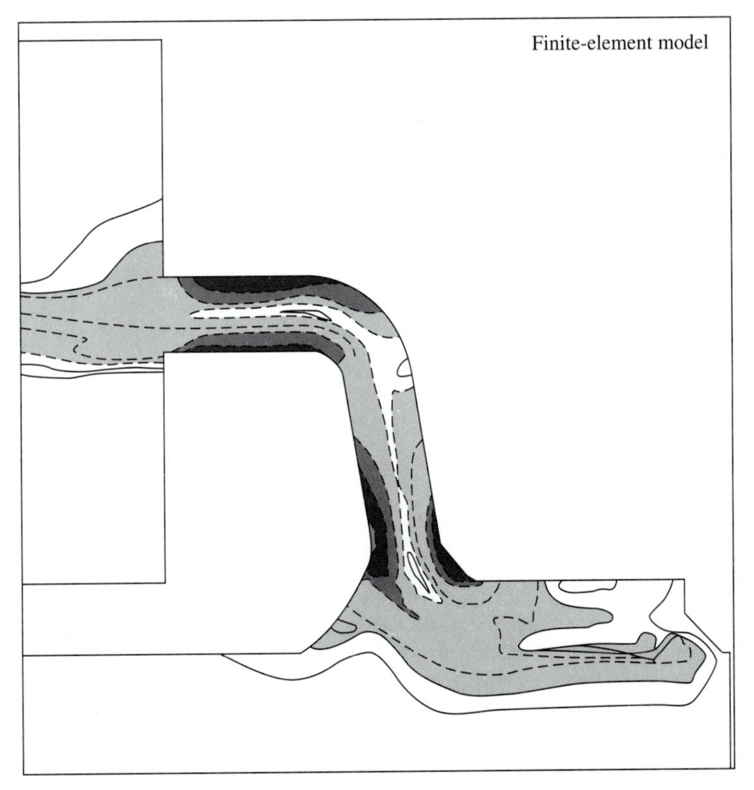

FIGURE 29
Finite element analysis (FEA) showing contours of constant stress in the plastic package lead and solder joint. The stress is greater where the shading is darker. (*Courtesy of Mitsubishi Electric Corp.*)

package TCE that is much closer to silicon's TCE (4 ppm/°C), than to that of the PCB. Solder fatigue[41, 42, 43] starts in the plastic deformation of the solder material and is followed by a tiny crack at the stress concentration point that develops along the crystal boundary into a breakage through accumulated temperature cycling operation. Finite-element analysis has become a powerful tool to characterize the thermal stresses on the solder joint, predicting[44] the stress concentration point and the location where the crack is likely to occur (Fig. 29).

Although higher and thinner leads using low-*E* materials have been successful[45, 46, 47] in settling the problem to some extent, improvements in both the single-chip and the next-level packagings are still required for a final solution.

10.6 SPECIAL PACKAGE CONSIDERATIONS

Various kinds of packaged-chip failures originate in the packaging. Figure 30 illustrates major package-induced failure modes for hermetic-ceramic and plastic packages. Table 4 lists the usual reliability tests performed to detect these failures.

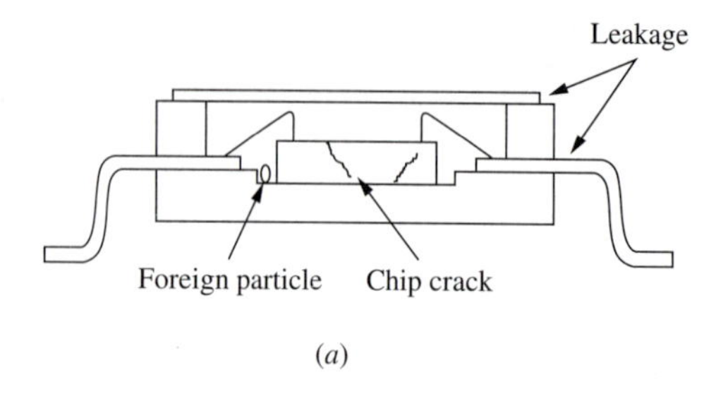

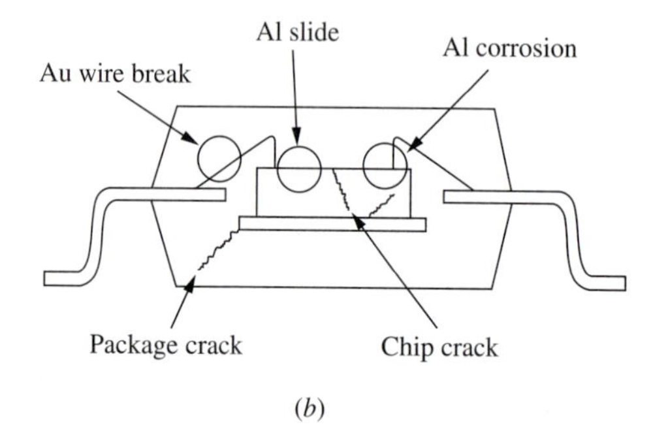

FIGURE 30
Major package-induced failure modes for (*a*) hermetic-ceramic and (*b*) plastic packages. Chip cracking from the back sometimes results from poor back grinding operation for both packages.

This section examines other aspects of packaging that must be addressed for a complete package-assembly methodology. Among them are the cleaning and assembly environment, package or lid sealing, corrosion, and α-particle protection.

10.6.1 Cleaning and Assembly Environment

The most critical cleaning step the chip undergoes is performed before bonding, encapsulation, or final lid sealing. This cleaning process must be chemically compatible with the chip metallurgy. Aluminum has a very narrow range of pH values in which its oxide protects it from corrosion in aqueous solution. Corrosion reactions, with metal dissolution, can occur in both basic and acidic solutions. There are two objectives in cleaning. One is to remove organic species that can affect bondability; an organic solvent is usually required for this. The other is to remove ionic species that can cause corrosion during the life of the device or, in an unusual instance, contribute to surface-charge accumulation. Water is a good solvent for ionic species.

Because environmental considerations have curtailed the use of powerful organic solvents, organic removal in the assembly process has become difficult. Today, new alternative chemicals are being used to some extent in place of chlorofluorocarbons. This difficulty and the ever-decreasing feature sizes on the ULSI chip, down to the submicrometer range, make environmental control in the assembly room a new issue that must be addressed. Typical assembly operations for ULSI devices are done in class 1000 to 10,000 cleanrooms. Otherwise, particles in the ambient that are not completely removed before molding (in plastic packages) or before lid seal-

TABLE 4

Reliability tests performed to detect package-induced failures

Test type	Typical conditions	Failure mode	Major causes	Operation to note
Temperature cycling	−65°C–150°C	Chip crack	TCE mismatch of components	Chip bond
High-temperature storage	150°C, 175°C	Wire lift-off	Undesirable intermetallics	Wire bond
Temperature cycling	−65°C–150°C		Poor bonding	
			Bonding pad contamination	
Temperature cycling	−65°C–150°C	Wire break	Poor bonding	
Vibration			Stress from the molding resin	
High-temperature, high-humidity storage	85° C, 85%	Al corrosion	Chip contamination	
High-temperature, high-humidity storage with bias	85°C, 85%, 7 V		Seal leakage	Encapsulation
Gross leak			Impurities from the molding resin	(mold, seal)
Pressure-cooker test (PCT)	130°C, 85%			
Pressure-cooker test with bias	130°C, 85%, 7 V			
Operating life	125°C, 7 V	Malfunction	TCE mismatch of components (stress from the molding resin)	Chip bond through encapsulation

ing (in ceramic packages) might degrade the reliability of the packaged device, most likely by hard failures caused by metal corrosion from particles with high levels of contaminants.

10.6.2 Package or Lid Sealing and Corrosion

The major objective of package sealing is to protect the device from external contaminants during its lifetime. Further, any contaminants present before sealing must be removed to an acceptable level before or during sealing. Ceramic packages may be lidded with polymers. However, since moisture usually penetrates organic sealants within a short period of time, almost all high-reliability applications have a hermetic seal made with glass or metal. Glass sealing for CERDIP is discussed in Section 10.4, and the process is essentially the same for lid sealing. Many of the metal alloys used for chip bonding are suitable for lid sealing. A leak-tight seal that excludes the external environment can be made without difficulty. The real difficulty has been freeing the package of contaminants, especially water, before sealing.[48, 49]

Problems can arise in a device in a low operating temperature, where water condenses inside the package. The following reactions can then lead to corrosion in the presence of a small halide contamination:

$$6HCl + 2Al \rightarrow 2AlCl_3 + 3H_2\uparrow$$

$$AlCl_3 + 3HOH \rightarrow Al(OH)_3 + 3HCl$$

$$2Al(OH)_3 + \text{Aging} \rightarrow Al_2O_3 + 3H_2O$$

Similar aluminum corrosion can occur in plastic packages, when water that penetrates the package condenses on the chip after a long period of time. In the presence of halide solubles from the molding compound, corrosion occurs particularly between the two adjacent and electrically biased lines:

$$Al + 4Cl^- \rightarrow AlCl_4^- + 3e^-$$

$$AlCl_4^- + 3H_2O \rightarrow Al(OH)_3 + 3H^+ + 4Cl^- \qquad \text{(Cathode line)}$$

$$H_2O + e^- \rightarrow \tfrac{1}{2}OH^- + H_2\uparrow$$

or

$$Na^+ + e^- \rightarrow Na, \qquad Na + H_2O \rightarrow Na^+ + OH^- + \tfrac{1}{2}H_2\uparrow$$

Then

$$Al + 3OH^- \rightarrow Al(OH)_3 + 3e^- \qquad \text{(Anode line)}$$

P_2O_5 in the chip passivation layer can accelerate the corrosion by forming H_3PO_5 in the presense of water[50]:

$$2Al + 6H^+ \rightarrow 2Al^{3+} + 3H_2\uparrow$$

$$2Al^{3+} + 3H_2O \rightarrow 2Al(OH)_3 + 6H^+$$

10.6.3 α-Particle Protection

Soft errors in memory circuits caused by α-particles emanating from packaging materials were first reported[51] in 1978; there have been many papers on the subject since then. The α-particles are emitted by the decay of uranium and thorium atoms contained as impurities in the packaging materials. Decreasing device design rules make circuits sensitive to this problem. Packaging materials, particularly molding compounds, have been pushed below the level of 0.001 to 0.01 α-particle/cm^2-h. Because α-particles have low penetrating power in solids, low-α materials, typically polyimides, have been used as α-absorbing coatings on silicon chips.[52, 53]

Although great progress has been made by reducing α-particle emission in current packaging materials, error correction is still required to control soft errors caused by α-particles.

10.7 OTHER ULSI PACKAGES

Sections 10.1 through 10.6 discuss standard assembly and package fabrication technologies, and they examine some design features that are mainly on standard single-chip packages. Other packages and their assembly techniques are described in this section.

10.7.1 Tape Carrier Package

Tape carrier package (TCP) is a generic term for packages in which a tape-automated bonding (TAB) technique is used for the pads-on-chip interconnects. The TAB technique, which started in gang-bonding, was originally used to improve the productivity of interconnects. TAB uses finely patterned thin metal, usually Cu foil plated with Au or Sn, in place of wires and connects the metal tips metallurgically to corresponding bumps (plated Au) formed on the Al pads on the chip. TAB has become a choice in finer-pitch interconnects for high-I/O ULSI devices because it enables finer pitch and longer span bondings than those achievable by wire bonding. TCPs are usually plastic-based.

Tape carrier

Figure 31[54] illustrates typical tape carriers used in TCP fabrication. The one-layer tape carrier is a thin Cu foil. It is cost effective but not preferred today; electrical testing of the packaged chip is difficult since all the Cu leads are mechanically connected as one body.

Two-layer tape that has a Cu layer, typically 25 μm thick, on a polyimide base tape, typically 75 μm thick, is manufactured in two ways—either Cu plating on the polyimide or polyimide casting on the Cu foil. A window in which to place the chip is formed, and along with other necessary holes, by etching away the polyimide tape, and the Cu layer is patterned into the designated shape by plating or etching

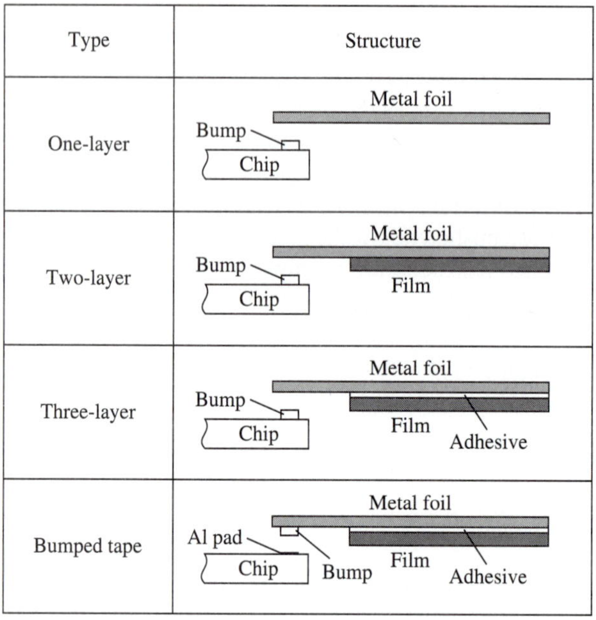

Type	Structure
One-layer	Metal foil; Bump; Chip
Two-layer	Metal foil; Bump; Film; Chip
Three-layer	Metal foil; Bump; Film; Adhesive; Chip
Bumped tape	Metal foil; Al pad; Chip; Bump; Film; Adhesive

FIGURE 31
Cross section of typical tape carriers used in TCP fabrication. (*After Clain, Ref. 54.*)

operations. Three-layer tape is similar to two-layer tape but has an additional adhesive layer between the Cu and the base plastic layers, enabling a choice of base plastics, such as glass-epoxy, BT resin, polyester, and the most prevailing polyimide resin.

Two-layer and three-layer tapes, particularly those with (0.5 μm-thick plating) Cu foils that are chemically Sn-plated are widely used in ULSI packaging. Bumped tape, which has Au bumps on the Cu foil innerlead tips, is also accepted to some extent, since it does not require wafer-bumping operations. Figure 32[55] illustrates the transfer-bumping operation used as a typical bumped-tape manufacturing method. Au bumps are formed by first performing plating operations on a substrate (typically glass-based). The bumps are then transferred to the aligned tips of the Cu foil (Sn-plated) through a bonding operation to formulate Au-Sn eutectic alloys. Bumped tape eliminates bumped chip availability problems.

Tape-automated bonding

Tape-automated bonding (also called innerlead bonding) is divided into two methods: bonding to the bumps-on-chip using one- to three-layer tapes, and bonding to Al pads-on-chip using a bumped tape. The former method is more popular in the industry today, therefore only this method is detailed here. In this method, bumps (25-μm-thick Au) are formed on the Al pads-on-chip and are attached to Sn-plated or Au-plated Cu foil tips.

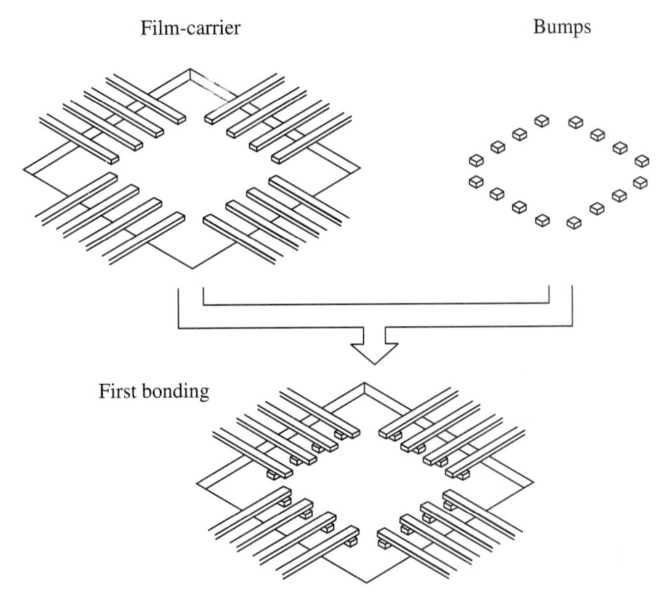

FIGURE 32
Illustration of the transfer-bumping operation. (*After Hatada, Ref. 55.*)

Figure 33[56] illustrates an example of the Au-bump-formation process that is performed at the wafer stage. First, a wafer covered with passivation glass except for the bond-pad areas is full-face sputtered by a combination of barrier metals such as Cr/Cu/Au, Ti/Ni/Au, or TiW/Au. Next, photolithography, in which 25- to 30-μm-thick photoresist materials are used, forms open windows at the bond pads where 20- to 25-μm-thick Au electroplating is performed, forming the Au bumps. Photoresist removal operations, and the following barrier-metal-etching operations, in which the formed Au bumps serve as masks to protect the barrier metals beneath the bumps, complete the bump formation.

Tape-automated bonding is usually a gang operation that connects all the bumps and corresponding foil tips at one time. Typical process conditions are thermocompression using heated (approximately 500°C) flat tools made of diamond and applying loads (approximately 10 kg/mm^2) for periods of time that range from 1 to 2 seconds. Figure 34 shows an example of a bonded chip at 110-μm pad spacing. Improved TAB techniques have been successful in reducing the spacing to less than 100 μm.

Tape-carrier package

Varieties of TCPs are used in industry today for ULSI devices. The package fabrication is usually done by plastic technologies—glob-top coating, or transfer molding that encapsulates the bonded chip with plastic resin materials. Figure 35 shows an example of a package in which the packaged chip can be electrically tested

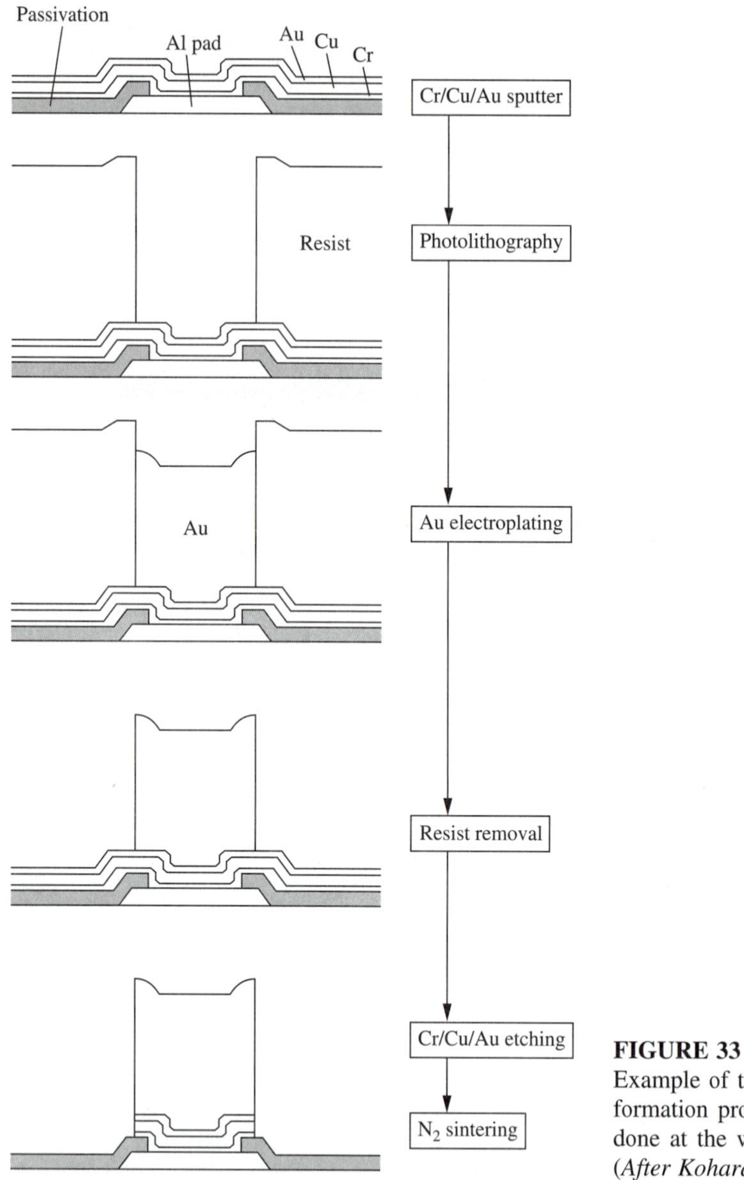

FIGURE 33
Example of the Au-bump-formation process that is done at the wafer stage. (*After Kohara, Ref. 56.*)

(including burn-in operations) using the larger test pads placed on the base polyimide tape that forms the opposite end of the bonded innerleads.[57] Posttesting and presoldering to the next-level packaging operation are the external lead-trim and form operations through which the package is formed into its final outer shape.

TCP has the potential to become a major choice in higher-I/O ULSI packaging since it enables one of the finest-pitch interconnects to be made. Obstacles to be

FIGURE 34
Scanning electron microscope photograph of an innerlead-bonded chip with 110-μm pad spacing. (*Courtesy of Mitsubishi Electric Corp.*)

surmounted are the relatively higher cost, the modest infrastructures involving test fixtures, and the next-level packaging technologies.

10.7.2 Flip-Chip Package

Flip-chip technologies fabricate bumps (typically Pb/Sn solders) on Al pads-on-chip and interconnect the bumps directly to the package media, which are usually ceramic- or plastic-based.

As Fig. 36[58] illustrates, in flip-chip technologies, the chip is face-down bonded to a package medium through the shortest path. These technologies can be applied not only to single-chip packaging, but also to higher or integrated levels of packaging in which the package media are larger and more sophisticated substrates that accomodate several chips to form larger functional units. Consequently, encapsulating flip-chip packages involves many variables, such as transfer molding, glob-top coating, and lid sealing, depending on the packaging level. Figure 36 shows a cross section in which the chip is attached to a ceramic substrate, yielding 528 pin interconnects. The flip-chip technique, using an area array, has the advantages of achieving the highest density of interconnection to the device and a very-low-inductance interconnection to the package. However, pretestability, postbonding visual inspection, and TCE matching to avoid solder bump fatigue are still challenges.

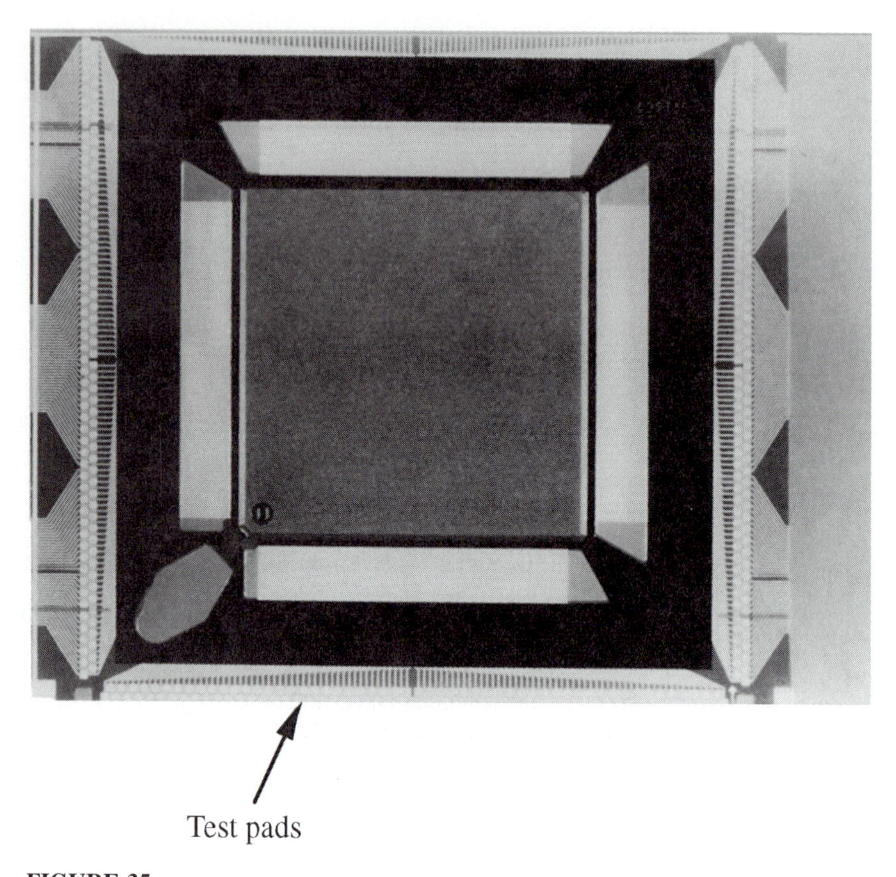

FIGURE 35
Example of a plastic TCP, in which the packaged chip can be electrically tested using the larger test pads placed on the base polyimide tape. (*Courtesy of Mitsubishi Electric Corp.*)

The solid, curved line in Fig. 37[59] shows that the original solder bump area is usually larger than that of barrier metals placed beneath. This area becomes closer in size to that of the barrier metals during the reflow-soldering operations. Thus, proper combinations of bump area, bump thickness, and barrier metal area must be selected to guarantee the interconnect integrities.

10.7.3 Ball-Grid Array Package

The recently developed ball-grid array (BGA) package is of interest in ULSI packaging. The BGA package is ceramic- or plastic-based and involves a variety of internal package structures. This package is now being assessed, standardized, and registered. This type of package features area-array external electrodes, usually Pb/Sn-

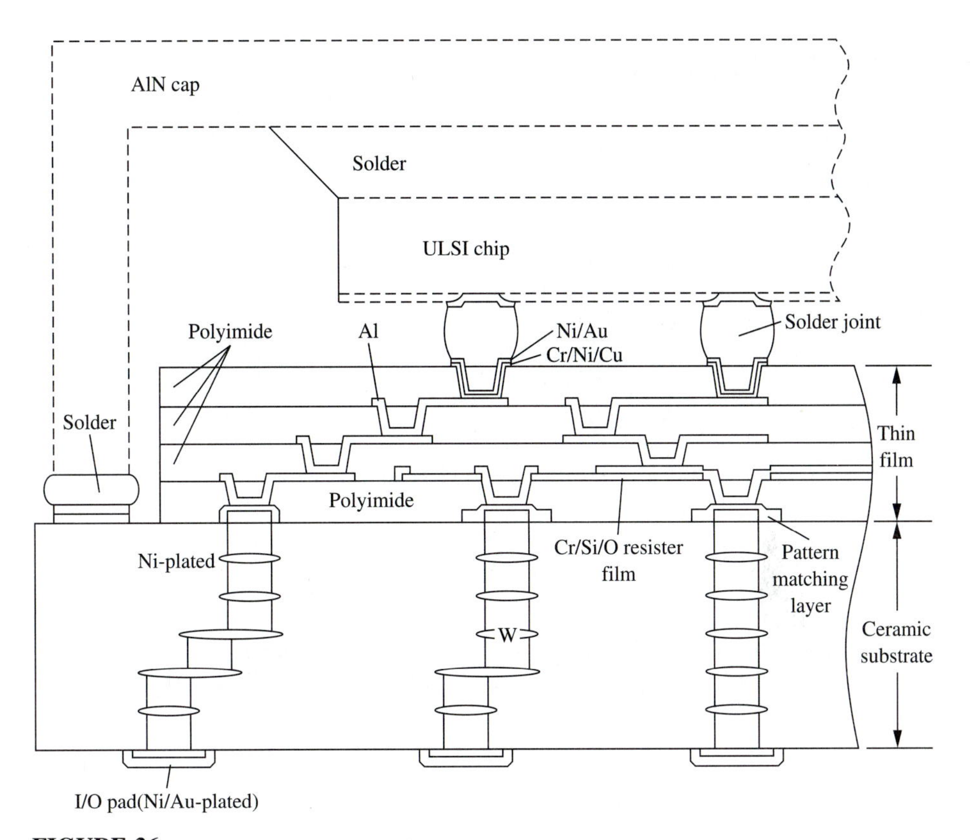

FIGURE 36
Cross section of a flip-chip package in which the chip is attached to a ceramic substrate and hermetically sealed. (*After Inoue, Ref. 58.*)

based bumps, placed on the package back surface at spaces ranging from 1 to 2 mm. The potential advantages are better external pin-count density, easier soldering to the PWB, lower inductance to the PWB, larger thermal paths to the package surroundings, and improved pretestability.

Figure 38[60] illustrates a simple BGA construction. An interconnected chip on a plastic resin substrate (typically BT resin) is transfer-molded, and solder balls are attached to the package back in post-molding operations. The figure shows thermal via holes placed in the substrate that provide shorter and significantly larger thermal paths from the chip to the mother PWB through the holes and the solder balls. More sophisticated BGA package structures have been proposed; these would contain multilayer substrates that have broader power planes, broader ground planes, or both for low-inductance and larger thermal-path connections.

Two popular choices for solder-bump formation are solder-ball attachment and solder-paste screen printing. Both methods would be followed by reflow operations to complete the metallurgical connections.

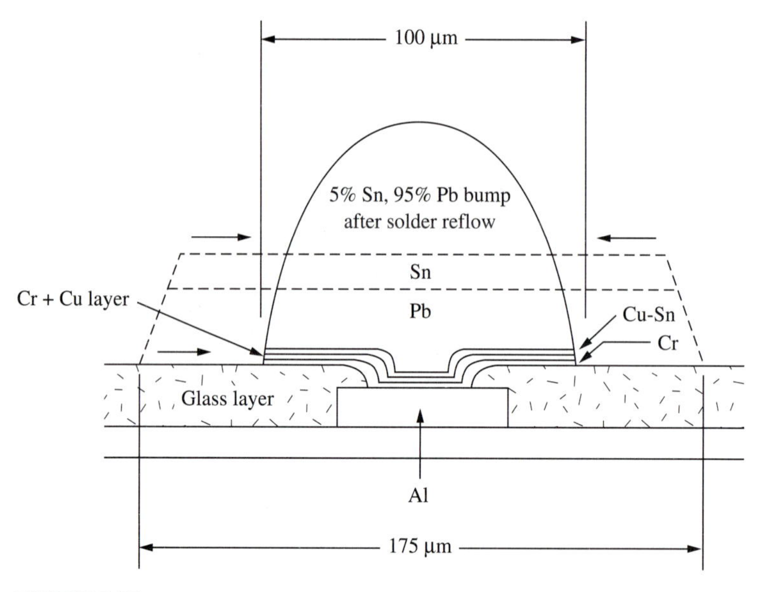

FIGURE 37
Cross section of a solder bump used in flip-chip technologies and the deformation caused by reflow-soldering operations. (*After Totta and Sopher, Ref. 59.*)

10.8 SUMMARY AND FUTURE TRENDS

Package pin count will undoubtedly continue to increase with increasing IC complexity. Alumina ceramic packages will continue to dominate a significant portion of the high-performance ULSI packaging technology until factors such as their high

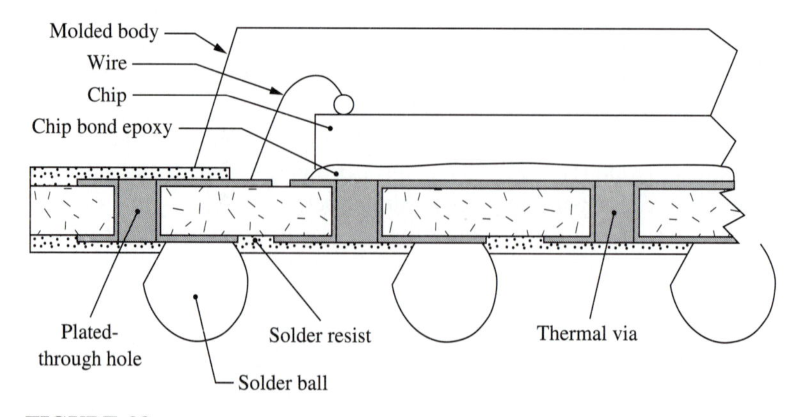

FIGURE 38
Cross section of a simple BGA package. The package is plastic-based and features area-array solder balls. (*After Sloan, Ref. 60.*)

dielectric constant, modest thermal conductivity, or cost force a change to other contending materials. Plastic packages, especially posttransfer-mold plastic packages, will dominate most of the ULSI packaging technology. Higher packaging density on the PWB level will continue to drive package design gradually but steadily toward smaller lead pitches, to approximately 0.3-mm spacings.

If chip integration keeps pace with decreasing feature size, today's interconnection technologies, especially wire bonding, will be challenged.[61, 62] So far, wire bonding has successfully maintained the relative proportions of chip size and interconnection area generally up to an I/O of approximately 300 and often up to an I/O of approximately 500 interconnects on a chip with the penalties of cost, yield, or some chip design restrictions. TAB and flip-chip techniques will be the candidates to replace wire bonding. However, it probably will be some time before appropriate infrastructures are available to replace wire bonding technologies. Interconnections of high I/O (> 500 per chip) will have to be the choice among the three techniques.

System design will increasingly depend on systematically optimizing the entire interconnection scheme to achieve the potential benefits of improved silicon capability. This optimization may lead to completely new requirements for assembly and packaging, but more than likely it will lead to sorting out the various existing assembly technologies and discovering the combinations that lend themselves to packaging-level integrations. The approach for various kinds of multichip modules (MCMs), shown in Fig. 39, verifies this trend.[63]

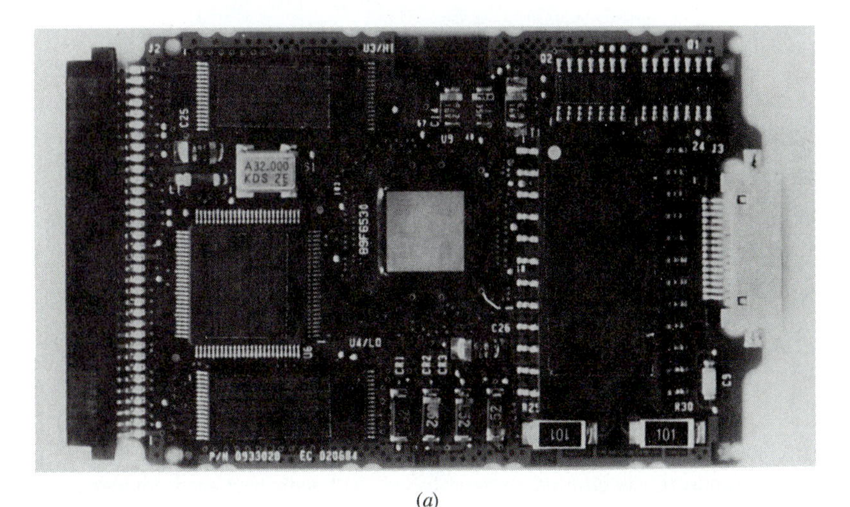

(*a*)

FIGURE 39
Three kinds of multichip modules. (*a*) MCM-Laminate (L) is PCB-based. Single-chip-packaged ULSI devices and other components, such as capacitors and resistors, are soldered onto the module. Sometimes bare chips can be used in lieu of packaged chips. Although the PWB is usually finely patterned, established surface-mount technology favors MCM-L growth. (*Courtesy of IBM Japan, LTD.*) (*continued*)

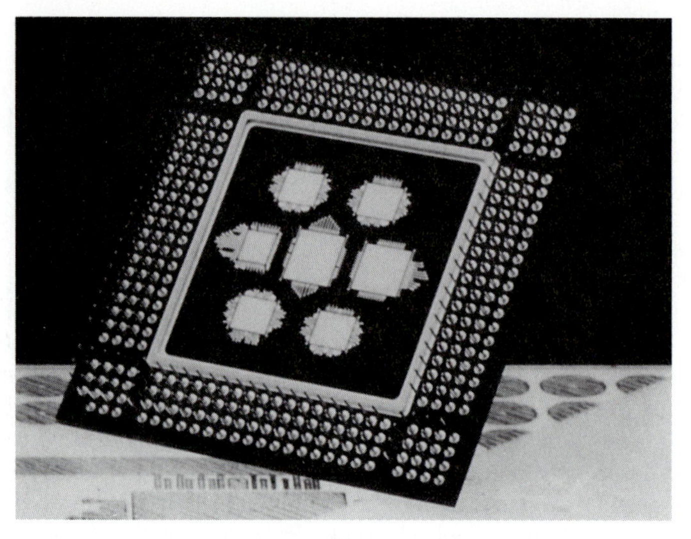

(*b*)

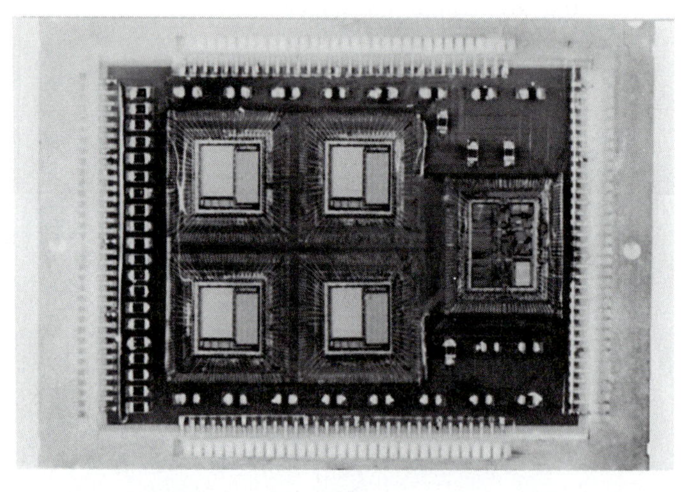

(*c*)

FIGURE 39 (*continued*)
(*b*) MCM-Ceramic (C) is co-fired ceramic-based. Although packaged devices can be placed on the substrate, MCM-C is more likely to have bare chips, since the ceramic provides excellent TCE matching and thermal conductivity. (*Courtesy of Kyocera Corp.*)
(*c*) MCM-Deposition (D) is usually ceramic- or silicon-based. Wiring patterns of the substrate are formed by deposition and achieve finer-features than MCM-L or MCM-C. Today, the use of this technology is limited, since the manufacturing cost is still high. (*Courtesy of Mitsubishi Electric Corp.*)

REFERENCES

1. J. S. Prokop and D. W. Williams, "Chip Carriers as a Means for High-Density Packaging," *IEEE Trans. Components, Hybrids, Manuf. Technol.,* **3,** 297 (Sept., 1978).
2. J. R. Howell, "Reliability Study of Plastic Encapsulated Copper Lead Frame Epoxy Die Attach Packaging System," *Proceedings of the International Reliability Physics Symposium,* 1981, p. 104.
3. B. Prince, *Semiconductor Memories,* 2nd ed., Wiley, New York, 1991, p. 209.
4. N. Miyazaki and Y. Hayashi, *Nikkei Microdevices,* 37 (Mar., 1992).
5. K. Ohtsuka, "Trend of ASIC Package Technology," *Semic. World,* 114, (Jun., 1991).
6. C. J. Bartlett, "Advanced Packaging for VLSI," *Sol. State Technol.,* 119 (Jun., 1986).
7. I. Anjoh, "Revolutions in Leadframe for Memory, LOC Structures for 16MEGDRAM," *Nikkei Microdevices,* 89 (Feb., 1991).
8. W. C. Ward, "Volume Production of Unique Plastic Surface-Mount Modules for the IBM 80-ns 1-Mbit DRAM Chip by Area Wire Bond Techniques," *Proceedings of the 38th Elect. Components and Technol. Conference,* 1988, p. 552.
9. N. Ueda, "Studies on a New LOC Structure Package," *Proceedings of the VLSI Packaging Workshop,* III-4, 1993.
10. T. Chiu, "High-Density Memory Packaging with Vertical Surface Mount Package (VPAK)," *Proceedings of the 1st VLSI Packaging Workshop,* 1992, p. 37.
11. B. Frayman, "Overmolded Plastic Pad Array Carriers (OMPAC): A Low-Cost, High-Interconnect-Density IC Packaging Solution for Consumer and Industrial Electronics," *Proceedings of the 41st Elect. Components and Technol. Conference,* May, 1991, p. 176.
12. R. R. Tummala, "Electronic Packaging in the 1990s—A Perspective from America," *IEEE Trans. Components, Hybrids, Manuf. Technol.,* **14(2),** 262 (Jun., 1991).
13. H. Wessely, "Electronic Packaging in the 1990s—A Perspective from Europe," *IEEE Trans. Components, Hybrids, Manuf. Technol.,* **14(2),** 272 (Jun., 1991).
14. T. Ohsaki, "Electronic Packaging in the 1990s—A Perspective from Asia," *IEEE Trans. Components, Hybrids, Manuf. Technol.,* **14(2),** 254 (Jun., 1991).
15. R. Iscoff, "Thin Outline Packages, Handle With Care!," *Semic. Intl.,* 78, (1992).
16. K. Fujita, "Development of Ultra-Thin Surface Mounting IC Package (0.8mm Thick)," *Proceedings of the 40th Elect. Components and Technol. Conference,* May, 1990.
17. M. Kakei, "Low Stress Molding Compounds for VLSI Devices," *Nikkei Microdevices,* 82 (1984).
18. C. E. T. White and J. Slatery, "An Update on Preforms," *Circuit Manuf.,* 78 (Mar., 1978).
19. *Bonding Tools and Production Accessories, Bonding Handbook and General Catalog,* Kulicke and Soffa Industries (1980).
20. S. Uchida, "Technology Trend for Leadframe Manufacturing," *Proceedings of the SEMI Tech. Symposium,* November, 1988, p. 423.
21. R. A. Garner and R. W. Nufer, "Properties of Multilayer Ceramic Green Sheets," *Sol. State Technol.,* 38 (May, 1974).
22. N. Mogi and H. Yasuda, "Development of High-Reliability Epoxy Molding Compounds for Surface-Mount Devices," *Proceedings of the 42nd Elect. Components and Technol. Conf.,* May, 1992, p. 1023.
23. K. Tomiyoshi, "The Technical Trend of Epoxy Molding Compounds for Thinner Package of Semiconductor Devices," *Proceedings of the SEMI Tech. Symposium,* Oct., 1990, p. 399.
24. T. Sudo, "Considerations on Package Design for High Speed and High Pin Count CMOS Devices," *Proceedings of the 39th Elect. Components and Technol. Conference,* 1989, p. 531.

25. H. R. Kaupp, "Characteristics of Microstrip Transmission Line," *IEEE Trans. on Computers,* **EC-16,** 185 (Apr., 1967).
26. R. K. Hoffmann, *Handbook of Microwave Integrated Circuits,* Artech House, Inc., Norwood, MA, 1987.
27. B. K. Bhattacharyya, "Ground Plane Design Parameters for CMOS VLSI Multilayer Packages," *Proceedings of the Technical Conference, Ninth Annual International Electronics Packaging Conference,* International Electronic Packaging Society, 1989, p. 659.
28. D. Mallik and B. K. Bhattacharyya, "High-Performance PQFP," *Proceedings of the 39th Elect. Components and Technol. Conference,* **VIII-1,** 1989, p. 493.
29. D. Mallik, "Multi-Layer Molded Plastic Package," *Proceedings of the 1989 Japan Intl. Elect. Manuf. Technol. Symp.,* **B4-2,** (Apr., 1989).
30. M. Aghazadeh and D. Mallik, "Thermal Characteristics of Single and Multilayer High Performance PQFP Package," *Proceedings of the 6th IEEE SEMI-THERM Symp.,* 33 (1990).
31. K. Manchester and D. Bird, "Thermal Resistance; A Reliability Consideration," *IEEE Trans. Components, Hybrids, Manuf. Technol.,* **31(4),** 550 (Dec., 1980).
32. P. G. Gabuzda, "Air Management System Yields High-Performance Cooling," *Electronic Packaging & Production,* 64 (May, 1988).
33. R. C. Chu, "Thermal Consideration and Techniques for Electronic Digital Computers," *Proceedings of the 9th IECP Symposium,* **5-3,** 1989, p. 3.
34. S. Kawai, "Structure Design of Plastic IC Packages," *Proceedings of the SEMI Tech. Symposium,* November, 1988, p. 349.
35. I. Fukuzawa, "Moisture Resistance Degradation of Plastic LSIs by Reflow Soldering," *Proceedings of the International Reliability Physics Symposium,* 1985, p. 192.
36. O. Nakagawa, "High Density Packaging Technology," *J. Electron. Matls.,* **18(5),** 633 (1989).
37. T. Nishioka, "Special Properties of Molding Compound for Surface Mounting Devices," *Proceedings of the 40th Elect. Components and Technol. Conference,* May, 1990, p. 625.
38. S. Ito, "Molding Compounds for Thin Surface Mount Packages and Large Chip Semiconductor Devices," *Proceedings of the 41st Elect. Components and Technol. Conference,* May, 1991, p. 190.
39. M. Kitano, "Analysis of Package Cracking during Reflow Soldering Process," *Proceedings of the International Reliability Physics Symposium,* 1988, p. 90.
40. J. H. Lau, *Solder Joint Reliability: Theory and Applications,* Van Nostrand Reinhold, New York, 1991.
41. J. H. Lau, and G. H. Barrett, "Stress and Deflection Analysis of Partially Routed Panels for Depanelization," *IEEE Trans. Components, Hybrids, Manuf. Technol.,* **10(3),** 411, (Sept., 1987).
42. H. D. Solomon, "Solder Joint Acceleration Factor," *J. Electronic Packaging, Trans. of ASME,* **113,** 186 (Jun., 1991).
43. H. D. Solomon, "Fatigue of 60/40 Solder," *IEEE Trans. Components, Hybrids, Manuf. Technol.,* **9,** 423 (Dec., 1986).
44. J. H. Lau, "Experimental and Analytical Studies of 208 Pin Fine Pitch Quad Flat Pack Solderjoint Reliability," *IEEE Trans. Components, Hybrids, Manuf. Technol.,* **9,** 122 (Jun., 1991).
45. J. H. Lau, "Solder Joint Reliability of Fine Pitch Surface Mount Technology Assemblies," *IEEE Trans. Components, Hybrids, Manuf. Technol.,* **13(3),** 534 (Sept., 1990).
46. J. H. Lau, "Thermal Stress Analysis of TAB Packages and Interconnections," *IEEE Trans. Components, Hybrids, Manuf. Technol.,* **13(1),** 182 (March, 1990).

47. J. H. Lau, "Thermal Stress Analysis of SMT PQFP Packages and Interconnections," *J. Electronic Packaging, Trans. of ASME,* **111,** 2 (Mar., 1989).
48. M. L. White, "Attaining Low Moisture Levels in Hermetic Packages," *Proceedings of the 20th International Reliability Physics Symposium,* 1982, p. 253.
49. M. L. White, "The Removal of Die Bond Epoxy Bleed Material by Oxygen Plasma," *Proceedings of the 32nd Elect. Components and Technol. Conf.,* 1982, p. 262.
50. W. Paulson and R. Kirk, "The Effect of Phosphorous Doped Passivation Glass on the Corrosion of Aluminum," *Proceedings of the International Reliability Physics Symposium,* 1974, p. 172.
51. T. C. May and M. H. Woods, "Alpha-Particle-Induced Soft Errors in Dynamic Memories," *IEEE Trans. Electron Dev.,* **26,** 2 (1979).
52. K. Yamaguchi and M. Igaroshi, "Screen Printing Grade Polyimide Paste for Alpha-Particle Protection," *Proceedings of the 36th Elect. Components and Technol. Conf.,* May, 1986, p. 340.
53. M. M. White, "The Use of Silicone RTV Rubber for Alpha Particle Protection on Silicon Integrated Circuits," *Proceedings of the International Reliability Physics Symposium,* 1981, p. 43.
54. R. L. Clain, "Beam Tape Carriers—A Design Guide," *Sol. State Technol.,* March, 1978, p. 53.
55. K. Hatada, "New Film Carrier Assembly Technology: Transferred Bump TAB," *IEEE Trans. Components, Hybrids, Manuf. Technol.,* **10(3),** 335 (1987).
56. M. Kohara, "New Development of Thin Plastic Package with High Terminal Counts," *IEEE Trans. Components, Hybrids, Manuf. Technol.,* **13(2),** 401 (1990).
57. T. Ueda, "High Pin Count and High Power Package," *IEEE Proceedings of the VLSI & GaAs Chip Packaging Workshop,* 1991, p. 13.
58. T. Inoue, "Micro Carrier for LSI Chip Used in the HITAC M-880 Processor Group," *Proceedings of the 41st Elect. Components and Technol. Conf.,* 1991, p. 704.
59. P. A. Totta and R. P. Sopher, "SLT Device Metallurgy and its Monolithic Extension," *IBM J. Res. Devel.,* **13,** 226 (1969).
60. J. Sloan, "Over Molded Pad Array Carrier (OMPAC): A New Kid on the Block," *Proceedings of the 1st VLSI Packaging Workshop,* Nov., 1992, p. 17.
61. K. Hatada, "Applications of New Assembly Method—Micro Bump Bonding Method," *Proceedings of the 1989 Japan Intl. Elect. Manuf. Technol. Symposium,* 1989, p. 47.
62. H. Fujimoto, "Bonding of Ultrafine Terminal-Pitch LSI by Micron Bump Bonding Method," *Proc. IMC1992,* 1992, p. 115.
63. K. Nakamura and Y. Nagahiro, "Emerging MCMs Behave Like System on Silicon," *Nikkei Microdevices,* 28 (April, 1993).

PROBLEMS

1. Estimate the necessary package pin count for a 40,000-gate, logic-gate-array chip using Rent's rule (Eq. 10.2). Assume that $\alpha = 1.5$, $\beta = 0.5$, and that the chip needs one pair of power/ground package pins per 10 signal pins (or signal terminals).

2. Suppose you have an Au-Al bond, and the metallurgical diffusion thickness (with Al as a solvent metal and Au as a diffusion metal) increased by 1 μm during a 150°C time-temperature storage operation. Calculate the period of time using Eqs. 10.3 and 10.4. Assume $D_o = 2.2 \times 10^{-4}$ (m^2/sec), $Q = 134$ (kJ/mol), and $R = 8.31$ (J/mol-K).

3. A high-temperature storage considerably accelerates Au-Al diffusion and affects the bond quality. Calculate the diffusion thickness increase caused by a 200°C, 60-day storage operation. Use the same assumptions as in Problem 2.

4. What chip size is required to place 352 pads (I/Os) on a pad-limited chip? Assume that the chip is square, the bonding-pad pitch is 120 μm, and the chip size is defined by the loci of bonding-pad centerlines. Use Fig. 16 to find the answer.

5. Derive Eq. 10.13. (Note that $l \gg a$).

6. Suppose you have a 5-volt, 1-nsec output-rise-time silicon chip with 68 signal leads and you want to package it so that inductive noise in the ground line is kept to 0.2 volt. You want to switch 18 lines simultaneously, and you want each package lead to have approximately 7 nH inductance. The output buffers draw 25 mA. How many package leads do you need for this device, assuming that there are eight power leads?

7. A high-power plastic packaged device that allows a maximum junction temperature of 85°C dissipates 10 watts in an ambient of 40°C. Calculate the required thermal resistance (θ_{ja}) of the package.

8. Discuss how the package described in Problem 7 affects the next-level (PWB) packaging and the final system construction.

CHAPTER 11

Wafer Fab Manufacturing Technology

T. F. Shao and F. C. Wang

11.1 WHAT IS MANUFACTURING?

Semiconductor manufacturing is often referred to as a *high-tech industry.* High-tech industries are typically characterized by a multidisciplinary approach and the quick adoption of advances in science and technology to commercial production. This chapter describes how technologies are implemented for manufacturing.

Manufacturing is one link in the iterative economic chain of providing goods for society. The following is a typical flow of goods:

> Customer's need → Conception of products → Product design →
> Product prototyping → Test marketing → Sales → Manufacturing →
> Customers → Payment → Customer satisfaction → More customer's needs →

In this economic chain, the key role of manufacturing is to produce products the customer wants. This role must be supported by the timely infusion of the necessary technologies to permit a manufacturer to continue serving its customers and make a profit.

11.1.1 Critical Success Factors

To serve the customers, the critical success factors considered important are these:

- *Quality:* Quality is only the first step in satisfying customers' demand.[1–6] Products must meet the customer's application requirements. These requirements typically are presented in the form of a product specification. To produce an acceptable product, the manufacturing process must have built-in quality control to ensure that the output meets the customer's requirements.

A modern integrated circuit is produced through hundreds of steps that involve sophisticated equipment and precise control either by an automatic or a manual discipline. Quality can be achieved, not by inspection and screening, but only through well-proven and controlled processes. The typical methodology for process control will be described later in this chapter.

- *Cost:* Given the same quality of product, customers invariably will compare the price of the product. Generally, the lower-priced, but equal-quality, product is bought. Thus, a manufacturer must constantly make sure that cost is low enough that the market price results in a profit.
- *Delivery:* Customers expect goods with the promised quality within the promised delivery time.[2] In many instances, quick delivery can result in premium prices or higher market share.

  Thus, the manufacturer must put in place the necessary infrastructures and training to produce integrated circuits within short cycle times and in the quantities that customers order. For commodity products such as DRAMs, the quantity can be in the millions; for prototype circuits, the quantity can be under 10.
- *Service:* Service is the "soft" side of hardware manufacturing in providing goods for the customers. The manufacturer must be able to provide services,[6] such as, how to select the right product for the customer's application; how to apply the product in the customer's application; how to troubleshoot; how to adjust delivery when customer's need for the product changes; etc. The customer service related to the product can become more and more important when the quality/cost/delivery characteristics of all manufacturers are similar.

11.1.2 Manufacturing Flow

An integrated circuit provides integration of active and passive components on a common substrate to perform desired functions. Typically, the structure is manufactured by a process involving photopatterning, film growth, implantation, diffusion and film etching, etc. A typical process flow is given in Table 1 to illustrate the process classifications. A measure of the complexity of the manufacturing flow is the number of photopatterning levels. It is not uncommon to see ICs made with more than 20 levels, and the total number of process steps can easily exceed 200.

11.1.3 Manufacturing Organization

The organization for manufacturing ICs is determined by the relation of manufacturing flow and its functional blocks as shown in Fig. 1. Based on the functional relationships, the goals of the total organization, and other factors such as the experience level of manpower, local perceptions, etc., many forms of organization can be structured. Two of these forms are known as classical organization, Fig. 2, and modular organization, Fig. 3. Each organizational structure has its pros and cons. The classical organization enhances professional development along its special func-

TABLE 1
Schematic illustration of a typical wafer fab manufacturing process flow

Process steps	Equipment used	Photo	Thin film	Ion implant	Diffusion	Etch
Wafer ID marking	Laser marker					
First oxidation	Furnace				X	
First nitride deposition	CVD furnace		X			
n-tank pattern	Coater/stepper/developer	X				
First nitride etch	Plasma etcher					X
n-tank implant	Ion implanter			X		
Stripe nitride	Hood				X	
p-tank pattern	Stepper	X				
Tank drive	Furnace					X
Second oxidation	Furnace					X
Second nitride deposition	CVD furnace		X			
Moat pattern	Coater/stepper/developer	X				
Second nitride etch	Plasma etcher					X
Field oxidation	Furnace				X	
Stripe nitride	Hood				X	
Gate oxidation	Furnace				X	
Poly 1 Deposition	CVD furnace		X			
⋮						
BPSG deposition	CVD reactor		X			
Reflow	Furnace				X	
Contact pattern	Coater/stepper/developer	X				
Contact etch	Plasma etcher					X
First metal deposition	Sputtering machine		X			
⋮						
Protective overcoat	CVD reactor		X			
Pad pattern	Coater/printer/developer	X				
Wafer test	Tester				X	

tional line. Experts can be developed in depth. The modular organization enhances the team effort by the process, equipment, and manufacturing in a given machine group such as photo, etch, diffusion, etc. Generally speaking, for any organizational structure, an interorganizational team is necessary to minimize the barriers that develop along the organizational lines and to enhance the cross-organizational teamwork.[2,3] The interplay and effectiveness of the organization cannot be rigidly prescribed. Management should comprehend the objectives and other background factors, then structure the organization appropriately.

11.1.4 The Cost Structure of Manufacturing

Semiconductor manufacturing is capital-intensive. Each wafer fab typically requires investment of several hundred million dollars. The physical building and facilities

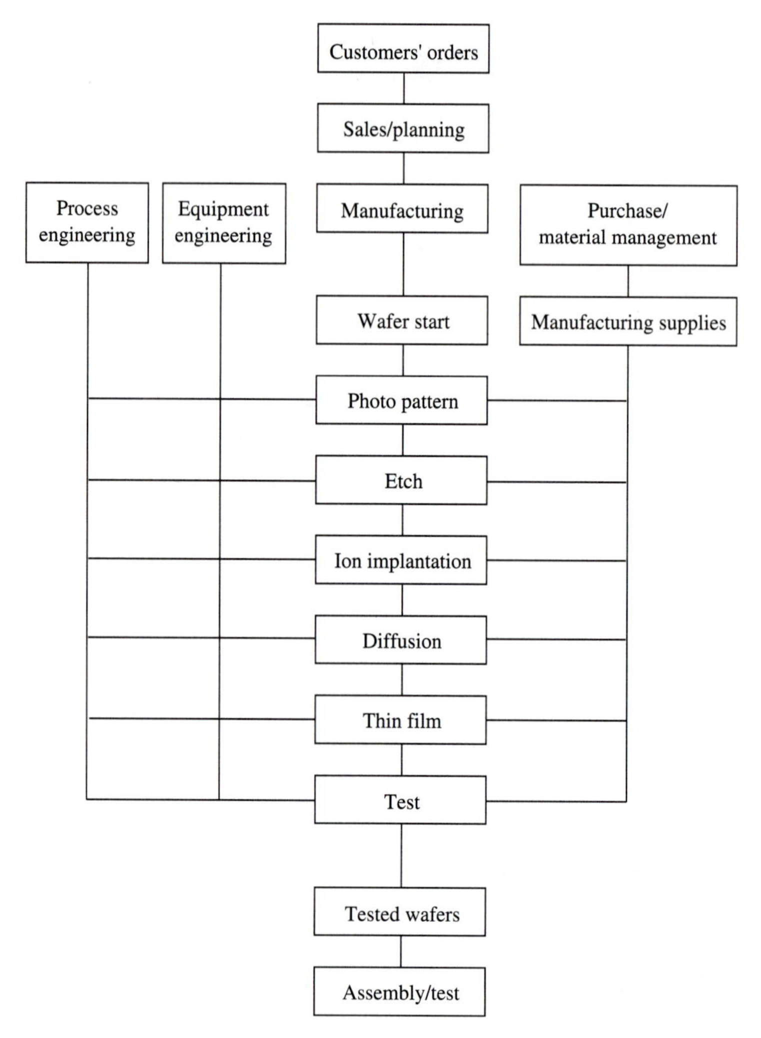

FIGURE 1
Manufacturing flow and functional block diagram.

can cost $100 million or more. Major process equipment may be priced on the order of $1 million. Beyond the capital cost, a representative monthly operating cost of a factory is several hundred thousand to more than $10 million consisting of material, labor, and overhead.

The function of manufacturing is to utilize the capital-intensive factory to produce the integrated circuits in a cost-effective manner. The speed with which a factory can move from start-up to full capacity is becoming increasingly important to meeting financial expectations and reducing risk of technology turnover.

How to manage a company for profit and growth is itself a subject requiring serious study by the management people.[1–3,5,6] For students starting on a career in

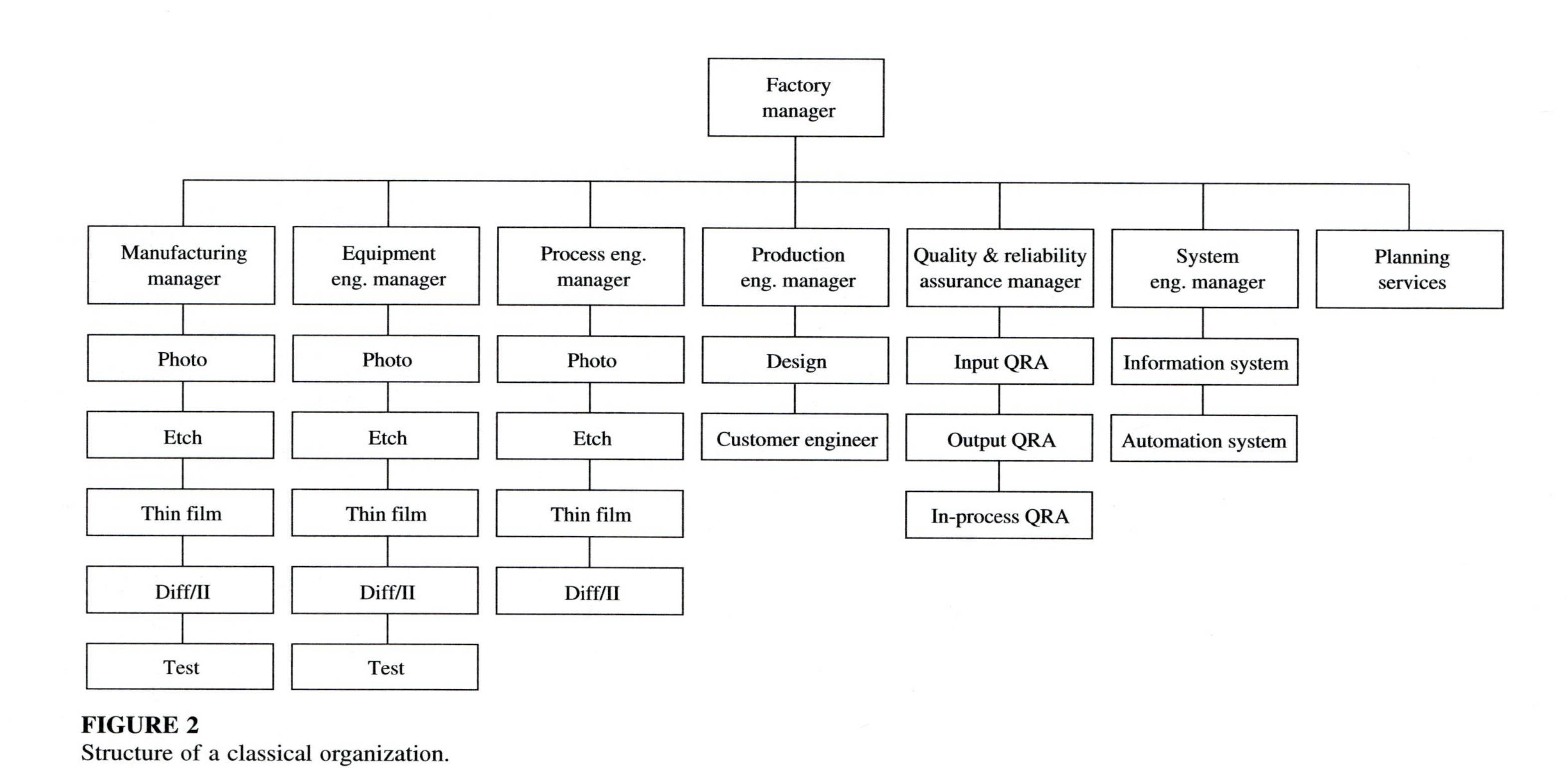

FIGURE 2
Structure of a classical organization.

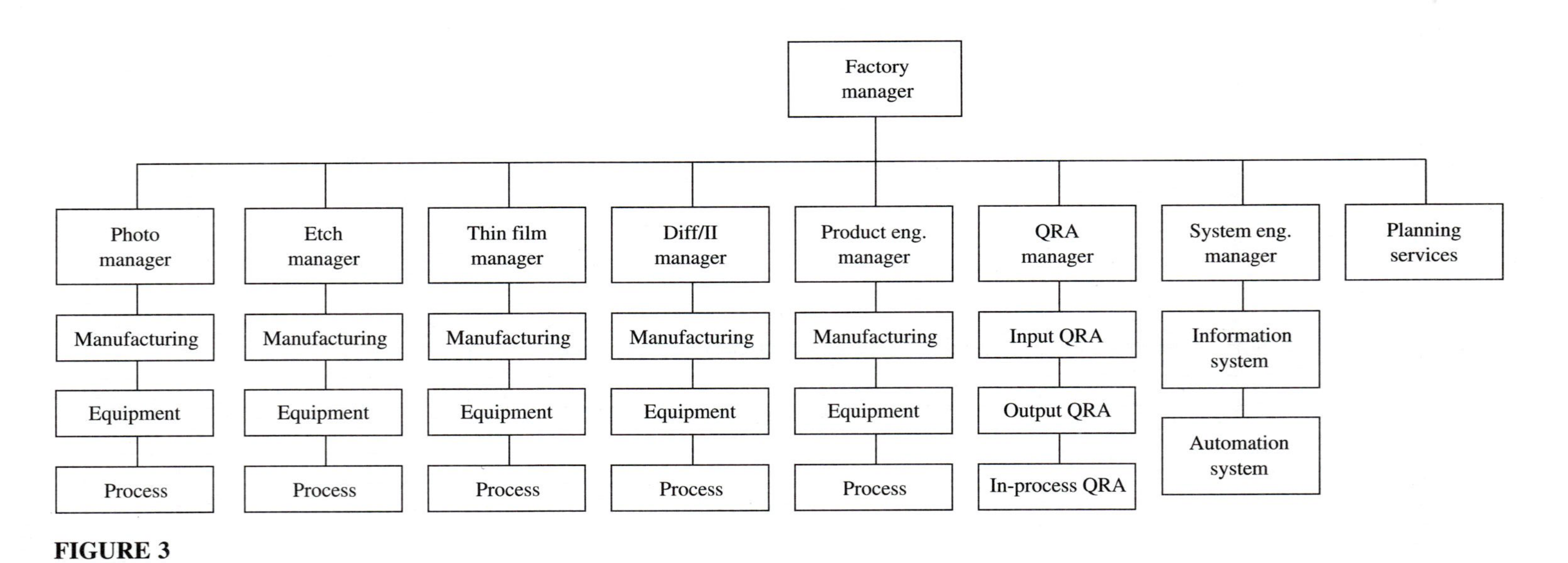

FIGURE 3
Structure of a modular organization.

engineering and manufacturing, it is important to know that each innovation or change must be reviewed with economic judgment before implementation in manufacturing.

Wafer fab manufacturing is very complicated with many dependent variables. A wafer fab commonly has several hundred pieces of equipment consisting of more than 30 different machine types. This equipment is operated continuously by several hundred operators in shifts around the clock. Under these conditions it takes months to complete one product cycle after hundreds of processing steps. Yet, the finished products must consistently meet customers' standards of quality and reliability. For commodity volume products, such as DRAMs in particular, several million units must be fabricated a month. To deal with such complexity, built-in manufacturability is most important, with sufficient margin at the start-up of a wafer fab and good process control to sustain volume production. The control and reduction of all variations[4,6] are emphasized for wafer fab manufacturing, which may not be the case for other chapters. To ensure that any subsequent changes enhance manufacturability but do not have a detrimental impact requires a control system for the improvement process. Those are the subjects to be considered in subsequent sections.

11.2 WAFER FAB MANUFACTURING CONSIDERATIONS

In this section, several common manufacturing considerations are introduced. These considerations are aimed at establishing the supporting foundation for designed-in manufacturability.

11.2.1 Typical Process Layout Considerations

It is difficult to discuss process layout without using an actual wafer fab design. However, several manufacturing concerns for layout follow:[2]

- *Relationship to utility connection:* Since the nature of each process group (diffusion, photo, etching, thin film) is different, the utility requirements are also different. For example, deionized water (DIW) and chemicals are used primarily by the wet-hood operation. Grouping similar processes in the process layout provides advantages by simplifying cost savings in acquiring supplies and by reducing maintenance efforts.
- *Wafer moving path:* Wafers move constantly during the entire manufacturing cycle from one equipment unit to another, from one process (area) to another process (area), and from a process to measurement/inspection. The moving paths based on the process flow (Fig. 1) should be planned carefully to minimize the movement of the wafers so as to reduce cycle time and defects.
- *Routing for manufacturing supplies:* In addition to the wafers that are being produced, other manufacturing supplies such as pilot/dummy wafers, quartzware, etc. also move constantly inside the wafer fab to support process needs. Routing considerations are similar to the foregoing.

- *Inventory storage:* Wafers waiting to be processed are stored somewhere inside the wafer fab. Sufficient storage space at appropriate locations must be provided in the layout to furnish easy access and clear segregation to avoid improper operation. Inventory storage of manufacturing supplies requires the same layout considerations.
- *Cross-contamination:* Processes that use little or no phosphorus and those that use much phosphorus should be segregated in the layout of equipment, process tools (wafer cassette, wafer transfer, etc.), and measurement instruments so as to avoid cross-contamination. Nonmetal and metal processes should be segregated as well.
- *Other considerations:* The layout must account for other manufacturing needs such as storage space for instructions (specifications, procedures, notes), housekeeping materials, maintenance parts/tools, etc. to ensure that the working area is clean and orderly.

Readers are encouraged to gain personal experience by studying actual wafer movement inside a wafer fab to appreciate the problems that can result from improper layout.

11.2.2 Automation

With the increased complexity of ICs and the sophistication of process equipment, automation is almost a necessity. The degree of automation is a trade-off of cost and benefit and is limited by the maturity of the automation hardware and software. Automation is also an important planning and control tool for wafer fab manufacturing. Automation can be classified into two categories:

- *Physical automation*

  Factory utility control: Temperature and humidity are usually controlled automatically (see Chapter 1 on cleanroom technology). The level of airborne particulates and the purity of chemicals are monitored, and equipment can be set to issue warning and shutdown signals.

  Wafer transfer: Wafers can be automatically transferred from one process equipment to another (e.g., cluster tools).

  Wafer processing: Wafers arriving at a given process equipment can be automatically loaded then unloaded after completion of processing. The process recipe for each process can be automatically downloaded from a central computer, which sets up the equipment for processing. Cross-checking by the computer can practically eliminate all errors that otherwise may be caused by human operator mistakes.

- *Information automation*

  Inventory control: All wafers or lots are given an identification and can be tracked throughout the processes. The movement of the inventory can be controlled by computer based on customer's need, equipment conditions, special instructions from engineering, etc.

Equipment control: The performance and maintenance status of the production equipment can be tracked by computer. This control will facilitate both production scheduling and equipment monitoring for capacity planning.
Cost collection: For purposes of financial decisions the cost of wafer processing can also be determined as each wafer passes through the process steps.
Technical data: At each process point, the input and output data, as well as real-time process conditions, can be collected and compared against the requirements (specifications). In case of a deviation, either automatic adjustments can be made or a signal sent as a warning or for shutdown. Technical data can be used not only for engineering purposes but also for process control.

11.2.3 Material Requirement

The quality and consistency of the materials greatly affect the yield and performance of the devices. Typically, materials can be separated according to whether they do or do not have direct contact with wafers during processing. Materials in contact include the wafers themselves, chemicals, gases, metal targets, and photo masks; these have more stringent requirements. Some materials do not have direct contact with wafers but still impact the process results, for example, quartz, wafer cassettes, cleanroom clothing, and cleanroom cleaning cloth for equipment. These critical materials should undergo incoming quality checks and specification controls to ensure that requirements for quality and consistency of supply are met. Other materials used inside the wafer fab, but not specifically mentioned here, are less critical except for considerations of minimal particle and contamination generation, cost, and supply convenience.

The material requirements for modern wafer fab manufacturing are extremely difficult to discuss because of the enormous variety of types of materials and broad ranges of specifications used for the IC industry. Instituting a requirement without understanding the exact process details and device parametric design is, in fact, dangerous. We will provide some basic material considerations in light of the importance of the subject to manufacturing success. Those generic requirements will attempt to balance both today's and near future advanced wafer fab needs over a broad scale of product varieties. The suppliers' capabilities are an integral part of material considerations. We do not want to make impractical demands and bear unnecessary high costs, but on the other hand, we do not want to compromise on actual process needs.

Direct material considerations

This section addresses wafers, chemicals, gases, metal targets, and photomasks. In Table 2, several common requirements[7–11] for 150- and 200-mm wafer processes are listed. ASTM[7] and SEMI[8] have on-going activities to discuss, upgrade, and standardize wafer requirements. For more details, please refer to ASTM/SEMI documents or attend meetings of the respective committees.

Unlike the wafer, there have been fewer attempts to standardize the chemical property requirements.[12] This is understandable since the effects of chemicals depend more on process sequence and the criticality of the process to device

TABLE 2
Typical wafer requirements for manufacturing

Item	Bare or EPI wafer	Purpose	Typical specification
General			
Diameter	B and E	Wafer size	150 or 200 ±0.5 mm
Orientation	B and E	Film growth and implantation	(100) ±1
Conductivity type	B and E	Type of doping for conductivity	P(boron), N(phosphorus)
Electrical			
Resistivity	B	Process and control device parameter	1–20 ohm-cm
	E		0.001–0.01 ohm-cm
Radial variation	B and E	Control of variation across wafer	< 15%
Minority carrier lifetime	B	Control of impurity concentrations	> 200 μsec
Mechanical			
Thickness	B and E	Mechanical strength and process equipment features	600–750 ±40 μm
Bow and wrap	B and E	Mechanical stress and photoprocess	> 60 μm
Global flatness	B and E	Control of thickness variation across water	< 20 μm
Site variation	B and E	Control of local thickness uniformity for photo exposure field	> 1 μm focal plane deviation Min. size 20 mm× 20 mm
Surface finish	B	Specify surface preparation technique for mechanical specifications and particle/contamination control	Chemical/mechanical polish
Backside finish	B and E	Particle and gettering control	Soft blasted/etched, oxide or poly-Si coated
Edge profile or chem/mech	B and E	Particle control and microcrack elimination	Edge beveled/etched polished
Structural			
Oxygen-induced stacking fault (OISF)	B	Minimize additional process-included defect generation	< 100 /cm^2
Dislocation	B	Same as above	< 50 etch pits/cm^2
Slip line	B	Same as above	None

TABLE 2 ***(continued)***
Typical wafer requirements for manufacturing

Item	Bare or EPI wafer	Purpose	Typical specification
Chemical			
Oxygen	B	Control of internal gettering effectiveness	20–32 ppma
Carbon concentration	B	Control of carbon-related nucleation for gettering	< 1 ppma
Surface metal concentration	B	Minimize variation of device electrical parameters related to surface contamination	$< 5 \times 10^{10}$ atoms/cm^2 for Fe, Ca, Na, Ni, Cr, Zn, Cu, Mg, etc.
Surface features			
Particle	B and E	Starting material defect control for yield consideration	< 0.16 /cm^2 @ 0.2 μm
Haze	B	Same as above	None detectable
Other surface defects (pits, dimples, cracks, chips, scratches, organic peels, contaminations, etc.)	B and E	Same as above	None visible

performance. For example, the importance of the impurity levels in chemicals varies according to whether the NH_4OH–H_2O_2–H_2O/HF-based or HCl–H_2O_2–H_2O-based wet cleanup is used as the last cleanup sequence for critical processes such as pre-gate oxidation. Yet, there is more demand for lower impurity levels for NH_4OH, H_2O_2, HF, or HCl than for H_2SO_4 or H_3PO_4 as a result of the process criticality itself (H_2SO_4-based photoresist or HNO_3-based quartz cleanups naturally are dirtier). In addition, not all impurities in the same chemical are important. Concerns about typical inorganic impurities (such as Al, Ca, Cu, Cr, Fe, Na, Ni, Mg, and Zn) are based on the known impact of these elements on device electrical performance [such as threshold voltage V_t shift, weak gate oxide integrity (GOI), and degraded minority carrier lifetime].[13] Because an element is not mentioned above does not imply that it has no effect on device performance or makes no contribution to manufacturing variations. The reader is advised to look into the contribution of specific elements to specific device parameters or under sporadic abnormal situations (e.g., excessive contamination). Refer to Chapter 2 for details on cleanup.

In Table 3, typical specifications for several elements of interest in some common chemicals are provided. The actual level of impurity for those elements, as well as the consistency of contamination from one lot to another, vary by orders of magnitude between different industrial suppliers. Readers must check the actual quality and consistency performance levels and cost with respect to individual process needs before selecting a particular vendor to supply a specific chemical. Impurities

TABLE 3
Typical chemical requirements for manufacturing

Impurity	LDL[†]	IPA[†]	Acetone	HF	NH_4OH	H_2O_2	HNO_3	HCl	H_2SO_4	H_3PO_4
Particles	> 0.5 μm	50/mL	20/mL	5/mL	30/mL	50/mL	50/mL	50/mL	30/mL	50/mL
					Parts per billion					
Chloride	< 1	50	50	10	200	100	100	—	100	1000
Nitrates	< 100	1000	—	10	—	2000	—	—	200	5000
Phosphates	< 10	50	50	5	100	200	50	50	500	—
Sulfates	< 100	500	—	10	500	200	100	50	—	10000
Aluminum	< 0.1	2	5	1	1	1	1	5	5	300
Calcium	< 0.1	1	0.5	1	3	3	10	5	5	500
Chromium	< 0.1	1	0.5	1	1	1	1	2	2	—
Copper	< 0.1	1	0.5	1	1	1	1	1	2	50
Iron	< 0.1	1	0.5	1	1	1	10	10	5	500
Magnesium	< 0.1	1	0.5	1	1	1	2	2	5	500
Nickel	< 0.5	1	0.5	1	1	1	1	2	2	50
Potassium	< 0.1	1	1	1	1	1	2	5	5	100
Sodium	< 0.1	5	5	1	1	5	5	10	5	100
Zinc	< 0.5	1	0.2	1	1	1	5	2	2	—

[†] LDL:lower detection limit
IPA: isopropyl alcohol

not listed in Table 3 are commonly below detection limits, however, detection limits may vary based on the vendor's analytic capability. Similarly, the particle content in chemicals varies greatly. This is related not only to the manufacturer's process capability and control of particles in chemicals but also to additional particle generation arising from packaging materials and transportation and to the sampling and measuring techniques used. The particle quality is typically guaranteed through extensive filtration in the distribution system rather than from direct control on incoming chemicals.

Gas requirements are identical to those for chemicals.[14] In Table 4 some typical purity specifications for common wafer fab gas supplies are listed. Specifications in Table 4 refer to values at the outlet of the wafer fab distribution systems after filtration. Once again, actual performance may be well below the specification value. Metallic impurities in gases are typically not specified and rely on the vendor's guarantee.

Specialty chemicals are primarily those used for photo and metal cleanup-related processes. A common requirement for photoresists, developers, SOG, and polymide is the transportation and storage temperature. The photosensitivity and lifetime of a photoresist varies with temperature, hence the need to maintain temperature of materials within specifications of 5 to 20°C for photoresist/developer and −20 to 10°C for SOG/polyimide. The specialty gases are primarily those used for deposition (SiH_4, DCS, PH_3, NH_3, HCl, etc.) and plasma etching (NF_3, CF_4, C_2F_6, BCl_3, HBr, etc.). Typically, common impurities such as N_2, O_2, H_2O, and THC (total hydrocarbon) are in the sub- to several-ppm range, as are metallic elements of concern. The required limits of the manufacturing by-products in a specific specialty gas—for example, $SiCl_4$, $SiHCl_3$, and SiH_3Cl in SiH_2Cl_2 (DCS)—can be handled only on a case-by-case basis. The interior coating materials for the cylinder, the connection valve assembly, and the vendor's handling and control procedures for cylinder preparation and gas filling are also areas of concerns in providing quality specialty gases. In most cases one would rely solely on the manufacturer's guarantee to meet the material requirements for specialty chemicals and gases routinely.

In Table 5 typical metal-target material requirements are listed. These are primarily guaranteed by the manufacturers at the time of delivery. As the materials are used, continuous cooperation with the manufacturer is necessary to understand the target degradation characteristics and target-to-target consistency.

Manufacturing considerations for photomasks emphasize defects and consistency. Defects and critical dimensions (CD) control within a mask and from photomask to photomask are important parameters, since they transfer to the patterned wafer every time it prints. Table 6 lists several material requirements for photomasks.

Deionized water (DIW) of high purity is actually an important material supply[15] but is often forgotten as a direct material, since it is generated at the wafer fab rather than being delivered by a vendor. DIW has been used extensively for all the wafer fab manufacturing process cleanup activities and for supply material preparation. In Table 7, typical DIW requirements are listed. The parameters in Table 7 should be monitored routinely at the DIW supply system and at the point of use.

All the requirements discussed in this section are for reference only. One should not apply them blindly.

TABLE 4
Typical gas requirements for manufacturing†

Gas	Purity (ppb) Ar	N_2	O_2	H_2	CO	CO_2	THC‡	H_2O	Particles #/FT^3 at $\geq 0.1\mu m$
Nitrogen	—	—	50	50	50	50	10	50	5
Oxygen	500	100	—	50	50	50	10	50	5
Hydrogen	10	20	10	—	50	50	10	50	5
Argon	—	50	50	50	50	10	50	50	5
Helium	—	50	50	50	50	50	10	50	5

† Specifications refer to values at the outlet of the distribution system after filtration
‡ THC: Total hydrocarbon

TABLE 5
Typical metal target requirements for manufacturing

Item	Purpose	Typical specification
General		
Dimension and shape	Match sputter machine configuration	Per sputter machine drawing for bonded target thickness and diameter
Bonding	Specify bonding requirements between target itself and backing plate for bonding area, backing plate materials, and minimum withstanding power	Per sputter machine specifications
Chemical		
Composition	Specify major constituents of target	Per process specification
Purity	Total purity level	> 99.99%
Total metallics	Control of metal impurity	< 100 ppm
Metallic impurity	Same as above	< Few ppm per each element
Other impurity	Control of nonmetallic impurity	< 1000 ppm per each element (H, N, O, C, etc.)
Visual defects		
Appearance	Quick check on surface feature nonuniformity	Uniform appearance
Stain, surface contamination	Quick check on surface quality	None visible
Chip, pit, crack, scratch, etc.	Same as above	None visible

TABLE 6
Typical photomask requirements for manufacturing

Item	Purpose	Typical specification
Reticle		
Reticle ID	Ensure proper device levels and test patterns are fabricated	Vendor code, device name, level code, part no., reticle lot no., serial no., alignment mark no.
Reticle material	Material specification	Cr/Cr_2O_3 or $Cr_2O_3/Cr/Cr_2O_3$ on quartz
Reduction ratio	Specify reduction ratio	Per stepper specification
Pattern defect	Define defect types (contaminant, opaque spot, hole, excess Cr, Cr line intrusion and scratch) and sizes that affect printing results	None $\geq 2\ \mu m$
Registration type	Specify needs for position accuracy	Position accuracy $< 0.5\ \mu m$
CD criteria	Control reticle CD variation within photomask and mask to mask	Per design CD tolerances
Tone	Specify tone convention	Normal: black Inverse: white
Alignment mark and chip coordinate	Align photomask to stepper	Per design and stepper specifications
Border line width	Define actual printing area on wafer	6 mm from edge
Pellicle		
Pellicle material	Specify pellicle source material	Per vendor specification
Printable particle	Define minimum defect size that affects printing	$\geq 40–60\ \mu m$

TABLE 7
Typical DIW requirements for manufacturing

Item	Typical specification	Item	Typical range
Resistivity	> 18 Megohms	Inorganics	
Particles	< 0.1/mL at $\geq 0.1\ \mu m$	Ions	< 0.1 ppb
Total organic carbon	< 10 ppb	(Cl, F, Br, NH_4, NO_2,	
Dissolved oxygen	< 20 ppb	NO_3, PO_4, SO_4, etc.)	
Silica	< 1 ppb	Metallics	
Bacteria	< 1 colony/mL	(Al, Ca, Cr, Cu, Fe, K, Li, Na, Ni, Pb, Zn, etc.)	< 0.05 ppb per element

Indirect material considerations

There are literally hundreds of materials used indirectly in a wafer fab. The criticality of those materials is reduced compared to that for direct materials. It is impossible to discuss them all in this chapter owing to their large quantity and wide range of variation. Readers should make selections based on their own needs and common cleanroom practices.

11.2.4 Specification Systems and Operation Procedures

To handle a complicated process and to control wafer fab manufacturing, maintenance of a stable operation is very important. Stability can be viewed from both the hardware (equipment/process) and the software (system/procedure/people) points of view. This section addresses the issue of systems and procedures.

System vs. people

What are the systems and procedures? Systems and procedures are ways to describe the sequences, requirements, and standards of how to do things.[16] For wafer fab operation with a high volume of output, everyone in the organization must know the detailed standards and what to do at all times. In this way mistakes are kept to an absolute minimum and deviations, if they occur, are realized early, preventing large losses.

There are two opposing schools of thought regarding systems and procedures. One believes that people's stability is not important so long as the systems and procedures are well-defined and implemented. This approach is possible for an operation with a limited number of products and rigorous management. The other school believes that the operation's stability comes from people's stability, because experience and judgment are required routinely and standards are sometimes difficult to define. Wafer fabs employing people with extensive experience and long tenure (10 years plus) find less need for detailed systems and procedures. Typically, wafer fab manufacturing requires both the well-defined system/procedure and experienced people to produce consistent, quality products; the system and people complement each other.

Specification system structure

Figure 4 shows the structure of a typical hierarchical specification system with examples. It starts with company policy and guidelines as the highest level and ends with standard operating procedures (SOPs) as the lowest level. The lower-level specifications must be in accord with higher-level specifications. For example, all equipment must have a calibration procedure in the manufacturing specification (lower-level specification) that is consonant with the calibration requirement in the quality specification (higher-level specification). Since the higher-level specifications provide the guiding principles for all operations, they can lead to an excellent company culture if they are properly established or be detrimental to the company's long-term existence if improperly set. Management must exercise extreme caution in defining such high-level specifications.

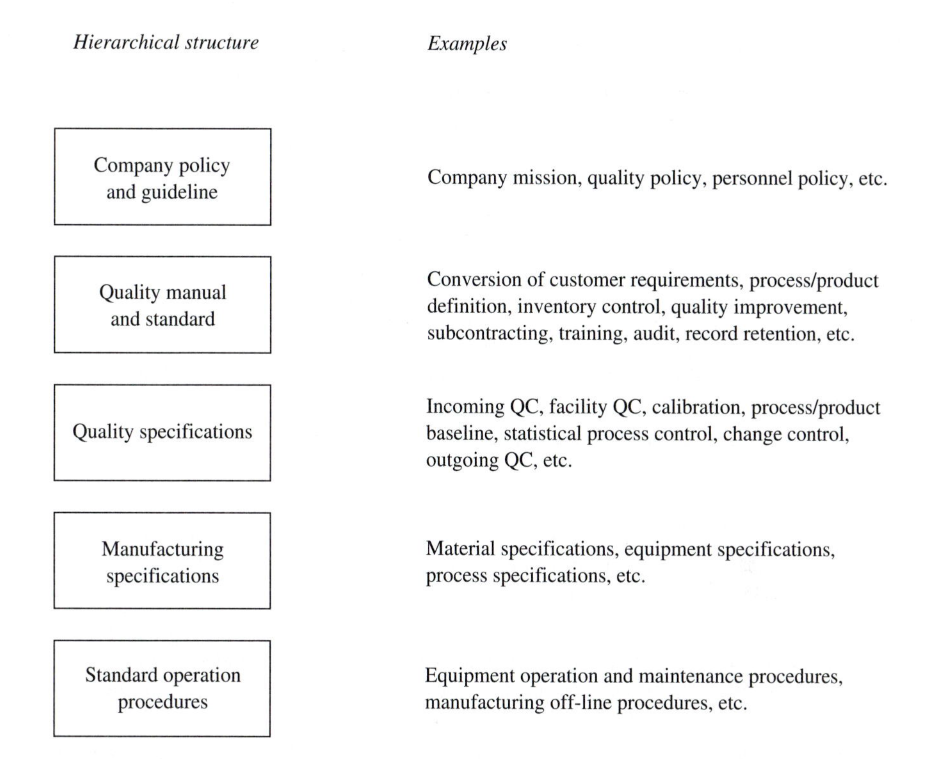

FIGURE 4
Illustration of a typical hierarchical specification system.

Along with a good system structure, proper system implementation requires a hierarchical sign-off control procedure. The sign-off not only acknowledges the specification but also provides a commitment from everyone to perform in agreed ways to carry out the requirements. For any high-level policy/guideline, top management is involved in the sign-off prior to the announcement to company employees. For other specifications, the sign-off follows the sequence: initiator, initiator's functional managers, other related functional managers, and the quality control (also quality and reliability assurance) manager. The QA manager is responsible for the specification activation. The fully signed-off and activated specification is then distributed to all involved personnel for immediate implementation.

Specification format

Since the specification/procedure defines the operational framework, its content needs to be detailed and clear so that it can be easily understood and followed.

Table 8 shows a typical formal content of a specification/procedure. Basically, it should describe in detail why (the purpose) the specification is needed and what will be done, how, by whom, and when. The scope of the specification and references should be given. Common words should be used for easy understanding and clarity because the specification is intended for all levels of the organization. An alternative format is to include a flow diagram for the cover page indicating the sequence of the

TABLE 8
Specification contents

	Format	Designating
Purpose	To state the purpose of the spec/procedure	Why
Scope	To define the applicability of the spec/procedure	—
References	To list specs/procedures related to this spec/procedure	—
Definition	To define terminologies to be used in the spec/procedure	—
Requirements	To present a detailed description of what, how, who, where, and when to do for the spec/procedure	What, how, who, where, when
Others	To indicate other requirements not included above, e.g., supporting documents/forms, record retention, etc.	What

procedures. Developing a flow diagram before writing the actual specification can be a good practice because the flow diagram clarifies the procedure.

Important systems

In Fig. 5 several important systems needs for wafer fab manufacturing are illustrated for the input, process, and output stages. For the input systems, the incoming and facility quality control ensure the proper incoming quality of all the materials and infrastructure environments (Sections 11.2.3 and 11.3.2). For the process-related systems, specifications for sustaining operations and process control are important. To sustain operations, one must identify and control the process/product baseline (Section 11.3.6), changes from the baseline (Section 11.4.3), and baseline abnor-

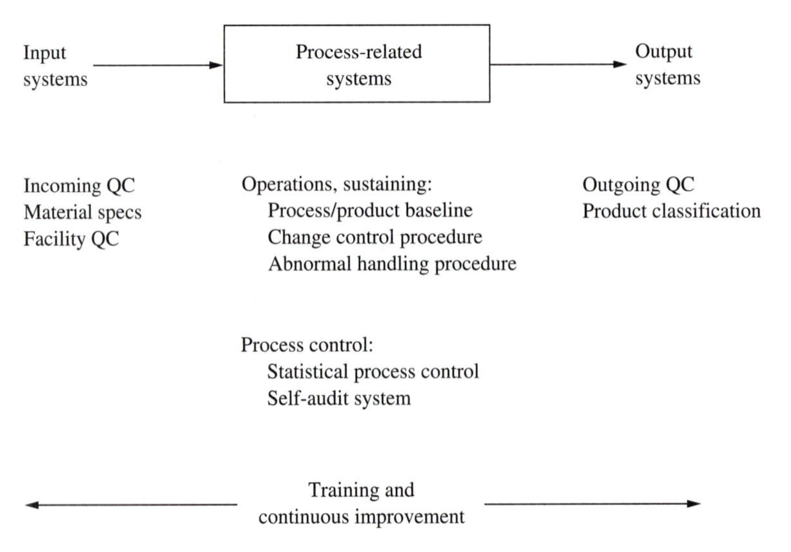

FIGURE 5
Schematic diagram for important operation specifications.

malities. For process control, a statistical process control (SPC)[17] self-audit program is commonly used to control drift in equipment and process parameters, as well as workers' adherence to procedure, in real time (Section 11.4.3). For output systems, the product classification and outgoing quality control are categorized for various customer needs and guarantee quality products for continuous customer satisfaction (Sections 11.3.5 and 11.4.3).

11.2.5 Metrology and Analytic Capability

Both measurement and analytic capability are required for wafer fab manufacturing. Inaccurate measurements, particularly of critical parameters, are detrimental, especially in high-volume production, and can have a large financial impact. Analytic capability, on the other hand, is like the eyes and ears of the wafer fab. It can serve either as a routine monitor to prevent problems or as a problem-solving tool. However, the establishment of those capabilities is a balance between capability and cost. In this section, several metrology and analytic capabilities for wafer fab manufacturing applications are recommended. If some of the suggested capabilities or, even better, more advanced analytic capabilities are available in the city where the wafer fab is located, the wafer fab can minimize its own cost yet have access to the capability. Sometimes, an independent source of analysis to verify internal analytical results is also locally available. This is the subject of infrastructure (Section 11.2.6).

Wafer fab metrology considerations

Accuracy and consistency are two key factors for manufacturing metrology considerations. Process control is not possible without accurate measurement. To achieve accuracy, absolute standards certified by a national body on calibration are required.[18] In Table 9, metrology needs along with the available absolute standards are provided for typical wafer fab manufacturing metrology applications.[18–21] For advanced processes, absolute standards are not available in several cases as shown in Table 9. If we use thickness measurement as an example, a gate oxide thickness of less than 100 Å with variations less than ±10 Å are commonly required. Absolute standards in this region (both thickness and variations) are not yet available. In this situation, a very thorough characterization is necessary using reference standards and other characterization means. For the gate oxide, the thickness is characterized, using gate oxide produced at the pilot line, with ellipsometer, transmission electron microscopy (TEM), and electrical measurements. Wafers with known thickness are then transferred to the wafer fab and used as the reference standards for calibration (or correlation) using similar characterization means. Once all the results are confirmed, the new ellipsometer, which will be used as a manufacturing metrology tool, is considered calibrated. It is, of course, preferable that this ellipsometer has already been calibrated against the available thicker absolute standards and the extrapolated results match the actual data using the thinner reference standards. This kind of approach, especially for metrology without an absolute standard, is essential for wafer fab start-up (Section 11.3). Where absolute standards are available, the metrology instruments must be calibrated against them directly.

TABLE 9
Typical wafer fab metrology instruments

Equipment	Purpose	Standard	Capability
Thickness			
Ellipsometer	To measure thickness and refractive index of oxide and nitride films	500 Å–1 μm	SiO_2: 10 Å–6 μm Si_3N_4: 10 Å–30 μm
Nanospec®	To measure thickness of poly Si, oxide, nitride films	500 Å–1 μm	25 Å–20 μm
Alpha step®	To measure film thickness	400 Å–25 μm	Low range < 13 μm High range > 13 μm–< 300 μm
Dimension			
SEM	To measure CD	0.5–5 μm at 1–20 K magnification	0.1–100 μm
Electrical			
Resistivity meter	To measure the film sheet resistance	0.1–150 ohm/sq cm	5 milli–5 meg ohm-cm
Carrier lifetime tester	To measure minority carrier lifetime in Si	Reference standard	1–5000 μsec
C-V (capacitance-voltage) tester	To measure device *C-V* characteristics such as gate-oxide V_T (threshold voltage)	1–500 pF	0–2000 pF
Particle/Defect			
Particle Counter	To measure particles on bare Si wafers	Reference standard	0.07–900 μm
	To inspect defects on patterned and nonpatterned wafers	Reference standard	0.2–10 μm
Defect detection and classification tool	To inspect defects on patterned wafer	Reference standard	0.15–10 μm Ability to classify defects
Inspection microscope	To inspect surface defects on grown/deposited films	Reference standard	0.15–100 μm
Others			
Stress measurement	To measure stress of films and finished wafers	Reference standard	Thickness: 400–1000 μm Bow/warp: ±350 μm
X-ray fluorescence	To measure boron and phosphorus content in PSG and BPSG film	Reference standard	B: 0–1%/1000 Å P: 0–1.5%/1000 Å
Fourier transform infrared spectrometer	To measure boron and phosphorus content in PSG and BPSG film	Reference standard	1–10 wt %
Stepper alignment inspection tool	To measure stepper alignment accuracy	Reference standard	0.5–20 μm

The consistency of the metrology instruments is the next key manufacturing consideration. To maintain consistency, periodic monitoring using the absolute standards (or reference standards) is necessary. Once the procedure and sampling frequencies are determined, one can control the consistency through statistical process control (SPC).[17] In Fig. 6 two charts are shown to demonstrate metrology accuracy using SPC. The example in Fig. 6*a* illustrates ellipsometer measurements with a thickness of 504 Å and variations of ±4 Å against an absolute standard of 506 Å. In Fig. 6*b* are measurements using the same ellipsometer with a 56-Å thickness and variations of ±4 Å using a reference (not absolute) standard of 56 Å. The variations shown in Fig. 6 arise primarily from the standard itself and from handling, i.e., placing the wafer on the stage, moving the stage to five preset locations, etc. The consistency of the metrology, however, can be maintained through SPC to keep manufacturing processes under control.

Other commonly needed analytic capabilities for wafer fab

Owing to process complexity (individual processes are dependent on each other) and long cycle time, various process monitors, as well as quick problem-solving capabilities, are essential when problems occur. The best approach is to equip the wafer fab with the analytic capabilities required most often.[11] Tables 10 and 11 list the commonly used analytic capabilities. Table 10 details the inspection needs for the wafer and reticle. The instruments listed in Table 10 are located inside the cleanroom so that the inspected wafer/reticle can return to production immediately without contamination. Of course, any inspection step can potentially introduce contamination, such as particles due to handling. Instruments listed in Table 11 are typically located outside the cleanroom even though samples may be taken from equipment inside the cleanroom. Sampling procedures and preparation are very important in advanced sub-ppb level analysis since any small handling error can introduce contamination and result in a large analytic error.

For problem solving and yield improvement,[9,11] the analytic capabilities in Tables 11 and 12 are needed. Instruments in Table 11 are used to identify potential contamination from the facility or wafer process. Instruments in Table 12 are tools for failure analysis of the electrical and physical properties of the finished or partially finished chips or wafers. Some of the most common applications using the analytic capabilities in Tables 11 and 12 are these:

1. Process control and improvement
    - Equipment/process release
    - DIW/chemical point-of-use monitor
    - B and P composition control in BPSG films
    - Cross-sectional process structural characterization
2. Problem solving
    - Facility quality problems
    - Abnormal process or low yield analysis
3. Incoming quality control (IQC)
    - IQC for direct materials (Section 11.2.3)
    - Evaluation and qualification of indirect materials (Section 11.2.3)

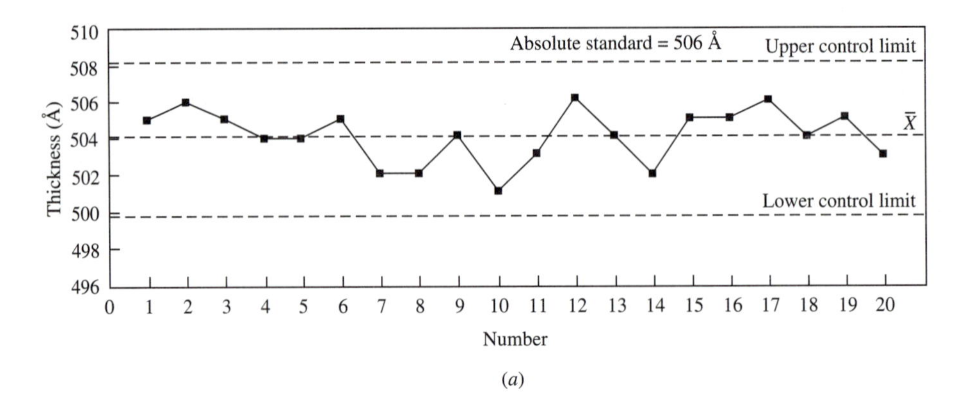

(*a*)

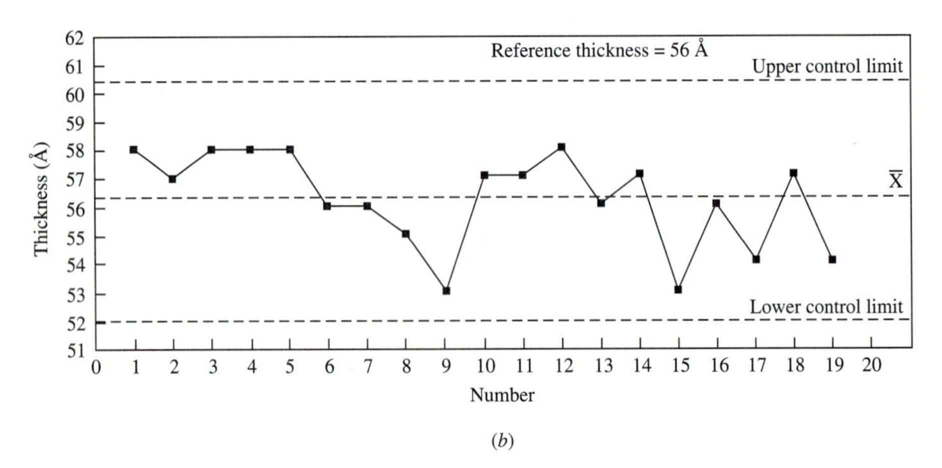

(*b*)

FIGURE 6
Ellipsometer metrology consistency indicated by SPC charts for (*a*) absolute standard with 506 Å and (*b*) a reference thickness of 56 Å.

Volume testing is a unique characteristic of these applications. Typically, hundreds or thousands of analyses are performed on a monthly basis; these are intended for everyday support of continuous manufacturing with quick turnaround time, sometimes in a matter of hours. Advance characterization techniques[9,11] in addition to those listed in Tables 11 and 12 are definitely helpful when the need occurs. Once again, it is preferable to have these available locally from outside services (Section 11.2.6).

11.2.6 Other Infrastructure Considerations

Even though a wafer fab may be equipped with many supporting facilities and analytic capabilities, it needs still other services because of the complexity and the nature of technological advancement as well as for daily manufacturing logistics. Many of these services will not be cost-effective if they are owned by one wafer

TABLE 10
Analytic capability for incoming wafer and reticle

Equipment	Purpose	Capability
Wafer		
Flatness tester and resistivity meter	To measure characteristics of the incoming wafer such as bow, warp, thickness, and resistivity	THK: 400–1000 μm Bow/warp: ±350 μm Resist.: 0.2–200 ohm-cm
Fourier transform infrared spectrometer	To measure the content of oxygen and carbon in silicon	O: 2–50 ppma C: 0.5–15 ppma
Particle counter	To measure particles on bare Si wafers	0.07–900 μm
Reticle		
Photomask defect inspection tool	To inspect defects on reticles	Defect size > 0.25 μm
Photomask particle and defect counter	To inspect particles on reticles	Particle size 0.5–60 μm

fab. It makes more sense to have those requirements satisfied by available local services, that is, the infrastructure. On the other hand, one wafer fab may not have enough work to support the service profitably. Wafer fabs and local infrastructure hence rely on each other. In any case, since wafer fabs, such as those producing advanced DRAM, are the driver, they play a leading role in local infrastructure development. Table 13 lists some of the basic infrastructure considerations required to support manufacturing. Of course, infrastructure availability is not simple because it is related to individual business needs and often is subject to local government policy and regulation. However, locations with wafer fabs should aim to provide the supporting infrastructure as suggested in Table 13.

11.3 MANUFACTURING START-UP TECHNOLOGY

The start-up of a new wafer fab is the most important, exciting, and work-intensive stage of a wafer fab history. It sets the foundation of the manufacturing facility. It also establishes the tone of future operations. A good start-up establishes positive reinforcement between manufacturability (process capability and people's practice) and continuous improvement. In this case, all resources contribute to capability improvement instead of troubleshooting of problems. Conversely, a poor start-up pushes the wafer fab into a negative cycle where one is constantly fighting to solve the remaining problems, which keep on manifesting themselves because of incomplete start-up checks. The start-up determines the fate of the wafer fab: either *designed-in manufacturability* or *designed-in problems.*

In this section, we introduce a start-up approach that has successfully brought in more than one advanced 0.5- and 0.8-μm wafer fab. Figure 7 is a schematic diagram

TABLE 11
Suggested wafer fab chemical analysis capability

Equipment	Purpose	Capability
Atomic Absorption spectrometer (AAS)	Detect cationic elements in liquid samples Analyze 17 common elements in samples (Al, As, B, Ca, Co, Cu, Cr, Fe, K, Mn, Na, Ni, P, Se, Si, V, and Zn)	Sub ppb to ppm
Inductively coupled plasma mass spectrometer (ICP/MS)	Determine 70% of the elements in periodic table in liquid samples Analyze more than 30 elements simultaneously	Sub ppb to ppm
Ion chromatograph (IC)	Detect trace anionic impurities in DIW or process water Analyze Cl, Br, NO_3, SO_4, PO_4, etc.	Sub ppb to ppm
Gas chromatograph		
TCD (Thermal conductivity detector)	Determine impurities in gases	ppm
DID (Discharge ionization detector)	Determine impurities in gases including H_2, O_2, N_2, CO, and CO_2, hydrocarbons, noble gases, and inorganic acid, ammonia vapors, etc.	ppb
Gas chromatograph mass spectrometer (GC/MS)	Identify all components in a gas system	0 to 650 amu
MSD (Mass detector)	Analyze gaseous and volatile liquid materials Provide qualitative information for unknown compounds	
Trace O_2 analyzer	Monitor trace oxygen content in gases	2 to 100 ppb
Trace moisture analyzer	Measure the trace moisture content in gases	10 ppb to 20 ppm
TOC analyzer	Detect total organic carbon Measure both the temperature and resistivity of DIW Monitor the organic contamination source in DIW system	Sub ppb to ppm
Titrators	Measure the components in liquid solution by titration method	ppm to at %
	Measure trace moisture content in organic solvents	ppm
	Check consistency of liquid acid mixtures	ppm to at %
Liquid particle counter	Measure particle counts both in DIW and chemicals (including corrosive solvents)	0.1 to 0.5 μm
Bacteria counter	Check total bacteria and particulate counts in DIW or liquids	Per DIW system specification

TABLE 12
Suggested wafer fab failure analysis (FA) capability

Equipment	Purpose	Capability
Physical FA		
Wet deprocessing (solvent/acid hoods and clean bench)	Strip device layer by layer Decorate defects for observation	Protective overcoat to substrate
Dry deprocessing (plasma etcher-RIE)	Remove certain device layers	Protective overcoat, metals, layers
Microsection station	Prepare device cross section of failed locations Device structural analysis	Identify failed locations
SEM and sputter coater	Detect microdefects, morphology, and dimensions	20 to 300 K magnification 150 Å resolution
Energy dispersive X-ray (EDX)	Determine material constituents qualitatively or quantitatively	Atomic %
Microscope	Observe gross defects, morphology, or structures	5–3500 magnification
Electrical FA		
Device tester	Test device AC/DC parameters, determine electrical failure phenomena Test the sensitivity of failure under various electrical operation conditions	Test pattern and algorithm generation Test full device operations and stress conditions
Bias and temperature test station	Test sensitivity of failure under bias and temperature stresses	0–150°C 0–50 V
Liquid crystal analysis	Identify failed location through hot spot Detect smaller DC, AC, and ICC device failures	Leakage > 100 μA
Emission microscope	Identify dielectric breakdown and transistor failure through detection of photoemission	Leakage > 100 μA
Laser spot size	Isolate metal and poly-Si lines for DC/AC testing	1 μm minimum laser spot size
Internal probe	Analysis of failure unit internal signal and waveform	

illustrating the major activities of wafer fab start-up. Some details of these activities are discussed in this section.

Because of the complexity of start-up, a very comprehensive project schedule and tracking plan must be developed first. Milestones and subtasks for each of the activities described in Fig. 7 should be identified. The interrelationships among activities also need to be comprehended to identify potential bottlenecks. In our experience, target time for starting wafer production is the major milestone. Activities

TABLE 13
Infrastructure considerations for manufacturing

Infrastructure service	Considerations
Laundry	Facility with DIW and cleanroom Low-Na and I-Cl raw materials (detergent) Thoroughly rinsed to remove all impurities Handling procedures for drying and packaging for particle control Anti-static bag
Quartz work	Facility with DIW and controlled environment Quartz cleanup capability Precision laser work, control, and measurement capabilities Stress annealing and inspection capabilities Established handling and control procedures, e.g., IQC, flame welding, outgoing inspection, and packing Experienced workers for precision high-temperature works
Calibration	Certification by national body Wide range of calibration capabilities Established procedures to ensure calibration accuracy and repeatability Issuance of proper certificate of calibration On time and fast turnaround Low cost Capability to develop and set up new calibration needs
Analytic service	Wide-range analytical availability to analyze physical, chemical, electrical, structural, and surface properties of materials Sufficient instrumentation with proper detection sensitivity Established procedures for calibration, sample preparation, and measurement accuracy and repeatability assurance Experienced analysts in different fields Reports issued with data analysis and conclusions On-time and fast turnaround Low cost Ability to develop and set up new analytical requirements
Chemical and gas supply (minimum requirements)	Facility with DIW, warehouse, and analytical lab Capability for purification, drum/cylinder cleaning, and drum/cylinder filling Capability for sampling and analysis by experienced analysts Established handling procedure, tracking, analysis, and outgoing control procedure Waste and used drum/cylinder treatment Ability for service, fast response, and problem solving
Cleanroon furniture	Supplier capability to clean up oil, grease, and fingerprints on finished furniture Good metal workmanship Supplier ability to pack cleaned furniture Electro-polished surface finishing preferred

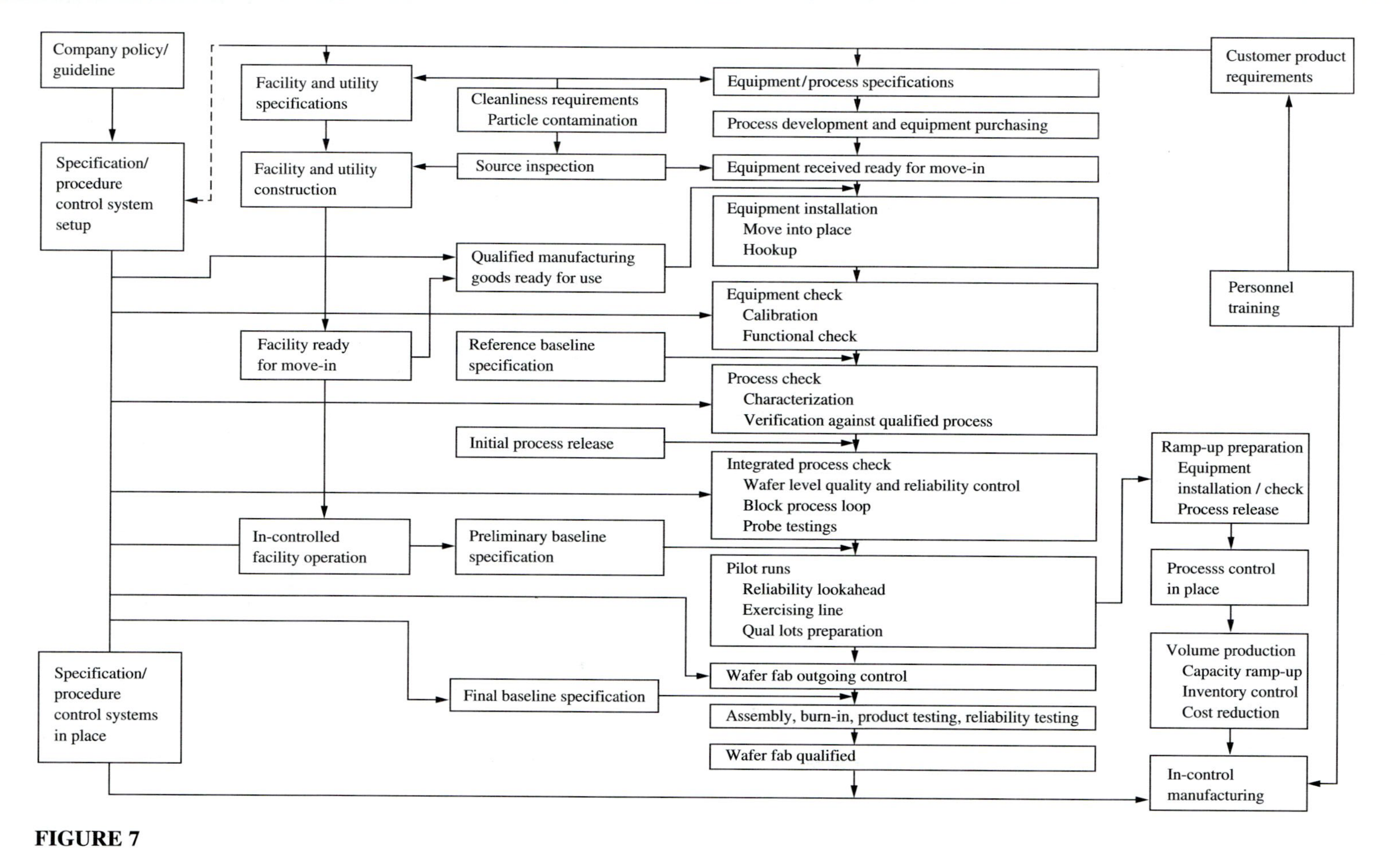

FIGURE 7
Wafer fab start-up activities.

required before the start, primarily facility construction and the acquisition of utilities and equipment, can be derived. Activities after the start of wafer production, such as the process release, product qualification, and volume ramp-up plan can also be determined.

11.3.1 Specification Generation

The first start-up activity, as shown in Fig. 7, is to generate various specifications from product to process to production, to meet the final manufacturing goal. This task is very important because all future activities are bound by these specifications. Our intention here is not to address the details of how the specifications should be derived. However, a generic methodology,[22] shown in Fig. 8, illustrates the process. For deployment, the characteristics of each item related to each block in Fig. 8 are decided first. Target values can then be determined based on their relationships. This process starts with understanding customer needs and continues until production requirements are defined. Details in other chapters of this book can be used for product, process, equipment, material, and facility considerations in the deployment process.

For a manufacturing wafer fab start-up, the more desirable situation is that only production planning (Fig. 8) is needed. However, owing to the short semiconductor product cycle time, sometimes it is inevitable that both process planning and production planning (Fig. 8) are required for start-up. In both cases, the following items should be given consideration in generating specifications:

- Process specification must accommodate design tolerances, demonstrated equipment performances, and manufacturing variations for good process margins.
- Equipment, material, and facility specifications should consider proven technology, industrial availability, and suppliers' continuous support and problem-solving capability to facilitate fast start-up and ramp-up with minimal resources.
- Connection and material specifications, and their quality standards with respect to equipment and facility, should be consistent so that they are not overspecified yet cause no loss of quality at the transition between the two.
- In all cases, the lessons learned from previous process and production experience should be incorporated as much as possible to eliminate costly known mistakes.

11.3.2 Utility Requirements and Control

For a start-up wafer fab, facility construction of the building and the utility systems (cleanroom, chemical, gases, and DIW distribution systems) starts first. A modern advanced wafer fab contains miles of pipes, thousands of valves, thousands of filters, and thousands of connections, weldings, and installations. All have high purity and cleanliness requirements.[23–25] The construction of high-purity utility systems can take over half a year. If work is not done carefully in terms of the system certification and cleanliness requirements, the results are either delayed completion or sacrificed system quality and reliability. In Chapter 1, details of the utility systems' design and specifications are discussed. In this section, we discuss the certification[26] and

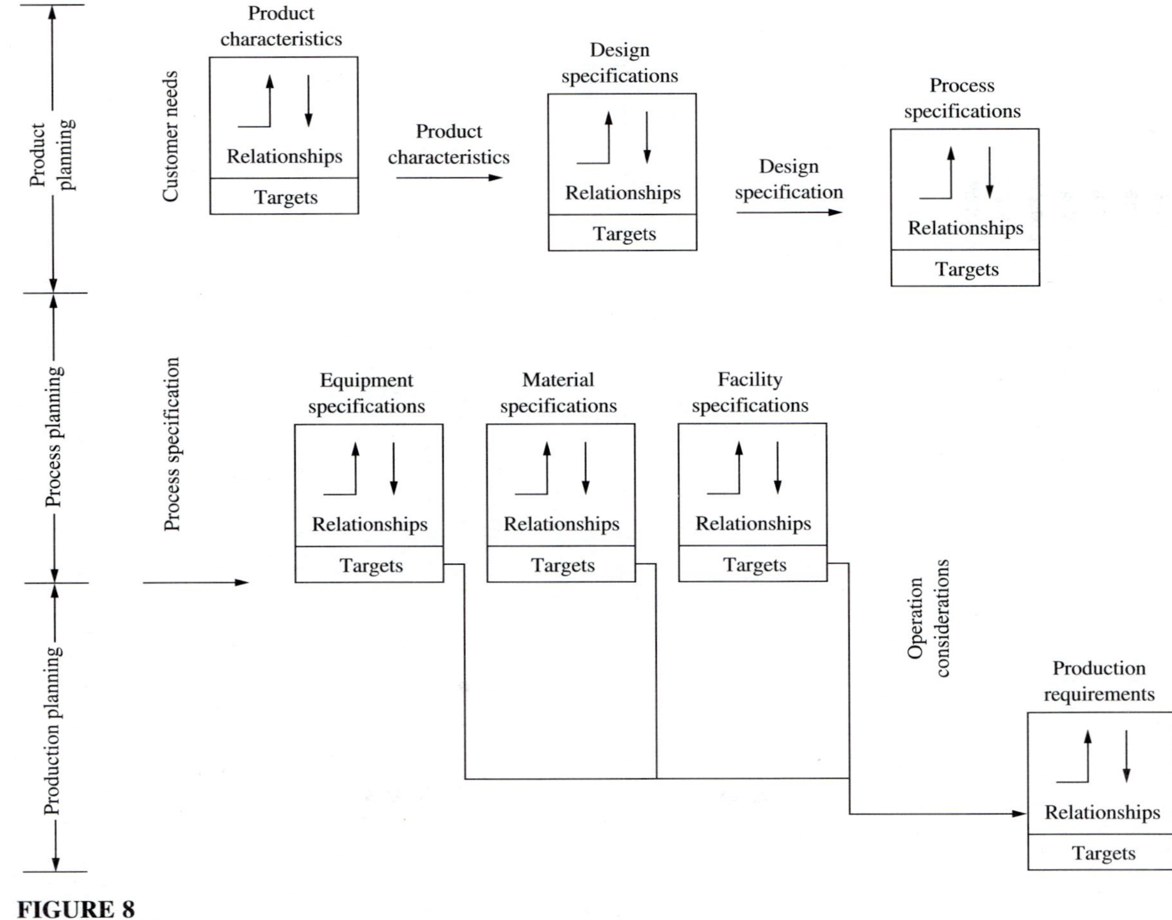

FIGURE 8
Deployment for specification generation—planning for quality.

control aspects of the utility systems. The suggestions in Section 11.3.1 are important to ensure fast system start-up and long-term stable system performance. Additional considerations are as follows:

- Detailed checklists for parts (pipe, valve, filter) and works (welding, pipe cleaning, filter installation) are prepared prior to construction based on experience and system requirements.
- Selection of parts and their material requirements is based on proven quality and reliability performance and is consistent with equipment specification.
- There is continuous worker training on procedures and cleanliness requirements.
- Routine and thorough checks by the construction teams using the prepared checklists and corrective actions are implemented on a real-time basis.

Utility systems certification

The certification of the utility systems ensures that the minimum process requirements of the utility supplies are met at time zero to eliminate supply variations at start-up. Certification will not be successful if the facility construction itself is poor. It is most important that certification be viewed only as the last stage of the construction quality control, whereas the main body of the quality control is actually done during the construction. If this is not the case, costly reworks, extended systems flushing (DIW and chemicals), and a delay of the supply operation are typical consequences. Another aspect is to certify each subunit of the system on-site on a real-time basis so that utility systems–related causes can be readily eliminated, which will avoid subsequent troubleshooting.

Figure 9 is a flow diagram illustrating the sequence for utility system certification. Two stages are involved: integrity tests and functional tests. The integrity tests are primarily to check that the system's mechanical strength under pressure, vacuum, and fluid conditions meets the system's mechanical design criteria. The functional tests are aimed at checking the system's quality performance in meeting system quality standards. In Table 14 are listed several parameters for each utility system. The actual acceptance value for each parameter is not listed. However, specifications discussed in Section 11.2.3 and Chapter 1 can be used for a particular application.

Maintaining facility systems stability

In addition to the mechanical and functional (quality) systems certification, the reliability of the systems' performances is also important. An unreliable system causes a great deal of supply fluctuation and a system supply shutdown in the worst case. Continuous monitoring of all utility systems ensures consistent system supplies for manufacturing. In today's wafer fab, the majority of the monitoring efforts are made automatic by using the continuous quality control (CQC) approach (Section 11.2.2). CQC consists of hardware (sensors and instruments) and software (data collection/analysis and control actions) allowing real-time and continuous monitoring, therefore minimizing human effort. The parameters not yet covered under CQC are marked with a "†" in Table 14. Therefore, they require periodic sampling and analysis by designated and trained personnel. In Fig. 10 an example of Fe impurity

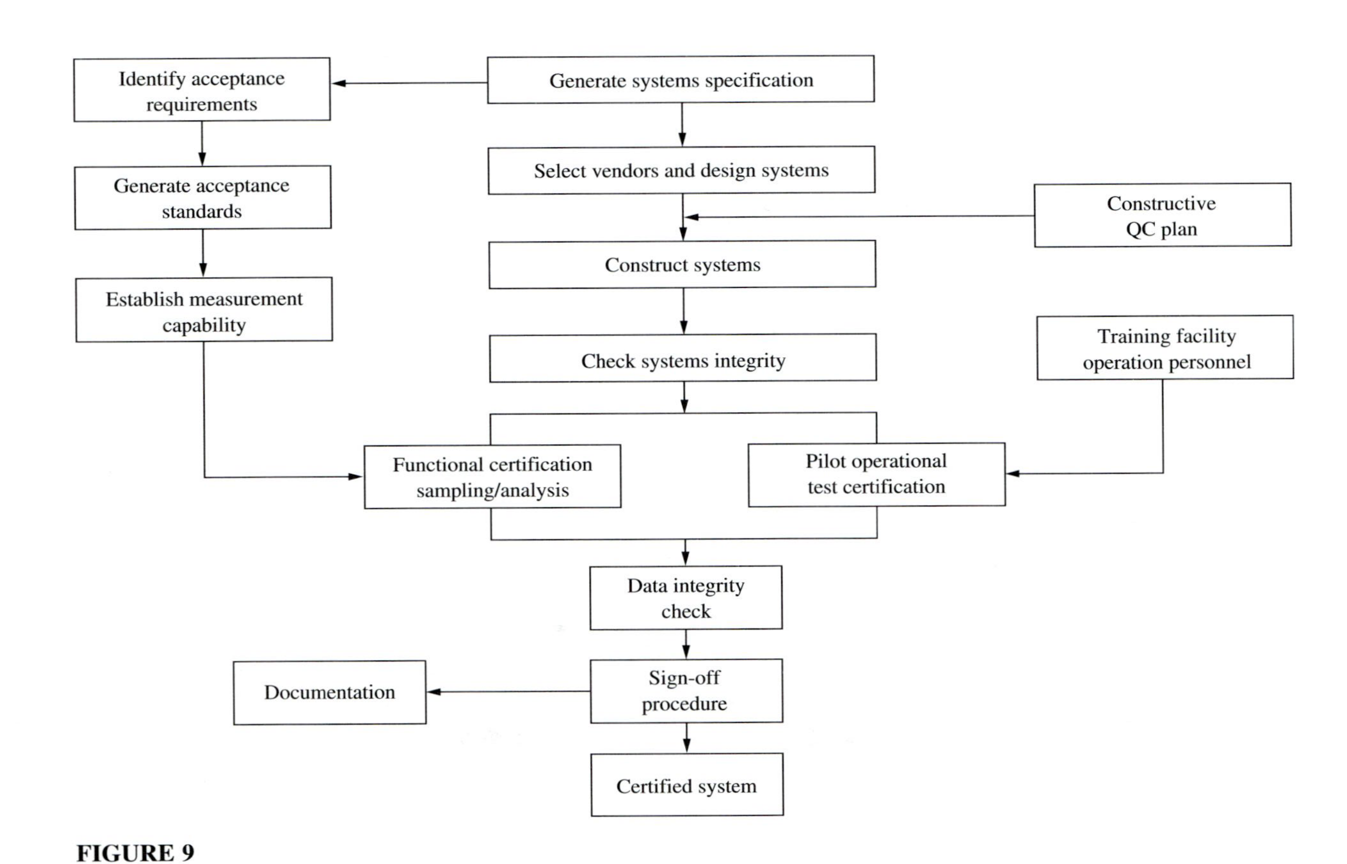

FIGURE 9
Schematic diagram of facility certification procedure.

TABLE 14
Parameters for facility system certification

		Gas			Chemical	
	Cleanroom	**Plant**	**Specialty**	**DIW**	**Acid**	**Solvent**
Integrity check	Filter grids and filter gross leak	Visual inspection Leak integrity, vacuum and pressure tests	Visual inspection Leak integrity, vacuum and pressure tests	Visual inspection Leak integrity, pressure decay with N_2	Visual inspection Leak integrity, pressure decay with N_2, DIW, and actual chemicals	Visual inspection Leak integrity, pressure decay with N_2
Functional certification	Clean classification Temperature Humidity Air flow velocity Air flow laminarity Vibration† Room pressure† Lighting level† Sound level† Air quality† Make-up air flow and recovery	Moisture Trace O_2 Particle† Impurities (Table 4)	Moisture‡ Trace O_2‡ Particle†	Temperature Pressure Resistivity D.O.§ TOC¶ Particle† Bacteria† Silica Impurities (Table 7)†	Particle using DIW Particle using actual chemical† Impurities (Table 3)†	Particle using N_2 Particle using actual chemical† Impurities (Table 3)†

† Parameters not monitored by CQC.
‡ Using plant gas, e.g., N_2.
§ D.O.: Dissolved oxygen
¶ TOC: Total organic content

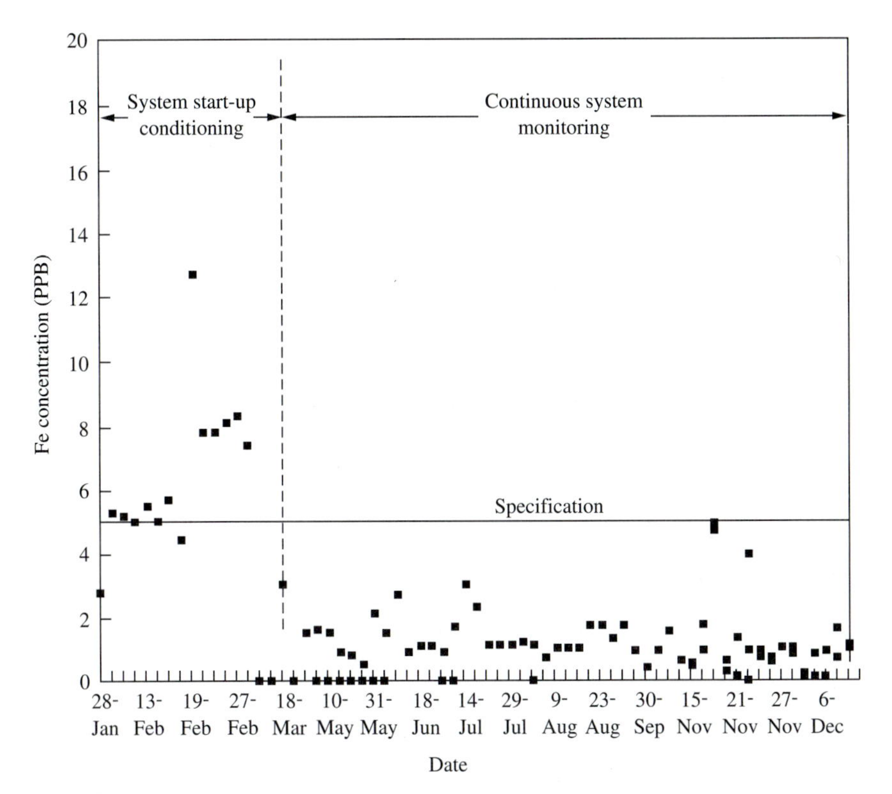

FIGURE 10
Measured level of Fe in HCl at the output filters of the chemical-dispensing system.

content in HCl at the point of delivery from the HCl distribution system is shown on a time scale. The system was under start-up conditioning before March. Leaching of Fe from the distribution system and contamination from HCl drums were observed. These problems were corrected and the system was certified on March 18 to meet the Fe specification of 5 ppb and maintained below 3 ppb. On November 17, a high Fe level of 5 ppb was again detected and traced back to contamination of the high-purity tubing connecting the chemical drum to the distribution system. The contamination was cleaned out, and system performance was restored. In other cases, one may not be able to resolve problems in facility systems with this efficiency and effectiveness. The effort, however, must be made to ensure that systems are under control for stable and consistent supplies.

11.3.3 Equipment Checks

The performance of all equipment is extremely important to a successful start-up. Equipment functional performances must first meet the targeted process requirements to demonstrate the capability. Subsequently, equipment stability and reliability are key to fast volume ramp-up and in-control manufacturing. To ensure good equipment performance, the following checks can be instituted.

Equipment source inspection

The concept of source inspection (inspection conducted at the vendor's site before shipment) is to eliminate problems prior to the equipment shipment. Table 15 suggests a checklist for source inspection. The purchasing specifications are the basis for the acceptance criteria for items 1–5 in Table 15. Items not meeting the acceptance criteria must be identified during source inspection and corrected prior to shipment.

Equipment cleanliness

Despite the high level of cleanliness (better than class 10 at 0.12 μm; see Chapter 1) required for today's advanced wafer fabs, equipment may carry gross contamination (dirt, grease, and fingerprints) into the cleanroom if it is not cleaned properly. Cleaning the equipment before moving it into the cleanroom is one way of eliminating contamination sources.

There are two main causes for equipment contamination: the equipment fabrication process and packaging. During the former, components and subassemblies can be contaminated by constant handling with bare hands, cutting/drilling/welding, or having dirty parts initially. To eliminate these contamination sources, a vendor must establish a good equipment-cleaning procedure, covering component and subassembly cleaning, handling parts with gloves, removing contaminants (dirt, drill/cutting debris) constantly, and continuously wiping down grease and oil. Subsequently, the exterior of the fabricated equipment should be wiped clean and checked for defects in workmanship, such as painting scratches, mechanical misalignment, and corrosion. A 5% solution of IPA (isopropyl alcohol) in DIW is commonly used for cleanup. For packaging, double or triple wrapping with precleaned plastic wrapping materials is used. Double (or triple) wrapping isolates outside contaminants (crate materials, dirt, dust) from the cleaned equipment. Equipment cleanliness is merely a good common sense practice rather than advanced technology. Most major vendors are equipped with Class 10,000 cleanrooms for assembly and have defined cleanliness and packaging procedures.

Equipment integrity

Once the equipment has been moved into its proper location inside the cleanroom, the next step is to assemble the equipment and make the utility connections. The equipment assembly includes mechanical assembly of the main body, leveling, connection to control units and vacuum pumps, and connection to stand-alone supply units (rf generator, ozone generator, cooling units, etc.) specific to that equipment.

Utility systems to be connected include electricity, chemicals, gases, DIW, process cooling water, plant vacuum, exhausts, and drain/waste-handling systems for processing and removal of the used/residue materials and by-products. The utility hookup should be consistent in materials and connector type between utility system and equipment, as shown in Table 16, to maintain high-purity supplies of chemicals, gases, and DIW.

The purpose of the equipment integrity check is to ensure the mechanical and point of use (POU) quality. The mechanical integrity check emphasizes visual

TABLE 15
Equipment source inspection checklist

Items	Description
1. Equipment specifications	Reconfirm specifications
2. Utility connection	Utility connection types and dimensions Piping and connector materials, e.g., PVDF for DIW piping Welding requirements Exact locations of the connection
3. Cleanliness requirements	Cleanliness procedure and setup Cleanliness inspection before packing Packing requirements
4. Safety issue	Common safety features, e.g., emergency shutoff device Specific safety features, e.g., laser safety interlock, specialty gas (SiH_4, etc.) safety interlock
5. Basic equipment function checks	Mechanical functions, e.g., stage accuracy, wafer transfer mechanisms, safety alarms, etc. Process functions, e.g., stepper lens performances, deposited film uniformity on pilot wafers, etc.
6. Calibration methodology	Vendor's calibration standards and procedures Vendor-provided certificates, as for mass flow controllers and thermocouples Understanding of total equipment on-site calibration requirements
7. Work out detailed equipment on-site checklist and schedule	List detailed equipment check items and estimate time required for scheduling
8. Confirm equipment shipping/arrival date and vendor on-site support resources	Confirm equipment actual shipping date to targeted shipping date Arrange for vendor on-site support items

inspection and pressure/vacuum tests that use pressurized inert gas and also tests lines and chambers for proper connection and sealing. The POU integrity check is to ensure that the supply quality is maintained from supply source to point of use and also for process quality assurance. Table 17 provides some POU quality requirements for gases, DIW, and chemicals.

Equipment functional check

The purposes of the equipment functional checks are to verify the equipment's ability to meet process margin requirements and to ensure the equipment's reliability and capability to sustain operation with the required consistency and capacity.

Before running the tests, the equipment needs to be calibrated (Section 11.2.5) so that the equipment parameters (temperature, gas flow, pressure, power, etc.) can be

TABLE 16
Typical materials and connectors used for equipment hookup

Utility system	Supply or drain	Materials	Connector type
Plant gas	S and D	316EL	VCR
(N_2, O_2, H_2, Ar, He)		316SS (nonprocess gas)	Swagelock (nonprocess gas)
Specialty gas	S and D	316EL	VCR
Hot/cold DIW	S	PVDF, PFA	Buttwelded union
	D	PP (cold), PVDF (hot)	Union
Chemicals			
Acid	S	PFA, PTFE	Teflon™ fittings
Solvent	S	316EL	VCR
Photoresist	S	304SS	—
Process cooling water	S and D	PVC	Swagelock/PVC
Plant vacuum	S	SS/PVC	Swagelock
Exhaust			
General and caustic	—	—	—
Acid	—	PP	Flange
Solvent	—	SS	Flange
Industrial waste	D	HDPE	Flange

controlled accurately. For new equipment, components such as thermocouples, mass flow controllers (MFC), pressure and vacuum gauges, etc. should be precalibrated at delivery and a certificate of calibration provided by the vendor. The whole equipment calibration is then performed on-site (using prearranged calibration standards) by the equipment vendor's field engineers. An example of equipment calibration is the stepper calibration for mechanical stages and optical/lens assembly.

Subsequent to calibration, the equipment is subject to verification for quality and reliability performances. Quality performance refers to equipment operation results associated with the process parameters at one instant of time. Reliability performance refers to failure characteristics of the equipment with time, which impacts the capacity and stability. With a deposition furnace as an example, illustrated in Table 18, film thickness and uniformity within a wafer, wafer to wafer, and run to run are parameters for quality performances. On the other hand, mean time between failures (MTBF), mean time to repair (MTTR), etc. are parameters for reliability performance. It is most important that equipment performance in both categories (quality and reliability) be verified to establish the foundation for good process margins and long-term manufacturability. In Table 19 several typical parameters are shown for functional verification of various types of equipment. Most of the reliability parameters for different types of equipment are similar to those listed in Table 18, and therefore are not repeated in Table 19.

Equipment operating procedure

Equipment operating procedure is not part of the equipment check but is a good practice to establish during the start-up stage. Clear operating procedures on equipment start-up and shutdown and loading and unloading procedures are minimal needs. Operating procedures for short-term (daily or weekly) and long-term (monthly or yearly) preventive maintenance[5,27] should be established after start-up

TABLE 17
Typical equipment point of use requirements

Utility system	Check item	Requirement	Check Location†
Plant gases	Leak check		
	Pressure decay	As required	POU
	Helium leak check	As required	POU
	Particle count	< 1 particle/ft^3; particle ≥ 0.1 μm	POD
	Moisture	< 50 ppb	POD
Specialty gases	Leak check		
	Pressure decay	As required	POU
	Helium leak check	As required	POU
	Particle count	< 1 particle/ft^3; particle ≥ 0.1 μm using N_2 or Ar	POD
	Moisture	< 50 ppb	POD
	Oxygen	< 50 ppb	POD
	THC	< 50 ppb	POD
Deionized water	Leak check	Visually inspect for leaks when equipment is turned on	POD
	Particle count	< 2 per mL; particles ≥ 0.085 μm	POD
	Bacteria	< 1 colony per liter	POU
	Impurities	< 0.05 ppb per metallic element	POU
Solvents	Leak check	As required	POD
	N_2 purge	As required	POD
	Particle count	< 20 mL > 0.2 μm	POD
	Impurities	< 1 ppb for Al, Ca, Cu, Fe, K, Ni, Cr, Zn	POU
Acids	Leak check	As required	POD
	DIW Flush	As required	POD
	Particle count	< 20 particles/mL > 0.2 μm	POD
	Impurities	< 1 ppb for Al, Ca, Cu, Fe, Na, K, Ni, Cr, Zn	POU

† POU: Point of use at the equipment
POD: Point of delivery to the equipment

but no later than the completion of the capacity ramp-up. These operating procedures are not only for the manufacturing operation but also for verification and improvement of the equipment availability for capacity planning and monitoring. An example is provided in Table 20.

11.3.4 Process Characterization

Process characterization at start-up is aimed at establishing actual process performances[10,11,28] against the design criteria and pilot production references. If the process step has never been studied owing to process changes or the use of new equipment, the process should be *characterized* to understand the key control parameters of the process requirements. Characterization also ensures that the individual process

TABLE 18
Typical equipment check parameters for a deposition furnace

Quality category	Reliability category
Temperature (set and ramp rate)	Equipment-dependent uptime (%)
Gas flow rate	MTBF (hrs) or MWBF (pieces)
Furnace pressure	MTTR (hrs)
Defect density (particle)	Component or subassembly
Film properties	Reliability, e.g., wafer transport mechanism
Deposition rate (Å/min)	
Thickness target	
Thickness and composition	Preventive maintenance (%)
Uniformity	MWBC (pieces)
Within wafers	Wafer boat
Wafer to wafer	Furnace tube
Run to run	
Resistivity	
Index of refraction	
Process yield loss (%)	
Max water scrap due to equipment (%)	
Furnace loading capacity & wafer size	
Maximum contamination level	

performances[10,11] meet design criteria for that process step and can be integrated[10,28] with other individual processes to meet overall device parameter requirements.[13] In this case, it is important that the process and equipment be characterized, as much as possible, at the vendor's location to minimize the unknowns during start-up. Most preferably, start-up process characterization should not be a *characterization* for new process/equipment but rather a *verification* of the process performances with respect to the process performances already achieved during pilot production. This is one approach for establishing a new wafer fab process baseline and fast start-up. Sometimes, the use of new equipment is inevitable so that running recipes for equipment parameters may not be transferable from the pilot line to production line. In this case, the performances of process output but not process recipe should still be maintained strictly compared to that at pilot running by fine-tuning equipment performances. Once again, this approach is to ensure no loss of process margin so that the equivalent pilot production device and yield can be achieved readily at the completion of start-up in the shortest time.

With this understanding of the characterization/verification of the start-up process, the release of a specific process, that is, the completion of its checks so that it can be moved to the production phase, becomes a matter of preparation and execution. In Fig. 11, a flow diagram illustrates the process release procedure. In it, the process release can be separated into two release steps: capability verification and stability check. The capability check is to verify that the process can perform the pilot production references (actual margin but not the spec) or the design criteria (process spec) if no references are available. The stability check, on the other hand, is to ensure that the process can be maintained without degradation over time after capability has been demonstrated.

TABLE 19

Typical equipment functional check parameters

Photo		Plasma		Metrology
Stepper	**Coater/Developer**	**Etchers**	**Ashers**	
1 Facility/safety check	1 Safety/facility check	1 Facility/safety check	1 Facility/safety check	1 Facility/safety check
2 Alignment (reticle/wafer)	2 Machine performance (uniformity, speed)	2 Gas check (MFC)	2 RF/temp control test	2 Repeatability and reproducibility
3 Illumination	3 EMO/interlock check	3 MFC check	3 Leak check	3 Accuracy
4 Stage/lens performance	4 Dispense/rinse cycle	4 Leveling	4 Strike plasma test	4 Correlation to pilot production
5 Focus/leveling	5 Wafer handing	5 Purge/temperature/ pressure check	5 Wafer handling	5 Calibration certificate
6 Wafer handling	6 Particle check	6 Calibration forward power	6 Particle check	6 EMO/interlock check reference
7 Particle check	7 Temperature/humidity	7 EMO/interlock check	7 Magnet check	7 Particle check
8 EMO/interlock check		8 Particle check	8 MFC check	8 Wafer handling
9 Process yield loss		9 Wafer handling		

Diffusion		Thin-film			
Furnace	**Wet**	**Sputter**	**CVD**	**Back grinder**	**Implant**
1 Facility/safety check	1 Facility/safety check	1 Facility/safety check	1 Facility/safety check	1 Facility/safety check	1 Facility/safety check
2 Gas check (MFC)	2 Host computer communications check	2 Robot indexer check	2 MFC/temp/rf calibration	2 Spindle angle	2 Beam energy capability /stability
3 External torch check	3 Leak check	3 Exhaust flow	3 Leak test	3 Chuck uniformity	3 Mass resolution
4 Leveling	4 Compressed air check	4 Target burn-in	4 Robot settings	4 Robot settings	4 High-voltage calibration
5 Heater baking	5 Wafer handling	5 Vacuum performance	5 Wafer handling	5 Wafer handling	5 Wafer handling
6 Temperature check	6 Particle check	6 EMO/interlock check	6 Particle check	6 Particle check	6 Particle check
7 EMO/interlock check	7 Filtration panel check	7 Particle check	7 EMO/interlock check	7 EMO/interlock check	7 Wafer cooling
8 Particle check	8 Running test	8 Wafer handling	8 Baratron gain/linearity		8 Beam Parallelism
9 Thermocouple calibration/check	9 POU certification		9 Load lock		9 Dose uniformity
10 Automechanism					10 EMO/interlock check

Product			
Tester	**Prober**	**Laser repair**	**Software Defect Analysis**
1 Facility/safety check	1 Facility/safety check	1 Facility/safety check	1 MTBF
2 Calibration	2 Alignment	2 Configuration check	2 Hardware linking
3 Host controller communications check	3 Chuck X-Y accuracy	3 Temperature check	
4 System burn-in	4 EMO/interlock check	4 EMO/interlock check	
5 EMO/interlock check	5 Particle check	5 Particle check	
6 Particle check	6 Wafer handling	6 Wafer handling	
7 Wafer handling			

TABLE 20
An equipment operating procedure (furnace tube change)

Item	Time (hr) Estimated	Actual
1. Process stop and heater off	6.0	6.0
Heater breaker off		
Cleaning of main valve	1	
Checking and cleaning of pumping parts for by-products	1	
Furnace cool down	6	
2. Tube/vacuum exhaust pull	4.0	2.0
Removal of shutter, nozzle, inner/outer tubes, thermocouple manifold	2	
Removal of vacuum piping	2	
3. Furnace/vacuum exhaust clean	3.0	2.5
4. Tube clean	0.0	0.0
5. Tube installation	4.0	2.7
6. Room temperature leak check	1.0	1.5
Performance of leak check for all seals	1	
Pumping and purging to degas		
7. Boat setup/registration	1.0	2.8
8. Heater on and heating up	2.0	1.0
9. High-temperature leak check	3.0	5.5
Process temperature leak check	0.5	
Pumping and purging to degas	2.5	
10. Temperature profile	8.0	7.0
11. Tube coating (process recipe with extended time)	3.0	3.0
12. Manufacturing test runs	7.0	7.0
Total time	42.0	41.0

In Table 21, a list of parameters for process checking is presented. The user decides which part of the list or whether additional specific parameters should be engaged in the user's own application. In Table 22, the process release steps just discussed are tabulated. A minimum of 25 tests reflects the statistical needs for meaningful control limit computation.[17] A solid process foundation can be established by combining the needs in Tables 21 and 22 and by thorough execution of the process release procedure shown in Fig. 11. As an example, process release data for the 4-Mbit DRAM trench capacitor critical dimension after etching are shown in Fig. 12 to illustrate the verification of the process capability and stability between transferred and received wafer fabs.

As a reminder, for manufacturing consideration, the process-related specifications and operating procedures (discussed in Section 11.2.4) also should be prepared in this stage to permit unambiguous execution by the manufacturing personnel.

11.3.5 Product Characterization

In the previous two paragraphs, we have discussed how to establish running and processing baselines for wafer fab equipment through characterization and verification

TABLE 21

Control parameters of process release

Photo			Plasma		Diffusion	
Stepper	**Resist**	**Coater/developer**	**Etcher**	**Asher**	**AP furnace**	**Wet**
1 Lens performance	1 Characterization	1 Particle test	1 Particle	1 Particle	1 Particle test	1 Chemical assay
Resolution		2 Thickness check	2 Etch uniformity	2 Ash rate	2 Oxide growth	2 Particle test
Mix and match		3 Uniformity check	3 Etch selectivity	3 Ash uniformity	Thickness	3 Etch rate
Depth of focus		Within wafer	4 Etch profile	4 Resist removal	Uniformity	4 MCLT†
2 Process CD		Wafer to wafer	5 Etch rate	capability	3 MCLT†	5 Water mark
Average		4 Resist type	6 Resist burning		4 GOI‡	6 Impurities
Uniformity (within			7 CD etch bias		5 Slipline	
wafer to wafer)			8 Corrosion		6 Bias temperature stress	
					7 Oxide growth during push	
					8 Doping uniformity for in-situ doped	
					9 Cross-sectional profile	
					10 Refractive index	
					11 Pinhole	

Thin-film					Product	Metrology
CVD-Oxide	**Scrubber**	**Implanter**	**CVD-W**	**Sputter**	**Tester**	
1 Particle test	1 Particle	1 Particle test	1 Particle test	1 Particle test	1 Calibration	1 Calibration
2 Thickness	removing	2 Sheet resistance	2 Resistivity	2 Reflectivity	2 Correlation	2 GRR§ study
nonuniformity	efficiency	uniformity	uniformity	Center to edge	to pilot reference wafer	
3 Stress		3 Metal	3 Reflectivity	3 Stress	Parametric	
4 Refractive index		contamination	Center to edge	4 Step coverage	Parameter	
5 Conformity		4 Implant angle	4 Step coverage	5 Resistivity	Laser	
6 Pinhole		Check	5 Mass gain	Uniformity	Softward defect analysis	
7 Peeling		5 GOI‡	6 Grain size	6 Grain size	Repair success rate	
characteristic		6 Implant species			Multiprobe	
8 Corrosion		check			Fail label	
		7 Dose and energy				

† MCLT: Minority carrier lifetime
‡ GOI: Gate oxide integrity
§ GRR: Gauge repeatability and reproducibility

TABLE 22
Process release steps

Purpose	Type of parameters	Material	Minimum # of tests
Capability	Parameters related to this process only	Bare pilot wafer	25
	Process integration parameters related to other processes	Preprocessed pilot wafers up to this process or	25
		Actual production lots	3–5
	One-time check parameters	Pilot or production wafers	3–5
Stability	Production control or monitoring parameters	Actual production wafers or production monitor wafers	25

of each process. With good preparation and execution, the approach has proven to be effective in our new advanced wafer fabs in fabricating good production units from the very beginning of the wafer fab start-up. The dependence of each process on the other (process integration) has been considered and checked through identified integration process parameters during the individual process releases. This approach minimizes the need to perform lengthy process integration checks using short-loop tests (combination of a few process steps but not the entire process flow) to verify the effectiveness of the process combination. It does not, however, eliminate the need for device/product characterization to ensure that not only process but also device margins are achieved. In addition, short-loop tests do serve other purposes—as a process monitor (e.g., gate oxide integrity) and as a device/product monitor (e.g., wafer-level reliability control)— and are commonly implemented on the manufacturing line for control. This paragraph will discuss several start-up device/product characterization considerations.

Parametric parameters

The parameteric parameters are those that are important to device performances.[13] With DRAM as an example, several parameters of interest to DRAM are the transistor threshold voltage (V_t), transistor GOI, cell dielectric integrity, cell leakage, and cell-to-cell isolation. In addition to these fundamental device parameters, there are also parametric parameters for process monitoring. Examples of parametric parameters are the electrical thickness of various dielectric layers (gate oxide and interleave isolation dielectrics) and the deviation of critical dimensions from design targets such as delta widths for gate and metal levels.

The approach/methodology for parametric parameter characterization is similar to that for the process release discussed in paragraph 11.3.4. It is equally important that the actual data from the pilot line or design criteria be clearly specified and collected. These data are compared with the data obtained from the start-up wafer fab. Often, fine-tuning of a few process parameters may be required to bring the parameter of concern back to the reference value if an actual shift is observed. The fine-tuning is aimed at adjusting device performance back to the designed margin. Even though poor performance margins may not impact the product salability,

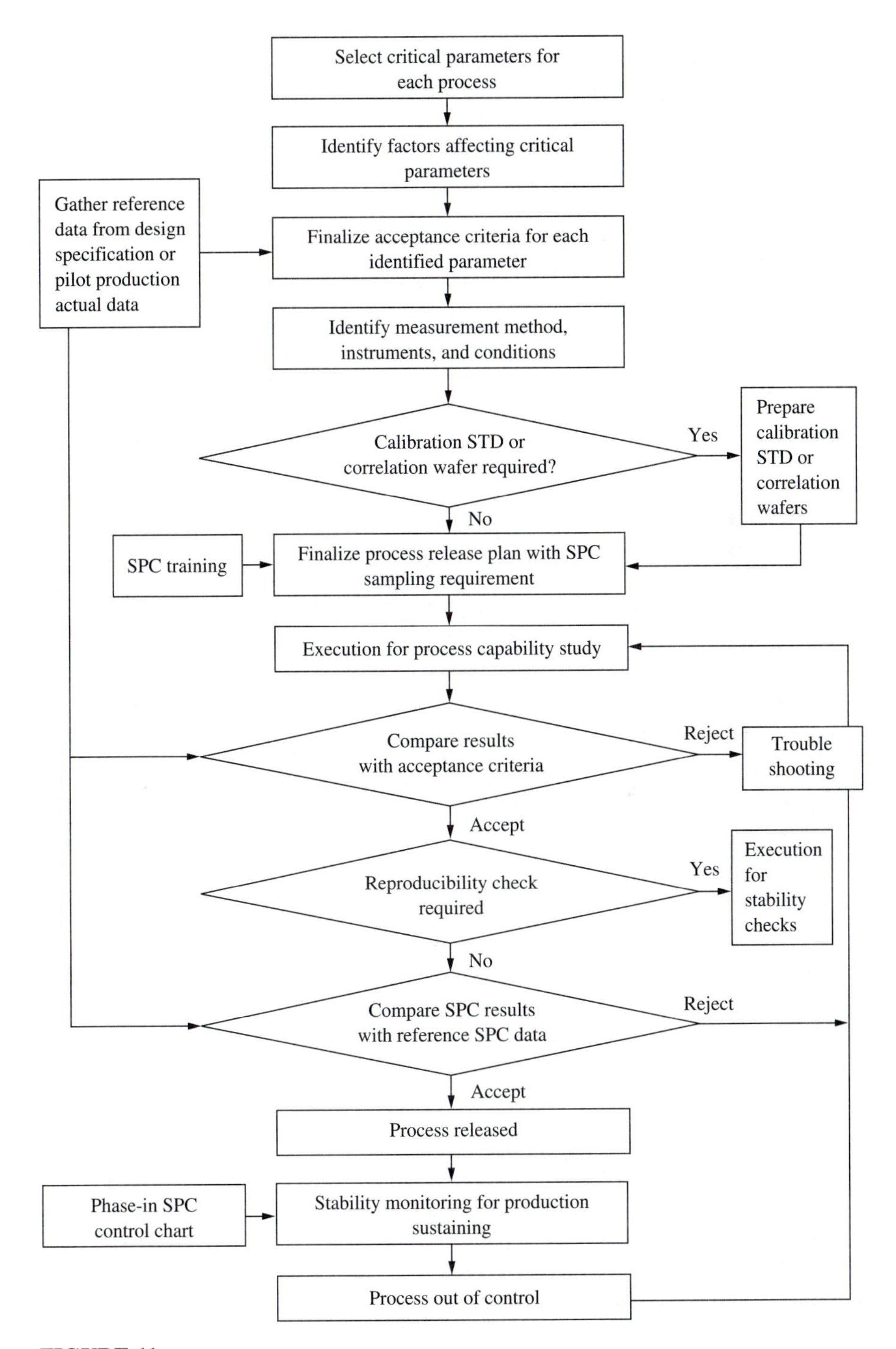

FIGURE 11
Schematic flow diagram for process release procedure.

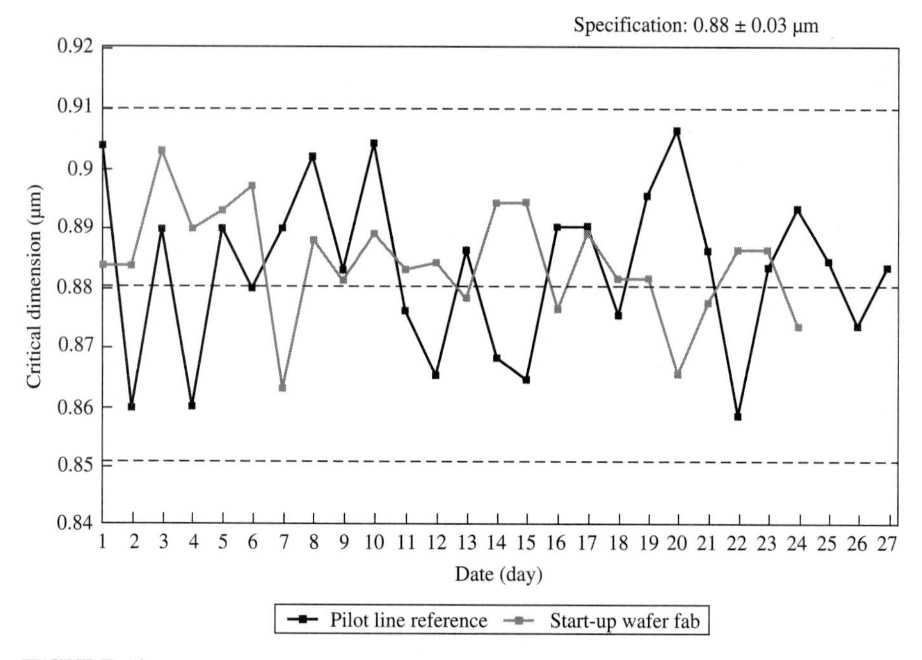

FIGURE 12
Verification of start-up 4-Mbit DRAM trench capacitor CD to reference performance.

maintaining a good margin is critical to guard against all variations in long-term volume manufacturability.[6] With DRAM as an example again, Fig. 13 illustrates a slightly worse device margin on 4-Mbit DRAM trench capacitance for the start-up wafer fab compared to that for the pilot reference. This variation indicates the need to fine-tune the capacitor dielectric thickness and critical dimensions.

Wafer-level reliability

In the preceding paragraph, we discussed the characterization of the functional parameters for the purpose of checking on the quality of the device at time zero. In this paragraph, we will consider aspects of the device reliability from the wafer-level point of view[29–31] for early detection of potential reliability problems. Wafer-level reliability control (WLRC) is implemented simply by physical or electrical means during the wafer fab processing. Some of the key factors for WLRC are these:

- Maximize in-situ reliability controls to replace product reliability control.
- Perform wafer-level reliability tests for fast process feedback, abnormality identification, and product classification.
- Develop device structure modeling to determine product life; investigate the feasibility of real-time test structures and wafer-level burn-in.

Typical WLRC tests and parameters[29–39] are summarized in Table 23. The selection of the WLRC parameters should be based on the historical problems, potential reliability concerns about the device of interest, and specific wafer fab process problems that affect product reliability.

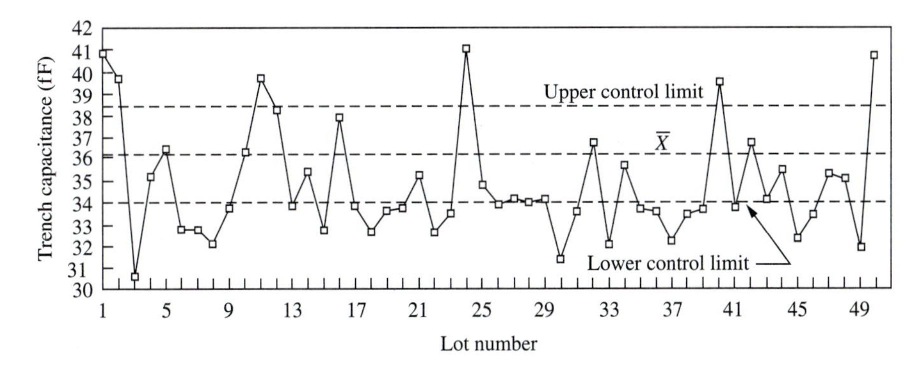

FIGURE 13
The slightly worse margin of the 4-Mbit DRAM trench capacitance for the start-up wafer fab requires process fine-tuning.

Probe testing

Wafer fab testing is not intended to replace final product testing after assembly but instead is aimed at studying the chip characteristics for process feedback and at classifying chips for various degrees of quality and reliability so as to simplify assembly/test operations for product sorting.

Typical wafer fab test programs include the following major categories:

- Parametric for testing the device parameters.
- Test of gross problems such as electrical continuity (open and short circuits).
- Test of various leakage mechanisms for device components (diode, transistor, capacitor, etc.), device isolation, and chip-level leakage in operating and standby modes.
- Test of chip functionality. The functionality test items, of course, are different for different devices. With DRAM as an example, functionality tests may include pause distribution, device timings (various circuit access times), test patterns (disturb, page, etc.), reliability in the form of voltage stress, and items specific to customer requirements.

Determining the test conditions for each of the test items is very important. The test must distinguish device quality levels but not eliminate good chips. Typically, the test conditions are decided on the basis of design, modeling, past experience, detailed correlation with packaged-product quality and reliability performances, and customer requirements. In many cases, a wafer fab test is conducted at higher than room temperatures (70 to 100°C) or at cold temperatures, because the typical guaranteed packaged-product temperature performance is in the range of −65 to 150°C or, a less stringent condition, −10 to 125°C. Unless there is a known low-temperature problem, wafer fab testing at high temperature is sufficient, because device performance is generally worse at high temperature.

One of the most important test parameters is, of course, the yield. Wafer fab test yield is not absolute but varies with different test items and conditions. For the same set of chips, a tighter test program may give a lower wafer fab yield but may provide

TABLE 23
Typical wafer level reliability control tests

Wafer level reliability test	Physical or electrical	Application (Refs. 30–39)	Judgment criteria
Metal			
Notch/void	P	EM induced by PO mechanical stress	% to minimum metal line width
Grain size	P	Sputter machine leak Al metal migration	Range of grain size
Si nodule	P	Metal (AlSi) sintering temperature control	Range of nodule size
Side hillock bridging	P	Sputter machine control	% to minimum metal line spacing No bridging
Step coverage	P	Control of via and contact metal deposition	Minimum of step coverage
Adhesion	P	Surface condition and contamination Film stress control	No metal lifting Range of stress
Corrosion	P	Post-metal-etch cleanup	No visual metal corrosion
Electromigration (EM)	E	Metal line EM	% Metal resistance change after electrical/temperature stress
Metal via and contact resistance	E	Via and contact EM Si Buildup at via and contact	% Via and contact resistance change after electrical/temperature stress
Protective overcoat (PO)			
Stress	P	EM induced by PO stress Metal peeling	Range of stress
Thickness	P	PO process control	Range of thickness
PO integrity	P	Corrosion due to moisture penetration	Minimum defect density (pinholes, etc.)
Wafer PCT (temperature and humidity)	P	Corrosion resistance	No corrosion

TABLE 23 *(continued)*

Typical wafer level reliability control tests

Wafer level reliability test	Physical or electrical	Application (Refs. 30–39)	Judgment criteria
Interlevel dielectrics			
Stress	P	Film cracking and delamination	Range of stresses
Thickness	P	Deposition process control Metal cracking relating to poor planarization	Range of thickness Reflow control
% phosphorus in PSG and BPSG	P	Phosphorus-induced corrosion	Maximum % of phosphorus
Wafer temperature cycle (TC) test	P	Interlevel strength	No peeling/delamination
Breakdown voltage cycle (VC) test	E	High breakdown strength	Minimum electrical field strength
Gate oxide integrity (GOI)			
Breakdown voltage	E	Gate process contamination, surface roughness, and dielectric quality control	Minimum electrical field strength
Device component			
Transistor channel hot carriers (CHC)	E	Short channel degradation	% Leakage current/transconductance increase after electrical/temperature stress
Diode leakage	E	Diode leakage kinetics	% Leakage current increase after electric/temperature stress

a higher final test yield after assembly, burn-in, and reliability checks. There are no set rules on whether a tighter or looser test program should be used except for the fine balance between the test and correlation with product performance, as well as a thorough understanding of the customer requirements.

In Fig. 14, four histograms show the distributions of test items in each category using DRAM devices. These items and a few others are used as the wafer fab outgoing quality control items for a commercially available product. With this kind of distribution, product classification becomes possible based on statistical variations. The quality and reliability performances of the sorted chips, using the distributions shown in Fig. 14 can then be correlated with those of the final packaged products. As a final-stage check for start-up at a wafer fab, the sorting criteria for quality and reliability classifications of a specific test item and distribution should be referenced to design criteria and/or pilot production data. If all checks are carried out thoroughly, the performance of the fabricated wafers at the start-up wafer fab are already within design tolerances or in line with the actual performance of the pilot production. Hence, the outgoing QC criteria from the reference pilot line wafer will be accurate (80 to 90%) indicators of the actual final quality and reliability performance of the new start-up. This is important for the fast qualification of the new wafer fab facility to produce and sell products.

11.3.6 Product Qualification

Finished wafers are subjected to next-stage processes, namely, assembly, burn-in, testing, and reliability checks. Some details of the assembly and the reliability are discussed in other chapters of this book and therefore are not covered here. However, we will briefly introduce the considerations for packaged product qualification with an emphasis on the wafer fab tests. The subject deserves some attention, because the successful completion of product qualification signals an important start-up milestone. Qualification is not an easy task and depends a great deal on product design, process margin, and the lessons learned (particularly from reliability-related problems) from past product generations and processes. The minimum qualification cycle time itself can take up to 2 months since several reliability tests require 1000 hours or more of test time. If the product fails, another attempt at qualification would easily delay the start-up milestone by 3 to 6 months, counting the time from wafer start to the completion of qualification tests. This time does not include the problem-solving efforts. In the late 1970s and early 1980s, product qualification took up to two years (as with 16K or 64K DRAMs), which is very expensive because of the large number of units needed, the operation cost, and loss of revenue. Products with designed-in reliability based on experience gained from past lessons learned and manufacturing start-up technology with thorough process/product checks (Section 11.3) are key to a robust qualification process. As a result, the qualification time has been reduced to less than 6 months (e.g., 4-Mbit DRAM) in the 1990s.

Internal product qualification

There are two test categories for qualification: a test for quality (the device is functional at time zero) and a test for reliability (the device is functional at a

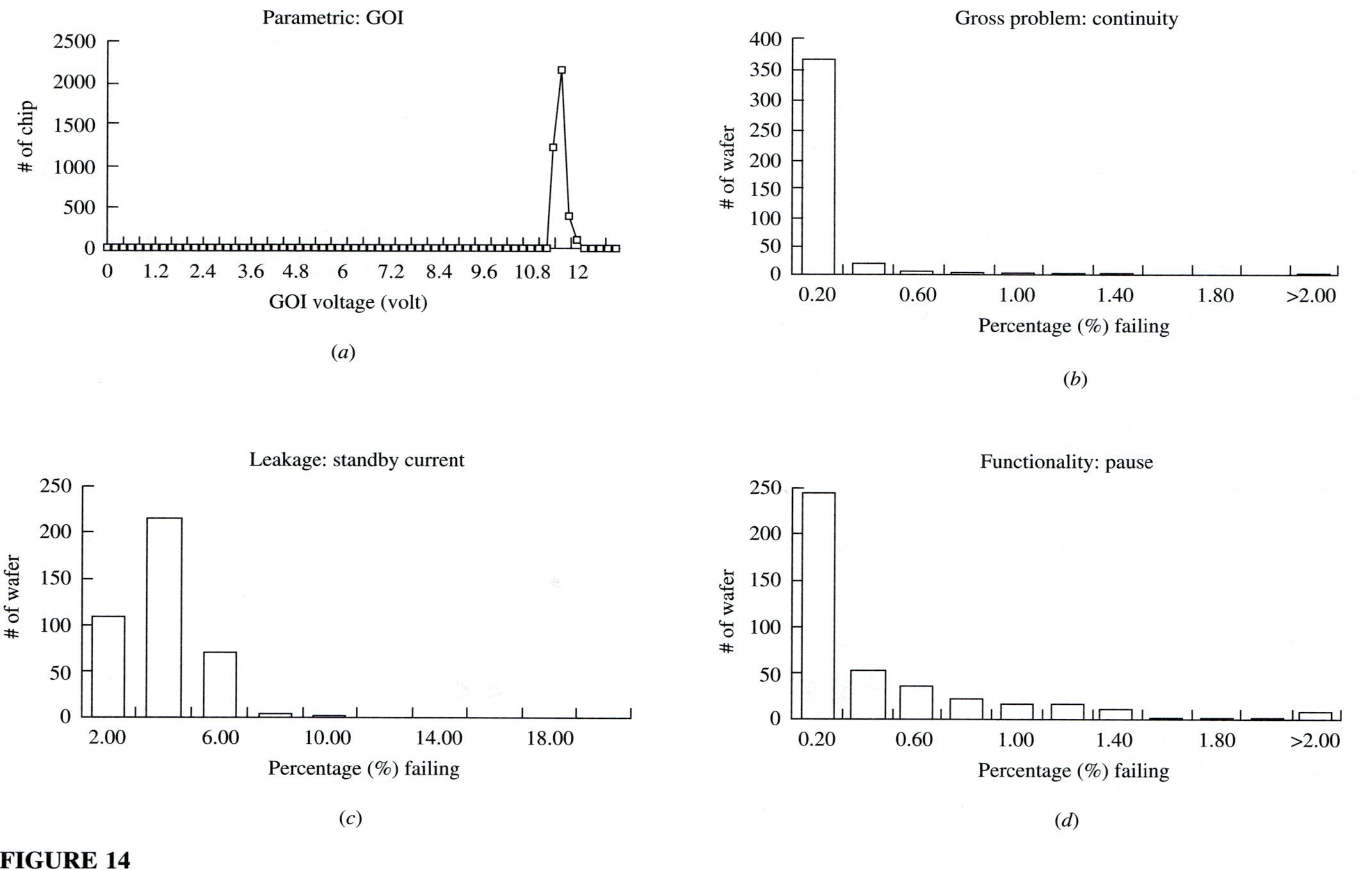

FIGURE 14
Examples for probe testing for (*a*) parametric, (*b*) gross problem, (*c*) leakage, and (*d*) functionality using DRAM devices and a sample size of 400 wafers from 20 lots.

future guaranteed product life). The tests of quality are tests against the product data sheet specifications. The tests themselves are not different from those discussed in Section 11.3.5, hence we shall not consider them further. The tests of reliability are summarized in Table 24. The exact stress conditions for each test may vary based on customer requirements and guaranteed product life. By knowing the customer requirements and the guaranteed product life, the acceptance criteria for each test can be generated based on a typical reliability failure rate (e.g., failure rate at a 60% confidence level, or FR60) calculation[40] with the stress conditions and a large enough sample size to judge qualification success. Last, in Table 25, a product qualification approach is provided. As one can see, good practice involves not only a few lots of snapshot testing (one can only do limited reliability tests because of time constraints) but also continuous monitoring of more lots over a period of time to ensure volume manufacturability.

Customer qualification

The ultimate goal of manufacturing, of course, is to sell the products that it produces. To do that, products need be qualified by customers first. There are two

TABLE 24
Typical tests of product reliability

Test	Purpose	Typical conditions
Dynamic operation life (op Life)	Test of product operation functionality under stress	−10°C, 125 to 150°C 5 to 8 V 500 to 2000 hours
Early failure rate (EFR)	Check product "infant mortality"	125°C, 7 to 8 V 20 to 200 hours
Humidity	Moisture test for mechanical strength and corrosion	65°C, 85°C 65%, 85% relative humidity for 168 to 1000 hours
Autoclave or pressure cooker test (PCT)	Test of mechanical strength and corrosion under severe conditions of total saturation and temperature stress	120 to 125°C 2 atm pressure with absolute 100% relative humidity for 96 to 240 hours
Temperature cycle (TC)	Test of mechanical strength	−65 to 0°/125 to 150°C 500 to 2000 cycles
Electrostatic damage (ESD) sensitivity	Test of ESD damage to products	MIL STD 883 D Method 3015.7
Soft-error rate (SER)	Test of radiation-induced soft failures	Test parameters per product data sheet, 1 μs/access timing, external checkerboard and inverse checkerboard data pattern
Package integrity (not discussed in this chapter)	Test of package-related reliability concerns	
Others		

TABLE 25
Product qualification approach

	Initial qualification		Volume ramp-up control	
Purpose	Capability demonstration		Manufacturability confirmation	
Tactics	Customer driven Process baseline establishment Assess intrinsic and early failure rate statistically Initial production start		Internally driven Process baseline sustaining Define product control range Final production release	
Source	Few lots with statistical validity		Every lot for a hundred lots or more	
Test	Highly accelerated stress margin test		Minimum lot coverage	
	Test	**Sample size**	**Test**	**Sample size**
	Op life	300–800	Op life	50–200
	EFR	500–2000	EFR	50–100
	Moisture	300–800	Moisture	50–100
	Thermal stress	300–800	Temp. cycle	50–100
	System soft error	200		
	ESD	20–50		
	Package integrity	200–300		
	Others	—		
Acceptance criteria	Customer requirements		Guaranteed product life	

possible levels of customer qualification: component and system. The exact customer qualification requirements are different for each customer. At the component level, however, qualification requirements are similar to those discussed in Section 11.3.5.6. Functional tests are based on the manufacturer's product data sheets, and reliability tests may include part or all of the tests listed in Table 24 as well as additional tests. The conditions for each test, as specified by each customer, may be different. However, they typically lie within those in Table 24. Hence, the component-level customer qualification is very similar to internal qualification from the test point of view. On the other hand, the customer system-level qualification is less known and predictable to manufacturers. For example, the same DRAM unit may produce quite different results in two different PC boards of the same type from two different system manufacturers. This outcome is possible because of the differences in the board design, materials used, processes, and system qualification test conditions. There are no general rules except to work with the customer and understand each specific customer requirement.

What would be an effective approach to deal with the whole range of test variations from many customers? One suggestion is simply to adopt the most stringent test conditions among all requirements for each test at the internal qualification stage. One can then work with the customers to have them accept the internal qualification results instead of the customer qualification at the component level. This is a win-win solution, because it reduces the qualification effort, cycle time, and cost for both the customer and manufacturer.

11.3.7 Process and Product Baseline

With all these efforts, we are finally in a position to establish the process and product baseline for the start-up wafer fab. Baseline is a document that describes all important factors of the process required to produce the product. This document is aimed at maintaining the process procedures and parameters and at controlling of future process changes of the product. Thus, the process and product baseline is extremely important and should be tightly controlled to ensure high-quality, high-reliability manufacturing. Table 26 is a summary of the typical contents of the baseline.

The bottom line in establishing the baseline is protecting our customers who will buy and use the products. Therefore, the baseline specification is customer auditable,

TABLE 26
Baseline specification description

Subject	Description
Header	Product name, device name, design revision, process flow, ID/revision, sign-off information
Purpose	Purpose of the baseline specification
Scope	Definition of applicability of the baseline
References	List of references to baseline specification
Change definition	Define changes that require notification and qualification approval, such as these: Major process steps Chip active circuit design change (photomask) Process flow changes Major process-technology-related equipment changes Change of direct materials used for processing (materials in contact with wafer during processing) Unspecified/unqualified reworks Changes of outgoing QC specifications
Notification and qualification approval	Specify in detail the requirements and procedures of notification and qualification approval for baseline changes defined in the previous entry
Baseline specification	Details of the baseline specifications Definition of major process steps Listing of process steps based on process flow with reference to process specification IDs and key parameter control ranges Photomask IDs and specifications Critical equipment related to key process technologies, with equipment vendor and model specified Critical materials such as wafer, chemicals, and gases with reference to each material specification and a brief description Allowed rework guidelines Test and qualification requirements Specification revision record
Others	If any

meaning that the customer can audit the wafer fab compliance based on this specification. Failures in baseline compliance found by a customer audit may result in the termination of the customer's order to purchase this product. Approval of the baseline document signals the commitment of the wafer fab to supply reliable, quality products in adequate quantities.

11.4 VOLUME RAMP-UP CONSIDERATIONS

Long-term success depends not only on a solid start-up foundation but also on the ability to produce products in volume. In Section 11.3, we discussed the foundation-building processes. In this section, we will address issues on volume production, which include equipment utilization, cycle time and inventory control, and sustaining production (baseline and change control).

The principal objective of a production wafer fab is to produce chips at a profit. For a new fab, the volume ramp-up is the key to profitability. Consider the following simple relationships. Conceptually, the profit can be expressed as

$$\begin{aligned} \text{Profit} &= \text{revenue} - \text{cost} \\ &= \sum(\text{output} \times \text{price} - \text{cost}) \\ &= \sum \text{output} \times \text{price} - \sum \text{cost} \end{aligned} \tag{11.1}$$

where $\text{Output} = (\text{wafer input}) \times (\text{process yield}) \times (\text{test yield})$

The wafer input rate is limited by the bottleneck equipment, that is,

$$\text{Wafer input} = (\text{bottleneck equipment throughput}) \times (\text{equipment utilization}) \tag{11.2}$$

Thus, the output or profit can be raised if the wafer input rate, process yield, and test yield can be increased. Of course, at the same time, the cost must be controlled and reduced without sacrificing quality/reliability.

11.4.1 Equipment Availability and Utilization

The wafer input rate can be increased for a given set of process equipment if more wafers can be processed through the bottleneck equipment. Typically, the utilization of the equipment must be increased. In this section, we examine the limitations of equipment in terms of production usage.

Process equipment, like any equipment, will experience trouble and require repair, resulting in the loss of production time. Process equipment also contains components that wear out, and equipment must be shut down periodically for preventive maintenance,[5,27] which also results in lost time.

A typical breakdown of the process equipment may be as follows:

Production	150 hours
Setup	20
Repair	5
Preventive maintenance (PM)	10
Idling	10
Total	195 hours, or 100%

Typical terminology for equipment used in wafer manufacturing can be expressed as follows:

$$\text{Availability} = \frac{\text{Time available for production}}{\text{Total time}} \tag{11.3}$$

$$= \frac{195 - (\text{repair} + \text{PM})}{\text{Total time}}$$

$$= \frac{180}{195} = 92.3\%$$

$$\text{Utilization} = \frac{\text{Time utilization for production}}{\text{Total time}} \tag{11.4}$$

$$= \frac{\text{Total time} - (\text{idling time available for production})}{\text{Total time}}$$

$$= \frac{150}{195} = 76.9\%$$

In this example, we see that even though availability for production is 92%, only 77% is actually utilized for production.

To improve utilization, one can break down the time loss further and then establish corresponding improvement actions (Section 11.5.1).

Time loss	Improvement actions
Repair	
Inexperienced equipment engineering staff	Increase training
Waiting for parts	Improve spare parts management
PM	
Inexperienced staff for maintenance	Increase training
	Reduce setup time
Idling	
Waiting for product	Increase inventory level as a buffer
No operator	Increase automation effort
	Cross-training of operators

11.4.2 Cycle Time and Inventory Control

In Section 11.4.1, we pointed out that profitability can be achieved by volume (output) ramp-up. The key factors listed were wafer input rate, process yield, and test yield. All of these are basically technical or engineering factors.

We will introduce in this section other nontechnical factors that are also key to successful manufacturing.[41]

- *Cycle time:* For a given process step, the cycle time is defined as the elapsed time from input to output. For a typical IC process flow, there may be tens to hundreds of process steps. The total cycle time is the sum of the cycle time of each step.
- *Work in process* (WIP) *inventory:* The quantity of wafers that reside in a given process step, whether actually in process or waiting to be processed, is called WIP.
- *Process speed:* Process speed is the rate of input being processed to the output (good units produced) plus the scrap due to defects.

These three parameters are related by the following equation:

$$\text{Processing speed} = \text{WIP/cycle time} \tag{11.5}$$

Ideally, the WIP can be reduced to the point that wafers are waiting to be processed, provided the process equipment never breaks down; the input rate of the wafer exactly equals output; supplies to the process step never run out; etc.

In real life the equipment does break down, the input rate of the wafer fluctuates, and some supplies can be interrupted. The result is that the WIP level will fluctuate above the ideal. In fact, one typical action to counter input rate fluctuation is to increase inventory level deliberately to ensure that the bottleneck equipment will never run out of production material.[41] The general relationship between inventory level and equipment utilization is shown in Fig. 15. Note that high utilization requires high inventory. However, if the fluctuations or variability (V) of operations (process, equipment, supplies, etc.) is reduced ($V_2 < V_1$), the curve of cycle time vs. utilization shifts to the right as shown in Fig. 15. By reducing variability (better process/equipment control, tight inventory movement control, etc.) the output can be increased and cycle time can be reduced. The reduction of manufacturing variability is thus of fundamental importance.[4,6,17,41]

High inventory, hence high cycle time [Eq. (11.5)], can enhance equipment utilization and increase factory output to generate more revenue. However, high inventory and cycle time result in some problems.[2,41]

- *Cost:* All inventory requires money and increases the cost of manufacturing.
- *Risk of defective inventory:* Figure 16 illustrates the inventory distribution of two cases, one with cycle time CT_1 and the other with cycle time CT_2. If a defect generated at point X_g is not detected until point X_d, the inventory at risk is the inventory distribution curve from X_g to X_d. Thus, high inventory or long cycle time results in an increased risk of accumulating defective inventory.
- *Slow response to customer demand:* For a commodity IC such as a standard memory device, the ICs can be made and sold without much consideration to cycle time, except for inventory cost. For ICs that are made to order in response to the

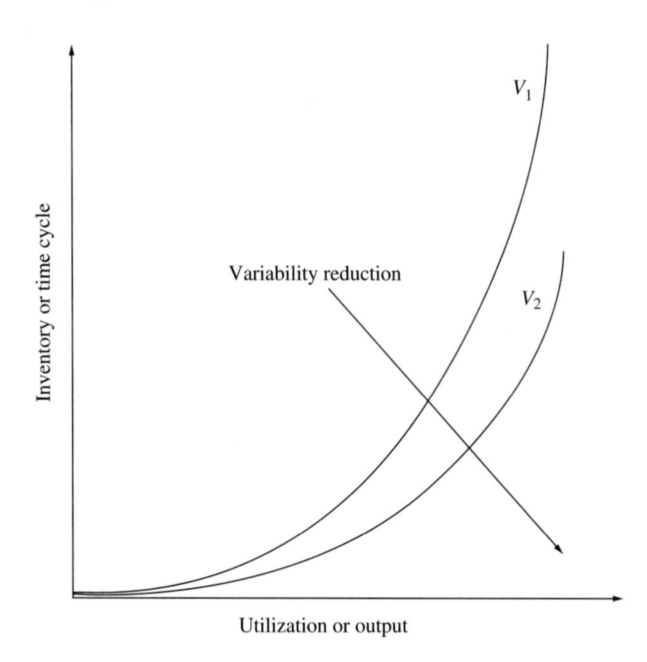

FIGURE 15
Relationships between cycle time and output.

quick pace of the market demand, which may be up or down, long cycle time results in long delivery time for the customer and potential excess inventory for the manufacturer.

Typical cycle time varies from 10 to 90 days. By going to all single-wafer processing, it has been reported that the cycle time of a complex CMOS IC can be less than 3 days![42]

11.4.3 Sustaining Production

We have repeatedly emphasized that minimizing variations is key to stable manufacturing.[4,6,17,41] Many of the wafer processes are interdependent and each may take a long time to complete, meaning that one may not know the exact impact of one change on others for a certain period of time. Typically, a large loss is incurred if one change adversely affects other processes. There are many painful real-life lessons where a large yield reduction (in excess of 10%) and loss of profit (in excess of $10M) were encountered when several changes or unknown variations were added, resulting in the loss of process margins. Perhaps the only way to deal with this situation is to control the variations proactively so that processes are kept stable to sustain continuous volume production. This situation is illustrated in Fig. 17 in terms of the cycle time needed for feedback (Section 11.4.1). It is obvious that only real-time control within an individual process provides the fast feedback needed for quick problem solving.

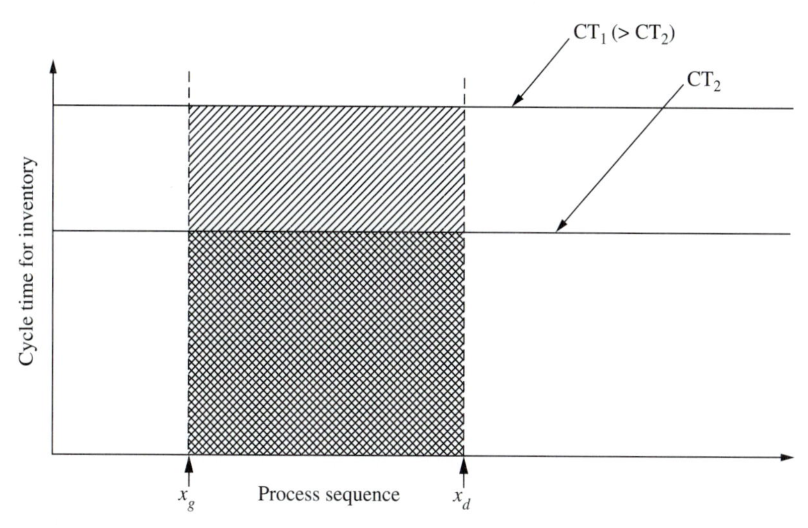

FIGURE 16
Risk for defective inventory (marked area) as affected by cycle time.

A scheme to achieve in-control manufacturing has been developed, as shown in Fig. 18. The concept is to implement designed-in capability and leading Q&R control parameters rather than to reject product based on lagging Q&R parameters. In addition to the use of the control strategy, as shown in Fig. 18, in-control manufacturing also depends on full compliance to procedures at all levels, which can only be ensured through repeated self-checkings.

Lagging parameters

Lagging parameters are those parameters that are the result of previous actions. They include assembly/test/reliability parameters (Table 24), wafer fab test probe parameters, and WLRC or short-loop process parameters (Table 23).

In all cases, the use of lagging parameters is less desirable for control because of the long feedback cycle time, as shown in Fig. 17, during which status information needed for diagnosis is lost. The lagging parameters are used to control the final net results, but they have more severe consequences (often, line shutdown) when an out-of-control situation occurs.

Concurrent quality (Q) and reliability (R) parameters

The concurrent parameters have real-time characteristics. They refer to property parameters such as film thickness (Table 21), process parameters themselves (Table 21), and equipment parameters (Table 19).

Concurrent Q&R parameters should be the main body of the process control and should be used heavily. One should lean toward using as many of the equipment parameters as possible for earliest problem detection. SPC is used to determine the out-of-control situation for each individual parameter.

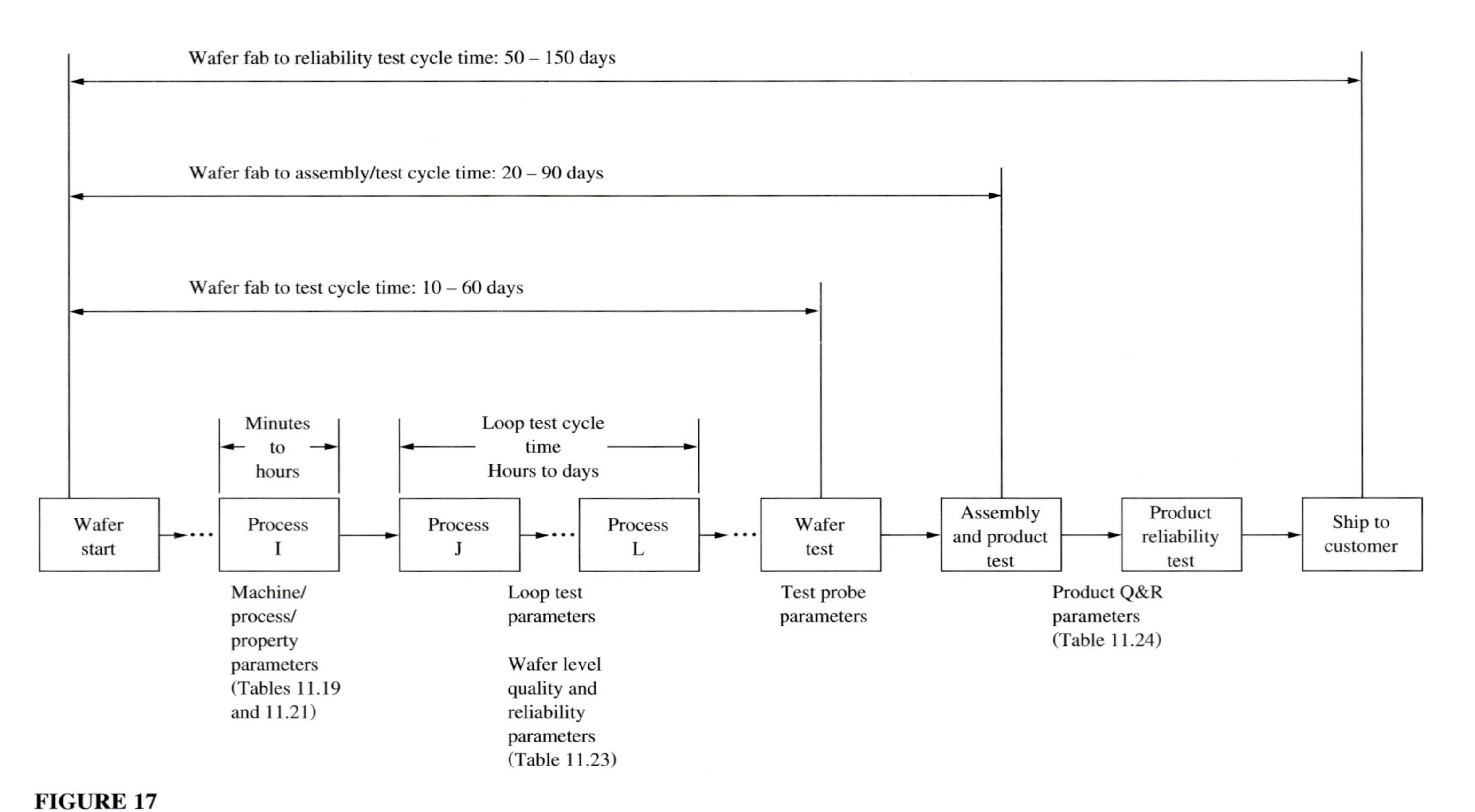

FIGURE 17
The long processing time indicates the importance of real-time control.

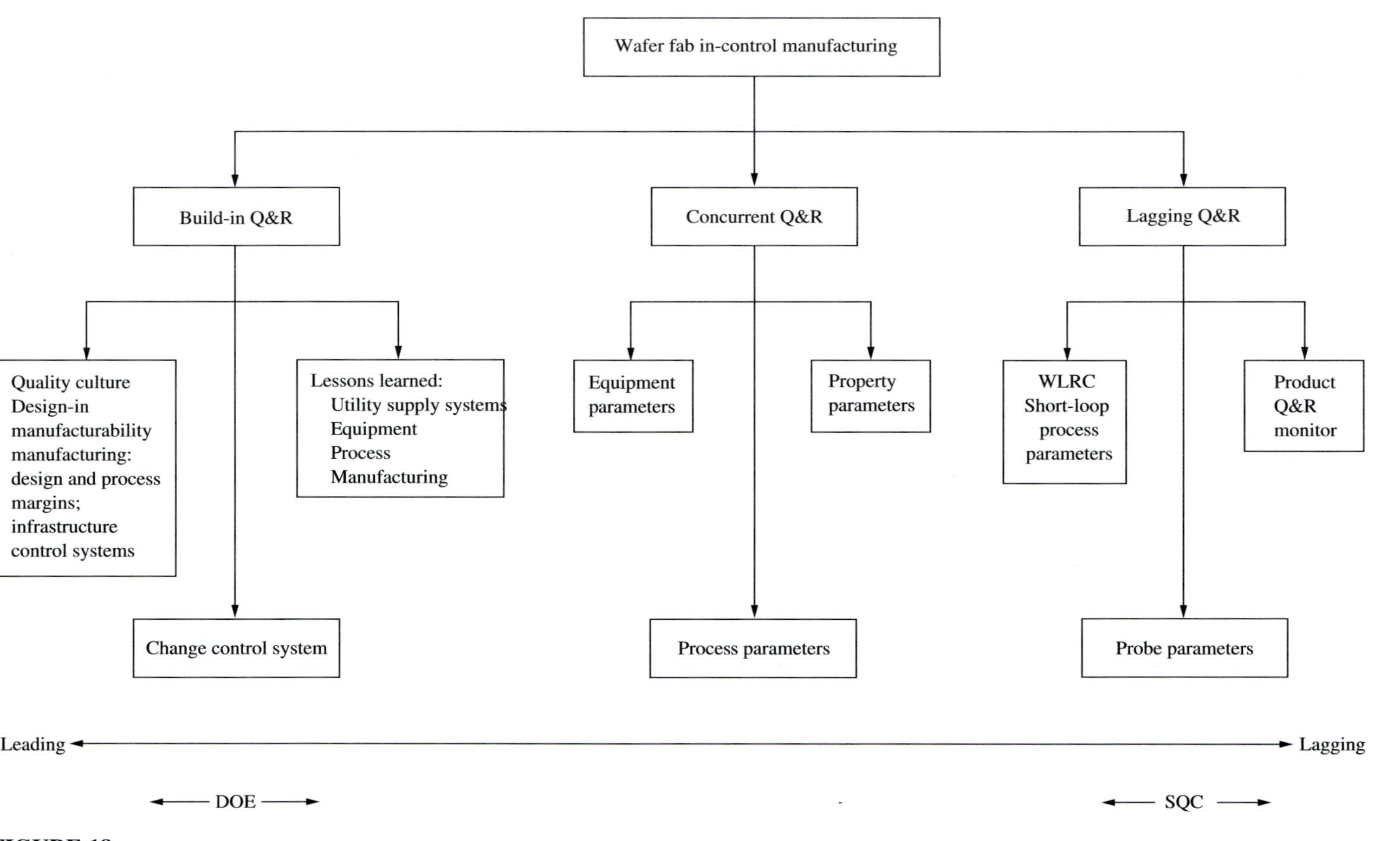

FIGURE 18
Scheme for an in-control wafer fab manufacturing strategy.

Build-in quality and reliability

The most powerful strategy is to *build* quality and reliability *in.* Control activities can be significantly relaxed if the process/product margins are so tolerant that variations cause little or no impact on the product quality and reliability. The build-in strategy can be further classified based on its impact. The first build-in category refers to newly established technology, infrastructure, practice, or methodology based on historical problem-solution lessons. The material requirements, facility systems start-up, and equipment cleanliness control are examples of build-in Q&R based on past lessons.

The second build-in category refers to control of change. We have repeatedly warned about the potentially adverse impact of changes on manufacturability. However, since changes are inevitable, we must learn how to control changes or make the best out of them. Indeed, a controlled change, when executed properly, improves process capability. The change control system of a wafer fab therefore has critical importance for continuous improvements in manufacturability. The system should utilize statistical methods on design of experiments (DOE)[4] and estimate the impact of the change on reliability. In Fig. 19 a flow diagram illustrates the approach for change control that builds upon the start-up process release, using the process release plan and actual data base as its foundation.

Quality culture and designed-in manufacturability are the most important build-in strategies. Regardless of the hardware setup, wafer fab will not be successful without a total quality culture[1,2,5] encompassing people who want to do the right things. This desire is something that no control systems can replace. The establishment of the quality culture reflects management's commitment to excellence and must be carefully thought out and implemented. On the other hand, the purpose of designed-in manufacturability is to develop proactive design-in activities for functionality, reliability, and infrastructure so as to establish the margins for simple and easy manufacturing prior to the wafer fab start-up. Designed-in manufacturability is desirable for optimum yield and quick volume ramp-up to achieve maximum financial results.

In summary, start-up emphasizes using build-in Q&R activities to establish the baseline foundation. Subsequent production emphasizes using concurrent Q&R parameters for sustaining the baseline and using the change control system for further continuous improvements. The strategy has produced profound results,[43] which have led to short- and long-term success in our advanced wafer fabs.

11.5 CONTINUOUS IMPROVEMENT

In previous sections, we focused on how to start up and establish capacity control to achieve volume wafer manufacturing. The emphasis has been to reduce variations of all kinds to sustain the baseline process for stability. Only in Section 11.4.3 was the change control system introduced (Fig. 19).

In this section, we introduce the concept of continuous improvement[1–3] through the constant use of problem-solving methodology.[4] The result is improvement in

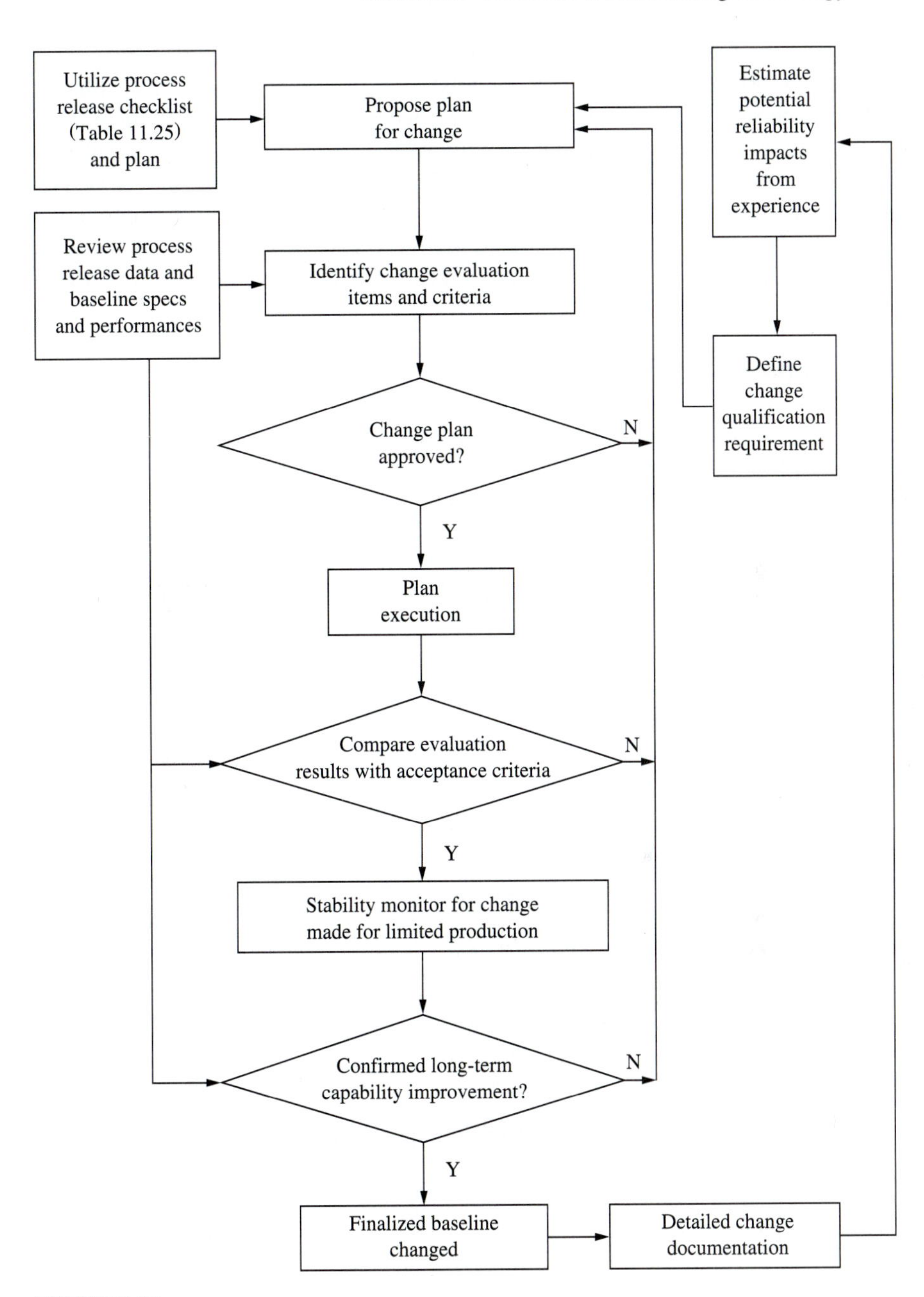

FIGURE 19
Flow diagram for change control.

yield, capacity, Q&R, and the cost/profit margin. Methodology for continuous improvement is different from the yield models[11,44] that are discussed in the technology books. Readers should realize that improvement opportunities are everywhere. Many small improvements can accumulate to a big gain.[3] The nature of stable manufacturing is not luck but hard work and good control. There are many potential >10% yield drops but very little opportunity for one 10% yield gain. This kind of improvement is often overlooked by most technologists yet has great impact on real-life manufacturing considerations. Continuous improvement is the key to operational excellence.

11.5.1 Improvement Process

Figure 20 shows a continuous improvement process using problem-solving methodology.[3,4] The methodology uses commonly accepted QC tools. Repeated, honest, and thorough use of the methodology is the secret of success. To illustrate the improvement process, we will discuss a case study using the problem-solving tool to improve the LPCVD nitride furnace utilization and thus to improve capacity. The case is chosen to be consistent with the manufacturing theme of the chapter.

CASE STUDY

Step 1: Select problem.

- Poor nitride furnace utilization

Step 2: Understand present situation.

- Target vs. actual utilization on monthly basis

	Target	**Actual**
MTBF (hr)	>300	360
Equipment utilization (%)	> 80	60

- Pareto analysis for downtime breakdown (Total: 288 hours): Fig. 21
- Major causes of downtime: 1. 144 hr, or 50%, due to maintenance
  2. 94 hr, or 34%, waiting for parts
- Breakdown of maintenance activities

Activity	**# of times performed**	**Total downtime (hr)**
Tube change	2	82
Boat change	3	33
Waiting for dummy preparation	3	29

- Details of tube change procedure: Table 20
- Details of boat change requirements:

Tube clean cycle	5 μm of nitride deposited
Boat change cycle	2 μm of nitride deposited
Dummy wafer reclean	1 μm of nitride deposited
Boat scrap frequency	After 5 chemical etches

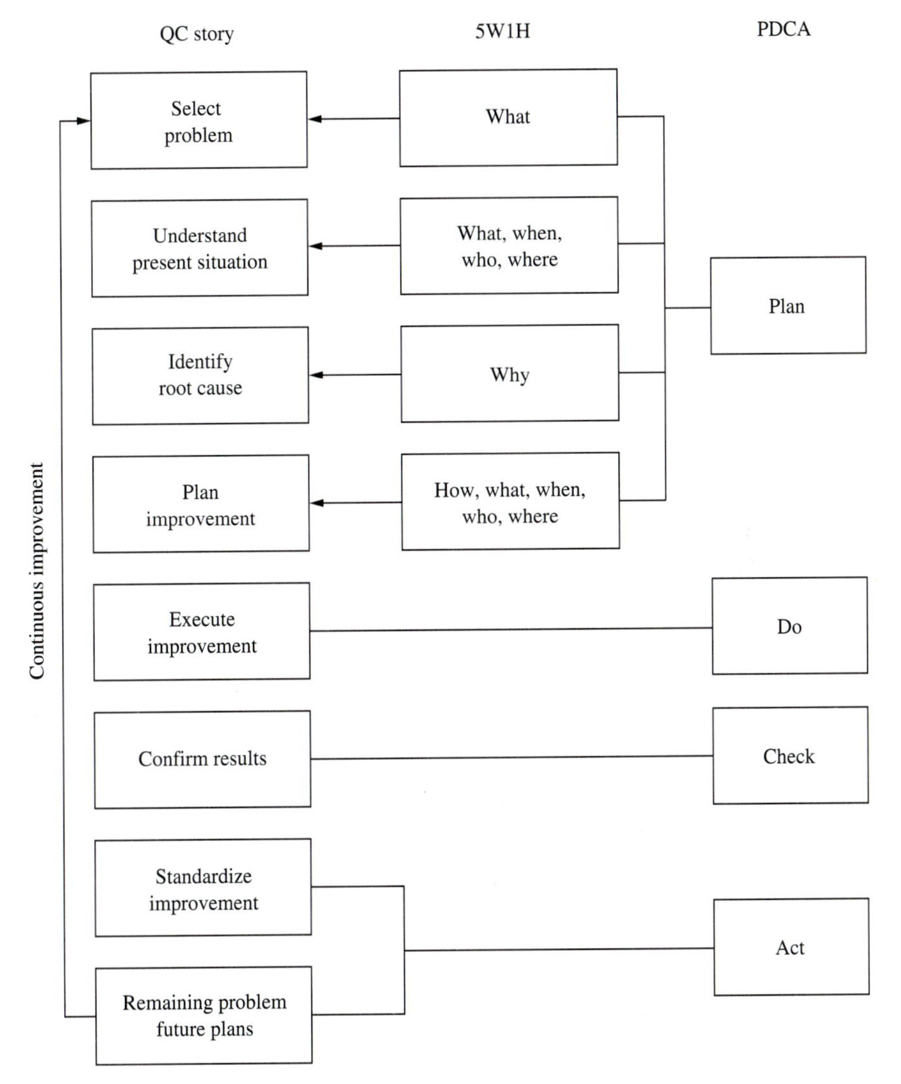

FIGURE 20
Continuous improvement through problem-solving methodologies.

Step 3: Identify root cause.

- Maintenance time loss

1. Poor PM procedure
2. Poor people skills in performing maintenance work such as proper sealing for leak-tight installation
3. Long boat setup/registration time due to alignment shift caused by change of quartz dimension with uncontrolled boat etching
4. Short tube/boat clean cycle

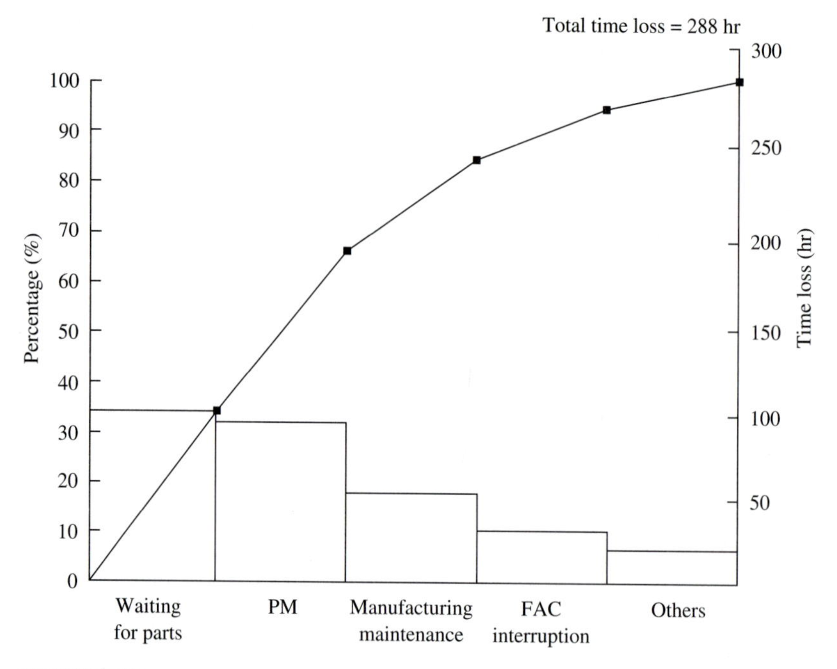

FIGURE 21
Pareto analysis on low furnace utilization.

5. Long waiting time for oxide dummy wafers used for nitride, due to poor peroxide furnace scheduling

- No spare parts, such as manifold, available
- Furnace interruption related to gas shutoff, due to improper alarm setting

Step 4: Plan improvement.

- Maintenance improvement

1. Revise tube change PM procedure based on actual experiences by equipment engineering.
2. Retrain furnace maintenance personnel.
3. Establish priority and communication systems for oxide furnace scheduling by manufacturing and planning.

- Setup of spare parts consignment and inventory management systems by purchasing/receiving
- False alarm elimination by facility engineering

Step 5: Execute improvement.

- Carry out plans outlined in step 4.

Step 6: Confirm results (3 months later).

- Target vs. actual utilization

	Target	Past actual	Current actual
MTBF (hr)	>300	360	480
Equipment utilization (%)	> 80	60	84

- Total downtime reduction: 172 hr

	Total downtime (hr)	
Improvement plan	Past	Current
Maintenance	144	99
Waiting for parts	98	0
Facility interruption	29	0
Other	17	17
Total:	288	116

- Maintenance improvement: 45-hr reduction in time lost

1. Revised PM procedure shown in Table 27
2. Confirmed maintenance activities

		Total downtime (hr)	
Activity	# of activity	Past	Current
Tube change	2	82	60
Boat change	3	33	33
Waiting for dummy preparation	3	29	6
	Total:	144	99

Step 7: Standardize improvement.
The PM procedure shown in Table 27 has been implemented for all nitride furnaces. Similar procedures are also applied to other LPCVD furnaces.

Step 8: Identify remaining problems/future plans.

- Large time loss due to boat change: 4.6% loss of utilization
- Other losses in Fig. 21

There are numerous wafer fab operations in which the improvement methodology as shown in the case study can be applied for areas such as process capability improvement; defect reduction; failure analysis; test yield improvement; and manufacturing improvements for wafer scrap, misoperation, and equipment capacity.

11.5.2 Supplier Improvement

A supplier improvement program[1–3] is also important to sustain and improve wafer fab manufacturing. Daily activities depend so much on suppliers for materials and services that they contribute significantly to improving or degrading wafer fab performance. Because a good supplier is hard to come by, owing to the high standards, particularly for the advanced wafer fabs, it is only through long-term cooperation with well-managed, capable vendors that wafer fab supplies can be guaranteed.

TABLE 27
Revised furnace tube change operating procedure

Item	Time (hr) Estimated	Actual
1. Pre-PM preparation	0	0
Prepare spare quartz, manifold, O-ring,		
Dummy wafers, etc.		
2. Process stop and heater off	5.0	5.0
Heater breaker off	—	
Clean main valve		
Check and clean pump parts for by-products		
Remove and clean vacuum exhausts		
3 hrs after heater off		
Furnace cool down		
3. Tube pull	2.0	2.0
Remove shutter, nozzle, inner/outer tubes,	—	
thermocouples, and manifold, and clean up		
4. Furnace clean	1.0	1.5
5. Tube/boat clean	0.0	0.0
6. Tube installation	2.5	2.5
Check tube for less than 5 times etching		
use previously prepared spare parts		
7. Room temperature leak check	1.5	1.5
Perform leak checks for all seals		
Pump and purge to degas		
8. Boat setup/registration	1.0	1.0
Check boat for less than 5 times etching		
Update records on part ID and etching time		
9. Heater on and heating up	1.0	1.0
10. High-temperature leak check	2.0	2.5
Process temperature leak check		
Pump and purge to degas		
11. Temperature profile and tube coating	7.0	7.0
12. Manufacturing test runs	6.0	6.0
Total time:	29.0	30.0

We shall not discuss details of the supplier improvement program. However, all of the principles discussed in this chapter are applicable to suppliers, and standards are just as stringent.

11.6
SUMMARY AND FUTURE TRENDS

In this chapter, we have discussed various aspects of wafer fab manufacturing. We have introduced a strategy for in-control manufacturing. We have discussed general manufacturing considerations for organization, layout, materials, systems and procedures, computer-integrated manufacturing, and cost structure. We have also described an approach/methodology for start-up and volume production. Lastly, we have illustrated a continuous improvement process with a case example. All of these are aimed at establishing wafer fab manufacturing to meet customer needs and generate satisfaction.

In practical situations, wafer fab must meet customer requirements and profitability. Automation in hardware to reduce variations in the production process and automation in information to manage the manufacturing dynamics in real time are an important approach to achieve this objective. Design-in manufacturability with sufficient margin so that variations can be easily brought into control is a definite trend. Overall, proactive design and implementation of activities upstream (toward input parameters) are essential to the success of wafer fab manufacturing, in contrast to the current practices of in-line control and screening of noncompliance in finished products downstream.

REFERENCES

1. W. E. Deming, *Out of the Crisis,* MIT-CAES, Cambridge, MA (1982).
2. R. T. Schonberger, *Japanese Manufacturing Techniques,* Collier Macmillan Publishers, London (1982).
3. M. Imai, *Kaizen,* Random House Business Division, New York (1986).
4. G. Taguchi, *Introduction to Quality Engineering,* NIPUB/Quality Resources, New York (1986).
5. H. Naguib, "The Implementation of Total Quality Management (TQM) in a Semiconductor Manufacturing Operation," IEEE Trans. Semic. *Manuf.* **6** (2), 156 (1993).
6. M. J. Harry and J. R. Lawson, *Six Sigma Producibility Analysis and Process Characterization,* Motorola University Press (1992).
7. The American Society for Testing and Materials (ASTM) Publication 25, Section 10: Electrical Insulation and Electronics, Philadelphia, PA (1993).
8. Semiconductor Equipment and Materials Institute (SEMI), Materials Volume, Mountain View, CA, 1993.
9. S. Wolf and R. N. Taubeer, *Silicon Processing,* Vol. 1, Lattice Press, Sunset Beach, CA (1987).
10. S. K. Ghandhi, *VLSI Fabrication Principles,* John Wiley & Sons, New York (1983).
11. S. M. Sze, Ed., *VLSI Technology,* 2nd ed., McGraw-Hill, New York (1988).
12. R. Iscott, "The Search for Ultrapure Chemicals," *Semic. Intl.,* 70 (July 1990).
13. S. M. Sze, *Physics of Semiconductor Devices,* 2nd ed., John Wiley & Sons, New York (1981).
14. D. Toy, "Gas Purity and Wafer Fabrication," *Semic. Intl.,* 114 (Apr. 1988).
15. R. Iscott, "Water Purity for the DRAM Generation," *Semic. Intl.,* 60 (Jan. 1991).
16. J. L. Lamprecht, "ISO9000 Implementation Strategies," *Quality,* 14 (Nov. 1991).
17. D. J. Wheeler and D. S. Chambers, *Understanding Statistical Process Control,* Statistical Process Controls, Inc., Knoxville, TN, 1986.
18. B. C. Belanger, "Traceability: An Evolving Concept," *ASTM Standardization News,* **8** (1), 22 (1980).
19. L. Peters, "Measuring Your Thinnest Films," Semic. Intl., 76 (June 1991).

20. M. T. Postek and D. C. Joy, "Submicrometer Microelectronics Dimensional Metrology: SEM," *J. Res. Natl. Bur. Stand.* (US), **92** (3), 205 (1987).
21. S. A. Rievi, "New Frontiers in the Metrology of Registration," *TI Tech. J.,* 76 (Jan.-Feb. 1990).
22. *Quality Function Development, Implementation Manual,* American Supplier Institute, Inc. (1989).
23. K. Kuhara, U. Katoh, and M. Shimbo, "Effects of Ultra-Clean Technology on Devices," *Ultra Clean Technol.,* **1** (2), 61 (1990).
24. Y. Motomura, "Evaluation of Piping Materials for Ultra Pure Water System—PVDF, PEEK," *Ultra Pure Technol.,* **1** (2), 86 (1990).
25. P. Burggraaf, "Valves Are Vital in UPDI Water Systems," *Semic. Intl.,* 112 (Sept. 1989).
26. D. C. Grant, S. VanDyke, and D. Wilkes, "Issues Involved in Qualifying Chemical Delivery Systems for Metallic Extractables," *Proceedings of the 39th Annual Technical Meeting,* Institute of Environmental Sciences, Mt. Prospect, IL, 1993, p. 200.
27. Y. Funaki, "Total Preventive Maintenance," *Semic. Intl.,* 162 (Sept. 1989).
28. S. Wolf, *Silicon Processing,* Vol. 2: Processing Integration, Lattice Press, Sunset Beach, CA (1990).
29. M. Davis and F. Haas, "In-Line Wafer Level Reliability Monitors," *Sol. State Technol.* 107 (May 1989).
30. C. L. Hong and D. L. Crook, "Breakdown Energy of Metal—A New Technique for Monitoring Reliability at the Wafer Level," *Proceedings of the 23rd IEEE International Reliability Physics Symposium,* 1985, p. 108.
31. S. K. Fan and J. W. McPherson, "A Wafer Level Corrosion Susceptibility Test for Multilayered Metallization," *Proceedings of the 26th IEEE International Reliability Physics Symposium,* 1988, p. 50.
32. J. W. McPherson, "Stress Dependent Activation Energy," *Proceedings of the 24th IEEE International Reliability Physics Symposium,* 1986, p. 12.
33. J. R. Black, "Electromigration—A Brief Survey and Some Recent Results," *IEEE Trans. Electron Dev.* **ED-16** (4), 338 (1969).
34. J. C. Ondrusek, C. F. Dunn, and J. W. McPherson, "Kinetics of Contact Wearout for Silicided ($TiSi_2$) and Non-Silicided Contacts," *Proceedings of the 25th IEEE International Reliability Physics Symposium,* 1987, p. 154.
35. S. C. Kolesar, "Principles of Corrosion," *Proceedings of the 12th IEEE International Reliability Physics Symposium,* 1974, p. 155.
36. J. W. McPherson and C. F. Dunn, "A Model for Stress-Induced Metal Notching and Voiding in VLSI Al-Si Metallization," *J. Vac. Sci. Technol. B,* **5,** 1321 (1987).
37. R. Moazzami and C. Hu, "Projecting Gate Oxide Reliability and Optimizing Reliability Screens," *IEEE Trans. Electron Dev.,* **37** (7), 1643 (1990).
38. J. W. McPherson and D. A. Baglee, "Acceleration Factors for Oxide Breakdown," *J. Electrochem. Soc.,* **132,** 1903 (1985).
39. E. H. Snow, A. S. Grove, D. E. Deal, and C. T. Sah, "Ion Transport Phenomena in Insulating Films," *J. Appl. Phys.,* **36,** 1664 (1965).
40. P. A. Tobias and D. Trindade, *Applied Reliability,* Van Nostrand Reinhold, New York, 1986.
41. E. M. Goldratt and J. Cox, *The Goal,* North River Press, Croton-on-Hudson, NY, 1986.
42. *Texas Instruments Corporate News,* June 28, 1993.
43. R. Iscoff, "SI Honors Top Fabs of 1994," *Semic. Intl.,* 76 (Apr. 1994).
44. SEMATECH, "Yield Models," Publication 91060594A-TR, Austin, TX (July 1991).

PROBLEMS

1. Describe key factors for customer satisfaction. How would those factors impact the success of wafer fab manufacturing?

2. Construct a complete manufacturing process flow following the example shown in Table 1 using a device of your choice.

3. List five to seven important material characteristics for wafer, chemicals, and gases and describe typical application ranges for each characteristic.

4. Describe parameters of concern for each of the utility systems (cleanroom, chemical, gas, and DIW) supplying the wafer fab. Specify those of particular interest to the point of delivery/use from each of the chemical, gas, and DIW systems.

5. Describe five or more of the process-checking parameters for releasing of the furnace, wet clean, stepper, plasma etching, CVD, ion implantation, and sputtering processes.

6. Describe the four major wafer fab test categories. Give an example of the test parameters for each test category using a device with which you are familiar. Also describe the purpose and test conditions for each of the five (op Life, EFR, humidity, PCT, and TC) product reliability tests.

7. Assume a wafer fab can produce 100 wafers a day. What are the wafer inventory levels if the cycle time to produce products A and B are 75 days and 10 days, respectively? By how much can the wafer inventory be reduced by reducing the product A cycle time to 60 days?

8. Prepare an example of your choice following the improvement process (Fig. 20) and case study (Section 11.5.1) discussed in this chapter.

CHAPTER 12

Reliability

J. T. Yue

12.1 INTRODUCTION

The reliability of integrated circuits involves many disciplines such as design, process, assembly, and test. Many delicate processing steps and manufacturing procedures are involved in producing a finished IC component. When the part fails in the field, a myriad of reasons could be the cause. Sophisticated and tedious analysis is required to get to the root cause of failure. From the customer perspective, what is really required of the product is continual functionality of the IC during the specified lifetime. Typically, manufacturers design an IC to last 10 years. Product reliability is ensured if each of the elements of the product is reliable for that length of time. There are three major elements of product reliability: design reliability, process reliability, and assembly reliability. If the reliability of each of these elements meets the required lifetime, then product reliability is ensured.

ULSI technology involves scaling of processes to the deep submicron (below approx. 0.35 μm) dimensions, as well as the additions of new process modules not used previously in the VLSI era. The reliability of each new process module, and how it interacts with other modules, will be critical to the final reliability of the entire process. Often, the scaling of a technology is performed without reducing the power supply voltage. Such an approach presents great challenge to the device engineers, reliability engineers, and process integration engineers. As a result, trade-offs are occasionally required among reliability, design, and process development. In the ULSI era, the concept of *design for reliability* is extremely important. Reliability must be designed in during all stages of IC development, including design, process development, and manufacturing. Therefore, process reliability is an extremely important subject.

First, basic process reliability wearout concepts will be reviewed, such as hot carrier injection, electromigration, stress migration, and oxide reliability. Then the effects of scaling on the reliability of these failure mechanisms will be discussed. Next, the relationship of dc to ac lifetime will be discussed. This topic is important since real-world reliability is based on time-varying, alternating stress conditions, not on constant dc stress conditions. Special reliability concerns arising from some of the recent ULSI processing approaches will be discussed, e.g., tungsten plugs and charge damage during manufacturing. The mathematics of reliability calculations and failure distributions will be included for completeness.

Integrated circuits are made of discrete components such as transistors, resistors, interconnections, dielectric thin films, and capacitors. These components are connected in a predetermined scheme (set by the circuit designers) and put together through a sophisticated sequence of process recipes (set by process integration). One way to measure the reliability of the entire IC process is to measure the reliability of each of the major process modules. Some of the key process modules from the reliability point of view are (1) transistor, (2) interconnection, and (3) gate dielectric.

After a transistor has met all of its designed performance, it has been shown that the transistor may not be able to maintain its performance throughout its intended lifetime. The degradation becomes more severe for shorter-channel devices. This degradation is called the *hot carrier injection* (HCI) *problem* and will be covered in Section 12.2. Similarly, a metal interconnection, either a series of vias or a single metal line, does not keep the same *resistance* value while current is continually passed through it. The resistance of the line will degrade by void formation in Al by a phenomenon called *electromigration* (EM). Even in the absence of current, it has been observed that an Al interconnection can suddenly open up as a result of mechanical stress relaxation, causing open circuits. This phenomenon is called *stress migration* (SM). Both EM and SM will be discussed in Sections 12.4 and 12.5. Gate oxide is an important element of the MOS transistor or of a storage capacitor in a DRAM circuit. This thin dielectric can break down, resulting in gate shorts, during a long application of electric field across it. This phenomenon is called *time-dependent dielectric breakdown* (TDDB) and is increasingly important to consider as gate oxide reaches the 100 Å level and below. The TDDB mechanism will be discussed in Section 12.5. The effect of device scaling on reliability is discussed in Section 12.6. The relationship of lifetimes between dc stress and ac stress is discussed in Section 12.7. Recent ULSI reliability concerns in tungsten plug and plasma damage are discussed in Section 12.8. Finally, the mathematics of failure distribution are briefly discussed in Section 12.9.

12.2 HOT CARRIER INJECTION

Hot carrier injection (HCI) is the phenomenon in which carriers (either electrons or holes), excited by the high electric field in the transistor channel region, are injected into the gate oxide of a MOS transistor, causing damage to the oxide/Si interface and/or trapped charges in the gate oxide. The carriers in the channel become overly excited (hence, *hot*) because they gain enough energy to cause impact ionization before they lose their energy by phonon scattering.[1] The oxide damage and trapped charges will cause electrical parameters of the transistor to drift.

Figure 1 illustrates the relationship of the carrier velocity of electrons and holes as a function of the channel longitudinal electric field. When the field exceeds 20 kV/cm, the carriers begin to lose energy by scattering with optical phonons, and their velocity saturates. Above 100 kV/cm, the carriers gain more energy from the field than they can lose by phonon or impurity scattering. Consequently, the energy of these carriers builds up. They are no longer in thermal equilibrium with the lattice and therefore are called *hot carriers*. When the excess energy exceeds a certain threshold, impact ionization is initiated. The electron-hole pairs generated in turn produce more pairs, and so on, resulting in the so-called avalanche multiplication process.

The channel electric field in the velocity-saturation region can be expressed[2] as

$$\mathscr{E}(x) = [V(x) - V_{\text{dsat}}]/l, \tag{12.1}$$

where V_{dsat} is the potential at the pinch-off point in the channel, x is taken to be zero at this point, and l is the effective length of the velocity-saturation region. The maximum electric field, $\mathscr{E}_{\text{max}}$, is found by replacing $V(x)$ by the drain voltage, V_{DS}, in Eq. (12.1). As the transistor channel decreases, V_{dsat} decreases and thus the $\mathscr{E}_m$ increases. HCI degradation is known to become worse in shorter channel devices.[3]

Figure 2 illustrates the impact ionization process in an NMOS transistor. For HCI degradation to occur, three events must happen: (1) there must be impact ionization, (2) the created electron (or hole) must be injected into the gate oxide (near the drain end), and (3) the injected carrier must be trapped near the oxide/Si interface or in the bulk oxide, or else break a chemical bond (e.g., Si-H).[2] Each of these events has a probability of occurrence.

$$\text{Event 1} \rightarrow \sim I_d \exp(-\phi_i/q\lambda\mathscr{E}_m) \tag{12.2}$$

$$\text{Event 2} \rightarrow \sim I_d \exp(-\phi_b/q\lambda\mathscr{E}_m) \tag{12.3}$$

$$\text{Event 3} \rightarrow \sim I_g \exp(-\phi_B/q\lambda\mathscr{E}_m) \tag{12.4}$$

The impact ionization energy in the silicon is ϕ_i, ϕ_b is the Si/SiO_2 barrier energy, and ϕ_B is the chemical-bond-breaking energy in the silicon dioxide (e.g., for Si–H). I_d is the drain current, I_g is the gate current, and $\mathscr{E}_m$ denotes the maximum electric field in Eq. (12.1).

The reliability consequence of these trapped charges (or damaged oxide/Si interface) is a degraded transistor. Parameters such as threshold voltage V_T, transconductance g_m, saturated drain current I_{dsat}, and linear drain current I_{dlin} are common parameters used to quantify the degree of HCI degradation. The HCI process also generates substrate currents I_{sub} as indicated in Fig. 2, which, if severe enough, can induce bipolar latch-up in circuits.[4]

12.2.1 NMOS and PMOS Transistor

Interface trap generation is believed to be the main degradation mechanism for nMOSFETs.[5–8] The mechanism of interface trap generation is not well understood and is still an active subject of research. It may involve hot electrons, hot holes,

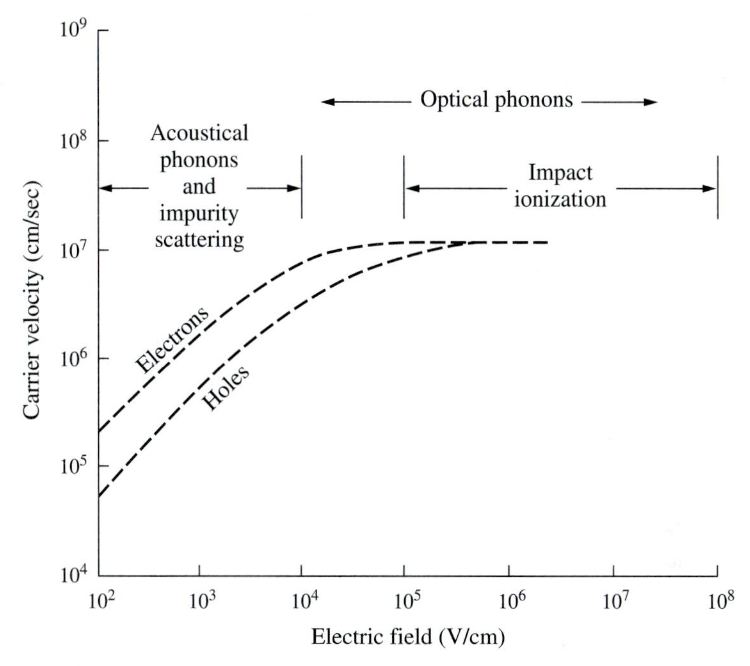

FIGURE 1
Effect of electric field on carrier velocity. (*After Sabnis, Ref. 1.*)

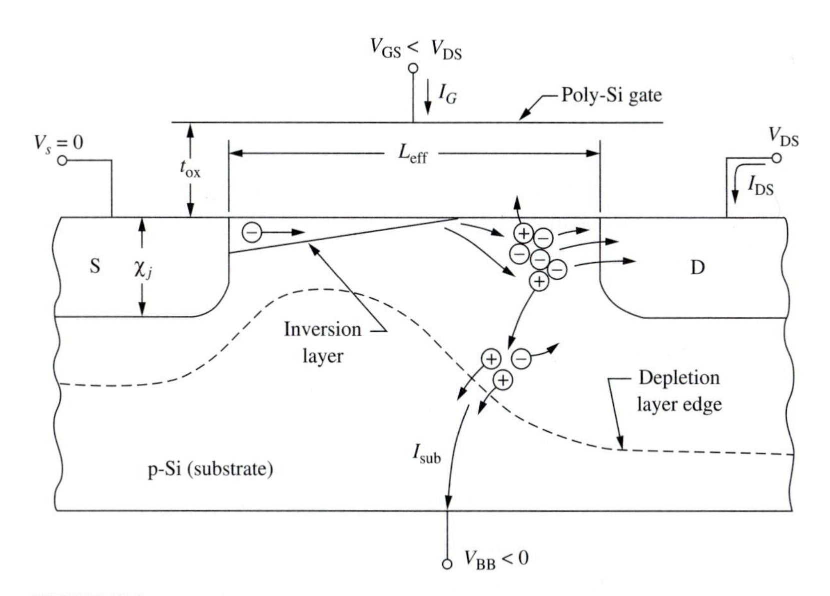

FIGURE 2
Impact ionization in saturation mode for NMOS shown as electron (−) and hole ⊕ generation in the transistor pinch-off region. Holes are swept to a p^- substrate by the electric field as substrate current, I_{sub}. (*After Sabnis, Ref. 1.*)

and/or hydrogen.[9] The simplest model is that hot carriers bombard the interface against a retarding electric field. The carriers break the Si–H bonds and cause a trap. This interface damage is shown in Fig. 3. A measure of the interface damage is ΔD_{it}, and it can be measured with the charge-pumping technique.[10]

The maximum interface damage has been shown to occur at the gate/drain bias condition at which the substrate current is a maximum. The gate : drain voltage ratio is typically about 1 : 2. An example of degraded *I-V* characteristics of an n-channel MOSFET is shown in Fig. 4*a*. Notice that the drain current is lower in the degraded curve than the fresh device curve. This decrease in the drain current is due to mobility degradation caused by interface trap generation.

The main degradation mechanism for pMOSFETs is believed to be trapped electrons in the gate oxide near the gate/drain interface.[4,11] In the half-micron or longer channel length transistor, the degradation seems to be greatest at a gate/drain bias condition at which the gate current is a maximum. The gate bias voltage is typically about $\frac{1}{4}$ of the drain bias voltage. However, at shorter channel lengths, degradation under maximum I_{sub} bias conditions could be larger.[12] Degradation parameters measured are similar to those for n-channel MOSFETs.

An example of degraded *I-V* characteristics of a p-channel MOSFET is shown in Fig. 4*b*. Notice the degradation of the drain current is an increase from the fresh device. The increase is due to channel shortening caused by electron trapping near the gate/drain region.[2,11] This trapping causes a decrease in $|V_T|$, an increase in subthreshold leakage current, and a decrease in the punch-through voltage.[3]

12.2.2 Equations of HCI

Degradation of the n-channel MOSFET transistor parameter under HCI stress has been shown[2] to follow the expression

$$\Delta p \sim \{(I_d/W)(I_{\mathrm{sub}}/I_d)^{\phi_{\mathrm{it}}/\phi_i} \cdot t\}^n \tag{12.5}$$

where Δp = change in a parameter, e.g., V_T, g_m, I_{dsat}, I_{dlin}
I_{sub} = maximum substrate current (A)
I_d = drain current at the maximum substrate current bias condition (A)
W = transistor width (meter)
ϕ_{it} = critical energy for device damage (eV)
ϕ_i = impact ionization energy (eV)
t = time (sec)
n = empirically fitted exponent

The maximum substrate current can be approximated by[2]

$$I_{\mathrm{sub}} \approx 2 I_d \exp(-\phi_i / q\lambda\mathscr{E}_m) \tag{12.6}$$

where λ = hot electron mean free path
$\mathscr{E}_m$ = maximum electric field

The lifetime τ has been shown[2] to follow the expression

$$\tau \approx (I_{\mathrm{sub}}/W)^{-\phi_{\mathrm{it}}/\phi_i} \tag{12.7}$$

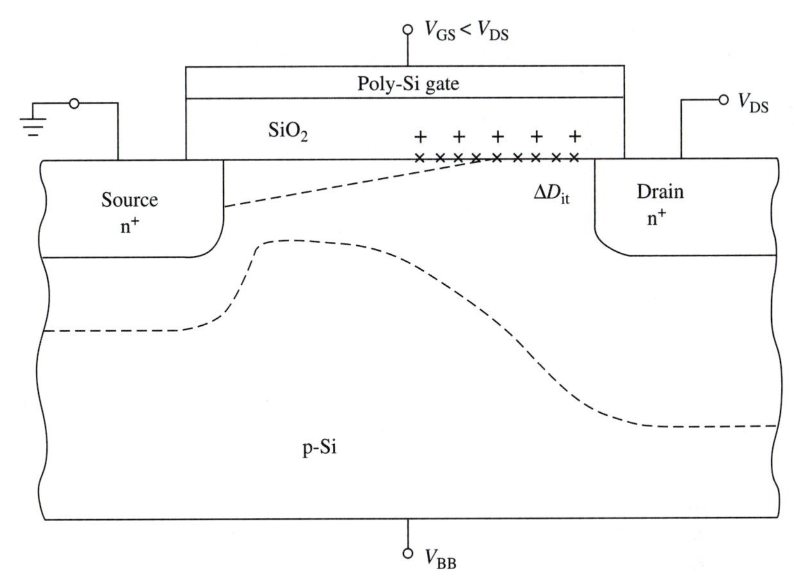

FIGURE 3
Diagram of trapped charges (+) and interface damage ΔD_{it}, or interface traps, during hot carrier injection in an NMOS transistor. Note that the trapped charges and interface damage are toward the drain end. (*After Sabnis, Ref. 1.*)

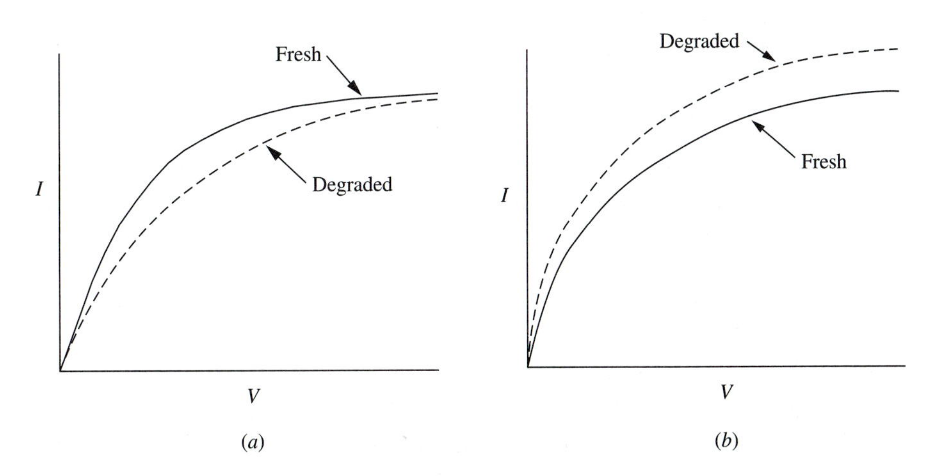

FIGURE 4
(*a*) Degradation of an *I-V* curve of an n-channel transistor after HCI stress.
(*b*) Enhancement of an *I-V* curve of a p-channel transistor after HCI stress.

where $\phi_{it}/\phi_i \approx 3$. The parameter τ is the dc lifetime of the device calculated at a given percentage of degradation of the parameter p. Typically a 10% in degradation of p has been used as a measure of τ, but this value has been chosen historically and has no physical meaning. The real value of Δp should be chosen from a circuit degradation viewpoint and could differ for different applications.

From an experimental point of view, Eqs. (12.5) and (12.7) can be expressed as in Eqs. (12.8) and (12.9) for NMOS, that is

$$\text{NMOS}: \quad \Delta p = At^n, \qquad n = 0.3 \text{ to } 0.7 \tag{12.8}$$

$$\tau = B(1/I_{\text{sub}})^m, \qquad m = 2.5 \text{ to } 3.5 \tag{12.9}$$

PMOS transistor HCI degradation is not so widely researched as NMOS, but recent studies have shown that from an empirical point of view, the HCI degradation can be expressed in a format similar to NMOS,[13,14] or

$$\text{PMOS}: \quad \Delta p = A't^{n'}, \qquad n' = 0.3 \text{ to } 0.7 \tag{12.10}$$

$$\tau = B'(1/I_{\text{gate}})^{m'}, \qquad m' = 2 \text{ to } 3. \tag{12.11}$$

Notice that the PMOS lifetime is now a function of the maximum gate current, I_{gate}. [However, for PMOS transistors at the quarter-micron channel length region, the lifetime has been shown to follow the maximum I_{sub} expression, similar to Eq. (12.9).[12]]

In Eqs. (12.8–12.11), the parameters n, m, n', and m' are empirically determined. Examples of Eqs. (12.9) and (12.11) are illustrated in Figs. 5*a* and 5*b* for NMOS and PMOS, respectively.

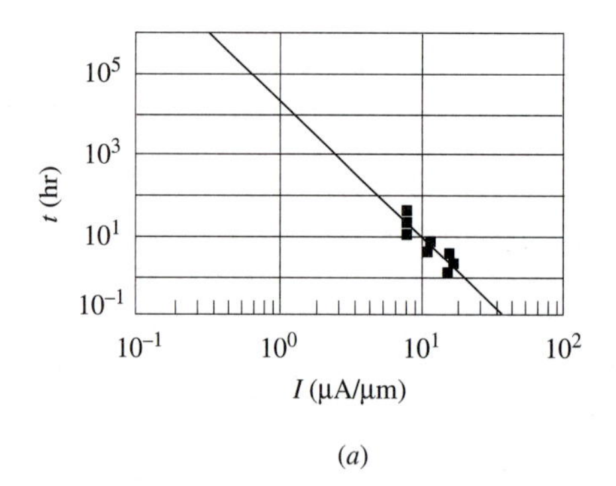

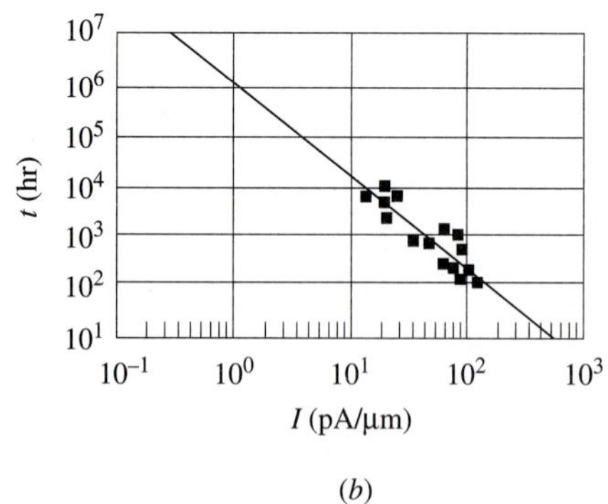

FIGURE 5
(*a*) Example of plotting lifetime data vs. I_{sub} for NMOS under several accelerated HCI stress conditions to project for lifetime to use condition. (*b*) Example of plotting lifetime data vs. I_{gate} for PMOS HCI under several accelerated stress conditions to project for lifetime to use condition. The regression fit of data in (*a*) is $\log t \text{ (hr)} = -3.34 \log I(\mu\text{A}/\mu\text{m}) + 4.3$. The regression fit of data in (*b*) is $\log t \text{ (hr)} = -1.81 \log I(\text{pA}/\mu\text{m}) + 5.97$.

12.2.3 HCI Summary

In this section, transistor degradation due to HCI effects has been discussed, and basic HCI equations have been given. It will be shown in a later section that HCI degradation is a limiting reliability factor to device scaling. The gate oxide quality has been reported to play a key role in HCI lifetime[15] and is expected to affect the parameters A and B in Eqs. (12.8) and (12.9). A more recently elucidated phenomenon related to HCI is hot electron–induced punch-through degradation.[16] This phenomenon is more pronounced in p-channel MOSFETs and could impose an even greater reliability limitation in the ULSI regime than the traditional HCI degradation. The key to HCI lifetime improvement is to reduce the maximum electric field through *drain engineering* and to produce trap-free, damage-free, and clean gate oxides.

12.3 ELECTROMIGRATION

Electromigration (EM) is the transport of metal ions through a conductor, and it results from the passage of direct electrical current. Electromigration is perhaps the most widely known and extensively researched failure mechanism of integrated circuits. Studies began in the 1950s with the development of the semiconductor.[17,18] With the discovery of crack formation in the aluminum conductor leading to failures in ICs in the late 1960s,[19] electromigration studies in thin Al films began in earnest. Although most metals will electromigrate, Al has by far been the metal of greatest research because of its common usage as the interconnect material of choice.

Electromigration is a combination of thermal and electrical effects on mass motion. The higher the temperature, the easier it is for the metal ions to electromigrate. For bulk metals, electromigration occurs at about three-fourths the melting temperature (T_m in K), whereas for a metallic polycrystalline thin film it occurs at about one-half of T_m. Thus, noticeable electromigration is observed at 423 K and above for Al.

The mass motion of Al ions is visualized in Fig. 6. At temperatures of about $\frac{1}{2}T_m$, the Al ions are sufficiently thermally agitated in the lattice potential well that there is a finite probability that some ions are situated at the top (or saddle point) of the potential well (dictated by Maxwellian statistics). These "activated" ions are essentially free of the lattice and are free to diffuse out of or fall back into the well. In the presence of an electric field, the activated ions experience electromotive forces $F_{\mathscr{E}}$ and F_p, where $F_{\mathscr{E}}$ is the force on the Al^+ ion due to the electric field and F_p is the *electron wind* force exerted on the Al^+ ion due to the momentum exchange of the electrons in the electric current.[20] F_p and $F_{\mathscr{E}}$ are opposite in direction and $F_p \gg F_{\mathscr{E}}$. Thus, a finite portion of the Al^+ ions are "pushed" in the direction of the electron current flow. Al ions move upstream while vacancies move downstream. Agglomerations of vacancies form voids, and large voids can grow completely across the width of the Al lines to cause circuit failures.

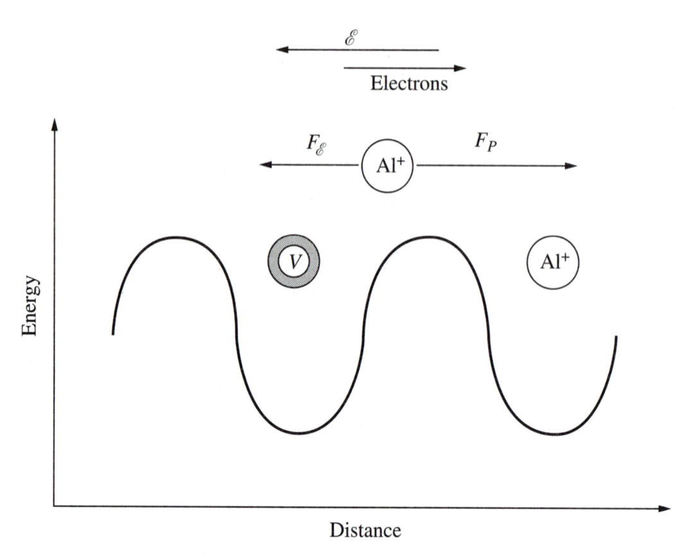

FIGURE 6
Schematic of a positive Al ion (Al^+) in a lattice potential well under opposing electric field ($F_{\mathscr{E}}$) and electron wind (F_p) forces. The letter V represents the vacancy left behind after the Al ion jump.

For Al to move from one well to another, a vacancy must be created, and an ion-vacancy jump process must also occur. The activation energy E_a for this sequence in pure, bulk Al is 1.48 eV.[21] The E_a for Al migration in a grain boundary has been found to be much lower, near 0.5 to 0.6 eV.[22] This difference can be explained by the fact that there is an ample supply of vacancies in grain boundaries and Al/oxide surfaces, so that the ion-vacancy jumping process is made much easier. Hence, electromigration in polycrystalline thin film is predominantly a grain-boundary, diffusion-driven mechanism.

The basic flux equation in electromigration, for an ideal metal in the absence of a temperature gradient, is given by[18,23]

$$F_m = ND_o/kT(Z^*q\mathscr{E})\exp(-E_a/kT) \tag{12.12}$$

where F_m = ion flux
N = density of atoms
D_o = diffusion coefficient
T = temperature in K
k = Boltzmann constant
Z^*q = effective ionic charge
$\mathscr{E}$ = electric field
E_a = activation energy

To have any void growth, there must be vacancy flux divergence, which is related to the ion flux divergence. This is dictated by the continuity equation,

$$dC_v/dt = -\nabla \cdot \boldsymbol{F}_v + (C_v - C_v^o)/\tau \tag{12.13}$$

where C_v = vacancy concentration
$\boldsymbol{F}_v$ = vacancy flux
C_v^o = thermal equilibrium vacancy concentration
τ = average lifetime of vacancy

Under a steady-state condition,

$$dC_v/dt = 0$$

so

$$\nabla \cdot \boldsymbol{F}_v = (C_v - C_v^o)/\tau \qquad (12.14)$$

If voids are to grow, $C_v \neq C_v^o$, thus $\nabla \cdot \boldsymbol{F}_v \neq 0$.

Figure 7 illustrates how grain boundary triple points and defects in the Al film can cause vacancy flux divergence, which leads to void and hillock formation. Temperature gradients and other structural inhomogeneities (e.g., irregular grain size) can also cause flux divergence.

Examples of electromigration voids in Al/0.5% Cu/1% Si films are shown in Fig. 8. Figure 8*a* shows a large void severing the entire width of the line, causing an electrical open. Figure 8*b*, however, shows that the large void did not cause an electrical open, as the underlying refractory metal barrier, Ti/TiN, is still intact. The temperature of test used for Fig. 8*b* was significantly less than one-half of the melting temperature of the refractory Ti/TiN barrier, so electromigration in the barrier was negligible. Thus, the shunting of the void prevented the line from becoming electrically open. The underlayer barrier of Al is one preventive measure for electromigration in ULSI processing. Other preventive measures will be discussed later.

12.3.1 Effect of Stress on Electromigration

The ion flux in Eq. (12.12) is an oversimplification. In reality, as Al ions move from one end of an interconnect to another under electromigration, a mechanical stress gradient is created in the line. This stress gradient will create an additional ion flux, opposite to the electron wind flux, due to the vacancy concentration difference

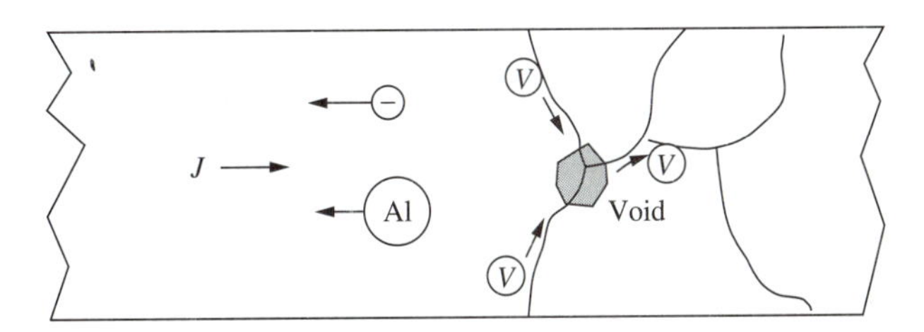

FIGURE 7
Schematic of void formation due to vacancy v flux divergence at grain boundary triple points. A triple point is defined when two or more grain boundaries intersect.

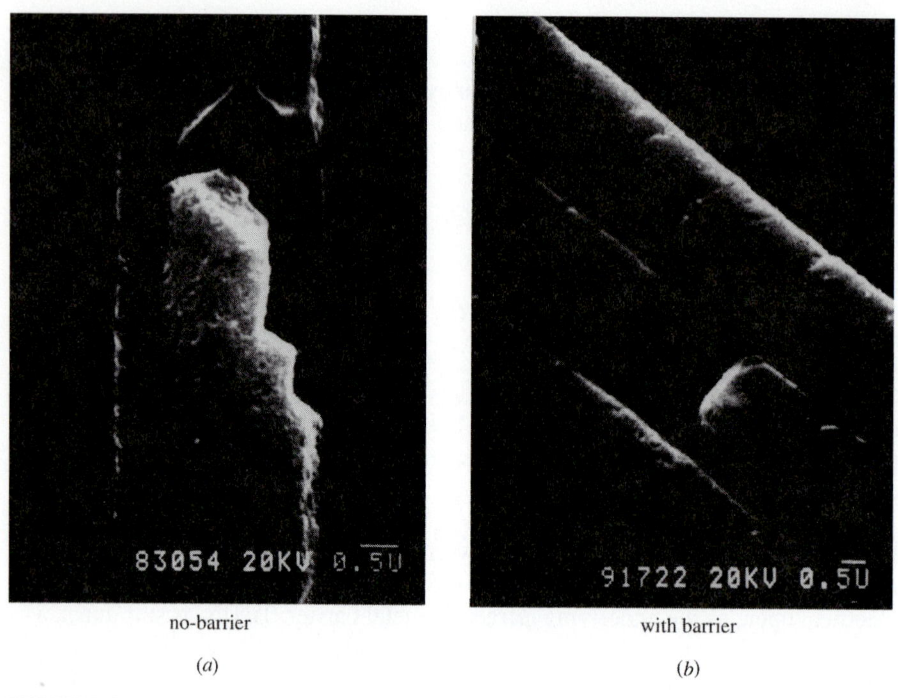

no-barrier (*a*)

with barrier (*b*)

FIGURE 8
(*a*) SEM micrograph of electromigration void in an Al–Cu–Si metal line without a barrier underlayer. (*b*) SEM micrograph of electromigration void in an Al–Cu–Si metal line with an underlayer barrier refractory metal showing that the barrier is still intact below the void.

between the cathode and anode ends. There are more vacancies in the cathode end than the anode end. The anode end of the interconnect becomes more compressively stressed than the cathode end.

The ion flux F_m, in the presence of the stress gradient, is given by[24]

$$F_m = ND/kT[Z^* q\mathscr{E} - \Omega(d\sigma_n/dx)] \tag{12.15}$$

where Ω = atomic volume
σ_n = stress normal to the grain boundary

The diffusion coefficient D is itself a function of σ_n such that it decreases exponentially with increasing stress. When a passivation overcoat is applied, the additional stress in the Al line results in a slower electromigration process. It has been shown that passivated Al lines have lifetimes that are about 10 times longer than nonpassivated lines.[25,26]

In Eq. (12.15), it is possible that the stress gradient term is large enough to equal the electron wind term. In that case then, $F_m = 0$ and

$$d\sigma_n/dx = Z^* q\mathscr{E}/\Omega \tag{12.16}$$

In general, Al films cannot support such a large stress gradient without growing hillocks or whiskers, which then relieve the stress. Hence, F_m rarely is zero. But by utilizing a pinhole-free passivation layer, hillock and whisker formation can be suppressed to minimize the electromigration damage.

From the drift velocity electromigration experiment, it has been shown that in very short strips (< 20 microns in length),[27] the stress gradient can be so high that Eq. (12.16) is satisfied. Notice that $d\sigma_n/dx$ in a strip is the stress at the anode end minus the stress at the cathode end divided by the length of the strip. Rewriting Eq. (12.16) for Δx, we have

$$\Delta x = \Delta\sigma_n \Omega / Z^* q\mathscr{E} \tag{12.17}$$

where Δx is the critical length, commonly referred to as the *Blech length.* For strip lengths less than Δx, electromigration damage cannot occur. Estimation of Δx from experimentally measured Z^* and typical stress values in Al films has resulted in general agreement with Eq. (12.17). An interesting concept in electromigration prevention is to design the interconnect lengths in segments of less than 20 microns at a time; however, this is impractical.

In multilevel interconnect processes, e.g., two to four layers of Al interconnects, Eq. (12.15) shows that the quality and stress values of the interlevel dielectric (ILD) films are important in determining the lifetime of the metal lines. In fact, the thermal expansion coefficients of ILD films and the thermal process history of the entire interconnect module are key parameters in determining the electromigration lifetime.

12.3.2 Electromigration of Lines

The reliability of a metal interconnect is most commonly described by a lifetime experiment on a set of lines to obtain the median time to failure (MTF). The data for the actual time to failure of each line are plotted on a lognormal graph, and the value of T_{50} (time for 50% of the lines to fail) is extracted along with the lognormal shape parameter (σ). The stress experiment involves stressing the lines at high current densities (usually from 10^6 to 3×10^6 A/cm^2) at temperatures ranging from 150 to 250°C. The failure criterion is typically an electrical open for nonbarrier conductors or about a 20% increase in line resistance for barrier metallization. The general MTF expression is

$$\text{MTF} = AJ^{-n} \exp(E_a/kT) \tag{12.18}$$

where A is a material constant based on the microstructure and geometric properties of the conductor, J is the current density, and E_a and kT have the previous definitions.

Equation (12.18) was first proposed by Black[28,29] and is based on a simple theoretical argument about the rate of mass transport occurring as a result of momentum exchange between the thermally activated ions and the electron wind. In that argument, the rate of mass transport, R, was proportional to the product of the momentum transferred in each collision ($q\mathscr{E}/v = q\rho J/v$) with the number of electrons available for collision ($N = J/q$), the effective cross section, and the activated Al ion density. If we assume R is inversely proportional to MTF, it then follows that MTF $\propto J^{-2}$ and $n = 2$ in Eq. (12.18). This argument is, however, not consistent with the ion flux divergence requirement. For Al lines, n has been found experimentally to range from 1 to 3 or even higher.

It has been argued that $n = 1$ in the absence of thermal gradients in the conductor and $n = 3$ when thermal gradients are taken into account.[30] A number of researchers have argued that conceptually $n = 2$ is correct.[31,32] Most recently, by taking into account the physical observation of void density and size during electromigration of a conductor line,[33] it was concluded that $n = 2$. In that work, the number of voids and hillocks formed per conductor length by electromigration was shown experimentally to be proportional to the current density. The average growth rate of each void was observed to be proportional to the current density. Since $(\mathrm{MTF})^{-1}$ is the product of (number of voids) $\times$ (void growth rate), it follows that $n = 2$.

Once T_{50} and σ are determined in the accelerated stress experiment, extrapolation to the circuit design condition gives the lifetime of the metal line. Typically, the electromigration goal is 10 years at the worst-case field usage condition for a 0.1% failure distribution ($T_{0.1\%}$) of the metal strips under test.

Equation (12.18) contains the geometry dependence in the parameter A. Simplistically, $A \approx wt$, where w is the width of the line and t is the thickness. Experimentally, it has been shown that the width dependence of MTF is a function of the ratio of the grain size d of the film and the width of the conductor w. As w/d decreases, the MTF will increase due to the *bamboo effect.* Figure 9 illustrates this effect for Al films without a barrier.[34] The minimum MTF generally occurs at about $w/d \approx 2$ to 4. As the grain size increases with respect to the line width, the interconnecting paths of the triple points parallel to the length of the conductor decrease and even vanish. So the probability of voids accumulation along the grain boundaries is reduced, thus retarding the electromigration process. Figure 10 shows a *bamboo* grain conductor versus a nonbamboo line. The final grain structure of the Al film not only is a function of the metal deposition conditions but is also strongly dependent on post-metallization annealing temperatures and the texture of the underlying surfaces.

In Al conductors with an underlayer barrier, the bamboo effect of MTF still holds and is shown in Fig. 11.[35] Although the Ti/Al–Cu/Ti layered film exhibits improved overall line MTF, it still shows a critical line width value at which the MTF is at a minimum. The line width dependence of MTF has significant implications from a design point of view and is usually characterized during the process development phase of any new metallization process.

12.3.3 Contact and Via Electromigration

Electromigration occurs not only in conductor lines but also in contacts and vias. Contact electromigration resembles the classic Al spiking phenomenon, except that the effect is activated by electric current. It is driven by the Si ion migration in the Al conductor in the direction of electron wind.[36] Figure 12 illustrates this process in an Al–Cu–Si conductor. As Si is driven from the anode near the contact by the electrons in the current, Si in the substrate is diffused into the Al to satisfy the Si solubility in the Al. Al atoms diffuse into the substrate to occupy the vacated Si positions. As time goes on, Al spiking can puncture the n^+ (or p^+) junction to cause junction leakage. The effect is particularly noticeable in shallow junction processes. The migrated Si

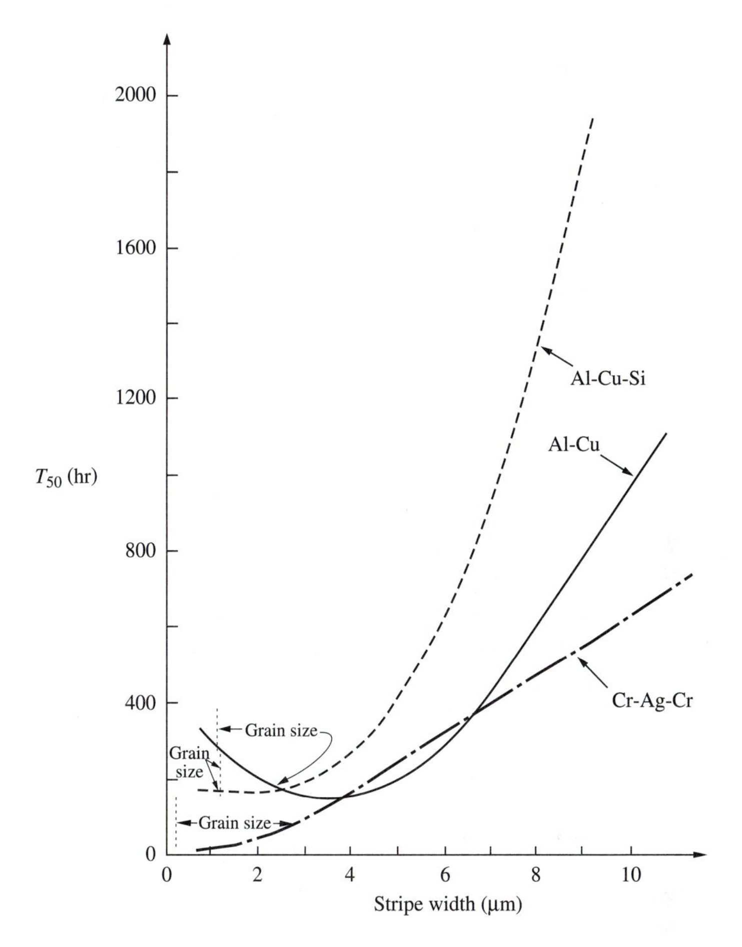

FIGURE 9
EM lifetime (T_{50}) as a function of Al line width for three different Al-based alloys. The *bamboo* effect is the increase in lifetime as the width decreases below a certain critical width. This width is typically about 2 to 4 times the average grain size in the metal film. (*After Scoggen et al., Ref. 34.* ©1975 IEEE.)

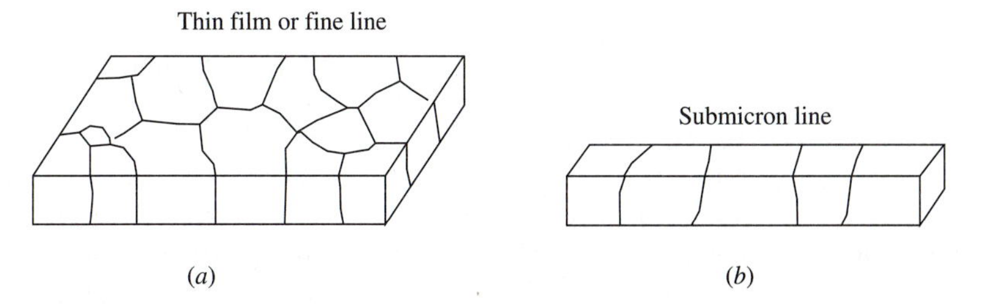

FIGURE 10
Sketch of (*a*) a nonbamboo structure in a wider line dominated by triple points, versus (*b*) a bamboo-like grain structure in a submicron line.

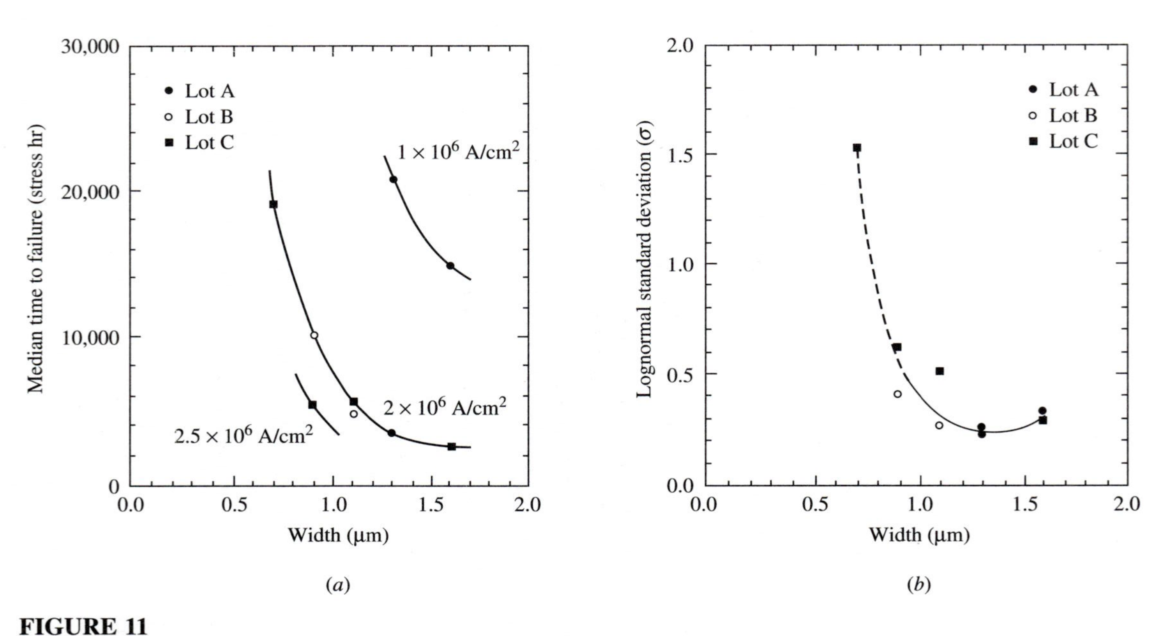

FIGURE 11
(*a*) EM lifetime data for three lots of Ti–Al–Ti metallization tested at three different current densities: 1×10^6, 2×10^6, and 2.5×10^6 A/cm^2. The characteristic *bamboo* effect is observed. (*b*) An increase in the lognormal σ with decreasing line width was measured for all three lots. *(After Rathore et al., Ref. 35.)*

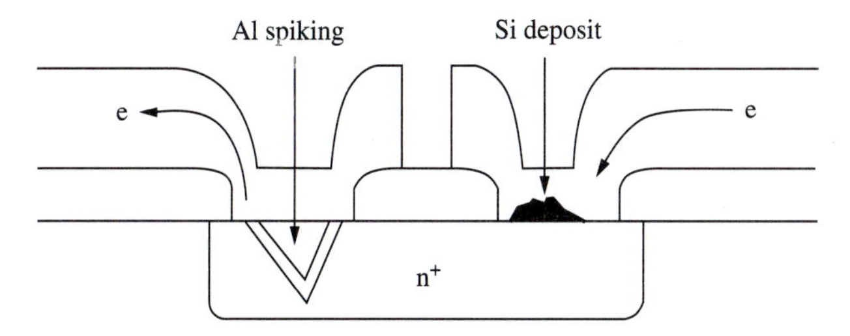

FIGURE 12
Diagram of the contact electromigration process without a barrier metal. Al spiking is shown on the left contact and Si pileup is shown in the right contact. Note the direction of the electron flow.

atoms can also deposit in the contact window in the anode side causing resistive or even open contacts.

The equation for contact electromigration (in the absence of barrier metal) has been shown to follow the empirical expression[36,37]

$$\mathrm{MTF} \propto x_j^2 (W^{\alpha}/I)^n \exp(E_a/kT) \tag{12.19}$$

where x_j is the junction depth, W is the contact window dimension, I is the current, and n and α are exponents. Unlike the line EM equation in Eq. (12.18), the current I is used in Eq. (12.19). It has been shown that n is a function of Joule heating and approaches 2 for lower current values. Since x_j is less than 1, MTF decreases with the square of the scaling dimensions.

To overcome this problem, a refractory barrier film has been developed that is impermeable to the diffusion of Si or Al. Most of the popular barrier systems are Ti/TiN and TiW barriers. However, even in these films contact EM can still occur if the barrier quality (porosity) is poor and if sufficient current is applied. For all practical purposes in today's field usage conditions, these barriers have solved the contact EM problem.

Note that poor step coverage of the Al film on the contact side walls can also lead to electromigration because the flux divergence is aggravated at those points. This effect is similar to the line electromigration effect discussed in Section 12.3.2 and should be distinguished from the contact electromigration mechanism.

Via electromigration became an issue in multilevel metallization technology. As one metal layer is connected to another metal layer through a via window, the current crowding at the via interface has been shown to exhibit a *liftoff* at the interface.[38,39] This can occur in Al–Al vias or even Al–barrier–Al vias. Figure 13 illustrates this failure mechanism. The driving force is the flux divergence at the interface due to grain size difference, which is aggravated by a very thin interfacial oxide film. This driving force can also be accentuated by the difference in the thickness of the metal layers. However, even with the best cleaning processes, via interfaces can still lift off during long-term current stressing. This failure mechanism should be characterized in a given process because it could be the lifetime-limiting factor and should be considered in the circuit design stage.

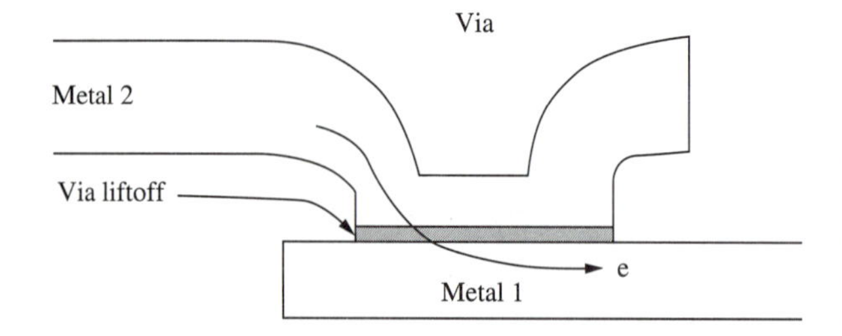

FIGURE 13
Diagram of a via EM failure mechanism due to via liftoff. The separation at the metal 2 and metal 1 interface is denoted by the shaded area.

12.3.4 Methods to Improve Electromigration Lifetime

To improve electromigration lifetime, one must reduce the magnitude of the flux, F_m, in Eq. (12.15) or the divergence of F_m. Examination of the variables that control F_m provides direction for improving electromigration lifetime. From Eq. (12.15) reduction in F_m can be achieved through the following:

1. Reduction in diffusivity of Al, D
2. Reduction in the driving force, $Z^* q\mathscr{E}$
3. Increase in the back pressure, $d\sigma_n/dx$

Reduction in the driving force $Z^* q\mathscr{E}$ is difficult and impractical, since $\mathscr{E}$ for ULSI devices is increasing owing to the circuit performance requirements ($\mathscr{E} = \rho J$). Reduction in the effective charge Z^* is difficult since this is an intrinsic property of the scattering process in Al.

Greater benefit can be achieved through the reduction in the diffusivity, D, of Al. The focus of research has been to reduce the grain-boundary diffusivity of Al by the addition of various solute atoms, most notably Cu. Cu is known to improve electromigration lifetime in Al strips.[40] Other solute atoms such as Mg, Mn, Ti, and Pd are also effective in improving EM lifetime.[41] The mechanism of diffusivity reduction is believed to be the retardation of atomic flux due to the segregation of these impurity atoms at the grain boundaries and their subsequent interaction with the migrating Al ions.[23] We can conceptualize this interaction as a modification of the potential well (Fig. 6) due to the presence of impurity atoms that reduce the number of "activated" Al ions in the grain boundaries. Thus, up to 4% Cu doping in Al has shown continued improvement in EM lifetime as well as in E_a.[42]

An interesting property of Cu is that its solubility in Al is only 0.05 at % at 230°C.[43] Typical doping levels in today's semiconductors are at 0.5%, 2%, or 4%. The additional Cu is then segregated at the grain boundary, and the greater the segregation of Cu, the more improvement in lifetime. However, values of Cu in Al $>$ 8 at % have been shown to reduce EM lifetime.[42] Cu also forms secondary compound, a theta phase particle with Al, $CuAl_2$, which is not necessarily situated in the Al grain but which serves as sources of Cu atoms to the Al grain boundary. This situation is particularly advantageous, because Cu also electromigrates, so a supply of Cu will keep grain-boundary diffusivity at a low level for a longer time.

Increasing the back-pressure term, $d\sigma_n/d\,x$, will reduce F_m. Earlier, from Eq. (12.17), we saw that short strips of conductor improve the back pressure and that for lengths below the Blech length of 20 microns, F_m goes to zero. However, designing interconnects with such short segments, interlaced with vias, is not a practical approach.

Efforts to reduce the flux inhomogeneity have been to modify the Al microstructure. Such microstructure modifications include the following:

1. Single-crystal Al stripes[44]
2. Conductor composed of chains of Al single grain[45]
3. Al films with large grain[46]
4. Ultra bamboo structures[47]

All these have shown improvements in EM lifetime. Ultra bamboo structures have shown improvement as much as 100 times.

Passivation overcoating, or ILD overcoating in multilevel metallization structures, modifies the conductor stress states and improves the lifetime.[23] Although the mechanism is not well understood, it is believed that the overcoating enhances the defect annihilation kinetics,[23] or increases the back pressure, or alters the surface diffusion kinetics. In fact, it has been shown that in narrow conductors, the surface diffusion mechanism contributes significantly to the EM drift velocity.[48]

Transition metal/barrier underlayering is becoming a preferred scheme in the ULSI process. These schemes have also shown marked conductor lifetime improvements ($\sim$ 10 times), e.g., TiN,[49] Cr,[50] and Ti–W.[51] The mechanism is believed to be the extra shunting provided by the underlayer, which allows voids to heal through localized joule heating. The barrier property improves the contact EM lifetime.[50] More exotic metal structures, such as Al metal interlayered with transition metal layers, have also demonstrated significant lifetime improvements ($\sim$ 50 to 100 times).[52] These sandwich structures of transition metals (Hf, Ti, Ta, or Cr) not only provide the shunting property like the single barrier layer just discussed but also modify the microstructure to cause a more bamboo-like conductor and prevent the linkage of voids between the top and the bottom Al–Cu layer.[53]

12.3.5 Electromigration Summary

Electromigration is a failure mechanism that has been widely studied in the last 30 years. The lifetime of any given interconnect technology needs to be evaluated separately because many processing variables determine the final lifetime value. Significant progress has been made to improve the electromigration lifetime, such as the use of Cu (or other transition metal) doping and barrier/refractory metal layering. However, the ever-increasing current density requirements in ULSI dictate that an even greater level of EM performance must be attained. Thus, research is continuing to reduce further the grain-boundary diffusivity in Al as well as the structural inhomogeneity. It appears that Al-doped thin films will still be the main interconnect at the 0.35-micron level, but below this level alternative thin films that have a better EM life, e.g., Cu or Au, may be considered. It remains to be seen how far below the 0.35 micron technology that Al-doped thin films, in combination with refractory layering, can continue to meet the high current density and reliability needs of ULSI.

12.4 STRESS MIGRATION

Stress migration (SM) is the phenomenon of metal voiding in conductor lines that are under tension in the absence of electrical current. The voids grow and in time can totally sever a conductor line to cause circuit failures. This reliability problem is a relatively recent phenomenon compared to electromigration or HCI but becomes a major reliability concern because of the initial lack of understanding of the failure mechanism. Figure 14 illustrates a typical metal void caused by stress migration.[54] It looks rather similar to a void caused by electromigration.

The stress migration problem was first discovered on 64K DRAMs employing Al–Si sputtered metallization for 2.5 to 3.5 micron line width passivated with PECVD SiN from various semiconductor suppliers.[55] As early as 1982, failures had occurred during mild thermal or electrical stress in vendor qualification tests or even in unstressed parts in storage.[56] Failure analysis showed metal opens in bit lines over steps as well as metal voids in flat topography.

Early phenomenological studies showed that the maximum SM failure was at 180°C. Failures were caused by N_2 contamination during sputtering,[57] by large silicon nodules in the Al,[58] and by topographical steps.[54] Metal voids had a strong functional dependence on the compressive stress of the passivation films, with the Al tensile stress varying with the thickness of the passivation layer.[54] A more rapid cooling rate from the passivation furnace produced fewer voids.[54] The stress-induced voids also degraded the electromigration lifetime.[54] Large silicon precipitates provided sources of vacancies.[58] The back-end assembly process with ceramic-glass-sealing temperatures at 450°C increased the metal void size and density.[59]

Experimental studies have shown that the SM lifetime on conductor lines tends to follow the expression

$$\tau \propto (\text{width})^m \times (\text{thickness})^n \qquad (12.20)$$

where m and n are experimentally determined constants that vary from 2 to 8.[60,61] The effect of Cu addition was studied, and it was shown that as little as 0.1% Cu added to Al–Si alloys substantially suppressed the open-failure rate.[60]

Stress-induced voids can be classified into two major categories, a wedge-shaped and a slit-shaped void. The wedge-shaped voids form at the edges of conductor lines and tend to peak in lines of 2 to 3 micron width. These can be observed even at the end of wafer processing. Slitlike voids are not easily observable after processing, but nearly all open-conductor failures have been caused by them.

12.4.1 Stress Migration Mechanism

Continuing research is devoted to further understanding the mechanisms of void growth and to fully explaining all the experimental data. However, based on the many studies thus far, some general understanding has been developed on the basic mechanisms of void formation and growth. Because of the importance of stress migration in ULSI, they will be discussed briefly.

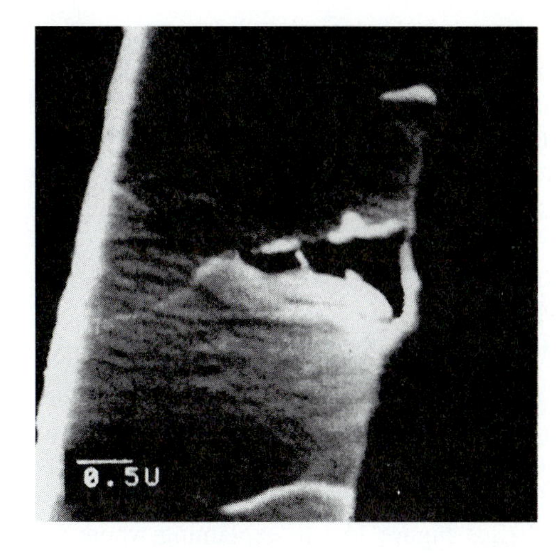

FIGURE 14
SEM micrograph of a stress-induced void in aluminum interconnect. Void was exposed after Ar back-sputter removal of passivation film. (*After Yue et al., Ref. 54.* ©1985 IEEE.)

It is generally believed that void nucleation and growth in Al conductors are a result of stress relaxation in a confined environment. The confined environment is provided by the surrounding passivation. Unpassivated Al lines cannot produce voids that cause failures because the free surfaces act as void sinks and prevent the accumulation of voids. The driving force is the tensile stress in the Al from the thermal expansion mismatch between the Al line and the surrounding passivation films and the Si substrate during the manufacturing process.

The stress in the confined Al line at the end of process, $\sigma_o(T)$, is given by[62]

$$\sigma_o(T) = f_c E(\alpha_{\text{Al}} - \alpha_{\text{Si}})(T_d - T) \tag{12.21}$$

where T is the temperature at the end of processing (such as room temperature), T_d is the passivation deposition temperature, E is the Young's modulus of Al, and α_{Al} and α_{Si} are the thermal expansion coefficients of Al and Si. The term f_c is a constraint factor that depends on the metal-insulator geometry and elasticity of each material. Note that at temperature T_d the stress in Al is zero by definition. An estimate of σ_o (room temperature) is 5×10^9 to 6×10^9 dynes/cm^2, by using $f_c \approx 1$, $E = 7 \times 10^{11}$ dynes/cm^2, $\alpha_{\text{Al}} = 23 \times 10^{-6}$/K, $\alpha_{\text{Si}} = 2.6 \times 10^{-6}$/K. This is comparable to or greater than the yield stress of Al film in confined Al lines.

The Al tries to relieve itself of this large tensile stress. The tensile stress results when the actual lattice spacing is greater than the thermal equilibrium spacing.[63] Since the Al is confined, it cannot relieve the stress by plastic deformation, although local internal plastic deformation could occur.[64] The Al tries to shrink but is prevented by the adhesion with the encapsulating, hard, rigid passivation. It is believed the major relaxation of stress occurs by diffusional or power-law creep, which causes void formation and growth. Thus, voids are created and grown as a by-product of stress relief. Smaller voids agglomerate into larger voids to reduce the total free energy of the system, and finally a void is large enough to sever a metal line.

Experimental data show that the maximum SM failure generally occurs in the temperature range between 150 and 200°C.[57,65] This can be explained by noting that in thermal creep the driving force due to stress decreases as temperature increases, but the vacancy diffusion rate increases exponentially with temperature. The product of these two driving forces gives a maximum somewhere between 150 and 200°C, with the peak temperature dependent on the thermal deposition conditions and the material properties of Al and the surrounding dielectrics.

A generalized power-law creep-rate model for SM is shown to be[65]

$$R(T) = C(T_o - T)^n \exp(-Q/kT) \tag{12.22}$$

where R = creep rate at temperature T
T_o = temperature of metal deposition
Q = activation energy associated with the diffusion process

Equation (12.22) can explain a maximum in R as a function of T, thus agreeing with the experimental data. By fitting Eq. (12.22) to a set of experimental data, Q was found to be 0.58 eV and $n = 2.33$.[65]

It has been observed that the void volume of an Al–Si conductor could be as large as 1 to 3% of the total volume.[54] (See Figure 15.) In explaining where the appreciable void volume comes from, we note that there are many sources of vacancies in a metal conductor. For example, supersaturation of vacancies by as much as 2.5% occurs in the metal conductor as it is pulled from a high-temperature passivation furnace at 400°C.[54] Si precipitation in Al during cooldown creates additional vacancies. Grain boundaries provide ample sources of vacancies. These vacancies can then agglomerate to form voids through such diffusional processes as Nabbarro-Herring creep (intragranular diffusion) or Coble creep (grain-boundary diffusion).[66] Recently it has been suggested that a H reaction[67] or even an Al–SiO_2 reaction[68] can also enhance void formation.

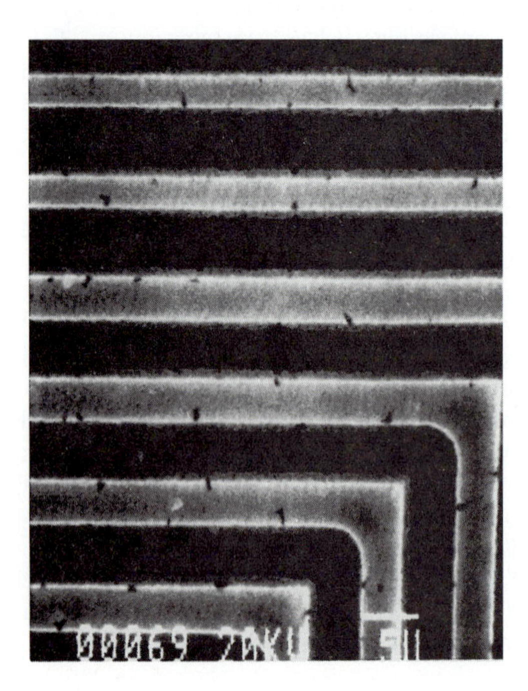

FIGURE 15
SEM micrograph of metal void density in aluminum interconnects, showing predominant location at line edges. (*After Yue et al., Ref. 54.* ©1985 IEEE.)

A further complexity to the void formation problem is that the encasing passivation of the conductor is not perfectly hard and could deform under various IC usage environments. It has been reported that localized passivation separation from the Al conductor does lead to void formation.[69] Although this adds an additional element of complication to the stress migration problem, deformation of the passivation should remain a second-order effect.

12.4.2 Dependence of SM on Al Aspect Ratio

It has been shown that the stress migration failure rate is a function of the Al line width and thickness, with an increasing failure rate for narrower and thinner metal lines. This experimental finding predicts a greater stress migration hazard in the ULSI regime. The solid lines in Figs. 16*a* and 16*b* show the SM failure times as a function of line thickness and line width, respectively.[70] This experimental behavior could be explained by taking into account void growth by grain-boundary diffusion.[71] In this diffusion theory, the factor n in Eq. (12.22) is 1. It is shown that

$$\begin{aligned} \tau &= Cw^3/\sigma \qquad \text{for } r < 1 \\ \tau &= Ct^3/\sigma \qquad \text{for } r \geq 1 \end{aligned} \tag{12.23}$$

where r is the ratio of thickness to width and σ is the longitudinal stress acting normal to the grain boundary. X-ray diffraction shows that σ is not a constant but is a function of w and t, decreasing with both w and t.[71] The fit of Eq. (12.23) to experiment is shown by the dotted lines in Figs. 16*a* and 16*b* after taking into account the experimentally determined σ values. Note that there is general agreement between the grain-boundary diffusional theory and experiment. The most accurate way to determine the stress values in a conductor line is by x-ray diffraction. Several researchers have used x-ray diffraction to understand stress relaxation in conductor lines as they are temperature-cycled.[62]

12.4.3 Testing for Stress Migration

Currently, IC manufacturers have no standard way of testing for SM. Typically, test structures of very long conductor lengths at a minimum line width of 100 thousand to 10 million microns are tested. These structures are annealed at temperatures ranging from 150 to 250°C, and line resistances are measured at periodic intervals. Failure criteria are arbitrarily set at some specific resistance increase, such as 20%. The lack of a standard testing procedure and test structure for SM could explain some of the variability reported in the stress migration failure rates. Consequently, there is a need to have a more standard SM testing methodology in the future.

12.4.4 Stress Migration Prevention and Summary

The power-law creep model or diffusional creep model explains many of the experimental behaviors of stress migration. However, a comprehensive model is still

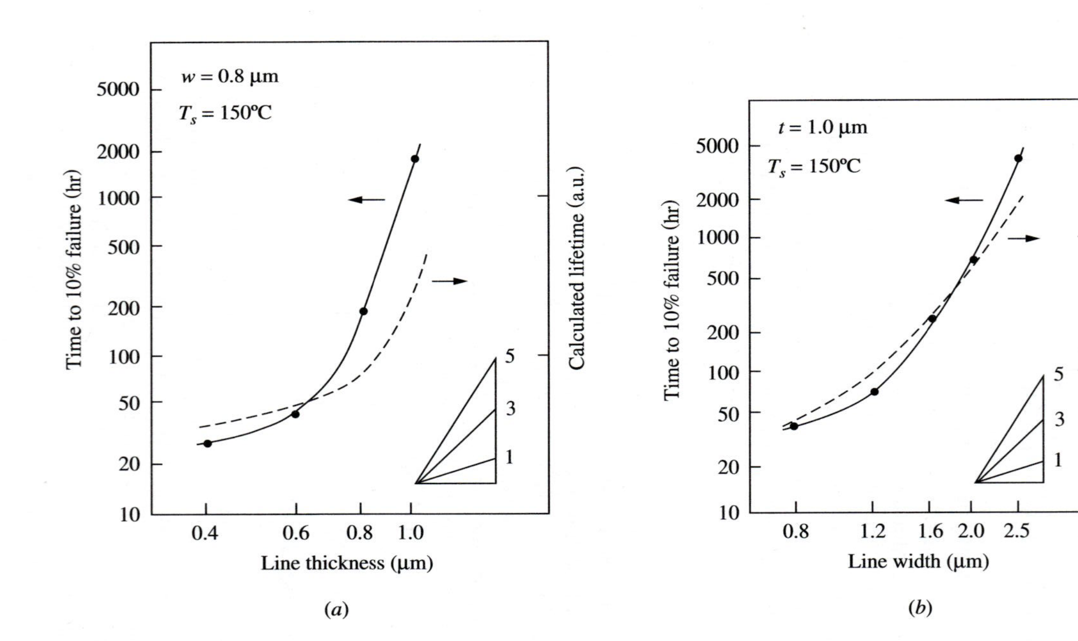

FIGURE 16
(*a*) Experimental 10% failure time (●) and calculated lifetime (- - -) of 0.8-micron wide Al lines plotted as a function of line thickness and aspect ratio. (*b*) Experimental 10% failure time (●) and calculated lifetime (- - -) of 1-micron thick Al lines plotted as a function of line width and aspect ratio. *(After Yagi et al., Ref. 70.)*

needed. The role of stress and stress relaxation has been identified to be very important in the formation and growth of voids. Based on this understanding, researchers have used alloy dopings, such as Cu, in Al to suppress the grain-boundary diffusion of stress migration. The nature of this suppression is similar to the Cu suppression in electromigration.

More recently, by using a refractory barrier or layered metallization scheme, failures due to the slit-type of voids were totally nullified as a result of the electrical shunting property of the refractory barrier. The refractory barrier is more resistant to stress migration than the Al conductor. Thus, for metal dimensions at the 0.5- to 0.35-micron level, it appears that the stacked Al conductor structure, with Cu doping, should be a sufficient safeguard against SM. For line widths below 0.35 microns, this approach may be inadequate against SM because of the dynamic interaction between SM voids and EM. Voids formed by stress relaxation could aggravate electromigration, and it is expected that this interaction will become worse as line widths become narrower. Further improvement could come from orientation control of Al films in ULSI, as a recent study has shown that void formation is accompanied by the minimization of surface free energy, which is achieved by creating the smallest-surface-energy plane, (111), of a face-centered cubic (fcc) crystal.[72]

12.5 OXIDE BREAKDOWN

Oxide breakdown is an important process reliability subject in the development of MOS and CMOS technologies. In the early days of DRAM devices, oxide breakdowns limited the yield and reliability of circuits. Therefore, understanding how oxides break down became important in MOS processing even before the studies of HCI, EM, or SM.

Oxide breakdown is described by the test methods used to cause the dielectric breakdown. In time-dependent dielectric breakdown (TDDB) a constant voltage is applied across the gate oxide at a given temperature. The leakage current across the gate oxide is monitored, and the time to breakdown is recorded when the current exceeds some value (e.g., 1 μA).

A second way of describing oxide breakdown is charge to breakdown, or Q_{bd}. In a Q_{bd} test, a fixed current in the Fowler-Nordheim (F-N) tunneling region is forced through the oxide and the voltage is monitored. When the voltage across the oxide suddenly decreases, the oxide is considered to have failed. The time to failure is designated by t_{bd}. Q_{bd} is the product of the fixed current and t_{bd}. The Q_{bd} test is important in nonvolatile memories, e.g., EPROMs and flash EPROMs, because the durability of tunnel oxides (typically 80 to 100 Å) used in these devices is characterized by Q_{bd}.

In the TDDB case, a relatively low electric field $\mathscr{E}$ is applied across the gate oxide (about 5 to 8 MV/cm). With TDDB, the times to failure could be in the thousands of hours. In the Q_{bd} case, a larger varying $\mathscr{E}$ field is applied across the thin oxide to achieve a constant tunneling current in the F-N region. The voltage usually is near the breakdown of the oxide, e.g., 9 to 10 MV/cm. With Q_{bd}, the time to failure is short, typically several minutes. Figures 17*a* and *b* compare the two measurement methods for TDDB and Q_{bd}.

12.5.1 Mechanisms of Oxide Breakdown

Oxide breakdown is generally believed to be caused by the positive charge buildup in the oxide near the injecting, cathode interface.[73] In oxide thicknesses greater than 120 Å, the source of these positive charges is traditionally believed to be due to impact ionization deep within the oxide[73] caused by the tunneling electrons as shown in Fig. 18*a*. Initially, the tunneling currents are extremely low. Positive charges drift back toward the cathode and are believed to be trapped at localized "weak spots."[73] These trapped positive charges lower the energy band and lead to further electron injection (Fig. 18*b*). The process leads to further impact ionization and positive charge trapping, resulting in a runaway process.[73]

The final high current I injected at localized spots at *positive charge-trapping* sites produces an I^2R heating sufficient to melt the SiO_2.[74] Before the catastrophic oxide rupture, oxide becomes leaky, and the leakage currents could cause circuits to be nonfunctional.[75]

Recent models of oxide breakdown in very thin oxides ($<$ 100 Å) and at power supply voltages less than 5 volts have suggested that another source of the positive charges could be *hot holes* created at the anode side.[76] These hot holes are created by electrons that have tunneled completely through the thin oxide. Upon reaching the anode side, these electrons transfer their gained kinetic energy to deep valence electron(s) through Auger-type interactions, thus creating hot holes in the valence band at the anode side. These holes subsequently are injected (tunneled) back into the oxide by the electric field and are then trapped by the same type of localized *weak spots* as in the thick oxide model. The trapped positive charges cause band lowering and therefore cause enhanced electron tunneling at the cathode. The process leads to thermal runaway, similar to that in the thick oxide model.

Note that both the thick and thin oxide breakdown models hinge on the concepts of positive charge trapping and thermal runaway. The difference is in the source of the positive charges. It is plausible, for very thin oxides, that the probability is increased for an electron to tunnel across the thin layer without resulting in impact ionization in the oxide. Perhaps the real situation is a combination of the two processes described above, and the transition from a thick oxide breakdown regime to the thin oxide breakdown regime is only a matter of degree.

Weak spots in the gate oxide are introduced during wafer manufacturing. They can be caused by process contaminations (causing localized oxide thinning), impurities from cleaning chemicals or dopants (e.g., mobile ions or metallics), process-induced damage (e.g., plasma or ion implantation), gate oxide thinning at the field oxide bird's beak, and high stress points due to the geometric layout.[75]

It is instructive to review the *I-V* curve of a capacitor as it is breaking down. Figure 19 shows typical leakage curves of three types of oxide failure as the voltage is increased, such as in a voltage ramp test. Curve *c* represents an intrinsically good gate oxide, as evidenced by the sudden increase in current beyond 10 MV/cm. This

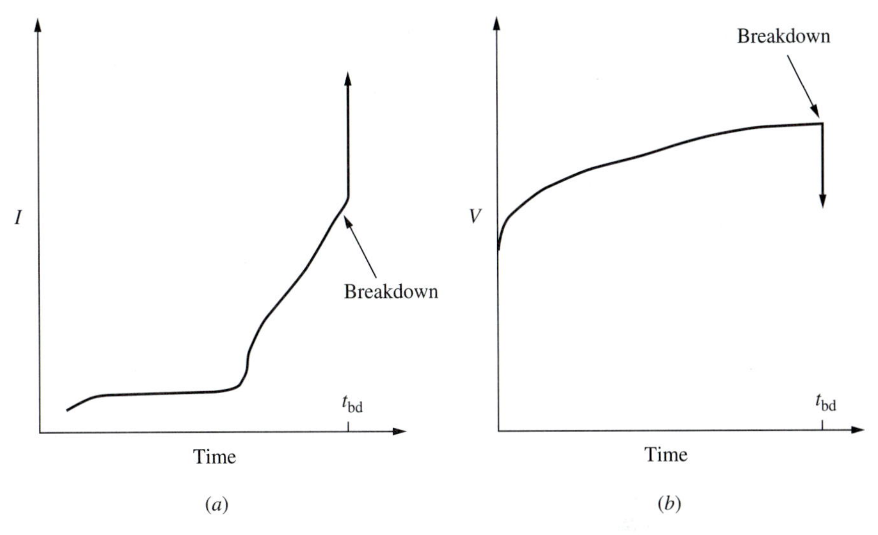

FIGURE 17
(*a*) In TDDB measurement, the leakage current i is monitored as a function of a fixed voltage. (*b*) In Q_{bd} measurement, the voltage is monitored as a fixed F-N tunneling current is applied across the oxide. The time to breakdown is denoted by t_{bd}.

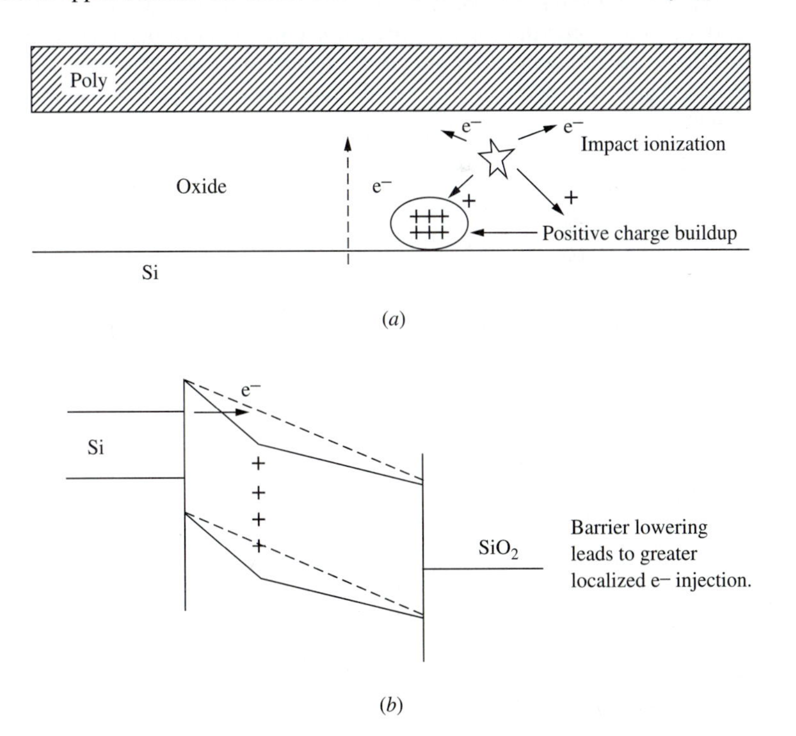

FIGURE 18
(*a*) A model for oxide breakdown mechanism. (*b*) Energy-band diagram showing a lowering of the injection barrier due to positive charge trapping in the oxide.

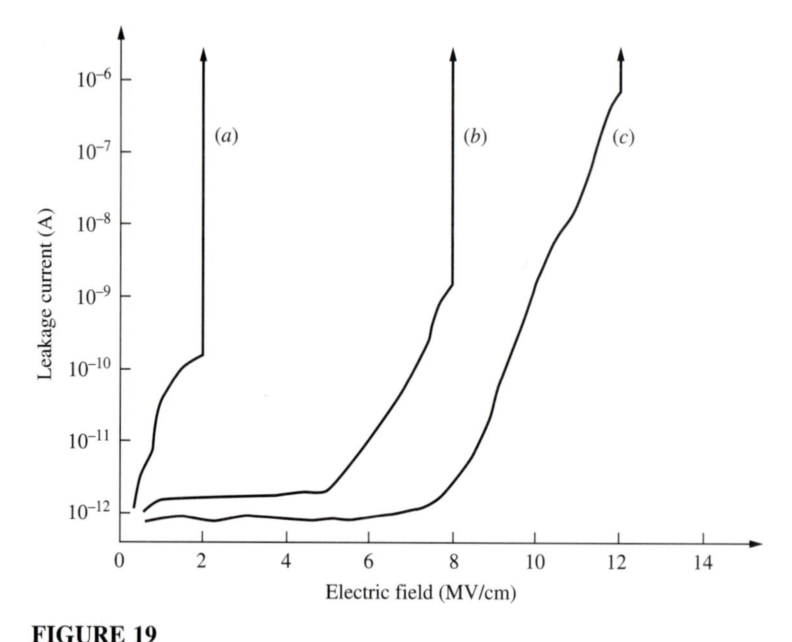

FIGURE 19
I-V curves of three types of oxide failures in a ramped voltage test. Curves (a), (b), and (c) represent Modes A, B, and C types of oxide defects. Mode A defect is usually defined for an electric breakdown value less than 2 MV/cm. Mode B is between 2 and 8 MV/cm. Mode C is greater than 8 MV/cm.

is known commonly as a Mode C type of defect (> 8 MV/cm). Curve *b* represents a Mode B defective oxide, as the rupture is about 7 MV/cm. Notice the leakage in curve *b* occurs earlier. This behavior is often exemplary of Mode B defects. This type of defect (between 2 and 8 MV/cm) poses a long-term reliability problem and such a capacitor could fail within the 10-year life of the circuit. Curve *a* represents a Mode A type of defect and is usually due to a gross oxide defect. Mode A oxide failures are usually tested out before the product is shipped to customers.

12.5.2 Lifetime Prediction of Oxide Breakdown

The lifetime prediction of oxide breakdown is based on empirical modeling of data. There are two popular models to determine the lifetime of oxides: the $\mathscr{E}$ model[77] and the $1/\mathscr{E}$ model.[78] Both appear to fit the experimental data at the high $\mathscr{E}$ fields at which laboratory experiments are obtained. Reliability prediction, however, involves extrapolation to lower electric field conditions. It is not clear at these lower fields (2.5 to 4 MV/cm) which model gives a better estimate of oxide lifetime since it has been extremely difficult to verify the long-term data.

The $\mathscr{E}$ model was obtained from analysis of TDDB failure data.[77,79] This analysis shows that the lifetime of oxide, t_1, stressed at an electric field, $\mathscr{E}_1$, is related to the lifetime of the same oxide, t_2, stressed at electric field, $\mathscr{E}_2$, by the relationship

$$t_1 = t_2 \exp\left[-\beta'(\mathscr{E}_1 - \mathscr{E}_2)\right] \tag{12.24}$$

where β' is the electric field acceleration factor. Since the experimental data are typically represented on a log base 10 scale, β is usually expressed in base 10 decades per MV/cm. Thus,

$$t_1 = t_2 10^{-\beta(\mathscr{E}_1 - \mathscr{E}_2)} \tag{12.25}$$

The relationship of oxide lifetimes at two different temperatures follows the Arrhenius relationship with an activation energy E_a of about 0.3 eV for intrinsic oxide failures. The combined acceleration AF due to voltage and temperature stresses can be calculated, by assuming β and E_a are independent of temperature and electric field, as

$$\mathrm{AF} = \mathrm{AF}_v \times \mathrm{AF}_T \tag{12.26}$$

where AF_v is the voltage acceleration factor and AF_T is the temperature acceleration factor. By using Eq. (12.26), the lifetime of an oxide at any field condition can be extrapolated from any accelerated TDDB tests.

Recent research has shown that in the low cumulative failure distribution there appears to be (1) a dependence of β on temperature, (2) a dependence of E_a on electric field at high temperatures (e.g., 150°C) and at higher fields (> 6 MV/cm),[79] and (3) no dependence of E_a on electric field at lower temperatures (e.g., 60°C) and fields (< 5 MV/cm).[80] This change suggests that the physical oxide breakdown mechanism at lower fields may be different from that at higher fields. The value of β has been measured to be about 5 decades per MV/cm at 60°C and the activation energy to be 0.75 eV at 4 to 5 MV/cm.[80]

The $1/\mathscr{E}$ model is based on the observation that oxide breakdown is a product of the Fowler-Nordheim tunneling currents and the positive charge generation.[81] Time to failure is expressed as

$$t_{\mathrm{bd}} \propto \exp(G/\mathscr{E}_{\mathrm{ox}}) \tag{12.27}$$

where G is about 320 MV/cm and $\mathscr{E}_{\mathrm{ox}}$ is the applied electric field across the oxide. From Eq. (12.27), it can be shown that

$$\beta = -\frac{d}{d\mathscr{E}_{\mathrm{ox}}}[\log_{10}(t_{\mathrm{bd}})] = \frac{G}{2.3\mathscr{E}_{\mathrm{ox}}^2}\left(\frac{\mathrm{decades}}{\mathrm{MV/cm}}\right) \tag{12.28}$$

The oxide lifetime can be modeled by the effective, localized oxide thinning, ΔX_{ox}, where $X_{\mathrm{eff}} = X_{\mathrm{ox}} - \Delta X_{\mathrm{ox}}$, and X_{ox} is the originally deposited oxide thickness. Thus,

$$t_{\mathrm{bd}} = \tau_0 e^{GX_{\mathrm{eff}}/V_{\mathrm{ox}}} \tag{12.29}$$

where τ_o is the intrinsic breakdown time under the applied voltage V_{ox}. In the $1/\mathscr{E}$ model, the oxide lifetime can be predicted once the X_{eff} distribution is determined experimentally.

It is believed that the $\mathscr{E}$ field model is more conservative in predicting oxide lifetime than the $1/\mathscr{E}$ field model.[80]

12.5.3 *V*-Ramp and *J*-Ramp Measurement Techniques

In a manufacturing environment, the TDDB test is too time-consuming to use as a process monitor. Instead, a stepped voltage test, the *V*-ramp test, is commonly used as shown in Fig. 20. This stepped voltage test produces a failure distribution that can be translated directly into that obtained by TDDB once a β is known.[82]

Another oxide monitor test used more recently is the current-ramping test, or the J^*t test.[83] In this test, the current is ramped until breakdown, and the integrated value of J^*t gives the total charge density to breakdown. This test is reported to reveal early oxide defects more distinctly than the *V*-ramp test.[83]

12.5.4 Oxide Breakdown Summary

Oxide breakdown is a complex mechanism. Both the $\mathscr{E}$ field and the $1/\mathscr{E}$ field models have been advanced to predict the oxide lifetime based on data from accelerated tests. It has been difficult to verify experimentally which one is more appropriate to project oxide lifetime at actual field stress conditions. As the oxide thickness becomes comparable to the mean free path of the tunneling electrons, it is suggested that new models are required, such as the hot hole model in Section 12.5.1. Furthermore, the activation energy of oxide defects is not well quantified and remains a challenge in estimating early defects. Therefore, it is expected that the understanding of oxide failure mechanisms in the 50 to 100 Å range will continue to be a major challenge.

12.6 EFFECT OF SCALING ON DEVICE RELIABILITY

The effect of device scaling on process reliability can be estimated from earlier discussions. Essentially, scaling leads to an increase in power density in ICs, which accelerates the wearout mechanisms described earlier. In the following discussion, the scaling factor will be defined by k, where $k > 1$. For example, a 20% shrinkage is denoted by $k = 1.20$.

12.6.1 Effect of Scaling on HCI

The HCI lifetime of a transistor is described by Eq. (12.7). The effect of scaling is to increase the substrate current as the channel length decreases. The increase can be described by rewriting Eq. (12.6) as

$$I_{\text{sub}} \propto \exp(-B/\mathscr{E}_m) \tag{12.30}$$

Then,

$$\frac{I_{\text{sub}}(\text{scaled})}{I_{\text{sub}}(\text{unscaled})} = \exp\left[-\frac{B}{\mathscr{E}_m}\left(\frac{1}{k} - 1\right)\right] \tag{12.31}$$

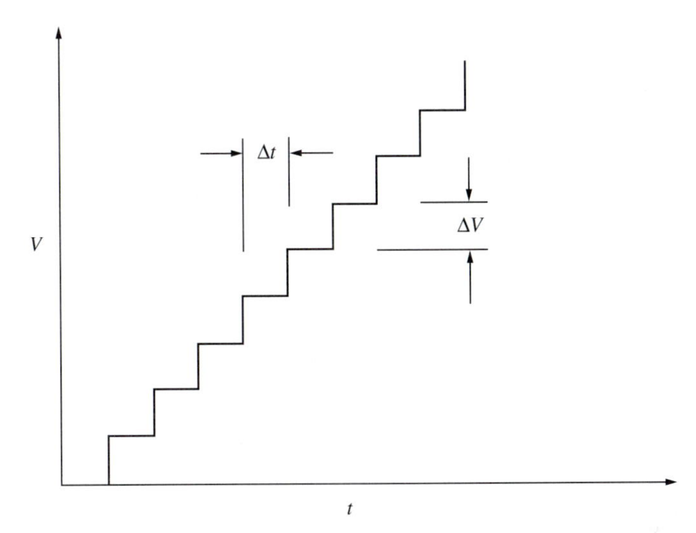

FIGURE 20
Voltage step waveform used in the V-ramp test.

where the increase in electric field is approximated by $k\mathscr{E}_m$. Therefore, the decrease in HCI lifetime is an exponential function of the scaling factor k, or

$$\Delta\tau \propto \exp\left[\frac{mB}{\mathscr{E}_m}\left(\frac{1}{k} - 1\right)\right] \tag{12.32}$$

12.6.2 Effect of Scaling on EM

The basic EM lifetime of a conductor line is described by Eq. (12.18). The current density J will be increased as a result of the increased current drive of scaled devices (for a fixed power supply) and the reduced interconnect cross section. To a first-order approximation, the current in the metal line is proportional to the transistor drive current, I_D, where

$$I_D \propto \frac{\epsilon}{t_{\mathrm{ox}}}\frac{w}{L} \tag{12.33}$$

where w is the width and L is the channel length of the transistor. In the case of a two-dimensional linear shrink, coupled with gate oxide shrink by the same shrink factor k, this expression becomes

$$I_D \propto \left(\frac{1}{1/k}\right)\left(\frac{1/k}{1/k}\right) \propto k \tag{12.34}$$

Since the current density is $J = I/A$, and $A \approx w \times L \propto k^{-2}$, it follows that

$$J \propto k^3 \tag{12.35}$$

Thus, EM lifetime will vary as

$$\text{MTF} \propto k^{-6} \tag{12.36}$$

if we assume $n = 2$ in Eq. (12.18). This sixth-power reduction in lifetime portends an ominous reliability warning for narrow interconnects. However, in reality the so-called bamboo effect shown in Fig. 11 mitigates this effect somewhat.

12.6.3 Effect of Scaling on SM

The effect of scaling on stress migration is estimated from Eq. (12.23), which agrees approximately with experimental data shown in Figs. 16*a* and *b*. If the thickness of the metal remains constant as the line width decreases, SM lifetime is expected to decrease as the inverse cube of the scaling factor, or

$$\Delta\tau \propto k^{-3} \tag{12.37}$$

The prediction shows that SM worsens for narrower interconnect line widths unless alternative prevention techniques are used, e.g., metal layering as discussed earlier.

12.6.4 Effect of Scaling on Oxide Reliability

The effect of scaling on oxide reliability is to increase the electric field across the oxide due to thinner oxides. From Eq. (12.24), the reduced lifetime, τ_s, due to the thinner oxide thickness can be described as

$$\frac{\tau_s}{\tau_u} = \exp[-\beta(k-1)] \tag{12.38}$$

where τ_s is the lifetime of the scaled oxide and τ_u is the lifetime of the unscaled oxide. The effect of scaling is to reduce the oxide reliability exponentially.

12.6.5 Scaling Summary

Scaling is a natural progression of cost improvement. The requirement of putting more functions and memory bits on a single chip dictates the need to go to smaller geometries. So far, this scaling has progressed without a change in the 5-volt power supply system requirement. This has resulted in increased reliability risks toward wearout mechanisms, e.g., HCI, EM, SM, and oxide reliability. However, as system manufacturers change the power supply to 3.3 V or even lower voltages, the reliability risks will be reduced.

12.7 RELATIONS BETWEEN DC AND AC LIFETIMES

Previous discussions have been based on dc stressing methods, primarily because dc measurements have been easier to perform than ac measurements. Since transistors, capacitors, and interconnect elements in ICs are operated in dynamic environments, it is important to understand how the dc lifetimes are related to ac lifetimes. It should be noted that researchers have not derived these relationships for HCI, EM, and TDDB in an entirely straightforward manner because various relaxation and switching effects in the frequency domain are encountered during ac stressing.

12.7.1 AC and DC HCI

The relationship between ac and dc HCI lifetimes can be illustrated by considering a CMOS inverter and the waveforms experienced by the n-channel pull-down transistor, shown in Fig. 21. As discussed in Section 12.2, the n-channel HCI degradation is greatest at the peak of the substrate current and is negligible when the I_{sub} is small. Figure 21 shows the short-duration pulses generated at I_{sub} peaks that cause HCI degradation in the ac case. The duration of these pulses is determined by the rise and fall times in V_D and V_G, and the pulses occur in intervals determined by the frequency of the clock.

$I_{sub}(t)$ will be a function of $V_G(t)$ and $V_D(t)$. It is natural to assume that in the ac (or dynamic) case, the net ac degradation is a summation of each small interval of dc stress defined by Eq. (12.9). Then the ac lifetimes of the n-channel transistor can be expressed as

$$\int_0^{t_{ac}} \left[\frac{I_{sub}}{W}\right]^m dt = B \tag{12.39}$$

where I_{sub}, B, and m have the same meaning as before and are obtained from dc HCI measurements. The term τ_{ac} is the ac lifetime, and W is the width of the transistor. Equation (12.39) is the so-called quasi-static model. Similar expressions can be derived for the p-channel transistor. The same arguments can be applied to other transistors in the circuits with their own $I_{sub}(t)$ or $I_{gate}(t)$ values.

It is now generally believed that the quasi-static model can basically predict the ac HCI lifetime based on dc HCI lifetime data for digital logic circuits.[84] Equation (12.39) has been verified for CMOS inverters and 101-stage ring oscillators over very long time periods and at 30 MHz.[85–87] For circuit designs susceptible to HCI degradation, circuit-reliability simulators, e.g., BERT (Berkeley Reliability Tools), can be used to simulate the degradation of circuit performance based on dc HCI degradations.[88,89]

There remain some subtle dynamic HCI effects that are subjects of on-going research. These include the following:

1. Hole and electron relaxation and neutralization effects from dynamic stressing[90]
2. Mechanical stress effects on HCI[91]
3. Enhanced AC degradation in the channel hot electron (CHE) region.[92]

The HCI degradation of n-channel and p-channel devices must be interpreted in light of degradation in the corresponding circuit blocks. When an IC parameter drifts beyond its specification limit, the IC is considered to have failed. Therefore, MOSFET parameter drifts must be correlated to the IC drift. For example, in DRAM, the hold

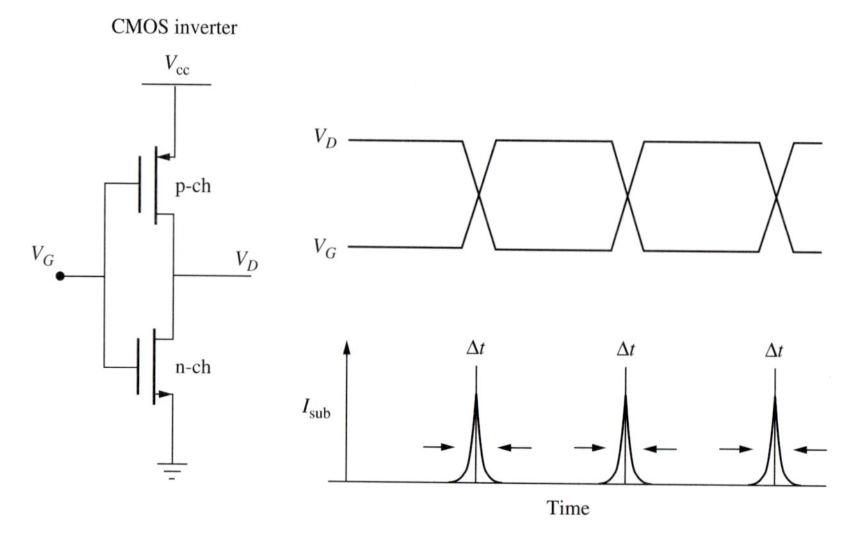

FIGURE 21
Example of an ac clock waveform in a CMOS inverter and corresponding substrate currents (I_{sub}) experienced by the n-channel pull-down transistor.

time, t_{hold}, and the minimum supply voltage of operation, $V_{DD,min}$, are typical parameters used to monitor the HCI drift. Bootstrapped nodes in DRAM are known to be vulnerable to HCI damage. Typically, a DRAM fails as a result of a sense amplifier mismatch, which is caused by slowing down of clocks, and leaky nodes, caused by HCI degradation.[1]

12.7.2 AC versus DC Electromigration

Conductor lines in ICs are stressed under either pulsed dc- or ac-current waveforms at various frequencies. The basic challenge in electromigration is to find the conductor lifetime under these stress conditions based on dc experiments. Hence, the question to address is this: Can the basic EM lifetime model based on dc experiments, described by Eq. (12.18), still adequately model the pulsed ac condition? If it can, then how should the current density J be defined in Eq. (12.18)?

To answer these questions, researchers have studied EM with square wave pulses of various duty cycles and frequencies. It is instructive to review the parameters of a general square wave pulse shown in Fig. 22. J_1 is the positive direction peak current; J_2 is the negative direction peak current; r is the duty ratio; τ is the period; and f is the frequency. Note that (1) for $r = 1$, the waveform becomes the simple dc continuous stress; (2) for $r < 1$ and $J_2 = 0$, the waveform is a unidirectional, pulsed dc (pdc) waveform; (3) for $r < 1$, $J_2 < 0$, the waveform is an ac bipolar, and (4) for $r = +0.5$ and $J_2 = -J_1$, the waveform becomes a perfectly symmetrical ac pulse.

In the case of a unidirectional dc pulse, it has been shown that at low frequencies ($f < 1$ kHz), the lifetime can be described by the on-time of the pulse.[93,94] The

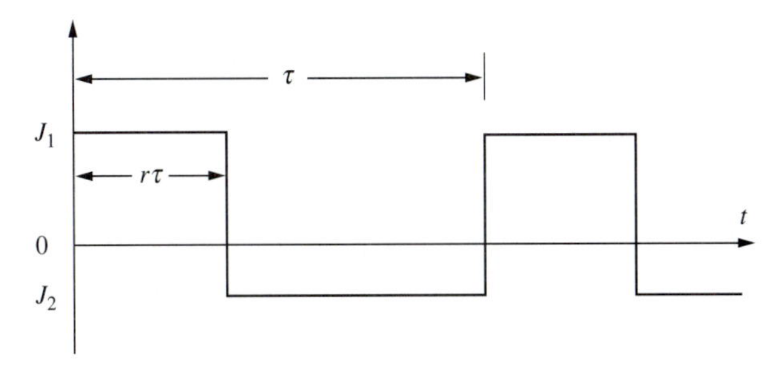

FIGURE 22
A general pulsed square waveform used in EM tests. J_1 is the positive peak current and J_2 is the negative current. The period is τ, and r is the duty cycle.

effective total time of stress is the sum of the times of each individual pulse. Since the on-time as well as the off-time of the pulse in this frequency range is longer than the thermal time constant of the Al ions, the vacancies can follow the on-off cycles of the electron wind. This phenomenon is described by

$$t_{p,\mathrm{dc}} = \frac{1}{r} t_{\mathrm{dc}} \qquad f < 1\ \mathrm{kHz} \tag{12.40}$$

Equation (12.40) is also the prediction of the root mean square (RMS) current density model.

At higher frequencies, it is expected that the electron wind will change direction at a rate faster than the response time of the individual atoms (or vacancies), so that the effective current density experienced by the ions or vacancies should be described by the time-average current density. This theory has generally been proven up to the low MHz range.[95–97] The time-average EM model can be described by

$$\mathrm{MTF} = \frac{A}{(J_{\mathrm{avg}})^n} \exp\left(\frac{E_a}{kT}\right) \tag{12.41}$$

where J_{avg} is the time-average current density of an arbitrary waveform. For the waveform shown in Fig. 22, with $n = 2$, Eq. (12.41) becomes

$$t_{p,\mathrm{dc}} = \frac{1}{r^2} t_{\mathrm{dc}} \tag{12.42}$$

This is the familiar $1/r^2$ time-average current model.

The shortcoming of the time-average current model is that there is a singularity for perfectly symmetrical wave pulses, i.e., $J_{\mathrm{avg}} = 0$. The physical implication of this is that the EM damage created by the electron wind in one direction cancels that of the other direction. For high frequency, this model is dangerous to use in the design rule, since it allows very high $\pm J_1$ peaks, which can cause high joule heating and temperature gradients. To overcome this, the vacancy relaxation model[98] and the average current recovery model[99] have been proposed. Furthermore, there is experimental evidence that the exponent for the duty factor r may be different from $n = 2$ in a generalized wave form such as shown in Fig.22.[99,100] Therefore, the time-average current model, although adequate for unidirectional pulsed dc waveforms, should be used carefully for ac electromigration design rules.

12.7.3 Dynamic Testing of Oxides

Historically, it has been postulated that ac and dc oxide lifetimes can be related by the duty factor. By taking into account the total on-time, the lifetime of a unipolar, pulsed dc stress, $t_{bd}(\text{pdc})$, can be related to the dc lifetime, $t_{bd}(\text{dc})$, by the expression

$$t_{bd}(\text{pdc}) = \frac{1}{r} t_{bd}(\text{dc}) \tag{12.43}$$

where r is the duty factor of the on-time and $t_{bd}(\text{dc})$ is the TDDB lifetime of the oxide discussed in Section 12.5.

Research has shown that Eq. (12.43) is correct at low frequencies, e.g., $<$ 200 Hz.[101] However, at higher frequencies, e.g., $>$ 10 kHz, $t_{bd}(\text{dc})$ is shown to be about four times longer than the lifetime predicted by Eq. (12.43) for 80 to 110 Å oxides.[101] This enhancement is attributed to the reduced hole trapping at the weak spots at the cathode side of the oxide.[101,102] Recall that hole trapping is responsible for oxide breakdown (Section 12.5.2). Thus, a reduction in hole trapping can lead to a longer lifetime. The reduced hole trapping is believed to arise from hole-trap relaxation during the off-time and has been found to have a time constant of 0.3 msec.[101] The charge to breakdown, Q_{bd}, is similarly increased during unipolar dc stress.

The lifetime under bipolar stressing is enhanced even more than in the unipolar case, that is, by about 40 to 100 times over that of the dc lifetime at frequencies ranging from 10 kHz to 4 MHz.[103] This further enhancement is believed to be due to the neutralizing effect of electron injection at both electrodes with the trapped holes near both electrodes.[103]

Experiments show that the electric field acceleration factor β, in the bipolar case, is the same as the dc stress value.[103] This fact allows the determination of the ac lifetime of oxides at low fields from a minimum amount of bipolar stress data at high fields.

12.8
SOME RECENT ULSI RELIABILITY CONCERNS

Newer materials and processing conditions have been developed at the submicron level. Some of these materials and processing conditions have caused reliability problems not observed previously. Two recent reliability problems will be discussed: (1) the effect of tungsten plug on electromigration of interconnect structures and (2) the effect of plasma damage on thin gate oxide reliability.

12.8.1 Tungsten Plug

The traditional approach to Al contact and via construction has led to poor Al step coverage in high-aspect-ratio contacts and vias. See Fig. 23 for an illustration. Two problems are due to poor Al step coverage: (1) Electromigration lifetime is reduced as a result of the thinned Al at the contact (via) walls, and (2) subsequent topography over the contact (via) structure is worsened. The first drawback limits the maximum current density allowed in a design. The second limits the compactness of layout rules on top of the contact (via), e.g., the use of stacked contacts. To overcome this problem of poor Al step coverage, tungsten plug processes have been developed to make the overlying topography smooth.[104] However, the interface between the W plug and the Al introduces a source of maximum Al ion flux divergence and causes shorter electromigration lifetimes than the traditional Al/Al via construction with good Al step coverage.[105] Figure 24 illustrates the problem. The upper conductor has a finite source of Al and, in the presence of a maximum flux divergence, Al atoms are quickly depleted near this W interface. The electromigration of Al can be delayed by adding 0.5 to 2% Cu alloys.[35,106] However, even Cu will electromigrate downstream,[106] so that eventually Al migration starts near the interface as the Al is depleted of Cu. It has been reported that via electromigration is the limiting factor (rather than line electromigration) in W plug interconnect technology.[106]

The challenge to researchers is to develop W plug interconnect schemes that will be reliable enough to handle the current density requirements of power-hungry logic circuits. An alternative via filling technique that uses very high Al sputtering temperatures ($> 500°C$) is also being developed. At this temperature, the Al flows into and completely fills the via. The advantage of this approach is that it removes the flux divergence problem at the W plug interface. But the reliability of these vias has not been demonstrated, and it is not clear how high an aspect ratio of vias this process can fill.

12.8.2 Plasma Damage

As gate oxide thicknesses become thinner (< 150 Å), they are more prone to plasma damage even under normal conditions of multilevel processing.[107] The symptoms of this damage range from outright oxide shorts to transistor threshold shifts, interface damage, and degraded transistor HCI performance.[108–110] These kinds of plasma damage have recently been reported in triple-level metal processing.[108,111] The cause has been attributed to the spatial nonuniformity in the plasma potential.[112,113] The nonuniform plasma causes an imbalance of ion and electron currents in the vicinity of the wafer. Currents are injected through the thin oxides as the wafer tries to balance out the inhomogeneous potential across the wafer surface.

In multilevel-metal schemes, many process steps potentially can cause plasma damage. These steps include interlayer dielectric (ILD) deposition and etching, contact and via etch, plasma resist stripping (or ashing), tungsten plug formation, aluminum plasma etch, and bond-pad etching. The behavior of plasma-damaged

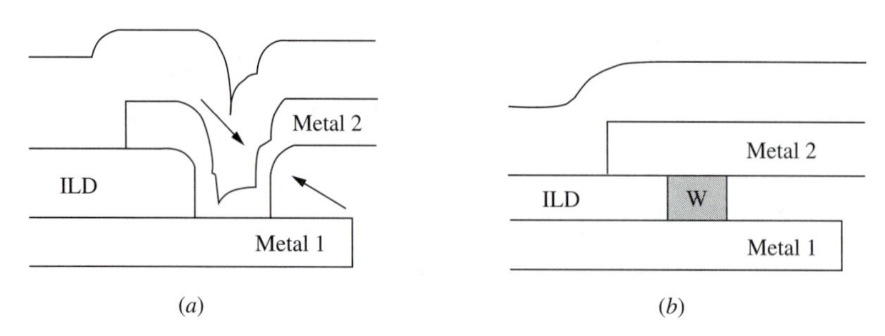

FIGURE 23
(*a*) Diagram of a traditional Al-to-Al via showing Al thinning, resulting in poor step coverage at the side wall (indicated by arrows); (*b*) W plug via showing practically no step coverage problem and good overlying metal topography.

transistors and capacitors at the end of the process can be simulated experimentally by injecting F-N currents into transistors or capacitors.[108,110] By studying the behavior of these injected structures, researchers have found that threshold shifts and degraded HCI performance are predominantly due to bulk trapping of electrons and holes. Damaged oxides cannot be annealed out totally, and the damage can reappear upon subsequent stressing of transistors. Transistors with large antenna ratios exhibit worse plasma damage, where the antenna ratio is defined as the total area of the conductor surface over the field oxide connected to a given transistor divided by the thin gate oxide area. By protecting the transistor gates with parallel diodes, the damage can be reduced.[108,111] Minimizing the plasma damage problem is expected to be a major challenge for the deep submicron technologies.

12.9 MATHEMATICS OF FAILURE DISTRIBUTION

The mathematics of failure distribution is important to process reliability because, to predict reliability from a limited number of samples, it is necessary to fit the experimental data to some failure distribution. It is the failure distribution that describes the failure characteristics of a specific failure mechanism. From the failure distribution,

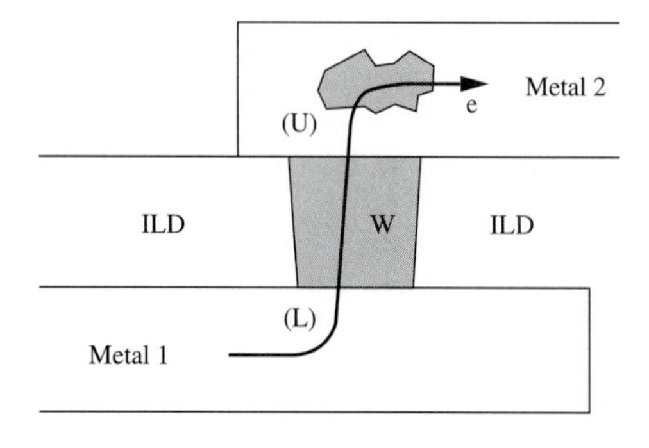

FIGURE 24
W plug electromigration. In the direction of electron flow, the upper (U) W/Al interface experiences a finite Al source. Voiding (shaded area) occurs rapidly as Al is swept downstream by the large flux divergence at the interface *u*.

one can then project the failure rate. If accelerated testing is used, e.g., in temperature, voltage, current, or humidity, then appropriate acceleration models must be used to project the reliability to field conditions.

In the foregoing discussions on HCI, EM, SM, and TDDB, failure data are commonly fitted to either the Weibull or lognormal failure distributions. Generally both models fit most sets of failure data equally well. The major difference between them is in the extrapolation beyond the range of the sample data. When projecting from a measured 5% cumulative failure data to, say, 0.01% of the failure population, the projection based on the lognormal model is usually more optimistic than the projection based on the Weibull distribution fit.[114] The question then arises as to which model should be used.

The Weibull model is based on the *extreme value distribution* that applies when many small defect sites compete with each other to be the one that causes the earliest time of failure.[114] Oxide failures fit this concept reasonably well, so Weibull models are commonly used for TDDB data. On the other hand, the lognormal distribution is based on a model for degradation processes,[114] which explains why it has been very successful in modeling failures due to chemical reactions, molecular diffusion, or migration, such as EM and SM. The general equations of failure distribution and the basic equations of Weibull and lognormal models will be summarized in the following sections.

12.9.1 Failure Rate Concepts

The cumulative distribution function (CDF) of failure is related to the probability density function (PDF) of failure by the relationship[114,115]

$$F(t) = \int_{-\infty}^{t} f(t')\,dt' \tag{12.44}$$

where $F(t)$ is the CDF and $f(t)$ is the PDF. The instantaneous failure rate is defined as

$$\lambda(t) = \frac{f(t)}{1 - F(t)} \tag{12.45}$$

Both the CDF, $[F(t)]$, and the PDF, $[f(t)]$, are commonly used terms in quantifying the reliability of a failure mechanism. For example, in an EM test of 20 samples, if 8 have failed after 1000 hours of stressing, then the corresponding CDF is 40% at 1000 hours. The instantaneous failure rate $\lambda(t)$ can be found by applying Eq. (12.45). The quantity $\lambda(t)$ is the failure rate of the survivors to time t in the very next instant following t. We define the units of failure rate to be 1 Failure Unit $\equiv$ 1 FIT $\equiv$ 1 failure in 10^9 device-hours.

12.9.2 Weibull Distribution

The CDF of the Weibull model is defined[114,115]

$$F(t) = 1 - \exp[-(t/c)^m] \qquad (\text{for } t > 0 \text{ only}) \tag{12.46}$$

and the PDF is defined as

$$f(t) = F'(t) = \frac{m}{t}\left(\frac{t}{c}\right)^m \exp\left[-\left(\frac{t}{c}\right)^m\right] \tag{12.47}$$

where m is the shape parameter, c is the characteristic lifetime parameter, and both m and c are positive.

The instantaneous failure rate can be written as

$$\lambda(t) = \frac{m}{t}\left(\frac{t}{c}\right)^m \tag{12.48}$$

From Eq. (12.46), the data of cumulative failure rate can be analyzed by using the expression

$$\ln\left\{\ln\left[\frac{1}{1 - F(t)}\right]\right\} = m \ln t - m \ln c \tag{12.49}$$

Therefore, by plotting the left-hand side of Eq. (12.49) against $\ln t$, the parameters m and c can be obtained. Standard Weibull plotting paper is available, as shown in Fig. 25; the ordinate is marked in both cumulative failure and in $\ln \ln[1/(1 - F)]$, and the abscissa is in $\ln t$.

For $m = 1$, $\lambda(t)$ becomes a constant ($\lambda = 1/c$), so the exponential distribution is a special case of the Weibull distribution when $m = 1$. Note the exponential distribution is defined to have $\lambda(t) =$ constant. For $m < 1$, $\lambda(t)$ decreases with time; and the Weibull distribution may be used to represent the early failure period (e.g., extrinsic TDDB failures or early device failures). For $m > 1$, $\lambda(t)$ increases with time; and the Weibull distribution may be used to represent the wearout period of a process or device (e.g., intrinsic TDDB failures).[114,115]

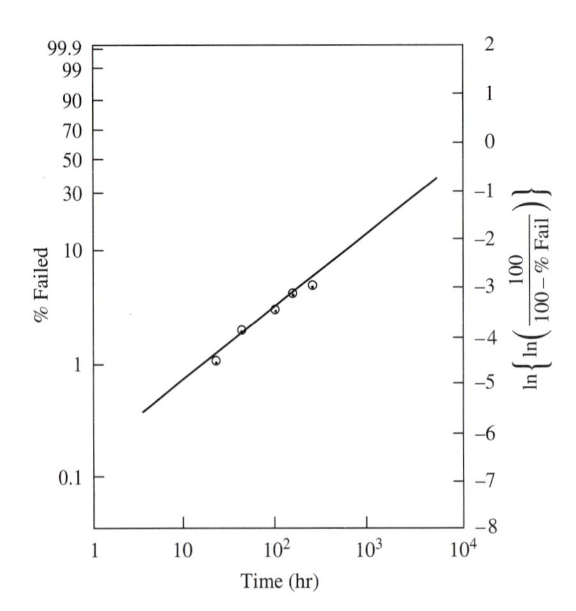

FIGURE 25
A typical Weibull plot. (*After* S.M. Sze, ed., *VLSI Technology,* 2nd ed., McGraw-Hill, 1988, reproduced with permission of McGraw-Hill, Inc.)

12.9.3 Lognormal Distribution Function

The PDF for the lognormal distribution is given by[114,115]

$$f(t) = \frac{1}{\sigma t\sqrt{2\pi}} \exp\left\{ -\left(\frac{1}{2\sigma^2}\right)\left[\ln\left(\frac{t}{T_{50}}\right)\right]^2 \right\} \tag{12.50}$$

where the shape parameter σ is the standard deviation of the logarithmic failure times, and T_{50} is the median time parameter in which 50% of the population has failed. The CDF, $[F(t)]$, is the integral of PDF from 0 to time t as prescribed in Eq. (12.44). Therefore,

$$F(t) = \Phi\left[\frac{\ln(t/T_{50})}{\sigma}\right] \tag{12.51}$$

where $\Phi(z)$ is the standard normal CDF function found in statistics books.[116] Equation (12.51) can be rewritten as

$$\ln t_f = \ln T_{50} + \sigma\Phi^{-1}[F(t_f)] \tag{12.52}$$

showing that in the lognormal distribution, the logarithm of failure times can be plotted as a linear function of $\Phi^{-1}(F) = z$, which is the inverse transformation of the CDF of the percentage of failures. Any standard linear regression program can be used to extract the slope, σ, and the intercept, $\ln T_{50}$. Standardized lognormal plotting paper is available, as shown in Fig. 26.

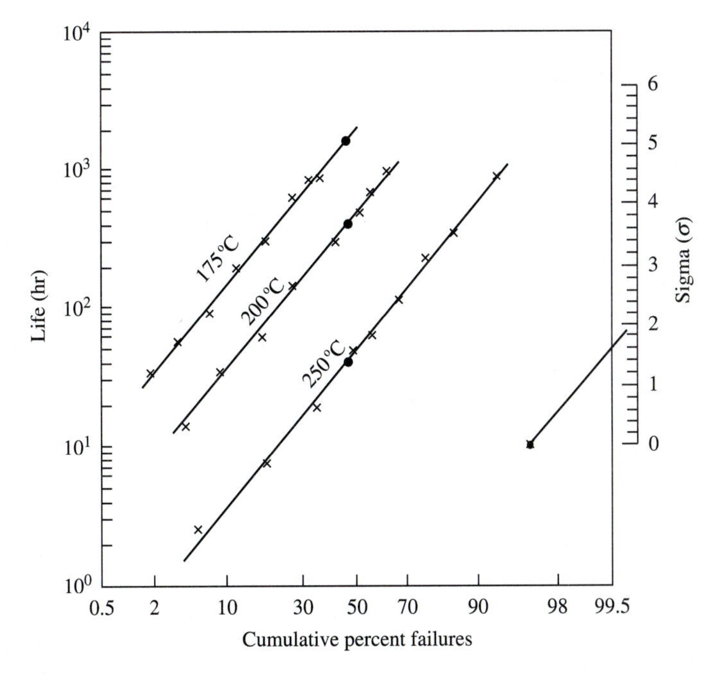

FIGURE 26
A typical lognormal plot. *(After Sze, Ref. 115.)*

Thus, in any lognormal analysis of failure rate data, the key parameters to obtain from the analysis are T_{50} and σ. Once they are extracted from the experimental data, the entire lognormal distribution function is determined and any time to failure at any percentage of cumulative failure distribution, t_f (%), can be obtained.

Commonly used lognormal properties are as follows:

$$T_{50} = t \exp[-\sigma \Phi^{-1} F(t)] \tag{12.53}$$

$$\sigma = \frac{\ln(t/T_{50})}{\Phi^{-1}[F(t)]} \cong \ln\left(\frac{T_{50}}{t_{16}}\right) \tag{12.54}$$

$$t = T_{50} \exp[\sigma \Phi^{-1} F(t)] \tag{12.55}$$

Note that the lognormal distribution function is closely related to the normal distribution function. If one takes the familiar normal distribution function $f(x)$ and substitutes into it $x = \ln(t)$, then Eq. (12.50) immediately follows. The lognormal distribution is simply the normal distribution of the logarithm in time.

The lognormal instantaneous failure rate has a wide variety of appearances depending on the shape parameter σ and is similar in appearance to the Weibull failure rate [Eq. (12.48)]. For $\sigma > 2$, the failure rate rapidly falls in time and can be used to model the early failures (similar to small values of m for the Weibull). For $\sigma < 1$, the failure rate increases in time, corresponding to the wearout failure rates (similar to large values of m in the Weibull).

12.10 SUMMARY AND FUTURE TRENDS

This chapter has emphasized process reliability subjects that are critical to new process technology development. Hot carrier injection (HCI), electromigration (EM), stress migration (SM), and oxide reliability are wearout failure mechanisms, yet they can be easily influenced by defect density and new process materials. Thus, in the development of any new process technology, the performance of these basic wearout mechanisms must be carefully studied. As shown in the chapter, scaling significantly degrades reliability. The traditional concept of a 10-year dc lifetime on test structures is no longer adequate with the push toward finer dimensions, higher current density and power, and more complex, multilevel processing. To calculate ac lifetimes accurately, dynamic degradation models and some circuit-level reliability simulators have been developed. The challenge remains for reliability engineers to define how much more performance can be squeezed out of a transistor, how much more current density can be pumped through an interconnect line or contact, or how much thinner gate oxide can be for a specific application without jeopardizing the 10-year field life goal. Other reliability parameters that are layout- and design-sensitive, e.g., latch-up, electrostatic discharge (ESD), and soft error, should also be studied during the early stage of technology development. The successful development of new technologies must include the identification and characterization of all these failure mechanisms and the subsequent process optimization to ensure the overall process reliability.

The task of the reliability engineer is still not completed when a technology is implemented into volume manufacturing. The engineer must still work with manufacturing engineers to maintain the same highly reliable process by ensuring that process drifts that could potentially damage reliability are monitored and corrected. Recent innovations in the IC industry in wafer level reliability (WLR)[117,118] and building in reliability (BIR)[119] are attempts to create these reliability monitoring systems. The idea in WLR is to monitor specific process reliability mechanisms inside the fab so as to detect quickly if any damaging process drifts have occurred. Or, as in BIR, the concept is to control the critical input parameters of all key process modules, at the earliest possible stage, to ensure that the process does not drift.

Finally, the element of manufacturing cost must be considered. Even after understanding the reliability of the process, difficult trade-offs are sometimes needed to reduce the cost of complex process modules at the expense of a lower, but adequate, reliability.

REFERENCES

1. A. G. Sabnis, "VLSI Reliability," in *VLSI Electronic Microstructure Science,* **22**, Academic Press, New York, 1990, Chapter 6, "Instabilities in I_c."
2. C. Hu, "Advanced MOS Device Physics," in *VLSI Electronic Microstructure Science,* **18**, Academic Press, New York, 1989, Chapter 3.
3. T. H. Ning, P. W. Cook, R. H. Dennard, C. M. Osburn, S. E. Schuster, and H. N. Yu, "1 μm MOSFET VLSI Technology, Part IV: Hot-Electron Design Constraints," *IEEE Trans. Electron Dev.,* **ED-26**, 346 (1979).
4. Y. W. Sing and B. Sudlow, "Modeling and VLSI Design Constraints of Substrate Current," *Tech. Dig. IEDM,* 732 (1980).
5. J. Y. Choi, P. K. Ko, C. Hu, and W. F. Scott, "Hot-Carrier-Induced Degradation of Metal-Oxide-Semiconductor Field-Effect Transistors: Oxide Charge versus Interface Traps," *J. Appl. Phys.,* **65**, 354 (1989).
6. S. K. Lai, "Interface Trap Generation in Silicon Dioxide When Electrons Are Captured by Trapped Holes," *J. Appl. Phys.,* **54**, 2540 (1983).
7. I. C. Chen, S. Holland, and C. Hu, "Electron-Trap Generation by Recombination of Electrons and Holes in SiO_2," *J. Appl. Phys.,* **61**, 4544 (1987).
8. E. Takeda, A. Shimizu, and T. Hagiwara, "Role of Hot-Hole Injection in Hot-Carrier Effects and the Small Degraded Channel Region in MOSFETs," *IEEE Electron Dev. Lett.,* **EDL-4**, 329 (1983).
9. C. Hu, S. Tam, F. C. Hsu, P. K. Ko, T. Y. Chan, and K. W. Terill, "Hot-Electron Induced MOSFET Degradation—Model, Monitor and Improvement," *IEEE Trans. Electron Dev.,* **ED-32**, 375 (1985).
10. P. Heremans, J. Witters, G. Groesneken, and H. E. Maes, "Analysis of the Charge Pumping Technique and Its Application for the Evaluation of MOSFET Degradation," *IEEE Trans. Electron Dev.,* **ED-36**, 1318 (1989).
11. M. Koyanagi, A. Lewis, R. Martin, T. Huang, and J. Chen, "Investigation and Reduction of Hot-Electron Punch Through (HEIP) Effect in Submicron PMOSFETs," *Tech. Dig. IEDM,* 722 (1986).
12. T. Tsuchiya, Y. Okazaki, M. Miyake, and T. Kobayashi, "New Hot-Carrier Degradation Mode and Lifetime Prediction Method in Quarter-Micron PMOSFET," *IEEE Trans. Electron Dev.,* **ED-39**, 404 (1992).

13. Q. Wang, M. Brox, W. H. Kraustschneider, and W. Weber, "Explanation and Model for the Logarithmic Time Dependence of p-MOSFET Degradation," *IEEE Electron Dev. Lett.,* **EDL-12**, 218 (1991).
14. T. C. Ong, P. K. Ko, and C. Hu, "Hot-Carrier Current Modelling and Device Degradation in Surface Channel PMOSFET," *IEEE Trans. Electron Dev.,* **ED-37**, 1658 (1990).
15. K. N. Quader, P. Fang, J. T. Yue, P. K. Ko, and C. Hu, "Simulation of CMOS Circuit Degradation Due to Hot-Carrier Effects," *Reliability Physics, 30th Annual Proceedings,* 1982, p. 16.
16. P. Fang, J. T. Yue, and D. Wollesen, "A Method to Project Hot-Carrier-Induced Punch Through Voltage Reduction for Deep Submicron LDD PMOS FETs at Room and Elevated Temperatures," *Reliability Physics, 30th Annual Proceedings,* 1982, p. 131.
17. H. Wever and W. Seith, "Neue Ergebnisse bei der Elecktrolyse Fester Metallischer Phasen," *Z. Elecktrochem.,* **59**, 942 (1955).
18. H. B. Huntington and A. R. Grone, "Current Induced Marker Motion in Gold Wires," *J. Phys. Chem. Solids,* **20**, 76 (1961).
19. I. A. Blech and H. Sello, "The Failure of Thin Aluminum Current-Carrying Strips on Oxidized Silicon," *Physics of Failures in Electronics* **5**, USAF-RADC Series, 496 (1966).
20. H. B. Huntington, "Electromigration in Metals," in A. S. Nowick and J. J. Burton, Eds., *Diffusion in Solids: Recent Development,* Academic Press, New York, 1975, p. 303.
21. J. Bass, "The Formation and Motion Energies of Vacancies in Aluminum," *Philos. Mag.,* **15**, 717 (1967).
22. J. R. Black, "Physics of Electromigration," *Reliability Physics, 12th Annual Proceedings,* 1974, p. 142.
23. T. Kwok and P. S. Ho, "Electromigration in Metallic Thin Films," in D. Grupta and P. S. Ho, Eds., *Diffusion Phenomena in Thin Films and Microelectronic Materials,* Noyes Publications, Park Ridge, NJ, 1988, p. 369.
24. K. N. Tu, J. W. Mayer, and L. C. Feldman, *Electronic Thin Film Science,* Macmillan, New York, 1993, Chapter 14.
25. J. R. Lloyd and P. M. Smith, "The Effect of Passivation Thickness on the Electromigration Lifetime of Al/Cu Thin Film Conductor," *J. Vac. Sci. Technol.,* **A1**, 455 (1983).
26. L. Yau, C. Hong, and D. Crook, "Passivation Material and Thickness Effects on the MTTF of Al-Si Metallization," *Reliability Physics, 23rd Annual Proceedings,* 1985, p. 115.
27. I. A. Blech, "Electromigration in Thin Aluminum Films on Titanium Nitride," *J. Appl. Phys.,* **47**, 1203 (1976).
28. J. R. Black, "A Brief Survey and Some Recent Results," *IEEE Trans. Electron Dev.,* **ED-16**, 338 (1969).
29. J. R. Black, "Electromigration Failure Modes in Aluminum Metallization for Semiconductor Devices," *Proc. IEEE,* **57**, 1587 (1969).
30. D. S. Chhabra, N. G. Ainslie, and D. Jepsen, "Theory of Failure in Thin Film Conductors," *Abstracts of the Electrochem. Soc.,* Spring Mtg. (1967).
31. J. R. Lloyd, "Electromigration Failure," *J. Appl. Phys.,* **69**, 7601 (1991).
32. M. Shatzkes, J. R. Lloyd, "A Model for Conductor Failure Considering Diffusion Concurrently with Electromigration Resulting in a Current Exponent of 2," *J. Appl. Phys.,* **59**, 3890 (1986).
33. K. Hinode, T. Furusawa, and Y. Homma, "Dependence of Electromigration Lifetime on the Square of Current Density," *Reliability Physics, 31st Annual Proceedings,* 1993, p. 319.
34. G. A. Scoggen, B. N. Agarawala, P. P. Peressini, and A. Brouillard, "Width Dependence of Electromigration Life," *Reliability Physics, 13th Annual Proceedings,* 1975, p. 151.

35. H. S. Rathore, R. G. Filippi, R. A. Wachnik, J. J. Estabil, and T. Kwok, "Electromigration and Current-Carrying Implications for Aluminum-Based Metallurgy with Tungsten Stud-Via Interconnections," *Proc. SPIE,* **1805**, 251 (1992).
36. P. A. Gargini, C. Tseng, and M. H. Woods, "Elimination of Silicon Electromigration in Contacts by the Use of an Interposed Barrier Metal," *Reliability Physics, 20th Annual Proceedings,* 1982, p. 66.
37. S. Vaidya and A. K. Sinha, "Electromigration Induced Leakage at Shallow Junction Contacts Metallized with Aluminum/Poly Silicon," *Reliability Physics, 20th Annual Proceedings,* 1982, p. 50.
38. N. D. Bui, V. H. Pham, J. T. Yue, and D. L. Wollesen, "A Via Failure Mode in Electromigration of Multilevel Interconnect," *Proceedings of the 7th International VLSI Multilevel Interconnection Conference,* 1990, p. 142.
39. T. Yamaha, M. Naitou, and T. Hotta, "Three Kinds of Via Electromigration Failure Modes in Multilevel Interconnections," *Reliability Physics, 30th Annual Proceedings,* 1992, p. 349.
40. I. Ames, F. M. d'Heurle, and R. Horstman, "Reduction of Electromigration in Aluminum Films by Copper Doping," *IBM J. Res. Devel.,* **4**, 461 (1970).
41. A. Gangulee and F. M. d'Heurle, "Effect of Alloy Additions on Electromigration Failures in Thin Aluminum Films," *Appl. Phys. Lett.,* **19**, 75 (1971).
42. A. J. Learn, "Electromigration Effects in Aluminum Alloy Metallization," *J. Electron. Matls.,* **3**, 531 (1974).
43. M. Hansen, in M. Hansen and K. Anderko, Eds., *Constitution of Binary Alloys,* McGraw-Hill, New York, 1958.
44. F. M. d'Heurle and I. Ames, "Electromigration in Single-Crystal Aluminum Film Conductors," *Appl. Phys. Lett.,* **16**, 80 (1970).
45. C. H. Herzig and W. Wiemann, "Experimental Determination of the Electrostatic Driving Force in the Electromigration of Tin in Gold and Gold-Tin Alloys," *Phys. Status Solidi,* **A26**, 459 (1974).
46. J. M. Pierce and M. E. Thomas, "Electromigration in Aluminum Conductors Which Are Chains of Single Crystals," *Appl. Phys. Lett.,* **39**, 165 (1981).
47. C. V. Thompson and H. Kahn, "Effects of Microstructure on Interconnect and Via Reliability: Multimodal Failure Statistics," *J. Electron. Matls.,* **22**, 581 (1993).
48. C. K. Hu, N. J. Mazzeo, and C. Stanis, "Electromigration in Two-Level Bamboo Grain Structure Al(Cu)/W Interconnections," *Mater. Chem. Phys.,* **35**, 95 (1993).
49. B. Grabe and H. U. Schreiber, "Lifetime and Drift Velocity Analysis for Electromigration in Sputtered Al Films, Multilayers and Alloys," *Solid State Electron.,* **26**, 1026 (1983).
50. E. Levine and J. Kitcher, "Electromigration Induced Damage and Structure Change in Cr-Al/Cu and Al/Cu Interconnection Lines," *Reliability Physics, 22nd Annual Proceedings,* 1984, p. 242.
51. L. J. Fried, J. Haves, J. S. Lechaton, J. S. Logan, G. Paal, and P. A. Totta, "A VLSI Bipolar Metallization Design with Three-Level Wiring and Area Array Solder Connections," *IBM J. Res. Devel.,* **26**, 362 (1982).
52. J. K. Howard, J. F. White, and P. S. Ho, "Intermetallic Compounds of Al and Transition Metals: Effect of Electromigration in 1–2 μm Wide Lines," *J. Appl. Phys.,* **49**, 4083 (1978).
53. T. Kwok, P. S. Ho, and H. C. W. Huang, "Summary Abstract: Electromigration Studies of Al-Intermetallic Structures," *J. Vac. Sci. Technol.* A, vol 2, 241 (1984).
54. J. T. Yue, W. P. Funsten, and R. V. Taylor, "Stress Induced Voids in Aluminum Interconnects during IC Processing," *Reliability Physics, 23rd Annual Proceedings,* 1985, p. 126.

55. J. Curry, G. Fitzgibbon, Y. Guan, R. Muollo, G. Nelson, and A. Thomas, "New Failure Mechanisms in Sputtered Aluminum-Silicon Films," *Reliability Physics, 22nd Annual Proceedings,* 1984, p. 6.
56. P. A. Totta, "Stress Induced Phenomena in Metallization: U.S. Perspective," in C. Y. Li, P. Totta, and P. Ho, Eds., *Stress Induced Phenomena in Metallization,* American Vacuum Society Series, **13** (AIP Conf. Proceedings No. 263), American Institute of Physics, Ithaca, NY, 1991, p. 1.
57. J. Klema, R. Pyle, and E. Domangue, "Reliability Implication of Nitrogen Contamination during Deposition of Sputtered Aluminum/Silicon Metal Films," *Reliability Physics, 22nd Annual Proceedings,* 1984, p. 1.
58. T. Turner and K. Wendel, "The Influence of Stress on Aluminum Conductor Life," *Reliability Physics, 23rd Annual Proceedings,* 1985, p. 142.
59. M. R. Lin and J. T. Yue, "Impact of Ceramic Packaging Anneal on the Reliability of Al Interconnects," *Reliability Physics, 24th Annual Proceedings,* 1985, p. 164.
60. S. Mayumi, T. Umemoto, M. Shishino, H. Nanatsue, S. Ueda, and M. Inoue, "The Effect of Cu Addition to the Al-Si Interconnects on Stress Induced Open-Circuit Failures," *Reliability Physics, 25th Annual Proceedings,* 1987, p. 15.
61. K. Hinode, N. Wada, T. Nishida, and K. Mukai, "Stress-Induced Grain Boundary Fractures in Al-Si Interconnects," *J. Vac. Sci. Technol.,* **85**, 518 (1987).
62. A. Tezaki, T. Mineta, and H. Egawa, "Measurement of Three Dimensional Stress and Modeling of Stress Induced Migration Failure in Aluminum Interconnects," *Reliability Physics, 28th Annual Proceedings,* 1990, p. 221.
63. K. N. Tu, J. W. Mayer, and L. C. Feldman, *Electronic Thin Film Science,* Macmillan, New York, 1993, Chapter 4.
64. P. A. Flinn, "Study by Wafer Curvature and X-Ray Diffraction Techniques," in *Stress Induced Phenomena in Metallization,* American Vacuum Society Series, **13**, 73 (1991).
65. J. W. McPherson and C. F. Dunn, "A Model for Stress-Induced Metal Notching and Voiding in Very Large-Scale-Integrated Al-Si (1%) Metallization," *J. Vac. Sci. Technol.,* **B5**, 1321 (1987).
66. I. Le May, *Principles of Mechanical Metallurgy,* Elsevier, New York, 1981.
67. P. A. Flinn and C. Chiang, "X-Ray Diffraction Determination of the Effects of Various Passivations on Stress in Metal Films and Patterned Lines," *J. Appl. Phys.,* **67**, 2927 (1990).
68. W. T. Tseng, "Interface Reaction Model for Process Voiding in Aluminum Conductor Lines," *Appl. Phys. Lett.,* **59**, 680 (1991).
69. Y. Sugano, S. Minegishi, H. Sumi, and M. Itabashi, "In Situ Observation and Formation Mechanism of Aluminum Voiding," *Reliability Physics, 26th Annual Proceedings,* 1988, p. 34.
70. H. Yagi, H. Niwa, T. Hosada, M. Inoue, H. Tsuchikawa, and M. Kato, "Analytical Calculation and Direct Measurement of Stress in an Aluminum Interconnect of Very Large Scale Integration," in C. Y. Li, P. Totta, and P. Ho, Eds. *Stress Induced Phenomena in Metallization,* American Vacuum Society Series, **13**, (AIP Conf. Proceedings No. 263), American Institute of Physics, Ithaca, NY, 1991, p. 44.
71. M. Kato, H. Niwa, H. Yagi, and H. Tsuchikawa, "Diffusional Relaxation and Void Growth in an Aluminum Interconnect of Very Large Scale Integration," *J. Appl. Phys.,* **68**, 334 (1990).
72. H. Keneko, M. Hasunuma, A. Sawabe, T. Kawanoue, Y. Kohanawa, S. Komatsu, and M. Miyauchi, "A Newly Developed Model for Stress Induced Slit-Like Voiding," *Reliability Physics, 28th Annual Proceedings,* 1990, p. 194.
73. I. C. Chen, S. Holland, and C. Hu, "Electrical Breakdown in Thin Gate and Tunneling Oxides," *IEEE Trans. Electron. Dev.,* **32**, 413 (1985).

74. D. R. Wolters and J. T. Van der Schoot, "Dielectric Breakdown in MOS Devices Part III: The Damage Leading to Breakdown," *Philips J. Res.*, **40**, 164 (1985).
75. A. G. Sabnis, "VLSI Reliability," in *VLSI Electronic Microstructure Science,* **22**, Academic Press, New York, 1990, Chapter 5.
76. K. Schuegraf and C. Hu, "Hole Injection Oxide Breakdown Model for Very Low Voltage Lifetime Extrapolations," *Reliability Physics, 31st Annual Proceedings,* 1993, p. 7.
77. D. Crook, "Method of Determining Reliability Screens for Time Dependent Dielectric Breakdown," *Reliability Physics, 17th Annual Proceedings,* 1979, p. 1.
78. I. C. Chen, S. Holland, and C. Hu, "A Quantitative Physical Model for Time-Dependent Breakdown in SiO_2," *Reliability Physics, 23rd Annual Proceedings,* 1985, p. 24.
79. J. W. McPherson and D. A. Baglee, "Acceleration Factors for Thin Gate Oxide Stressing," *Reliability Physics, 23rd Annual Proceedings,* 1985, p. 1.
80. K. C. Boyko and D. L. Gerlach, "Time Dependent Dielectric Breakdown of 210 Å Oxides," *Reliability Physics, 27th Annual Proceedings,* 1989, p. 1.
81. J. C. Lee, I. C. Chen, and C. Hu, "Modeling and Characterization of Gate Oxide Reliability," *IEEE Trans. Electron Dev.,* **35**, 2268 (1988).
82. A. Berman, "Time-Zero Dielectric Reliability Test by a Reamp Method," *Reliability Physics, 19th Annual Proceedings,* 1981, p. 204.
83. D. Crook, "Detecting Oxide Quality Problems Using JT Testing," *Reliability Physics, 29th Annual Proceedings,* 1991, p. 337.
84. W. Weber, M. Brox, T. Kunemune, H. Muhlhoff, and D. Schmitt-Landsiedel, "Dynamic Degradation in MOSFET's—Part II: Application in Circuit Environment," *IEEE Trans. Electron. Dev.,* **ED-38**, 1859 (1991).
85. P. M. Lee, P. K. Ko, and C. Hu, "Relating CMOS Inverter Lifetime to DC Hot-Carrier Lifetime of NMOSFET's," *IEEE Electron. Dev. Lett.,* **EDL-11**, 39 (1990).
86. K. N. Quader, P. Fang, J. T. Yue, P. K. Ko, and C. Hu, "Simulation of CMOS Circuit Degradation Due to Hot-Carrier Effects," *Reliability Physics, 30th Annual Proceedings,* 1982, p. 16.
87. P. Fang, R. Rakkhit, and J. T. Yue, "An Investigation of Hot Carrier Effects in Submicron CMOS Integrated Circuits," *Microelectron. Reliab.,* **33**, 1713 (1993).
88. "BERT-Berkeley Reliability Tools," University of California-Berkeley Memorandum No.UCB/ERL M91/107 (1991).
89. K. N. Quader, P. Fang, J. T. Yue, P. K. Ko, and C. Hu, "Hot-Carrier Design Rules for Translating Device Degradation to CMOS Digital Circuit Degradation," *IEEE Trans. Electron. Dev.,* **41**, 681 (1994).
90. M. Brox and W. Weber, "Dynamic Degradation in MOSFET's—Part I: The Physical Effects," *IEEE Trans. Electron. Dev.,* **ED-38**, 1852 (1991).
91. A. Hamada and E. Takada, "AC Hot-Carrier Effect under Mechanical Stress," *1992 Symposium on VLSI Technology—Digest of Technical Papers,* 1992, p. 19.
92. K. Mistry and B. Doyle, "The Role of Electron Trap Creation in Enhanced Hot-Carrier Degradation during AC Stress," *IEEE Electron. Dev. Lett.,* **EDL-11**, 267 (1990).
93. F. M. d'Heurle, "Electromigration and Failure in Electronics: An Introduction," *Proc. IEEE,* **59**, 1409 (1971).
94. J. S. Suehle and H. A. Schafft, "The Electromigration Damage Response Time and Implications for DC and Pulsed Characterizations," *Reliability Physics, 27th Annual Proceedings,* 1989, p. 229.
95. J. A. Maiz, "Characterization of Electromigration Under Bidirectional (BC) and Pulsed Unidirectional (PDC) Currents," *Reliability Physics, 27th Annual Proceedings,* 1989, p. 220.
96. J. M. Towner and E. P. van der Van, "Aluminum Electromigration under Pulsed DC Conditions," *Reliability Physics, 21st Annual Proceedings,* 1983, p. 36.

97. J. J. Clement, "Vacancy Supersaturation Model for Electromigration Failure under DC and Pulsed DC Stress," *J. Appl. Phys.,* **71**, 4264 (1992).
98. B. K. Liew, N. W. Cheung, and C. Hu, "Projecting Interconnect Electromigration Lifetime for Arbitrary Current Waveforms," *IEEE Trans. Electron. Dev.,* **37**, 1343 (1990).
99. L. M. Ting, J. S. May, W. R. Hunter, and J. W. McPherson, "AC Electromigration Characterization and Modeling of Multilayered Interconnects," *Reliability Physics, 31st Annual Proceedings,* 1993, p. 311.
100. K. Hatanaka, T. Noguchi, and K. Maneguchi, "A Generalized Lifetime Model for Electromigration under Pulsed DC/AC Stress Conditions," *Proc. 9th Symposium on VLSI Technology—Digest of Technical Papers,* 1989, p. 19.
101. Y. Fong, I. C. Chen, S. Holland, J. Lee, and C. Yu, "Dynamic Stressing of Thin Oxides," *Tech. Dig. IEDM,* 664 (1986).
102. M. S. Liang, S. Haddad, W. Cox, and S. Cagnina, "Degradation of Very Thin Gate Oxide MOS Devices Under Dynamic High Field/Current Stress," *Tech. Dig. IEDM,* 394 (1986).
103. E. Rosenbaum and C. Hu, "High-Frequency Time-Dependent Breakdown of SiO_2," *IEEE Electron. Dev. Lett.,* **12**, 267 (1991).
104. Workshops on "Tungsten and Other Refractory Metals for ULSI Applications," *Materials Research Society Conference Proceedings,* 1986–1990.
105. J. Tao, K. K. Young, N. W. Cheung, and C. Hu, "Comparison of Electromigration Reliability of Tungsten and Aluminum Vias under DC and Time-Varying Current Stressing," *Reliability Physics, 30th Annual Proceedings,* 1992, p. 338.
106. J. J. Estabil, H. S. Rathore, and E. N. Levine, "Electromigration Improvements with Titanium Underlay and Overlay in Al(Cu) Metallurgy," *Proceedings of the 8th International VLSI Multilevel Interconnection Conference,* 1991, p. 242.
107. F. Shone, K. Wu, J. Shaw, E. Hokelet, S. Mittal, and A. Haranahalli, "Gate Oxide Charging and Its Elimination for Metal Antenna Capacitor and Transistor in VLSI CMOS Double Layer Metal Technology," *Proc. 9th Symposium on VLSI Technology—Digest of Technical Papers,* 1989, p. 73.
108. Y. H. Lee, L. Yau, R. Chau, E. Hansen, B. Sabi, S. Hui, P. Moon, and G. Vandentop, "Correlation of Plasma Process Induced Charging with Fowler-Nordheim Stress in p- and n-Channel Transistors," *Tech. Dig. IEDM,* 65 (1992).
109. H. Shin, C. C. King, T. Horiuchi, and C. Hu, "Thin Oxide Charging Current during Plasma Etching of Aluminum," *IEEE Electron. Dev. Letts.,* **13**, 404 (1991).
110. H. Shin, C. C. King, and C. Hu, "Thin Oxide Damage by Plasma Etching and Ashing Processess," *Reliability Physics, 30th Annual Proceedings,* 1992, p. 37.
111. R. Rakkhit, F. P. Heiler, P. Fang, and C. Sander, "Process Induced Oxide Damage and Its Implications to Device Reliability of Submicron Transistors," *Reliability Physics, 31st Annual Proceedings,* 1993, p. 293.
112. S. Fang and J. McVittie, "Thin Oxide Damage from Gate Charging during Plasma Processing," *IEEE Electron. Dev. Letts.,* **13**, 288 (1992).
113. S. Fang and J. McVittie, "A Mechanism for Gate Oxide Damage in Nonuniform Plasma," *Reliability Physics, 31st Annual Proceedings,* 1993, p. 13.
114. P. A. Tobias and D. Trindade, *Applied Reliability,* Van Nostrand Reinhold, New York, 1986.
115. S. M. Sze, ed., *VLSI Technology,* 2nd ed., McGraw-Hill, 1988.
116. P. A. Tobias and D. Trindade, *Applied Reliability,* Van Nostrand Reinhold, New York, 1986, Chapter 5, Table 5.1.
117. T. A. Dillon, W. A. Miller, D. G. Pierce, and E. S. Snyder, "Wafer Level Reliability," *Microelectronic Manufacturing and Reliability, SPIE,* **1802**, 144 (1992).

118. JEDEC task group on Wafer Level Reliability, JEDEC Committee JC14.2, *JEDEC Standard Docs.* **33** (1993) and *JEDEC Standard Docs.* **35** (1992).
119. H. A. Schafft, D. A. Baglee, and P. E. Kennedy, "Building-In Reliability: Making It Work," *Reliability Physics, 29th Annual Proceedings,* 1991, p. 1.

PROBLEMS

1. **Hot Carrier Injection—Part I:** A group of NMOS transistors is stressed under HCI tests. The transistors are stressed under 3 V_{ds} values (V_{ds} = 7 V, 6.6 V, and 6.2 V) at the maximum I_{sub} conditions. The average percentages of degradation in time of the three stressed groups of transistors are given below:

$\Delta I_{dsat}/I_{dsat}$	1.4 hr	4 hr	10 hr	40 hr
6.2 v	0.43%	0.5%	0.68%	1.0%
6.6 v	0.62%	0.8%	1.0%	1.4%
7.0 v	1.15%	1.3%	1.9%	2.8%

Find the lifetime (in hours) to reach 10 percent I_{dsat} degradation for each V_{ds} stress group. Use Eq. (12.8). (*Hint*: Use the power law regression analysis, $\Delta p/p = At^b$).

2. **Hot Carrier Injection—Part II:** In Problem 1, the I_{sub} values for each of the V_{ds} groups were 1, 2.4, and 6 μA/μm for the V_{ds} values of 6.2, 6.6, and 7 V, respectively. (*a*) Using Eq. (12.9), project for the HCI lifetime of this group of transistors at V_{ds} = 5.5 V, if the I_{sub} value is 0.2 μA/μm at that bias. (*Hint*: Use power law regression analysis, $\tau = AI_{sub}^m$). (*b*) What is the slope m? (*c*) Is this lifetime acceptable from a customer point of view?

3. **EM—Part I:** In an electromigration experiment, 13 samples were stressed to failure under the condition of 5×10^6 A/cm² at an oven temperature of 175°C. The failure times were found at times = 1, 1.2, 1.8, 2, 2.05, 2.2, 2.3, 2.3, 2.8, 3, 3.2, 4.7, and 5.2 hr. Use the lognormal statistics to estimate the T_{50} and σ of this experiment. [*Hint*: Use Eq. (12.54) and a lognormal graphing plot such as Fig. 12.26; assume the percent cumulative failure rate = $(i - 0.3)/(N + 0.4)$, where i is the ith failure in a total population of N parts.]

4. **EM—Part II:** Typical electromigration studies of a new metal interconnect system involve the measurement of kinetic parameters n and E_a. The study is often performed with a "three-cell" stress experiment. Suppose one of the cells was at the condition stipulated in Problem 3, with the T_{50} solution of Problem 3; the other two conditions and T_{50} results are given below:

	225°C	175°C
4.1×10^6 A/cm²	9.6 hr	
5×10^6 A/cm²	2.6 hr	T_{50} (Prob. 3)

Using the fact that the effective temperatures (T_{eff}) of the three cells were 325, 295, and 259°C as a result of joule heating, find the best set of n and E_a that fits the experimental data. (*Hint*: solve for E_a first, then use that E_a to normalize the data to the same temperature before solving for n.)

5. Use the T_{50} found in Problem 3 to find the lifetime at 0.1% of the lognormal failure distribution at a field condition of 100°C junction temperature and a current density of 1×10^6 A/cm^2. [*Hint*: The T_{eff} is 259°C. Assume $T(0.1\%) = 0.177\, T(50\%)$.]

6. **Oxide Reliability:** In a TDDB experiment of 100 Å oxide, the time required to get 1% of the capacitors to fail at 9.5 MV/cm of electric field stress was 3.15×10^3 sec. The time to get 1% of another similar set of capacitors to fail at 8.5 MV/cm was 3.15×10^5 sec. Suppose it is desired to analyze these data with the $\mathscr{E}$ field model [Eq. (12.25)]. (*a*) Find the field acceleration factor beta. (*b*) Find the 1% lifetime of the 100 Å oxide at 5.5 volts.

7. Suppose it is desired to analyze the TDDB data in Problem 6 with the $1/\mathscr{E}$ model. (*a*) Find the 1% lifetime of the 100 Å oxide using Eq. (12.27). (*b*) Which model predicts a shorter lifetime, the $\mathscr{E}$ or $1/\mathscr{E}$ model?

8. Often in lognormal analysis it is useful to know the relationship between the 0.1% cumulative failure rate ($T_{0.1}$) and the median failure rate (T_{50}). In Problem 5, it was assumed that $T_{0.1} = 0.177\, T_{50}$. In general, $T_{0.1}$ is related to T_{50} through the σ of the lognormal distribution. Use Eq. (12.55) and the property of inverse function of Standard Normal CDF to show the relationship of $T_{0.1}$ and T_{50} as a function of σ.

APPENDIX A

Properties of Si at 300 K

Properties	Si
Atoms/cm^3	5.0×10^{22}
Atomic weight	28.09
Breakdown field (V/cm)	$\sim 3 \times 10^5$
Crystal structure	Diamond
Density (g/cm^3)	2.33
Dielectric constant	11.9
Effective density of states in conduction band, N_C (cm^{-3})	2.8×10^{19}
Effective density of states in valence band, N_V (cm^{-3})	1.04×10^{19}
Effective mass, m^*/m_0:	
Electrons	$m_1^* = 0.91$, $m_t^* = 0.19$
Holes	$m_{1h}^* = 0.15$, $m_{hh}^* = 0.53$
Electron affinity (V)	4.05
Energy gap (eV)	1.12
Heat capacity (J/mol-K)	20.07
Index of refraction	3.42
Intrinsic carrier concentration (cm^{-3})	1.02×10^{10}
Intrinsic Debye length (μm)	24
Intrinsic resistivity (Ω-cm)	3.16×10^5
Lattice constant (Å)	5.435
Linear coefficient of thermal expansion, $\Delta L/L\Delta T$ ($°\text{C}^{-1}$)	2.59×10^{-6}
Melting point (°C)	1412
Minority-carrier lifetime (s)	2.5×10^{-3}

Properties	**Si**
Mobility (drift) (cm^2/V-s):	
μ_n (electrons)	1450
μ_p (holes)	505
Optical phonon energy (eV)	0.063
Phonon mean free path λ_0 (Å):	
Electrons	76
Holes	55
Poisson's ratio, (100) orientation	0.28
Specific heat (J/g-°C)	0.7
Thermal conductivity (W/cm-°C)	1.5
Thermal diffusivity (cm^2/s)	0.9
Vapor pressure (Pa)	1 at 1650°C, 10^{-6} at 900°C
Young's modulus (GPa)	130

APPENDIX B

List of Symbols

Symbol	Description	Units
a	Lattice constant	Å
B	Magnetic induction	Wb/m^2
c	Speed of light in a vacuum	cm/s
C	Capacitance	F
D	Electric displacement	C/cm^2
D	Diffusion coefficient	cm^2/s
E	Energy	eV
E_f	Fermi energy level	eV
E_g	Energy bandgap	eV
$\mathscr{E}$	Electric field	V/cm
$\mathscr{E}_m$	Maximum field	V/cm
f	Frequency	Hz
h	Planck's constant	J-s
$h\nu$	Photon energy	eV
I	Current	A
J	Current density	A/cm^2
k	Boltzmann constant	J/K
kT	Thermal energy	eV
L	Length	cm or μm
m_0	Electron rest mass	kg
m^*	Effective mass	kg
n	Refractive index	
n	Density of free electrons	cm^{-3}
n_i	Intrinsic density	cm^{-3}
N	Doping concentration	cm^{-3}
N_A	Acceptor impurity density	cm^{-3}

Symbol	Description	Units
N_D	Donor impurity density	cm^{-3}
p	Density of free holes	cm^{-3}
P	Pressure	N/m^2
q	Magnitude of electronic charge	C
Q_{it}	Interface-trap density	$charges/cm^2$
R	Resistance	Ω
t	Time	s
T	Absolute temperature	K
v	Carrier velocity	cm/s
v_s	Saturation velocity	cm/s
V	Voltage	V
V_{bi}	Built-in potential	V
V_B	Breakdown voltage	V
W	Thickness	cm or μm
x	x direction	
∇T	Temperature gradient	K/cm
ϵ_0	Permittivity in vacuum	F/cm
ϵ_s	Semiconductor permittivity	F/cm
ϵ_i	Insulator permittivity	F/cm
ϵ_s/ϵ_0 or ϵ_i/ϵ_0	Dielectric constant	
τ	Lifetime or decay time	s
θ	Angle	rad
λ	Wavelength	μm or Å
ν	Frequency of light	Hz
μ_0	Permeability in vacuum	H/cm
μ_n	Electron mobility	$cm^2/V\text{-}s$
μ_p	Hole mobility	$cm^2/V\text{-}s$
ω	Angular frequency ($2\pi f$ or $2\pi\nu$)	rad/s
Ω	Ohm	Ω

APPENDIX C

International System of Units

Quantity	Unit	Abbreviation	Units
Length	meter	m	
Mass	kilogram	kg	
Time	second	s	
Temperature	kelvin	K	
Current	ampere	A	
Frequency	hertz	Hz	1/s
Force	newton	N	kg-m/s^2
Pressure	pascal	Pa	N/m^2
Energy	joule	J	N-m
Power	watt	W	J/s
Electric charge	coulomb	C	A-s
Potential	volt	V	J/C
Conductance	siemens	S	A/V
Resistance	ohm	Ω	V/A
Capacitance	farad	F	C/V
Magnetic flux	weber	Wb	V-s
Magnetic induction	tesla	T	Wb/m^2
Inductance	henry	H	Wb/A

APPENDIX D

Physical Constants

Quantity	Symbol	Value
Ångstrom unit	Å	1 Å = 10^{-1} nm = 10^{-4} μm = 10^{-8} cm = 10^{-10} m
Avogadro constant	N_{Avo}	6.02204×10^{23} mol^{-1}
Bohr radius	a_B	0.52917 Å
Boltzmann constant	k	1.38066×10^{-23} J/K (R/N_{Avo})
Elementary charge	q	1.60218×10^{-19} C
Electron rest mass	m_0	0.91095×10^{-30} kg
Electron volt	eV	1 eV = 1.60218×10^{-19} J = 23.053 kcal/mol
Gas constant	R	1.98719 cal mol^{-1}K^{-1}
Permeability in vacuum	μ_0	1.25663×10^{-8} H/cm(4×10^{-9})
Permittivity in vacuum	ϵ_0	8.85418×10^{-14} F/cm($1/\mu_0 c^2$)
Planck constant	h	6.62617×10^{-34} J-s
Reduced Planck constant	$\hbar$	1.05458×10^{-34} J-s ($h/2\pi$)
Proton rest mass	M_p	1.67264×10^{-27} kg
Speed of light in vacuum	c	2.99792×10^{10} cm/s
Standard atmosphere		1.01325×10^{5} N/m^2
Thermal voltage at 300 K	kT/q	0.0259 V
Wavelength of 1-eV quantum	λ	1.23977 μm

INDEX

ACKNOWLEDGMENTS

Chapter 1

Figure 10, p. 20, from Suzuki, "Supercleanroom" (in Japanese), *Hitachi Hyoron* (Commentary), **68** (9), 737, 1986, reprinted with permission from Hitachi, Ltd.
Figure 14, p. 37, reprinted with permission from Murata Machinery, Ltd.
Figure 16, p. 40; Fig. 19, p. 43; Table 9, p. 44, reprinted with permission from Christ AG. Switzerland.
Figure 18, p. 42, reprinted with permission from Asahi Chemical Industry Co., Ltd.
Figure 24, p. 51, from M. Nakamura, T. Ohmi, and K. Kawada, "All Metal and Oxygen Passivation Tubing Technology for Ultra Clean Gas Delivery System," *Institute of Environmental Sciences Technical Meeting,* San Diego, 1991, p. 605, reprinted with permission from the Institute of Environmental Sciences.
Table 10, p. 45, reprinted with permission from Merck-Kanto Advanced Chemicals Ltd.
Table 11, p. 48, and Table 12, p. 51, reprinted with permission from BOC Gases, Murray Hill, NJ.

Chapter 4

Figure 6, p. 152; Fig. 7, p. 154; Fig. 38, p. 187, © 1990 IEEE. Reprinted with permission from The Institute of Electrical and Electronics Engineers, Inc.
Figure 9, p. 157; Table 2, p. 195, reprinted with permission from C. W. Osburn.
Figure 11, p. 160; Fig. 12, p. 161; Fig. 15, p. 165; Fig. 16, p. 166; Fig. 17, p. 166; Fig. 28, p. 178; Fig. 31, p. 181; Fig. 33, p. 183, reprinted with permission from Academic Press, Inc.
Figure 18, p. 167; Fig. 32, p. 182; Fig. 44, p. 194, reprinted with permission from Elsevier Science Publishers BV.
Figure 19, p. 168; Fig. 20, p. 169; Fig. 23, p. 172, reprinted with permission from Materials Research Society.
Figure 21, p. 170, reprinted with permission from the American Institute of Physics.
Figure 22, p. 170, © 1991 IEEE. Reprinted with permission from The Institute of Electrical and Electronics Engineers, Inc.
Figure 26, p. 175; Fig. 27, p. 176, © 1992 IEEE. Reprinted with permission from The Institute of Electrical and Electronics Engineers, Inc.
Figure 34, p. 185; Fig. 45, p. 196, reprinted with permission from The Electrochemical Society.
Figure 36, p. 186, reprinted with permission from *Journal of Electronic Materials,* Vol. 18 (1989), p. 143, a publication of The Minerals, Metals & Materials Society, Warrendale, PA 15086.

Chapter 6

Table 1, p. 273, reprinted with permission from the April 1993 issue of *Solid State Technology,* © 1993 PennWell Publishing Co.
Figure 4, p. 277, reprinted with permission from the April 1977 issue of *Solid State Technology,* © 1977 PennWell Publishing Co.

Figure 5, p. 278, reprinted with permission from the August 1977 issue of *Solid State Technology,* © 1977 PennWell Publishing Co.

Figure 6, p. 279, reprinted with permission from the August 1980 issue of *Solid State Technology,* © 1980 PennWell Publishing Co.

Figure 7, p. 280; Fig. 8, p. 281; Fig. 9, p. 283; Fig. 12, p. 287; Fig. 13, p. 287; Fig. 14, p. 289; Table 2, p. 290; Fig. 22, p. 302, reprinted with permission from Society of Photo-Optical Instrumentation Engineers, Bellingham, WA.

Figure 10, p. 284; Fig. 11, p. 285, © 1982 IEEE. Reprinted with permission from The Institute of Electrical and Electronics Engineers, Inc.

Figure 15, p. 291, reprinted with permission from *Solid State Technology,* © 1974 PennWell Publishing Co.

Figure 16, p. 292, © 1975 IEEE. Reprinted with permission from The Institute of Electrical and Electronics Engineers, Inc.

Figure 24*a,* p. 304; Fig. 24*b,* p. 304; Fig. 25, p. 306; Fig. 26, p. 309; Fig. 27, p. 312; Fig. 29*b,* p. 314; Fig. 30, p. 319; Fig. 31, p. 320; Fig. 32, p. 322, reprinted with permission from the American Institute of Physics.

Table 8, p. 317, reprinted with permission from the February 1986 issue of *Solid State Technology,* © 1986 PennWell Publishing Co.

Chapter 7

Figure 3, p. 335, © 1992 by International Business Machines Corporation, reprinted with permission.

Figure 4, p. 336; Fig. 6, p. 339; Fig. 7, p. 340; Fig. 19, p. 351; Fig. 20, p. 352, reprinted with permission from the American Institute of Physics.

Figure 5, p. 337; Fig. 16, p. 347; Fig. 18, p. 349; Fig. 22, p. 356; Fig. 23, p. 358, reprinted by permission of the publisher, The Electrochemical Society, Inc.

Figure 8, p. 340, reprinted from D. L. Flamm and V. M. Donnelly, "The Design of Plasma Etchants," *Plasma Chem. Plasma Process.,* **1,** 317 (1981), with permission from Plenum Publishing Group.

Table 2, p. 350, reprinted with permission of *Semiconductor International* magazine, 1993, © by Cahners Publishing Co.

Table 3, p. 354, reprinted with permission of *Semiconductor International* magazine, 1992, © by Cahners Publishing Co.

Figure 24, p. 359, reprinted with permission from *Japanese Journal of Applied Physics.*

Chapter 9

Figure 1, p. 473, © 1991 IEEE. Reprinted with permission from The Institute of Electrical and Electronics Engineers, Inc.

Figure 2, p. 475; Fig. 3, p. 476; Fig. 5, p. 481; Fig. 16, p. 498, © 1980 IEEE. Reprinted with permission from The Institute of Electrical and Electronics Engineers, Inc.

Figure 6, p. 482, © 1992 IEEE. Reprinted with permission from The Institute of Electrical and Electronics Engineers, Inc.

Figure 7, p. 484; Fig. 8, p. 486; Fig. 23, p. 509; Fig. 29, p. 516, © 1987 IEEE. Reprinted with permission from The Institute of Electrical and Electronics Engineers, Inc.

Figure 9, p. 487; Fig. 10, p. 488, © 1988 IEEE. Reprinted with permission from The Institute of Electrical and Electronics Engineers, Inc.

Figure 12, p. 491; Fig. 27, p. 515; Fig. 28, p. 516, © 1989 IEEE. Reprinted with permission from The Institute of Electrical and Electronics Engineers, Inc.

Figure 15, p. 497, © 1986 IEEE. Reprinted with permission from The Institute of Electrical and Electronics Engineers, Inc.
Figure 24, p. 511, © 1993 IEEE. Reprinted with permission from The Institute of Electrical and Electronics Engineers, Inc.
Figure 30, p. 518, © 1985, 1989 IEEE. Reprinted with permission from The Institute of Electrical and Electronics Engineers, Inc.

Chapter 12

Figure 1, p. 659; Fig. 2, p. 659; Fig. 3, p. 661, reprinted with permission from Academic Press, Inc.
Figure 9, p. 669, © 1975 IEEE. Reprinted with permission from The Institute of Electrical and Electronics Engineers, Inc.
Figure 11, p. 670, reprinted with permission from H. S. Rathore.
Figure 14, p. 675; Fig. 15, p. 676, © 1985 IEEE. Reprinted with permission from The Institute of Electrical and Electronics Engineers, Inc.
Figure 16, p. 678, reprinted with permission from the American Institute of Physics and H. Yagi.
Figure 25, p. 694; Fig. 26, p. 695, reprinted with permission from McGraw-Hill, Inc.